11th TSME-International Conference on Mechanical Engineering (TSME-ICoME 2020)

IOP Conference Series: Materials Science and Engineering
Volume 1137

Ubon Ratchathani, Thailand
1-4 December 2020

Editors:

Kulachate Pianthong
Anchasa Pramuanjaroenkij
Nattadon Pannucharoenwong
Komsan Rattanakijsuntorn

ISBN: 978-1-7138-4616-1
ISSN: 1757-8981

Printed from e-media with permission by:

Curran Associates, Inc.
57 Morehouse Lane
Red Hook, NY 12571

Some format issues inherent in the e-media version may also appear in this print version.

Institute of Physics
Dirac House, Temple Back
Bristol BS1 6BE UK

Phone: 44 1 17 929 7481
Fax: 44 1 17 920 0979

techtracking@iop.org

Additional copies of this publication are available from:

Curran Associates, Inc.
57 Morehouse Lane
Red Hook, NY 12571 USA
Phone: 845-758-0400
Fax: 845-758-2633
Email: curran@proceedings.com
Web: www.proceedings.com

TABLE OF CONTENTS

IOP Conf. Series: Materials Science and Engineering 1137 (2021) 011001 doi:10.1088/1757-899X/1137/1/011001

PREFACE

The 11th TSME International Conference on Mechanical Engineering (TSME-ICoME 2020) was hosted by Department of Mechanical Engineering, Ubon Ratchathani University, Ubon Ratchathani, Thailand. The conference was held at Sunee Grand Hotel & Convention Center, Ubon Ratchathani, Thailand, during December 1 – 4, 2020. The TSME-ICoME is annually co-organized by Thai Society of Mechanical Engineers (TSME) to promote an academic platform where scientists, engineers and researchers from around the world can meet and discuss about significant issues regarding the mechanical engineering.

The scientific program covered a wide range of topics, including Alternative Energy and Combustion (AEC), Automotive, Aerospace and Marine Engineering (AME), Applied Mechanics, Materials and Manufacturing (AMM), Biomechanics and Bioengineering (BME), Computation and Simulation Techniques (CST), Dynamic Systems, Robotics and Controls (DRC), Engineering Education (EDU), Energy Technology and Management (ETM), and Thermal System and Fluid Mechanics (TSF)

This year conference brought together more than 200 delegates to exchange knowledge by in-depth discussions and to promote international and multi-disciplinary collaborations. The theme of the TSME-ICoME 2020 was "Challenging Trends in Mechanical Engineering" consisting of mechanical research fields, industrial panel sessions, keynote speeches, papers and poster presentation. The hybrid presentation system, on-site and online, was adopted according to the COVID-19 pandemic precautions.

We would like to thank the program chairs, conference staff, reviewers, members of the TSME and TSME board committees for their valuable work and time. Additionally, we would also like to thank the editing team members from TSME and Ubon Ratchathani University for screening, preparing, and editing all manuscripts submitted for the TSME-ICoME 2020 (IOP Conference Series: Materials Science and Engineering) publication.

We are grateful to all those who have contributed to a success of TSME-ICoME 2020. We hope that all participants and other interested readers receive benefit scientifically from the TSME-ICOME 2020 proceeding and find it stimulating to read.

Finally, we are looking forward to meeting you and discussing valuable research in the next 12th TSME International Conference on Mechanical Engineering.

With our warmest regards,

The TSME-ICoME 2020 Editors
Prof. Dr. Kulachate Pianthong
Assoc. Prof. Dr. Anchasa Pramuanjaroenkij
Asst. Prof. Dr. Nattadon Pannucharoenwong
Dr. Komsan Rattanakijsuntorn

IOP Conf. Series: Materials Science and Engineering 1137 (2021) 011001 doi:10.1088/1757-899X/1137/1/011001

Address at the 11th Thai Society of Mechanical Engineers International Conference on Mechanical Engineering (TSME-ICoME 2020)

In 2010, the first International Conference on Mechanical Engineering (ICoME) was organized by TSME and, since then, the conference was held every year. The conference was also supported by the American Society of Mechanical Engineers (ASME), the Japan Society of Mechanical Engineers (JSME) and the United Kingdom's Institute of Mechanical Engineers (IMechE). This year, the new cooperation with the Korean Society of Mechanical Engineers (KSME) was established.

The aim of the International Conference on Mechanical Engineering (ICoME) is to build a sustainable network of scientific students, researchers, and other groups of academic institution. Every year, participants will be given an opportunity to present their research work to the community, to engage in discussion with professional in the field and to develop new collaboration among mechanical departments from several universities.

The TSME-ICoME has been organized and popular for 10 years. In this honor, I would like to express my congratulation and thankfulness to Ubon Ratchathani University for the dedication to organize the 11th TSME-ICoME (TSME-ICoME 2020) which was held at Sunee Grand Hotel & Convention Center, Ubon Ratchathani, Thailand, during December 1 – 4, 2020. On behalf of the Thai Society of Mechanical Engineers, I would like to thank all supporting team and staff for your hard working along with the TSME board committees. I hope the conference is successful for all purposes and believe that this conference will gather researchers in the mechanical engineering.

Prof. Dr. Phadungsak Rattanadecho

President of Thai Society of Mechanical Engineers (TSME)

IOP Conf. Series: Materials Science and Engineering 1137 (2021) 011001 doi:10.1088/1757-899X/1137/1/011001

Committee

- Honorary Chair
  Prof. Dr. Phadungsak Rattanadecho
  President, Thai Society of Mechanical Engineers (TSME)

- International Advisory Board
  - Prof. Dr. Phadungsak Rattanadecho, President of Thai Society of Mechanical Engineers (TSME)
  - Prof. Dr. Ashwani K. Gupta, University of Maryland, USA
  - Prof. Dr. Ning Zhu Shizuoka, Institute of Science and Technology, JAPAN
  - Prof. Dr. Katsunori Hanamura, Tokyo Institute of Technology, JAPAN
  - Prof. Dr. Jens Nørkær Sørensen, Technical University of Denmark, DENMARK
  - Prof. Dr. Ashim K. Datta, Cornell University, USA
  - Prof. Dr. Kambiz Vafai, University of California, USA
  - Prof. Dr. Naoki Maruyama, Mie University, JAPAN
  - Assoc. Prof. Dr.Seokwoo Jeon, KAIST, REPUBLIC OF KOREA
  - Prof. Dr. Elmi Abu Bakar, Universiti Sains Malaysia, MALAYSIA
  - Prof. Dr. Pham Tuan Anh, Ho Chi Minh City University of Technology, VIETNAM
  - Assoc. Prof. Dr.lman K., Reksowardojo Institute of Technology, Bandung, INDONESIA
  - Prof. Dr. Le Anh Tuan, Hanoi University of Science and Technology, VIETNAM
  - Prof. Dr. Ocktaeck Lim, University of Ulsan, REPUBLIC OF KOREA
  - Prof. Dr. Huynh Thanh Cong, Ho Chi Minh City University of Technology, VIETNAM
  - Prof. Dr. Hendri Syamsudin, Institut Teknologi Bandung, INDONESIA
  - Prof. Dr. Pham Huu Tuyen, Hanoi University of Science and Technology, VIETNAM
  - Prof. Dr. Hironori Saitoh, Sojo University, JAPAN
  - Prof. Dr. Jau-Huai Lu, National Chung Hsing University, Republic of China
  - Dr. Norazila Binti Othman, Mechanical Engineering, Universiti Teknologi Malaysia

- Local Organizing Committee, Ubon Ratchathani University
  - Asst. Prof. Dr. Mongkol Pusayatanont - Conference Advisor
  - Asst. Prof. Dr. Prachasanti Thaiyasuit - Conference Chair
  - Assoc. Prof. Dr. Thanarath Sriveerakul - Associate Conference Chair
  - Asst. Prof. Dr.-Ing. Nantawatana Weerayuth - Subcommittee Chair of Venue and Formalities
  - Asst. Prof. Dr. Ratchada Sopakayang - Subcommittee Chair of Academic Affairs
  - Asst. Prof. Dr. Wirapan Seehanam - Subcommittee Chair of Public Relations
  - Assoc. Prof. Dr. Kulachate Pianthong - Subcommittee Chair of Funding and Special Events
  - Assoc. Prof. Dr. Umphisak Teeboonma - Subcommittee Chair of Finance and Registration
  - Assoc. Prof. Dr. Chawalit Thinvovgpituk - Committee
  - Assoc. Prof. Dr. Anirut Matthujak - Committee

  - Assoc. Prof. Dr. Lt. JG.Somya Poonaya - Committee
  - Asst. Prof. Dr. Adun Janyalertadun - Committee
  - Asst. Prof. Dr. Apinunt Namkhat - Committee
  - Asst. Prof. Rattapong Patikanung - Committee
  - Dr. Komsan Rattanakijsuntorn - Committee
  - Dr. Patinya Samanuhut - Committee
  - Chakrit Pongarm - Committee
  - Songsupa Pumchumpol - Committee
  - Bongkot Juntamas - Committee
  - Krittaya Chaiyot - Committee
  - Nitikorn Promduang - Committee
  - Paryat Samanmit - Committee
  - Pakawat Naosopa - Committee
  - Samak Intha - Committee
  - Suk Dampatik - Committee
  - Watcharaporn Chantharakan - Committee
  - Jirapreeya Ok-oon - Committee
  - Narongrit Majan – Committee

- Local Scientific Committee
  - Prof. Dr. Sumrerng Jugjai, King Mongkut's University of Technology Thonburi - Alternative Energy and Combustion (AEC)
  - AM. Prof. Dr. Noppol Hanklar, Navaminda Kasatriyadhiraj, Royal Thai Air Force Academy - Automotive, Aerospace and Marine Engineering (AME)
  - Assoc. Prof. Dr. Vitoon Uthaisangsuk, King Mongkut's University of Technology Thonburi - Applied Mechanics, Materials and Manufacturing (AMM)
  - Assoc. Prof. Dr. Jackrit Suthakorn, Mahidol University - Biomechanics and Bioengineering (BME)
  - Prof. Dr. Phadungsak Rattanadecho, Thammasat University - Computation and Simulation Techniques (CST)
  - Prof. Dr. Suwat Kuntanapreeda, King Mongkut's University of Technology North Bangkok - Dynamic Systems, Robotics and Controls (DRC)
  - Assoc. Prof. Dr. Ratchaphon Suntivarakorn, Khon Kaen University - Engineering Education (EDU)
  - Assoc. Prof. Dr. Jaruwat Charoensuk, King Mongkut's Institute of Technology Ladkrabang - Energy Technology and Management (ETM)
  - Prof. Dr. Somchai Wongwises, King Mongkut's University of Technology Thonburi - Thermal System and Fluid Mechanic (TSF)

- Thai Society of Mechanical Engineers (TSME) Organizing Committee
  - Assoc. Prof. Dr. Worawut Wisutmethangoon - Senior Advisor
  - Assoc. Prof. Dr. Withaya Yongchareon - Senior Advisor
  - Prof. Dr. Sumrerng Jugjai - Senior Advisor
  - Asst. Prof. Dr. Chinda Charoenphonphanich - Senior Advisor
  - Manit Koothanapath - Adviser
  - Yuthana Charoenwong - Adviser
  - Dr. Payung Sukdasawit - Adviser
  - Apichit Lumlertpongpana - Adviser
  - Assoc. Prof. Dr. Yossapong Laoonual - Adviser

IOP Conf. Series: Materials Science and Engineering 1137 (2021) 011001 doi:10.1088/1757-899X/1137/1/011001

- Assoc. Prof. Dr. Pongpan Kaewtatip - Adviser
- Assoc. Prof. Dr. Perapong Tekasakul - Adviser
- Jiraboon Vityakraisign - Adviser
- Maj.Gen.Chukiat Chuaypetch - Adviser
- Tinakon Pongluxna - Adviser
- Prof. Dr. Kulachate Pianthong - Vice President
- Assoc. Prof. Dr. Jaruwat Charoensuk - Vice President
- Asst. Prof. Dr. Pornthip Keangin - Secretary
- Asst. Prof. Dr.Warakom Nerdnoi - Treasurer
- Asst. Prof. Dr. Nattadon Pannucharoenwong - Receptionist
- Dr. Nopparat Khamporn - Registrar
- Dr. Montana Rungsiyopas - Public Relations
- A.M. Prof.Dr.Noppol Hanklar - Committee
- Assoc. Prof. Dr. Smith Eiamsa-ard - Committee
- Assoc. Prof. Dr. Preechar Karin - Committee
- Prof. Dr. Tanakorn Wongwuttanasatian - Committee
- Dr. Sopida Sungsoontorn - Committee
- Assoc. Prof. Dr. Kanit Wattanavichien - Committee
- Assoc. Prof. Dr. Amornrat Kaewpradap - Committee
- Assoc. Prof. Dr. Anchasa Pramuanjaroenkij - Committee
- Dr. Samroeng Netpu - Committee
- Dr. Taperit Thongshoob - Committee
- Asst. Prof. Dr. Wirachai Roynarin - Committee
- Asst. Prof. Dr. Nattawoot Depaiwa - Committee

IOP Conf. Series: Materials Science and Engineering 1137 (2021) 011002 doi:10.1088/1757-899X/1137/1/011002

Peer review declaration

All papers published in this volume of *IOP Conference Series: Materials Science and Engineering* have been peer reviewed through processes administered by the Editors. Reviews were conducted by expert referees to the professional and scientific standards expected of a proceedings journal published by IOP Publishing.

- **Type of peer review: Double-blind**

- **Describe criteria used by Reviewers when accepting/declining papers. Was there the opportunity to resubmit articles after revisions? The TSME ICOME applied the peer review criteria by which 2 out of 3 reviewers have to accept a manuscript and, then, the manuscript will be accepted. On the other hand, if 2 out of 3 reviewers reject a manuscript, the manuscript will be declined. There was the opportunity to resubmit the revision or revised manuscripts if the reviewers required the revision. The reviewer arrangement was decided by the assigned chair of each topic. We had 100 reviewers from various professional fields in Mechanical Engineering and each manuscript was assigned to the reviewers in the field by the chair of the field.**

- **Conference submission management system:** We use our own submission system which is an in-house system developed by our society; http://www.tsme.org/submission/icome2020/. We have been using this system for 5 years in the series of our conference. The example pictures of the submission management system are as following;

IOP Conf. Series: Materials Science and Engineering 1137 (2021) 011002 doi:10.1088/1757-899X/1137/1/011002

TSME
ICOME 2020
Challenging Trends in Mechanical Engineering

• **Home** • **Create New Account** • **Online submission** • **Login/Registrations**

Announcement

Dear all online presenters, please check your email concerning the Online Presentation Instruction !!! To remind you again, On this Monday, November 30th (1.00PM-2.00 PM), there will be the online-test session. The online test-session is only provided to test the online system. Full-time presentations and final draft of PPT from each presenter are not required on the test-session. Please check the information in your mail box and follow step by step.

Details for Payment via Bank Transfer
Bank Name: KRUNG THAI BANK
Account No.: 693-0-33388-3
Account Name: The Thai Society of Mechanical Engineers (TSME) Conference
Swift Code (for international wire transfer only): KRTHTHBK
Reference/Payment details: [Your Paper ID: ICOME-2020-XXXX]

http://www.icome.tsme.org/icome2020/
facebook:TSME-ICoME2020
tsme-icome2020@ubu.ac.th
Tel: +66 45 353323
Fax: +66 45 353308

TSME
ICOME 2020
Challenging Trends in Mechanical Engineering

IOP Conf. Series: Materials Science and Engineering 1137 (2021) 011002 doi:10.1088/1757-899X/1137/1/011002

Name	Date
OFF Abstract Submission Deadline:	10-August-2020 23:59 UTC+07:00
ON Abstract Acceptance Notification Date :	20-August-2020 23:59 UTC+07:00
ON Full Manuscript Submission Deadline :	30-September-2020 23:59 UTC+07:00
ON Full Manuscript Acceptance Notification Date : 1-10 Nov 2020	10-November-2020 23:59 UTC+07:00
ON Camera-Ready Manuscript Submission Deadline :	17-November-2020 23:59 UTC+07:00
OFF Early-bird Registration Deadline : 1-20 Nov 2020	20-November-2020 23:59 UTC+07:00
ON Conference Dates: December : 1-4 Dec 2020	04-December-2020 00:01 UTC+07:00

Browser Support

The numbers in the table specify the first browser version that fully supports the attribute.

Attribute			
required	5.0	10.0	4.0
Download	Download	Download	Download

facebook:TSME-ICoME2020
tsme-icome2020@ubu.ac.th
Tel: +66 45 353323
Fax: +66 45 353308

TSME-ICoME 2020 Committee
Department of Mechanical Engineering,
Ubon Ratchathani University
85 Sathonlamark Rd. Warin Chamrap, Ubon Ratchathani 34190, Thailand

- **Number of submissions received: 147**

IOP Conf. Series: Materials Science and Engineering 1137 (2021) 011002 doi:10.1088/1757-899X/1137/1/011002

- **Number of submissions sent for review: 146**
- **Number of submissions accepted: 139**
- **Acceptance Rate (Number of Submissions Accepted / Number of Submissions Received X 100): 94.5%**
- **Average number of reviews per paper: 3**
- **Total number of reviewers involved: 100 reviewers**
- **Any additional info on review process (ie plagiarism check system):**
  We do check a template (format) of the manuscripts according to the conference requirement, typo and grammatical errors. The plagiarism is checked by using the program; "Turnitin", before submitting them to IOP. We do not allow any manuscripts to have an individual plagiarism more than 15%. However, we do request the author to conform the general rules of the organization and the IOP publication before submitting the manuscript to IOP.
- **Contact person for queries: Assoc.Prof.Anchasa Pramuanjaroenkij, Ph.D. (anchasa@gmail.com)**

Improvement of a hot water producing system in a chili sauce

Warachit Phayom[1,*] and Jakkaphan Bamrungcharoen[1]

[1]Energy Engineering, Faculty of Technology, Udon Thani Rajabhat University, Udon Thani, Thailand 41000

* Corresponding Author: warachit.ph@udru.ac.th

Abstract
In Udon Thani Province (Thailand), Mahachai Food is a small-sized food processor that manufactures and distributes chili sauce to the local market. The manufacturer is faced with various problems in its production facility. A significant problem was in its hot water system. The system required workers to transfer hot water by buckets. Also, its energy costs were quite high. Thus, this research aims to upgrade the manufacturer's hot water production system. In doing so, the investigator designed vessels and LPG stoves that preheated the water by solar flat plate collectors and implemented a water transfer system using a pipe system. The results are as follows: 1) Water transfer required only one-third of the original time. This reduced the need for workers from four persons to just one. It also helped increase safety in the workplace since the potential of human exposure to hot water was greatly reduced. 2) The efficiency of the hot water producing system was increased. 3) The cost of energy-producing hot water was reduced.

Keywords: Chili Sauce Manufacturing Facility, Hot Water Production System, Improving the Performance, Thermal Efficiency, Solar flat plate collector.

1. Introduction

Udon Thani province is located in the north-east region of Thailand where its economy is growing and expanding rapidly. Udon Thani has the 3rd highest economic growth in the north-east region presently [1]. The most potential and funded industry in Udon Thani is the food industry [2]. However, most food industry is a small-sized food factory. Also, Mahachai Food (hereinafter called the 'manufacturer') is a small-sized food processor that manufactures and distributes chili sauce to the local market. The production capacity is approximately 10-12 cycles per month. Each cycle is produced 5000-6000 bottles of 750 milliliters (4000-4500 liters). The main ingredients of the production are chilies, flour, and hot water. The most important matter in the manufacturing process was producing hot water. There were many problems in producing hot water as follows:

1. Heating water resulted in high energy losses. The total produced hot water was approximately 4000-4500 liters per cycle. So, it required up to 80-90 kilograms of LPG gas. This was due to using ordinary cooking vessels along a single wall with no insulation. Additionally, they had no lids which increases heat loss. There were two-vessel sets for switching. The first set consisted of two 200 liters vessels and another set had one vessel of 280 liters, as can be seen in Figure 1.

(a) (b)

Figure 1. The stoves and cooking vessels used by the manufacturer
(a) A set of two 200-liters vessels (b) A set of a 280-liter vessel

2. Each cycle, the manufacturer would boil water about 20-23 rounds. During boiling water, there was much wasted time. Waiting for enough water for every vessel before transferring material to the tank used for mixing chili sauce because boiling time caused long delays. To boil the water took approximately 35 minutes and transferring it approximately 20 minutes in each round. Because the manufacturer had 2 sets of chili sauce mixing tanks, water was transferred almost all the time.

3. The process of transferring hot water from the vessels to the chili sauce mixing tanks required four workers; 1 person for scooping hot water from the vessels into 10-liters buckets, 2 people for carrying the buckets to the chili sauce tank with a distance of about 10 meters, and another one for pouring hot water from the buckets into the chili sauce tank. In each chili sauce mixing round needs to use 150 liters of water at a temperature of 90 degrees Celsius and about 10 liters of other ingredients to fill in a 280-liter mixing tank. The worker who scooping hot water into the 3 liters of a bowl had to make about 15-20 buckets. Then the workers who carries the buckets had to finish about 10 times per round. This process was a very unsafe workplace practice. As an aside, [3] reported human skin pain receptors are located at approximately 0.1 mm depth and the pain temperature threshold is 44.8°C. A portage of hot water was very dangerous. That also slows the manufacturing process (Figure 2).

Figure 2. Manual workers were used to transferring hot water from the vessels to the chili sauce tank

These manufacturing problems contributed to energy losses and increased manufacturing expenses. Thus, to improve manufacturing capacity, this research aimed to improve the manufacturer's hot water production system by increasing the overall thermal efficiency and alleviating the need for the manual worker, thus reducing costs and expenses for energy while improving workplace safety.

2. Materials and Methods

For this study, the plant layout of the manufacturer was improved. The original layout of the plant had a fairly distributed work area (Figure 3a) and the working area was still being used ineffectively. Besides, the distance between the boiling water area and the sauce mixing area was more than 10 meters away and the boiling water area was outdoor where there might be ruined by the sunshine, wind, and rain. Therefore, plant layouts of the machinery, water transfer system, and the water heating system equipment were designed as shown in Figure 3b.

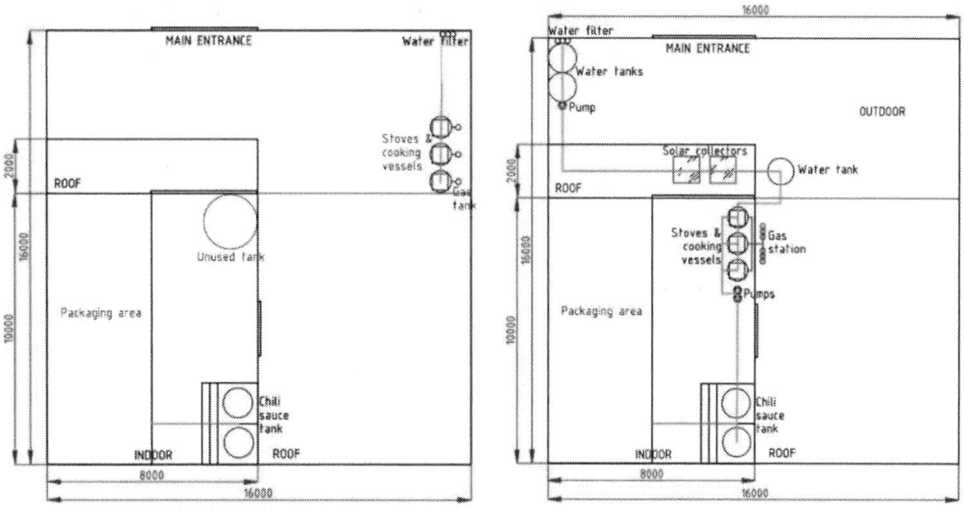

(a) (b)

Figure 3. Plant layout (unit: millimeter) (a) the original layout (b) the new layout

The details of the new system: water storage, pipelines, and pumps replaced human workers in the water transfer system. The proper location of the hot water pipelines was important. It should not obstruct the walkways used by workers. Also, it should be as short as possible to reduce heat losses. However, the pipe must not be covered so that it could be removed and easily cleaned. The stations of all equipment should be stable, durable, support their function, and maintenance. Moreover, the cooking vessels, the stoves, and the gas station would be moved to indoor and put next to the mixing sauce area. The gas stove should use low conductivity materials and insulations [4], such as refractories bricks, as much as possible. The gas stove had a width of 60 centimeters on each side and a height of 40 centimeters. Its wall was made of refractory bricks, the outside was covered with stainless steel, and the 4 pillars were made of stainless steel. Additionally, the station where gas was located should have connections for multiple tanks. The new vessels were made of stainless steel with double insulation. Their maximum safe volume was 250 liters. The chili sauce mixing tank can receive 150 liters per batch, as shown in Figure 4.

The installation of a solar water preheating system can significantly improve the system. It can help reduce fuel energy used to heat water [5, 6]. The parts of a solar water preheating system consisted of 2 flat plate solar collectors with a total net area of 2 square meters, which was a recessed 500 litre-vessel with insulation as shown in Figure 5. The solar water preheating system was designed by the water flowing from the inlet into the outlet with a series flow pattern to increase the water temperature. The flat plate solar collector was covered with transparent glass and its absorber plate which was painted black to absorb the maximum amount of solar irradiance. While the underneath and side of the collector were insulated by glass wool to decrease the heat loss by conducting through the absorber plate. Besides, pressure relief valves, pressure, and temperature gauges have been installed. It was considered stable safe, and economical.

(a) (b) (c)

Figure 4. The new system (a) Stoves (b) Gas station (c) Vessels

Figure 5. Solar water preheating system (a) Flat plate solar collectors and (b) A recessed 500 litre-vessel with insulation

Data was collected on both original and new systems for analysis gas consumption (Q_{LPG}, kJ per day) useful energy (Q_{useful}, kJ per day) heat transfer rate to water within the solar collectors. ($\dot{Q}$, Watts) overall thermal efficiency of the system (η_{th}) thermal efficiency of the Flat plate solar collector (η_{FP}) breakeven analysis and payback period.

3. Results and Discussions

Examination and analysis could be divided into two parts. First, the manufacturer's original system was examined. Then, the ability to perform or the performance of the researcher's proposed system was determined. The following was found:

3.1 Performance of the manufacturer's previous system

The original system required four workers, one working on transferring water into and out of the vessels, two workers transferring water from the water heating area to the chili sauce mixing tank, located 10 meters away. One worker controlled the overall manufacturing processes. Collectively, their wages were 1,200 THB per day at least (300 THB/day/person).

The manufacturer's original process consisted of multiple processing lines, each with its vessel, gas stove, and a 48 kg gas tank. One line had a vessel with a volume of 280 liters, and the other two had volumes of 200 liters/vessel.

According to the energy analysis using LPG to heat water, it was found that the average daily volume of water heated was 4.52 cubic meters (4,520 liters). This required 83.57 kg of LPG or 4,119,831.04 kJ or 2 gas tanks a day. The useful energy transferred to the water heating system to increase the average water temperature at the initial state of 25.60°C to the average water temperature at the final state of

101.50°C or the average temperature difference (ΔT) of 75.90°C was 1,398,553.66 kJ. Overall, the thermal efficiency of the system was 34.02%.

Heat losses account for 65.98% of the energy used. On this basis, the system required 912.31 kJ per liter of water for heating for THB 0.39 per liter of hot water or THB 0.30 per bottle of chili sauce. The total daily (cycle) expenses for heating water were THB 1,778.44, representing a loss of THB 1,173.45. At the time of the analysis, the LPG price was THB 21.28 per kg [7]. Moreover, the increased temperature that the water experienced during heating was 75.90 degrees Celsius. The capacity for hot water production was only 3.89 liters per minute. See Table 1.

Table 1. Analysis of energy use in heating water with the original system

No.	V_W (m^3/day)	m_{LPG} (kg/day)	ΔT (°C)	Q_{LPG} (kJ/day)	Q_{useful} (kJ/day)	η_{th} (%)	Time (min)	Capacity (L/min)
1	4.72	94.19	79.70	4,643,190.24	1,526,369.43	32.87	1,142.00	4.13
2	4.38	78.86	75.84	3,887,482.56	1,358,465.72	34.94	1,214.00	3.61
3	4.45	77.67	72.17	3,828,820.32	1,310,825.83	34.24	1,132.00	3.93
Avg	4.52	83.57	75.90	4,119,831.04	1,398,553.66	34.02	1,162.67	3.89

3.2 Performance of the new design

After the hot water system was improved, only one worker was needed to control the system. The process tasks included turning valves on and off to produce the desired flow.

As shown in Table 2, it was found that only 4.02 cubic meters (4,020 liters) of water were successfully heated. This was because there were unexpected obstacles in the process. Other system components failed many times, especially in the chili sauce mixing part. Thus, the volume of heated water appears to be as small as the same process duration was needed. However, the amount of LPG gas used daily was only 45.75 kg or 2,255,292.00 kJ or only a gas tank. The total amount of energy transferred into the water to increase the average water temperature at the initial state of 25.79°C to the average water temperature at the final state of 92.28°C or the average temperature difference (ΔT) of 66.49°C was 1,060,717.92 kJ at an overall thermal efficiency of 47.12%. The new system produced 10.24 liters of hot water per minute at a daily energy cost of THB 973.56. Furthermore, the rate of gas used was 560.43 kJ per liter of water for THB 0.24 per liter. This was a 38.46% savings over the manufacturer's original system.

Table 2. Analysis of energy use in heating water with the new system

No.	V_W (m^3/day)	m_{LPG} (kg/day)	ΔT (°C)	Q_{LPG} (kJ/day)	Q_{useful} (kJ/day)	η_{th} (%)	Time (min)	Capacity (L/min)
1	4.41	44.69	49.75	2,203,038.24	1,078,714.68	48.96	387.00	11.39
2	4.26	48.46	59.36	2,388,884.16	1,056,495.14	44.23	401.00	10.63
3	3.40	44.10	73.62	2,173,953.60	1,046,943.95	48.16	391.00	8.70
Avg	4.02	45.75	66.49	2,255,292.00	1,060,717.92	47.12	393.00	10.24

For the water preheating system by solar flat plate collectors, the collectors were faced with the south direction and the tilt angle of 25 degrees [8, 9]. The experiments were done during the rainy season when the solar irradiance was unlikely to be high. Figures 6 shows that the highest solar irradiance during the examination was only 586 Watts per square meter.

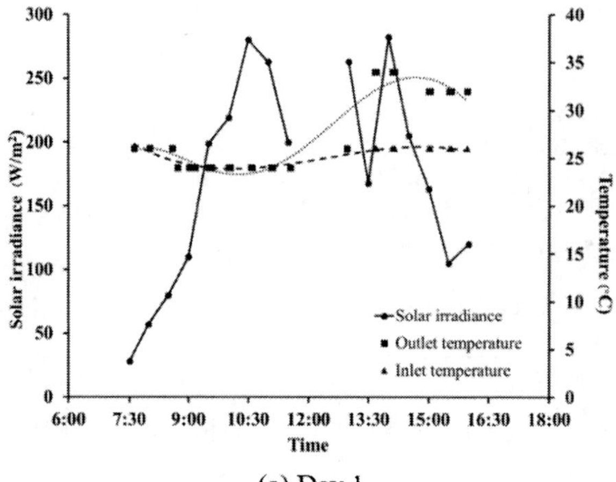

(a) Day 1

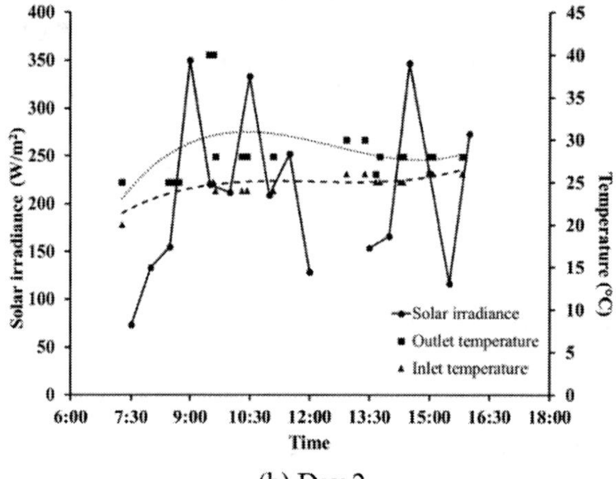

(b) Day 2

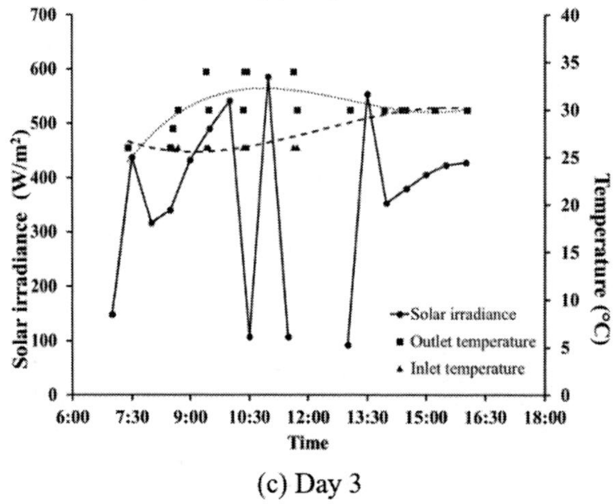

(c) Day 3

Figure 6. The solar irradiance and the water temperatures at the entrance and exit of the solar collectors

The average air temperature ranged from 29.53 to 34.05 degrees Celsius. The average relative humidity of air was 65.57% to 80.59%. The average wind speed varied from 0.30 to 0.53 km per hour. Examination of water mass flow rate through the flat plate collectors revealed a flow rate was 5 liters per minute or 300 liters per hour. This is the rate at which warm water could flow into the receiving vessel without overflowing. It was found that the difference between inlet and outlet temperatures of water was averaged as 3.07 degrees Celsius. Furthermore, the outlet temperature of the water could be as high as 40 degrees Celsius, even when the solar irradiance was low. Heat transferred to the collectors was calculated as 1,069.6 Watts. The thermal efficiency of solar flat plate collectors was 91.26%. However, if the manufacturer needs a greater volume of warm water, it is suggested that more flat plate collectors should be installed. This is because collectors in a series connection will help increase the water temperature. Nevertheless, it should be noted that such an installation will experience an increased pressure drop within the system. To address this pressure drop [10, 11], it is suggested that the installation be made up of two flat plate collectors along with additional warm water heaters.

3.3 Break-even analysis and payback period

Break-even analysis used the principle of [12] with the straight-line method, whilst the payback period followed the one of [13, 14]. Fixed costs included depreciation costs, interest expense, salary, and shipping costs. Variable costs included costs of LPG gas, water, electricity, maintenance costs, ingredient, and packaging. The results found that the break-even point was around 240,769 bottles per year. That means, if normal production is 60,000 bottles per month, the manufacturer must produce 4 months to break-even. The energy costs related to hot water production were cut by approximately half. The original system produced water for THB 0.39 per liter while the new system reduced this to THB 0.24 per liter at a cost savings of THB 6,750 per month. The new system also reduced wages to THB 13,500 per month. Lastly, the total capital expenditure on the new system was THB 1,000,000, and it would take 49 months to reach the payback period.

4. Conclusion

The improvement of the hot water production system at this chili sauce factory increased the manufacturer's efficiency in the following ways:

1. The newly designed hot water production system as well as the water piping and pumping system to supply the chili sauce vessel, clearly reduced the time needed for water transfer. Whilst the previous system took 20 minutes to transfer water from the vessel and then to the chili sauce mixing tank, the new system could transfer the same volume of water within 5 minutes. Furthermore, the developed system also helped to promote safety in the workplace and reduce the labor required from four workers to just one.

2. The developed hot water production system helped increase the heat efficiency of the boiler by 12.79%. Also, it helped increase the capacity for hot water production by a factor of three. It was also better in that it allowed for transfer water to the sauce mixing tank as one volume rather than in many small buckets.

3. The economic benefits from energy and labor costs when comparing before and after improvement systems were 13,500 baht per month, resulting in a payback period of 49 months. It should be noted that if the manufacturer increases its production, the variable costs will be reduced and the payback period will be sooner.

Although the manufacturer has significantly reduced the costs, the addition of a solar hot water production system can also be done. Besides, the coming automation systems or the internet of things (IoT) will help to make the operation more efficient and environmentally friendly.

Acknowledgments

This project was funded by the Thailand Research Fund (TRF): RDG6050144. The authors are grateful to Krisda Buarapa for supporting the experiments and Asst.Prof.Nutthanuch Mekara and Prof.Dr.Jeffrey C. Nash of Udon Thani Rajabhat University for their continued interest and encouragement.

IOP Conf. Series: Materials Science and Engineering 1137 (2021) 012001 doi:10.1088/1757-899X/1137/1/012001

5. References

[1] Marome W and Pholcharoen T 2019 *Environment Asia* **12**(3) 1-11
[2] Udon Thani Industry Office, Ministry of Industry 2020 http://www.industry.go.th/udonthani/index.php/component/k2/item/10552?rss_id=12206, accessed 1 December 2020
[3] Log T 2018 *Int J Environ Res Public Health* **15** 808-821
[4] Nilrat M Suthummanon S and Taweekun J 2017 *NUEJ* **12**(1) 115-30
[5] Bhowmik H and Amin R 2017 *Energy Rep* **3** 119-23
[6] Kalogirou S A 2014 *Solar energy engineering: processes and systems* (Waltham, MA: Elsevier Inc.) pp 14-15
[7] Energy Policy and Planning Office, Ministry of Energy 2019 *Energy statistics of Thailand 2019* (Nakhon Pathom, Thailand: Energy Policy and Planning Office, Ministry of Energy) p 204
[8] Stanciu C and Stanciu D 2014 *Energ Convers Manage* **81** 133-43
[9] Jamil Ahmad M and Tiwari G N 2009 *Open Renew Energ J* **2** 19-24
[10] Hassan J M Abdul-Ghafour Q J and Mohammed M F 2015 *J.Eng* **21**(5) 55-71
[11] Zelzouli K Guizani A Sebai R and Kerkeni C 2012 *Engineering-PRC* **4** 881-93
[12] Hunt D and Wilson D 2016 *Farm power and machinery management* (Ames, Illinois: Waveland Press) p 212
[13] Ardalan K 2012 *JEFE* **11** 10-16
[14] Umar A B 2010 *AMA-AGR MECH ASIA AF* **41**(4) 82-6

Influences of alternative diesel fuels on tailpipe emissions and particulate matter under the highway fuel economy test cycle

Phanuwat Boontatao and Sathaporn Chuepeng*

ATAE Research Unit, Department of Mechanical Engineering,
Faculty of Engineering at Sriracha, Kasetsart University,
199 Sukhumvit Road, Chonburi 20230, Thailand

* Corresponding Author: schuepeng@eng.src.ku.ac.th

Abstract
Diesel engine vehicles are widely used worldwide in community and goods transport sectors due to their engine's high efficiency and durability. Despite continuous improvements in various aspects, the combustion of compression ignition in conventional diesel engine is yet an important source of pollution and particulate matter that causes issues to human health and the environment. The objective of this research work is to study the use of conventional diesel (B7), biodiesel (B10) and premium diesel (DHi) fuels that affects to the release of exhaust gas and particulate matter emissions from tailpipes during a Highway Fuel Economy Test (HWFET) cycle. The experimental results have shown that the combustion of B7 and B10 reduced particulate matter by 0.38% and 51.0%, reduced total unburned hydrocarbon by 25.3% and 28.3%, reduced carbon monoxide by 43.3% and 64.0%, and reduced nitric oxide by 6.0% and 16.5%, respectively compared to DHi baseline on mass basis. The vehicle running on B10 releases a higher particle concentration with a smaller size compared to DHi.

Keywords: Biodiesel, Diesel, Emission, HWFET, Particulate matter.

1. Introduction

Among various atmospheric pollutants, there are major sources associated with human respiratory diseases. A number of studies founded that biomass burning and secondary particulate matter are as a result of the combined reaction of automotive exhaust and ammonia from agricultural fertilizers [1]. Diesel engine vehicles are widely used worldwide in community and goods transportation sectors due to their engine's high efficiency and durability. Statistically, numbers of diesel vehicles registered throughout the world increase year by year. The most pollutants in diesel engines are nitrogen oxides (NO_x) and particulate matter (PM). The main cause of the pollutant is from the incomplete combustion of the diesel engine. The pollutants will affect the lower respiratory tract when inhaled causing illness and affect the environment [2].

Armas et al. (2013) [3] studied the potential of two alternative fuels (biodiesel and gas-to-liquid (GTL) synthetic diesel), and diesel under the New European Driving Cycle (NEDC) that significantly affects the formation of pollutant emissions. The experimental results have shown that total unburned hydrocarbon (HC) and carbon monoxide (CO) were reduced with alternative fuels, especially during the first urban cycle. These trends suggest that the composition of alternative fuels favors a cleaner combustion. In the case of NO_x emissions, similar results were observed between diesel and GTL; a slight decrease was obtained with biodiesel during most of NEDC, except in the last part of the cycle.

Macor et al. (2011) [4] assessed the two Euro-3 commercial trucks fueled with a 30% v/v biodiesel/diesel blended fuel (B30) and pure diesel fuel in laboratory under the standard driving conditions (Urban Driving Cycle, UDC, and Extra Urban Driving Cycle, EUDC) and the Common Artemis Driving Cycles (CADC) "URBAN" test cycle. The core of this paper is the measurement of regulated emissions (CO, NO_x, HC and PM). Among the results from these tests, the substitution of diesel with B30 blend causes an increase in CO by 20% and HC by 20%. These increments were particularly evident in the UDC cycle. The NO_x emissions showed higher level during URBAN cycle when driving condition was heavier with more acceleration phases. The PM produced by B30 was nearly 30% lower for both vehicles.

Lapuerta et al. (2015) [5] tested a new advanced biofuel-FAGE (fatty acid formal glycerol ester) produced from crude glycerol blended with diesel fuel (20% FAGE content) in an automotive engine following the New European Driving Cycle (NEDC). FAGE blend and the reference fuel showed similar NO_x emissions, independent on the engine temperature while a substantial reduction (approx. 20-40%, with the highest reduction at warm engine temperature) of soot and particulate matter was observed for the FAGE blend. The FAGE bonded oxygen, higher than that of common biodiesel fuels, favors a complete combustion, therefore reducing these emissions.

Gao et al. (2019) [6] tested for the fuel consumption and exhaust emissions of a Euro-6 compliant light-duty diesel vehicle in Worldwide Harmonized Light Vehicles Test Cycles (WLTC) on a chassis dynamometer. The results showed that the exhaust emissions in the first 300 s dominated the accumulative emissions of the whole cycle; they were 63.2%, 55.4% and 56.0% of the total emissions for HC, NO_x and CO, respectively. In the last 200 s of the cycle, increasing NO_x emissions resulted from the high in-cylinder temperature despite of high catalyst efficiency.

The purpose of this paper is to investigate the use of conventional diesel (B7), biodiesel (B10) and premium diesel (DHi) fuels in a common rail diesel engine vehicle. These fuels have been widely used in the local market of Thailand. The trial was conducted under the Highway Fuel Economy Test (HWFET) cycle. By this scheme, the gaseous emissions, particulate matter and particle number-size distribution will be quantified and presented.

2. Materials and Method

2.1 Vehicle engine and equipment

The experimental tests were carried out using a 4-stroke, turbocharged diesel engine (Isuzu, Model 4JK1-TC), equipped with a common rail direct injection system. Its main specifications are shown in Table 1 and Figure 1 show the vehicle and equipment arrangement for the test.

Table 1. Specifications of the test engine

Item	Specification
Fuel type	Diesel
Number of cylinder	4
Displaced volume	2,499 cm^3
Maximum torque	280 Nm @ 1,800-2,200 rpm
Maximum power	85 kW @ 3,600 rpm
Bore · stroke	87.4 mm · 95.4 mm
Compression ratio	18.3

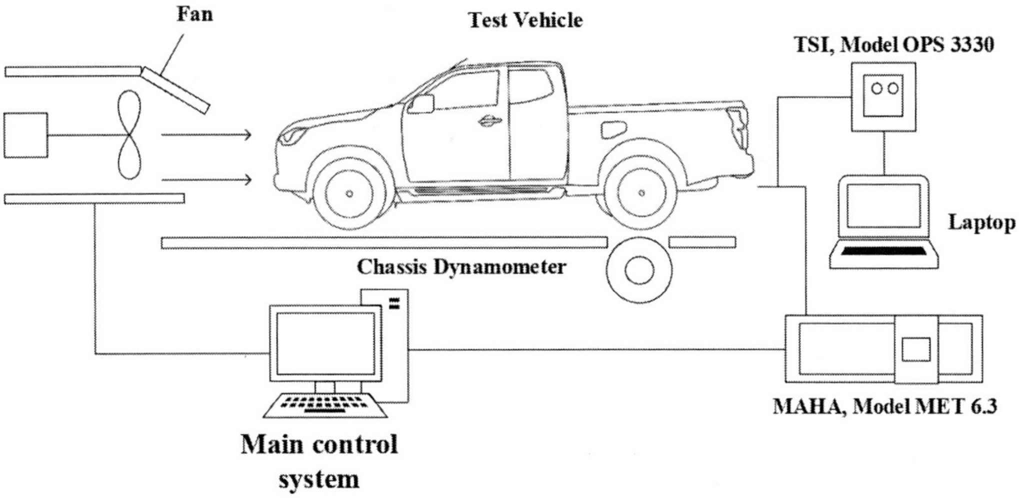

Figure 1. Schematic diagram of vehicle test system

The test of exhaust gas and particulate matter during Highway Fuel Economy Test (HWFET) cycle was based on chassis dynamometer (Maha, Model LPS3000). Tailpipe exhaust emissions such as CO, NO, HC, and smoke opacity were measured by emission tester (MAHA, Model MET 6.3). Smoke opacity was then converted to PM emission by calculation regarding the scheme explained by AVL in [7]. Meanwhile, particle number concentration and size distribution were measured by optical particle sizer (TSI, Model OPS 3330) with the measuring range between 300 and 10,000 nm.

2.2 Test conditions

The experimental procedure was accomplished based on Highway Fuel Economy Test (HWFET or HFET) cycle, a chassis dynamometer driving schedule developed by the U.S. Environmental Protection Agency (U.S. EPA) for the determination of fuel economy of light duty vehicles [8]. Additionally, the HWFET cycle simulates freeway travel and makes no complete stops until the end of the test cycle; details are shown in Table 2 [9]. The tests on each type of fuels were triplicate and replacement of the fuel filter during fuel changing was always accomplished.

Table 2. U.S. EPA HWFET cycle details [9]

Driving schedule attributes	HWFET
Top speed	96.6 km/h
Average speed	77.7 km/h
Maximum acceleration	5.1 km/h/s
Distance covered	16.6 km
Time elapsed	12.75 min
Individual full stop	0
Percentage of time stopped	0

2.3 Test fuels

In this research work, three types of fuel were used in the test: conventional diesel (B7), biodiesel (B10) and premium diesel (DHi). Selected properties of the test fuels are enumerated in Table 3.

IOP Conf. Series: Materials Science and Engineering 1137 (2021) 012002 doi:10.1088/1757-899X/1137/1/012002

Table 3. Fuel properties

Properties	Premium diesel (DHi)	Conventional diesel (B7)	Biodiesel (B10)
Density at 15°C (g/mL)	0.8217	0.8328	0.8339
Viscosity at 40°C (mm^2/s)	3.475	2.882	2.912
Cetane index	61	56	56.2
Flash point (°C)	97	64	62
Water and sediment (% v/v)	0.1	0.02	0.02
Lower heating value (MJ/kg)	45.45	45.02	44.72

3. Results and Discussion

3.1 Vehicle performance

Figure 2 shows the profiles of actual vehicle speed registered during the test of HWFET cycle with different fuels. These results were controlled to be similar with the cycle pattern for all fuels.

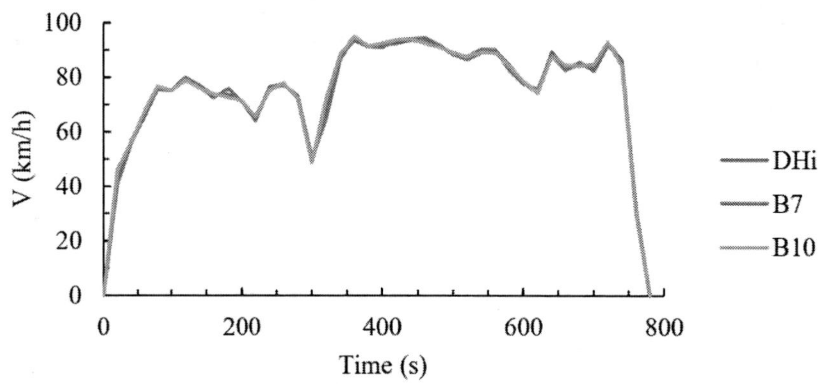

Figure 2. Actual vehicle speed under HWFET cycle

3.2 Exhaust emissions

The exhaust gas emissions described in this section are associated with HC, CO, and NO.

3.2.1 Total unburned hydrocarbon

Figure 3 shows time evolution of HC concentration during HWFET with the three fuels tested. It is clearly shown that the combustion for all the three fuels in this vehicle engine emitted very low HC concentration throughout the cycle. The results showed that the HC concentration characteristics of the are similar for all fuels. The maximum HC concentrations for DHi, B7 and B10 are by 12, 9 and 11 ppm, respectively. On mass basis in the unit of g/km, the accumulative HC emission during HWFET cycle from the vehicle running on the test fuels are in the order of B10<B7<DHi. Compared to DHi baseline, the combustion of B7 and B10 reduced the HC emissions by 25.3% and 28.3%, respectively. The presence of oxygen in the biodiesel molecule promotes a more complete combustion and, consequently, lower HC emissions [3].

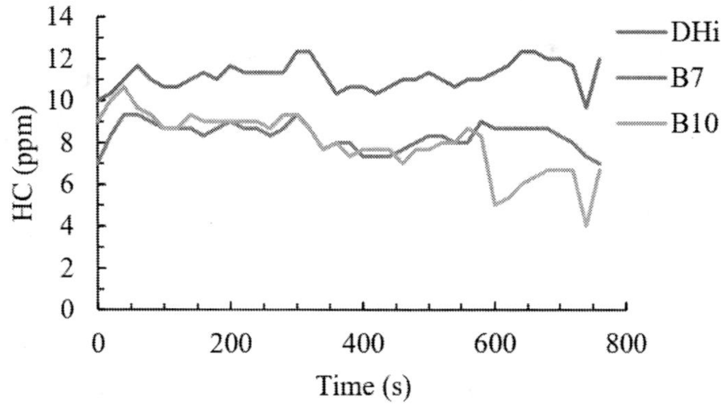

Figure 3. Time evolution of HC concentration during HWFET with the three fuels tested

3.2.2 Carbon monoxide

Carbon monoxide is an incomplete combustion exhaust gas and also depends on other factors such as speed, injection pressure and injection timing, etc.; the key factor is mainly associated with the rich mixture between fuel and air [10]. Figure 4 depicts the time evolution of CO concentration during HWFET with the three fuels tested. In term of CO concentration, the experimental results show similar trends of HC emissions. There are three common spikes of the CO emitted at 190, 310, and 650 s where the engine was suddenly accelerated to higher velocity. At these points, the engine has to be put in more fuel that generates greater fuel-to-air ratios. The maximum CO concentrations for DHi, B7 and B10 are by 0.836, 0.632 and 0.239 %, respectively. On mass basis in the unit of g/km, the accumulative CO emissions during HWFET cycle from the vehicle running on the test fuels are in the order of B10<B7<DHi, similarly to that of the HC release. When comparing to the DHi baseline, the combustion of B7 and B10 reduced the CO emission by 43.3% and 64.0%, respectively. Fueling the engine with B10 at high oxygen content may shift the fuel and air combination toward leaner mixture, and hence lower CO emission.

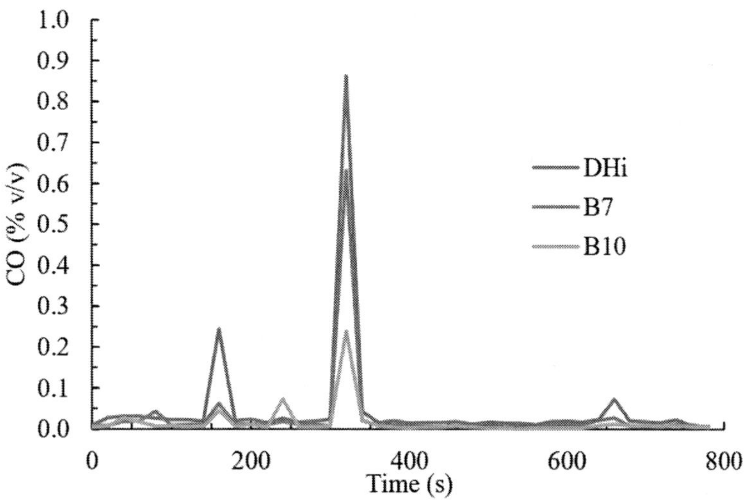

Figure 4. Time evolution of CO concentration during HWFET with the three fuels tested

3.2.3 Nitric oxide

Typically, nitric oxide is formed by the in-cylinder combustion at high temperature and oxygen concentration, following Zeldovich mechanism [10]. Figure 5 shows time evolution of NO concentration during HWFET with the three fuels tested that shows to affect the NO concentration.

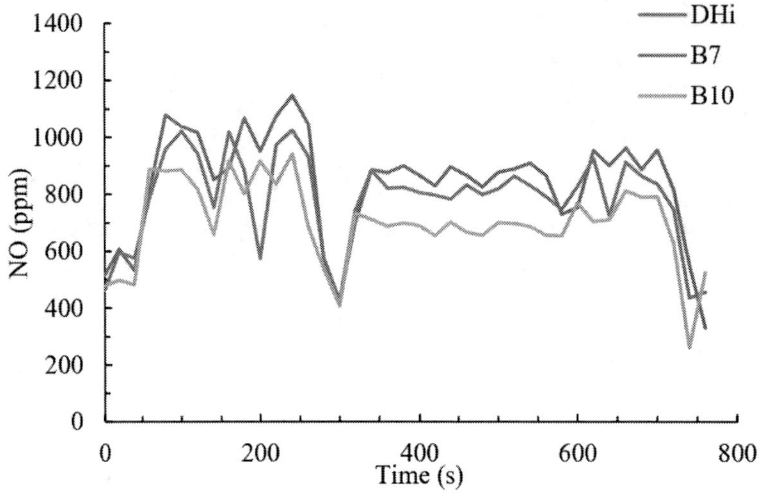

Figure 5. Time evolution of NO concentration during HWFET with the three fuels tested

In Figure 5, the trends of the results of NO concentrations are similar but differ in amount. On mass basis in the unit of g/km, the accumulative NO emission during HWFET cycle from the vehicle running on the test fuels are in the order of B10<B7<DHi, similarly to those of CO and HC emissions. When comparing to the DHi baseline, the combustion of B7 and B10 reduced the NO emission by 6.0% and 16.5%, respectively. Due to higher cetane number of DHi, the ignition delay is shorter compared to the others and, therefore the fraction of injected fuel is burnt in pre-mixed combustion regime. The pre-mixed combustion is associated with higher pressures, pressure gradients and temperatures in the chamber; all of which increase NO formation [11].

3.3. Smoke opacity and particulate matter

Figure 6 shows time evolution of smoke opacity during HWFET for the three fuels tested. There are three common spikes of the smoke emitted at first acceleration, 190, 310, and 650 s where the engine was suddenly accelerated to higher velocity, presumably high load occurrence. At these points, the engine has been supplied by lots of fuel amounts; these fuels may not be completely burned within limited time. The maximum smoke opacity for DHi, B7 and B10 are by 0.73, 0.67 and 0.30 m^{-1}, respectively. On mass basis in the unit of g/km after conversion from the unit of m^{-1}, the accumulative PM emission during HWFET cycle from the vehicle running on the test fuels are in the order of B10<B7<DHi, similarly to those of previous gaseous emissions. Compared to the DHi baseline, the combustion of B7 and B10 lowered the PM emission by 0.38% and 51.0%, respectively. PM formation, caused by high temperature decomposition, mainly takes place in the fuel-rich zone at high temperature and pressure, especially within the core region of each fuel spray. Increasing oxygen concentration in the biodiesel-diesel blended mixture enhances a soot oxidation process. This reduces local fuel-rich regions and limits soot nucleation early in the formation process, thus reducing PM emissions and smoke opacity [12].

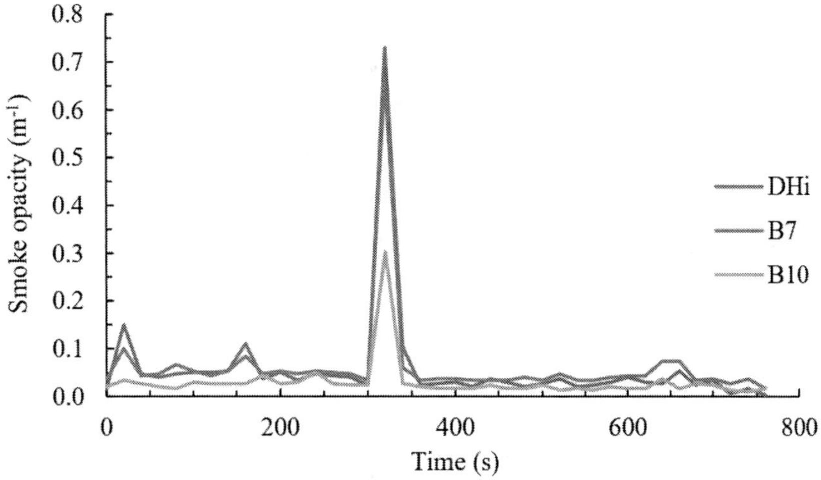

Figure 6. Time evolution of smoke opacity during HWFET with the three fuels tested

Figure 7 evidently shows the emissions of particulate matter and nitric oxide from the combustion of DHi, B7 and B10 during HWFET cycle. The B10 combustion confirms the lower PM level than B7 and DHi, respectively. The use of biodiesel for this case of engine with the HWFET cycle seems to simultaneously reduce PM and NO emissions.

Figure 8 shows particle number concentration-size distribution during HWFET cycle. From the combustion of all the fuels tested, the particles show a greater concentration in smaller size for all fuels but differ in concentration amounts. The total particle number concentrations for DHi, B7 and B10 were by 17,552, 16,901 and 38,085 cm^{-3} respectively. The count mean diameters for DHi, B7 and B10 were by 380, 508 and 430 nm, respectively. In summary, the greater amount of biodiesel blended fuel such as B10 results in higher particle concentration with smaller size compared to DHi.

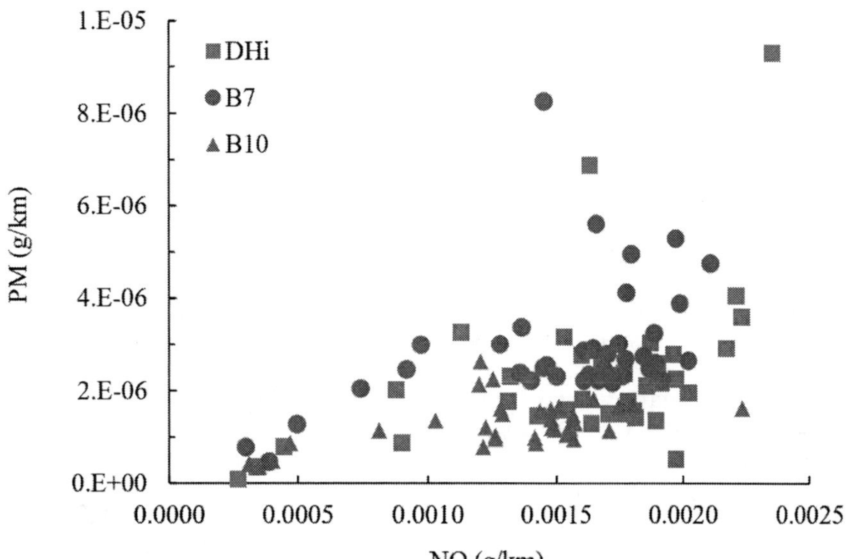

Figure 7. Particulate matter and nitric oxide emissions

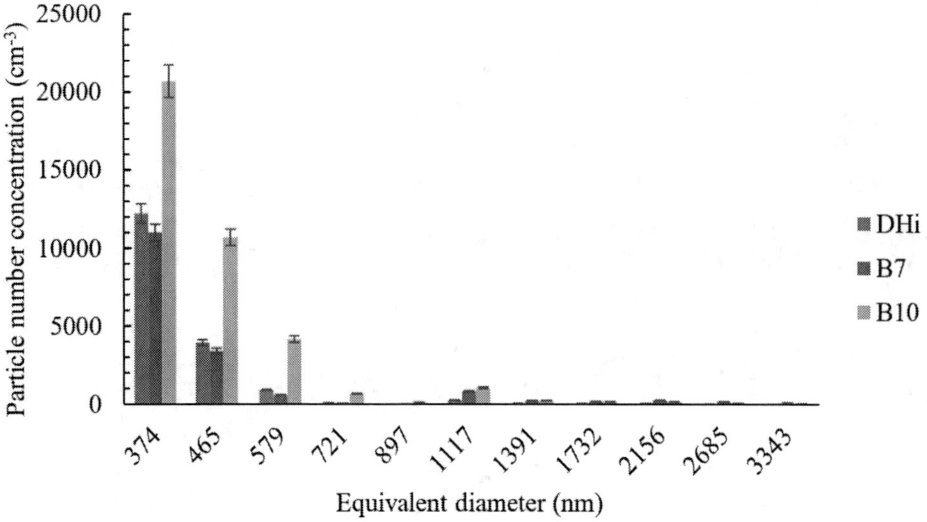

Figure 8. Particle number concentration-size distribution during HWFET

4. Conclusion

This work studies and compares the release of exhaust gases and particulate matter from diesel vehicle tailpipe fueled with B7, B10 and DHi during HWFET cycle. HC, CO, NO and PM were reduced with alternative biodiesel blended fuels, prominently for B10. Although oxygen content in biodiesel partly presented in the blends promotes more complete combustion by reducing these emissions, the combustion temperature dominates NO reduction among oxidizer-rich mixture. Compared to DHi baseline on mass basis, the combustion of B7 and B10 reduced all emissions: by 25.3% and 28.3% for HC, by 43.3% and 64.0% for CO, by 6.0% and 16.5% for NO, and by 0.38% and 51.0% for PM, respectively. The use of biodiesel for this engine running on the HWFET cycle can reduce PM and NO simultaneously. The total particle number concentrations for all fuels are in the range of 16,901 to 38,085 cm^{-3} while the count mean diameters are in the range of 380 to 508 nm. The greater amount of biodiesel blended fuel such as B10 results in higher particle concentration with smaller size compared to DHi.

Acknowledgments

Kasetsart University Sriracha Campus is acknowledged for the research funding (Contract No. 4/2557).

References

[1] Rivas I Beddows D C S Amato F Green D C Järvi L Hueglin C Reche C Timonen H Fuller G W Niemi J V Pérez N Aurela M Hopke P K Alastuey A Kulmala M Harrison R M Querol X and Kelly F J 2020 *Environment International* **135** 105345

[2] Kim D Chen Z Zhou L F and Huang S X 2018 *Chronic Diseases and Translational Medicine* **4**(2) 75-94

[3] Armas O García-Contreras R and Ramos Á 2013 *Applied Energy* **107** 183-190

[4] Macor A Avella F Faedo D 2011 *Applied Energy* **88** 4989-5001

[5] Lapuerta M Rodríguez-Fernández J and García-Contreras R 2015 *Energy* **93** 579-568

[6] Gao J Chen H Li Y Chen J Zhang Y Dave K and Huang Y 2019 *Energy Conversion and Management* **196** 605-613

[7] AVL 2005 *Smoke value measurement with the filter-paper-method* (Graz: AVL List GmbH)

[8] U.S. EPA 2018 *Code of Federation Regulations, Title 40 Protection of Environment, Part 600, Subpart B* (Washington: U.S. Government Publishing Office)

[9] Mersky A C and Samaras C 2016 *Transportation Research Part* C **65** 31-48

[10] Heywood J B 1988 *Internal Combustion Engine Fundamentals* (Singapore: McGraw-Hill)
[11] Gill S S Tsolakis A Dearn K D and Rodríguez-Fernández J 2011 *Progress in Energy and Combustion Science* **37** 503-523
[12] Giakoumis E G Rakopoulos C D Dimaratos A M and Rakopoulos D C 2012 *Progress in Energy and Combustion Science* **38** 691-715

Thermal performance of heat pump dryer using R32 as refrigerant.

P. Ponwapee, P. Somsila, U. Teeboonma, A. Namkhat, and S. Pumchumpol*

Department of Mechanical Engineering, Faculty of Engineering,
Ubon Ratchathani University, Ubon Ratchathani, Thailand 34190

* Corresponding Author: songsupa.p@ubu.ac.th

Abstract
The objective of this research was to study the performance of heat pump dryer using refrigerant R32 and ginger was selected as testing material. The experiments were conducted under conditions; drying temperature of 45, 50, and 55°C and air velocity of 1.0, 1.5 and 2.0 ms-1. Drying rate, the specific moisture extraction rate, the specific energy consumption, and the coefficient of performance of heat pump were tested. The results revealed that the increment of drying temperature or air velocity resulted in increment of drying rate, power of heat pump dryer, and specific moisture extraction rate whilst specific energy consumption decreased. Additionally, it was revealed that drying temperature or air velocity did not affect the coefficient of performance of the heat pump.

Keywords: Coefficient, Drying, Heat pump

1. Introduction

Advantages of reducing moisture in agricultural products are weight reduction, extended shelf life and convenience for transportation. From the past to the present, there have been many kinds of drying technologies such as drying in the sun, superheated steam drying, infrared drying, vacuum drying, heat pump drying and etc. It is well known that the drying process has a relatively high energy consumption [1]. Presently, most researchers found that heat pump drying technology has highly efficient energy consumption and high product quality [2-4]. However, refrigerants of commercial heat pump dryer are categorized in high global warming potential, GPW, such as R22, R134a, R407C and R410a are having GPW 1,810, 1,430, 1,774 and 2,090, respectively [5]. Recently, refrigerant R32 is a lower GWP refrigerant, GWP is 675, which is an alternative refrigerant that is used as a replacement for high GWP refrigerants. Many researchers studied alternative refrigerants for use in heat pump dryers. In et al. [6] were to compare the study of refrigerants R410a, R32 and L41b for heat pump dryers. They found that the average coefficient of performance, COP, of R32 as a refrigerant for heat pump dryers is higher than R410a at about 5%. Abdullah et al. [7] were to study and test an alternative refrigerant for replacing of refrigerant R410a and they found that a suitable refrigerant for replacing of refrigerant R410a is refrigerant R32. Akhilesh et al. [8] were to develop mathematical model to study performance of heat pump dryers for selecting alternative refrigerants which replace refrigerant R134a. Refrigerants R290, R600a, R32, R152a and R123yf were used in the mathematical model of this study. Results of the study revealed that refrigerant R32 for heat pump dryers had the highest dehydration removal rate.

Dehydration removal rate of heat pump using refrigerant R32 was higher than R134a at 8%, however, COP of heat pump dryer using R152a as refrigerant is the highest. However, past research on performance of heat pump dryers using R32 as refrigerant were still relatively few. Therefore, the objective of this research is to study and test the performance of a heat pump dryer using R32 as refrigerant.

2. Methodology

2.1 Experimental Equipment

This research aimed to develop and test of heat pump dryer using R32 as a refrigerant. The maximum refrigerating effect of the heat pump dryer was 2.64 kW. The main equipment of the heat pump dryer consists of 0.75 kW compressor, 0.05 kW ventilation fan, air damper (3), evaporator (4), external blower, internal condenser (5) and external condenser (6). Schematic diagram of the heat pump dryer shown at Figure 1.

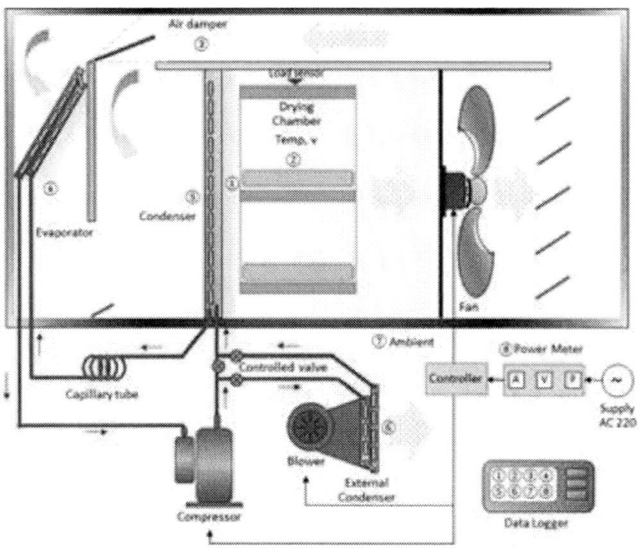

Figure 1. Schematic diagram of heat pump dryer.

The heat pump dryer operation was started by blower (6) discharged drying air throughout the condenser (3). Drying air was heated by refrigerant in an internal condenser. At the meantime, refrigerant at the condenser outlet would be condensed to saturated liquid. Therefore, drying air flowed through the drying chamber for transferring heat to products. Suddenly, products were heated up when receiving heat from hot drying air and water in products were evaporated to drying air at the drying chamber outlet. Drying air at drying chamber outlet were humidified, whereas, drying air temperature at drying chamber outlet were decreased. Some of the humid drying air was recirculated throughout the evaporator. Moisture in humid air was condensed by evaporator. Refrigerant in the evaporator would be heated to vapor. The refrigerant vapor at the evaporator outlet was sucked and discharged to the condenser by compressor. Meanwhile, dried air at the evaporator outlet flowed and heated at the condenser before entering the drying chamber. The process of heating and dehumidifying the drying air takes place on a cyclic until achieving required products moisture level. Drying air temperature would be controlled by an external condenser equipped at the outside drying chamber and refrigerant cycle would be controlled by 3 solenoid valves. Heating process of drying air by heat pump cycle, solenoid valve was opened to control the refrigerant flow into the internal condenser, whereas, solenoid valve

IOP Conf. Series: Materials Science and Engineering 1137 (2021) 012003 doi:10.1088/1757-899X/1137/1/012003

was closed and refrigerant from the compressor would pass to the external condenser when the temperature inside the drying chamber reaches a certain level.

2.2 Experimental setup

Refrigerant R32 was used for testing of heat pump dryer. The drying air temperature was 45, 50 and 55°C and drying air velocity were 1.0, 1.5 and 2.0 ms-1, respectively. The ginger was dried for a total 4 hours. Initial moisture and weight of ginger were controlled. Temperature, pressure of heat pump cycle, the weight of the ginger, electrical power and energy were recorded by the data recorders. The initial and final moisture of ginger were calculated by AOAC standard. Drying air temperature and velocity had been controlled at starting stage. The drying process was operated until equilibrium stage by investigated on every recorded data. At equilibrium stage, gingers were put in drying chamber and the experiment was conducted continuously for 4 hours.

2.3 Data analysis

The experimental data are used to analyze various parameters. By the last condition used in comparative analysis such as moisture ratio; MR, drying rate; DR, specific moisture extraction rate; SMER, specific energy consumption; SEC and coefficient of performance of heat pump; COPh. The result was calculated by the following equations;

$$MR = \frac{M_t}{M_{in}} \quad (1)$$

$$DR = \frac{m_i - m_f}{T} \quad (2)$$

$$SMER = \frac{m_i - m_f}{E} \quad (3)$$

$$SEC = \frac{3.6E}{m_i - m_f} \quad (4)$$

$$COP_h = \frac{Q_c}{P_c} \quad (5)$$

Where;

M_{in} = Initial moisture (%dry basis)
M_t = Moisture (%dry basis)
M_i = Initial weight (kg)
M_f = Final weight (kg)
E = Electrical energy consumption (kWh)
Q_c = Heat at condenser (kW)
P_c = Electrical power of compressor (kW)

3. Result and Discussion

The experiments on drying ginger by heat pump dryer using R32 refrigerant were controlled drying air temperature of 45, 50 and 55°C and air velocity of 1.0, 1.5 and 2.0 ms-1, respectively. Results are shown as follows;

Figure 2 shows changing in moisture ratio at each experimental condition. Considering the conditions from the humidity ratio of 0.2 was found that either increasing the drying temperature or drying air velocity would affect the rate of reduction of the moisture content increased. In other words, the drying rate increased as shown in Figure 3. The factor contributing when the drying air temperature increases, it increases the heat transfer. As the temperature difference of the hot air and the product increases the product received more heat, the result being that the evaporation of water from the product increased accordingly. Therefore, drying air temperature increases would increase the drying rate. Similarly, increasing the drying air velocity would result in greater heat transfer to the product. Due to heat transfer is primarily a form of convection transfer. For this reason, when the drying air velocity increases, resulting in a heat transfer coefficient increased. Therefore, the product receives more heat. As a result, the drying rate increases as well. Figure 3, if considered under constant drying air temperature conditions, it was found that when the drying air velocity increases approximately 50% would affect in an average increase of drying rate of approximately 10%. Similarly, if the drying air velocity remains to be found that when the drying air temperature increased by an average of 10%, the drying rate increased by 10% as well. Therefore, the drying air temperature had a greater effect on the drying rate than the drying air velocity.

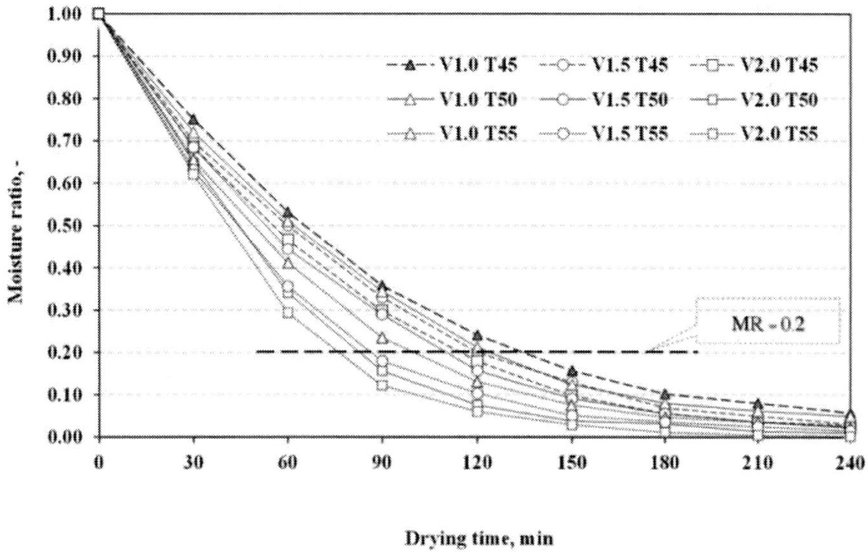

Figure 2. Moisture ratio at each of the experimental condition.

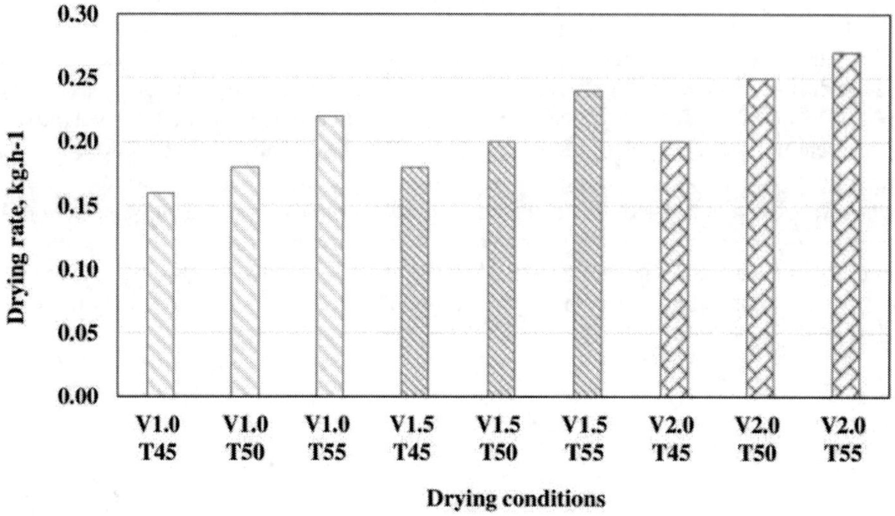

Figure 3. Drying rate at each of the experimental condition under moisture ratio 0.2

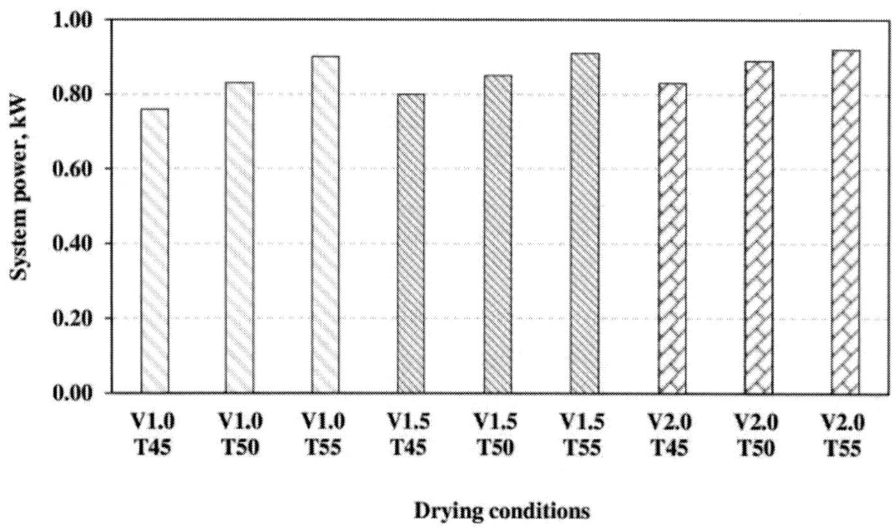

Figure 4. Electrical power at each of the experimental condition.

Figure 4 shows the electrical power of the heat pump dryer at each of the experimental condition. The result revealed that an increase in either drying air velocity or drying air temperature would increase the system's power. Due to the increase of drying air temperature, the compressor had to pressurize the refrigerant to a sufficiently high pressure and corresponding to the saturated vapor temperature to achieve the desired drying temperature. At meantime, it would result in increased heat load on the evaporator. The system had to inject a higher amount of refrigerant. Therefore, the system required more electrical power. In addition, Figure 4 indicated that a 10% increase in drying air temperature results in a 5-7% increase in electrical power, while a 50% increase in drying air velocity results in a 3- 5%, which was found that drying air temperature had a greater effect on electrical power of the heat pump dryer than drying air velocity.

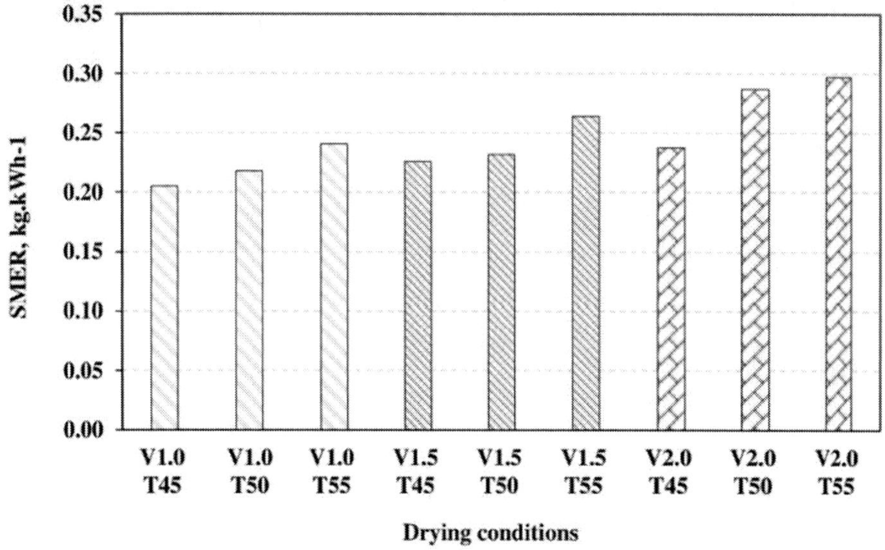

Figure 5. Specific moisture extraction rate at each of the experimental condition

Figure 5 shows the specific moisture extraction rate of the heat pump dryer at each of the experimental condition, an increase in either drying air velocity or drying air temperature would result in an increased specific moisture extraction rate, resulting in a more efficient energy consumption consistent with the specific energy consumption presented in Figure 6. The result in Figure 6 is similar to that of Figure 5, which could be explained by Figures 3 and 4 is that the drying increment was greater than the electrical energy increase. As a result, the specific moisture extraction rate would increase while the energy consumption decreased.

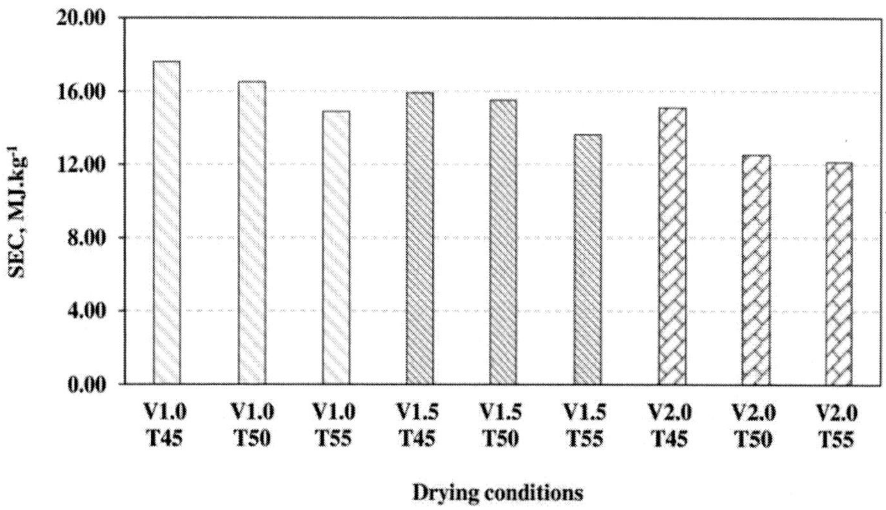

Figure 6. Specific energy consumption at each of experimental condition

Figure 7 shown coefficients of performance the heat pump dryer at each of the experimental conditions. Result found that the change in coefficient of performance of the heat pump dryer had little change. According to the definition, coefficient of performance of heat pump is equal to the total heat reject from condenser divided by power input. Both of these energies vary in approximately the same ratio. Therefore, drying air temperature and drying air velocity had very little effect on the coefficient of performance of the heat pump dryer.

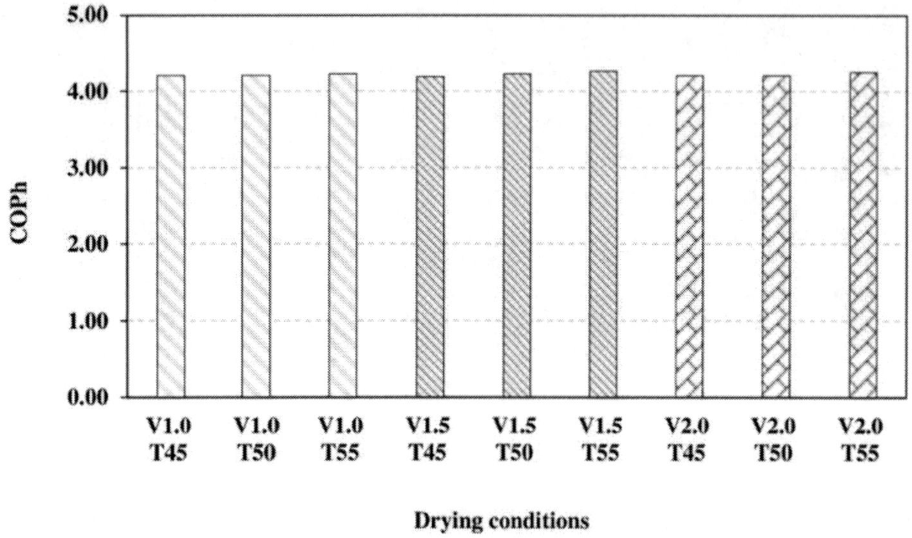

Figure 7. Coefficient of performance at each of experimental condition.

4. Conclusion

Refrigerant R32 was used as the refrigerant for the heat pump dryer and ginger was selected for this experiment. The result revealed that either higher drying air temperature or higher drying air velocity was effective in increase in the drying rate, and the electrical power of the compressor and specific moisture extraction rate, whereas, specific energy consumption was decreased. Moreover, either drying air temperature or drying air velocity had a little effect on the coefficient of performance of the heat pump dryer.

Acknowledgments

The authors would like to express their sincere thanks to Faculty of Engineering, Ubon Ratchathani University for financial support.

References

[1] Huelsz G Urbiola-Soto L López-Alquicira F Rechtman R and Hernández-Cruz G 2013 *Drying Technology* **31** 576-586
[2] Erbay Z and Hepbasli A 2017*Energy* **127** 502-515
[3] Niklas B Thomas A Wilhelm T Marcos B Andreas M and Jürgen K 2019 *Applied Thermal Engineering* **161** 1-14
[4] Te G W Butterfield A Caldwell D Crook A and Winkleman A 2017 *Applied Thermal Engineering* **124** 170-17
[5] Jing W and Zhou M.W 2020 *Applied Thermal Engineering* **174** 1-11
[6] In S Cho K Lim B Kim H and Youn B 2014 *Applied Thermal Engineering* **72** 315-322

[7] Abdullah A Radia E Yunho H Vikrant A and Reinhard R 2015 *International Journal of Refrigeration* **60** 106-117
[8] Akhilesh S Jahar S and Rashmi R S 2019 *Food and Bioproducts Processing* **117** 1-13

Physical and chemical characteristics of carbonized corncob through hydrothermal and pyrolysis conversion

Nongnoot Srilek[1,*] **and Pruk Aggarangsi**[1,2]

[1]Department of Mechanical Engineering, Faculty of Engineering, Chiang Mai University, Chiang Mai Thailand 50200
[2]Energy Research and Department Institute-Nakornping, Chiang Mai University, Chiang Mai Thailand 50100

* Corresponding Author: k_nongnoot@hotmail.com

Abstract
Thailand and south-east Asia have been struggled with agriculture biomass residues management for decades where open burning causes severe air pollution affecting millions of people every year. Thermal carbonization is one of the most cost-effective and environmentally friendly techniques which converted the low value residues such as straw, corn stovers and corncob into a more valuable material such as solid biofuel, chemical adsorbent feed stocks. This study aims to explore crucial characteristics of carbonized corncob including fibre composition, BET surface area and FT-IR spectrums. The characteristics can be used to identify the potential of the material for further surface enhancement or surface activation to produce the bio-based activated carbon. The carbonization process in this work includes mild temperature hydrothermal technique and high-temperature pyrolysis process. The hydrothermal carbonization temperature is set at 250°C in a pressurized reactor where a pyrolysis is operated at 480°C and 380°C at atmospheric pressure. The BET surface area of "hydrochared" corncob derived from hydrothermal is 11.53 compared with 16.13 m^2 g^{-1} of the raw materials. The pyrolized biochar at 480°C and 380°C yields surface area of 7.66 and 6.12 m^2 g^{-1}. The oxygenated functional groups on char surface and BET surface area are also compared to provide baseline for carbon activation.

Keywords: Hydrothermal carbonization, Hydrochar, Corncob, Bio-based activated carbon, Pyrolysis, Biochar

1. Introduction

A pollution is a major problem in Thailand. It has a huge effect on a living quality, long term health care and tourism economy, and has been recognized as the national agenda. The open burning of agricultural wastes, especially rice and corn residues, is a leading cause of pollution in a particular place, especially in the northern part of Thailand. Corn or maize are the annul crops mainly producing the animal feeds. Corn faming and harvesting take 3-4 months for each period. The first period or rainy season are cultivated during March to June and August to October. The second period or summer are cultivated during November to February. The conversion of low-value residue biomass has been a tremendous challenge for researchers around the world to effective management. It also converts biomass residues such as corn, an immense amount, to higher value materials. Corn residues are lignocellulose materials which major constituents including cellulose, hemicellulose, lignin and extracted material. These fibres are main compositions and contents can be largely different according to the type and age of the plant, atmosphere and geographic. A cellulose, $(C_6H_{10}O_5)_n$, is a linear homopolymer that is composed of glucose monomer ($C_6H_{12}O_6$). Hemicellulose, $(C_5H_{10}O_5)_n$, is a copolymer that includes various monomer units. The lower degree polymerization of hemicellulose has an effect on a smaller crystalline structure. As such, it is easier to decompose than cellulose. Lignin, on the other hand, is the polyphenol polymer with a three-dimensional mesh of aromatic unit including many hydroxyl and polar groups. Thus, it is insoluble in any solvent [1], which leads to various difficulties limiting options of conversions. The chemical constituents of cellulose are 43.2 % C, 5.7 % H and 51.1 % O and 66.0 % C, 6.0 % H and 26.9 % O of lignin composition [2]. Hemicellulose, cellulose and lignin constituents in lignocellulose material can be decomposed at 180-290, 240-350 and 280-500°C respectively [3].

Pyrolysis process is a non-combustion thermochemical conversion that generally can be classified into slow, fast and flash process (see Table 1). The temperature ranges of pyrolysis depend on the expected products. Solid carbonized products as residues from pyrolysis are often called biochar or char derived from the biological feedstock. Solid carbonized products as residues from pyrolysis are often called biochar or char derived from biological feedstock. Biochar consists of carbon that is molar hydrogen/molar carbon (H/C) ratio of biochar. It was found that it is related to the degree of charring. We can also see that it contains a maximum H/C of the molar ratio of 0.7 and carbon content of at least 10% are fundamental specification [4].

Table 1. Reaction temperature, heating rate and the holding time for thermochemical conversion.

Thermochemical conversion	Temperature (°C)	Heating rate (°C/min)	Holding time	Ref.
Slow pyrolysis	400 – 600	10 – 30° C/minute	minutes -days	[5],[6]
Fast pyrolysis	400 – 600	> 100 °C/minute	~ 1 second	[5],[6]
Hydrothermal carbonization	180 – 250	5 – 10 °C/minute	1 –12 hours	[5]

Similar to pyrolysis, a hydrothermal conversion technique is also called a non-combustion thermochemical conversion. This technique is similar to the pyrolysis, however, there is a difference in terms of the process. Its process is conducted under a pressure and typically has a high humidity content. The hydrothermal conversion can be divided into three types: carbonization, liquefaction and gasification [7] (see Figure 1). The HTC is the thermochemical conversion of biomass and its products are aimed for carbon-rich solid material. An HTC reaction process can normally be conducted in a pressurized vessel reactor with a liquid medium which can be vaporized into steam. Thus, it can lead to a decrease in the reaction pressure and it usually consists of the external heating. For example, a condition is placed where the hydrothermal temperature is 180 - 240°C and the pressure is at 1 - 3 MPa (10 - 30 bars). If the generated pressure is contained and controlled, water still maintains in liquid form. It is the sub-critical water according to water-steam thermodynamic properties [8]. The hydrothermal products is often called hydrochar or hydrothermal char which underlined for wet process. The four major fibre constituent including cellulose, hemicellulose, lignin and extracted material were extracted at different condition of hot compressed water through hydrolysis reaction. Generally, hydrochar has a higher H/C molar ratio and lower aromatic structure than the biochar, and it has less fused aromatic structures [5]. The hydrothermal refinery can also develop more oxygen-functional groups (OFGs) on the char's surface [9]. Likewise, typical the oxygen/carbon and hydrogen/carbon molar ratio of the hydrothermal char is considered higher when compared to the pyrolytic char. Nevertheless, the most previous reports in hydrothermal carbonization extend the potential of corncob as solid fuel application [10-11].

Activated carbon is considered as a highly porous substance which the carbonized char has undergone the activation so that it should be treated chemically or physically to develop pores structure or interconnected pores inside the carbon structure. The commonly used feedstocks for traditional activated carbon production include wood, coal, petroleum residues, peat, lignite and polymers. Those kinds of products are very expensive and non-renewable [12]. Activated carbon is well known for its adsorption properties and is used in many applications and has a relate high market value. Hence, the bio-based activated carbon is always expected target materials produced from renewable material. The in-depth characteristics can have an increase in the correct upgrading. Therefore, the purpose of this study aims to explore the crucial characteristics including physical and chemical of carbonized corncob derived from hydrothermal and pyrolysis process that can be used to identify the potential of the renewable material for further surface enhancement.

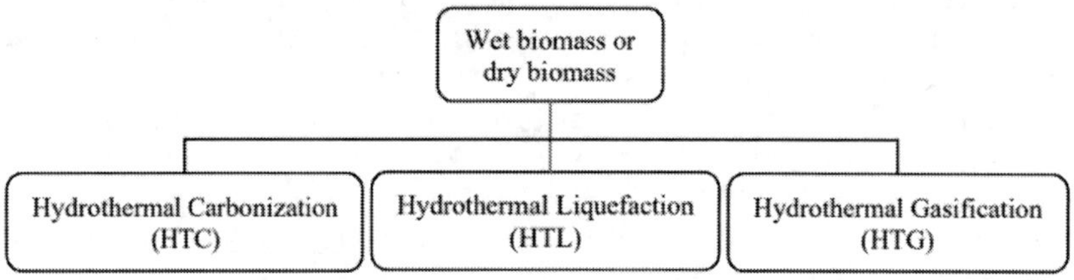

Figure 1. Hydrothermal classification [7]

2. Materials and methods

2.1 Materials

Corncob, is one of the immense amounts of agricultural wastes in Northern Thailand. It can be classified into 2 types as corn for consumer purpose and maize for animal feed purpose which has more amount than sweet corn. Corncob is used as feedstocks (see Figure 2). The input material is processed for size reduction and sieved with 2 - 0.84 mm. The prepared material is then dispersed in deionized water (DI) for the hydrothermal process.

Figure 2. Corncob

2.2 Experiment

The slow pyrolysis conversion includes laboratory and mobile plant scale and yields the biochar. Laboratory scale pyrolysis is normally heated at 10°C/minute, 15 minutes of holding time and 1.5 l/minute of the N_2 flow rate of the rate (see Thermal System Research Laboratory, Chiang Mai University for more detail). The reaction temperature is performed at 480°C [13] and the previous study reported in the intersection of TGA and DTGA plot of corn stover at 420-470°C [14]. The mobile pyrolysis plant is operated in 2°C/minute with 190 minutes of holding time, as shown in Figure 3. As can be seen from Figure 4, the hydrochar is produced from 10 litres pressurized fixed vessel reactor at 250°C with the suspension of sieved corncob and deionized water (1:20 w/w). Consistent with the literature, this research found that the higher temperature tended to decrease a specific surface and produced a maximum S_{BET} of 8.033 $m^2 g^{-1}$ at 250°C, 20 min of residence time [8]. The holding time and heating rates are 100 minutes and 5 °C/minute respectively. The hydrochar was rinsed by DI and dried in oven at 105°C, 24 hours. The processed chars are collected tightly for in the characterising step.

2.3 Physical and Chemical Characterization

Fibre analysis, Brunauer-Emmett-Teller surface area (BET surface area, S_{BET}) and Fourier Transform Infrared (FT-IR) analysis are characterized for feedstock and carbonized feedstock. The fibre analysis and BET surface area are defined as physical characters. Analysing of fibre using a detergent method or is so-called the Neutral Detergent Fibre and the Acid Detergent Fibre and Acid Detergent Lignin were applied. Regarding the surface area analyzer, the model quanta chrome with model autosorb 1 MP was also employed to determine the BET surface area (S_{BET}). The S_{BET} determination was operated with N_2 gas as adsorbate at 0.05-0.35 relative vapor pressure (P/P_0) with the temperature of the outgas at 150°C, and the outgas time spent was four hours. The operating time was at 46.8-57.9 minutes. The FT-IR is a crucial qualitative chemical characterization that has an estimation OFGs on the surface material.

Figure 3. Mobile Pyrolysis Plant (Supported by ERDI-CMU).

Figure 4. Hydrothermal fixed vessel reactor (Supported by ERDI-CMU).

3. Results and discussion

The four samples of Corncob with 2 - 0.84 mm, Corncob, Biochar-480" (pyrolyzed from the laboratory scale reactor), Biochar-380 (pyrolyzed from the Mobile Pyrolysis Plant) and "Hydrochar-250" (produced through hydrothermal conversion). They were then tested.

The physical appearance of Hydrochar-250 is slightly dark brown comparing to the corncob sample with a yellowish colour. The Biochar-480 and Biochar-380 are black as provided in Figure 5.

(a) (b) (c) (d)

Figure 5. (a) Corncob (b) Hydrochar-250, hydrothermal conversion at 250°C
(c) Biochar-480, laboratory pyrolysis reactor at 480°C
(d) Biochar-380, Mobile pyrolysis Plant at 380°C

3.1 Fibre analysis

The determination of fibre constituents results in % of NDF, ADF and ADL. The interpretation of these compositions is the % of cellulose, hemicellulose and lignin as illustrated in Table 2.

Table 2. Fibre constituents

Sample	%			%		
	NDF[a]	ADF[b]	ADL[c]	Cellulose	Hemicellulose	Lignin
Corncob	85.09	44.30	7.22	37.15	40.73	7.22
Biochar-480	77.65	74.95	89.36	N/A	2.69	89.36
Biochar-380	80.33	74.89	83.48	N/A	2.51	78.78
Hydrochar-250	91.91	56.38	10.26	46.12	35.23	10.26

[a]NDF defined Neutral Detergent Fibre (cellulose, hemicellulose, lignin)
[b]ADF defined Acid Detergent Fibre (cellulose, lignin)
[c]ADL defined Acid Detergent Lignin (lignin)

The current study found that the fibre constituents of feedstocks corncob are cellulose (37.15%), hemicellulose (40.73%) and lignin (7.22%) respectively. The composition of holocellulose includes cellulose and hemicellulose as a majority. Regarding fibre, the results indicate that the content of the hemicellulose is close to the cellulose. Biochar-480 and Bichar-380 that produced from pyrolysis have 89.36% and 78.78% of the lignin. However, both biochar differs in terms of the reaction temperature and can result in the retain lignin contents. The absolute decomposition of cellulose was found to be pyrolysis process of Biochar-480 and Biochar-380. The highest lignin with 89.36% produced from higher temperature pyrolysis process at 480°C is fully carbonized. The results of the detected hemicellulose through both pyrolysis process did not show a significant increase when compared to the lignin of Hydrochar-250. It shows a slight increase compared to the origin of raw material. Nevertheless, the holocellulose of Hydrochar-250 still has a high portion. The hydrothermal conversion, cellulose significantly started a hydrolysis at 200°C while hemicellulose fraction decomposed about 180°C and lignin that thermally more stable started a degradation at 280°C [15]. The study also suggested that the decomposition temperature should be under a hydrothermal of three components following the previous study illustrated the strong adaptation and optimization requirement. For this 250°C the hydrothermal temperature, hemicellulose constituent decrease compared to origin of raw material during a slight increase of the lignin slightly increase. Although the findings from the fibre analysis indicate the better degradation for holocellulose (cellulose and hemicellulose), this study set out with the aim of the degradation of hemicellulose component to emerge from the analysis.

3.2 BET surface area (S_{BET}*)*

S_{BET} analysis plays a crucial role in physical adsorption aspect of gases and vapours on porous carbon shown via actual adsorption isotherm [16]. The BET analysis directly measures surface area and pore size distribution. The S_{BET} of both higher temperature pyrolysis (Biochar-480 and Biochar-380) are lower than the hydrochar, Hydrochar-250, while higher in BET C-constant (see Table 3).

Table 3. S_{BET}, BET C-constant and fibre composition.

Sample	S_{BET} (m^2 g^{-1})	BET C-constant	Cellulose (%)	Hemicellulose (%)	Lignin (%)
Corncob	16.13	2.49	37.15	40.73	7.22
Biochar-480	7.66	6.46	N/A	2.69	89.36
Biochar-380	6.12	8.11	N/A	2.51	78.78
Hydrochar-250	11.53	3.98	46.12	35.23	10.26

The hydrothermal carbonized of Palm Empty fruit bunches (EFB) developed S_{BET} of 6.079, 8.033 and 2.043 m^2 g^{-1} with 150, 250 and 350°C of reaction temperature and 20 minutes of the reaction time. The results compared with the EFB shown 1.478 m^2 g^{-1} of S_{BET} [7]. The S_{BET} of corn stover's carbonized char derived at 550°C, 15 mins pyrolysis and 250°C, 4 hours hydrothermal are 12 and 4 m^2 g^{-1} respectively [17]. As can be seen from Table 3, there are the different trend in this study. Notably, the characters of surface material still combined the point of view of physical properties (fibre contents and S_{BET}) and chemical properties.

3.3 FT-IR Characteristics

FT-IR technique is crucial chemical characterization and aims to find the OFGs on material surface. The acidic carbon-oxygen surface groups of activated carbons make the carbon surface hydrophilic behaviour and polar in character [18]. The vibration of the functional group in the material that absorbed the infrared radiation at a specific frequency will be detected. These FT-IR spectrums plot the relation of wavenumber (cm^{-1}) and % transmittance.

From the FT-IR spectrums in Figure 6, it represents the distinction of corncob and biochar. Vibrational band of Biochar-480 and Biochar-380 were quite flat and is in the range of 3500-1500 cm^{-1} especially Biochar-480 corresponding with the higher reaction temperature (O-H stretching representing the Hydroxyl, 3200 cm^{-1}). The spectra range of 1300-500 cm^{-1} indicates the aromatic chemical skeleton and its result relate to 89.36% and 78.78% lignin in Biochar-480 and Biochar-380 respectively differ vividly from corncob. The band of 1205 and 1103 cm^{-1} indicated the functional group of C-O stretching and it was found the vibration in lignin. The range was at 875-850 cm^{-1} to testify the aromatic region.

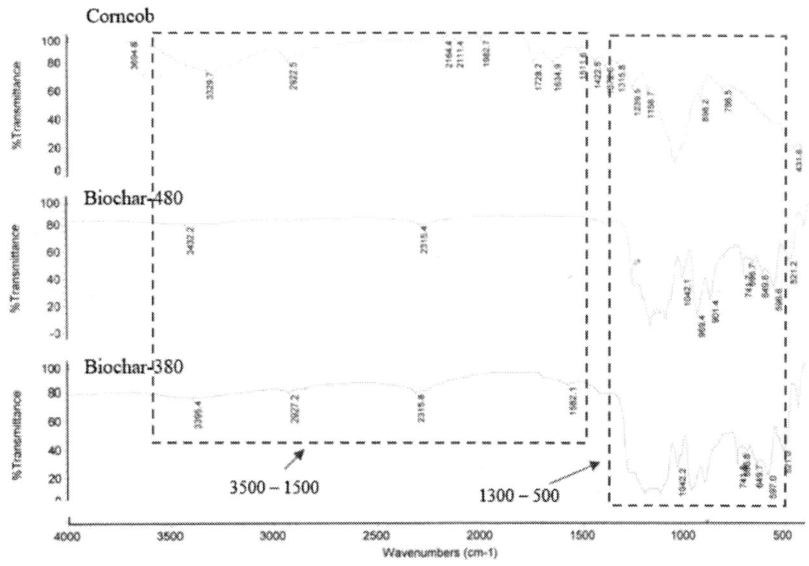

Figure 6. FT-IR spectrums of Corncob, Biochar-480 and Biochar-380

FT-IR spectra of Hydrochar-250 represents in Figure 7 was very similar to the spectra of Corncob given in this study. The region at 1725 cm^{-1} of C=O stretching functional group represents the vibration band in hemicellulose. This can be found in corncob which can decreases in the intensity of the Hydrochar-250. The FT-IR result of pure cellulose also indicated the higher C=O compound containing while hemicellulose contained the O-H and C-O composition [19]. The interpretation of this result can be verified by 35.23% of hemicellulose in Hydrochar-250 and was found to be lower than the 40.73% of hemicellulose in Corncob. It can thus be suggested that can enhance the increasing of the oxygenated surface functional group.

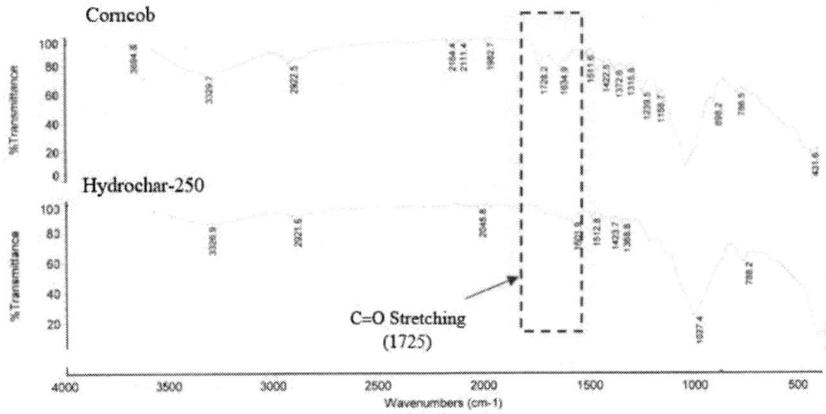

Figure 7. FT-IR spectrums of Corncob and Hydrochar-250

3.4 Discussion

The comparison of FT-IR spectrums of 4 samples involved Corncob, Biochar-480, Biochar-380 and Hydrochar-250. As shown in Figure 8, we can see the critical disparity of carbonized corncob material from the thermochemical biomass conversion process are the pyrolysis and hydrothermal carbonization. Figure 8 presents the blue dash line as it showed the C=C stretching and C-H bend functional group interpreted as the aromatic skeletal in lignin which has less polar properties. Oxygenated functional groups (OFGs) on surface material has the polar behaviour can be represented in following bands with red dash line: 3200 cm^{-1} (O-H stretching representing the hydroxyl, carbonyl groups (acid properties), alcohol (cellulose) or phenol (lignin)), 1725 cm^{-1} (C-O stretching representing the vibration in hemicellulose) found in the origin of raw material. Also, it showed a decrease of the intensity of the Hydrochar-250 but has less in Biochar-480 and Biochar-380 corresponding to fibre content results, 1240 cm^{-1} (C-O stretching representing the hemicellulose esters), 1205 cm^{-1} (C-O vibration in lignin). Besides, it has a more distinct peak of both biochar with higher lignin content and polymerized new aromatic portion.

The biochar through both high-temperature pyrolysis has a high lignin content and it demonstrated complete carbonized char with the S_{BET} range of 7.66 and 6.12 $m^2 g^{-1}$ for laboratory scale and mobile plant respectively. The hydrothermal char is partially carbonized with the S_{BET} of 11.53 $m^2 g^{-1}$, resulting in the degrading of the hemicellulose. Therefore, this is a critical point for optimization the degradation of hemicellulose component. The quantity of oxygen component on the surface of the carbon material is related to the increase in hydrophilic behaviour. The aromatic skeletal of both biochars demonstrated new polymerized aromatic structure.

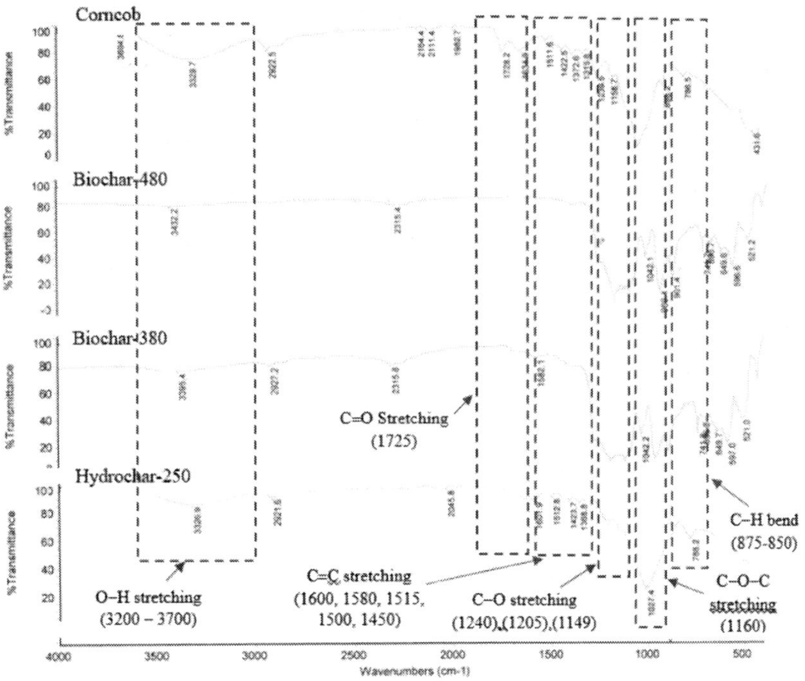

Figure 8. FT-IR spectrums of Corncob, Biochar-480, Biochar-380 and Hydrochar-250

4. Conclusion

A limitation of this study is that physical and chemical characteristics of carbonized corncob in producing precursor material for the bio-based activated carbon was an uncontrolled factor. The evidence from this study suggested that the identification of the potential of the material for surface enhancement should be taken into consideration. The findings of this study suggest that the analytical method and the optimization process includes the fibre compositions, S_{BET} and FT-IR. The oxygenated functional surface groups (OFGs) of biochar and hydrochar has also analysed to provide the baseline for carbon activation. The adsorption ability does not only relate to the surface area defined as physical property but it also strongly involves the OFGs defined as chemical property which prominent electrostatic interaction in nature. The higher OFGs and higher S_{BET} of hydrothermal char show significant results than biochar. This study set out with the aim of optimizing the hydrothermal conversion and analysing the crucial physical and chemical characteristics with a providing of the baseline for carbon surface activation of the renewable material.

Acknowledgments

This project received financial support from The Energy Conservation Fund (ENCON Fund), 2018. The experimental equipment and technical support from the Energy Research and Development Institute-Nakornping, Chiang Mai University (ERDI-CMU) and Thermal System Research laboratory, Chiang Mai University.

References

[1] Chen H 2014 *Biotechnology of Lignocellulose: Theory and Practice* (Dordrecht : Springer Netherlands) pp 25-71
[2] Dinjus E, Kruse A and Tröger N 2011 *Chem. Eng. Technol.* **34** 2037-2043
[3] Dimitriadis A and Bezergianni S 2017 *Renew. Sust. Energ. Rev.* **68** 113-125

[4] Version 2.1 of the IBI Biochar Standards International Biochar Initiative 2015 https://biochar-international.org/characterizationstandard/, accessed on 1 December 2020
[5] Kambo H S and Dutta A 2015 *Renew. Sust. Energ. Rev.* **45** 359-378
[6] Qian K Kumar A Zhang H Bellmer D and Huhnke R 2015 *Renew. Sust. Energ. Rev.*
[7] Srilek N and Aggarangsi P 2020 *IOP Conf. Ser.: Earth Environ. Sci.* **495** 012026
42 1055-1064
[8] Parshetti G K Hoekman S K and Balasubramanian R 2013 *Bioresource. technol.* **135** 683-689
[9] Liu Z Zhang F S and Wu J 2010 *Fuel.* **89** 510-514
[10] Arellano O Flores M Guerra J Hidalgo A Rojas D and Strubinger A 2016 *Chem. Eng. Trans.* **20** 235-240
[11] Nakason K Panyapinyopol B Kanokkantapong V Viriya-empikul N Kraithong W and Pavasant P 2018 *Biomass. Convers. Biorefin.* **8** 199-210
[12] Chen Y Zhu Y Wang Z Li Y Wang L Ding L and et al. 2011 *Ad. Colloid. Interfac.* **163** 39-52
[13] Srilek N and Aggarangsi P 2019 *Int.J.Eng.Adv.Technol.* **8** 1089-1098
[14] Kumar A Wang L Dzenis Y A Jones D D and Hanna M A 2008 *Biomass. Bioenerg.* **32** 460-467
[15] Funke A and Ziegler F 2010 *Biofuel. Bioprod. Bior.* **4** 160-177
[16] Wang T Zhai Y Zhu Y Li C and Zeng G 2018 *Renew. Sust. Energ. Rev.* **90** 223-247
[17] Sohi S P Krull E Lopez-Capel E and Bol R 2010 *Adv. Agron.* **105** 47-82
[18] Zhang L Wang Q Wang B Yang G Lucia L A and Chen J 2015 *Energ. Fuel.* **105** 872-876
[19] Yang H Yan R Chen H Lee D H Zheng C 2007 *Fuel.* **86** 1781-1788

Design of small wind turbines to produce electricity for low wind speed in Thailand

Suradet Panyawong and Woradej Manosroi*

Department of Mechanical Engineering, Faculty of Engineering,
Chiang Mai University, Chiang Mai, Thailand 50200

* Corresponding Author: woradej.manosroi@gmail.com

Abstract
This work studied the design of small Vertical Axis Wind Turbine (VAWT) S Shape and Leaf Shape which are appropriate for low wind speed in Thailand. The theory of Overlaps Ratio (β) , End Plates and Blade Areas which effect to wind turbines operation were analysed. The turbines were tested in a wind tunnel and connected to permanent magnet generator under wind speed conditions close to the real wind speed. The starting up wind speed and Power Coefficient of Leaf Shape were 0.81 m/s and 0.011 and the starting up wind speed and Power Coefficient of S Shape were 0.63 m/s and 0.038, respectively. The results showed that the S-Shape performance was better than Leaf Shape. The results from this study can be applied for developing and improving the small turbine for the application at the low wind speed area in Thailand

Keywords: Vertical Axis Wind Turbine, Permanent Magnet Generator, Low Wind Speed in Thailand

1. Introduction
Nowadays, global warming crisis such as greenhouse effect, melting ice at the north and the south pole since the use of energy from fossil such as petroleum, coal, natural gas that affected to environment and in the future, all of these energies will be exhausted. Therefore, the research of renewable energy such as solar energy, hydropower energy, and wind energy were made to solve this problem. Wind energy is alternative energy because of the cleanliness and no pollution. The wind turbine is the device which converted the mechanical energy from the wind to electrical energy. Most of research studied on large Horizontal Wind Turbines (HAWT) which require high wind starting up speed for the operation, but wind speed is low in the most part Thailand especially in the northern Thailand. Another wind turbine that can handle low wind speed is a Vertical Axis Wind Turbine (VAWT). It can also receive the wind in all directions. Thus, Vertical Axis Wind Turbine Savonius type was selected for this study [1]

2. Researching and Methods

2.1 Wind Speed data

Abdul Latif Manganhar et al. [2] performed a design and comparison of vertical axis wind turbine test between 2 types which were Three Blade Savonius and gap between blades (WAG-RH, Wind Accelerating and Guiding Rotor House) by testing inside the wind tunnel and testing outside area. The results showed that the wind turbine Power Coefficient was 0.125 while wind turbines with a gap between blades Power Coefficient was 0.218 by tested at the same at wind speed 5 m/s. It was founded that the gap between of blade influenced to operation of wind turbine with increased efficiency and the use of a program model could help improve the performance by improving the relevant parameters as well.

Jae-Hoonlee et al. [3] analyzed twist angle of Savonius wind turbine by test Tip Speed Ratio at angles of 0, 45, 90 and 135 degrees. The numerical method of the air flow through the wind turbine at each twisted angle was performed to find out which parameters involved, to construct and to test the turbine in wind tunnels. The wind tunnel tests were, then, used to compare and to analyse the results. From the numerical flow simulation and the wind tunnel flow test found that the highest Power Coefficient was 25.5 percent at the angle 45 degree and the lowest at the shape was at a twist angle of 135 degree.

Abdullah Al-Faruk and Ahmad Sharifian [4] studied the operation of a two blade Savonius wind turbine by hot air from utilizing industrial waste heat. The test was derived from the exhaust of jet engine installed in the laboratory and could adjust the wind speed and temperature and installed the turbine in the wind tunnel. Numerical method on air flow in conjunction with the actual test was performed. The actual test results and simulation tests were compared and analyzed to determine the parameters and to improve wind turbine performance. The results showed that the relevant parameters were propeller gap and propeller torsion angle, with the appropriate values after improvement, the gap between the blades was 16 mm and the propeller torsion angle was 195 degrees and the Power Coefficient was increased 24.12 %.

Keum Soo Jeon et al. [5] studied the shape of vertical axis wind turbine which affected performance of wind turbine by using a two blade Savonius wind turbine with a twist angle of 180 degrees. The top and bottom of the wind turbines were in total of 4 types ordered from small area to large area and tested in a wind tunnel. The Power Coefficient data were recorded. The results showed that the highest area and Power Coefficient was type 4 or circular flap which up to 36% compared to type 1 (without a top cover). The research team concluded that the top end plate reduced wasteful airflow deflection and increase torque.

Jean Luc Menet and Nachida Bourabaa [6] studied the increase performance of Savonius wind turbine by using numerical methods for selected and adjusted relate parameter. After improving relate parameter, this research suggested ones to use two blade Savonius wind turbine by putting top and bottom end plate. The height of the wind turbine should be twice of its diameter and the overlap ratio must be between 0.15 and 0.3.

From literature review, the concept was applied in this research. The design new turbine with end plate top and bottom which had difference area with the same height with the overlap ratio between 0.15 and 0.3 of wind turbine blade and tested the turbine in wind tunnel and compared result between 2 turbine types.

2.2 Wind Speed data

According to the wind speed data from Chiang Mai Meteorological Station in Thailand, the maximum and minimum average wind speed in past 10 years of Chiang Mai were in May and December consequently. Table 1 and 2 show the maximum, minimum and average wind speed of Chiang Mai, Thailand in the past 10 years (2008 to 2018) in May and December (Source: Chiang Mai Meteorological Station).

Table 1. The maximum and minimum average wind speed in the past ten years in May

Maximum Wind Speed (m/s)	Minimum Wind Speed (m/s)	Average Wind Speed (m/s)
17.505	3.089	7.650

Table 2. The maximum and minimum average wind speed in the past ten years in December

Maximum Wind Speed (m/s)	Minimum Wind Speed (m/s)	Average Wind Speed (m/s)
10.773	2.052	4.542

The wind energy assessment in Chiang Mai was analysed by the wind power divided by the number of seasons in Thailand (3 seasons- Winter Summer and Rainy). The calculation showed that the highest and lowest wind power were in summer and winter as shown in Figure 1.

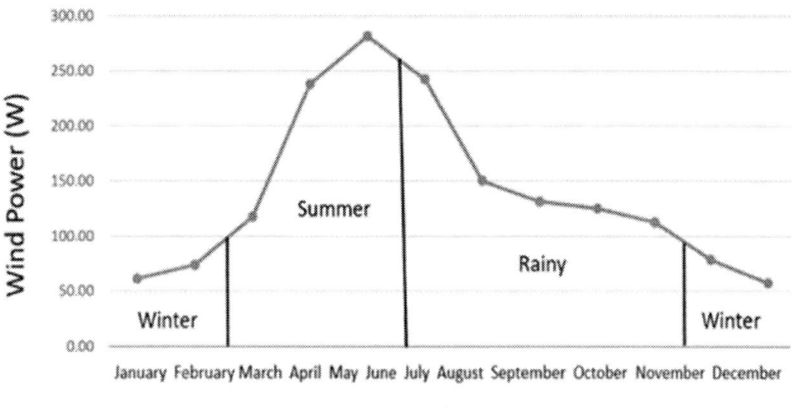

Figure 1. Wind power assessment in Chiang Mai province by season.

The real wind speed was measured for collecting the wind speed data for designing the wind turbine. The wind speed data from meteorological stations cannot be used for designing wind turbines because the measurement of maximum wind speed of each day was not suitable for estimating the wind energy in each season. The wind speed was measured by Anemometer as shown in Figure 2.

IOP Conf. Series: Materials Science and Engineering 1137 (2021) 012005 doi:10.1088/1757-899X/1137/1/012005

Figure 2. Anemometer and device for record wind speed data.

The wind speed data were collected from December to January since 18th December to 17th January (30 days) which was in the winter because the lowest wind speed is in the winter and this study wanted to test the wind turbine at the lowest wind speed in the winter. The average wind speed from the collected data was 1.15 m/s as shown in Figure 3.

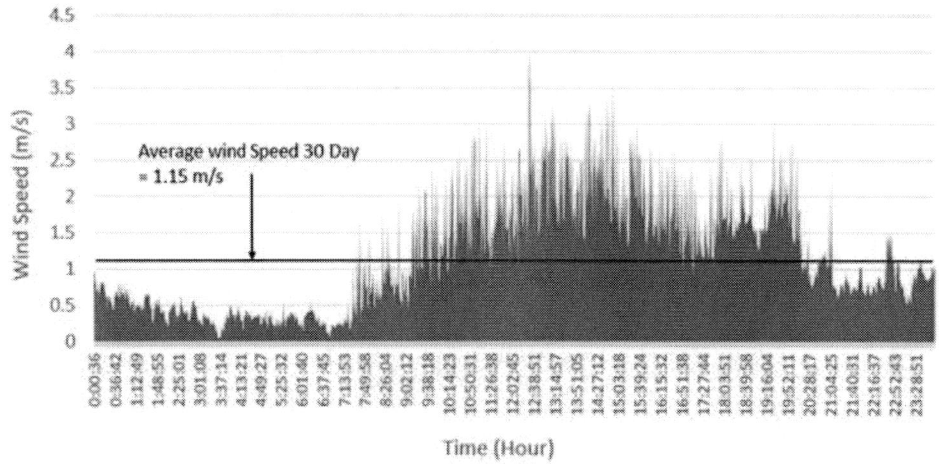

Figure 3. Average wind speed data from the Anemometer

2.3 System Design

The design of wind turbine begins by selected shapes from research of interest, Leaf-Shape wind turbines [7] and S-Shape wind turbines [8] and overlap theory were used for the redesigning [9] then adding End Plate of the wind turbine at the end for increasing the efficiency [5] as shown in equation (1).

$$\beta = \frac{e}{d} \tag{1}$$

When β is Aspect Ratio
e is Overlap Distance (m)
d is Chord Length Rotor (m)

Figure 4. Displays the vertical axis wind turbine from the designing by using overlap theory and average wind speed data from the measurement.

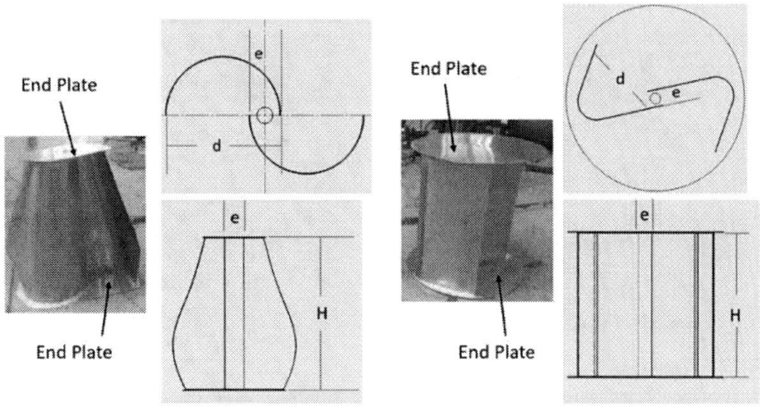

(a) (b)

Figure 4. (a) Leaf Shape wind turbine and schematic diagram of design (b) S Shape wind turbine and design schematic diagram

The input power can be calculated by the Equation (2) [10].

$$P_a = \frac{1}{2}\rho A v^3 \tag{2}$$

Where P_a is Wind Power (W)
A is Blade Swept Area (m^2)
ρ is Density of Air (1.225 Kg/m^3)
v is Wind Speed (m/s)

The output power (electrical power) can be calculated by Equation (3) [11].

$$P_e = IV \tag{3}$$

where P_e is Electrical Power (W)
I is Current (A)
V is Voltage (V)

The parameter indicates the efficiency of wind turbine is Tip Speed Ratio (TSR) (**λ**). Tip Speed Ratio increase with the increasing of efficiency which can be calculated by Equation (4) [12]

IOP Conf. Series: Materials Science and Engineering 1137 (2021) 012005 doi:10.1088/1757-899X/1137/1/012005

$$\lambda = \frac{(RPM \times \pi \times D)/60}{v} \tag{4}$$

Where

λ	is Tip Speed Ratio (TSR)
RPM	is Round Per Minute
D	is Diameter of Wind Turbine (m)
v	is Wind Speed (m/s)

The efficiency of wind turbine can be determined by Equation (5) [13].

$$Cp = \frac{P_e}{P_a} \tag{5}$$

Where C_P is Power Coefficient

Table 3. Data of system wind turbine test

Device	Specification
Leaf Shape Wind Turbine	Turbine Diameter(D) = 0.25 m, High(H) = 0.3 m, e = 0.04 m, Overlap Ratio(β) = 0.276, Area of Turbine = 0.0589 m^2, Material = Aluminum
S Shape Wind Turbine	Turbine Diameter(D) = 0.3 m, High(H) = 0.3 m, e = 0.04 m, Overlap Ratio(β) = 0.253, Area of Turbine = 0.09 m^2, Material = Aluminum
Generator	Permanent Magnet Generator: Power = 10 W, Voltage = 12V
Load	Light Bulb: 5 W, 12V

3. Experimental

Schematic of wind turbine system test diagram in wind tunnel is shown in Figure 5 and component of measuring tools and wind turbine testing presents in Figure 6.

IOP Conf. Series: Materials Science and Engineering 1137 (2021) 012005 doi:10.1088/1757-899X/1137/1/012005

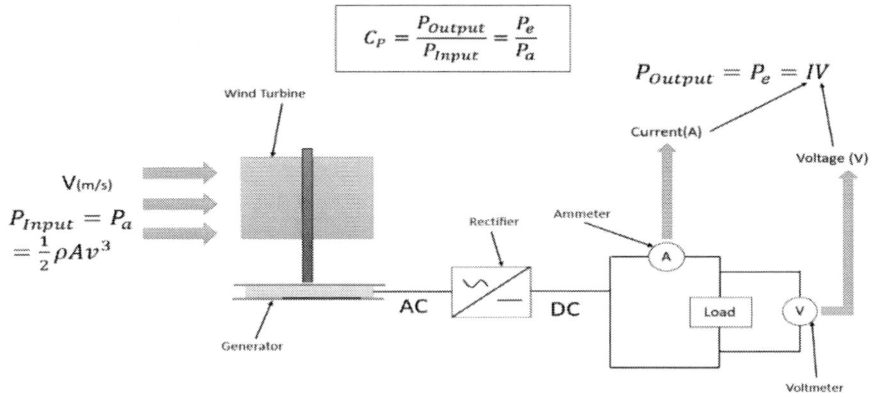

Figure 5. Wind turbine test system in wind tunnel

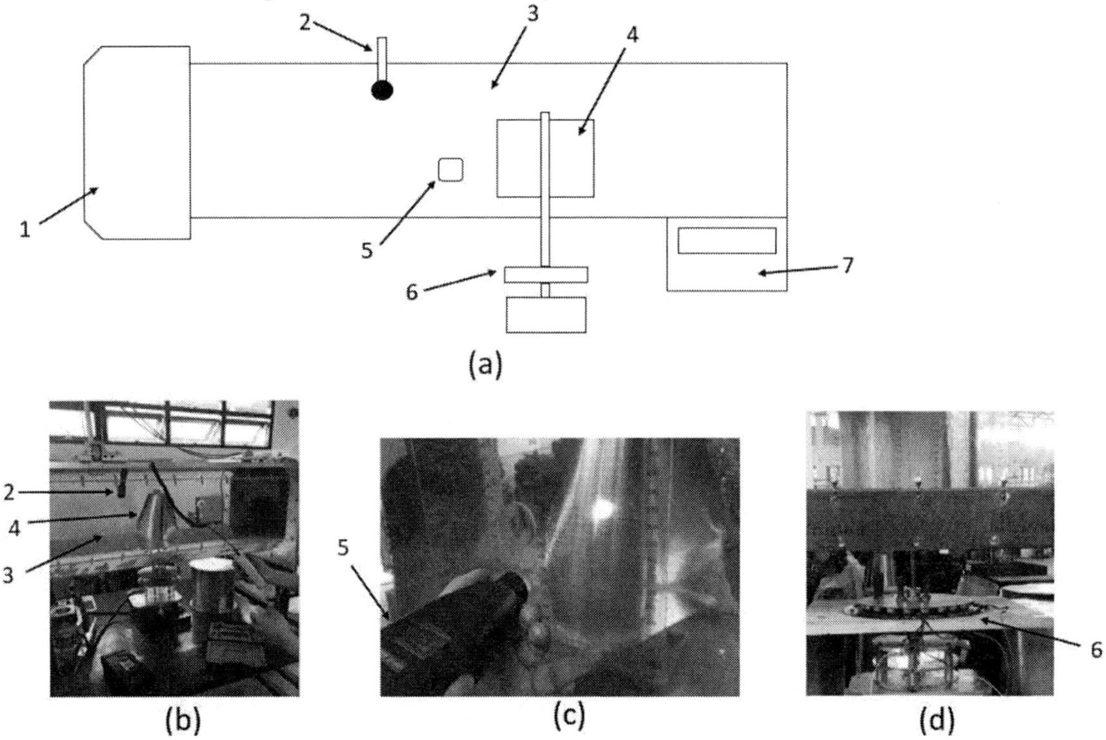

Figure 6. (a) Schematic diagram of the experiment, (b) Measuring wind speed (c) RPM speed measuring, (d) Generator and device during the electric measurement. The device and measuring tools shown as follows 1. Wind Tunnel Fan 2. Anemometer 3. Wind Tunnel 4. Wind Turbine 5. Tachometer 6. Generator and Electric Measuring Tools 7. Control Panel of Wind Speed Adjust

Figure 5 presented the relationship between equation and actual test by the experiment. The experiment started by turning on the wind tunnel fan. The wind turbine power can be calculated by Equation (2) which is input power. The anemometer measured and recorded the wind speed data. The turbine speeds (RPM) were measured and recorded by Tachometer for determining the Tip Speed Ratio (TSR) which can be calculated by Equation (4). The wind turbine transfer force was transferred from shaft to generator and spin to produce the electric power. The electrical power was measured by Voltmeter and Ammeter. The Output Power can be determined by Equation (3). The tests were repeated by adjusting wind speed in the wind tunnel started from 0 m/s then increased gradually to 6 m/s.

4. Result and Discussion

The results showed that the starting up wind speed and average starting up wind turbine speed of S Shape were 0.63 m/s and 48.31 RPM respectively and the average starting up wind speed and starting wind turbine speed of Leaf Shape were 0.81 m/s and 28.32 RPM consequently. Figure 7 displays Tip Speed Ratio (TSR) of S Shape and Leaf Shape wind turbine.

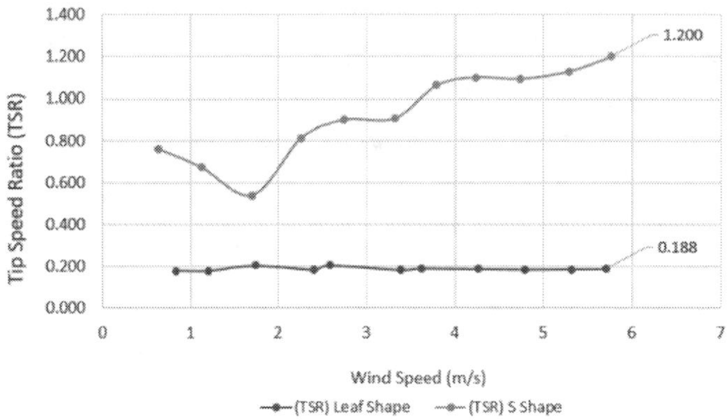

Figure 7. Tip Speed Ratio of S Shape and Leaf Shape.

According to Figure 7, Tip Speed Ratio of S Shape were higher than Leaf Shape wind turbine also output power of S Shape were higher than Leaf Shape as shown in Figure 8. The Power Coefficient of S Shape was higher than Leaf Shape as displayed in Figure 9.

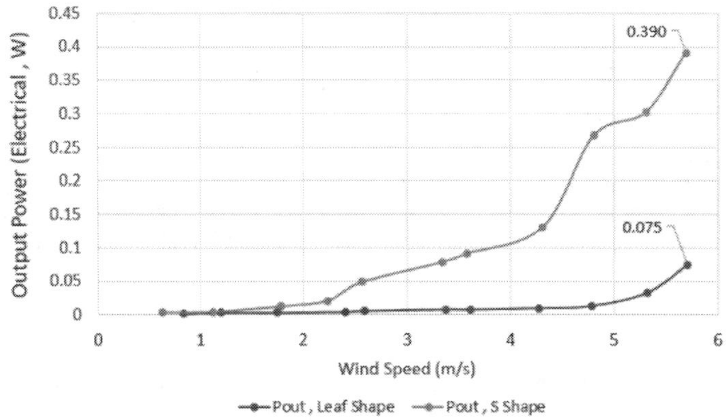

Figure 8. Output power of S Shape and Leaf Shape.

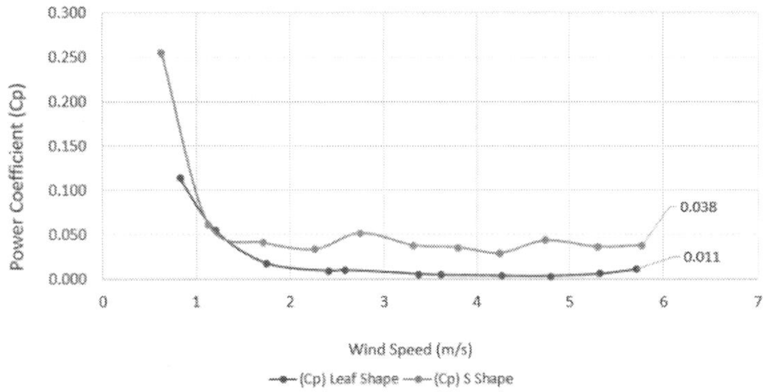

Figure 9. Power Coefficient of S Shape and Leaf Shape.

At wind speed range 3 to 6 m/s, Tip Speed Ratios of the S Shape wind turbine were higher than Leaf Shape wind turbine as displayed in Figure 7. The output power increased with the increasing of wind velocity as well as Tip Speed Ratio as shown in Figure 8. For the Leaf Shape wind turbine, Tip Speed Ratio was almost constant because the output power and Power Coefficient were lower than the S Shape wind turbine due to the S Shape wind turbine Overlap Ratio was lower than Leaf shape wind turbine as shown in Table 3. The starting up speed of S Shape was lower than Leaf Shape because the Chord Length of S Shape was longer than Leaf Shape. Thus, S Shape wind turbine blade area and swept area were more than Leaf Shape wind turbine. The Power Coefficient of S Shape wind turbine was more than Leaf Shape wind turbine due to S Shape wind turbine plate area end plate and wind catchment area were more than Leaf Shape wind turbine.

5. Conclusion.

From this study S Shape wind turbine starting up speed was lower than Leaf Shape wind turbine. The starting up speed of S Shape and Leaf Shape were 0.63 and 0.81 m/s and starting of turbine speed S Shape and Leaf Shape were 48.31 RPM and 28.32 RPM respectively. The S Shape wind turbine Tip Speed Ratio increased with the increasing of wind speed, but the Tip Speed Ratio of Leaf Shape was almost constant because S Shape wind turbine Power Coefficient was higher than Leaf Shape wind turbine. The Power Coefficient of the wind turbine increased with increasing of Overlap Ratio and End Plate. The wind turbine performance increased with the increasing of the Chord Length Rotor, Swept Area and End Plate Area. It can be concluded that, there is a possibility for the operation of S Shape wind turbine at low wind speed area in winter season in northern part area of Thailand.

Acknowledgments

This research was funded by Chiang Mai University.

References

[1] Casini M 2016 *J. Clean Energy* **4**(1) 56-65
[2] Manganhar A L Rajpar A H Luhur M R Samo S R and Manganhar M 2019 *J. Renew. Energy* **136** 512-520
[3] Lee J H Lee Y T and Lim H C 2016 *J. Renew. Energy* **89** 231-244
[4] Al-Faruk A and Sharifian A 2016 *Alex. Eng. J.* **55**(3) 2055-2064
[5] Jeon K S Jeong J I Pan J K and Ryu K W 2015 *J. Renew. Energy* **79** 167-176
[6] Menet J L and Bourabaa N 2004 Increase in the Savonius Rotors Efficiency via a Parametric Investigation (Valenciennes: university of valenciennes)
[7] Rehman S M Mohammed S F and Patka M A 2020 *Int. Research J. Eng. Technol.* **7**(7) 4692-4695
[8] Ramadan A Yousef K Said M and Mohamed M H 2018 *Energy J.* **151** 839-853
[9] Zemamou M Aggour M and Toumi A P 2017 *Proc. Int. Conf. on Power and Engineering System Engineering* (Berlin) vol 141 p 386

[10] Mahmoud N H El Haroun A A Wabha E and Nasef M H 2012 *Alex. Eng. J.* **51**(1) 19-25
[11] Abraham J P Plourde B D Mowry G S Minkowycz W J and Sparrow E M 2012 *J. Renewable Sustainable Energy* **4** 042703
[12] Bakırcı M and Yilmaz S 2018 *Eng. Sci.* **21**(6) 1128-1142
[13] Saeed H A H Elmekawy A M N and Kassab S Z 2019 *Alex. Eng. J.* **58**(2) 429-441

Analyzing esterification kinetics of model bio-oil with ethanol catalyzed by a green catalyst

Prapaporn Prasertpong[1,*], Nakorn Tippayawong[2]

[1]Department of Mechanical Engineering, Faculty of Engineering,
Rajamangala University of Technology Thanyaburi, Pathum Thani Thailand 12110
[2]Department of Mechanical Engineering, Faculty of Engineering,
Chiang Mai University, Chiang Mai Thailand 50200

* Corresponding Author: prapaporn_pr@rmutt.ac.th

Abstract
Fast pyrolysis of biomass is a potential method for the production of bio-oils as a renewable liquid fuel. Bio-oils contain a large number of organic compounds, particularly carboxylic acids, which cause corrosiveness and instability that are troublesome for storage and for the transportation of fuel. Upgrading is an important process to transform bio-oils into high-quality biofuels. Contained organic acids are treated with alcohol to convert them into esters by esterification which may be considered as a good intermediate bio-oil upgrading method. This work presents a kinetic study for 12-tungstosilicic acid (a type of heteropoly acid, HPA) catalyzed esterification of a model bio-oil with ethanol. The main compound of the agro-residue-derived bio-oil was illustrated by oleic acid. The kinetic study was investigated under the molar ratio between ethanol to oleic acid of 9.11:1 and catalyst loading of 10 wt% condition with the varied reaction temperature in the range of 35-75°C. The result showed that a pseudo-first-order was suitable to represent the kinetic model for esterification of the model bio-oil. By fitting between the experimental results and kinetic model, the activation energy (Ea) and the pre-exponential factor (A) were found to be 31.5 kJ/mol and 432 min^{-1}, respectively. This comparatively low activation energy equated to other catalytic esterification proved that the HPA catalyst (12-tungstosilicic acid) considered here was an assured green catalyst for esterification in the bio-oil upgrading process.

Keywords: Bioenergy, Bio-oil upgrading, Catalytic esterification, Chemical conversion, Renewable energy

1. Introduction
In recent times, global attention is attracted to biomass as an important source for alternative renewable energy, particularly in developing countries [1] due to the advantages of its profusion, reproducibility, economic potential, and environmental friendliness [2]. Jatropha curcas is an energy plant that is widely cultivated and considered as one of the inherent feedstocks for the production of biodiesel by transesterification of jatropha oil obtained from seed extraction [3-4]. This generated waste from the

jatropha seed oil extraction process, jatropha cake, can be further transformed into biofuel as a value-added product [5-6]. Fast pyrolysis is one of the potential techniques for transforming such biomass waste into liquid fuel or bio-oil through a thermochemical process [7-9]. In general, bio-oils consist of water and various oxygenated compounds including acids, alcohols, aldehydes, ketones, and phenols [10-11], which are responsible for unappropriated properties of bio-oil that limited its economical utilization as the fuel in ether stationary and transportation applications including low heating capacity, high water, and oxygen content, instability, high viscosity and corrosiveness [12-13]. The composed acid compounds in bio-oil were considered as the inclined reactive compound, which can cause the viscosity change and instability in bio-oil [14]. Moreover, a large amount of the contained acid compound is responsible for the corrosiveness of bio-oil that limits the direct use of bio-oil in storage and transportation of fuel without upgrading [15]. One of the effective bio-oil upgrading techniques is esterification, which can transform contained acid in bio-oil into ester by treating this compound with alcohol that can reduce the acidity and improve the heating capacity of bio-oil [16]. For esterification investigation, the bio-oil derived from different biomass feedstock such as mallee eucalypts [17], silver grass [18], and risk husks [19] was represented by acetic acid, which was found as the major acid component composed in bio-oil. However, oleic acid was found to be the major acid component composed in bio-oil derived from jatropha cake [20-21] which was rare and few works of literature have reported about it.

Mineral acid catalysts are widely studied and applied for the esterification reaction in the industrial sector due to their high activity and selectivity [22]. However, the equipment corrosion and wastes of the toxic hazardous are still considered as a concerning problem for mineral acid catalyst utilization [23]. In recent times, heteropoly acid (HPA) has gained attention as an interesting and effective catalyst group for esterification through its advantages of low toxicity, volatility, and corrosiveness, while remaining a selectivity and high activity from the high Brønsted acid strength, which can provide fast reaction under mild conditions [24]. The successful HPA catalyzed esterification was reported from Nowakowski et al. [25] for bio-oil upgrading with n-butanol, Prasertpong et al. for the upgrading of model bio-oil with ethanol using acetic acid [22] and oleic acid [26] as the model bio-oil.

The kinetic of esterification catalyzed by several catalysts were reported. For HPA catalyzed esterification, a kinetic study was reported for tungstophosphoric acid [27], tungstophosphoric acid/SBA-15 [28], tungstophosphoric acid/MCM-41 [5] catalyst for biodiesel production, while the tungstosilicic acid catalyzed esterification was reported for esterification of model bio-oil represented by acetic acid [25]. The pseudo-first-order kinetic model was proposed by Jiang et al. [29] as a suitable model for describing the kinetic for a sulfonated cation exchange resin catalyzed esterification reaction, as well as for a kinetic parameter investigation. Moreover, the evaluation of a kinetic parameter using the fitting of a proposed kinetic model and experimental results were reported by Ilgen [30] for the kinetic study of Amberlyst 46 catalyzed esterification of oleic acid with methanol.

2. Materials and Methods

2.1 Materials

In this work, model bio-oil was represented by oleic acid, which was found as the major component composed in bio-oil derived from the jatropha cake fast pyrolysis process provided by TISTR [21]. Oleic acid (purity $\geq$ 99%) and 12-tungstosilicic acid (H4SiW12O40•nH2O) were supplied from QRec Co., Ltd, and Sigma Aldrich Co LLC, respectively. Ethanol (purity $\geq$ 99.5%), pure potassium hydroxide (KOH), isopropanol (purity $\geq$ 99.8%), and toluene (purity $\geq$ 99.5%) were provided by RCI Labscan Co., Ltd. The solution of phenolphthalein in ethanol (0.1 M) was purchased from Vechavit Part., Ltd. All chemicals were of analytical reagent grade and were utilized as a received condition without further purification.

2.2 Esterification experiments

The esterification reactions were performed in a 250 mL three-necked flask batch reactor connected with thermocouple and refluxed condenser under 500 rpm continuous mechanical stirring and atmospheric pressure. The initial reacting mixture for the typical experimental run was included oleic acid 40 g, a certain amount of 12-tungstosilicic acid, and ethanol to obtain the catalyst loading of 10 wt% and the molar ratio between ethanol to oleic acid of 9.11:1 condition, which was selected according to the previous work antecedent to this work [26]. A constant temperature oil bath equipped with a temperature control system was used for heating the whole reaction system to maintain the desired reaction temperature in a range of 35-75 °C to obtain sufficient experimental data for the analysis of the kinetic profile [31]. The reaction time was started when the reaction temperature reached the expected value. The samples of reacting mixture were taken every 15 minutes for the first hour of reaction time, then every 1 hour for the last 3 hours to observe the change in conversion of oleic acid during the reaction.

2.3 Analytical methods

The collected samples were analyzed with an acid-base color-indicator titration method according to ASTM D974-04 to determine the acid value (AV) [32]. The approximate 0.1 g of sample was dissolved in 100 mL of 100: 99: 1 solution of toluene, isopropanol, and deionized water, and 0.5 mL of the 0.1M phenolphthalein indicator solution, then titrated with 0.1 M KOH standard solution until the color change of the sample mixture was observed. The acid value of the sample was calculated by Eq 1, where AV is the acid value of the sample (mg of KOH/ g), V is the used volume of KOH solution (mL), M is the concentration of the KOH solution (M), MW_{KOH} is the molecular weight of KOH (56.1 g/mol), and W_{sample} is the weight of the sample (g).

$$AV = \frac{V \times M \times MW_{KOH}}{W_{sample}} \tag{1}$$

After the determination of the acid value of the samples, the oleic acid conversion was calculated from Eq (2), where AV_i is the acid value of the initial sample and AV_t is the acid value of the taken sample, respectively.

$$Conversion = \frac{AV_i - AV_t}{AV_i} \times 100\% \tag{2}$$

3. Results and discussion

The HPA catalyzed esterification of oleic acid with ethanol was considered as a reversible reaction. The main products produced from this reaction are ethyl oleate and water, which can be illustrated with Eq. (3)

$$\underset{Oleic\cdot\ acid}{RCOOH(A)} + \underset{Ethanol}{C_2H_5OH(B)} \cdot \underset{k_2}{\overset{k_1}{\rightleftharpoons}} \underset{Ethyl\cdot\ oleate}{RCOOC_2H_5(C)} + \underset{Water}{H_2O(D)} \tag{3}$$

The liquid phase pseudo-second-order kinetic can be used to describe this reaction in the forward and reverse direction [33-34]. The overall reaction rate of this reaction can be represented as Eq. (4) [27, 30, 34], where [A], [B], [C], [D] were the concentration of oleic acid, ethanol, ethyl oleate, and water respectively; X, Y, Z, W denote their reaction order. The reaction rate constant in the forward and reverse direction was denoted by k_1 and k_2, respectively.

$$-\frac{d[A]}{dt} = k_1[A]^X[B]^Y - k_2[C]^Z[D]^W \tag{4}$$

In this work, the molar ratio between ethanol to oleic acid of 9.11:1 was used, which was over the balance value and indicated the higher concentrations of the initial reactant compared with other components. Thus, $k_1[B]^Y$ can be supposed as the constant [25, 35]. Besides, this esterification reaction was considered as the irreversible reaction since the excess alcohol can push the reaction equilibrium in the forward direction to form the ethyl oleate, which causes the much higher value of the kinetic constant in the forward direction (k_1) comparing with the kinetic constant in the reverse direction (k_2) [25, 36]. The reaction rate of this reaction was simplified into Eq. (5), which can be expressed with the pseudo-first-order kinetic model, where $k= k_1[B]^Y$ [27, 30, 34].

$$-\frac{d[A]}{dt} = k[A]^n \tag{5}$$

When the initial concentration and conversion of oleic acid are defined as $[A_0]$ and X, respectively, then the concentration of oleic acid can be defined as Eq. 6 [37]

$$[A] = [A_0](1-X) \tag{6}$$

Eq. 5 can be transformed into Eq. 7 [37-38]

$$-\frac{d[A]}{dt} = -\frac{d\{[A_0](1-X)\}}{dt} = [A_0]\frac{dX}{dt} = k[A]^n \tag{7}$$

Then:

$$\frac{dX}{dt} = \frac{k}{[A_0]}\{[A_0](1-X)\}^n \tag{8}$$

$$\frac{1}{(1-X)^n}dX = \frac{k}{[A_0]^{(1-n)}}dt \tag{9}$$

When the reaction order (n) = 1, Eq. 9 was transformed into Eq. 10 [25, 30, 39]

$$\frac{1}{(1-X)}dX = -kdt \tag{10}$$

After the Integrating of Eq. 10, the pseudo-second-order kinetic equation in terms of acid conversion was obtained as illustrated in Eq. 11.

$$\ln(1-X) = -kt \tag{11}$$

The obtained experimental data for esterification of oleic acid at various reaction temperatures were fitted with Eq. 11 as displayed in Figure 1 and the reaction rate constant (k) for the respective reaction temperature was calculated from the slope of the line. The summary of the obtained reaction rate constants and their particular coefficient of determination (R^2) were represented in Table 1.

The reaction rate constant, summarized in Table 1, tended to increase with the increased reaction temperature, which implied that the reaction temperature can increase the rate of the reaction. Since the esterification is an endothermic reaction, the increase of reaction temperature was led to improve molecular motion speed and the mass transfer rate of the reaction system, which can be accelerated the rate of reaction [18, 40]. This result was consistent with the report of Yu et al. [35] for SO_4^{2-}/ZrO_2 catalyzed wasted cooking oil upgrading, Prasertpong et. al. [25] for HPA catalyzed bio-oil model compound upgrading as well as various studies for biodiesel production [34, 39, 41].

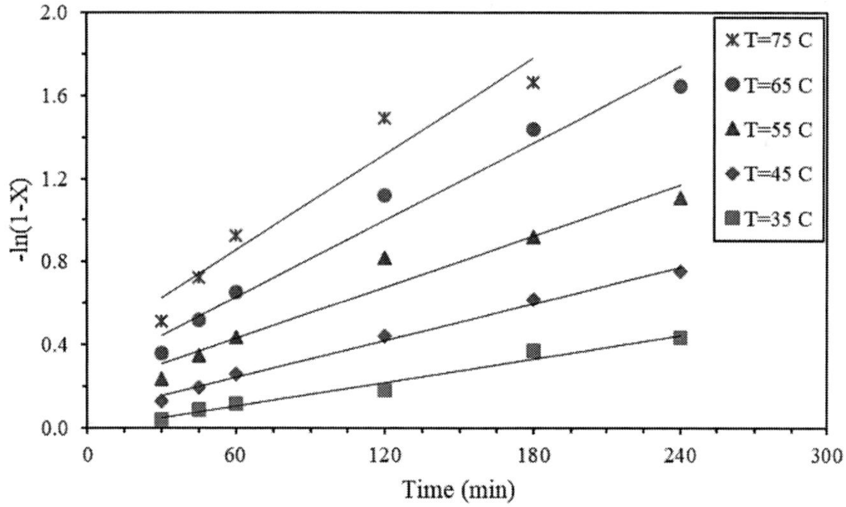

Figure 1. The kinetic modeling curve for the HPA catalyzed esterification reaction under different reaction temperatures.

Table 1. The reaction rate constants (k) and coefficient of determination (R^2) for esterification under different reaction temperatures.

T (°C)	k (min^{-1})	R^2
35	0.0019	0.9754
45	0.0030	0.9929
55	0.0041	0.9523
65	0.0062	0.9734
75	0.0077	0.9364

A good correlation between the experimental data and the calculated value was observed from Table 1 with the above 0.93 of the coefficients of determination (R^2) [34]. These results proved that the pseudo-first-order kinetic model had justified describing the esterification reaction catalyzed by the HPA catalyst. The literature was also reported the pseudo-first-order kinetic model for bio-oil upgrading with the catalytic esterification [25] and biodiesel production via Amberlyst 46 [30], F^--SO_4^{2-}/MWCNTs [34], HPA catalyst [27, 42], and TPA_3/SBA-15 [28] catalyzed esterification.

After the reaction rate constants under different reaction temperatures were determined. The Arrhenius equation (Eq. 12) was applied to model the relation between the reaction temperature and the rate coefficient, which can be used for kinetic parameter calculation, where Ea, A, R, and T were the activation energy (J/mol), pre-exponential factor (min^{-1}), gas constant (J/K mol) and reaction temperature (K), respectively. The logarithms of the reaction rate constant (ln k) were plotted versus 1/T as illustrated in Figure 2, while the value of Ea and A can be determined from the slope and interception on the y-axis of this plot [18, 25].

$$-\ln k = \frac{E_a}{RT} - \ln A \tag{12}$$

The linear relationship between the logarithms of the reaction rate constant and the reverse of temperature demonstrated the ratification of the Arrhenius equation to explain the esterification reaction catalyzed by the HPA catalyst with the R^2 value of 0.99 [34]. The activation energy and pre-exponential factor for this reaction were calculated and compared with the result obtained from the literature, which was summarized in Table 2.

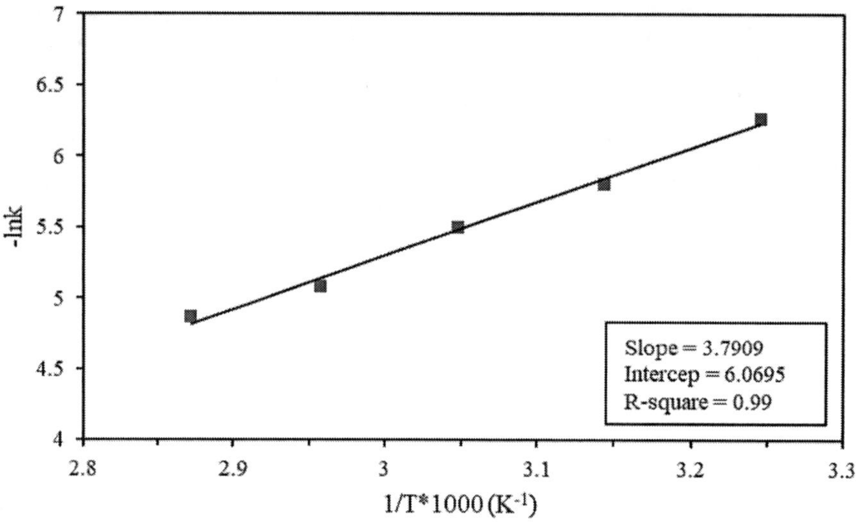

Figure 2. Arrhenius plot for HPA catalyzed esterification reaction.

The pre-exponential factor of 432 min^{-1} and apparent activation energy of 31.5 kJ/mol and were gained for the HPA catalyzed esterification of oleic acid as the model bio-oil with ethanol. This activation energy related to the HPA catalyst was observed to be lowest comparing with the results obtained from the other comparable esterification reaction catalyzed by ether homogeneous and heterogeneous catalysts shown in Table 2. This low activation energy implied that this reaction with the HPA catalyst required lower energy to make the reaction occurred. Hence, the HPA catalyst can be considered as a promising catalyst with a high catalytic performance for bio-oil upgrading via esterification due to its relatively low activation energy. The higher frequency of collusions between the reactant molecules of the esterification reaction catalyzed by the HPA catalyst was indicated by the higher pre-exponential factor related to the HPA catalyst comparing with other catalysts.

Table 2. Comparison of kinetic study for catalytically esterification reaction.

Catalyst	Ea (kJ/mol)	A (min^{-1})	Reference
H_2SO_4	51.6	-	[43]
H_2SO_4	46.0	1.8×10^5	[44]
TPA3/SBA-15	44.6	12.7	[28]
F^--SO_4^{2-}/MWCNTs	32.7	-	[34]
zinc acetate	32.6	120	[36]
HPA; tungstosilicic acid	31.5	432	This work

4. Conclusion

The kinetic behavior of HPA (tungstosilicic acid) catalyzed esterification of oleic acid as the model bio-oil with ethanol was investigated in the batch condition under atmospheric pressure. The results showed that tungstosilicic acid is an effective catalyst for the esterification process. The kinetic of this esterification reaction was appropriately described with the pseudo-first-order kinetic model. The experimental data were fitted with the kinetic model to obtain the pre-exponential factor (A) of 432 min^{-1} and the apparent activation energy (Ea) of 31.5 kJ/ mol for this esterification reaction. The tungstosilicic acid was expected to be an assuring catalyst for upgrading of bio-oil with the esterification process with the comparatively low activation energy comparing with other related catalytically esterification reaction.

Acknowledgments
Financial support of this work was partially provided by Thailand Science Research and Innovation, Royal Golden Jubilee program, Chiang Mai University, and Rajamangala University of Technology Thanyaburi.

References

[1] Akinrinola F S Darvell L I Jones J M Williams A and Fuwape J A 2014 *Energy Fuels* **28**(6) 3821-3832
[2] Thapa S Indrawan N and Bhoi P R 2018 *Environ. Technol. Innov.* **9** 210-219
[3] Odetoye T E Abu Bakar M S and Titiloye J O 2019 *Nigerian Journal of Technological Development* **16**(2) 71-77
[4] Pambudi N A Laukkanen T Syamsiro M and Indra MG 2017 *J. Energy Inst.* **90**(5) 672-679
[5] Patel H Rajai V Das P Charola S Mudgal A and Maiti S 2018 *Renew. Energy* **122** 310-322
[6] Primandari S R P Islam A K M A Yaakob Z and Chakrabarty S 2018 *Jatropha curcas L. biomass waste and its utilization Advances in biofuels and bioenergy* ed M Nageswara-Rao and J Soneji (London: IntechOpen) chapter 14 pp 273–282
[7] Bridgwater A V 2013 *Fast pyrolysis of biomass for the production of liquids Biomass Combustion Science, Technology and Engineering* ed L Rosendahl (Cambridge: Woodhead Publishing) chapter 7 pp 130-171
[8] Khuenkaeo N and Tippayawong N 2020 *Chem. Eng. Commun.* **207**(2) 153-160
[9] Jaroenkhasemmeesuk C Diego M E Tippayawong N Ingham D B and Pourkashanian M 2018 *Int. J. Energy Res.* **42**(9) 2983-2996
]10 [Sanghuang J Alwi S R W Nawi W N R M Manan Z A and Klemes J J 2019 *Chem .Eng .Trans.* **72** 139-144
[11] Zhang Q Xu Y Li Y Wang T Zhang Q and Ma L 2015 *Appl. Energy* **160** 633-640
[12] Bridgwater A V 2012 *Biomass Bioenergy* **38** 68-94
[13] Sundqvist T Oasmaa A and Koskinen A 2015 *Energy Fuel* **29** 2527-2534
[14] Hu X Wang Y Mourant D Gunawan R Lievens C and Chaiwat W 2013 *AIChE J.* **59**(3) 888-900
[15] Sharifzadeh M Sadeqzadeh M Guoa M Borhani T N Konda N V S N M Garcia M C Wang L Hallett J and Shah N 2019 *Prog. Energy Combust. Sci.* **71** 1-80
[16] Mei D Guo D Wang C Dai P Du J and Wang J 2020 *J. Energy Inst.* **93**(4) 1382-1389
[17] Hu X Lievens C Mourant D Wang Y Wu L and Gunawan R 2013 *Appl. Energy* **111** 94-103
[18] Liu Y Li Z Leahy J J and Kwapinski W 2015 *Energy & Fuel* **29**(6) 3691-3698
[19] Zhang Q Chang J Wang T and Xu Y 2006 *Energy & Fuels* **20**(6) 2717-2720
[20] Kim S W Koo B S Ryu J W Lee J S Kim C J and Lee D H 2013 *Fuel Process. Technol.* **108** 118-124
[21] Prasertpong P Jaroenkhasemmeesuk CTippayawong N and Thanmongkhon Y 2017 *CMU J. Nat. Sci.* **16**(2) 135-144
[22] Prasertpong P Jaroenkhasemmeesuk C Regalbuto JR Lipp J and Tippayawong N 2020 *Energy Reports* **6** 1-9

[23] Zhang L Xian M He Y Li L Yang J and Yu S 2009 *Bioresour. Technol.* **100**(19) 4368-4373
[24] Cardoso A L Augusti R and Silva M J D 2008 *J. Am. Oil. Chem. Soc.* **85**(6) 555-560
[25] Prasertpong P and Tippayawong 2019 *Energy procedia* **160** 253-259
[26] Prasertpong P and Tippayawong T 2020 *Energy Source Part A* DOI:10.1080/15567036.2020.1746867
[27] Talebian-Kiakalaieh A Amin N A S Zarei A and Noshadi I 2013 *Appl. Energy* **120** 283-292
[28] Brahmkhatri V and Patel A 2011 *Appl. Catal. A-Gen.* **403**(1-2) 161-172
[29] Jiang Y Lu J Sun K Ma L and Ding J 2013 *Energy Convers. Manageme.* **76** 980-985
[30] Ilgen O 2014 *Fuel Process. Technol.* **124** 134-139
[31] Fauzi A H M Amin N A S and Mat R 2014 *Applied Energy* **114** 809-818
[32] ASTM Standard D974-04 2004 *Standard Test Method for Acid and Base Number by Color-Indicator Titration* (Pennsylvania :ASTM International)
[33] Lieu T Yusup S and Moniruzzaman M 2016 *Bioresour. Technol.* **211** 248-256
[34] Shu Q Zou W He J Lesmana H Zhang C Zou L and Wang Y 2019 *Renew. Energy* **135** 836-845
[35] Yu S Wu S Lia L and Gea X 2020 *Fuel* **276** 118019
[36] Song C Qi Y Deng T Hou X and Qin Z 2010 *Renew. Energy* **35**(3) 625-628
[37] Heynderickx P M Chaemchuen S and Verpoort F 2020 *Data Brief* **31** 106027
[38] Ni Z Zhai Y Li F Wang H Yang K Wang B and Chen Y 2020 *Renew. Energy* **147** 719-729
[39] Tang Z E Lim S Pang Y L Shuit S H and Ong H C 2020 *Renew. Energy* **158** 91-102
[40] Özbay N Oktar N and Tapan N A 2008 *Fuel* **87**(10-11) 1789-1798
[41] Alenezi R Leeke G A Winterbottom J M Santos R C D and Khan A R 2010 *Energy Convers. Manage.* **51**(5) 1055-1059
[42] Gaurav A Dumas S Mai C T Q and Ng F T T 2019 *Green Energy Environ.* **4**(3) 328-341
[43] Mazubert A Taylor C Aubin J and Poux M 2014 *Bioresour. Technol.* **161** 270-279
[44] Chai M Tu Q Lu M and Jeffrey Yang Y 2014 *Fuel Process. Technol.* **125** 106-113

Investigation on continuous FAME production using rotor reactor with bumpy number variation under theoretical methanol to oil molar ratio condition

Prachasanti Thaiyasuit[1]*, Wuttisan Khiowthong[1], and Wirapan Seehanam[1]

[1]Department of Mechanical Engineering, Faculty of Engineering,
Ubon Ratchathani University, Warin Chamrap, Ubonratchathani, Thailand 34190

* Corresponding Author: prachasanti.t@ubu.ac.th

Abstract

Continuous methyl ester production process from waste cooking oil (WCO) was studied by using a transesterification process at room temperature. The aim of this research was to investigate the optimized production conditions and specific energy consumption (SEW*h/kg of neat biodiesel) of the bumpy surface rotor reactor. There are 6.9%, 13.8%, and 27.6% of area fractions (AF) respectively. The reaction was under the condition being theoretical methanol to oil molar ratio and the reactor volume was 0.5L. At 2,000-4,000rpm of rotor speed, 1.0-1.5%w/w of KOH and 1.132-5.110L/min of precursor flow rate are used. It was found that the highest FAME content (fatty acid methyl ester in neat biodiesel, %) being 98.6%. It was achieved by using AF 27.6% rotor reactor under experimental conditions including 3,000rpm, 1.50%w/w of KOH and 2.027L/min of precursor flow rate. In addition, this condition gave SEC of 12.5W*h/kg. It was lower than that of orifice hydrodynamic cavitation reactor, power ultrasonic reactor, and mechanical stirring reactor around 93.2%, 95.0%, and 97.5% respectively. The properties of produced FAME met both EN 14214 and ASTM D 6751 standards.

Keywords: Continuous biodiesel production, Bumpy surface rotor reactor, Specific energy consumption, Retention time

1. Introduction

Information Methyl ester or Biodiesel is an alternative fuel to compensate diesel. Biodiesel can be produced from edible oil such as palm, coconut, soybean, sunflower oil or animal oil. In addition, it can be produced from non-edible oil such as used vegetable oil, Jatropha oil, and it also contains oil from rubber granules [1, 2]. About 95% of biodiesel production in the world is derived from edible oil. The price of oils for edible was the main cost of biodiesel production and made the cost increase by 1.5-2.0 times the diesel fuel price [3, 4].

Currently, the raw material cost is a major part of the total biodiesel production cost which is about 75-78% [5]. Biodiesel production on a commercial scale was necessary to reduce raw material cost. For this reason, used oils such as waste cooking oils have become more attractive to the alternative raw material. Therefore, the energy efficiency and raw material cost were major contributors to decrease the total cost of biodiesel production [6, 7].

It was estimated that the WCO amount in Thailand had 74 million L/year approximately [8]. After the suspension was filtered out of WCO and passed oils to the dehumidification process that the treated oils had properties nearly the pure refined vegetable oils since it was low in percentage of free fatty acid (%FFA). It could be used to produce biodiesel by transesterification process [9, 10].

The optimized production conditions for mechanical stirring (MS) including 6:1 of MeOH to oil molar ratio, 1.0 %wt. of KOH, 60°C of reaction temperature, and 60 min of reaction time [11, 12]. The research by Ji, J. [1] showed the comparison to the biodiesel productions among MS, hydrodynamic cavitation (HC), and power ultrasonic cavitation (UC) which the optimized condition including 45°C of reaction temperature, using 6:1 of MeOH to oil and 1% w/w of KOH as a catalyst. MS operated at 900rpm, PU used the power150W at 19.7MHz, and the single-edged plate HC worked at 0.7MPa. They compared the SEC that sorted by MS, PU, and HC at 500 W*h/kg, 250 W*h/kg, and 183W*h/kg respectively. HC had a slightly lower reaction rate than PU while MS had the lowest reaction rate. The production of biodiesel by the MS via transesterification process was interrupted with the reaction rate limited by mass transfer of the oil and alcohol [13]. They could not be mixed homogeneously and the process was a reverse reaction as Equation 1 which had a limit for reversing by the discontinuity in the product circulation. It didn't benefit from continuous production.

$$TGR + 3MeOH \overset{Catalyst}{\rightarrow} GL + 3FAME \quad (1)$$

Where a mole of triglyceride (TGR) reacted with 3 moles of the methanol (MeOH) by using KOH as a catalyst that changed to a mole of glycerol (GL) and 3 moles of the fatty acid methyl ester (FAME).

HC was started with the cavitation phenomenon caused by the transition of status from liquid to vapor. Because the cavitation beginning point pressure was less than the vapor pressure over the surface of that substance. In order to the cavitation could be achieved by using the reactor at an approximately constant temperature within the precursor flow. Cavitation was similar to boiling which it was caused by a change in pressure that was controlled by the fluid dynamics, but it was not due to the change in temperature [14, 15]. The intensification technology, such as a bumpy surface rotor reactor (BSRR) could be increased the mass transfer rate between oil and alcohol at room temperature. It has some challenges for this research that still could be reduced the MeOH quantity, the SEC, and retention time respectively.

In this study, the WCO methyl ester was produced by using the BSRR via transesterification process and using theoretical MeOH to oil molar ratio. The optimized production conditions were investigated by studying the variables including rotor speed, [KOH] %w/w, precursor flow rate, and the investigated parameter was the SEC. The FAME content was compared when using a reactor with rotors AF 6.9%, 13.8%, and 27.6% that the properties of biodiesel produced met both of EN 14214 and ASTM D 6751 standards.

2. Materials and methods

2.1 Material

The waste cooking oil was bought from a local fried chicken shop nearby Ubon Ratchathani University, Thailand. The suspensions particles in WCO were removed by filtration through a filter 30 sieve and the moisture was removed by boiling at 110°C which treated WCO has 1.89% FFA. It was packed in a closed container when cooled down at 30-40°C. The WCO properties were summarized in Table 3. Methanol used anhydrous grade with 99.9% purity. Potassium hydroxide was used with industrial grade 90% purity.

2.2 Methods

The experimental equipment and setup show in Figure 1 that consisted of a precursor storage tank with a 40L capacity (1). The storage tank inlet was connected to the WCO feeder (2) and feeder of MeOH mixed with KOH (3). The precursor flowed pass to inlet valve (4) through pump model PMD-331B (5). The precursor flow rate into BSRR could be adjusted by opened ball valve (6) 16.5-25.5° to vary the flow rate from 1.132-5.110L/min. Within BSRR (7), it consisted of three type rotor AF 6.9%, 13.8%, and 27.6% of area fractions that relationship was shown in Equation 2, and equipment detail was shown in Figure 2. Inside the reactor had a gap between the rotor surface and the contact surface jacket within 4.5mm. At the bottom of the reactor was connected with a ball valve (8) for storing product samples. The inner temperature was measured by thermocouple Type K 0-399°C (9). The reactor was driven by Induction Motor 5HP model SF-JR (10). The speed control device was the inverter model VB5N 43P7G-S / 45P5P (11). The energy consumption was measured with LCD Watt-Hour 3 phase Energy Meter model DTM 024 (12).

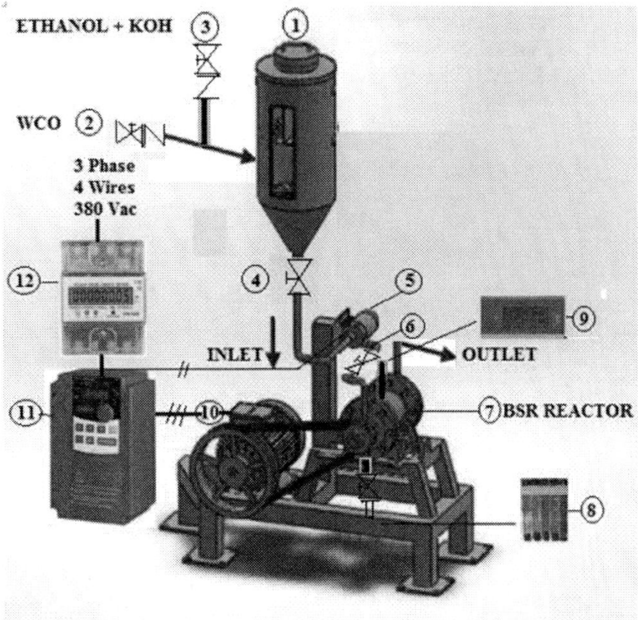

Figure 1. The Bumpy Surface Rotor Reactor (BSRR) with 0.5L capacity

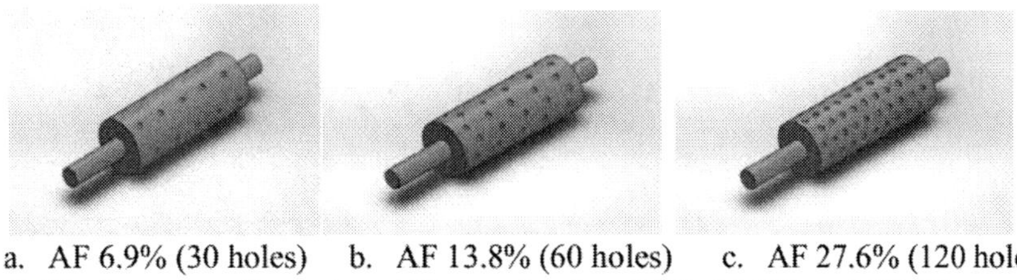

a. AF 6.9% (30 holes) b. AF 13.8% (60 holes) c. AF 27.6% (120 holes)
Figure 2. The bumpy surface rotors with area fraction 6.9%, 13.8%, and 27.6% respectively

$$Area\ Fraction = \frac{A_{Holes}}{A_{Rotor\ Surface}} \times 100 \quad (2)$$

Where Area fraction (%) is the percentage of the total area of holes A_{Holes} per the surface area of the rotor $A_{Rotor\ Surface}$

IOP Conf. Series: Materials Science and Engineering 1137 (2021) 012007 doi:10.1088/1757-899X/1137/1/012007

The properties of WCO and biodiesel were checked under standards that shows in Table 3. The percentage of FAME was evaluated according to EN14214 under EN14103. Regarding with gas chromatography (GC) analysis by CLARUS 680 GC equipped with CLARUS SQ 8 CMS which was shown in Figure 3.

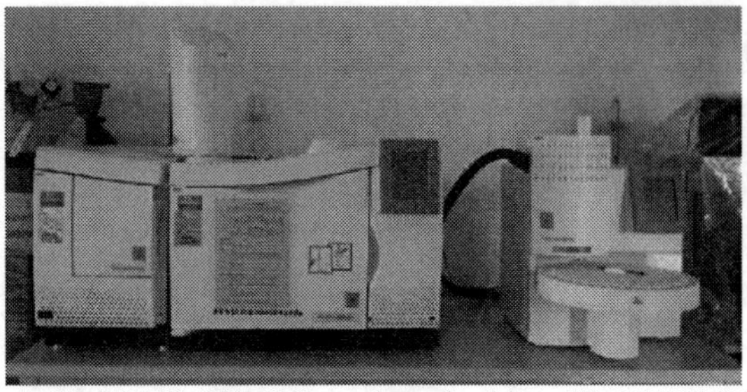

Figure 3. A CLARUS 680 GC equipped with a CLARUS SQ 8 C MS

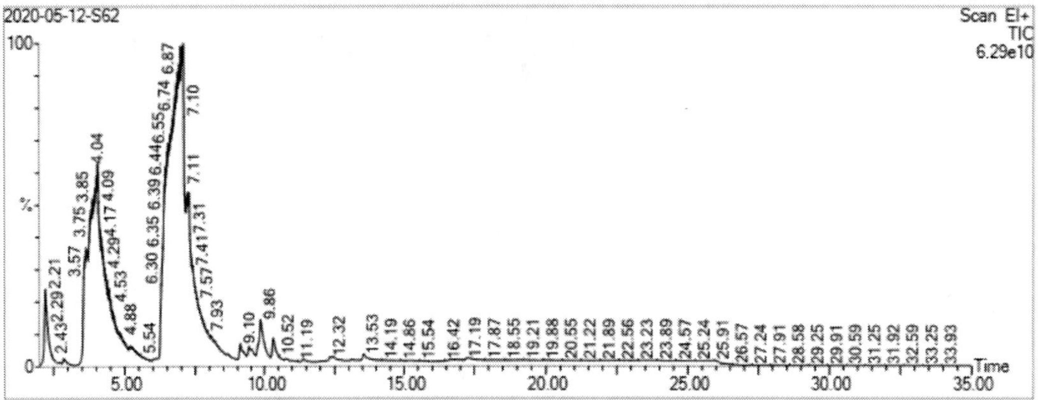

Figure 4. The GC graph of biodiesel was used to analyze for the summation of % FAME

The gas chromatography graph (GC graph) of neat biodiesel was analyzed by using the area under curve (AUC) method as show in Figure 4 which %FAME was investigated by using the gas chromatography technique. The FAME content in each experiment was calculated by the following expression as shown in Equation 3.

$$FAME\ content = \frac{A_{FAME}}{\sum A} \times 100 \quad (3)$$

Where *FAME content* (%) was the percentage of FAME in neat biodiesel which A_{FAME} (%) was the area under the methyl ester curve, $\sum A$ (%) was the total area under the curve acquired by GC.

2.3 Experiment

The experiment included two parts. In the first part, the methanol to oil molar ratio was set at 3:1 with a constant speed of 3000rpm. The BSRR with rotor AF 6.9%, 13.8%, and 27.6% were investigated by comparison the effect to FAME content. In addition, the variables were checked by chose the optimal variables that were flow rate (1.132, 2.027, 3.000, 4.128, and 5.110L/min) and [KOH] (1.00, 1.25, and 1.50%w/w).

The second part was an experiment to find a suitable speed by using the [KOH] %w/w and the flow rate from the first part with theoretical molar ratio. The sampling collection time was 1min for all two parts of the experiment. The optimal variable of the best result was set up to repeat three times. The electrical energy consumption and mass of neat biodiesel were recorded and used to calculate parameter SEC as shows in Equation 4.

The specific energy consumption, SEC (W*h /kg) was calculated by the following expression.

$$SEC = \frac{EC}{m_{NB}} \tag{4}$$

Where energy consumption, EC (W*h) was the total energy used to produce a net mass of neat biodiesel, m_{NB} (kg).

3. Results and discussion

3.1 Effect of KOH concentration

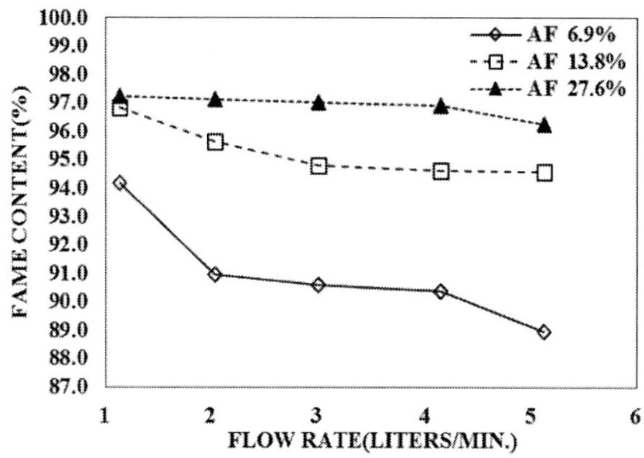

Figure 5. Effect of flow rate on FAME content with different area fraction: [KOH] 1.00%w/w

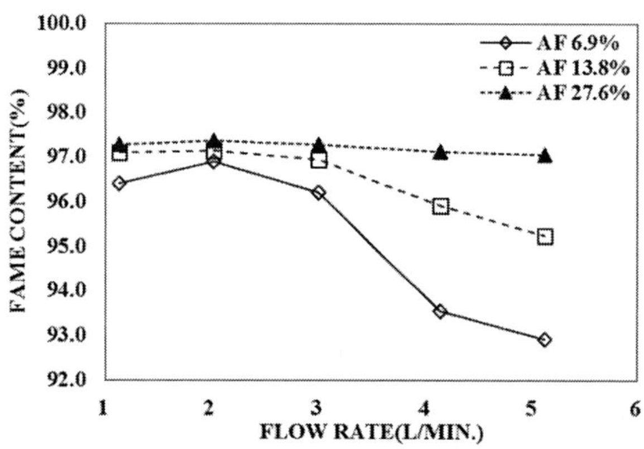

Figure 6. Effect of flow rate on FAME content with different area fraction: [KOH] 1.25%w/w

Figures 5 to 7 show the [KOH] %w/w affected the conversion of triglycerides to methyl esters with various flow rates at different area fraction. They show the influence of [KOH] %w/w on the evolution of ester content with varied flow rate as shown in Figure 5, 6, and 7 by increasing of FAME content varied on the increasing of AF. The best results were reached with further increased in [KOH] %w/w and the optimal was [KOH] 1.50%w/w. All of the experiment results FAME content were greater than 96.5% which the best FAME content of rotor AF 6.9%, 13.8%, and 27.6% were 97.8%, 98.2%, and 98.6% respectively with a flow rate 2.027L/min as shown in Figure 7.

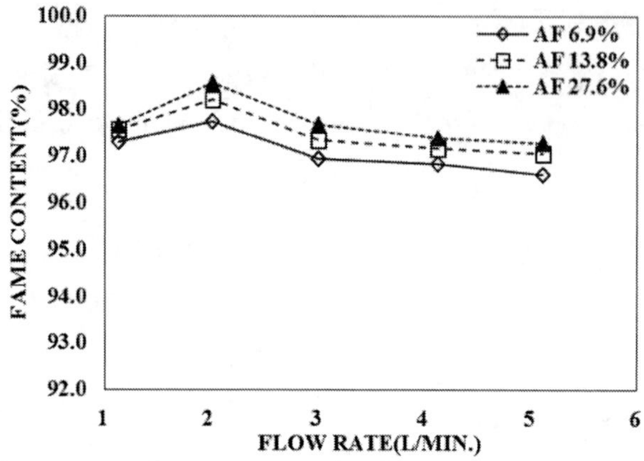

Figure 7. Effect of flow rate on FAME content with different area fraction: [KOH] 1.50%w/w

3.2 Effect of rotor speed

Figure 8 shows the rotor speed affected the FAME content to vary on the increasing of AF. It shows the influence of speed on the FAME content that it increased when speed was increased. The experiment gave the best results at 3000rpm that FAME content was 98.6% by rotor AF 27.6% and [KOH] 1.50%w/w.

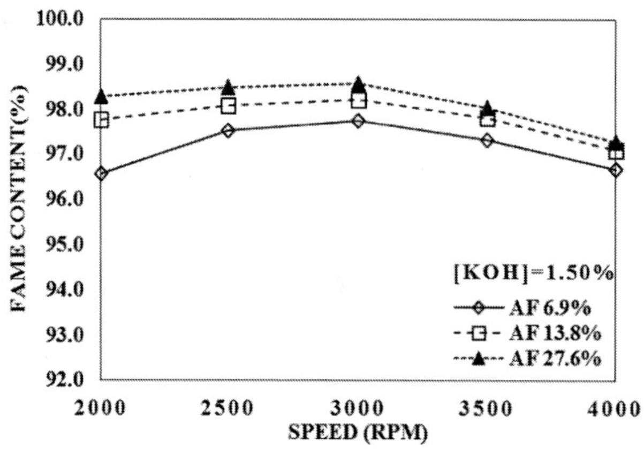

Figure 8. Effect of speed on FAME content with different area fraction

3.3 Effect of rotor speed on temperature

Figure 9 shows the increased speed affected the increasing of temperature which varied on each AF. When AF was constant, the influence of speed increased in the range of 2000-4000rpm and affected the temperature to increase 2°C. All temperatures were recorded after operated for 1 min at initial temperature of 30°C without an external heat source. When the rotor speed was constant and AF was increased while temperature was increasing in range 1-4°C.

IOP Conf. Series: Materials Science and Engineering 1137 (2021) 012007 doi:10.1088/1757-899X/1137/1/012007

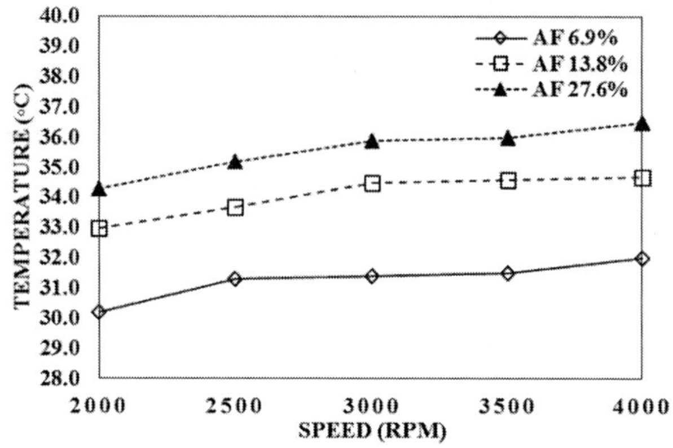

Figure 9. Effect of speed on temperature with different area fraction without an external heat source

3.4 Specific energy consumption

Table 1 shows the increasing of SEC when speed was increased and AF was constant. At the same speed shows the decreasing of SEC since m_{NB} increased when AF was increased. The best results at the optimal conditions showed that SEC with AF 27.6% was 12.50 W*h/kg.

Table 1. Effect of speed on SEC with MeOH to oil ratio 3:1 and [KOH] 1.50%w/w by different AF

Speed (rpm)	AF 6.9%			AF 13.8%			AF 27.6%		
	EC W∗h	m_{NB} kg	SEC W∗h/kg	EC W∗h	m_{NB} kg	SEC W∗h/kg	EC W∗h	m_{NB} kg	SEC W∗h/kg
2000	10	1.35	7.41	10	1.50	6.67	10	1.55	6.45
2500	15	1.50	10.00	15	1.55	9.68	15	1.55	9.68
3000	20	1.50	13.33	20	1.55	12.90	20	1.60	12.50
3500	25	1.50	16.66	25	1.55	16.13	25	1.60	15.63
4000	30	1.50	20.00	30	1.55	19.35	30	1.60	18.75

Table 2. Comparison of BSRR performance with different reactors [1]

Description	Reactor Type			
	BSRR	HC	PU	MS
SEC (W*h/kg)	12.5	183	250	500
Methanol to Oil by mol.	3:1	6:1	6:1	6:1
Temp. °C	Room Temp.	45	45	60
KOH wt.%	1.50	1.00	1.00	1.00
Time (min.)	Retention time	Reaction Time		
	0.25	30.0	10.0	60.0

3.5 Performance

The optimal conditions gave SEC of 12.5W*h/kg which less than HC reactor, PU reactor, and MS reactor to 93.2%, 95.0%, and 97.5%. The retention time was 0.25min which less than HC, PU, and MS reactor to 29.8 min, 9.8 min, and 59.8 min respectively as shown in Table 2. The produced FAME physicochemical properties are met both ASTM D6751 and EN 14214 standards as shown in Table 3.

Table 3. Physicochemical properties of WCO and biodiesel in comparison to biodiesel standard

Properties	WCO	WCO biodiesel	ASTM D6751	EN 14214	Standards /Method
Density (kg/m^3, 15°C)	915.0	880.0	-	860-900	ASTM D1298
Kinematic viscosity (cSt,40°C)	40.50	3.52	1.9-6.0	3.50-5.00	ASTM D445
Sulfur content (mg/kg)	10.50	10.00	15 max	10 max	ASTM D2622
High heating value (MJ/kg)	38.65	39.74	-	-	ASTM D240
FAME content	-	98.6	-	96.5 min	EN 14103

3.6 The comparison of production conditions

In early studies of the FAME production compared to the production conditions studied as shown in Table 4 found that BSRR can be operated at room temperature. This reactor didn't want to use a reactant preheating as the conventional method, it used the lowest methanol to oil molar ratio only 3:1, using 1.50%w/w of KOH, 3000rpm of rotor speed, and 2.027L/min of flow rate. This continuous production used the SEC only 12.5W*h/kg of neat biodiesel and conversed high FAME content 98.6%.

Table 4. The FAME production conditions summary by using WCO with methanol as reactant

Researchers	Catalyst (%w/w)	alc./Oil Molar Ratio	Reaction Condition	FAME Content (%)	Other details
Chen G [16]	CH_3ONa, 0.75 NAOH, 0.75	6:1	65°C 3min 600rpm	97.9 96.2	MC 1200W FFA<1% MC 750W FFA<1%
Azcan N [17]	CH_3ONa ,1.00	6:1	60°C 5min	98.8	MC Power not available
Barnard T M [18]	KOH, 1.00	6:1	50°C 0.5min.	97.9 2.0 L/min	MC 1.6kW 0.11% FFA Con. Pro. with Teflon tubing coil 8cm ID
Maghami M [19]	KOH, 1.00	6:1	55°C 30min	79.9	100% UC 400W 20kHz
Pukale D D [20]	K_3PO_4, 3.00	6:1	50°C 90min	92.0	50% UC 375W 22 kHz
Anuar M R [21]	Hydrotalcite ,0.08	15:1	57°C 1min	76.5	55% UC 400W 20kHz
Maddikeri G L [22]	CH_3OK, 1.00	12:1	40°C 30min	90.0	60% UC 450W 22 kHz
Ghayal D [6]	KOH, 1.00	6:1	60°C10 min	95.0	HC 7.5kW Cap. 10L
Maddikeri G L [23]	CH_3OK, 3.00	12:1	40°C30 min	89.2 90.0 70.0	HC 3 bar Cap.15L UC 60%450W 22kHz MS 1000rpm
Chuah L F [24]	KOH,1.00	6:1	60°C15 min 60°C90 min	98.0 97.0	HC4kW 2bar50L MS
Chen K S [25]	CH_3ONa, 0.75	15:1	65°C 3 min	97.9	MW 1.2kW FFA<1%
Thaiyasuit P	KOH,1.50	3:1	Room temp. Continuous Production	98.6 2.027 L/min	BSR Reactor Cap0.5L speed 3000rpm, SEC12.5W*h/kg

4. Conclusion

In summary, a transesterification process with BSRR was applied to carry out the experiment to produce high purity biodiesel with WCO which having a low %FFA. The continuous production condition is an efficient method that saved a lot of energy, time, and cost.

The optimum conditions for continuous production of WCO methyl ester under BSRR conditions and theoretical molar ratio of 3:1 were as follows; the first independent variable was the alkaline catalysed concentration of KOH %w/w which affected the conversion of TGR to FAME when [KOH] %w/w was constant, increasing of FAME content varied on the increasing of AF. When AF was constant, the best results were reached with a further increased in [KOH] %w/w and the optimum was [KOH] of 1.50%w/w. The second variable was the rotor speed that affected the FAME content when the speed was constant and AF was increased and affected the FAME content to be increased. The best result was a rotor speed of 3000 rpm at the precursor flow rate of 2.027L/min with 0.25 min of retention time. The highest FAME content of rotor AF 6.9%, AF 13.8%, and AF 27.6% under the achieved optimum condition were 97.8%, 98.2%, and 98.6% respectively.

When BSRR was operated at room temperature that HC was started with the cavitation caused by the transition of precursor flowed. The status of methanol in the mixture of KOH and WCO was changed from liquid to vapour. At the beginning point of cavitation, the pressure in the reactor was less than the vapour pressure over the surface of that substance. The cavitation could be achieved by using the BSRR which was caused by a change in pressure at an approximately constant temperature with the changing in the narrow range. After operating the reactor for 1 min at an initial temperature of 30°C when speed was constant and the AF increased, the result was showed that the temperature increased in the narrow range of 1-4°C. For all of the AF when the speed was changed to increase in the range of 2000rpm to 4000rpm which affected the temperature to be increased in the range of 2°C. Therefore, this reactor was not required to use reactant preheating as the conventional method.

The SEC was a parameter that increased when AF was constant and rotor speed was increased. EC was dependent on the rotor speed when the rotor speed was constant and AF was increased in which affected the neat biodiesel mass to increase, but EC was constant with the rotor speed, so SEC was decreased. The optimum conditions showed that SEC with AF 27.6% was 12.50 W*h/kg with EC of 20 W*h but could be produced neat biodiesel 1.6kg in 1 min. Finally, the performance comparison of SEC with the other reactor found that BSRR was less than the HC reactor, PU reactor, and MS reactor to 93.2%, 95.0%, and 97.5%. The retention time was 0.25min which less than HC, PU, and MS reactor to 29.8 min, 9.8 min, and 59.8 min, respectively. The results were showed that the FAME content increased due to an increase in the AF that the most optimized BSRR was achieved by using AF 27.6% (120 holes) rotor reactor. It was found that the highest FAME content at 98.6% was met both ASTM D6751 and EN 14214 standards.

Acknowledgments

Thank you to the Department of Mechanical Engineering, Faculty of Engineering, Ubon Ratchathani University that supports biodiesel plant to be research site, and thank you to the Scientific Instrument Center that supports the examination of %FAME with GC/MS and biodiesel physicochemical property.

References

[1] Ji J Wang J Li Y Yu Y and Xu Z 2006 *Ultrasonics* **44** 411
[2] Thaiyasuit P Pianthong K and Worapun I 2012 *Journal of Oleo Science* **61** 81
[3] Tan K Lee K and Mohamed A 2011 *Energy* **36** 2085
[4] Maddikeri G L Pandit A B and Gogate P R 2012 *Ind. Eng. Chem. Res.* **51** 14610
[5] Zhang Y Ghayal D M A Mclean D D and Kates M 2003 *Bioresource Technology* **90** 229
[6] Ghayal D Pandit A B and Rathod V K 2013 *Ultrason. Sonochem.* **20** 322
[7] Klemes I and Huisingh D 2008 *J. Clean. Prod.* **16** 159
[8] Punsawat T 2007 *Biodiesel: EPPO Thailand* (Bangkok: Graphical Company Limited)
[9] Enweremadu C C and Mbarawa M M 2009 *Renew Sustain Energy* **13** 2205

[10] Corro G Tellez N Jimenez T Tapia A Banuelos F and Vazquez C O 2011 *Catal Today* **166** 116
[11] Phan A N and Phan T M 2008 *Fuel.* **87** 3490
[12] Dyneshwar G Aniruddha B P and Viranda K R 2013 *Ultrasonics Sonochemistry* **20** 322
[13] Zheyan Q Lina Z and Laurence W 2010 *Chemical Engineering and Processing:Process Intensification* **49** 323
[14] Arojo S and Benito Y 2008 *Ultrasonics Sonochemistry* **15** 204
[15] Franc J P and Michel J M 2004 Fundamentals of Cavitation (Netherlands: Springer).
[16] Chen G Shan R Shi J and Yan B 2014 *Bioresour. Technol.* **171** 428
[17] Azcan N and Yilmaz O 2013 *Fuel* **104** 31
[18] Barnard T M Leadbeater N E Boucher M B Stencel M and Wilhite B A 2007 *Energy & Fuels* **21** 1777
[19] Maghami M Sadrameli S M and Ghobadian B 2015 *J. Appl.Therm. Eng.* **75** 575
[20] Pukale D D Maddikeri G L Gogate P R Pandit A B and Pratap A P 2015 *Ultrason. Sonochem.* **22** 278
[21] Anuar M R and Abdullah A Z 2016 *Appl. Catal. A General.* **514** 214
[22] Maddikeri G L Pandit A B and Gogate P R 2013 *Fuel Process.Technol.* **116** 241
[23] Maddikeri G L Gogate P R and Pandit A B 2014 *Fuel* **137** 285
[24] Chuah L F Yusup S Aziz A R A Bokhari A and Abdullah M Z 2016a *J. Clean. Prod.* **112** 4505
[25] Chen K S Lin Y C Hsu K H and Wang H K 2012 *Energy* **38** 151

Design and fabrication of a solar-dish concentrator with 2-axis solar tracking system

Pongnarin Savangvong[1], Banterng Silpsakoolsook[2], and Sutapat Kwankoameng[1,*]

[1]Department of Mechanical Engineering, School of Engineering, King Mongkut's Institute of Technology Ladkrabang, Bangkok, 10520, Thailand
[2]Department of Chemistry, Faculty of Science, Siam University, Bangkok, 10160, Thailand

* Corresponding Author: sutapat.kw@kmitl.ac.th

Abstract
This paper presents an economical parabolic solar collector including design analysis, fabrication, and testing of the solar concentrator with a dual-axis solar tracking system. The design goals of the solar dish are high concentrating temperature and power density but low weight and cost. To achieve a high-efficiency dish, minimizing loss is important therefore the design focuses on high reflectability, low heat absorption with an optimum dish profile. The dish is a parabolic shape with two layers of different materials. Three reflector materials, an emergency blanket, PET-Mirror, and Mirror are introduced on the top layer as the solar reflectance. The second layer, fiberglass that is lightweight, easy to shape up in various features, and cheap, is used as a dish structure. The dish is installed with a 2-axis solar tracking system and supported by a steel column. The ¼ inch copper tube is coiled as a receiver. The receiver is placed at the focal point of the solar dish. To evaluate the heat at the focal point, water is pumped through the receiver coil to absorb the heat from the focal point. The heat output of the solar dish is calculated from the temperature difference of the water inlet and outlet of the receiver coil. The temperature at the focal point was presented along with the heat power output, and solar dish efficiency.

Keywords: Solar concentrator, Parabolic dish, Solar tracking

1. Introduction

Global warming is a serious environmental problem because it wildly affects the climate around the world. Causing many parties to take this issue seriously. The rise in global temperatures is caused by ozone in the atmosphere, which reflects some of the solar radiation, being destroyed by CFCs, NOx, CO_2 [1]. One of the main sources of these gases is energy production from non-renewable fossil fuels. However, we cannot cut down on the amount of energy we use because the various economic activities tend to require more energy. Therefore, the sustained solution to energy demand and global warming is to find clean energy sources that can be circulated back and forth unlimitedly.

Solar energy is the most potent and abundant permanent renewable energy source. The amount of sunlight that is continuously hitting Earth's atmosphere is about 1.75×105 TW, around 60% or 1.05×105

TW transmission through atmospheric reaching the Earth's surface continuously [2].

There are two main categories of technology to utilized solar radiation: solar thermal and solar photovoltaics (PV). For solar thermal, solar radiation is converted to heat by thermal collectors, while PV produces directly electricity. The advantages of solar thermal energy over PV are higher efficiencies, lower investment costs, and solar thermal can store thermal energy efficiently and may use other fuels such as natural gas or biogas as a backup to ensure continuous operation [3]. The solar collector working as a heat exchanger that harvests the solar radiation and converts it to heat, then transfer to a fluid flowing through the collector [4]. The solar collector technologies can be divided by tracking type as non-tracking, single-axis tracking system, dual-axis tracking system as in figure 1[5]. The dual-axis tracking system, which concise of a linear Fresnel reflector, solar tower receiver, and parabolic dish concentrator (PDCs), provides the highest efficiency [6].

The PDCs are a compact focal point that can achieve very high concentration ratios up to 1000. At temperatures exceeding 1,000°C, they can produce power efficiently by utilizing high energy conversion cycles [7]. The PDCs can couple with various applications such as water heating receiver, Stirling engine, etc. [8]. The PDCs that are coupled with the Stirling engine with the typical cycle efficiency of ~38% may behave annual solar-electric efficiency of ~25% [6].

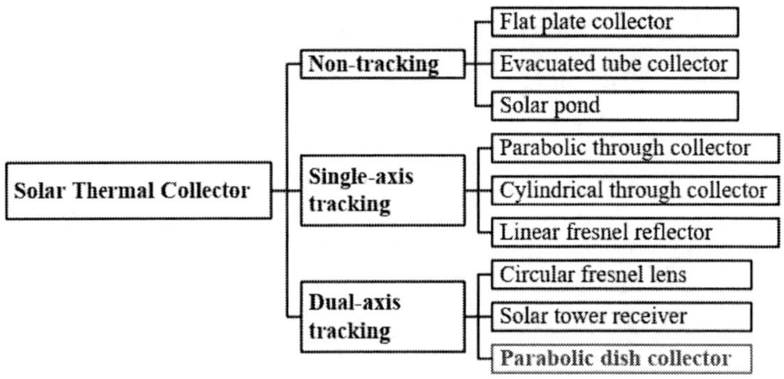

Figure 1. Type of solar collector.

For the study and experiment of PDCs, many researchers study the operation of the system by building and experimenting. Bianchini et al were presented the performance assessment of a solar parabolic dish for domestic use. They evaluate the performance of a small-scale commercial parabolic dish of 11.5 kW and shown the result in different environmental conditions [9]. Pavlovic et al. have presented the experiment of a prototype solar dish collector. A mathematical model was used to evaluate the system performance of the collector in many operating cases [10]. Mavire and Taole studied the thermal performance of the cavity receiver for parabolic dish concentrator namely SK-14. Different types of receivers and focal lengths of parabolic dishes were also tested with the experimental setup [11]. Hijazi et al. were presented the mechanical design of a low-cost parabolic solar dish concentrator. One of the highlights of their study is the optimum size of the reflector to be cut to reduce both cutting cost and sheets cost [12].

Several related papers on the parabolic concentrator dish were shown that the solar dishes are high-cost devices and very heavy [9, 10, 12]. They need very large nearly perfect curve mirrors to effectively concentrate solar radiation. In order to build a dish prototype for study and experiment requires a considerable amount of budget for the whole dish structure and reflective material. Therefore, this research is to investigate the feasibility of fabricating a prototype of the solar dish system at the lowest cost and weight, while maintaining at high the efficiency of the solar collector system.

2. Background and theory

2.1 The sun

This section provides the background theory, radiation from the sun, and its position observation from the earth, to understand the physics of solar radiation for the analysis of solar energy systems.

2.1.1 Heat and radiation from the sun

The sun is a 13.9 × 105 km diameter sphere composed of many layers of gases. At its surface, the fusion reaction is continuously reacted and releases the energy of 3.8×1023 kW in form of an electromagnetic wave, only a tiny fraction approximately 1.7x1014 kW, is intercepted by the earth which is located approximately 150 million km from the sun [1]. The sun radiates energy to the earth is referred to as solar radiation. As the solar electromagnetic energy passes through the atmosphere of the earth, while the solar energy levels are around 1 kW/m2 when it reaches the surface of the earth [2]. The amount of solar energy per unit area received is normally define as irradiance, qs which varies with geographical position on the earth, the orientation of the collector, meteorological conditions, and the time of day [3].

For concentrating systems, the net solar energy, Qs is proportional to the collector area Aa and the proportionality of the solar energy falling on the reflector, qs (W/m2)[3]. The thermal efficiency is defined as the ratio of the useful energy delivered to the energy incident on the collector aperture. Therefore, the radiation falling on the receiver q0 is a function of the optical efficiency, which accounts for all errors [3].

2.1.2 The Sun Position

According to the earth move around the sun and rotates daily about its polar axis, it is important for the parabolic dish concentrator system to mathematically calculate the position of the sun [2]. The coordinates of the location of the sun at any instant of time can be expressed by two variables, Azimuth and Elevator angle as shown in Figure 2. Both angles are the function of the site location and time as Equations 1 and 2 respectively [2].

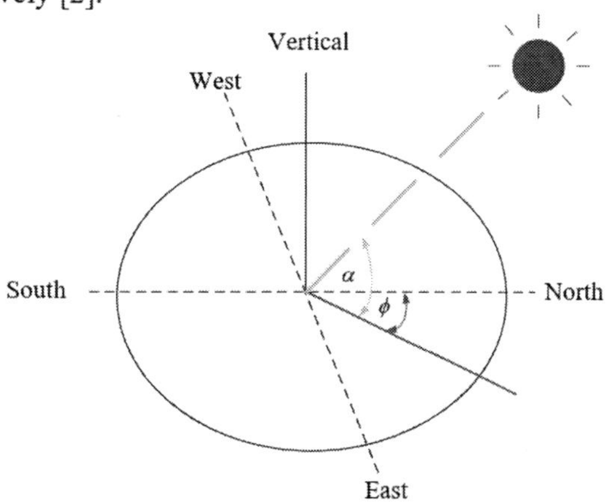

Figure 2. Azimuth and elevator angle.

$$\phi = \cos^{-1}\left(\frac{\sin(\delta)\cos(\varphi) - \cos(\delta)\sin(\varphi)\cos(\omega)}{\cos(\alpha)}\right) \tag{1}$$

$$\alpha = \sin^{-1}(\sin\delta\sin\varphi + \cos\delta\cos\varphi\cos\omega) \tag{2}$$

2.2 Collector

Solar thermal collectors are classified into different tracking structural designs and working principles as shown in figure 1. The parabolic dish concentrator (PDC) is a single focal point type so that it provides the highest temperature compare to other collector structure. The detail of the PDC and the thermal analysis were presented in this section.

2.2.1 Parabolic dish

The equation of a parabolic in the x-y coordinate and focal point is given in Equations 3 and 4 respectively[1].

$$y = \frac{x^2}{4f} \tag{3}$$

$$f = \frac{D^2}{16d} \tag{4}$$

Where f is the focal point of the dish, D is the diameter of the dish and d is the depth of the dish as detail in Figure 3. The rim angle ϕ_r is the angle between the sunray and a line from the focus to the concentrator dish edge. The rim angle affects the radiation ray of the sun and the manufacturing of the parabolic dish. The rim angle of a parabolic can be written as Equation 5.

$$\phi_r = \tan^{-1}\left(\frac{D/2}{f-d}\right) = \tan^{-1}\left(\frac{4fD/2}{4f^2-(D/2)^2}\right) \tag{5}$$

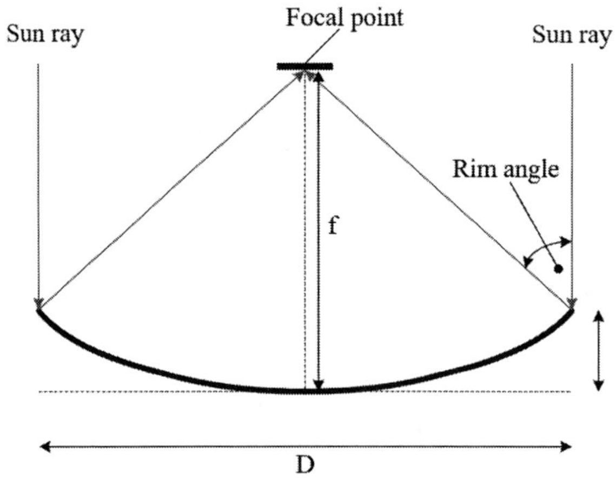

Figure 3. Schematic of the parabolic curve.

Another parameter that indicates the solar dish is the concentrating ratio. Due to the flux density of solar radiation is estimated at not over 1000 W/m2. Therefore, it needs to concentrate to obtain higher solar radiation. So that the solar concentration ratio is an important concept for the solar collector. The concentration ratio, C is defined as the ratio of the aperture area to the receiver area as written in Equation 6.

$$C = \frac{A_a}{A_r} \tag{6}$$

2.2.2 Thermal analysis

As mention in section 2.1.1 that the amount of heat obtained from the dish is the radiation intensity (qs) of light that hits the light multiplied by the aperture area (Aa) as in Equations 7 and 8. To analyze the concentrator efficiency, we consider the heat loss, (Q_{loss}) from the receiver to ambient temperature as follows.

$$Q_s = q_s A_a \quad (7)$$

$$q_0 = n_0 q_s = \frac{n_0 Q_s}{A_a} \quad (8)$$

$$Q_{loss} = Q_s - Q \quad (9)$$

$$Q_{loss} = U_r A_r (T_r - T_0) \quad (10)$$

Where U_r is the overall heat transfer coefficient and A_r is the receiver area.

From equation 10, if all heat loss to the ambient then the maximum temperature occurs at the reciever[3] and the temperature can be write in dimensionless as equation 11.

$$\theta_{max} = \frac{T_{r,max}}{T_0} = 1 + \frac{Q_s}{U_r A_r T_0} = 1 + \frac{q_r A_a}{\eta_0 U_r A_r T_0} = 1 + \frac{q_r C}{\eta_0 U_r T_0} \quad (11)$$

The collector efficiency is

$$\eta_c = \frac{Q}{Q_s} = 1 - \frac{\theta - 1}{\theta_{max} - 1} \quad (12)$$

3. Design and Manufacture

3.1 Design and Manufacturing the Parabolic Dish

This section provides the design information and fabrication technique of parabolic solar dishes. The solar dish collector designed with a diameter of 1 meter, and a rim angle of 45°, and other specifications are shown in Table 1. The diameter of 1 meter was constrained by the size of 1.2mx2.4m of the emergency blanket.

Table 1. The design parameter of a parabolic dish.

Specification	Value
Diameter	1 m
Rim Angle	45°
Depth	0.01 m
Focal point	0.06 m
Aperture area	0.785 m^2
Receiver area	0.00785 m^2
Concentrating ratio	100

3.1.1 Parabolic mold

In order to fabricate a fiberglass parabolic dish, it is necessary to create a parabolic mold. In this research, a highly reflective and flexible emergency blanket was applied as a mold. The emergency blanket was stretched by the wood frame as shown in Figure 4 (left). The stretched blanket was glued in a circle with a diameter of 1 meter onto the PVC sheet.

To forming a parabolic shape of the blanket, a pressurized technique was used. Air was compressed slowly into the cavity between the PVC sheet and the blanket to push the blanket to be inflated until achieving the design curve as Figure 4 (right). Note that, parabolic moles with different depths can be easily created by increasing or decreasing the amount of compressed air.

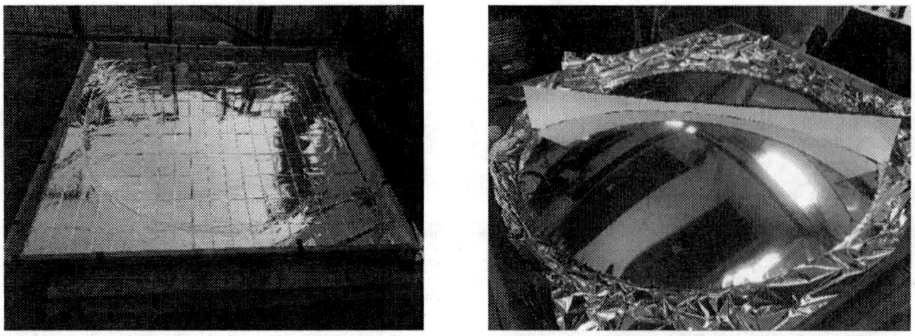

Emergency blanket glued on PVC sheet Parabolic mold from the blanket

Figure 4. The emergency blanket is a mold of the parabolic dish.

3.1.2 Fiberglass and resin

Once the parabolic mold is obtained, the fiberglass sheet no.450 was used to strengthen the resin no. CPS-515 PTW. As shown in Figure 5 (left), the fiberglass sheet was placed on the emergency blanket mold, and Figure 5 (right) was shown the resin is poured over the fiberglass to achieve the desired parabolic shape.

Figure 5. The fiberglass and resin dish.

3.2 The reflector

The reflector material should be excellent reflective to reflect a direct beam efficiently into the receiver. However, the material must be available, low cost, and quality. In this paper, three reflector materials, emergency blankets, PET-mirror, and mirrors were studied.

Emergency Blanket PET-Mirror Mirror

Figure 6. Three types of the reflector dish

Three resin dishes obtained from the casting process in section 3.1 are attached to three reflective materials: emergency blanket, PET-mirror, and mirror. Dish 1 is a cast resin plate with an existing emergency blanket as a reflective material. Dish 2 and 3 are attached PET-mirror and mirrors over the emergency blanket, respectively as in Figure 7. The data of the 3 dishes are shown in Table 2

3.3 The tracking system

3.3.1 Tracking structure

The parabolic concentrator is a point-focus optical system, It is, therefore, necessary to have a dual-axis solar tracking system for the highest efficiency. In this research, three dishes with a diameter of 1 m need to test simultaneously to maintain the same condition. Thus, a two-axis sun tracking system as show in Figure 7 was built. The structure is built to be lightweight and easy to transport. Hence, the elevator and azimuth can be disassembled.

Table 2. Data of each dishes.

Data	Dish 1	Dish 2	Dish 3
Diameter(m)	1	1	1
Reflector material	Emergency blanket	PET-Mirror	Mirror
Mass(Kg)	4.04	4.82	6.31
Cost(THB)	550	1,250	2,325

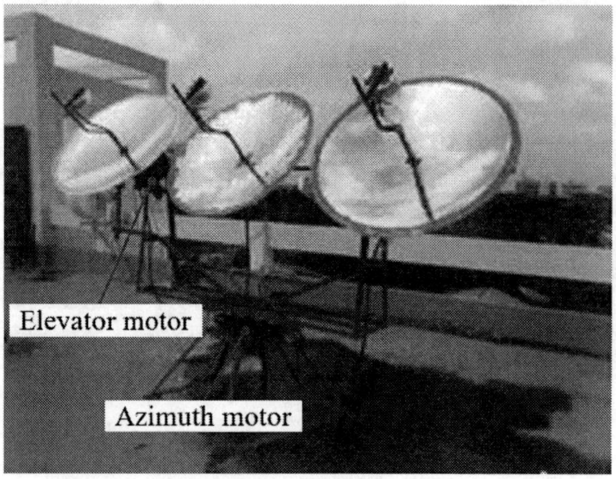

Figure 7. The structure of the tracking system.

3.3.2 Tracking programming

To control the tracking system, the Arduino Mega board was used along with a real-time clock (RTC) to calculate the real-time position (azimuth and elevator angle) of the sun. Two absolute encoders were us to measure the real value of azimuth and elevator angle. The program in the Arduino board is comparing both angles from calculation and encoder to drive azimuth and elevator motor to track the sun as a closed-loop system.

4. Experimental setup

The experimental setup was place Department of Mechanical Engineering, Faculty of Engineering, King Mongkut's Institute of Technology Ladkrabang on 14th September 2020. In order to investigate the performance of the solar dish with different reflector material, the experiment was set up as shown in Figure 8-9. The water pump was used to pump water pass through the receiver to absorb heat. The rotameter with a range of 0.4-2.4 LPM was used to indicate the flow rate of water through the receiver of each dish.

To test all dishes with the same condition, all three dishes were placed on the 2-axis tracking system as in Figure 8. The tracking control box calculates and tracks the sun's position every minute to turn the dish to face the sun. The 12VDC battery was used as the power source of the tracing system.

In each dish, three K-type thermocouples were used. Two of them were used to measure the temperature of the water inlet and outlet of the receiver, respectively. Another one was placed at the focal point of the dish as shown in figure 9 to measure the temperature of the focal point. Nine thermocouples were used for three dishes. All of the thermocouples were wired to the Wisco AI 210 data locker to collect the value of all thermocouple temperature. The useful power output from the receiver or receiver power is given by Equation 13 [4].

$$Q_{heat} = mc_p (T_{out} - T_{in}) \tag{13}$$

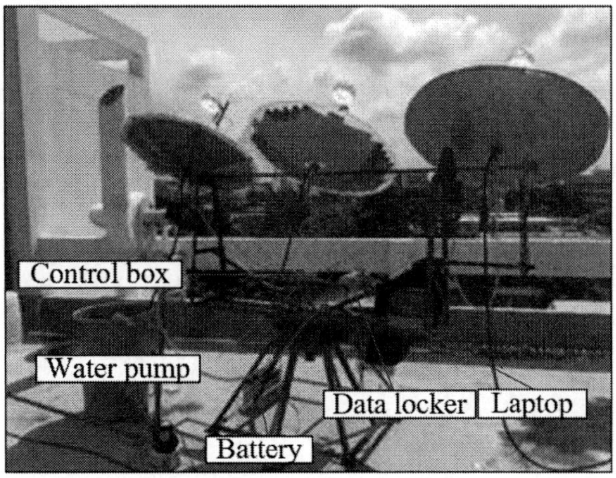

Figure 8. Experimental Setup.

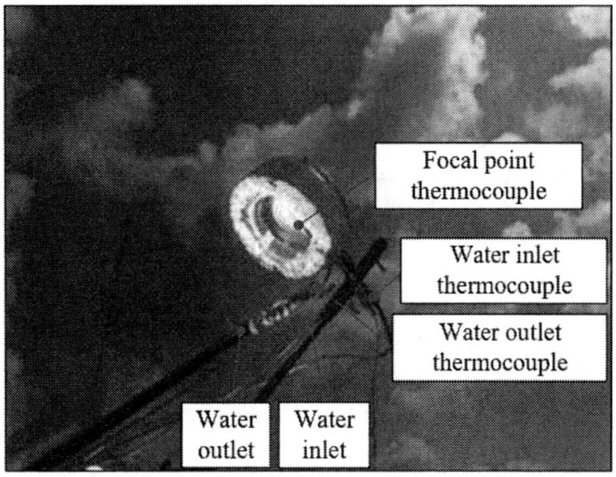

Figure 9. Thermocouple at the focal point.

5. Result and discussion

The efficiency of three solar parabolic dishes with a diameter of 1 meter and a rim angle of 45° was tested. The reflector material of each dish was made from an emergency blanket, PET-Mirror, and Mirror, respectively. The temperature of the focal point of each dish was shown in Figure 10.

In Figure 10, during the experiment period from 9.00 am-3.00 pm, the temperature at the focal point of all dish was ranged between 200 and 875°C depend on the weather condition. The maximum temperature of dish 1, 2, and 3 are 817°C, 724°C, and 874°C, respectively. The average temperature of dish 1, 2, and 3 are 582°C, 506°C, and 643°C, respectively.

Figure 11 shows the temperature of the water outlet from the receiver of each dish. The maximum water outlet temperature of dish 1, 2, and 3 are 42.4°C, 41.5°C, and 43.4°C, respectively. The average water outlet temperature of dish 1, 2, and 3 are 38.1°C, 37.1°C, and 39.0°C, respectively. The heat power of dish 1, 2, and 3 are 389W, 361W, and 420W, respectively.

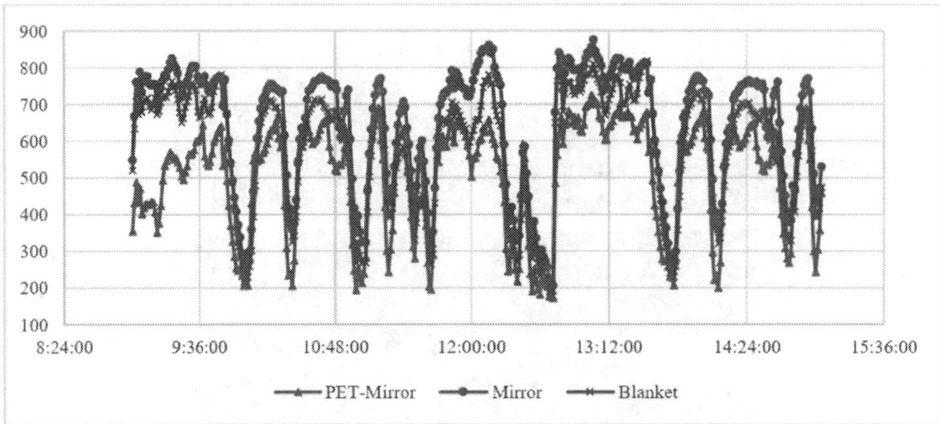

Figure 10. The focal point temperature of each dish.

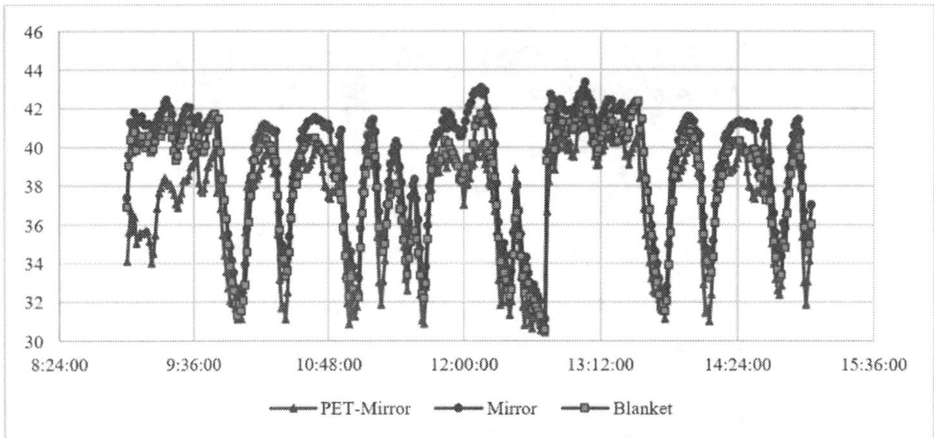

Figure 11. The temperature of the water outlet from the receiver of each dish.

Table 2. was shown the data and experimental results of the three dishes. Besides the maximum temperature and heat power, other interesting ratios are mass and cost of each per its heat power. The result was shown that dish 1 with an emergency blanket has the lowest mass per heat power and cost per heat power of 1.04 x10-3 k/W and 1.41 THB/W, respectively.

Table 3. Summarize experimental results of each dishes.

Data	Dish 1	Dish 2	Dish 3
Diameter (m)	1	1	1
Reflector material	Emergency blanket	PET-Mirror	Mirror
Mass (Kg)	4.04	4.82	6.31
Cost (THB)	550	1,250	2,325
Maximum focal point temperature (C)	817	724	874
Maximum heat power (W)	389	361	420
Mass per heat power (x10^{-3}k/W)	1.04	1.34	1.50
Cost per heat power (THB/W)	1.41	3.46	5.54

6. Conclusion

First of all, this paper presents the parabolic dish fabrication technique. Forming a resin parabolic dish from an emergency blanket mold is a flexible technique. This technique allows the fabrication of a dish of various sizes and rim angles at which low cost. Therefore, it is ideal to create a parabolic dish for research purposes.

The performance of three solar parabolic dishes with different reflector material (Emergency blanket, PET-Mirror, and Mirror) were investigated by experimental test. The dish was designed with has a diameter, focal point, and rim angle of 1m, 0.6m, and 45° respectively. The dual-axis tracking system was used to turn the tested dish to the sun so that the solar dish can be performed at the highest efficiency.

The experimental result was shown that the maximum temperature of the dish with an emergency blanket, PET-Mirror, and Mirror are 817°C, 724°C, and 874°C, respectively. The heat power of the dish with an emergency blanket, PET-Mirror, and Mirror are 389W, 361W, and 420W, respectively.

In addition, other parameters that were present are the mass of each dish per its heat power and fabrication cost per its heat power. The result was shown that the dish with an emergency blanket has the lowest mass per heat power and cost per heat power of 1.04 x10-3 k/W and 1.41 THB/W, respectively. This shows that the fabricating technique is worthwhile in terms of efficiency and price.

Acknowledgments

The financial support of this research is from the Research and Researcher for Industry (RRi) and Part Rich Precision Co., Ltd is gratefully acknowledged.

References

[1] Kalogirou S A 2004 *Progress in Energy and Combustion Science* **30** 231-295
[2] Goswami D Y 2015 *Principles of Solar Engineering 3rd ed* (CRC Press)
[3] Faraz T 2012 *Proc. 2nd International Conference on the Developments in Renewable Energy Technology* 1-5
[4] Gorjian S Ebadi H Calise F Shukla A and Ingrao C 2020 *Energy Conversion and Management* **222** 113246
[5] Suman S Khan M K and Pathak M 2015 *Renewable and Sustainable Energy Reviews* **49** 192-210
[6] He Y L Qiu Y Wang K Yuan F Wang W Q and Li M J 2020 *Energy* **198** 117373
[7] Barlev D Vidu R and Stroeve P 2011 *Solar Energy Materials and Solar Cells* **95** 2703-2725
[8] Loni R Askari E Areh A Ghobadian B Kasaeian A B Gorjian S and Najafi G 2020 *Renewable Energy* **145** 783-804
[9] Bianchini A Guzzini A Pellegrini M and Saccani C 2019 *Renewable Energy* **133** 382-392
[10] Pavlovic S Daabo A M Bellos E Stefanovic V Mahmoud S and Al-Dadah R K 2017 *Journal of Cleaner Production* **150** 75-92
[11] Mawire A and Taole S H 2014 *Energy for Sustainable Development* **19** 162-169
[12] Hijazi H Mokhiamar O and Elsamni O 2016 *Alexandria Engineering Journal* **55** 1-11
[13] Prinsloo G and Dobson R 2015 *Solar Tracking* (South Africa: Prinsloo, Dobson)
[14] Venkatachalam T and Cheralathan M 2019 *Renewable Energy* **139** 573-581

Results of the measurement of SOFC fuel cell stacks under pressure conditions

H.P. Berg[1],*, R. Dückershoff[2], M. Vorpahl[1], and A. Himmelberg[1]

[1]BTU Cottbus-Senftenberg, Chair of Combustion Engines and Flight Propulsion, Siemens-Halske-Ring 14, D-03046 Cottbus, Germany
[2]University of Applied Sciences Mittelhessen (THM), Friedberg, Department of Mechanical Engineering, Mechatronics and Material Technology, ZEuUS, Wilhelm-Leuschner-Str. 13, 61169 Friedberg, Germany

* Corresponding Author: Peter.Berg@b-tu.de

Abstract
Results from the measurement of SOFC fuel cell stacks under pressure conditions are presented. As part of a measurement campaign, the operation of a stack system is investigated, particularly under the operating conditions of a recuperated micro gas turbine. Above all, the performance and effectiveness of selected stack types at various operating pressures and operating temperatures are measured. With the test facility set up for such investigations, cell systems could be examined under atmospheric conditions and with pressures of up to 5 bar.

It is shown that in operating conditions under pressure, the output of the fuel cell systems are improving. From a gauge pressure of 4 bar, the performance curve is flattened and higher pressures only produced a marginal increase in performance. Furthermore, the cells tested at overpressure show a steady-state behavior more quickly under load change requirements than in atmospheric operation. This means that more flexible operating modes with faster response behavior can be realized. By choosing a suitable operating temperature, the efficiency of the system is further increasing. Care was taken to select the operating conditions of the cell systems so that coking can be prevented.

Finally, a statement can be made about the pressure dependence of the fuel conversion rate. Parasitic reactions at the anode can be related to the power output. The tests carried out have shown that the high-temperature fuel cell is a promising service provider of the future. A combination of SOFC high-temperature fuel cells and micro gas turbines in one machine could, in addition to the internal provision of the required pressure, also lead to better dynamics of the entire system and increase the energy yield from the primary energy source.

Keywords: Fuel Cell, MGT-SOFC, Pressurized System, Measurement data

1. Introduction

In a fuel cell, the chemical energy of the fuel is not converted into electrical energy by releasing heat, but directly into electricity through an electrochemical process. This so-called cold combustion has the advantage that losses are significantly lower, as the diversion of energy conversion via heat generation and subsequent work output to generate electrical energy is replaced by such a direct process with lower heat loss. Highly efficient fuel cells are e.g., SOFC (solid oxide fuel cells). These are high-temperature fuel cells which use membranes made of yttrium-stabilised zirconium oxide as a solid electrolyte. This type of fuel cells shows an electrical efficiency of up to more than 50% at operating temperatures around 700°C under atmospheric conditions.

Simulations show that an increase in the operating pressure of SOFC fuel cells can increase cell performance. As a result, hybrid energy converters can be developed which make better use of primary energy on the basis of micro gas turbines, whose combustion chamber has been replaced by an SOFC fuel cell stack grape. By reusing the fuel cell exhaust air with an afterburner chamber for the conversion of residual fuel in a turbine, the electrical yield and the electrical efficiency of the entire system can be further increased [1-5].

Such an increase in efficiency can be used in hybrid systems consisting of micro gas turbines and high temperature fuel cells (MGT-SOFC) to develop small, decentralized energy converters with high electrical efficiency [6]. In this context, electrical efficiencies in continuous operation of 65% and more can be achieved in optimal utilization of the primary energy input. At the BTU Cottbus-Senftenberg, Germany, a research project on the development of such efficient energy converters is currently being completed under the development name TurboFuelCell [1, 6].

In order to demonstrate the increase in efficiency, a test facility has been developed at the LS VFA to investigate SOFC fuel cells under pressure conditions. Individual fuel cell stacks can be supplied with fuel and the load behavior at different operating temperatures and pressure ranges can be investigated. In this investigation, measured values and analyses up to a maximum operating pressure of 5 bar absolute are presented on the response behavior, power output and efficiency of the fuel cell conversion. The investigated systems are operated with pure hydrogen.

2. Determination of performance and efficiency of fuel cells

In an SOFC fuel cell, the chemical energy of the fuel is directly converted into electricity by electrochemical processes. Such a process, called "cold combustion", offers the advantage that losses are significantly lower. For their electrochemical reactions, such fuel cells use hydrogen as fuel, which is supplied to them either directly or as a fossil pre-product. In addition to hydrogen, air is supplied as a second operating medium (oxidizer), from which the oxygen is to react with the hydrogen. Both reaction partners are separated from each other by means of an ion-conducting material, the electrolyte. Yttrium-stabilized zirconium oxide in the form of a solid oxide membrane is used in SOFC systems. This membrane is permeable to oxygen ions O^{2-}. In the process, the O^{2-} ions migrate through the membrane after absorbing electrons on the cathode side and react on a catalyst layer of nickel with electron-donating H^{+} ions on the anode side to form water. The electron migration can be used as electric current via the connected electrodes [7-9].

Anode:
$$2\,H_2 + 2\,O^{2\ominus} \rightleftharpoons 2\,H_2O_g + 4\,e^{\ominus}$$
$$CO + O^{2\ominus} \rightleftharpoons CO_2 + 2\,e^{\ominus} \qquad E_0 = 0\ V \qquad (1)$$

Cathode:
$$O2 + 4\,e^{\ominus} \rightleftharpoons 2\,O^{2\ominus} \qquad E_0 = 1.23\ V \qquad (2)$$

$$2\,H_2 + O_2 \rightleftharpoons 2\,H_2O$$

In the process, the O^{2-} ions migrate through the membrane after absorbing electrons on the cathode side and react on a catalyst layer of nickel with electron-donating H^+ ions on the anode side to form water. The electron migration can be used as electric current via the connected electrodes [7-9].

These normal potentials are only valid for standard conditions (T=25 °C and p=101325 Pa). The difference between the normal potentials of both electrodes is called reversible cell voltage or open-circuit voltage.

$$\Delta E_0 = E_{0,Cathode} - E_{0,Anode} \tag{3}$$

$$\Delta E_0 = 1.23V \tag{4}$$

The reversible cell voltage can be used to determine the GIBBS free reaction enthalpy (ΔG_0). This indicates how large the useful energy is related to the mole of fuel gas.

$$\Delta G_0 = -z \cdot F \cdot \Delta E_0 \tag{5}$$

The reaction determines the number of electrons involved (z). For the reaction in fuel cells, this has a value of 2, resulting in the following free enthalpy

$$\Delta G_0 = -237.13\, kJ/mol \tag{6}$$

Conversely, this equation can also be used to calculate the reversible cell voltage E_0 using the Nernst equation [7].

$$E_0 = \frac{-\Delta G_0}{-z \cdot F} \tag{7}$$

With the Faraday constant F. Since the electrical voltage of a single cell with ΔE_0=1.23 V is very low and not sufficient for technical applications, the operating voltage can be increased by connecting several fuel cells in series. In the stacks investigated here, 30 cells are connected in series.

An ideal voltage is only present in no-load operation. The theoretical value is never reached, since voltage losses occur, e.g., in case of reaction inhibitions and insufficient gas diffusion. Therefore, values between 0.9V and 1V are more likely to be measured under no-load conditions. Furthermore, voltage losses are caused by activation losses, whereby H_2 molecules give off electrons at the catalyst and form protons. Similarly, ohmic losses are caused by the voltage drop at the membrane or the electrodes on the fuel cell. In the diffusion-controlled area, a further voltage loss is caused by the drop in concentration shortly before the gas escapes from the cell area.

2.1 Fuel cell efficiency [7]

The efficiency of the fuel cell η_{FC} shows the ratio of supplied chemical energy and generated electrical energy, which summarizes the losses inside the fuel cell.

$$\eta_{FC} = \frac{electrical\ power}{chemical\ performance\ in\ used\ hydrogen} \tag{8}$$

This can be represented in an efficiency chain, so that the internal losses can be more specifically attributed to their origins and the Fuel Utilization of Hydrogen in the Fuel Cell F_U:

$$\eta_{FC} = \eta_{SOFC}^{rev} \cdot \eta_U \cdot \eta_I \cdot \eta_H \cdot F_U \tag{9}$$

A theoretical efficiency η_{th} can be defined by analysing the fuel cell reaction:

$$2H_2(g) + O_2(g) \rightleftharpoons 2H_2O(l) \tag{10}$$

Since a free enthalpy of the reaction of $\Delta G_0 = -237.13\, kJ/mol$ and an enthalpy under standard conditions of $\Delta H = 285.83\, kJ/mol$ can be determined, regarding the theoretical efficiency follows:

$$\eta_{th} = \frac{\Delta G_0}{\Delta H} = \frac{237.13kJ/mol}{285.83kJ/mol} = 82.96\% \tag{11}$$

These 82.96% represent the theoretical maximum which can be achieved with the reacting substances H_2 and O_2. By means of the voltage efficiency η_U the internal losses due to electro catalysis, electrolyte and cell design are shown. This can be calculated by means of the ratio of clamp voltage to open-circuit voltage:

$$\eta_U = \frac{U_{cell}}{E_0} \tag{12}$$

U_{cell} is the voltage applied to the cell at the current operating point and always follows the model of the characteristic curve. E_0 is the electromotive force, which depends, among other things, on the temperature and the activity of the oxidized and reduced form. This can be represented, among other things, with its dependence on pressure, numbers of electrons transferred during the reaction z and the Faraday constant F:

$$E_0(p) = E_0(p_0) - \frac{\Delta v \cdot R \cdot T}{z \cdot F} \cdot ln\left(\frac{p}{p_0}\right) \tag{13}$$

Here, Δv refers to the molten change of the gas molecules (Δv =-1.5). Higher partial pressures thus lead to higher stresses and cause an increase in performance. The dwell time in the cells required for improved diffusion is also increased for the same mass flow.

The current efficiency or also FARADAY efficiency (η_I) indicates the ratio of the theoretical maximum gas volume V_{theo}, which can be determined from the measured current, to the actually consumed gas volume $V_{transformed}$:

$$\eta_I = \frac{V_{theo}}{V_{transformed}} \tag{14}$$

If one now sets the theoretically possible current in relation to the actual current efficiency, one obtains the current efficiency η_I

$$\eta_I = \frac{I}{I_{theo}} = \frac{I}{z \cdot F \cdot \dot{n}_{H_2}} \tag{15}$$

Voltage efficiency (η_U) and current efficiency (η_I) depend on the cells used and indicate how efficiently they convert the chemical energy supplied to them into electrical energy. Thus they form the internal or also electrical efficiency (η_{el}):

$$\eta_{el} = \eta_I \cdot \eta_U \tag{16}$$

The heating value efficiency (η_H) determines the ratio of the heating value of the usable material to the heating value of the supplied mixture:

$$\eta_H = \frac{H_{H_2} \cdot x_{H_2}}{\sum_i (H_i \cdot x_i)} \tag{17}$$

Finally, there is the Fuel Utilization (F_U). This indicates how high the amount of hydrogen used in electricity generation is compared to the amount of hydrogen supplied and is expressed as the ratio of the hydrogen used to the hydrogen supplied.

$$F_U = \frac{\dot{n}_{H_2,use}}{\dot{n}_{H_2,supplied}} \tag{18}$$

This is needed to evaluate the practical application in order to prevent depletion at the electrodes. If the fuel utilization is 100%, the oxygen partial pressure rises so high that the cell voltage collapses. Since the anode usually contains part nickel, the risk of nickel oxide formation would increase as a result. This would lead to the formation of inactive areas, which would only become available again with a subsequent reduction of the cell with a sufficiently high hydrogen concentration. As only pure hydrogen was used for the experiments, the efficiency chain can be further shortened:

$$\eta_{FC} = \eta_{th} \cdot \eta_{el} \cdot F_U \tag{19}$$

In the investigations, the electricity efficiency was not taken into account, but the converted fuel flow was determined from the electrical power achieved. The hydrogen used for the calculation of the fuel utilization was calculated using the actual electricity flow, so that parasitic reactions are neglected. Therefore, electrical efficiency and fuel utilisation were combined into one factor

$$\hat{F}_U = \eta_{el} \cdot F_U \tag{20}$$

Thus, the mass flows for the determination of the fuel utilization in a 30 cell-stack are as follows:

$$\dot{n}_{H_{2,use}} = \frac{30 \cdot I}{2 \cdot F} \tag{21}$$

$$\dot{n}_{H_{2,supplied}} = \frac{\dot{m}_{H_{2,norm}} \cdot p_{norm}}{R_{S,H_2} \cdot T_{norm} \cdot M_{H_2}} \tag{22}$$

2.2 Determination of performance

Fuel Utilization is also needed to determine the chemical performance of the hydrogen consumed:

$$P_{chem,use} = \dot{m}_{H_{2,supplied}} \cdot H_{S,H_2} \cdot \hat{F}_U \tag{23}$$

For a fuel cell, a characteristic curve can be drawn with which the electrical voltage falling across the current and the resulting power can be determined. Since the active area of the fuel cell also has a significant effect on the power, it is common practice to specify the voltage not via the current but via the current density as shown in Figure 1.

Since the electrical power of the fuel cell is defined as the product of current and voltage, the characteristic curve results in a point of maximum power. However, since the efficiency of the fuel cell decreases with increasing current, the areas of highest efficiency are, however, in the partial load area. The power characteristics of the fuel cell can also be determined from the characteristic curve. Thus, the point of intersection of the voltage line with the Y-axis, or voltage axis, results in the open-circuit voltage of the fuel cell. This is also called OCV (open-circuit voltage) and is lower than the theoretically achievable, reversible cell voltage due to losses at the cathode. In fuel cells, the electrical voltage does not occur directly, but takes some time to build up in the cell.

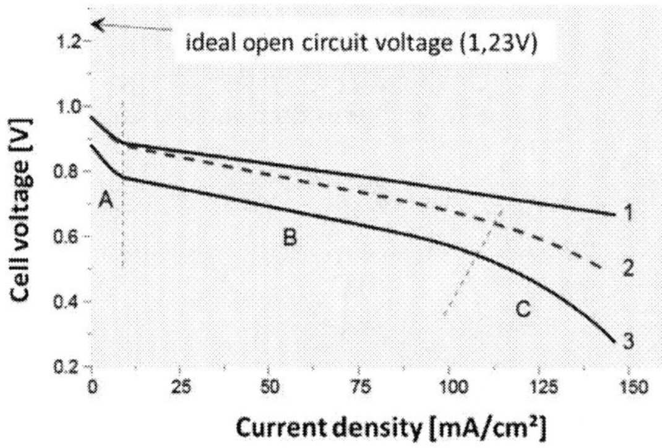

Figure 1. Current-Voltage characteristic of a fuel-cell [7]

3. Test facility [10]

At the VFA chair of the BTU Cottbus-Senftenberg, a test facility has been set up with which fuel cell stacks can be investigated under various pressure conditions and temperature influences. All operating parameters of the stack could be monitored and the respective test object could be supplied with different operating media.

For the SOFC fuel cells under investigation, an external heat source had to be integrated into a pressure vessel, with which a stack could be heated to an operating temperature of approx. 800°C. For this purpose, a commercially available ceramic furnace was used. The furnace chamber had to be insulated against the pressure vessel to prevent excess temperature on its outer wall. To monitor the temperature, various thermocouples were installed in fuel gas lines, air inlets and outlets and gas chambers. This enabled the heat distribution of the stack to be resolved in three dimensions. The control of the gas flows was realized by means of mass flow controllers, which can realize a minimum flow rate of 0.7 (ml_N)/min. Afterwards the gases are led to the test stand by means of a hose system. There, the gases are transported in scale-free pipes into the pressure chamber and further into the interior of the furnace. There they are preheated to the required inlet temperature in a 2 m long pipe bend inside the furnace and then fed to the fuel cell stack. Since the fuel cell membranes are sensitive to pressure surges, pressure sensors were installed on the anode and cathode sides. This allowed a maximum permitted pressure difference of 50 mbar to be regulated and monitored, so that the stack could be tested under constant conditions without damage. Figure 2 shows the realized test facility for the determination of pressurized SOFC Fuel Cells.

An important component of the test stand is the outer pressure vessel. This enables tests to be carried out at increased pressure. The boiler pressure is built up by blowing in approx. 300 l_N/min of air and flushed for heat dissipation. A manometer and pressure sensors are used to monitor the boiler pressure. To monitor the performance development of the stack, all 30 cells of a stack were equipped with voltage sensors. This allowed a good overview of damage to individual cells during the tests. Furthermore, these always provided information about the current performance point of the stack. The regulation and control of the test stand was carried out with a LabView application. SOFC stacks from the Chinese manufacturer SOFCMAN were used for the tests. Their characteristic curves were to be validated and extended for a future characteristic map.

The current density (current intensity divided by active cell area) is often used in the presentation of results. This allows cell efficiencies of different manufacturers to be displayed. In this comparison only the stack type SOFCMAN, 64cm², 30 cells [11] was used. Therefore, the current intensity was used for comparison and classification of the measuring points.

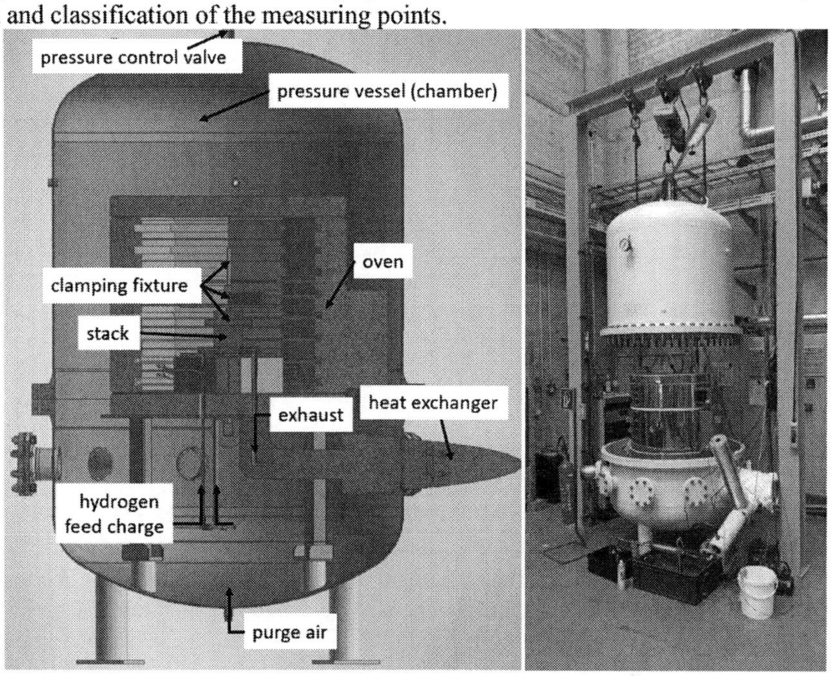

Figure 2. Test Facility to determine pressurized SOFC-Fuel Cells.

3.1 Measurement and data aquisition

In the test facility, the SOFC fuel cell from the company SOFCman [11] was heated to an operating temperature of 700°C and the working pressure in the pressure vessel was set to operating pressure. The stack behaviour was measured in the plant at absolute operating pressures between 1bar and 5bar. After reducing the fuel cell stack with pure hydrogen, the SOFC fuel cell system is operated with the hydrogen flow for power operation on the anode side and a suitable air flow on the cathode side. The pressure in the flow is thus set to operating pressure. To operate the stack, a current intensity was set depending on the fuel flow and the power output of the stack was recorded. Under these conditions, stack temperatures, pressure difference and stack current voltage were measured. In addition, the cell current voltage was measured for each cell of the used 30-cell stack. The fuel utilisation was calculated from the electrical results and the efficiency of the stack power was determined. The data obtained at different stack loads were analysed and graphically processed. By adjusting the operating temperature and pressure, an attempt was made to optimise the performance of the system. When using pure hydrogen as the fuel system, no further analysis of the exhaust gas composition is required, because the fuel utilisation can be calculated from the power and the heat output of the stack. This will be carried out in subsequent work when the fuel cells are operated with synthesis gas instead of hydrogen.

4. Test results

To compare the performance of the fuel cells, performance data are analyzed in particular. Thus, the increase in stack performance due to the increased operating pressure can be clearly documented. Furthermore, SOFC stacks under pressurized atmosphere show an increased response behaviour after load changes. A stationary operating state under correspondingly maximum power output is set up much faster. This is particularly advantageous when using the systems under load change conditions. Furthermore, the efficiency of the cell systems must be assessed in particular. Here, an improvement in the utilization of the primary energy input can be seen above all through the improved conversion of the fuel.

Figure 3 now shows the performance curves of the examined fuel cells under different load requirements as a function of the operating overpressure. In relation to a power output of about 620 W maximum power under atmospheric conditions, power increases of more than 60 W/stack from 30 cells can be seen.

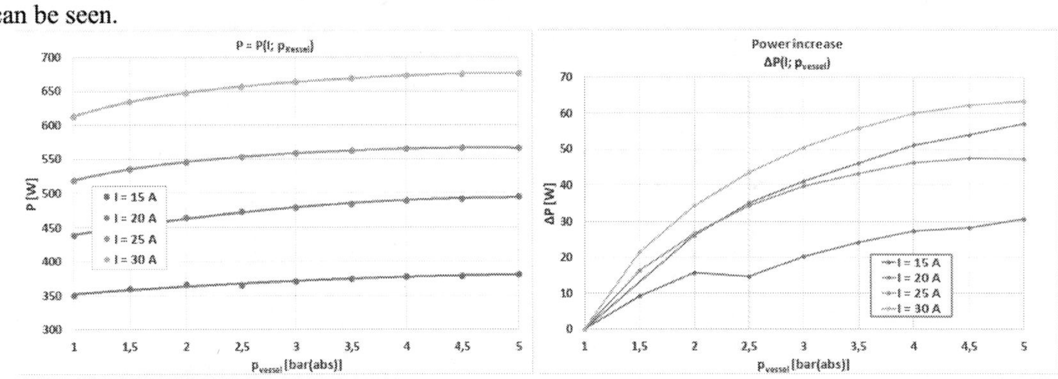

Figure 3. Performance curves for different load requirements [10].

Figure 4a shows the increase in electrical efficiency at maximum stack power. The result is related to the total energy content (calorific value) of the supplied hydrogen fuel (blue). The green curve shows the efficiency related only to the fuel converted with the fuel utilization. The second graph (4b) shows the increase in efficiency when the operating pressure is increased to 3.5 bar (g). Thereby the power increases almost linearly with increase. If the demanded power is reduced, the decrease in stack power is greater than the increase with an increase in power demand. This is due to the fact that the stack has a higher operating temperature at higher power and that a cooling down at load reduction does not succeed completely to the equilibrium temperature at load increase. Thus, a hysteresis loop is formed in the power characteristic curve. This appears to be less pronounced at higher pressure.

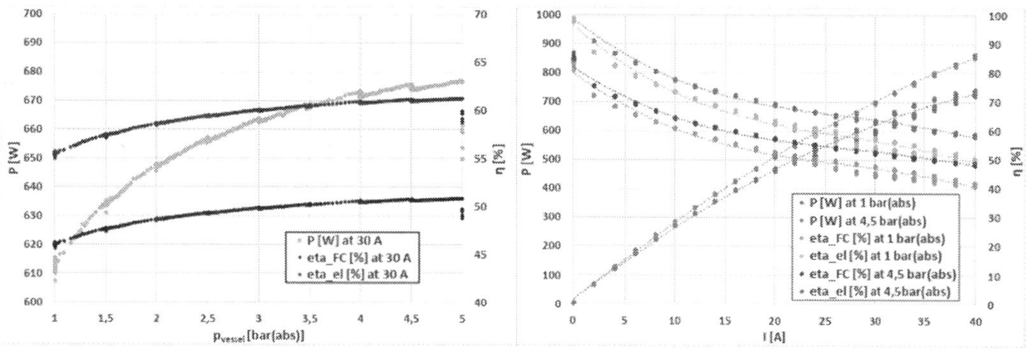

Figure 4. a) el. Efficiency and power increase at increased operating pressure, b) efficiency at increased stack load [10]

Efficiencies also decrease with increasing power demand. An atmospherically operated stack at full power demand (40 A), for example, shows a drop in efficiency to just over 40% compared to an idling load of more than 80%. At higher pressures, on the other hand, an electrical efficiency of almost 98 % can be determined with a very low load demand of almost 0 A. At maximum load, this is still measured at just over 50%. The fuel cell efficiency even reaches about 59%. Even when looking at the efficiency, a hysteresis between increasing load decrease and decreasing load demand can be recognized. Here too, this is less pronounced at higher pressures.

Furthermore, the response behaviour of the stacks to load changes will be investigated. When considering the voltage curve after the load requirement has been increased by increasing the stack current intensity, it can be seen that the voltage response reaches a maximum faster at the considered higher operating pressure and thus a stationary operation is ensured. If one compares the temporal tensions with the tensions reached after 15 minutes, it becomes clear that especially with higher load requirements the new operating point needs longer time to reach the maximum possible power, expressed by the highest voltage. Tests with higher operating pressure show that the lower voltage at the beginning of the load requirement does not deviate so much from the final voltage. In addition, the equilibrium voltage is usually reached earlier than 15 minutes after the start of the test. (Figure 5)

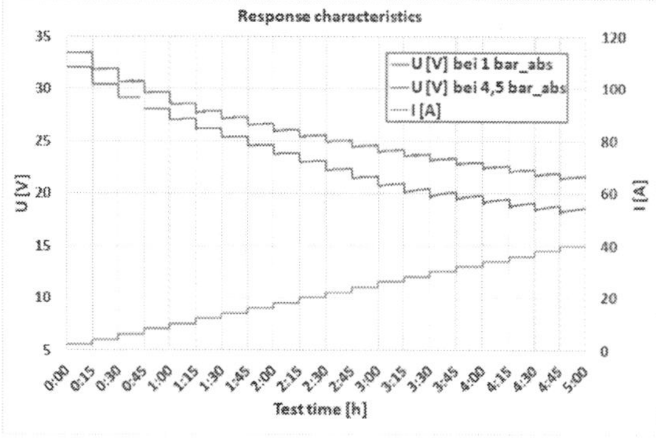

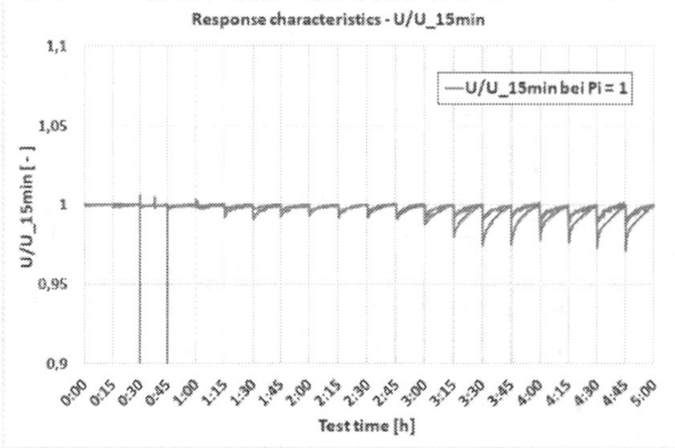

Figure 5. Comparison of the response behaviour of SOFC fuel cell stacks at different operating pressures [10].

Finally, the average increase of the operating voltage at load demand in the form of demanded stack current will be considered. The increase in pressure may well reach an increase of more than 11%. This increase decreases with the pressure increase, so that at a limit pressure no improvement of the stack performance is to be expected. (Figure 6)

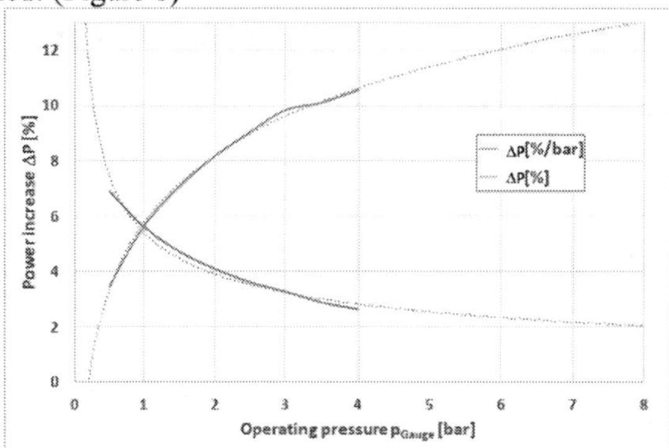

Figure 6. Increase of the operating voltage for the same load requirements at increased operating pressure compared to atmospheric test conditions (load point=30A)

5. Analysis

Within the scope of the investigations, operating data of SOFC fuel cells at atmospheric operating conditions and at higher pressures were compared. Various partial load cases were also investigated and the influence of the operating pressure on the performance of such systems was shown.

Pressurized fuel cells provide higher performance at the same fuel mass flow compared to the same cells operated only under atmospheric pressure conditions. An increase in stack performance of up to 11% at maximum cell load was observed. This increase is particularly noticeable at high loads. Similarly, higher performance increases can also be achieved at lower load requirements. For commercial applications an optimum between stack performance (cost factor) and pressure charging can thus be found. A significant increase can only be achieved up to operating pressures of about 8 bar. At higher pressure the curve also flattens out noticeably.

The response behaviour and regulation of the fuel cell stack during pressurized charging is still much faster than under atmospheric operating conditions. A stationary operating condition can be reached faster. This is particularly advantageous for applications to compensate for load changes. Likewise, if the load is increased, the initially achieved cell voltage is higher compared to the maximum achievable voltage after a 15-minute settling period. Thus, the difference between the minimum and maximum voltage after a load change is significantly lower. Furthermore, a hysteresis of the power development between power increase and power reduction at the same operating point can be seen. The temperature dependence of the power output of a cell stack plays a particularly important role here. In the case of a load reduction, the stacks must first cool down to a temperature relevant to the operating point in order to achieve optimum power output. This requires a longer start-up phase. This hysteresis and also the start-up time can be significantly reduced by pressure charging. This reduces the temperature dependency, especially with load changes.

Furthermore, the electrical efficiency for stack yield can be increased by up to 10% over the entire power range. In lower partial load ranges, it is evident that the pressure increase can even result in a higher efficiency increase. Thus, the selection of stack packs also requires consideration of the number of stacks and stack load and the resulting stack wear.

The pressure ranges in which an increase in performance is significant correspond to the efficient operating pressures of recuperated micro gas turbines. This demonstrates the compatibility of the two energy conversion systems, micro gas turbine and SOFC fuel cell. This can be used in hybrid systems to significantly increase the electrical energy yield from primary energy systems. In the reference [1], it was shown that an electrical efficiency of up to 65% can be achieved.

Finally, it should be noted that the tests have not yet allowed any conclusions to be drawn regarding degradation/longevity. Long-term tests are planned to be carried out for this purpose. These will be carried out in subsequent projects after the expansion of the test facility to continuous operation. For this purpose, further tests will be performed to investigate the temperature dependence of the power output.

6. Conclusion

A system for the investigation of pressure charged SOFC fuel cells has been introduced. Results from atmospheric and pressurized fuel cell tests were presented. It was shown that pressure charging can increase the performance of the investigated stacks by up to 11% and increase the efficiency by more than 10%. Furthermore, the improvement of the response behavior of a cell stack to load changes under pressure-charged operating atmosphere was determined and the reduced temperature dependence of the power output was shown. Especially in hybrid systems with recuperated micro gas turbines operating in a similar pressure range, pressure charging can be used to significantly increase the primary energy utilization.

References

[1] Berg H P Himmelberg A Lehmann M Dückershoff R and Neumann M *IOP Conf. Series: Materials Science and Engineering* **297** (2018) 012004
[2] Berg H P Kleissl M Himmelberg A Lehmann M Prechavut N and Vorpahl M *IOP Conf. Series: Materials Science and Engineering* **501** (2019) 012007
[3] Berg H P and Krienke C 2015 *12. Magdeburger Maschinenbautage* CD-ROM, UB Magdeburg
[4] Berg H P Dückershoff R Lehmann M and Prechavut N 2017 *12th European Conference on Turbomachinery Fluid dynamics & Thermodynamics E* (Stockholm, Sweden) ETC2017-266
[5] Henke M Willich C Westner C Leucht F Leibinger R Kallo J and Friedrich K A 2012 Electrochimica Acta **66** 158-163
[6] Berg H P 2020 https://www.b-tu.de/T-Cell/, *Die T-Cell als Energieerzeuger der Zukunft*, accessed 14 Sepetember 2020
[7] Kurzweil P 2016 *Brennstoffzellentechnik* (Wiesbaden: Springer Fachmedien Wiesbaden GmbH)
[8] Baehr H D and Kabelac S 2016 Thermodynamik (Springer)
[9] Droste-Franke B 2009 Die Technologien Brennstoffzelle und Virtuelles Kraftwerk und ihre Bewertung im Hinblick auf eine zukunftsfähige Energieversorgung *Brennstoffzellen und Virtuelle Kraftwerke* (Ethics of Science and Technology Assessment, Vol. 36) ed Gethmann CF (Berlin: Springer) chapter 3 pp 43–133
[10] Vorpahl M 2020 *Versuchsdurchführung sowie Auswertung und wissenschaftliche Analyse der experimentell erhobenen Messdaten druckaufgeladener Hochtemperaturbrennstoffzellen* Master Thesis BTU Cottbus-Senftenberg, Germany
[11] SOFCMAN 2016 http://www.sofcman.com/, 2017 Ningbo SOFCMAN Energy (Zhejiang, P.R. China)

Takeoff Performance Analysis of a Light Amphibious Airplane

Sinchai Chinvorarat[1,*], Boonchai Watjatrakul[1], Pongsak Nimdum[1], Teerawat Sangpet[1], and Pumyos Vallikul[1]

[1]Department of Mechanical and Aerospace Engineering
King Mongkut's University of Technology North Bangkok
1518 Pracharaj 1, Bangkok, 10800, Thailand

*Corresponding author: sinchai.c@eng.kmutnb.ac.th

Abstract
The performance analysis of a light amphibious airplane during takeoff from the water surface has been presented in this paper. The airplane is one of the light amphibious airplane series built in Thailand with a bi-wing configuration. The conceptual design aims to build a short takeoff and landing (STOL) amphibious airplane. The mathematical model cooperated with several complicated parameters is presented and solved by MATLAB. The water-drag resistance of the airplane is estimated from typical seaplane hull NASA technical information, which enhances the correctness of the model at takeoff configuration. The simulation reveals that the seaplane can successfully takeoff from calm water in 24.5 seconds at the speed of 54 KCAS and an approximate distance of 890 feet with the AOA of 10 degrees.
Keywords: takeoff performance analysis, water-drag resistance, light amphibious airplane

1. Introduction

The "NAX" project initiated by the Royal Thai Navy research team has been started since the year 2002, aiming to design and build a light, versatile amphibious airplane (seaplane) and exploiting several missions such as coastal patrol, search and rescue, fire extinguish including any others mission assigned by military and civilian purposes. The initial version of the airplane, NAX-3, has been designed from the concept of the EDRA Aeronautic Super Pétrel, an amphibious pusher configuration biplane. The NAX-3 has demonstrated its satisfaction flying qualities in both performance and safety, but the total weight of the airplane is above 650 kilograms, which is the number classified as a light sport airplane category stated by the Civil Aviation Authorities of Thailand (CAAT).

The new design of the NAX-4 project has started with the concept of modifying an amphibian pusher bi-wing configuration. The main seaplane structure is mostly made from composite, gains benefits in lightweight control, and enhances the strength of structure with aluminum alloy as a structural bone. By the retractable landing gears, the seaplane can takeoff and landing on a short airfield or water runway. A more powerful 115 Hp, Rotax 914 piston engine with three blades constant pitch propeller is installed as a "pusher" bi-wing configuration. The seaplane basic design performance is shown in Table 1.

Table 1. "NAX-4 light amphibious airplane

Specifications:		
Crew	1	person
Capacity	1	person
Length	19.88	ft
Upper Wingspan	32.94	ft
Upper Wing area	97.52	ft^2
Lower Wingspan	18.77	ft
Lower Wing area	52.96	ft^2
Cabin width	4.12	ft
Empty weight	880	lbs
Max fuel capacity	40	gallon
Max takeoff weight	1,430	lbs
Performance:		
Maximum speed	179.81	mph
Cruise speed	141.47	mph
Stall speed	53.98	mph
Endurance	5	Hrs
Rate of Climb	863.75	fpm
Take off Distance (Runway)	650	ft
Take off Distance (Water)	890	ft
Landing Distance (Runway)	600	ft
Landing Distance (Water)	700	ft

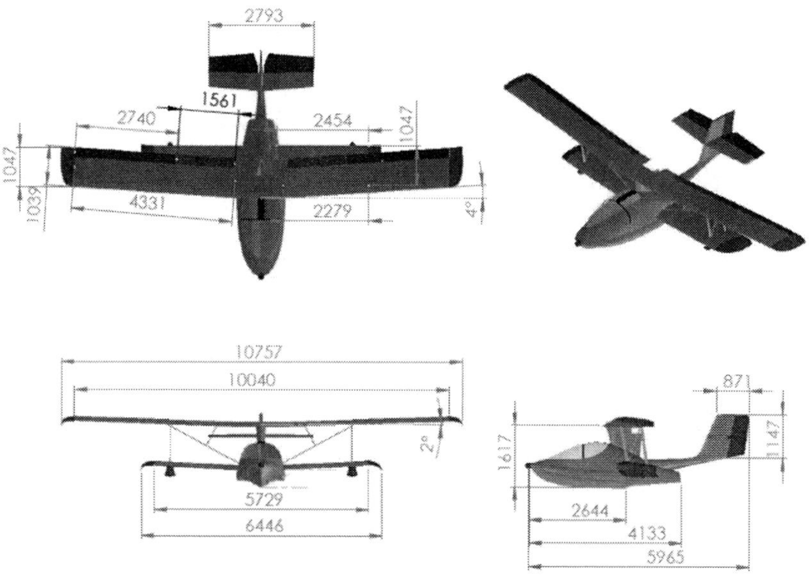

Figure 1. "NAX-4 amphibious light airplane" general dimension.

Since the seaplane can operate from the water surface, it's motion is distinguished from conventional land airplane [1], [2]. When the seaplane left the water surface, it operates like other fixed-wing airplanes but draggier than a landplane due to its hull design and many numbers on posts that support wings and undercarriage [3], [4]. The determination of the airborne performance follows the procedure typically determine the airborne of the landplane. The motions of seaplanes on the water, however, introduce additional complications. The seaplane goes through a transition process from a state of static buoyancy to dynamic planning [5]. The seaplane must be designed to smoothly transferable in four phases, i.e., displacement phase, hump or ploughing phase, planning or on the step phase, and airborne or liftoff flight.

This paper determines the analysis of the NAX-4 amphibious light airplane in takeoff configuration incorporate with essential and imperative parameters that help to understand the characteristics of the seaplane in detail. The results from the modeling analysis will be validated and verified through the actual flight test.

Abbreviations	
F_b	Buoyancy force
$C_\Delta = \frac{F_b}{\rho_W g B^3}$	Buoyancy force coefficient
$C_v = \frac{v}{\sqrt{gB}}$	Speed coefficient
C_R	Resistance coefficient
α_{trim}	Trim angle
R	Water resistance
R_{Froude}	Froude water resistance
v	Linear velocity of the airplane
α	Incidence angle
α_{trim}	Trim angle
m	Mass
T	Thrust
g	Gravity

2. Seaplane Motions in the Water

When the seaplane is taxying with an idle RPM engine, it is considered as a displacement phase [6]. Normally the taxi speed is about 6-8 knots so that the propeller will not pick up water spray that causes serious erosion problems of the propeller blade. Since the seaplane has no brakes, it is very important to taxying at low speed to avoid any collision in a confined area. When the power is increased significantly above idling, the seaplane will usually become nose-up or ploughing position. For seaplanes that have a high thrust-line will tend to nose down upon applying power, it is imperative that the elevator should be in the full up position. The ploughing position is about the combination of high velocity streamline flows through the elevator and the hydrodynamic force exerts on the immersed part of the hull. However, hydrodynamic or water drag becomes the major part of the force resisting acceleration. This resistance reached its maximum at a speed of about 30 knots, just before the hull is placed into the planing phase. The trim angle needs to be slowly increasing approximately two more degrees helping the hull to ride on the step. The point of the highest resistance is called the "hump." After the hump is passed and the seaplane is traveling on the step, and water resistance R continue to decrease.

The planning or step position is attained by advancing the throttle to full power, and the elevator is slightly up. As the seaplane accelerates, it will gradually assume a nose-high pitch attitude raising the hull up to the step area. As a result of aerodynamic and hydrodynamic lifting, the seaplane is raised high in the water, allowing the hull to ride on top of the water wave. The airborne is merely transferring lift force from the hull to the wing by applying back elevator pressure resulting in the seaplane lifting off the water and becoming airborne. To avoid a rhythmic pitching motion caused by dynamic instability in forces along the hull, so-called "porpoising" while on the step, it is crucial to maintain the proper pitch angles during the takeoff run. Too much back-elevator pressure during planning will force the aft part of the hull more profound into the water, creating a strong resistance and retarding takeoff. On the contrary, insufficient bake elevator pressure will cause the hull to remain in the water, which results in excessive water drag. During the seaplane acceleration, the water is increasing sprayed upward, outward, and rearward from the bow and will be directed into the propeller, causing erosion of the blades as well.

IOP Conf. Series: Materials Science and Engineering 1137 (2021) 012010 doi:10.1088/1757-899X/1137/1/012010

Figure 2. displacement phase, hump or ploughing phase, planing or on the step phase.

2.1 The Hydrodynamic Force, Resistance, and Drag Force

For the seaplane operating near the water surface, the hydrodynamic force is affected by gravity acting on the wave and by surface tension [7]. By dimensionless analysis, the force acting for the seaplane body can be expressed by:

$$F_H = k\frac{1}{2}\rho_W v^2 L^2 \, Re\left(\frac{\rho v L}{\mu}\right) Fr\left(\frac{v}{\sqrt{gL}}\right) We\left(\frac{\rho v^2 L}{\sigma}\right) \tag{1}$$

where k is a constant coefficient, ρ_W is water density, v is the seaplane velocity, and L is the submerged length of the seaplane in the water. Equation (1) illustrates that the hydrodynamic force depends on the function of three characteristic dimensionless ratios, the Reynolds Number, the Froude Number, and the Weber Number lied in order. The Reynolds Number indicates the ratio of dynamic forces to viscous forces, the Froude Number is the ratio of dynamic forces to gravity force, and the Weber Number is the ratio of dynamic forces to surface tension forces. In the model of the seaplane framework, the effects of Webber number are confined to low speeds and are considered constant, which equals to unity.

The total resistance of a seaplane consists of (i) hull form water drag, which includes wave and eddy drag, (ii) parasitic drag, and (iii) induced drag. The hull drag portion of the hydrodynamic force while the seaplane is moving on the water surface can be determined from the resistance coefficient C_R as:

$$C_R = \frac{R}{\rho_W g B^3} \tag{2}$$

where R is water resistance, and B is the seaplane cabin width.

The parasitic drag and induced drag can also be determined through a drag coefficient C_D using a quadratic polynomial as:

$$C_D = C_{Dmin} + \frac{C_L^2}{\pi.AR.e} \tag{3}$$

where C_{Dmin} is the minimum drag coefficient, AR is the reference aspect ratio, and e is Oswald efficiency.

These resistances directly affect the takeoff safety and performance of the seaplane and yet are taking into account in the design and production phase.

3. Seaplane Takeoff Estimation

The seaplane takeoff from the water surface is more complicated than conventional airplanes due to the complexity of water resistance and aerodynamic drag. The seaplanes are generally very draggy due to their shape of the hull and associated support structure, generating lots of drag. However, the shape of the seaplane is dictated by hydrodynamic design and operation, and yet it is practically impossible to reduce its drag. Drag can be minimized by other means such as using aerodynamic strut profile, keep landing gear behind a smooth door compartment, or making fuselage and main parts of the seaplane by composite material (to reduce weights and number of rivets), etc.

To be able to lift off from the water surface safely and efficiently, several parameters have to be considered as follow,

Buoyancy force coefficient:

$$C_\Delta = \frac{F_b}{\rho_W g B^3} = \frac{\Delta}{\rho_W g B^3} \tag{4}$$

where, $F_b = \rho_W g \forall = \Delta$ is the Buoyancy force.

the maximum Buoyancy force coefficient (when the seaplane is stationary) is denoted as:

$$C_{\Delta_0} = \frac{W}{\rho_W g B^3} \tag{5}$$

and speed coefficient as a function of the seaplane speed is denoted by:

$$C_v = \frac{v}{\sqrt{gB}} \tag{6}$$

To estimate the water-resistance of the seaplane more effectively, the water resistance coefficient C_R as a function of speed coefficient C_v can be obtained from the experimental model towed in the water tank [8]. However, NACA TN-2481 [5] provides the graph of C_R and C_v for planing-tail seaplane model and can be the empirical curve fitting of a Gaussian function as:

$$C_R = 0.05789e^{\left(-\left(\frac{(C_v-1.907)}{0.7561}\right)^2\right)} + 0.0273e^{\left(-\left(\frac{(C_v-1.347)}{0.2536}\right)^2\right)} - 0.3322e^{\left(-\left(\frac{(C_v-23.41)}{11.74}\right)^2\right)} + 0.07924e^{\left(-\left(\frac{(C_v-3.227)}{1.951}\right)^2\right)} \tag{7}$$

For reduction of water resistance with loading, the following expression is given:

$$C_{R\Delta} = C_R.\left(\frac{C_\Delta}{C_{\Delta_0}}\right) \tag{8}$$

and water resistance is given by:

$$R = C_{R\Delta}\rho_W g B^3 \tag{9}$$

Also, the trim angle curve can be approximated by the equation:

$$\alpha_{trim} = \alpha_1 + \frac{(\alpha_2-\alpha_1)}{2} * \left(1 + tanh(aC_v + b)\right) \tag{10}$$

where $a = \frac{5.294}{C_{v2}-C_{v1}}$ $\quad b = -(2.647 + AC_{v1})$

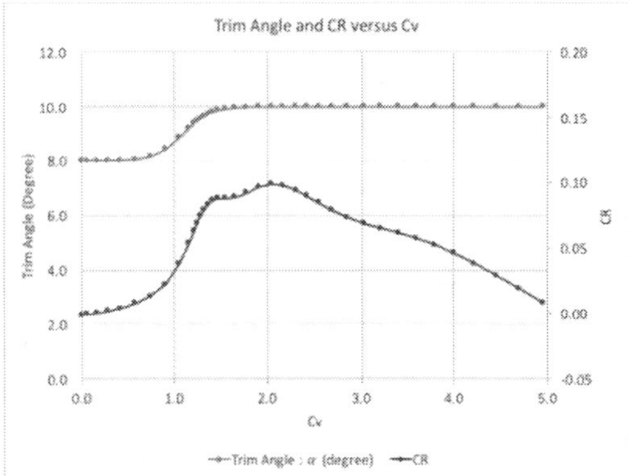

Figure 3. Trim angle and water-resistance as a function of speed coefficient.

From Figure 3, the "hump" or the peak of C_R, the value indicates the maximum water resistance resulting from the seaplane reaches its hull speed [9]. Practically, it means the seaplane is on the step and ready for the pilot to rotate the seaplane. The trim angle indicates the orientation of the seaplane on the water surface during takeoff. It is very important for takeoff analysis because it directly affects the angle of attack (AOA) of the seaplane. By adding the trim angle to the AOA, the seaplane will merely on the step at a certain C_v value at the planing stage and ready to airborne [10, 11].

The friction resistance of submerged surfaces, while the hull is in the planning stage, by William Froude is introduced by:

$$R_{Froude} = f.S_W.v^n \tag{11}$$

where f is the coefficient of friction resistance depends on the hull surface, S_W is the wetted area, n is constant depend on surface quality, usually close to 2.

The thrust of the seaplane as the functions of power and velocity for fixed-pitch propeller can be approximated by a cubic spline method as the form [12, 13]:

$$T(v) = Av^3 + Bv^2 + Cv + D \tag{12}$$

when A, B, C, D are constants and can be determined from the following matrix equation

$$\begin{pmatrix} 0 & 0 & 0 & 1 \\ v_C^3 & v_C^2 & v_C & 1 \\ 3v_C^2 & 2v_C & 1 & 0 \\ v_H^3 & v_H^2 & v_H & 1 \end{pmatrix} \begin{bmatrix} A \\ B \\ C \\ D \end{bmatrix} = \begin{bmatrix} T_{static} \\ T_C \\ -n_P.P_{BHP}/v_C^2 \\ T_H \end{bmatrix}$$

where n_P is the propeller coefficient, v_H is maximum level airspeed, v_C is cruise airspeed, T_{static} is static thrust at zero airspeeds, P_{BHP} is engine brake horsepower, T_H is thrust at maximum airspeed, T_C is thrust at cruise airspeed.

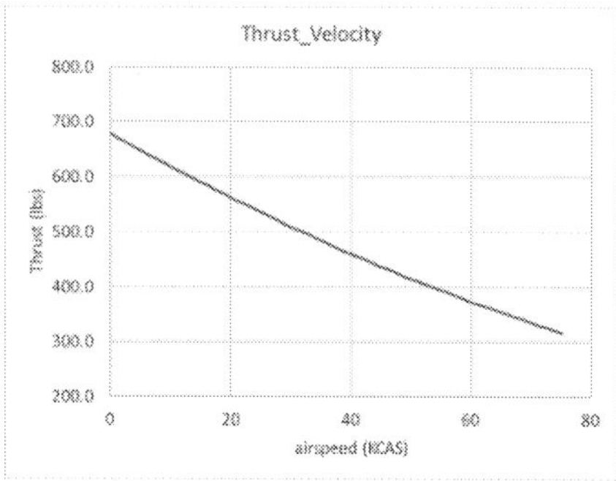

Figure 4. Thrust-velocity for fixed-pitch propeller.

Equation of motion for 2-D [14] taking off from water of seaplane in the longitudinal axis is expressed as:

$$T.r - D - R - R_{Froud} = ma \tag{13}$$

where r is ramp up the factor of a pilot to slowly increase power from takeoff, which is given by the expression:

$$r = \begin{cases} 0.1 + 0.9\,(t/10) & if\ t < 10\ sec \\ 1 & if\ t \geq 10\ sec \end{cases}$$

4. Simulation Results

The necessary information of the NAX-4 light amphibious airplane [15, 16] for 2-D takeoff estimation under the calm water condition is given in Table 2.

Table 2. Technical information of NAX-4.

Fuselage Beam	B	4.27	ft
Fuselage Water Length	*L*	13.55	ft
Wing Span	*b*	32.94	ft
Wing Area	*S*	152.85	ft^2
Wing Aspect Ratio	*AR*	7.1	
Wing AOI wrt Keel	*iw*	4	degrees
Gross Weight	*W*	1430	Lbf
Rotax 914 UL Engine Power	*P*	115	BHP
Propeller Diameter	*Dp*	6	ft
Spinner Frontal Area	*Aspinner*	5	ft^2
Maximum Speed @5000 ft	*Vmax*	151	KCAS
Propeller Efficiency	η	0.82	
Thrust Function @ Max Power	*T*	1972.32	Lbf

The simulation shows that the seaplane starts increasing power gradually from idle to full power in 10 seconds. The seaplane starts to pick up the speed, and also the aerodynamic drag, water drag, and lift force start increasing at different rates. However, the water resistance that is the function of the speed starts growing exponentially and reach the top at hull speed in 12 seconds, making the seaplane to be the planning stage and begin to get on the step at this period. Once the seaplane is on the step, the hydrodynamic resistance, R_{Froude}, starts its value due to the submerged part of the hull while the water resistance reduces to "zero" at the speed of 36 KCAS. The pilot can rotate the seaplane at this speed but must be aware of the rear shifting of the center of buoyancy (CB) that could make the seaplane enters porpoising if too much rotation. The seaplane will lift off the water when the displacement load becomes "zero" at the speed of 54 KCAS within 24.5 seconds, and the total distance is 893 feet.

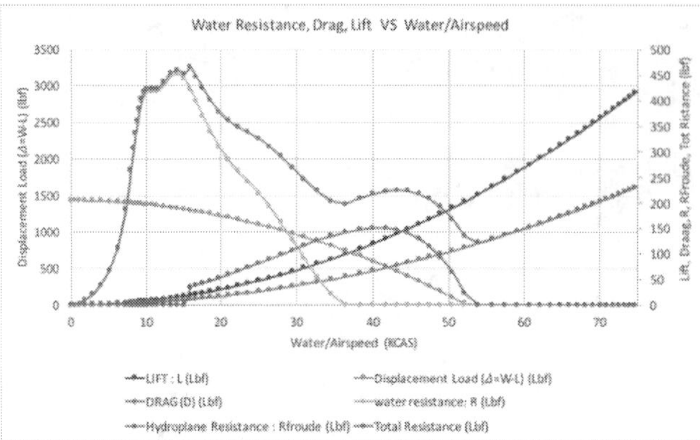

Figure 5. Lift, Drag, Water resistance, Hydroplane resistance, Displacement Load versus Water/Airspeed.

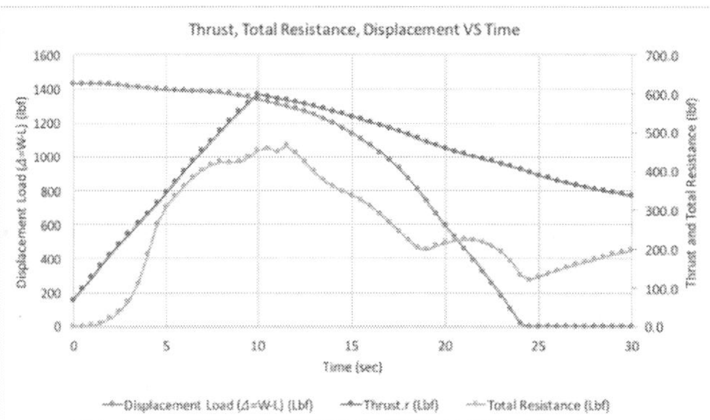

Figure 6. Displacement Load, Thrust, Total Resistance versus Time.

The takeoff configuration of the seaplane has been analyzed using ANSYS- fluent in VOF module. The 3-D model of NAX-4 is built without engine and propeller to study the seaplane behavior at the planning stage. The FEM mesh is tetrahedrons with 538,282 elements and 104,052 nodes. The multiphase – VOF CDF model is analytically selected along with a realizable k-epsilon model. The couple solution technique is chosen with a flow courant number equal to 0.5, and the surface tension between two media, air and water, is set to be 0.75. The AOA of the seaplane is 10 degrees, and the airspeed is 30 meters/sec.

The simulation shows that the lift coefficient C_L converges to the value of 1.007 while the drag coefficient C_D converges to 0.08, making the aerodynamic lift equal to 1,515 lb_f at liftoff. One can visualize some turbulence after the upper wing that creates drag to the seaplane and the surface shear (red arrows) along the body surface. This turbulence affects the stabilizer fins and elevator for maintaining the trim angle, which the pilot needs to be aware of while lifting off the seaplane.

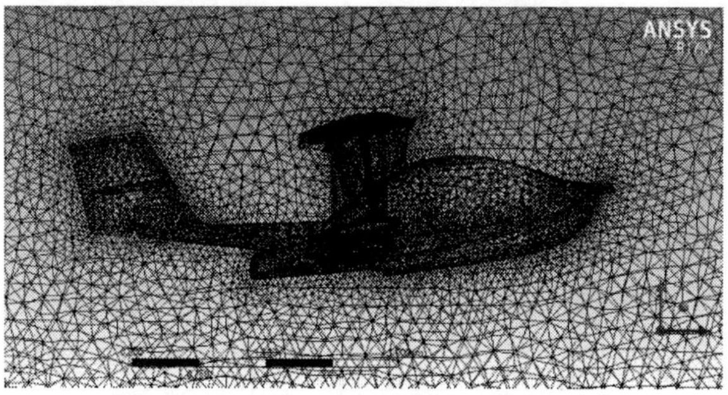

Figure 7. NAX-4 FEM mesh model.

Figure 8. NAX-4 while on the planing stage.

Figure 9. NAX-4 two-phases VOF model on the planing stage.

5. Conclusions

A light amphibious airplane takes off modeling and analysis are presented in this paper. The motion of the seaplane on the water surface consists of four different phases as displacement phase, hump or ploughing phase, planning or on the step phase, and airborne or liftoff flight. The water resistance is considered as the most complicated resistance in addition to aerodynamics drag directly involving in the seaplane takeoff configuration. The water resistance coefficient as the function of the speed coefficient and trim angle are two key parameters to determine the successful takeoff performance of the seaplane. For our NAX-4 light amphibious airplane, the 2-D takeoff simulation and the two-phases VOF model in ANSYS-fluent at the planning phase show the corresponding hydrodynamics and aerodynamics parameters. The seaplane can successfully takeoff from calm water in 24.5 seconds at the speed of 54 KCAS and an approximate distance of 890 feet with the AOA of 10 degrees. However, the NAX-4 light amphibious airplane is currently on the maintenance schedule; therefore, all the seaplane takeoff parameters will be verified through an actual water takeoff flight once the NAX-4 is ready to fly in the near future.

Acknowledgments

The authors would like to thank Thailand Science Research and Innovation (TSRI), and Department of Mechanical and Aerospace Engineering, KMUTNB for the research cooperation and financial support through the project. A special acknowledgment goes to Royal Thai Navy (RTN) for invaluable information and research collaboration throughout this project.

References

[1] Walter S 1924 Static Stability of Seaplane Floats and Hulls *NACA TN-183*

[2] Perring W G A and Glauert H 1932 Stability on the Water of a Seaplane in the Planing Condition *Reports and Memoranda* Aeronautical Research Committee

[3] Parkinson J 1948 Takeoff Performance of Light Twin-float Seaplanes *TN 1524* Langley Memorial Aeronautical Laboratory, Langley Field, Va, USA

[4] Hugli W and Axt, W 1951 Hydrodynamic Investigation of a Series of Hull Models Suitable for Small Flying Boats and Amphibians *NACA TN 2503*, Stevens Institute of Technology, Washington, USA

[5] Suydam H B 1952 Hydrodynamic Characteristics of a Low-Drag Planing-Tail Flying- Boat Hull *NACA TN 2481*

[6] Ludwig P and Tietjens O G 1957 *Fundamentals of Hydro- and Aeromechanics*, Dover Publications

[7] River P 1958 Hydrodynamics Manual *Naval Air Test Center*, Flight Test Division, Maryland

[8] Dyck V 1989 Seaplanes and the Towing Tank, Advanced Marine Vehicles Conference *AIAA Paper* 1533

[9] Cook M V 1994 The theory of the longitudinal static stability of the hang-glider *Aeronautical Journal* (98)978

[10] Hancock G J 1995 *An introduction to the flight dynamics of rigid aeroplanes* Hemel Hempstead: Ellis Horwood Ltd

[11] Biran A 2003 *Ship Hydrostatics and Stability* Butterworth-Heinemann

[12] FAA-H-8083-23 2004 *Seaplane, Skiplane, and Float/Ski Equipped Helicopter Operations Handbook* U.S. Department of Transportation, Federal Aviation Administration, Flight Standards Service

[13] Gudmundsson S 2014 *General Aviation Aircraft Design: Applied Methods and Procedures* Elsevier Inc., Waltham, MA, USA

[14] Seth A and Liem R 2018 Takeoff analysis of amphibious aircraft with implementation of a hydrofoil *Structures Congress (Structure18)*, Songdo Convensia, Inchoen, Korea

[15] Chinvorarat S, Watjatrakul B, Nimdum P, Sangpet T, Soontornpasatch T, and Vallikul P 2019 Static testing for composite wing of a two-seater amphibious airplane *IOP Publishing: Material Science and Engineering*, doi:10.1088/1757-899X/501/1/012026

[16] Chinvorarat S, Watjatrakul B, Nimdum P, Sangpet T, Soontornpasatch T, and Vallikul P 2020 Six-degree of freedom mathematical dynamic model of a light sport aircraft *IOP Publishing: Material Science and Engineering*, doi:10.1088/1757-899X/886/1/012011

Investigation of loose contact oxidation kinetic analysis on diesel particulate filter's wall surface using non-isothermal TGA technique

Swe Zin Win[1,*], Preechar Karin[1], Watanyoo Phairote[1], Nuwong Chollacoop[2], Peerawat Saisirirat[2], and Katsunori Hanamura[3]

[1]Department of Mechanical Engineering, School of Engineering,
King Mongkut's Institute of Technology Ladkrabang, Bangkok 10520, Thailand
[2]National Metal and Materials Technology Center, National Science and Technology Development Agency, Pathum Thani 12120, Thailand
[3]School of Engineering, Tokyo Institute of Technology, Tokyo 152-8550, Japan

* Corresponding Author: ms.swezinwin96@gmail.com

Abstract
Carbonaceous soot particles emitted from diesel engines are dangerous to human health because of its carcinogenic activity and can penetrate into lungs. In order to control soot emissions, diesel particulate filters (DPFs) are widely used to meet progessively stringent regulations of vehicle emissions limits. In this article, the characterization of conventional DPF microstructure was briefly investigated by scanning electron microscopy (SEM) addition with energy dispersive X-ray analysis (EDX) and X-ray fluorescence analysis (XRF). Non-isothermal oxidation mechanism of soot on conventional DPF was carried out from thermogravimetric analysis (TGA) under air atmosphere with the flow rate of 40ml/min at the heating rates of 5, 10, and 15°C/min. The kinetic parameters were calculated by three non-isothermal model-free methods. The average value of apparent activation energy obtained by Flynn-Wall-Ozawa (FWO), Kissinger-Akahira-Sunose (KAS) and Kissinger (K) were 77.02, 65.29 and 70.48 kJ/mol, respectively at different heating rates. The average results got from this work indicate that non-isothermal experiments were quite low compared to kinetics parameters estimated by the isothermal method from previous research.

Keywords: Particulate matter, Diesel particulate filter, Soot oxidation, Thermogravimetric analysis, Non-isothermal method

1. Introduction

Compression ignition (diesel) engines are the entire power source in heavy-duty trucks and off-road vehicles as well as their usage is rapidly increasing in passenger cars owing to offering high fuel efficiency. Anyway, the tremendous amount of particulate matter (PM) or soot emissions from diesel exhaust system cause serious health issues and environmental consequences, resulting in the increment of strict emission regulations [1]. The compositions of diesel particulate matter include mainly of soot (solid carbon), agglomerated carbonaceous compounds, ash and volatile components. However, each of PM composition from diesel engine mainly depends on the engine operating conditions and properties of fuel [2]. Since the substantial knowledge of PM formation and oxidation processes have been improved from previous research, it can support a dramatic development of controlling diesel exhaust emissions in aftertreatment technologies [3, 4].

Currently, installing the diesel particulate filter (DPF) in exhaust pipe system is one of the trending technologies of reducing PM from diesel engines. The working principle of DPF filtration system is carried out through a physical trapping mechanism of PMs in the honey-comb shaped structure with alternating channels which the particles are trapped and accumulated on the wall surface. Hence, the cleaned exhaust gas is released into the atmosphere [5]. The trapped PMs are needed to be thermally regenerated in order to burn off the soot cake layer from the filter. The removal of particles in active (periodical) regeneration, along with the increasing of the temperature inside a DPF, or passive (continuous) regeneration, through the normal activity of DPF filtration mechanism. During the active regeneration, the temperature rise is occurred by post-injection of a certain amount of fuel to burn out the DPF upstream itself and heat the gas temperature inside the filter. This type of regeneration is compulsory, when the temperature inside DPF stream is not adequate to make sure a passive regeneration [6].

Some of the materials applied in fabricating of porous ceramics DPFs involve cordierite, silicon carbide (SiC), silicon nitride, mullite and aluminium titanite [7]. Currently, cordierite and SiC are popular and generally used in the market. Cordierite materials are normally made up of using the sources from talc, alumina and clay to fabricate cordierite all round many steps of reaction [8]. Cordierite DPF is popular by its ceramic structure and not only has a low coefficient of thermal expansion but also has a good thermal shock resistance [8].

Besides the influence of mechanical properties, the PM trapping and soot burning (oxidation) of diesel particles emission is also the important role of DPFs. Previous research has been reported that the methods for calculating the reaction kinetics of soot oxidation by isothermal thermogravimetric analysis (TGA) and temperature program oxidation (TPO) with conventional cordierite DPF powers [9]. It can be seen that the microstructure and oxidation properties of carbon black (CB) are the acceptable sample to simulate instead of real diesel soot particles [10]. In this work, soot burning (oxidation) mechanism of CB on DPF powders was studied by thermogravimetric analysis using the non-isothermal technique. The complete kinetics estimation was carried out by three different model-free methods. The morphology characteristics and chemical and elemental composition of conventional DPF structure were also briefly determined by SEM-EDX and XRF analysis.

2. Experimental Setup and Methodology

2.1. Physicochemical characterization of conventional diesel particulate filter (DPF)

A conventional diesel particulate filter (DPF) sample was investigated in this study as displayed in Figure 1. The morphological information of DPF sample was determined by SEM micrographs. A HITACHI SU5000 model of FESEM in the resolution of 30kV providing 10 to 600,000x magnification with BSE detector was carried out to focus on the sample's microstructure surface. The identification of the elemental composition of a DPF sample was investigated by using EDX detector together with SEM which can observe the periodic elements from the range between Be to Pu. X-Ray fluorescence analysis

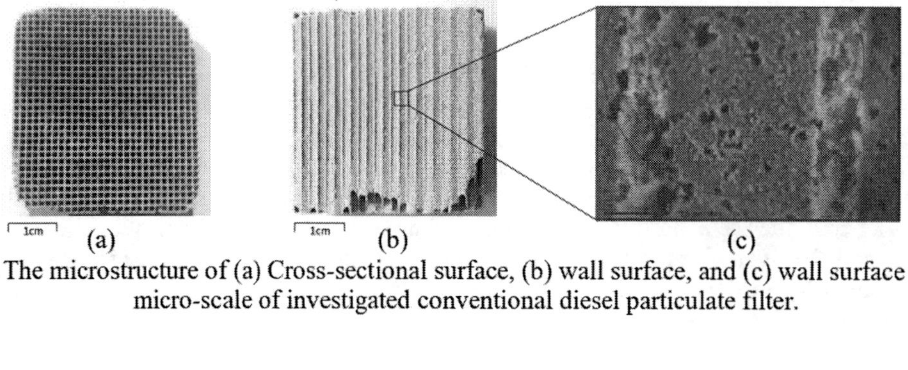

Figure 1. The microstructure of (a) Cross-sectional surface, (b) wall surface, and (c) wall surface in micro-scale of investigated conventional diesel particulate filter.

was also applied to make sure the detail chemical composition of a sample which can detect elements in wider ranges than EDS analysis.

The investigation of soot (CB) oxidation mechanism on DPF's surface was carried out with thermogravimetric analyzer (NETZSCH) by mixing of DPF sample in powder condition and CB N330 as in a ratio of 1:1 referring to the loose contact method for the laboratory-scale study described in the literature [11]. For non-isothermal technique, the sample weight of approximately 2-5 mg with 0.1% accuracy in mass balance was placed into the aluminum dioxide (Al_2O_2) crucible pan and heated at a temperature from 30 to 900 °C accompanied by three various rates of heating in 5, 10 and 15 °C/min undergoing air atmosphere with a flow rate of 40ml/min. Thermal kinetics of soot (CB) oxidation on DPF's surface was studied by characterization of mass conversion patterns from experimental TGA data.

2.2. Kinetic estimations of non-isothermal analysis

Equation (1) represents the soot (CB) combustion reaction which can be observed during the thermogravimetric experiment. Based on the basic principle of chemical kinetics reaction [12], the oxidation behavior with non-isothermal technique is described by Equation (2).

$$C + O_2 \rightarrow CO_2 \tag{1}$$

$$-\frac{d[C]}{dt} = k\,[C]^n[O_2]^m \tag{2}$$

where C is mass of carbon black, t is reaction time, k is the constant of chemical reaction rate, m and n are the orders of reaction in regards to C and O_2, respectively. The reaction order n is presumed to be 1 when the whole internal C surface is fully diffused by oxygen molecules and the rate of reaction is approximately proportional to the concentration of C [13, 14]. Hence, the oxygen limitation of CB particles diffusion in the internal surface can be negligible and thus the order of m becomes zero. Referred to the Arrhenius Equation [12], the rate of a chemical reaction can be illustrated as:

$$k = Ae^{\frac{-E_a}{RT}} \tag{3}$$

where A is the pre-exponential factor, E_a is the apparent activation energy in kJ/mol, R is the value of an ideal gas constant in 8.314 J/mol K, T is the kelvin temperature. The kinetics equation can be obtained by substitution of Equation (2) into Equation (3):

$$-\frac{d[C]}{dt} = Aexp\,[-\frac{E_a}{RT}][C] \tag{4}$$

For non-isothermal experimental condition, the sample is heated at a linear rate, $\beta = \frac{dT}{dt}$, Equation (4) can be rewritten as:

$$-\frac{d[C]}{dT} = \frac{A}{\beta}exp\,[-\frac{E_a}{RT}][C] \tag{5}$$

For estimating solid-state kinetics parameters form thermogravimetric analysis, both model-fitting and model-free (isoconversional) methods can be used [15]. For non-isothermal studies, the model-fitting approach requires to extract the specified value of kinetic parameters for the overall process while in fitting different models to conversion and temperature curves. However, this method may lead to obtaining unpredictable values of Arrhenius parameters [16]. On the other hand, model-free methods allow the estimation of activation energy for an independent model at specific degrees of conversion with a sequence of evaluation data using various linear heating rates [16]. In this paper, the kinetics parameters such as apparent activation energy and pre-exponential factor were estimated by model free methods of Flynn-Wall-Ozawa (FWO), Kissinger-Akahira-Sunose (KAS) and Kissinger (K).

The Flynn-Wall-Ozawa, obtained from the integral iso-conversional method [17] is derived applying Doyle's approximation [18] for the integral to transform into new equation.

$$\ln\beta_i = \ln\left(\frac{A_\alpha E_\alpha}{Rg(\alpha)}\right) - 5.331 - 1.052\,\frac{E_\alpha}{RT_{\alpha i}} \tag{6}$$

In this Equation, $g(\alpha)$ is used as the integral conversion form of reaction model which is equal to $-\ln(1-\alpha)$ for first order reaction [19]. FWO method enables the estimation of apparent activation energy (E_a) from a graph of heating rates ($\ln\beta_i$) verse the inverse temperature ($1/T_{\alpha i}$) which represents the linear reaction with an associated value of conversion at several heating rates. The slope and intercept values of liner straight lines from the plot can be calculated the activation energies and pre-exponential factors, respectively.

The Kissinger-Akahira-Sunose (KAS) method [20] applying with Coats-Redfern temperature integral approximation [21] can be improved to transform into iso-conversional approaches.

$$\ln\left(\frac{\beta_i}{T_{\alpha i}^2}\right) = \ln\left(\frac{A_\alpha R}{E_\alpha\, g(\alpha)}\right) - \frac{E_\alpha}{RT_{\alpha i}} \tag{7}$$

The apparent activation energy (E_α) can be evaluated from the value of slope by a graph of ln ($\beta_i/T_{\alpha i}^2$) versus $1/T_{\alpha i}$ for the variable fractional conversion values which make up the linear lines to the plot. The value of the pre-exponential factor (A_α) can be derived from the intercept of those straight lines.

The Kissinger (K), also called maximum rate method provides the kinetic parameters of a solid-state reaction which does not need to know the reaction mechanism. Kissinger method is invented for the model-free approach of non-isothermal technique but does not estimate apparent activation energy for each conversion [22].

$$ln\left(\frac{\beta_i}{T_m^2}\right) = ln\left(\frac{AR}{E}\right) - \frac{E}{RT_m} \tag{8}$$

The constant value of E_α and A were calculated depending on equation (8) by the plotting of ln (β_i/T_m^2) versus $1/T_m$, where T_m means the temperature of the maximum weight loss peaks from DTG curves.

3. Results and Discussion

3.1. Surface Topography and characterization of DPF structure

Figure 2 presents the characterization of surface topography of DPF sample using SEM. The left image indicates SEM image with 300x magnification while the right image indicates 1000x magnification. It can be observed that the widely distributed pores are different in size and non-uniform in shape. As the microstructure of pore surfaces are strongly connected to the efficiency of surface filtration and soot burning mechanism [23], the result of opening in porosity was one of the advantageous properties of cordierite DPF. It was also found that some of the micro-cracks can be developed on DPF's wall surface during the high-temperature fabrication processes because the material of cordierite has low heat capacity [8].

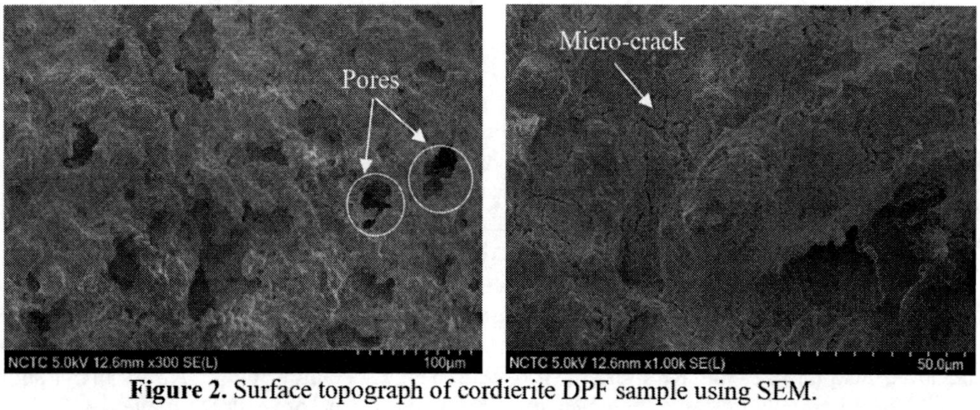

Figure 2. Surface topograph of cordierite DPF sample using SEM.

The elemental characterization of a DPF sample was described by applying EDX detector together with SEM in a small area of interest on the sample as presented in Figure 3. The energy of the emitted X-ray during the process can be detected the specific element which has a unique atomic structure. As reported by EDX analysis, the sample involves mainly of oxygen (O), silicon (S), aluminium (Al) and another trace element in impurity. Another elemental analysis results from XRF technique were expressed in figure 4. Form these results, we have found that both experiments observed the same elements expect the highly reactive non-metal in XRF analysis. On the other hands, XRF technique can even detect very low weight percent of elements like Ti, Fe, Ni and Pt. These elements data assure that DPF sample has the ongoing ceramics form to be cordierite structure which is generally produced from low-cost natural clay materials such as talc and kaolin [8].

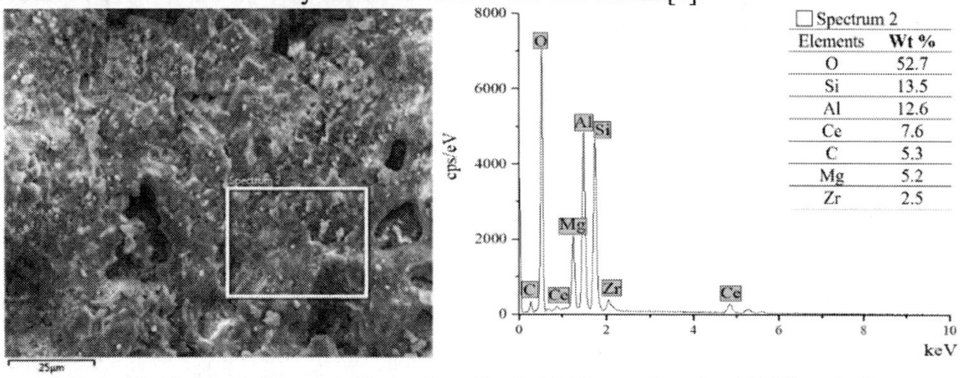

Figure 3. Elemental composition of cordierite DPF sample using EDX analysis.

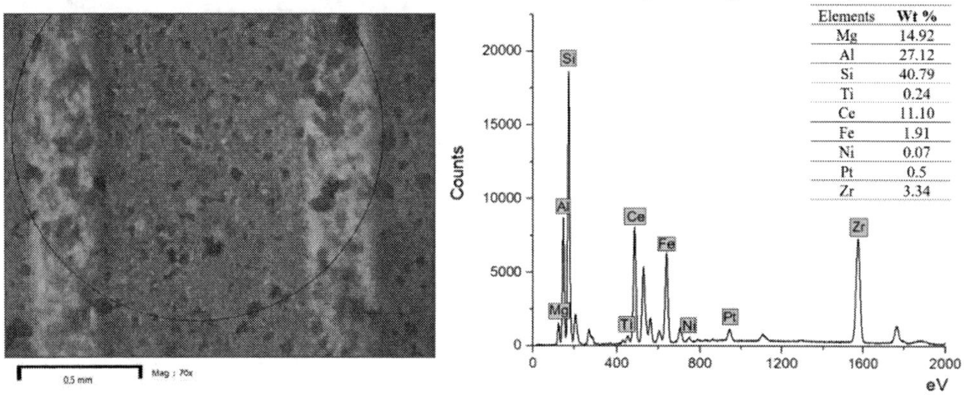

Figure 4. Elemental composition of cordierite DPF sample using XRF analysis.

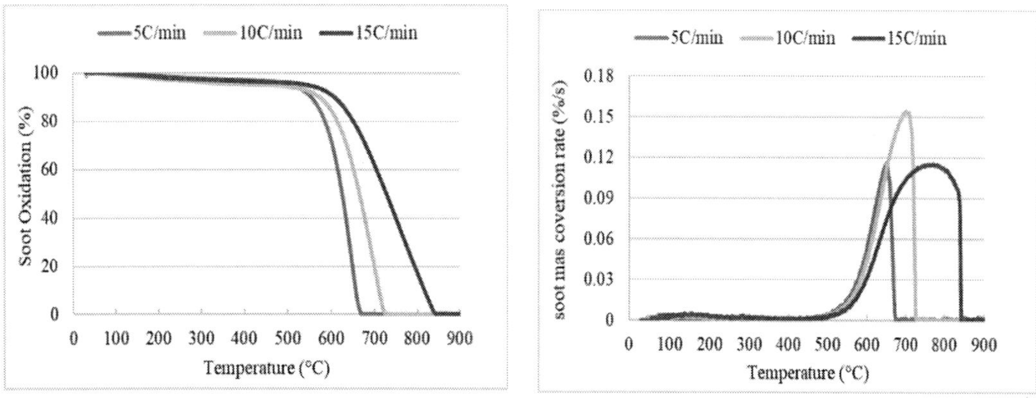

(a) (b)

Figure 5. (a) TG and (b) DTG curves for soot oxidation process of DPF sample with the non-isothermal method at three different heating rates.

3.2. Non-isothermal soot oxidation reactions

Figures 5 (a) and (b) show the thermo gravimetric (TG) and derivative thermo gravimetry (DTG) curves of loose contact in soot burning (oxidation) process of CB with DPF powder under pure air atmosphere which were estimated at various heating rates of 5, 10, and 15°C/min, respectively. All of the TG curves were asymmetric as observed in figure 5 and the individual conversions were obtained at higher temperatures. This means that the maximum degradation points of TG and DTG are slightly moved towards the increasing of heating rate. The maximum soot mass reaction rate at 5 and 10°C/min occurred around the temperature between 600-700°C with well-defined peak while the value at 15°C/min was occurred nearby the temperature about 600-900°C with mildly shoulder peak. It can also be observed that the peaks on DTG curve become more widen with higher heating rate. Due to heat transfer limitations, the higher the heating rate, the larger the reaction time becomes.

3.3. Evaluation of kinetics parameter

The results got from TGA non-isothermal analysis were analysed by three model-free methods using Flynn-Wall-Ozawa (FWO), Kissinger-Akahira-Sunose (KAS) and Kissinger (K) methods to calculate the kinetics parameters such as the apparent activation energy (E_a) and pre-exponential factor (A) as shown in Figure 6,7 and 8. For all the soot oxidation value of 20-80%, the liner plots obtained by FWO, KAS, and K methods get in the coefficient of regression (r^2) higher than 0.9. Figure 9 shows the variation of apparent activation energy (E_a) depends on the soot oxidation value of 20-80% for the non-isothermal process of CB oxidation on DPF derived by model-free methods. It can be found that the data

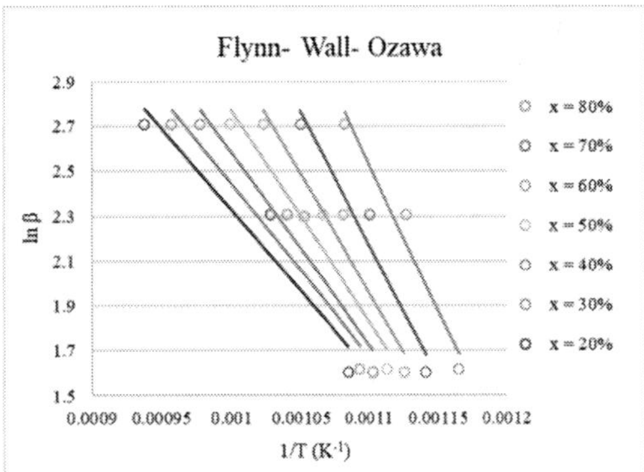

Figure 6. FWO plot for different values of soot oxidation at three heating rates.

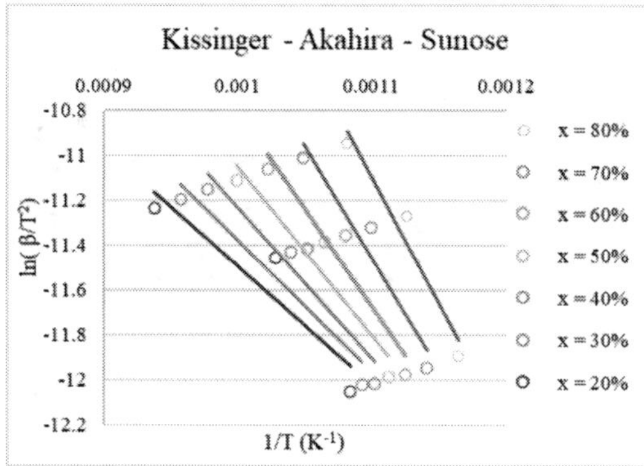

Figure 7. KAS plot for different values of soot oxidation at three heating rates.

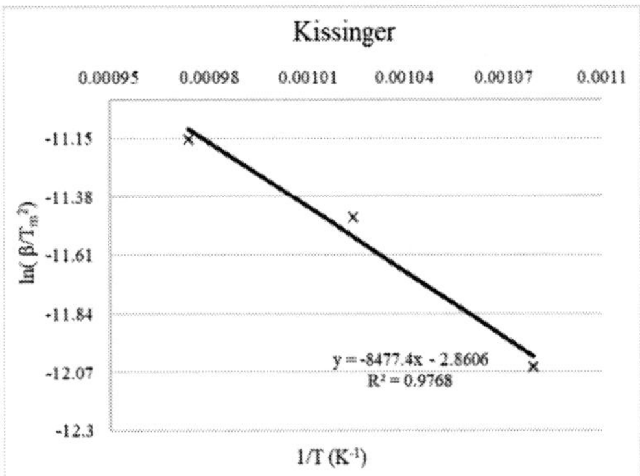

Figure 8. Kissinger plot at three heating rates.

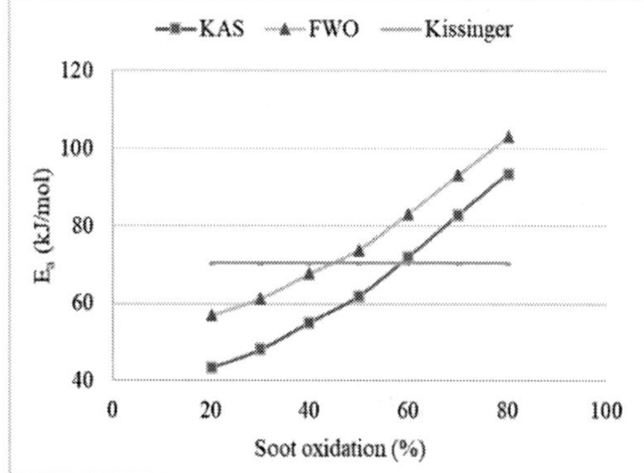

Figure 9. Dependence of activation energy of soot oxidation (%) derived by FWO, KAS and K methods.

of activation energy have a rising trend with increasing of soot oxidation value. The average values of activation energy and pre-exponential factor resulted from each method are described in table 1. The average activation energy value obtained from FWO and K methods were 77.02 and 70.48 kJ/mol, respectively as these values are greater than the value calculated by KAS method which was 65.29 kJ/mol. Previous research reported that the activation energy for isothermal soot oxidation on DPF powders was covered in the range from 109 to 131 kJ/mol [9]. While comparing with calculated E_a value under isothermal technique, the result got under the non-isothermal condition in this work is quite low. However, the E_a calculated from non-isothermal condition takes the mean value over the unpredictable range of rising temperature while the value estimated from isothermal condition takes the average value over the range of temperature chose for the analysis. One of the researcher groups has been suggested that the result of the analysis data generated from kinetics model under isothermal and non-isothermal experiments cannot be relevantly compared [24].

Table 1. Calculated kinetics parameters of soot oxidation by three model-free methods.

Model-free (isoconversional) methods	Apparent Activation Energy, E_a (kJ/mol)	Pre-exponential Factor, A (min^{-1})
Flynn-Wall-Ozawa (average)	77.02	1.69E + 04
Kissinger-Akahira-Sunose (average)	65.29	1.76E + 03
Kissinger	70.48	4.85E + 02

4. Conclusion

The physicochemical analysis of pure DPF sample and soot oxidation mechanisms of loose contact on the surface of DPF were investigated in this work. The morphology of DPF structure was observed that the various pores are different in size and irregular in shape. The morphology and microstructure of the cordierite DPF sample were successfully observed by the elemental analysis using SEM-EDX and XRF technique. Non-isothermal thermogravimetric analysis was implemented at three different heating rates of 5, 10, and 15 °C/min undergoing pure air gas flow rate of 40 ml/min to study the estimation of kinetic parameters. From thermogravimetric soot oxidation graph, it has been observed that heating rate could affect TG curve locations, maximum soot mass conversion rate and position of the maximum temperature peak. The activation energy values derived from FWO and KAS methods were significantly increased as the soot oxidation was risen to 80 %. In Kissinger method, the activation energy value was constant at the various conversions. This work can be summarized that the average values of apparent activation energy of loose contact soot oxidation on DPF using three model-free methods were in the range between 65.29 – 77.02 kJ/mol.

Acknowledgments
The authors are grateful for research funding from National Research Council of Thailand (Diesel Engine's Particulate Matters Reduction using Ethanol- Biodiesel- Diesel Blends and Particulate Filter 398/2563).

References

[1] Englert N 2004 *Toxicol. Lett.* **149** 235-242
[2] Karin P Oki H Hanamura K and Charoenphonphanich C 2012 *J. Res. Appl. Mech. Eng.* **1**(2) 3-8
[3] Mohankumar S and Senthilkumar P 2017 *Renew. Sust. Energ. Rev.* **80** 1277-1238
[4] Burtscher H 2005 *J. Aerosol. Sci.* **36** 896-932
[5] Hanamura K Karin P Cui L Rubio P Tsuruta T Tanaka T and Suzuki T 2009 *Int. J. Engine. Res.* **10**(5) 305- 321
[6] Kumar P A Tanwar M D Bensaid S Russo N and Fino D 2012 *Chem. Eng. J.* **207-208** 258-266
[7] Alder J 2005 *Int. J. Appl. Ceram. Technol.* **2** 429-439

IOP Conf. Series: Materials Science and Engineering 1137 (2021) 012011 doi:10.1088/1757-899X/1137/1/012011

[8] Cutler W A 2008 *28th Int. Conf. on Advanced Ceramics and Composites A: Ceramics Engineering and Science Proceedings* chapher 61 pp 421-430

[9] Karin P Boonsakda J Siricholathum K Saenkhumvong E Charoenphonphanich C and Hanamura K 2017 *Int. J. Automol. Technol.* **18**(1) 31–40

[10] Atribak I Bueno-Lopezv A and Garcia-Garcia A 2010 *Combust. Flame.* **157** 2086-2094

[11] Fino D Bensaid S Piumetti M and Russo N 2016 *Appl. Catal. A-Gen.* **509** 75-96

[12] Espenson J H 1995 *Chemical Kinetics and Reaction Mechanisms* Second Edition McGraw-Hill Inc.

[13] Stanmore B R Brilhac J F and Gilot P 2001 *Carbon* **39** 2247-2268

[14] Bojova M N Decarne C Abi-Aad E Pryakhin A N Lunin V V and Aboukais A 2005 *Thermochimica Acta.* **428** 165-171

[15] Khawam A 2007 *Doctoral Thesis* (Iowa: University of Iowa)

[16] Sergey V and Charles A W 1999 *Thermochimica Acta.* **340-341** 53-68

[17] Flynn J and Wall L 1966 *J. Polym. Sci. Pol. Phys.* **4** 323-328

[18] Doyle C D 1961 *J. Appl. Polym. Sci.* **5** 285-292

[19] Jankovie B 2008 *Chem. Eng.* **139** 128-135

[20] Akahira T and Sunose T 1971 *Res. Rep. Chiba Inst. Technol.* **16** 22-31

[21] Coats A W and Redfern J P 1964 *Nature.* **201** 68

[22] Kissinger H 1956 *J. Res. Natl. Bur. Stand. (U. S.)* **57** 217-221

[23] Karin P Cui L Rubio P Tsuruta T and Hanamura K 2009 *SAE. Int. J. Fuels. Lubr.* **2**(1) 661-669

[24] Vyazovkin S and Wight C A 1997 *J. Phys. Chem. A.* **101** 8279-8284

IOP Conf. Series: Materials Science and Engineering 1137 (2021) 012012 doi:10.1088/1757-899X/1137/1/012012

Design for crash safety of electric heavy quadricycle structure

Suphanut Kongwat[1], Suchanet Jaroenjittakam[1], Siravit Chaianan[1], Intuch Atchariyauthen[1], and Pattaramon Jongpradist[1,*]

[1]Department of Mechanical Engineering, Faculty of Engineering, King Mongkut's University of Technology Thonburi, Bangkok Thailand 10140

* Corresponding Author: pattaramon.jon@mail.kmutt.ac.th

Abstract

An electric small four-wheeler, categorized by the European Union as L7e so-called heavy quadricycle or microcar, is one of the solutions to eco-friendly and sustainable mobility for personal transport. Nonetheless, quadricycles typically do not offer the equivalent passive safety as larger passenger car models, in case of accidents, owing to the lack of energy absorption in the vehicle's structure. This paper presents a proposed heavy quadricycle structure using plain weave carbon fibre-reinforced polymer in the passenger cell and aluminium alloy 6061-T6 in the crumple zone. The behaviors of electric heavy quadricycles under impact are simulated in accordance with the test guidelines of European New Car Assessment Programme using a non-linear finite element analysis via LS-DYNA to examine structural crashworthiness and characteristics. A full-frontal crash with a rigid wall at 30 km/h to 50 km/h shows that the front crush box, longerons and subframes in the crumple zone can efficiently absorb energy from a frontal crash up to 54.8%. The maximum damage in the structure occurs at the joints of the A-pillars and side beams to the front panel. Whilst the occupant safety space is safe under side collision by a 1350-kg moving deformable barrier at 50 km/h speed. However, the quadricycle tends to experience overturns from a side crash due to the vehicle's lightweight and high center of gravity. Finally, suggestions for crashworthiness improvements to the composite passenger cell are provided based on the attained damage mechanism under a crash.

Keywords: Heavy quadricycle, Composite material, Crash safety, Energy absorption

1. Introduction

International Energy Agency (IEA) [1] forecasted greenhouse gas (GHG) emissions from various sectors would increase continuously from 2013 to 2050. Without any measures to reduce GHG emissions, the earth's average temperature will rise 6 degrees Celsius by the end of the century. The electric vehicle industry is one of the technology sectors aiming towards the reduction of COx emissions and pollution problems in big cities. Electric vehicles produce fewer direct and life-cycle emissions, that contribute to climate change. than conventional vehicles. Battery electric vehicles produce no GHG emissions at the tailpipe but cannot cause emissions from the carbon intensity of the electricity used to recharge and the vehicle's consumption.

The development of small and lightweight electric vehicles will improve energy consumption efficiency and reduce battery size. An electric small four-wheeler categorized by European Union as

L7e so-called heavy quadricycle is one of the solutions to eco-friendly and sustainable mobility for personal transport. The statistics of quadricycle registrations in the United Kingdom [2] between 2010 and 2018 show that quadricycles have become significantly popular, especially in London. In Europe the motorized quadricycles market is set grow at a CAGR of 5% between 2019 and 2029 [3]. This is mainly due to the car's flexibility in traveling and parking in the city and urban zones, low-cost maintenance and its advantages in reducing city pollution. With the rising trend of heavy quadricycles, the possibility of accidents from collisions is also increased. In the year 2014 and 2016, European New Car Assessment Programme (Euro NCAP) tested crash safety of several quadricycles in L7e class [4]. The Euro NCAP aims to assess the crashworthiness of quadricycles for passenger safety like several regulations of other vehicles such as bus rollover protection [5, 6]. The evaluation showed that the heavy quadricycles are more hazardous, and prone to occupant injury during a crash than other larger passenger car models. They typically do not offer equivalent passive safety owing to the lack of energy absorption of the vehicle structure and its crash incompatibility [7].

Fiber reinforced polymers are frequently used in structures where high strength-to-weight ratio and stiffness or rigidity are required. Liu et al. [8] analyzed the strength of car frames designed by using glass fiber and carbon fiber reinforced polymer (CFRP) under pole side crash according to the Euro NCAP criteria. It was found that the CFRP structure underwent less deformation and damage under crash. Stein et al. [9] designed a model of heavy quadricycle called Epsilon with CFRP space-frame made of woven carbon fibers square cross-section with low-density polyethylene foam core. The car structure was shown to pass all 6 crash tests following Euro NCAP testing. Moreover, the structural improvement of an L7e car based on Kriescher et al. [10] is unique in adding ring structure around the base of the passenger cell to help absorb and distribute energy when collisions occur.

This paper proposes a design of lightweight heavy quadricycle body made of plain weave CFRP in the passenger compartment and aluminium alloy 6061-T6 in the crumple zone. The structural behaviors of electric heavy quadricycles under front and side impact are then analyzed in accordance with the test guidelines of Euro NCAP via LS-DYNA finite element simulation to examine crashworthiness and crash characteristics.

2. Materials and Methods

2.1 Body design

The specifications for L7e require that the vehicle's unladen weight, excluding battery mass, must not be more than 450 kg for transport of passengers. The maximum power does not exceed 15 kW with a maximum speed not less than 90 km/h.

The baseline quadricycle is 1.38 m wide, 2.27 m long and 1.26 m high. The structure is separated into 2 main parts, i.e., the front crumple zone made of aluminum alloy 6061-T6 with thickness of 2 mm and the passenger cell part made of T700SC plain weave fabric with thickness of 5 mm. Figure 1 shows the quadricycle components and the finite element model. The front crumple zone comprises of longeron and subframe parts while the passenger cell includes A-pillars, B-pillars, roof frame, front panel, rear panel, floor, and side beam. All joints are assumed to be perfectly joined. To avert galvanic corrosion between carbon composite and aluminium alloy, application of anodizing as a pretreatment process to generate oxide coatings is necessary.

The whole structure is meshed by using a 4-node Belytshko-Tsay element with six degree-of-freedom (3 translations and 3 rotations) at each node. The element size is set to 10 mm. The model consists of 103,795 elements and 103,034 nodes. The mass of other components not included in this FE model such as, battery, motor and control system, as well as passengers and passenger seats, is applied to the model by adding uniformly distributed mass elements of 300 kg to the car body. The total weight of the baseline quadricycle is 386 kg.

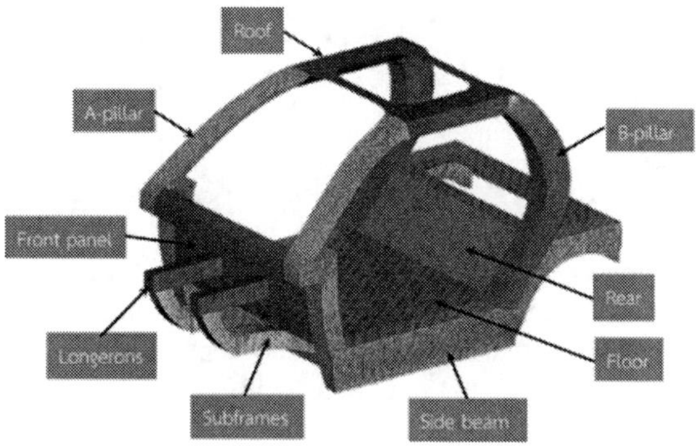

Figure 1. Baseline quadricycle components and meshing

2.2 Material model

Table 1 and 2 provide the material properties of Al 6061-T6 and CFRP T700SC, respectively. The material model for aluminium alloy is implemented by MAT24 Piecewise Linear Plasticity model. The CFRP is defined by MAT54 Enhanced Composite Damage model able to simulate damage progression in dynamic simulation. The material orientation is specified along the horizontal direction in the plane of the member. Material properties and model formulations are validated with simple experimental results from Osborne et al. [11].

Table 1. Material properties of Al 6061-T6 [12]

Property	Magnitude
Modulus of Elasticity (GPa)	68.9
Poisson's Ratio	0.33
Shear Modulus (GPa)	26
Ultimate Tensile Strength (MPa)	310
Tensile Yield Strength (MPa)	276
Shear Strength (MPa)	207

Table 2. Material properties of CFRP T700SC 12k/2510 plain weave fabric [11]

Property	Magnitude
Axial Young's modulus (GPa)	55.9
Transverse Young's modulus (GPa)	54.4
Shear Modulus (GPa)	4.2
Axial tensile strength (MPa)	910.1
Axial compressive strength (MPa)	910.1
Transverse tensile strength (MPa)	772.2
Transverse compressive strength (MPa)	703.3
Shear Strength (MPa)	131
Axial tensile failure strain (mm/mm)	0.0164
Axial compressive failure strain (mm/mm)	-0.0127
Transverse failure strain (mm/mm)	0.014
Shear failure strain (mm/mm)	0.03

3. Structural crashworthiness analysis

Crashworthiness of the quadricycle structure under full frontal and side collision is analyzed in this section.

3.1 Full frontal crash analysis

Full frontal crash analysis is performed with varying initial velocities of 30 km/h, 40 km/h and 50 km/h to a rigid wall as shown in Figure 2. To facilitate the analyses, the wheels are not modelled. Instead, the rear part of the car is constrained from translation in y-direction. Automatic surface-to-surface contact is assigned between the car front to the rigid wall and automatic single surface is specified to the entire quadricycle model such that when the structure deforms, penetration of its own parts does not occur. The residual space or the safety zone of the occupant is measured from the front of the vehicle to the driver's position, i.e., 640 mm from the front of the structure and 215 mm from the front panel.

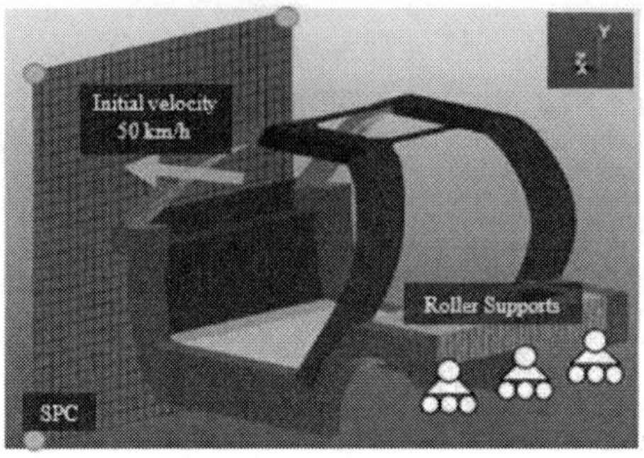

Figure 2. Model set up for full frontal crash test

Figure 3a shows damage of the structure at the end of front crash simulations with different impact speeds. When the crash speeds are 30 km/h and 40 km/h, the crumple zone can efficiently absorb all impact energy and the passenger compartment is intact. When the crash speed is at 50 km/h, the crumple zone is extremely damaged and not able to absorb all impact energy. In this case, damage of the passenger cell occurs at areas where the front panels connected to A-pillar and side beams (Figure 3b). Table 3 listed the energy absorption of the crumple zone and specific energy absorption for each part (shown as numbers inside the brackets) at varying impact speeds. The crumple zone can absorb most of the impact energy when the crash is at a lower speed of less than 40 km/h. Although the longerons seem to absorb slightly higher energy than the subframe, when the parts are compared in terms of specific energy absorption, it is obvious that the EA of the longerons is more efficient.

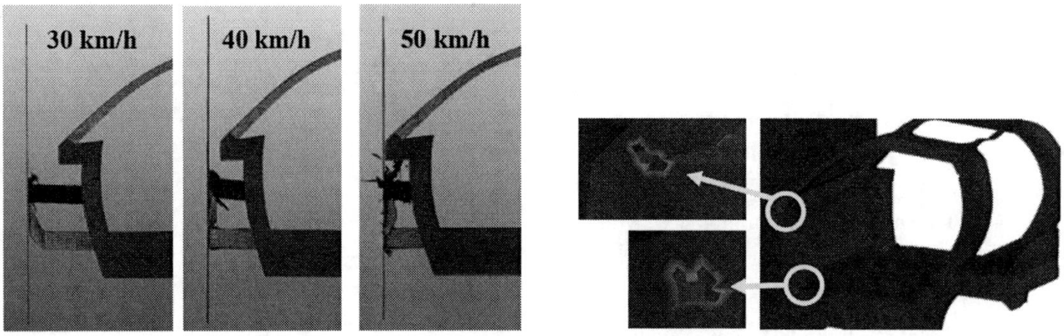

(a) (b)

Figure 3. Structural damage from full frontal crash (a) damage of crumple zone at different speeds (b) damage of passenger cell at 50 km/h speed

Table 3. Energy absorption of the crumple zone

Impact velocity	Impact energy (kJ)	Energy absorption, EA (kJ) [Specific energy absorption, kJ/kg]			Percent of EA
		Longerons	Subframe	Crumple zone	
30 km/h	12.1	5.0 [0.46]	3.8 [0.12]	8.8	72.5 %
40 km/h	22.0	8.0 [0.74]	7.5 [0.24]	15.5	70.0 %
50 km/h	33.8	10.6 [0.98]	7.9 [0.25]	18.3	54.8 %

An example of energy absorption distribution after frontal crash at 50 km/h is displayed in Figure 4a. Most of the impact energy dissipates to four main parts, i.e., subframe, longerons, front panel and side beam. This show that the impact force effectively transmits from front of the car to the base of the passenger cell. The passenger compartment is re-analysed based on the remaining impact energy when the crumple zone is fully damaged or when the crash speed greater than 50 km/h. It is found that when the crash speed reaches 80 km/h, the plastic hinges at A-pillars are fully formed and damages at B-pillars initiate. This causes the front structure to intrude into the residual space and the occupant safety is compromised. Figure 4b illustrates clearances from the rigid components of the front structure when the crash speed is 50 km/h and 80 km/h. It can be noticed that the A-pillars intrude into the safety zone by 98 mm at the occupant's chest and 52-mm intrusion is found at the location of passenger's head. Further investigation by including occupant dummies into the FE model can predict the passenger's injury severity from both the crash deceleration and direct impact of these rigid parts to the occupant.

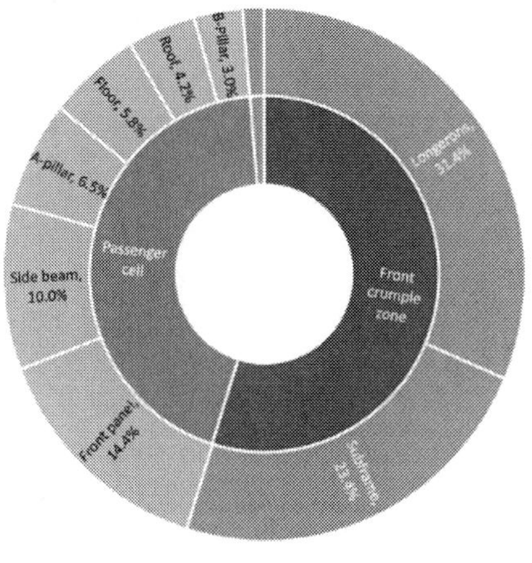

(a)

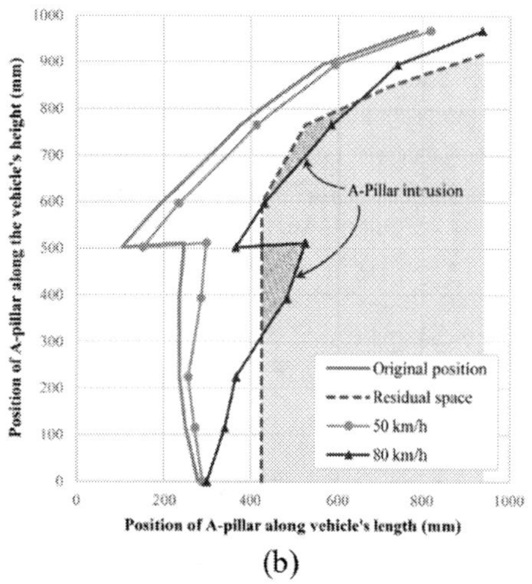

(b)

Figure 4. (a) Energy absorption distribution under frontal crash (b) A-pillar deformation after crash

3.2 Side impact analysis

Euro NCAP side impact analysis is conducted by collision of a 1350-kg mobile deformable barrier (MDB) to the side of the heavy quadricycle at 50 km/h. The MDB is used to represent another passenger car crash into the vehicle of interest. The location of impact corresponds to the location of the center of gravity (CG) along the length of the quadricycle. In this case, the car's wheels need to be included into the FE model. The front wheels are constrained to the longerons while the rear wheels constrain to the rear part of the car. Automatic surface-to-surface contacts are assigned between the MDB and the quadricycle side and between the vehicle wheels to the floor as shown in Figure 5. Static and dynamic friction coefficients between the wheels and the floors are 0.85 and 0.80, respectively.

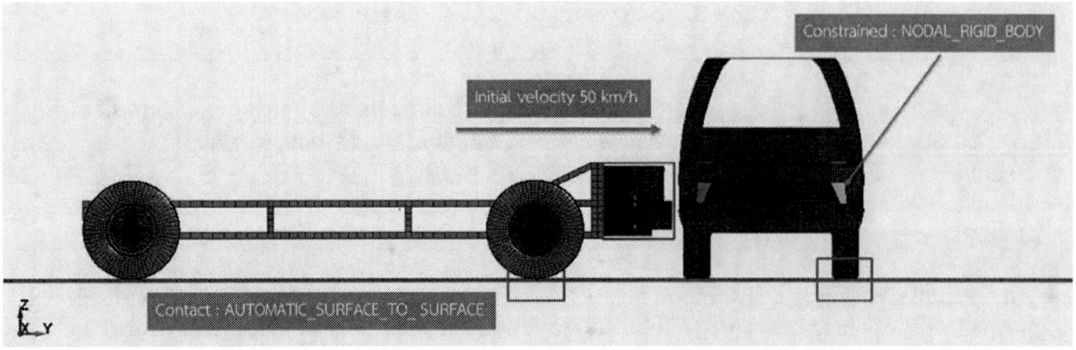

Figure 5. Side impact analysis by mobile deformable barrier

The CG of the baseline quadricycle is 599 mm from the floor when the ground clearance is at a typical value of 180 mm while the CG of the MDB is 500 mm. It is found that at this height, overturning of the quadricycle may occur when it is crashed by the MDB trolley due to its small weight and its incompatibility with other passenger cars.

The baseline model is then modified by lowering the ground clearance of the quadricycle to 150 and 130 mm. The quadricycle is analysed at crash speeds of 50 km/h and 80 km/h. It is found that the car with the ground clearance of 150 mm does not overturn when the MDB speed is 50 km/h but still overturns at high-speed crash of 80 km/h. Nevertheless, the car does not experience overturning during crash when the ground clearance is lowered to 130 mm. Damage from side crash simulation at 50 km/h of the modified model is shown in Figure 6. The car structure can well-resist energy from side crash of another passenger car. The honeycomb profile impactor is completely damaged and absorb most of the crash energy. The quadricycle is slightly damaged at the joints of near-side beam with pillars while the passenger cell is still unharmed. Thus, the occupant's residual space is safe.

The thickness of the car frame is decreased to 3 mm to study the energy absorption of structural parts under severe damage. Damages are noticed to first occur on the side beam at the same locations as in the previous cases and afterward B-pillar collapses and the occupant's safety space is intruded. It is observed that, in this case, the impact energy can mostly be absorbed by the side beam and pillars. Energy dissipation to other parts via the structural parts is not as efficient as in the case of frontal crash. Redesign of the base structure with crossbeams to dissipate the side impact energy is therefore recommended.

Figure 6. Damage from side crash test at 50 km/h

4. Conclusions

This study presents finite element analysis of a heavy electric quadricycle with a CFRP passenger cell and aluminium crumple zone under frontal and side crash. It is shown that the front crumple zone can efficiently absorb impact energy from frontal crash and the car structure can efficiently distribute energy and guarantee safety of the occupant under Euro-NCAP front crash test with a rigid wall. When the baseline structure is side-crashed by an MDB representing another car, quadricycle is prone to overturn due to the light vehicle's weight and high center of gravity. Lowering ground clearance can be an option to avoid overturning of the quadricycle. When overturning does not occur, the baseline structure is proved to be safe under side impact test up to 80 km/h. However, if the quadricycle is severely crashed, the side structure cannot efficiently dissipate the impact energy to other structural parts and thus supplementary energy dissipated members should be added to improve crashworthiness of the structure under side crash. Experiments and further investigation on the actual behavior of the CFRP components under crash scenario should also be examined in the future work.

Acknowledgments

The authors wishing to acknowledge financial supports from Mitsui Sumitomo Institute Foundation Grant 2020.

References

[1] Beyond one million electric cars 2016 *Global EV Outlook* (Paris: International Energy Agency Publications) 7
[2] Statista Number of licensed quadricycle microcars in the United Kingdom (UK) between the 1st quarter of 2010 and the 3rd quarter of 2018 https://www.statista.com/statistics/310552/number-of-quadricycle-microcars-in-the-united-kingdom/, accessed on 12 March 2019
[3] Persistence Market Research 2019 *Europe Market study on Motorized Quadricycles*
[4] European New Car Assessment Program Quadricycles' Tests *https://www.euroncap.com/en/vehicle-safety/safety-campaigns/2016-quadricycles-tests/*, accessed on 14 September 2018
[5] Kongwat S Jongpradist P and Hasegawa H 2020 *Int. J. Automot. Technol.* **21**(4) 981-991
[6] Kunakorn-ong P and Jongpradist P 2020 *Int. J. Automot. Technol.* **21**(1) 215-225
[7] Sadeghipour E Lienkamp M and Duddeck F 2014 Crash Compatibility of Microcars: A Study on Current Test Approaches *Crash Tech 2014* (Munich: Germany)
[8] Lui Q Lin Y Zong Z Sun G and Li Q 2013 *Compos. Struc.* **97** 231-238
[9] Stein J Matheis R Faßbender S and Seidel K 2017 *Small Electric Vehicle Concept Epsilon with Innovative Lightweight Design Body* (Aachen: CORDIS EU Research Results) FP7-SST-2013-RTD-1
[10] Kriescher M Scheibe S Rieger C Straßburger P and Herreweghe C V 2019 Development of the Safe Light Regional Vehicle (SLRV) Vehicle Concept within the DLR Next Generation Car (NGC) project *32nd Electric Vehicle Symposium* (Lyon: France)
[11] Osborne M Wade B and Feraboli P 2012 *Simulating laminated composites using LS-DYNA material model MAT54 part I: [0] and [90] ply single-element investigation* 3-33
[12] Singhatanadgid P 2016 *Mechanics of Composite Materials* (Bangkok: Chulalongkorn University Press)

Analysis of aluminum front bumper beam for heavy quadricycles under impact loading

Kiadtikun Phanmueang[1], Pattaramon Jongpradist[1] and Kitchanon Ruangjirakit[1,*]

[1]Department of Mechanical Engineering, Faculty of Engineering, King Mongkut's University of Technology Thonburi, Bangkok Thailand 10140

* Corresponding Author: kitchanon.rua@mail.kmutt.ac.th

Abstract

A heavy quadricycle is a small four-wheeled vehicle suitable for urban mobility owing to its compact, lightweight and energy-efficient advantages. However, such small vehicles have a comparatively short crumple zone leading to a higher risk of occupant injury in case of frontal collision. In this paper, the structural performance and energy absorption of aluminum alloy AA6063-T6 thin-walled front bumper beams for heavy quadricycles are studied. Mechanical characteristics of the beams with simple cross-sectional profiles, i.e., square, rectangle and double-cell rectangle as well as a commercially available bumper beam of a heavy quadricycle under cylindrical-rod impact loading are investigated through finite element analysis using LS-DYNA. A parametric study is also performed to examine the effects of open and closed sections, different cross-sectional shapes and various fillet radii on impact responses of the bumper beam including energy absorption, specific energy absorption, peak force, and crush force efficiency. The results show that although section profiles with sharp corners possess higher energy absorption, a larger fillet radius could mitigate the unfavourable peak force from impact up to 40%. The partition plate of the double-cell rectangle section induces a much higher peak force compared to others. In addition, when comparing an open-section beam with closed-section beams, it is found that an open-cell hat-shape model with fillet radius of 8 mm is the most efficient design. The crush force efficiency and peak force of the section are 32% higher and 17% lower, respectively, when compared to the closed-cell section with fillet radius of 8 mm.

Keywords: Bumper beam, Frontal impact, Energy absorption, Heavy quadricycle

1. Introduction

At present, one of the greatest challenges faced by automotive industry is to provide safer vehicles with high energy efficiency at a competitive cost [1]. The use of high-performance lightweight materials for structural members offers great potential for increasing vehicle efficiency while crashworthiness requirements are also satisfied. A heavy quadricycle, categorized by the European Union as L7e, is a small four-wheeled vehicle that has an unladen mass equal to or less than 450 kg without battery, a maximum net motor power not exceeding 15 kW and a maximum speed not less than 90 km/h [2]. Quadricycles are suitable for urban mobility owing to their compact, lightweight and energy-efficient

IOP Conf. Series: Materials Science and Engineering 1137 (2021) 012013 doi:10.1088/1757-899X/1137/1/012013

advantages. However, such vehicles have a comparatively short crumple zone leading to a higher risk of occupant injury in case of frontal collision.

Front crumple zone is mostly designed by mounting thin-walled stiff bumper beam to crash boxes so as to absorb and dissipate kinetic energy in accidental impacts. Crash energy must be absorbed in a controlled manner with limiting forces exerted on occupants and unstable behaviors such as buckling or catastrophic failures be avoided [3]. The energy absorption capacity and crashworthiness performance of the bumper and crash boxes have received extensive attention in the engineering fields [4]. To achieve higher energy absorption efficiency, researchers proposed various methods to achieve the better design of the members [5-8]. Energy dissipation mechanisms for diverse design of crumple zone are also substantially explored using basic theory of deformation and observation of beam under indentation-mode impact loading [9-14]. Based on the Technique for Order of Preference by Similarity to Ideal Solution (TOPSIS) used in bumper beam profile selection process for small road vehicle, it is found that the double hat profile fits the bumper beam for a small car. The selected concept can be strengthened by adding ribs or increasing the thickness of the bumper beam to make the best geometry according to defined product design specifications [15-16]. Dynamic tests on a double hat profile and a simple hat profile under impact showed that the double hat profile exhibits better stress distribution and lower deflection after impact [17-18] which benefits both passenger and pedestrian protection [19].

This paper aims to apply finite element analysis to investigate the energy absorbing behavior and failure mechanism of the aluminum alloy AA6063-T6 thin-walled front bumper beam for heavy quadricycles under impact loading. A parametric study is also performed to examine the effects of open and closed sections, different cross-sectional shapes and various fillet radii on impact responses of the bumper beam.

2. Bumper Beam Model

2.1 Material Properties

The bumper beam in this study is made of aluminum AA6063-T6. The material mechanical properties as listed in table 1 are obtained from uniaxial tensile tests according to the ASTM E8M-04 standard [20] using three specimens. The tensile tests are conducted using Zwick Roell HB250 universal testing machine. The loading speed is set constant at 1 mm/min. The true stress-strain curve as illustrated in Figure 1 is assigned to the bumper material model in the finite element simulation.

Table 1. Material properties of AA6063-T6 obtained from the uniaxial tensile test.

Material	E (GPa)	v [14]	ρ (kg/m^3) [14]	σ_y (MPa)	σ_u (MPa)	ε_{max}
AA6063-T6	64 ± 2.72	0.33	2700	197.72 ± 2.52	214 ± 2.88	0.0991 ± 1.25

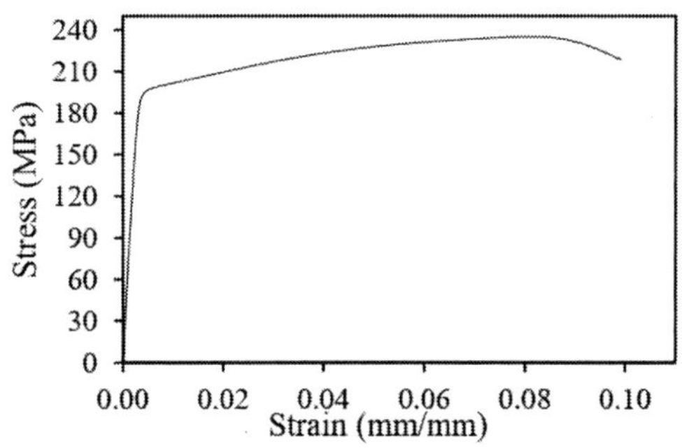

Figure 1. True stress-strain curve of AA6063-T6.

2.2 Bumper Beam Profiles

In order to assess the most suitable profile for the bumper beam of an L7e quadricycle, four aluminum profiles, which are commercially available profile, rectangular section, double-cell rectangular section and square section, are proposed and their dimensions are shown in Table 2. The length of the bumper beam should not exceed the vehicle width of 1,500 mm. In this study, the bumper beam length is 720 mm taken from a commercial model of heavy quadricycle. To facilitate result comparison of energy absorbing capacity of different beam sections, the mass of all beam profiles is controlled at 0.6 kg by changing the beam thickness.

Table 2. The geometry of the bumper beam cross section for the impact loading test.

Section profile / Details	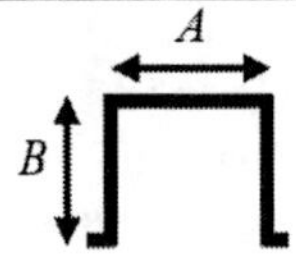	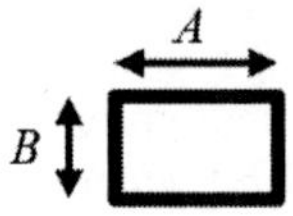	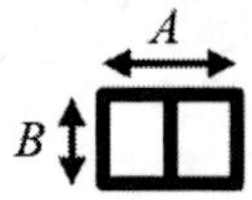	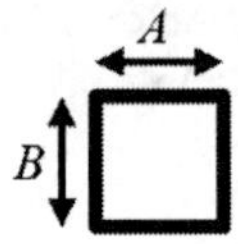
Model	Commercial	Rectangle	Double-cell rectangle	Square
Weight (kg)		0.608		
Length (mm)		720		
A (mm)	50	48	40	38
B (mm)	50	30	26	38
Thickness (mm)	1.89	2.00	1.97	2.00

3. Cylindrical-rod Impact Simulation

The cylindrical-rod impact loading tests are simulated using non-linear explicit finite element code in LS-DYNA. The bumper beam of length 720 mm is supported by two cylindrical rods with a span of 620 mm at the locations of the crash boxes (Figure 2a). The beam is discretized using Belytschko-Tsay 4-node shell elements. The 20-mm diameter cylindrical rod and supports are modelled using 8-node solid elements. An automatic single surface contact algorithm is applied to model the contact of the beam to itself. To account for the contact between the impactor (master) and the beam (slave), and between the beam (master) and the supports (slave), automatic node-to-surface contacts are defined. The friction coefficient for all contact pairs is set to be 0.3 [21]. The MAT_024 Piecewise Linear Plasticity Material model is assigned to the beam material using the properties obtained from the test results and the stress-strain curve as shown in Figure 1 and MAT_020 Rigid is used to model the cylindrical-rod and supports. The cylindrical rod of mass 25 kg is assigned with an initial impact velocity of 4.71 m/s (17 km/h), which provides an initial impact energy of 279 J. Mesh sensitivity is performed by comparing the simulated average force of models with element sizes from 0.5 mm to 5 mm as shown in Figure 2(b). The resulting average impact force obtained from mesh sizes of 0.5, 1 and 2 mm are relatively similar. Therefore, 2-mm element size was chosen to reduce computation time without compromising result accuracy. Strain rate effect is not considered in the present numerical model since the aluminum alloy 6063-T6 is strain-rate insensitive during low strain-rate impact between 110 s^{-1} and 850 s^{-1} [22].

IOP Conf. Series: Materials Science and Engineering 1137 (2021) 012013 doi:10.1088/1757-899X/1137/1/012013

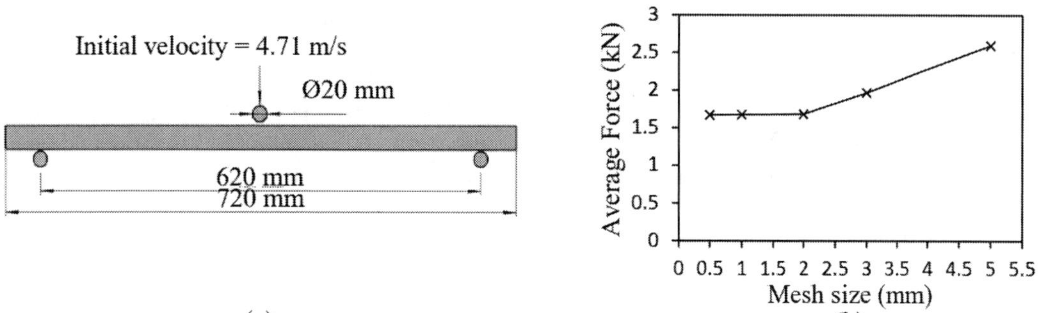

(a) (b)

Figure 2. (a) Cylindrical-rod impact modelling. (b) Mesh convergence study.

4. Simulation Results and Discussion

4.1. Effects of Bumper Beam Profiles

The deformed shapes and effective plastic strain of the bumper beams under cylindrical-rod impact loading are illustrated in Figure 3. After impact, all beam sections exhibit similar fracture failure starting from the top surfaces underneath the point of impact and propagating to the sides of the beams. Force-displacement curves of each cross-section profile under impact are compared in Figure 4 and similar trend can be clearly seen. Peak forces are observed at the beginning stage of the impact when failure at the corners of the top surface first occurs. Then, the forces drastically drop afterward and continuously decrease when the beam's side walls damage causing large deformations to the beam. The double-cell rectangular profile shows the highest peak force (F_{max}) of 11.3 kN due to the stiffened vertical wall inside the cross-section. The commercially available hat-section gives a noticeably lower peak force compared to that of other profiles owing to the deformation mechanism of the open-section in which its vertical walls could undergo larger horizontal deformation and plastic strain to absorb the first impact force.

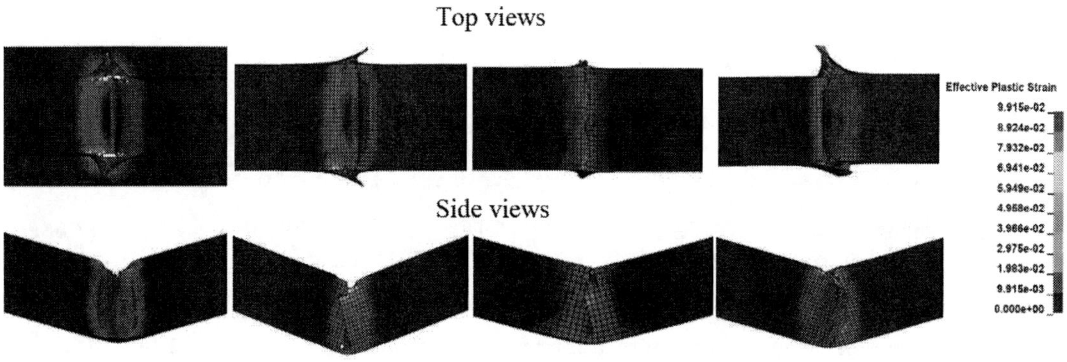

(a) Commercial (b) Rectangle (c) Double-cell rectangle (d) Square

Figure 3. Effective plastic strain of the bumper beam under cylindrical-rod impact.

Results on energy absorption (EA), specific energy absorption (SEA), crush force efficiency (CFE), average force (F_{avg}) and maximum deformation (d_{max}) are listed in Table 3. The average force is computed by averaging the force over the entire distance of the impact event. Crush force efficiency is defined as the ratio of the average force to the peak force. Thus, a preferable profile for a bumper is one with a low peak force to mitigate the injury risk of the occupant during crash, a high SEA for mass and cost efficiency, and CFE should be close to 1 where the impact force is well transferred to different parts of the structure. The commercial open-cell model is found to exhibit the highest SEA of 202.8 J/kg. However, the double-cell rectangular section gives the highest CFE among other sections. Figure 5 illustrates result comparison of the rectangular, double-cell rectangular, and square sections with the commercially available baseline model. It is found that the closed-section bumper beams have lower EA and SEA than those of baseline model by 30-40%. The double-cell rectangle although provides a higher CFE and F_{avg}, its F_{max} is somewhat dangerously higher than other models. Therefore, considering the above discussion, the optimal shape for a heavy quadricycle is the commercial section as it exhibits the highest EA and SEA and less peak force compared to other sections.

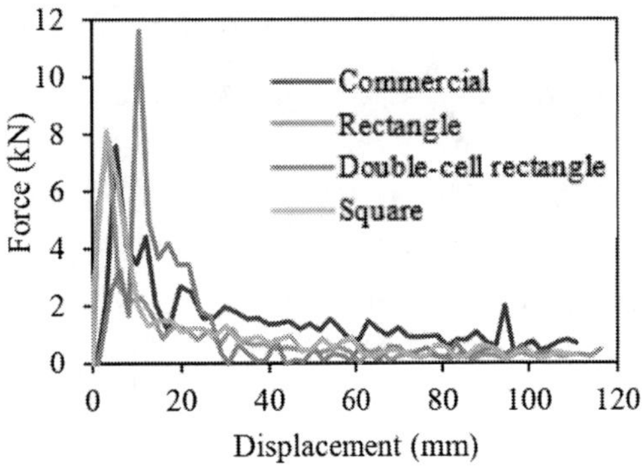

Figure 4. Force-displacement curves.

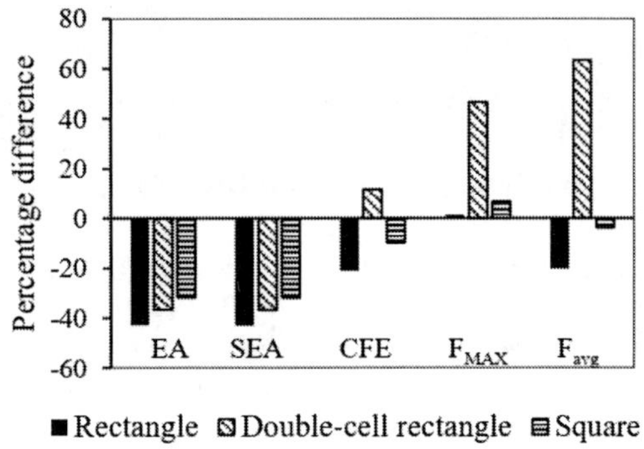

Figure 5. Section performance compared with baseline.

Table 3. Energy absorption capability of different bumper beam cross sections.

Cross section	Commercial	Rectangle	Double-cell rectangle	Square
EA (J)	123.3	70.6	78.2	84.1
SEA (J/kg)	202.8	115.9	128.3	138.1
CFE	0.205	0.161	0.228	0.184
F_{max} (kN)	7.7	7.8	11.3	8.2
F_{avg} (kN)	1.6	1.3	2.6	1.5
d_{max} (mm)	109.77	116.16	107.58	112.68

4.2 Effects of Fillet Radius

The failure mechanism simulation results of the bumper beams under rod impact testing have shown that the crack initiates from the top corners of the beams and propagating to the side walls. Therefore, adding rounded fillets at corners will help improve the load carrying and energy absorption capability of the bumper beam. Therefore, the effects of fillet radius for both open-cell and closed-cell commercially available sections with the fillet radii of 0, 2, 5, 8 mm are investigated.

The force-displacement results of various fillet radii for both open-cell and closed-cell cross sections are illustrated in Figure 6. The energy absorption performance comparison with the baseline model is also shown in Figure 7. It is noticed that, for open-cell sections, the size of fillet radius largely affects the section performance. The beam section with a larger fillet radius has a distinguishably less peak force and thus higher CFE and F_{avg}. The most preferable open-cell section is one with fillet radius of 8 mm in which the F_{max} is reduced by 40% compared to the baseline model while CFE and F_{avg} are improved by 100.9% and 20.5%, respectively. However, for closed-cell sections, the advantages of fillet radius are not as evident. Force-displacement curves for closed-cell section with different fillet radii are similar. Although the closed-cell sections with fillet radius of 8 mm show a lower F_{max} compared to the open-cell and the baseline sections, the SEA and F_{avg} are also reduced due to the characteristics of the closed-cell bumper damage during impact. In addition, the performance of all closed-cell sections is found to be less favorable when compared with those of open-cell sections.

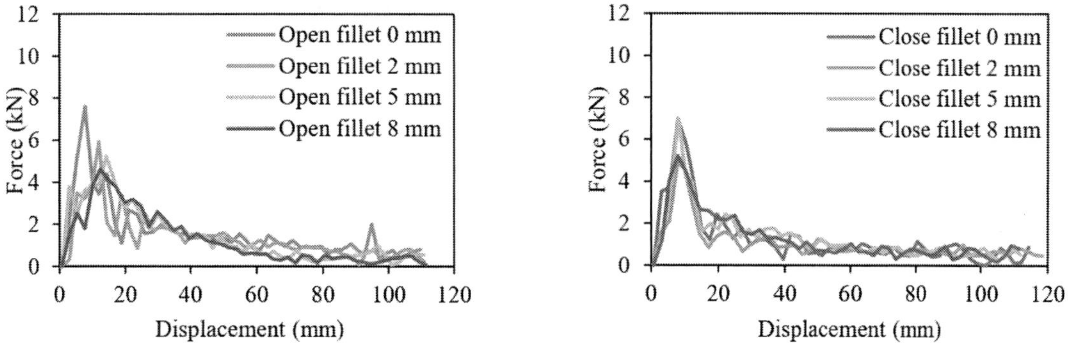

Figure 6. Force-displacement curves of (a) open-cell (b) closed-cell sections with different fillet radii.

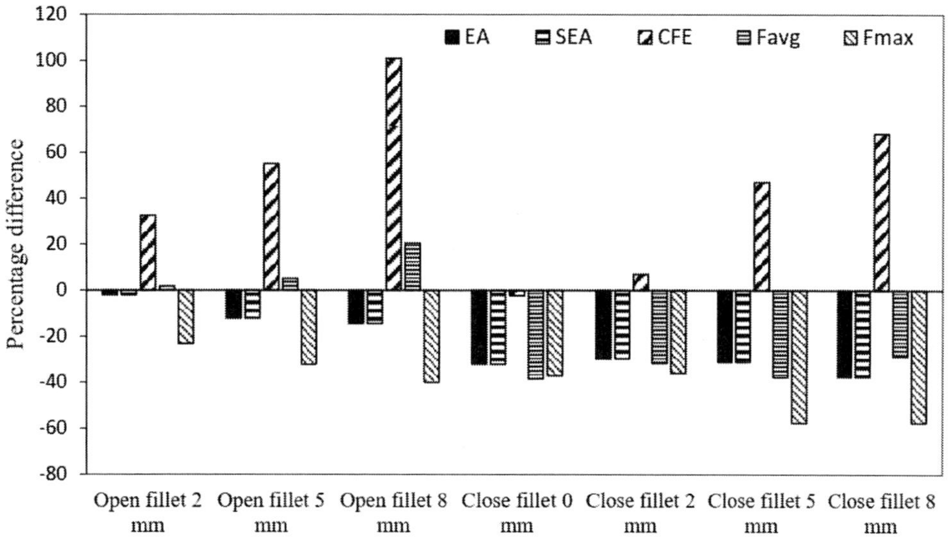

Figure 7. Performance comparison of commercial section with various fillet radii to the baseline model.

5. Conclusion

The behavior and performance of the heavy quadricycle bumper beam under the cylindrical-rod impact loading with various section profiles and fillet radii are examined. Among the closed-cell sections, the double-cell rectangular bumper beam provides the highest CFE and F_{avg}. However, the open-cell commercial section gives a much higher SEA than other beam profiles with the lowest F_{max}. The analyses of fillet size for open-cell and closed-cell sections indicate that the use of larger fillet radii up to 8 mm can improve the performance of the open-cell sections while the effect of fillet radius to the closed-cell sections are not as prominent. The open-cell hat-shape model with fillet radius of 8 mm is found to be the most efficient design as it exhibits the highest CFE and the lowest F_{max}.

This model should be examined further under other impact and crash loads in order to use efficiently for a heavy quadricycle.

References

[1] Chandrakant R S and Ajit L S 2018 *J. Inst. Eng. India Ser.* **99** 599-606
[2] Campos J A L Robled A S Vilán J A Nieto P J G and Cordero J B 2015 *Materials* **8** 6893-6908
[3] Duarte I Vesenjak M Krstulović-Opara L and Ren Z 2015 *Compos. Struct.* **124** 128-139
[4] Lu G and Yu T 2003 *Energy Absorption of Structures and Materials* (Cambridge: CRC Press–Woodhead Pub)
[5] Zhang Z Liu S and Tang Z 2009 *Thin Wall Struct.* **47** 868-878
[6] Zhang X Zhang H and Wang Z 2016 *Int. J. Mech. Sci.* **106** 107-116
[7] Baroutaji A Morris E and Olabi A G 2014 *Thin Walled Struct.* **82** 262-277
[8] Baroutaji A Gilchrist M D Smyth D and Olabi A G 2015 *Int. J. Impact Eng.* **82** 74-88
[9] Xiang X M Lu G Li Z X and Lv Y 2017 *Int. J. Impact Eng.* **110** 138-148
[10] Zhang X and Zhang H 2018 *Int. J. Impact Eng.* **112** 165-179
[11] Chen W 2001 *Int. J. Solids Struct.* **38** 7919-7944
[12] Huang Z and Zhang X 2018 *Int. J. Mech. Sci.* **144** 461-479
[13] Chen W Wierzbicki T and Santosa S 2002 *Acta Mech.* **153** 183-206
[14] Huanga Z and Zhang X 2019 *Thin-Walled Struct.* **137** 231-250
[15] Davoodi M M Sapuan S M Ahmad D Aidy A Khalina A and Jonoobi M 2011 *Materials & Design* **32** 4857-4865
[16] Yantao C Lingyu S and Lijun L 2019 *Journal of Cleaner Production* **241** 118316
[17] Rao S S Viswatej K and Adinarayana S 2016 *IJETT.* **37** 2
[18] Qi C Sun Y and Yang S. A 2018 *Thin Walled Struct.* **129** 327-341
[19] Yang S Sun Y and Qi C 2020 *Int. J. Mech. Sci.* **165** 105210
[20] ASTM International *ASTM E8M-04 2004 Standard test methods for tension testing of metallic materials* (West Conshohocken, Pennsylvania: ASTM)
[21] Hallquist J O 2006 *LS-DYNA theory manual* (Livermore Software Technology Corp.) pp 25-31
[22] Singh N K Singha M K Cadoni E and Gupta N K 2012 *Advanced Materials Research* **548** 169-173

IOP Conf. Series: Materials Science and Engineering 1137 (2021) 012014 doi:10.1088/1757-899X/1137/1/012014

A comparative study of equivalent circuit models for a Li-ion battery pack of an electric Tuk-Tuk

Natcha Rajchapanupat[1] and Poowanart Poramapojana[1,*]

[1]Department of Mechanical Engineering, Faculty of Engineering, Kasetsart University, Bangkok 10900, Thailand

* Corresponding Author: fengpap@ku.ac.th

Abstract

This paper presents developments and validations of Lithium Nickel Manganese Cobalt Oxide (NMC) battery cell and battery pack models. A 7.4 kWh NMC battery pack is installed on an electric Tuk-Tuk equipped with a 3 kW motor. The battery pack is configured of four cells connected in parallel and 20 sets of the four cells connected in series. Based on the first-order and the second-order equivalent circuit models, battery cell and battery pack models are developed in Matlab/Simulink environment. For parameter identifications of the cell model, pulse discharge tests are performed at room temperature with 10% State of Charge (SOC) interval. The lumped heat capacitance method is applied to the battery models. The battery cell model is validated with SOC, cell center temperatures and terminal voltage using the experimental data of constant current discharge tests. With the battery pack configuration, the battery cell model is scaled up to match the battery pack voltage and its capacity. The dynamics of the battery pack models are validated with the vehicle testing data using the no-load, wide-open throttle and real road tests. The simulation results show good agreement with the test data. Using the developed battery models, vehicle performance and its energy consumption can be improved by optimizing battery cell and pack configurations.

Keywords: Li-ion battery, Battery pack models, Electric Tuk-Tuk

1. Introduction

Tuk-Tuk is a unique three-wheeled vehicle name in Thailand. The vehicles have been widely used in many countries. Within the constraint of cost, electric Tuk-Tuks generally are installed with lead-acid batteries. With the advancement of Lithium-ion battery technology and electric powertrain, their performance and energy efficiency can be improved explicitly. The optimization of the powertrain and battery pack configuration can be achieved using accurate battery and powertrain models in the design processes [1-5].

Many battery models have been proposed to capture behaviors of battery cells. To understand electrochemical reactions in the cell, electrochemical models can achieve high accuracy. However, the models are not designed for battery management systems in electric vehicles because of their complex equations and a large number of unknown parameters [6]. For real-time and control-oriented purposes, equivalent circuit models are mostly used to manage vehicle power and battery control units. The equivalent circuit models have been developed. For example, the first-order resistance-capacitance

(RC) and second-order RC (Dual polarization model) [7-8]. Comparisons of twelve equivalent circuit models for LiNMC and Lithium Iron Phosphate (LiFePO4) cells were studied [9]. The simulation results of the study show the first-order RC model is preferred due to its accuracy and simple to estimate cell voltage and SOC.

This paper presents effectiveness comparisons of the first-order model and the second-order model from the cell model to the battery pack model. The remainder of this paper is organized as follows. In Section 2, the experimental of the battery cell testing and the on-road battery pack testing are described. Section 3 demonstrates the battery cell model developments and validations. Section 4 presents the battery pack model developments and validations. Finally, Section 5 concludes the present work.

2. Experimental

2.1 Battery Cell Testing

A diagram of a battery test system used in this present work is shown in Figure 1. The test bench consists of a battery cell, an ITECH IT8511A electronic DC load, an IT6722A DC power supply and a Graphtec GL840 data logger. The cell voltage, current, ambient temperature and cell center temperature are recorded by GL840 data logger with the sampling time of 1 second. In addition, Type K thermocouples are used in the tests. Two types of cell tests are performed at room temperature including constant current (CC) discharge and pulse current discharge tests. The CC discharge tests are conducted to determine the Open Circuit Voltage (OCV) test and to validate the cell model. The pulse current discharge tests are performed to identify battery cell parameters for the models. Moreover, the experimental data is recorded every 0.1 second for the pulse test.

For CC discharge tests, the cell is discharged from the fully-charged state to the cut-off voltage of 3.0 V. For each test, the cell is charged using a constant-current constant-voltage protocol. The cell is charged with a 1/3 C-rate constant current step until the cell voltage equals to 4.2 V. Then, the cell is charged by a constant voltage step until the cut-off current reaches 1/20 C-rate. After the cell is fully charged, one hour rest before performing the CC discharge test is used to make the cell reaches the thermal equilibrium state. For the battery parameters identification, details of the testing protocol are explained in [10].

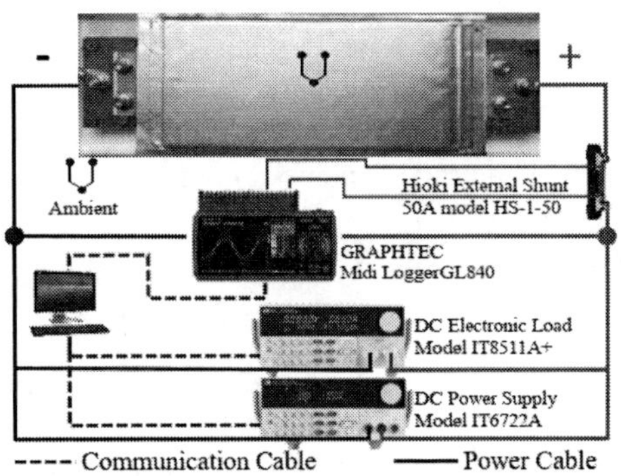

Figure 1. The battery cell test bench

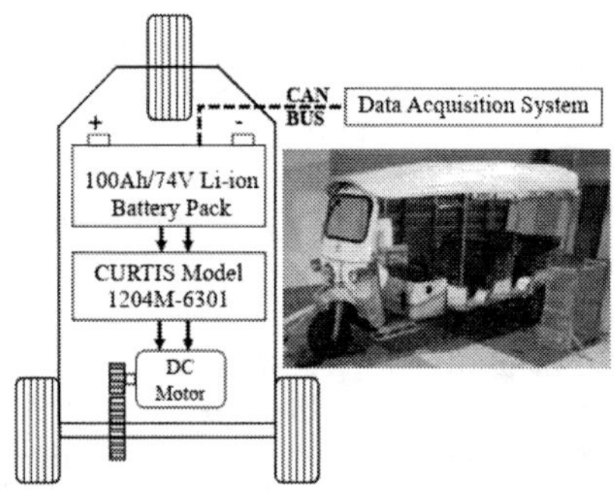

Figure 2. The electric Tuk-Tuk configuration

2.2 On-road Battery Pack Testing

The electric Tuk-Tuk configuration used in the present work is shown in Figure 2. The vehicle is propelled by a 3 kW motor. A 1204M Curtis motor controller is energised by a 7.4 kWh (74 VDC, 100Ah) battery pack. To record battery cell voltage in the pack, a data acquisition system is used to receive data through the Control Area Network (CAN) bus interface.

To evaluate the effectiveness of the battery pack model, three test conditions are performed to capture the dynamic behavior of the battery pack. First of all, a non-road load test is performed to simulate the steady-state response of the system. The vehicle is lifted their rear wheels. The vehicle speed is increased with an incremental of 100 rpm and maintain at the speed for a few seconds. Secondly, a Wide-Open Throttle (WOT) test is conducted to simulate a step response of the system. The vehicle is driven with the full pedal position until the vehicle reaches the maximum speed. Finally, a real-world road test is performed on a local road for 6 minutes.

3 Battery Cell Model Validation

3.1 Equivalent Circuit Models

The resistance-capacitance network-based equivalent circuit models are shown in Figure 3. The cell terminal voltage $V(t)$ is determined using equation (1) where $V_{ocv}(SOC)$ is the open-circuit voltage (OCV) as a function of battery State of Charge (SOC). $R_0(SOC)$, $R_1(SOC)$ and $R_2(SOC)$ are the resistors. $C_1(SOC)$ and $C_2(SOC)$ are the capacitors. $V_1(t)$ is the voltage drop across $R_1(SOC)$ and $C_1(SOC)$. $V_2(t)$ is the voltage drop across $R_2(SOC)$ and $C_2(SOC)$. $I(t)$ is the current.

$$V(t) = V_{ocv}(SOC) - V_1(t) - V_2(t) - I(t)R_0(SOC) \tag{1}$$

The SOC can be evaluated using equation (2) where SOC_0 is the initial SOC and Q_{batt} is battery cell capacity in the unit of Ah.

$$SOC = SOC_0 - \frac{1}{3600*Q_{batt}} \int I\,dt \tag{2}$$

The rate of $V_1(t)$ and $V_2(t)$ are expressed in equation (2) and (3). By taking time integral to the equations, $V_1(t)$ and $V_2(t)$ can be obtained.

$$\frac{dV_1}{dt} = -\frac{V_1(t)}{R_1(SOC)C_1(SOC)} + \frac{I(t)}{C_1(SOC)} \tag{3}$$

$$\frac{dV_2}{dt} = -\frac{V_2(t)}{R_2(SOC)C_2(SOC)} + \frac{I(t)}{C_2(SOC)} \tag{4}$$

To estimate cell surface temperature, the energy conservation with a lumped-capacitance method [11] is shown in equation (5) where m is the cell mass, c_p is the specific heat capacity, T is the cell surface temperature, h is the heat transfer coefficient, A_s is the cell surface area. T_{amb} is the ambient temperature and $\dot{Q}$ is the convective heat generation rate. In this present work, $\dot{Q}$ is estimated using Equation (6). The reversible heat is neglected to simplify the problem [12-13].

$$mc_p\frac{dT}{dt} = hA_s(T - T_{amb}) + \dot{Q} \tag{5}$$

$$\dot{Q} = I^2(R_0 + R_1 + R_2) \tag{6}$$

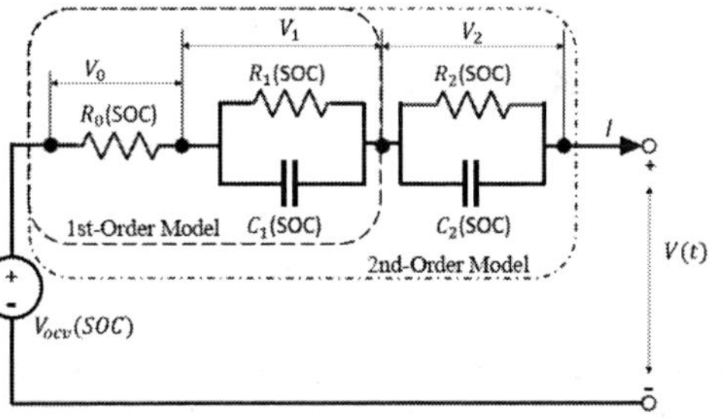

Figure 3. The resistance-capacitance network-based equivalent circuit models

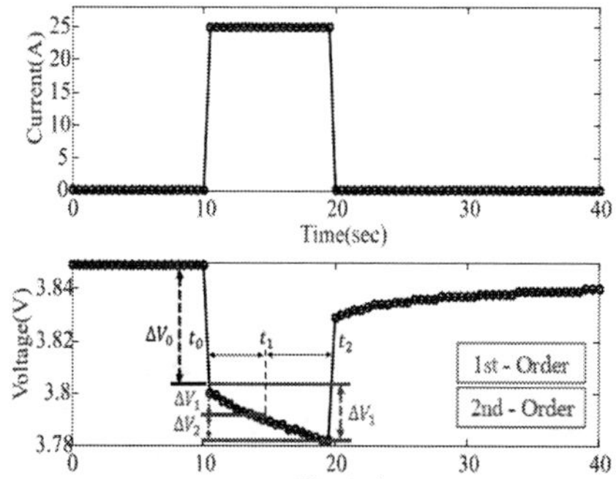

Figure 4. The pulse current discharge test for the parameters identification at 70% SOC

3.2 Parameter Identification for The Battery Cell

Based on the models shown in figure 3, the OCV, resistors and capacitors can be evaluated using the experimental data. To find the OCV equation as a function of SOC as shown in Equation (7), MATLAB curve fitting toolbox is used to obtain constants by comparing with the C/20 constant current discharge test. d_0, d_1, d_2, d_3, d_4 and d_5 are constants.

$$V_{OCV} = d_o + d_1(soc) + d_2(soc)^2 + d_3(soc)^3 + d_4 e^{-d_5(soc)} \tag{7}$$

The resistors and the capacitors are estimated from the pulse current discharge tests. Figure 4 demonstrates the pulse current discharge test for the parameters identification at 70% SOC. At the time t_0, $R_0(soc)$ of both models is determined using Equation (8) where ΔV_0 is the voltage drop when the pulse discharge is applied. $R_1(SOC)$ can be evaluated using equation (9). For the first-order model, ΔV_1 is the voltage drop from t_0 to t_2 whereas ΔV_1 for the second-order model is defined from t_0 to t_2. t_1 is the half time of t_0 and t_2. t_2 is the end time of the pulse discharge test. Time constants $\tau_1(SOC)$ and $\tau_2(SOC)$ of the first-order and the second-order can be estimated using Equation (10). $C_1(SOC)$ can be obtained from the definition of a time constant as shown in Equation (11). Figure 5 shows the OCV, resistors and capacitors of the models.

$$R_0(soc) = \frac{\Delta V_0}{I} \tag{8}$$

$$R_1(soc) = \frac{\Delta V_1}{I} \tag{9}$$

$$V_1(t) = IR_1\left(1 - e^{-\frac{t-t_0}{\tau_1}}\right) \tag{10}$$

$$\tau_1(SOC) = R_1(SOC)C_1(SOC) \tag{11}$$

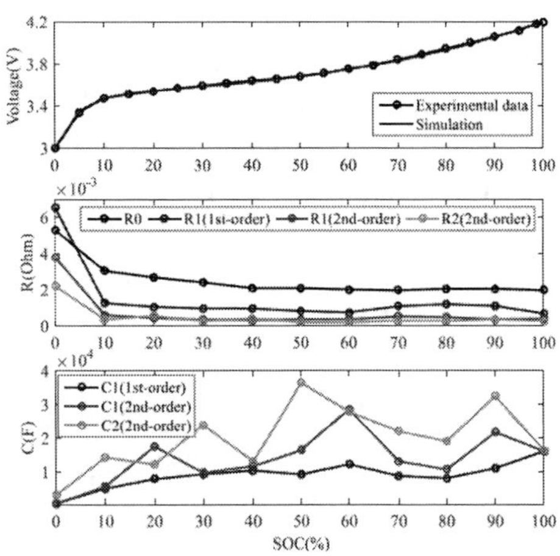

Figure 5. The open-circuit voltage, resistors and capacitors of the models

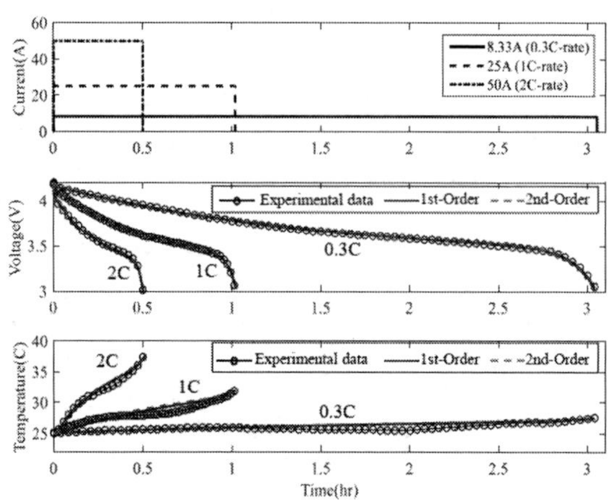

Figure 6. Cell model validations at room temperature

3.3 Cell Model Validations

Validation of discharge voltage at room temperature is shown in Figure 6. Noted that the cell was discharged using N3300A Agilent DC electronic load for 2C discharge [10]. The discharge current is the input of the models. The cell voltage and cell temperatures are compared with the experimental results. For validations of cell voltage, the simulation results show good agreement with the experimental data. For the first-order model, the maximum errors are 0.026 V, 0.027 V and 0.034 V for

IOP Conf. Series: Materials Science and Engineering 1137 (2021) 012014 doi:10.1088/1757-899X/1137/1/012014

C/3, 1C, and 2C-rates respectively. The maximum errors of the second-order model are 0.021 V, 0.024 V and 0.023 V for C/3, 1C, and 2C-rates respectively.

For cell temperatures validations, the thermal properties of the models are listed as follows. The cell mass is 0.585 kg and the specific heat capacity is 950 kJ/kg·K [10]. The convective heat transfer coefficients during the experimental validation are 3.382 W/m^{2}K, 5 W/m^{2}K, and 11.5 W/m^{2}K for 1/3C, 1C, and 2C-rates respectively. For validation of the cell center temperatures, the simulation results show that the lumped-capacitance model of both models follows the experimental results. For the first-order model, the maximum errors are 4.34% (1.11°C), 4.90% (1.39°C) and 3.15% (1.03°C) for 1/3C, 1C, and 2C-rates respectively. For the second-order model, the maximum errors are 4.24% (1.08°C), 4.04% (1.14°C) and 3.03% (0.99°C) for 1/3C, 1C, and 2C-rates respectively. The errors of the temperature validation can be reduced by including the reversible heat into the heat generation equation.

4 Battery Pack Model Validation

This section presents the effectiveness of the battery pack models for the first-order and the second-order models.

4.1 Battery Pack Model

The battery pack configuration for the first-order model is shown in Figure 7. A group of four cells are connected in parallel to meet the required capacity. 20 groups of the cells are connected in series to meet the working voltage. Assuming that 80 cells in the pack have the same characteristics and capacity [14]. The resistance and the capacitance of the battery pack can be calculated using Equation (12) and (13) where R_{0_pack} is the resistance of the pack, R_{0_cell} is the resistance of the cell, n_s is the number of the cell in series, n_p is the number of the cell in parallel, C_{1_pack} is the capacitance of the battery pack, and C_{1_cell} is the capacitance of the battery cell. R_{1_pack} and R_{2_pack} are determined in the same manner of R_{0_pack}. The battery pack model in MATLAB / Simulink environment is shown in Figure 8.

$$R_{0\_pack} = \left(\frac{n_s}{n_p}\right) R_{0\_cell} \tag{12}$$

$$C_{1\_pack} = \left(\frac{n_p}{n_s}\right) C_{1\_cell} \tag{13}$$

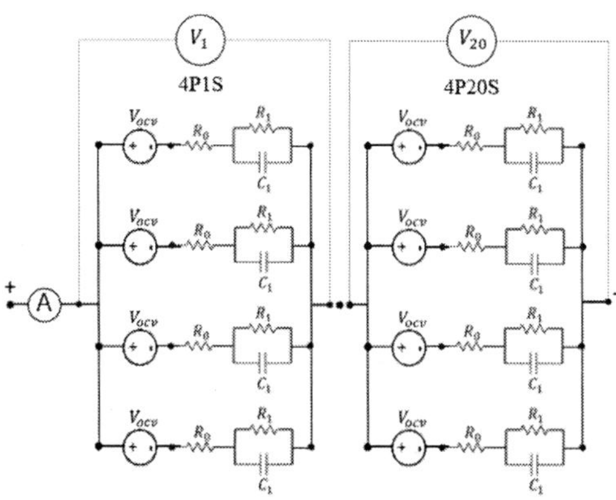

Figure 7. Battery pack configuration for the first-order model

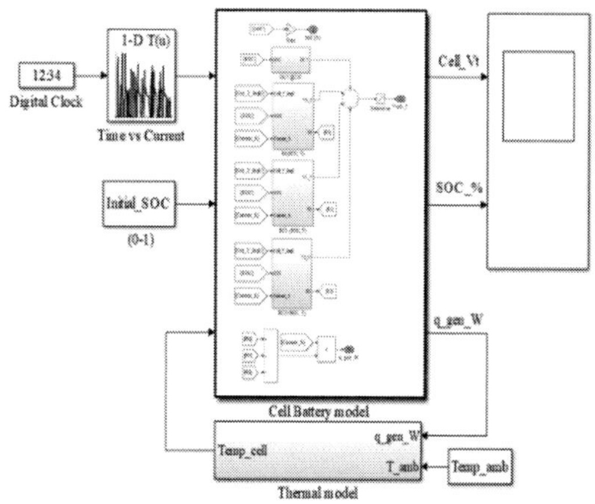

Figure 8. Battery pack model in MATLAB / Simulink environment

4.2 Comparisons of Battery Pack Model Validations

Figure 9 shows current, voltage and SOC of the cells in the battery pack for the non-road load test. For the first-order model, the maximum errors are 0.74 mV, 1.46 mV and 1.66 mV for the average Root-Mean-Squared Error (RMSE), the minimum and the maximum cell voltage in the pack respectively. For the second-order model, the maximum errors are 0.74 mV, 1.29 mV and 1.81 mV for the average RMSE, the minimum and the maximum cell voltage in the pack respectively.

Figure 10 shows current, voltage and SOC of the cells in the battery pack for the Wide-Open Throttle test. For the first-order model, the maximum errors are 1.33 mV, 2.32 mV and 3.21 mV for the average RMSE, the minimum and the maximum cell voltage in the pack respectively. For the second-order model, the maximum errors are 1.18 mV, 2.38 mV and 3.23 mV for the average RMSE, the minimum and the maximum cell voltage in the pack respectively.

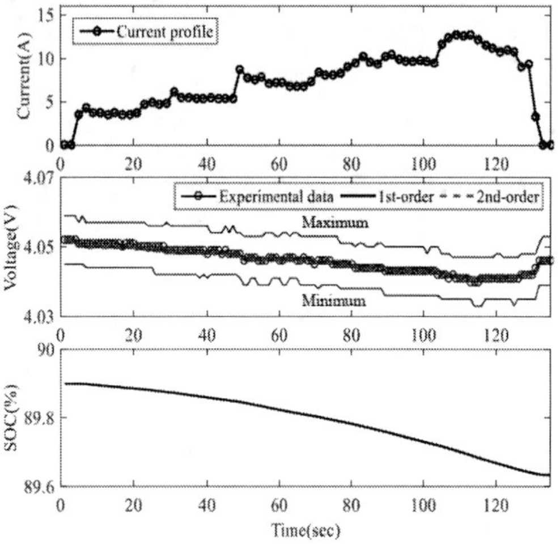

Figure 9. Current, voltage and SOC of the cells in the battery pack for the non-road load test.

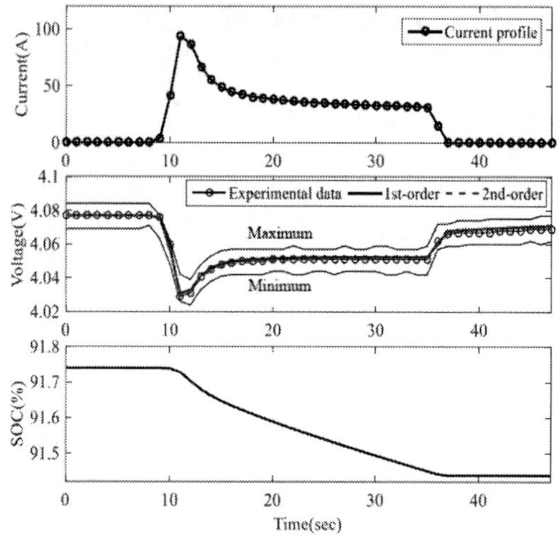

Figure 10. Current, voltage and SOC of the cells in the battery pack for the Wide-Open Throttle test.

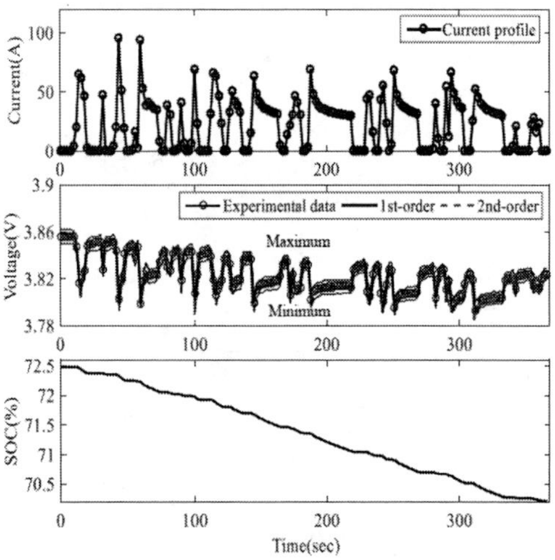

Figure 11. Current, voltage and SOC of the cells in the battery pack for the real-world road test.

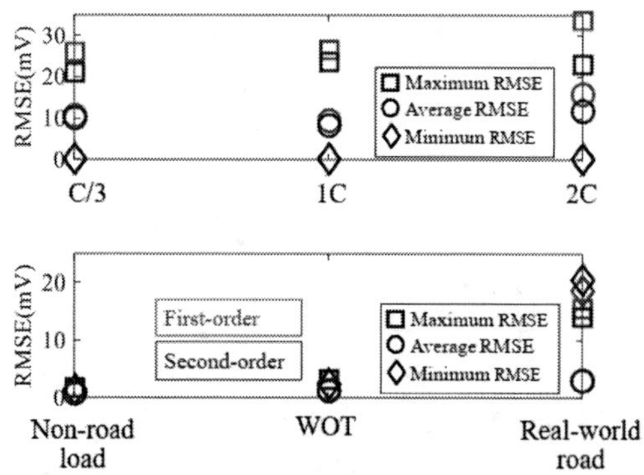

Figure 12. Model validation performance for the battery cell and the on-road battery pack tests.

Figure 11 shows current, voltage and SOC of the cells in the battery pack for the real-world road test. For the first-order model, the maximum errors are 3.00 mV, 18.66 mV and 15.50 mV for the average RMSE, the minimum and the maximum cell voltage in the pack respectively. For the second-order model, the maximum errors are 3.10 mV, 20.48 mV and 14.14 mV for the average RMSE, the minimum and the maximum cell voltage in the pack respectively.

Figure 12 shows the model validation performance for the battery cell and the on-road battery pack tests. The average RMSE, the minimum and the maximum of the cell voltage between the experimental data and the simulation results are summarized. The RMSE of the battery cell validation is increased when the discharge current is increased. For the on-road battery pack validation, the RMSE is high when the discharge current has highly fluctuated.

5 Conclusion

Based on the resistance-capacitance network, the first-order and the second-order equivalent circuit models are developed for an NCM cathode pouch cell and a battery pack. For the cell model, the parameters identification of the models is obtained from pulse current discharge tests. For cell validations of both models, the cell voltage and temperatures at three discharge rates show good agreement between the experimental data and the simulation results. The second-order model demonstrates better dynamic behavior of the cell. The single-cell model is used to estimate a group of four cells connected in parallel in the battery pack. A non-road load test, a Wide-Open Throttle (WOT) test and a real-world road test are conducted to capture steady-state response, step response and real-road drive respectively. The simulation results of the battery pack model validations show good agreement with the experimental data for both models.

Acknowledgments

The authors are also grateful to Beta Energy Solution Company Limited for the electric Tuk-Tuk and an on-board data acquisition system.

References

[1] Lee D Rousseau A and Rask E 2014 *SAE Technical Paper* 2014-01-1809
[2] Li Y Agashe P Ge Z and Chen B 2013 *SAE International Journal of Alternative Powertrains* **2** 241-251
[3] Rajendran A V Hegde B Ahmed Q and Rizzoni G 2017 *IEEE Conference on Control Technology and Applications (CCTA)* pp 261-266
[4] Ritter A Elbert P and Onder C 2016 *IFAC-PapersOnLine* **84-377** 49
[5] Ma Y Li B Li G Zhang J and Chen H 2017 *IEEE/CAA Journal of Automatica Sinica* **204-195** 4
[6] Xiong R He H Guo H and Ding Y 2011 *Procedia Engineering* **2874-2869** 15
[7] Yao L W Aziz J A Kong P Y and Idris N R N 2013 *IECON 39-2013th Annual Conference of the IEEE Industrial Electronics Society* pp 1734-1729
[8] Zhang L Peng H Ning Z Zhongqiang M and Sun C-Y 2017 *Applied Sciences* **1002** 7
[9] Hu X Li S and Peng H 2012 *Journal of Power Sources* **367-359** 198
[10] Theinglim K and Poramapojana P 2019 *SAE Technical Paper* 2019-01-2261
[11] Huria T Ceraolo M Gazzarri J and Jackey R 2012 *IEEE International Electric Vehicle Conference* pp 8-1
[12] Kim Y Mohan S Siegel J B Stefanopoulou A G and Ding Y 2013 *52nd IEEE Conference on Decision and Control* pp 5685-5680
[13] Muratori M Canova M Guezennec Y and Rizzoni G 2010 *IFAC Proceedings* **197-192** 43
[14] Lee J Ahn J and Lee B K 2017 *IEEE Energy Conversion Congress and Exposition (ECCE)* pp 5247-5242

A fuzzy controller for two energy storage management system in electric bicycle

P. Pera [1], M. Masomtob[2], W. Intano[1] and A. Kaewpradap[1*]

[1]Department of Mechanical Engineering, King Mongkut's University of Technology Thonburi, Bangkok, 10140, Thailand
[2]National Energy and Technology Center (ENTEC), National Science and Technology Development Agency (NSTDA), Pathum Thani, 12120, Thailand

* Corresponding Author: amornrat.kae@mail.kmutt.ac.th

Abstract
This paper presents an energy management system for an electric bicycle that runs in local community in Thailand. Fuzzy logic technique is applied in this system to control supercapacitor and battery management according to the demand of load power. The load of the electric bicycle depends on speed of each driving cycle. Mamdani-based fuzzy logic controller is selected to control the ratio of output between supercapacitor and lead acid battery. Fuzzy logic controller aims to reduce battery usage during peak current occured from the electric bicycle and passes speed bump in order to extend the lead acid battery life time.

Keywords: Fuzzy logic control, Energy management system, Speed bump, Electric bicycle.

1. Introduction
In the last few years, Plug-in Hybrid Electric Vehicle (PHEV) and Battery Electric Vehicle (BEV) in Thailand has become more popular due to its zero emission [1]. The number of registered cars accumulatively increase from 7705 to 156,038 cars during of 2015-2019 [2]. Electric bicycle is one type of EV that has become more popular to use in city area. It is used to replace motorcycle and bicycle for people who travel less than 20 km. People prefer to use electric bicycle because no driving license needed, ecofriendly, easy to maintenance and the price is cheaper than an internal combustion motorcycle. For the electric bicycle to become cheap and easy to maintenance lead acid battery is used as energy storage. Government launches many policies such as zero tax for Electric Vehicle (EV) manufactured in Thailand, 100,000 baht subsidy for users who buy EV [3], and support to industries and researches to create more EV.

One of the most important research topics is the development of energy management systems for battery. Energy management system is one of the most essential systems in EV because battery is the most important part to make EV running. The better energy management system could extend the driving range of the EV. Due to the limitation of the battery Hybrid Energy Storage System (HESS) as shown in Figure 1. This system is applied to the system to help increase driving range and extend battery lifetime [4].

The previous research of HESS is used to manage two energy sources i.e., fuel and battery. Since EV has been more popular the manufacturers and researchers want to change from fuel to battery and other types of storage system. The problem of pure EV is car acceleration required high power battery causes insufficient energy for the battery to distribute power to motor. On the other hand, only the battery is not enough for the EV. As the result of, the present studies try to develop a HESS which uses supercapacitor to improve insufficient energy [5]. Supercapacitor is a new type of energy storage that can distribute high power with a rapid charge and discharge rated [6]. HESS between supercapacitor and battery is importance to use logic control and design the power distribution at different situation [5]. There are many types of logic control for HESS. Simple frequency decoupling technique which separates power demand by using low pass and high pass filter. Supercapacitor is applied for transients and peak loads [7]. The load can be classified by low and high frequency. The control method for energy management system in solar panel –battery super capacitor hybrid energy system in solar car is utilized by if-clause algorithm [8]. Fuzzy is a Linguistic language that represents the human words or feeling into value that computer understands. Therefore, fuzzy is not commonly use with HESS. The research about Fuzzy usually uses with air-condition and electric water heater [9][10]. It is found that the fuzzy algorithm can be applied for energy management for EV. Thus, this research focuses on fuzzy logic in HESS in EV of supercapacitor and battery.

Electric bicycle is used in as a parameter to build fuzzy logic control. However, this logic control can be applied for EV in the future. The influence factors in fuzzy logic control are the following: First, the effect of the ratio between battery and supercapacitor. Then, the speed differentiation. This research is to study on experiment in MATLAB Simulink. Bangkok Driving Cycle and speed bump to evaluate the fuzzy logic control algorithm[11]. For extension of the lead acid battery lifetime.

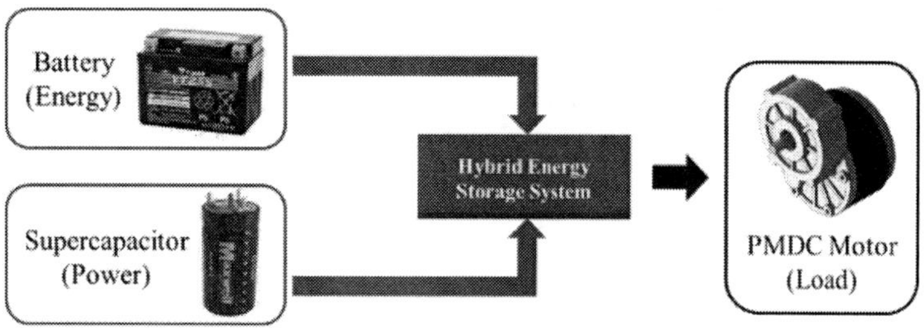

Figure 1. Hybrid Energy Storage Management System.

2. Nomenclature

A	Frontal area of vehicle (m^2)
α	Gain
C_d	Drag coefficient
F	Force (N)
F_t	Tractive force (N)
g	Gravity force (N)
m	Mass (kg)
P	Electric power (W)
ρ_{air}	Air density (kg/m^3)
R_a	Aerodynamic drag resistance (N)

R_i	Inertial resistance (N)
R_g	Grade resistance (N)
R_r	Rolling resistance (N)
μ	Rolling resistance coefficient
v	Speed vehicles (m/s)

3. Methodology

3.1 Hybrid Energy Storage System (HESS)
HESS in this system uses fuzzy logic control to evaluate the amount of the power load from the electric bike. Battery and Supercapacitor are input for fuzzy that can distribute energy depend on load. Power load is input for fuzzy but demand energy.

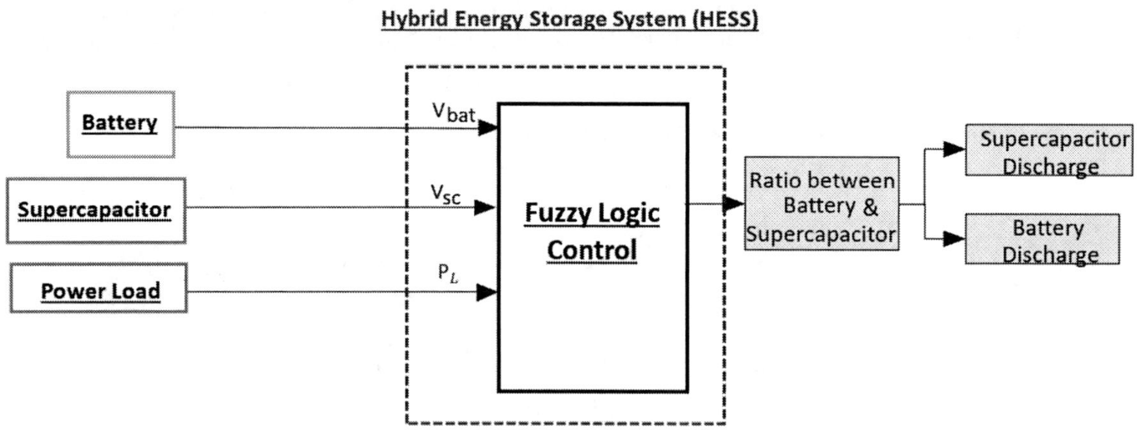

Figure 2. Overview of HESS system.

3.2 Energy consumption [12]
The total energy consumption can be derived by total force and vehicle speed as shown in Equation (1)

$$P = F \cdot v \tag{1}$$

The electric power can be obtained from by total resistance force and vehicle speed as illustrated in Equation (2)

$$P = (\mathrm{R_a} + \mathrm{R_r} + R_g + R_i) \cdot v \tag{2}$$

The energy consumption in Equation (1) can be calculated by total force as shown in Equation (3)

$$F = R_a + R_r + R_g + R_i \tag{3}$$

The total force act on bicycle is shown in Figure (3).

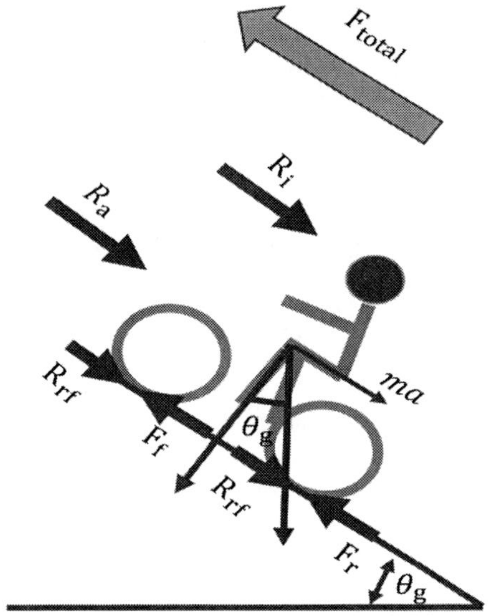

Figure 3. Forces acting on bicycle.

Aerodynamic drag resistance is the air resistance during the vehicle moving which is calculated by the Equation (4). The 1.1 of drag coefficient (C_d) is applied with 1.1839 kg/m^3 of air density in this experiment.

$$R_a = \frac{1}{2} C_d \cdot \rho_{air} \cdot A \cdot v^2 \tag{4}$$

Rolling resistance is the resistance occurring on the wheel turns on the surface which related to the weight of the vehicle as shown in Equation (5). The 0.04 of rolling resistance coefficient tire turns on asphalt road, 90 kg of mass for electric bike and driver [8]

$$R_r = \mu \cdot m \cdot g \cdot cos\theta \tag{5}$$

Grade resistance is the resistance that occurs when the vehicle travels up or down the slope. As shown in Equation (6). In this experiment the vehicle runs on plain path which have less gradient even the speed reduces. The road grade (θ) is almost zero degree thus the grade resistance is neglected.

$$R_g = m \cdot g \cdot sin\theta \tag{6}$$

3.3 Fuzzy logic control

The first step of Fuzzy Logic Control (FLC) is to transform linguistic language into computer language as an input for FLC and make a decision with its own algorithm that the user designs to get the desirable output, for example in the Figure 4 the temperature control of the water. Normal control can only separate on and off the heater as 1 and 0. However FLC separates the input into more detail i.e., cold, cool, nominal, warm and hot. Then substitute all this feeling into the computer language.

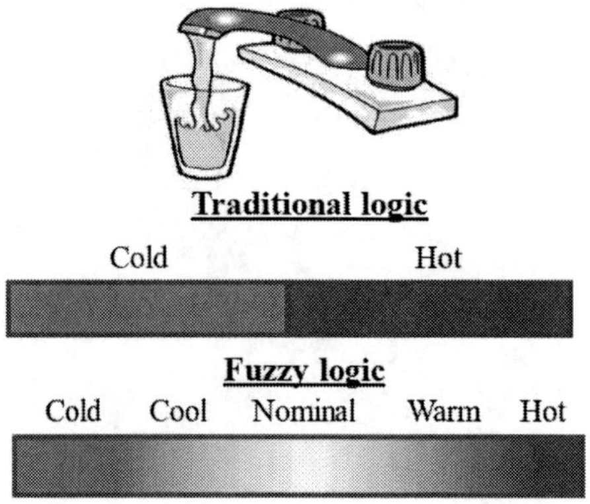

Figure 4. Comparison of water temperature control[13].

For the algorithm of design part, the all-output responses from each input are needed to know for design of the control algorithm based on real data as shown in Figure 5. However, the unknown output result from FLC algorithm by using the function design fuzzy in MATLAB then importing the fuzzy file into the control algorithm.

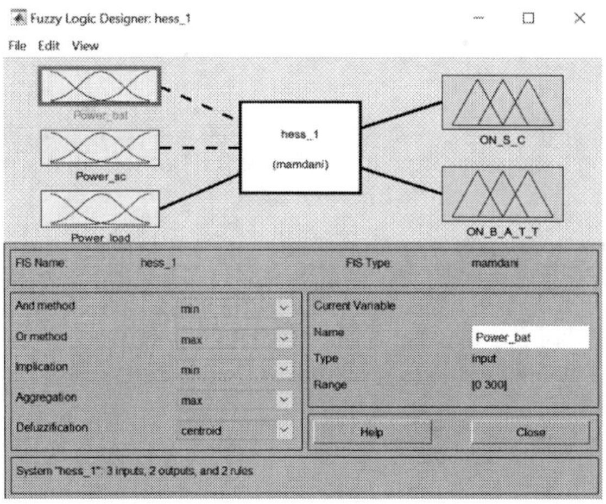

Figure 5 Fuzzy logic control design in MATLAB.

3.4 Bangkok driving cycle

Bangkok driving cycle is the standard for test of the vehicle fuel consumption and power consumption in Thailand. Bangkok driving cycle consists of 6 phases as shown in Figure 6 and 7. As the previous research, this driving cycle was applied for study of motorcycle. Thus, the Bangkok driving cycle is expected to apply for the experiment of electric bicycle. Following the Bangkok driving cycle the speed is calibrated by gain of 0.3 to comply with electric bicycle. The speed is used to calculate the power consumption of electric bicycle for this driving cycle based on the theory of bicycle model.

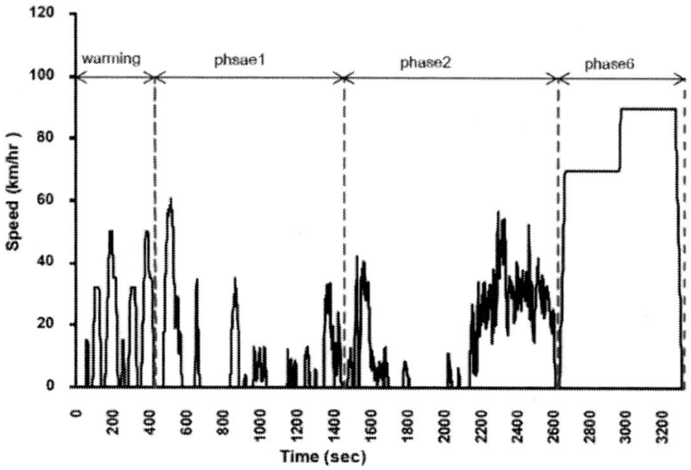

Figure 6. Bangkok driving cycle phase 1,2,6 [11].

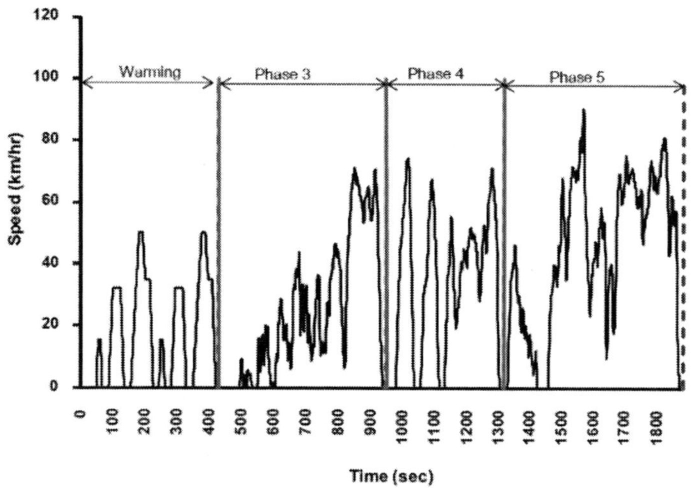

Figure 7. Bangkok driving cycle phase 3,4,5 [11].

The relation of speed according to the Bangkok driving cycle is shown in Figure 8. In case of driving cycle is applied without supercapacitor, the battery has to supply all the power for the load.

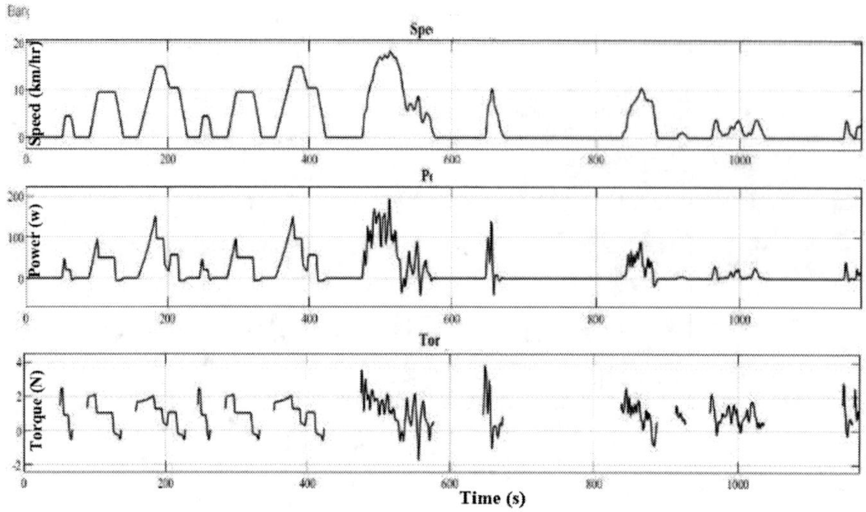

Figure 8. Power consumption on Bangkok driving cycle.

Table 1. The simulation parameters for electric bicycle.

Parameter	Value	Unit
Cross Section Area (a)	0.8×0.5×1.6	m^2
Air Density (ρ)	1.1839 at 25°C	
Drag Coefficient (C_d)	1.1	
Mass (m)	90	kg
Gravity Force (g)	9.81	m^2/s
Rolling Friction Force (R_f)	0.018 at dry asphalt	
Slope (Δ)	0	$degree$
Angle	$\frac{\Delta}{180 \cdot \pi}$	rad

3.5 Specification of electric bicycle

Table 2. Specification of electric bicycle.

	Specifications
Motor	DC Gear Motor SNK 24 V 250 W
Battery	Battery 6 lead acid Yuasa YTZ5s 2 series then 3 parallel 24 V
Supercapacitor	350 F, Nominal voltage 2.7 V, 6 series

Speed bump characteristic is collected by the experiment with variable parameter in this study. The illustration of actual experiment is represented in Figure 9.

IOP Conf. Series: Materials Science and Engineering 1137 (2021) 012015 doi:10.1088/1757-899X/1137/1/012015

Figure 9. Actual experiment of speed bump.

3.6 Fuzzy algorithm

The fuzzy algorithm is designed base on real experiment. The maximum power load, steady load battery and supercapacitor characteristic are used to design the fuzzy input and output as shown in Figure 10 (a) and Figure 10(b). There are two rules in this fuzzy algorithm. In the first condition, if the power load is steady with the power between 50 to 200 w then the battery is on. In the second, when the power load is 125 w to 300 w then battery and supercapacitor is turned on. As show in Figure 10(c). The ratio of turned on and turned off supercapacitor and battery for switch control is represented in Figure 11.

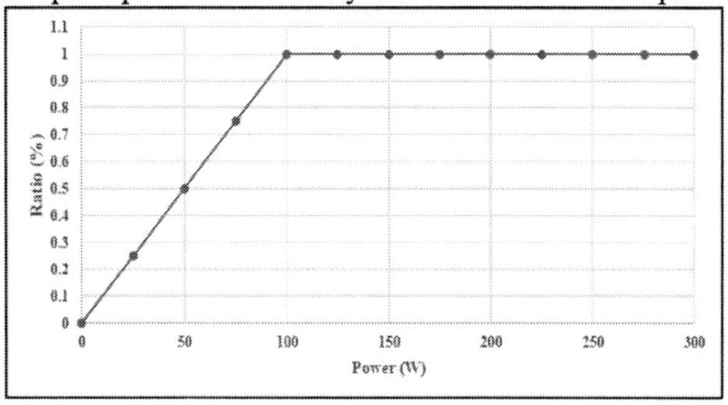

Figure 10 (a) Power ratio of supercapacitor.

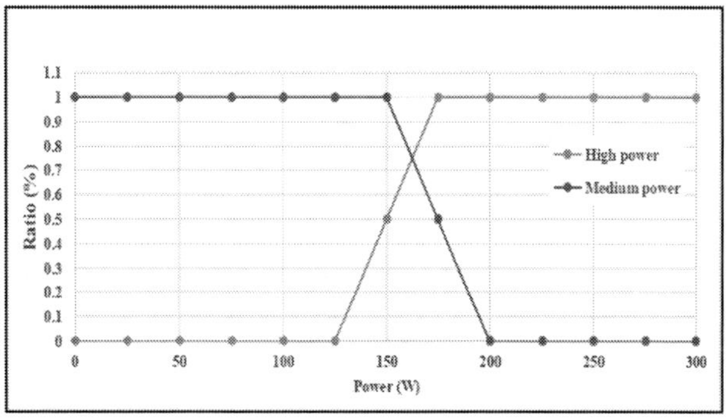

Figure 10 (b) Power ratio of battery.

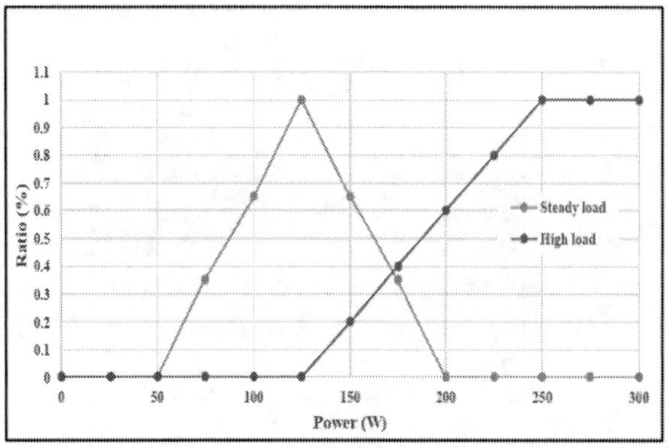

Figure 10 (c) Relation between power and power load ratio for steady and high loads.

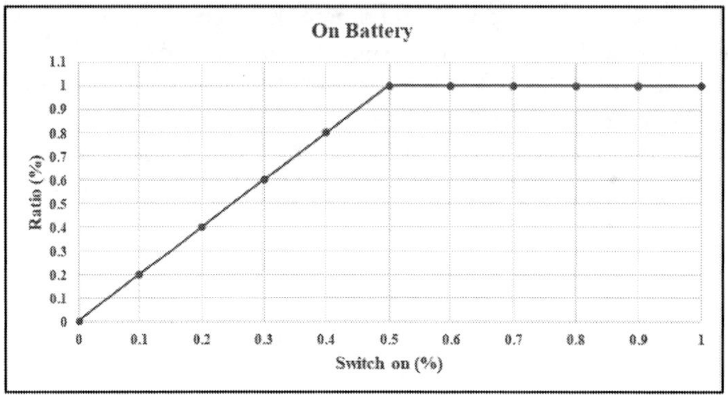

Figure 11(a) Relations between battery ratio and turn-on switch.

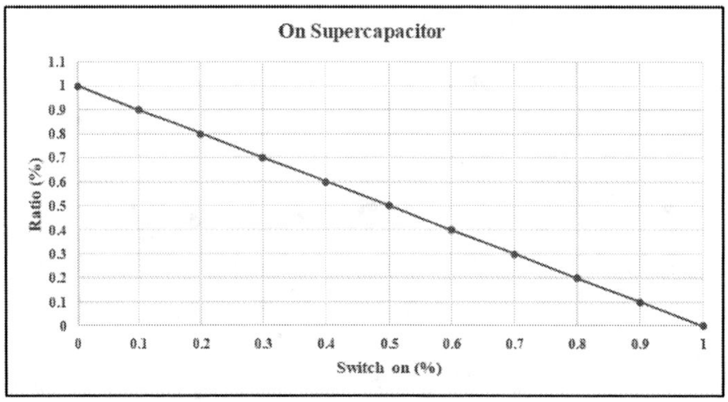

Figure 11(b) Relations between supercapacitor ratio and turn-on switch.

4. Result

4.1 Bangkok driving cycle analysis

Bangkok driving cycle with FLC is shown in Figure 12. Supercapacitor is used to decrease the power discharge from the battery and provide the energy for meeting the power requirement. FLC is used to decide if the power require is more than 150 w. Then, FLC will sent signal to switch that control the output discharge of the supercapacitor as present in Figure 8. The status of the on and off switch that controls with FLC is shown in Figure 13.

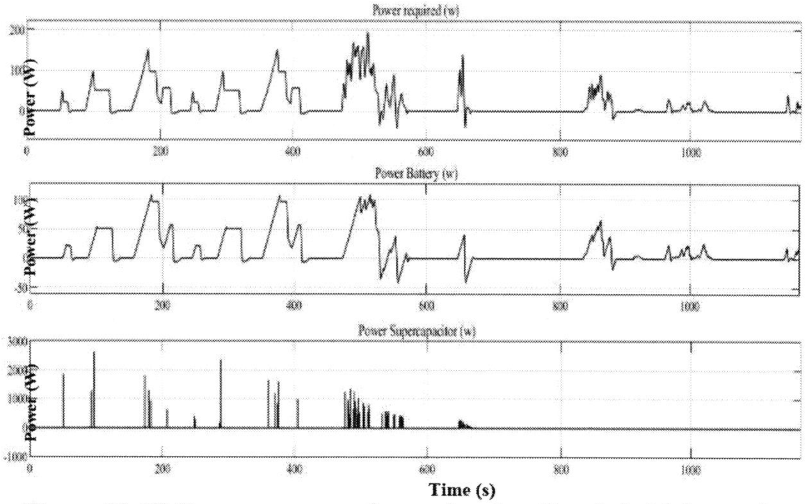

Figure 12. FLC power consumption analyzed on Bangkok driving cycle.

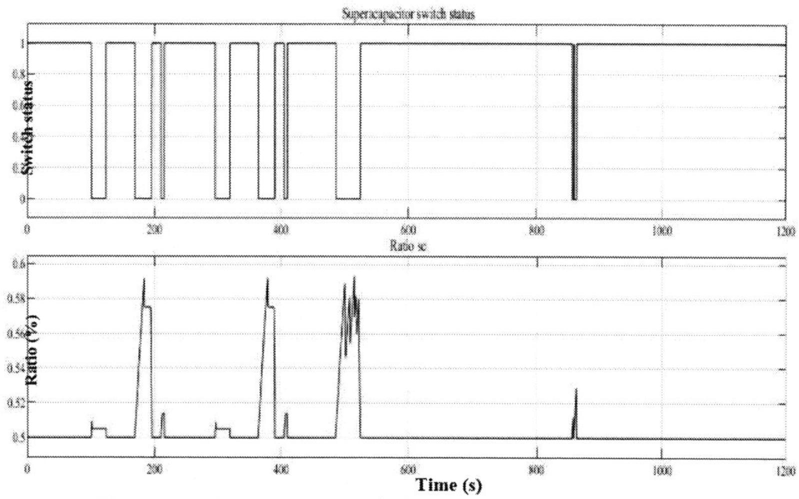

Figure 13. FLC contact switch on Bangkok driving cycle.

4.2 Speed bump

In Thailand, there is a speed bump which is not concerned to the Bangkok driving cycle. Speed bump is usually found in the city area. The results show that at speed of 17.5 km/hr is observed at distance 10 m of brake distance. The average power for this speed bump is shown in Figure 14. This power is put into the simulation to test the power consumption of HESS.

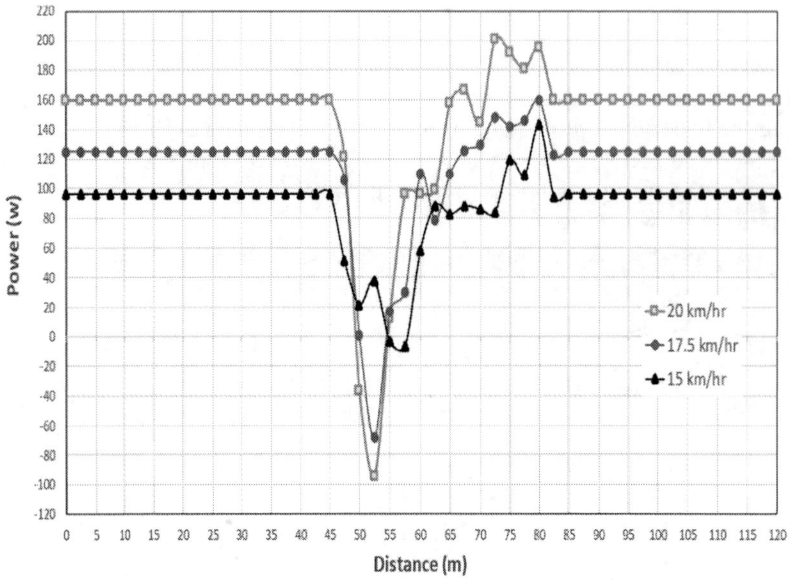

Figure 14. Power consumption on speed bump for speed of 15,17.5 and 20 km/hr.

The result of power consumption using FLC with the is shown in Figure 15 and the on off switch which controls with FLC the status of the switch is shown in Figure 16

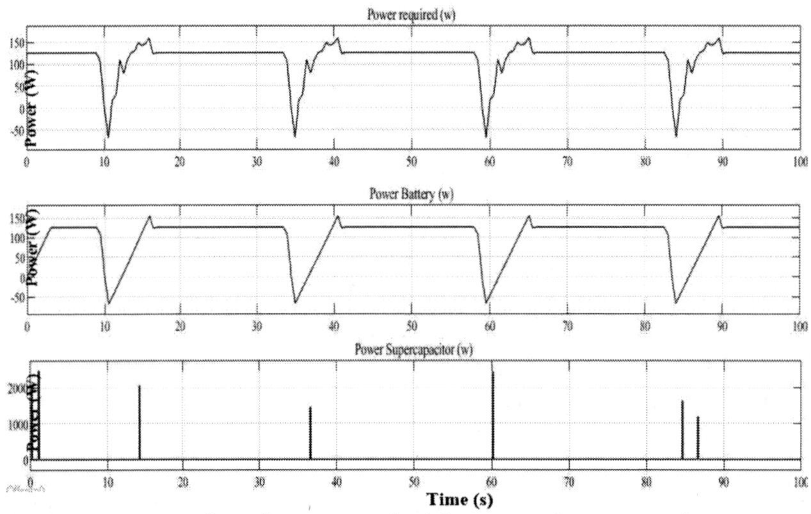

Figure 15. Power consumption of power required, battery and supercapacitor on speed bumps.

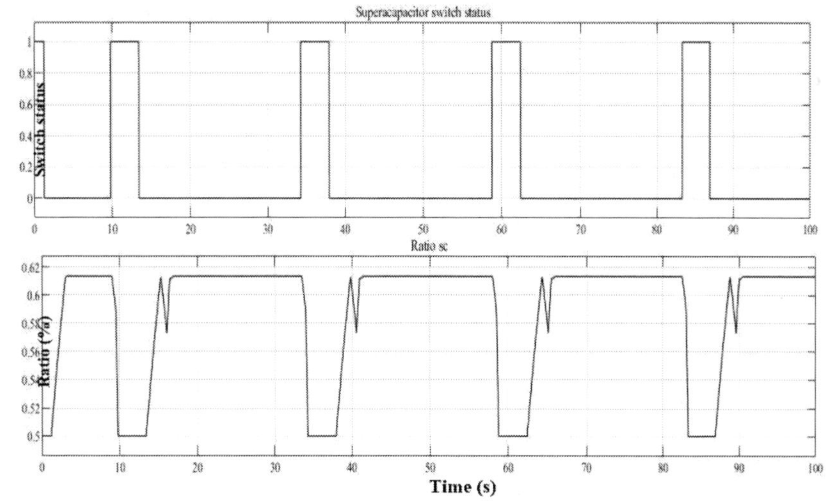

Figure 16. 4 Relations between power ratio and switch status.

5. Conclusion

Supercapacitors could be applied to reduce the power consumption of the battery. The HESS is used to manage the energy between the battery and supercapacitor. The energy management could be controlled automatically with FLC. Speed bump show high amount of impact in power consumption than the Bangkok driving cycle because of its characteristic of speed bump force the driver to break and accelerate again. The power consumption is increased by the speed bump.

The usage of supercapacitor can extend the lead acid battery lifetime because the peak load is distributing with the supercapacitor with its advantage of fast charge, discharge and long-life cycle time. So, it is suitable to use supercapacitor to help the lead acid battery. From the result speed bump is one of the most interesting parameters in EV HESS. So, in future work, the author will study effect of speed bump and other type of speed reducers in Thailand.

Acknowledgements

The author would like to acknowledge for personal financial support from Thailand Advance Institute of Science and Technology-Tokyo Institute of Technology (TAIST-Tokyo Tech), National Energy and Technology Center (ENTEC) and King Mongkut's University of Technology Thonburi for supporting facilities.

Reference

[1] Intano W Kaewpradap A Hirai S and Masomtob M 2020 *J. Res. Appl. Mech. Eng.* **8** 11-21
[2] Laoonual Y 2019 *Next Generation Mobility Technology*
[3] eppo 2020 www.eppo.go.th/index.php/en/ *2020*
[4] Hatami A Tousi M R Bayat P and Bayat P 2015 *Proc. ICEE 2015 - Proc. 23rd Iran. Conf. Electr. Eng.* **10** 1498-503
[5] Sun L Zhang N Awadallah M and Walker P 2017 *Proc. - 2017 IEEE Int. Conf. Mechatronics* 434-439
[6] Rashidi S Esfahani J A and Hormozi F 2020 *Ref. Modul. Mater. Sci. Mater. Eng.* 1-12
[7] Singh A and Pattnaik S 2017 *Proc. 1st IEEE Int. Conf. Power Electron. Intell. Control Energy Syst.*
[8] Wu B Zhuo F Long F Gu W Qing Y and Liu Y 2011 *Proc. 8th Int. Conf. Power Electron. - ECCE Asia* **2** 1682-1687
[9] Attia A H Rezeka S F and Saleh A M 2015 *Alexandria Eng. J.* **54** 395-403
[10] Nehrir M H LaMeres B J and Gerez V 1999 *IEEE Eng. Soc. Winter Meet.* **1** 433-436

IOP Conf. Series: Materials Science and Engineering 1137 (2021) 012015 doi:10.1088/1757-899X/1137/1/012015

[11] Not D 2013 *Investigation of Motorcycle Emissions Tested by European Standard Driving Cycle and Bangkok Driving Cycle* 1–6
[12] Wong J Y 2002 *Theory of ground vehicles* vol 216 (John Wiley & Sons, Inc)
[13] Singhala P Shah D and Patel B 2014 *Int. J. Instrum. Control Syst.* **4** 1-10

Three-dimensional effects on the performance of multi-level overtopping wave energy converter

S. Jungrungruengtaworn[1,*], N. Thaweewat[1] and B.S. Hyun[2]

[1]Department of Maritime Engineering, Faculty of International Maritime Studies, Kasetsart University, Chonburi, Thailand 20230
[2]Department of Naval Architecture and Ocean Systems Engineering, Korea Maritime and Ocean University, Busan, Korea 49112

* Corresponding Author: sirirat.jun@ku.th

Abstract
The performance of a multi-level overtopping wave energy converter has been numerically investigated in a three-dimensional wave tank. The device has three extra slots and is positioned on a breakwater. The span length of the breakwater is the main parameter associated with the three-dimensional effects on the hydraulic efficiency of the wave energy device. It has been shown that the device with a finite span yields a lower captured crest energy due to waves falling-down the edges during the run-up process. The result also implies that as the span of the device is increased, the efficiency tends to increase further to resemble the two-dimensional device. The three-dimensional mechanism has a significant influence on the potential energy in the water being stored by the higher reservoirs, while that of lower ones have relatively smaller effects. In addition, a relation between hydraulic efficiencies of two and three-dimensional devices has been proposed.

1. Introduction
Nowadays, the carbon dioxide (CO2) emission from combustion engines in transportation and industrial sectors causes the air pollution crisis in many cities such as Delhi, Seoul, Chiang Mai, and central and eastern China [1-4]. Subsequently, researchers have been seeking for methods in order to resolve the crisis. As a result, the development of renewable energy technology is considered as an option to alleviate the air pollution problem as well as energy shortage and crisis. The utilization of renewable energy seemingly contributes to the reduction of greenhouse emission level [5].

Harvesting renewable energy is significantly advantageous since it generally replenishes. However, the utilization of renewable energy is currently minor compared to total energy consumption. Particularly, the hydropower holds the largest share of worldwide renewable energy utilization with increasing tendency every year, while the electricity generation from ocean renewable energy is relatively small [6]. Nevertheless, ocean energy resources are considerably huge and predictable. Scientists and engineers have consequently been working to innovate ocean energy technology, e.g., tidal current turbines, offshore wind turbines, and wave energy converters.

Wave energy converters (WECs) are various in design and can be categorized according to their principles in the energy capture process, e.g., oscillating device, oscillating water column (OWC), and overtopping wave energy converter (OWEC). The overtopping devices collect and accumulate wave energy in the form of the potential energy of stored water volume in a reservoir which has the mean water level higher than the sea level. The stored water is simultaneously released back to the sea through a tube or channel in which a low head turbine is typically installed to extract the energy.

Optimization of geometric design yielding a high-volume flow rate into the reservoir has been one of the main interests for the OWEC studies. The OWEC originally possesses only one reservoir. This results in a relatively low efficiency since the ocean conditions are noticeably variable and the single-level device could deliver its best efficiency only at a specifically designed condition. An example of the single-level OWEC is the Wave Dragon [7-9]. Generally, the hydraulic efficiency of an OWEC, based on the single-level design, can be 20-30% approximately. A multi-level configuration has therefore been proposed for the performance enhancement in a wider range of tidal and wave conditions.

Several experimental and numerical investigations [10-14] have been performed to better understand the unsteady hydrodynamic mechanisms in wave overtopping behavior. It is difficult to capture all physical aspects of the flow by performing only experiments. Therefore, numerical simulations are mainly used to preliminarily study and efficiently find out the optimal design. Experimental and numerical studies concerning OWECs have mainly been conducted based on two-dimensional flow. In the present study, numerical simulations of multi-level OWEC integrated with breakwaters are performed in a three-dimensional wave tank. This study investigates the influence of the breakwater span length on hydraulic efficiency. The results obtained from this study are compared with that of a two-dimensional OWEC. Flow physics and overtopping mechanisms have been additionally discussed.

2. Materials and Methods

The numerical tool used to investigate the wave overtopping behavior of a multi-level OWEC device is a commercial CFD solver ANSYS FLUENT V17. The governing equations are the continuity and the Reynolds Average Navier-Stokes (RANS) equations for viscous, incompressible, three-dimensional flow. The standard $k - \varepsilon$ turbulence model is applied for turbulence closure while the free surface is tracked and located using the two-phase volume of fluid (VOF) method proposed by [15].

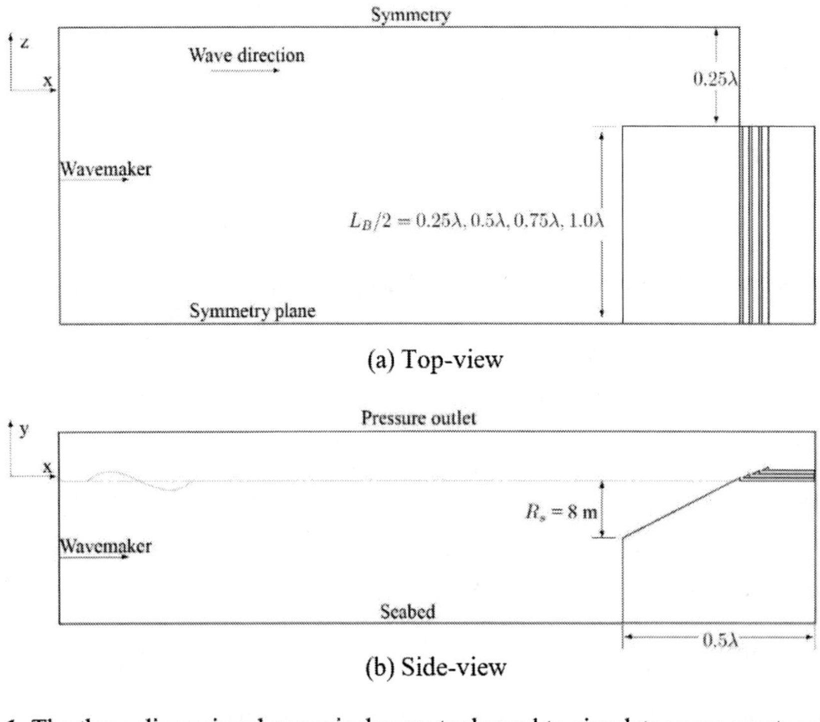

Figure 1. The three-dimensional numerical wave tank used to simulate wave overtopping behavior

2.1 Numerical Wave Tank

The wave overtopping behavior is simulated in a three-dimensional numerical wave tank as shown in Figure 1, which has a water depth of $d = 20$ m and a tank length of 300 m, approximately corresponding to 0.36λ and 5.4λ respectively, where λ is the wavelength. Wall conditions, combined with a dynamic mesh model, is utilized on the left boundary, which acts as a piston-type wavemaker. As for the right boundary, the pressure outlet combines with the open channel flow condition is applied. In order to reduce the domain size and number of meshes, the symmetric plane is utilized. The first-order linear wave is generated as an incident wave. In order to prevent an impulsive movement at the beginning, the kinematics of the piston type wavemaker $x(t)$ is defined by the following:

$$x(t) = \frac{x_0}{2}\left(1 - e^{\frac{-5t}{2T}}\right)\sin \omega t \quad (1)$$

where x_0 is the maximum displacement of the wavemaker, yielding the desired wave height, and ω is the angular frequency relevant to the wave period.

2.2 Wave Energy device and Ocean conditions

The sectional layout is based on a multi-level OWEC used by [10] integrated with breakwater, which is schematically illustrated in Figure 1 and 2. The device consists of a ramp of slope $S = 1/2$, together with extra slots at different levels which are connected to corresponding reservoirs. The width of these slots is identical as $w = 0.4$ m. The crest height of the lowest reservoir is $R_{c1} = 0.5$ m, while that of the higher ones have a constant increment of 0.5 m. The crest of the main reservoir is, therefore, $R_{c4} = 2.0$ m.

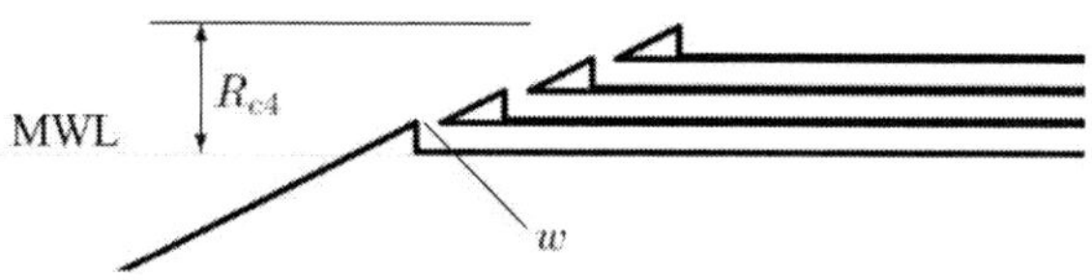

Figure 2. Sectional layouts of the OWEC device. The incident wave direction is from left to right

In order to study the influence of span length L_B on the performance of the device, the parameter is varied within the range of $0.5\lambda \leq L_B \leq 2.0\lambda$ with a 0.5λ increment. The generated incident wave height $H = 2$ m, together with wave period $T = 6$ s is utilized following [10]. This combination yields the wave steepness $H/\lambda = 0.036$, the relative wave height $H/d = 0.1$, and the relative slot width, $w/R_{c4} = 0.2$.

2.3 Discretization of Numerical Model

A grid system of the computational domain has been generated with adaptive refinement. A dense, more refined mesh is applied and generated in regions of interest which are the vicinity of the free surface and the OWEC device, especially near the slots and the tip of the device.

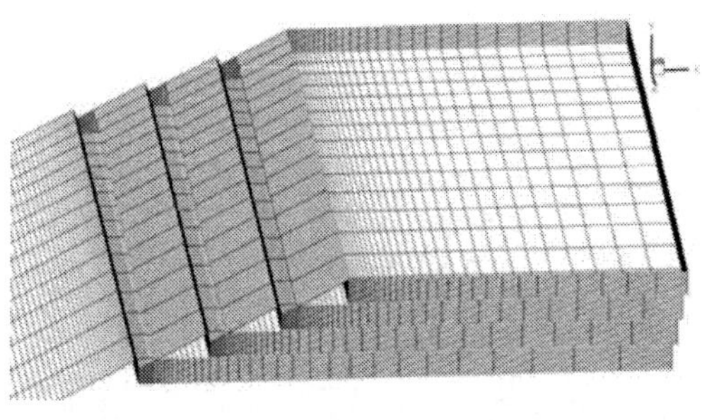

Figure 3. The Surface mesh of the multi-level OWEC device

The mesh size is dependent on wave height and length. The number of cells per wave height is 20, in combination with 100 cells per wavelength. The current discretization is finer than the suggestion from [16]. The time step size is $dt = T/6000$. It has been found that the larger time step of $dt = T/3000$ yields a slight deviation in the simulated overtopping discharge compared to $dt = T/6000$ case.

2.4 Wave Energy Conversion Performance

In order to generate electricity from wave energy using overtopping principle, the wave energy is firstly captured and collected in potential form in a reservoir. The potential energy of stored water is then converted to mechanical energy via a low-head turbine and consequently transformed into electricity by a generator. These energy conversion processes correspond to hydraulic, reservoir, turbine and generator efficiencies respectively. In this study, the hydraulic efficiency η is used to present the performance of the wave energy device. This performance indicator is also known as the capture width ratio (CWR).

$$\eta = \frac{P_{\text{crest}}}{L_B \cdot P_{\text{wave}}} = \frac{\sum_{i=1}^{n} Q_i \rho g R_{c,i}}{L_B \cdot \frac{1}{16}\rho g H^2 c\left[1+\frac{2kd}{\sinh(2kd)}\right]} \tag{2}$$

where P_{crest} is the rate of overtopping energy captured by the device, P_{wave} is the wave energy per unit width, Q_i is the time-averaged overtopping flow rate of the corresponding i^{th} reservoir, ρ is the water density, g is the gravitational acceleration, $R_{c,i}$ is the crest height of the ith reservoir, H is the incident wave height, c is the phase velocity, k is the wave number, and d is the water depth. The parameter L_B is the span length of the wave energy device which is the independent variable of primary interest.

3. Results and Discussion

The hydraulic efficiency of the three-dimensional device with different dimensionless spans is compared with a two-dimensional device in Figure 4. Note that the two-dimensional device has an infinite span $L_B/\lambda = \infty$. Clearly, the three-dimensional effects play an important role in the wave energy capture mechanisms. The device with the shortest span yields the lowest relative efficiency, while increasing the span length yields greater performance. Another increase in span length resulted in a further performance enhancement, which tends to converge with a device with an infinite span, i.e., the two-dimensional device, as presented by the overall hydraulic efficiency in Figure 4(a).

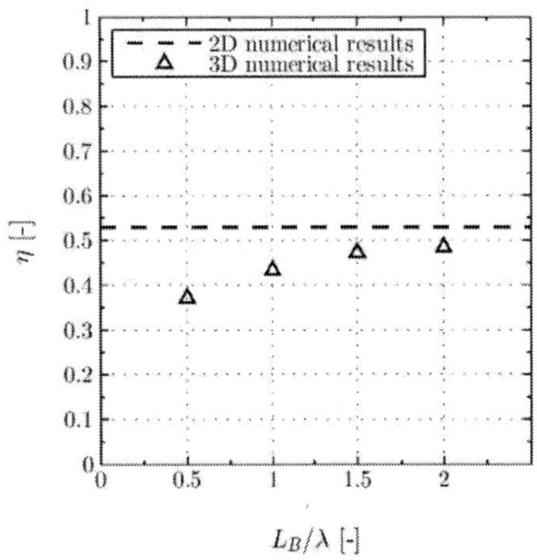

(a) Overall hydraulic efficiency

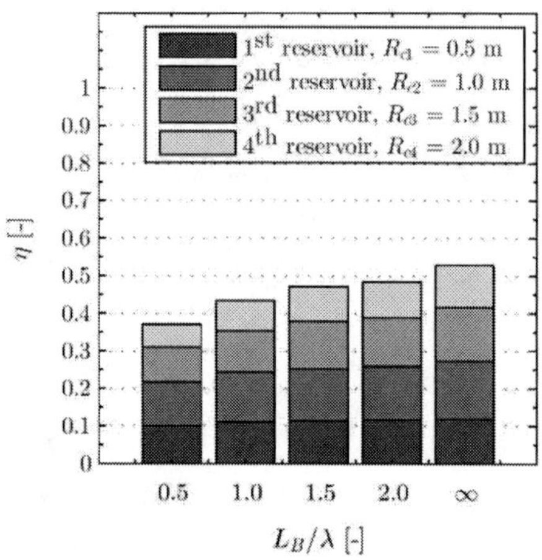

(b) Partial hydraulic efficiencies

Figure 4. Hydraulic efficiency η as a function of dimensionless span L_B/λ

This mechanism is analogous to the difference between two and three-dimensional wings, as a wider aspect ratio yields a greater lift coefficient, a shorter one results in a lower lift coefficient due to three-dimensional vortices generated at wing tips. Similarly, the three-dimensional OWEC model has a finite span possessing tips, which leads to waves falling-off the edges during the run-up process as seen in Figure 5. As a result, a wave at the tip of the device cannot run-up as high as a wave near the centerline. This consequently leads to a drastic decrease in partial efficiency of the main reservoir, whereas the effects are in-determinant for reservoirs with a relatively low crest level. It is subsequently suggested to install a guide vane at the tips of the device, which could prevent or reduce the three-dimensional effects and increase the effective span of the device, similar to what winglets achieve on an aircraft.

(a) $L_B = 0.5\lambda$

(b) $L_B = 1.0\lambda$

(c) $L_B = 1.5\lambda$

(d) $L_B = 2.0\lambda$

Figure 5. Top-view snapshots of wave flow into reservoirs at time t, $t + 0.2T$, $t + 0.4T$, $t + 0.6T$, $t + 0.8T$ respectively from left to right. The wave direction is from left to right

Figure 4(b) shows the partial hydraulic efficiencies for different reservoir levels. When the span length is increased, the enhancement in partial hydraulic efficiency varies depending on the level of the reservoirs. Within the parametric range considered, the span length may have no significant influence on the potential energy stored in the lowest reservoir, since the partial hydraulic efficiency seems to change unnoticeably. However, the potential energy stored in the main reservoir, i.e., the highest one, is seemingly more sensitive to the change in span length.

In addition to the numerical results, an approximation used to predict the relationship between two- and three-dimensional partial hydraulic efficiencies of this multi-level OWEC device has been proposed.

The three-dimensional partial efficiency η_i^{3D} can be expressed as a function of the dimensionless span L_B/λ and the two-dimensional partial efficiency η_i^{2D} as:

$$\eta_i^{3D} = \frac{L_B/\lambda}{L_B/\lambda + Z_i} \cdot \eta_i^{2D} \tag{3}$$

where Z_i is a factor used to fit the curve. This factor could be a function of wave steepness, water depth, or the geometry of the device such as slope ratio, slot width, or the crest height of the ith reservoir. In figure 6, the partial hydraulic efficiencies are plotted along with the prediction corresponding to equation (3). The values of Z_1, Z_2, Z_3 and Z_4 are 0.0757, 0.1635, 0.2440 and 0.3923 respectively. It should be noted that the smaller value of Z_i gives a more rapid convergence to the efficiency of the device with an infinite span. The proposed formula, combined with an appropriate value of Z_i appears to sufficiently predict the convergent trend of hydraulic efficiency in spite of a slight deviation.

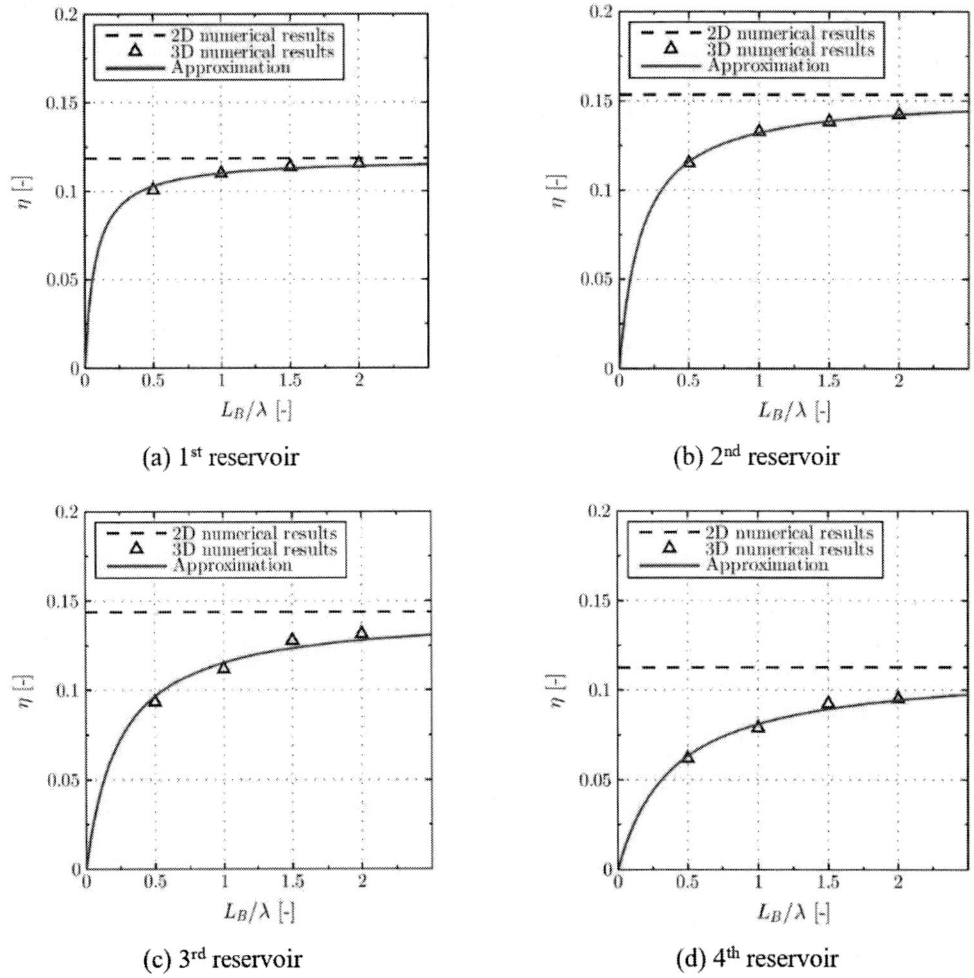

(a) 1st reservoir

(b) 2nd reservoir

(c) 3rd reservoir

(d) 4th reservoir

Figure 6. Partial hydraulic efficiencies obtained using two and three-dimensional models. The solid line represents the approximation of partial hydraulic efficiency η_i^{3D} of the device with a finite span as a function of dimensionless span L_B/λ

4. Conclusion

This study aims at conversion and utilization of wave energy using the overtopping principle. A numerical model for three-dimensional flow was used to investigate the effects of span length of multi-level overtopping wave energy devices on energy captured. The hydraulic efficiency of the three-dimensional device is compared with the baseline two-dimensional device. A convergent trend is found for the relation between span length and hydraulic efficiency, both overall and partially. As for the device with a relatively narrow span, an increase in span length gives a huge enhancement in the hydraulic efficiency. However, the hydraulic efficiency is slightly affected by the change in span length when the device is relatively wide. Moreover, a formula used to predict the relation between efficiencies of two and three-dimensional devices has been proposed which appears to be valid with acceptable accuracy.

References

[1] Kim M J 2019 *Heliyon* **5**(12) e02953
[2] Moran J NaSuwan C and Poocharoen O O 2019 *Environ. Sci. Policy* **97** 1-15
[3] Sharma N Taneja S Sagar V and Bhatt A 2018 *Procedia Comput. Sci.* **132** 1077-85
[4] Zheng H and Xu L 2020 *Resour. Conserv. Recycl.* **155** 104661
[5] Alvarez-Herranz A Balsalobre-Lorente D Shahbaz M and Cantos J M 2017 *Energy Policy* **105** 386–97
[6] Shadman M Silva C Faller D Wu Z de Freitas Assad L Landau L Levi C and Estefen S 2019 *Energies* **12**(19) 3658
[7] Kofoed J P and Frigaard P 1999 *Evaluation of hydraulic response of the Wave Dragon* (Aalborg: Danish Wave Energy Program, Aalborg University)
[8] Kofoed J P Frigaard P Friis-Madsen E and Sørensen H C 2006 *Renew. Energy* **31**(2) 181-189
[9] Tedd J and Kofoed J P 2009 *Renew. Energy* **34**(3) 711-717
[10] Jungrungruengtaworn S and Hyun B S 2017 *Int. J. Nav. Archit. Ocean Eng.* **9**(6) 668-676
[11] Jungrungruengtaworn S and Hyun B S 2018 *Proc. OCEANS - MTS/IEEE Kobe Techno-Oceans* (Kobe, Japan)
[12] Jungrungruengtaworn S Reabroy R Thaweewat N and Hyun B S 2020 *Ocean Syst. Eng.* **10**(4) 359-371
[13] Liu Z Han Z Shi H and Yang W 2018 *Int. J. Nav. Archit. Ocean Eng.* **10**(5) 651-659
[14] Han Z Liu Z and Shi H 2018 *Ocean Eng.* **150** 94-101
[15] Hirt C W and Nichols B D 1981 *J. Comput. Phys.* **39** 201-225
[16] Connell K O and Cashman A 2016 *Proc. Int. Conf. on Electrical, Electronics, and Optimization Techniques* (Chennai, India) pp 3008-3012

Fuzzy-based potential field collision avoidance technique for unmanned surface vehicles

Kantapon Tanakitkorn[1,*]

[1]Faculty of International Maritime Studies, Kasetsart University, Si-Racha, Chonburi, Thailand

*Corresponding Author: kantapon.ta@ku.th

Abstract
Collision avoidance is a crucial part of the autonomous operation of an unmanned surface vehicle (USV), especially in a dynamic environment. This work presents a fuzzy-based potential field algorithm for collision prevention when a USV encounters an obstacle, either moving or static. Two types of repulsive potential functions are employed. The Gaussian repulsive potential is applied to repel the own ship (OS) away from the target ship (TS). On the other hand, the vortex potential function is used to smoothly deviate the heading of the OS towards the side that quickly resolves the collision. The fuzzy membership functions and rules are imposed to classify the encounter condition and apply the appropriate potential functions accordingly. Head-on and crossing encounter scenarios are simulated to demonstrate the performance of the proposed algorithm. The simulation results have shown that, when the collision is detected, the algorithm is effective in generating the vector field that directs the OS away from the collision path. Once the collision is resolved, the algorithm lets the OS continue the original course.

Keywords: Collision avoidance, USV, Fuzzy, Potential field, APF

1. Introduction

The development of collision avoidance systems not only can potentially help human navigators to react promptly and make a better judgement but also is a key component toward autonomous navigation of unmanned surface vehicles (USVs). An overview of the collision avoidance techniques for both manned and unmanned vessels has been presented by [1]. A potential roadmap to transform a manned ship into a fully autonomous ship was also discussed in this work.

In the literature, the collision avoidance can broadly be classified into global methods and local methods. The global methods seek for an optimal path that is collision-free and leads a USV from the start location the desired destination [2,3,4]. The main drawback of this framework is that it needs complete knowledge of the environment. If anything changes, the global pathfinder needs to update the map and replan the path accordingly. Since an intensive computational power is required for updating the map and replanning the path, the global method may not be suitable for real-time collision avoidance, especially in a critical encounter scenario.

Artificial potential field (APF) is widely used for a local collision avoidance [5,6]. In this regard, the ship under our control is typically referred to as the own ship (OS); the other ship that the OS would potentially make physical contact with at a certain time in the future is called the target ship (TS). The principle of the APF is to consider the OS as a particle under the influence of the artificial force field. The goal destination is virtually assigned with the long-range attractive potential that pulls the OS toward to it, whereas the obstacle is assigned with the virtual short-range repulsive potential that repels the OS away. The collision-free path is then established by following the gradient of the resulting vector field. However, the APF approach does not guarantee a feasible solution because the gradient of the APF may lead the OS to the local trap instead of the destination [7]. A few methods to overcome the local trap of the APF were discussed in [8,9,10]. As an alternative to the traditional potential functions, [11] has suggested using a vortex function to evade the obstacle and a dipole function to make the OS converges to the desired path. A simulation study was presented to demonstrate the applicability of the concept.

Instead of blindly repelling the OS away from the TS, the velocity obstacle (VO) concept is adopted to evaluate the collision risk between the two vessels and execute the evasive action only when it is needed. In the VO concept, a collision is detected when the following conditions are met simultaneously: 1) the relative distance between the vehicles is smaller than a predefined threshold, and 2) the projection of the relative velocity vector from the OS intercepts with the safety circle of the TS. The collision can be resolved by changing speed and course of the OS so the relative velocity vector is pointing outside the collision cone [12, 13, 14].

Fuzzy logic is another framework that is employed by many researchers for developing marine collision avoidance algorithms [15, 16]. Mostly, the relative distance and bearing between the two ship are used to describe the encounter condition. However, more membership functions may be developed to accurately described the encounter condition (i.e., head-on, crossing, or overtaking) as well as the role of the OS (i.e., stand-on or give-way) [17, 18]. Then, the if-then rules are applied to interpret the linguistic variables and determine the deviation angle of the OS. The COLREGs rules [19] may be incorporated into the fuzzy inference system so the evasive action of the OS complies with the marine rules of the road.

This work presents a collision avoidance algorithm that combines the APF with the fuzzy logic. Section 2 explains the repulsive potential functions for collision prevention. The fuzzy membership functions and the inference system are developed in Section 3. This is followed by the simulation results in Section 4 and conclusion in Section 5.

2. Repulsive Potential Functions

Two repulsive potential functions are employed for collision avoidance. The first function is the Gaussian repulsive potential that repels the OS away from the TS. (This function is regarded as the STAY CLEAR action in Section 3.) The advantageous of this function is that its magnitude and range can be controlled more effectively when compared to the conventional FIRAS function [9]. The Gaussian repulsive potential and its derivative are as follows [10]:

$$U_g = c_g \exp\left(\frac{-\|d\|}{l_g}\right), \tag{1}$$

$$\nabla U_g = -\frac{c_g}{l_g}\exp\left(\frac{-\|d\|}{l_g}\right)\left[\frac{\partial d}{\partial x}, \frac{\partial d}{\partial y}\right]^{\mathrm{T}}. \tag{2}$$

Here, the vector d denotes the relative distance between the two vessels. The magnitude and effective range of the Gaussian potential function is defined by c_g and l_g, respectively.

The second function is the vortex repulsive potential. Depending on the encounter condition, the vortex function generates either clockwise or counter-clockwise spinning force field that deviates the heading of the OS away from the collision path. (This function is regarded as the VORTEX_CW and VORTEX_CCW actions in Section 3.) The vortex potential function and its derivative are modified from [11,20] as presented below:

$$U_v = c_v \exp\left(\frac{-||d||}{l_v}\right), \tag{3}$$

$$\nabla U_v = -\tau \frac{c_v}{l_v} \exp\left(\frac{-||d||}{l_v}\right) \left[-\frac{\partial d}{\partial y}, \frac{\partial d}{\partial x}\right]^{\mathrm{T}}, \tag{4}$$

where $\boldsymbol{c_v}$ and $\boldsymbol{l_v}$ are used for controlling the strength and the effective range of the vortex function. The parameter $\boldsymbol{\tau}$ is set to either $\mathbf{-1}$ for a clockwise vortex or $\mathbf{1}$ for a counter-clockwise vortex.

3. Fuzzy logic decision-making

The fuzzy logic framework helps to design which potential function should be used when the collision is detected in different encounter conditions; otherwise, it lets the OS keep course and speed when the path is collision-free. To this end, the encounter situation between the two vessels is characterised by two membership functions: 1) relative distance and 2) encounter angle. (As illustrated in Figure 1.)

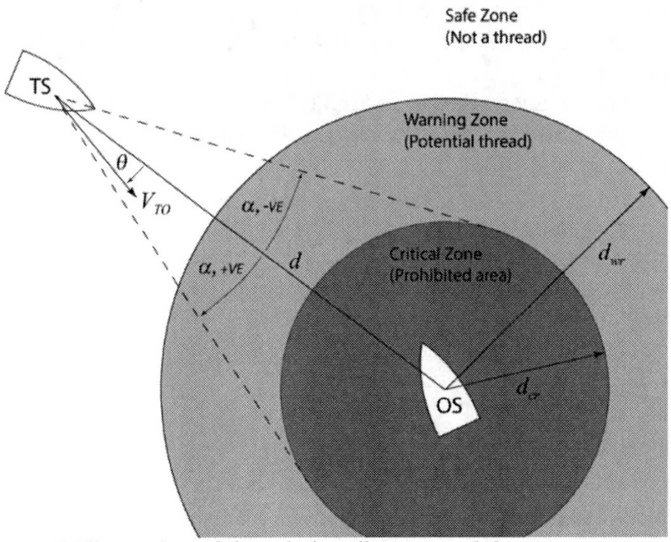

Figure 1. Illustration of the relative distance and the encounter angle for describing the encounter condition between the two vessels

3.1 Relative distance membership function

Two concentric circles are located over the OS and divide Two concentric circles are located over the OS and divide the space into three zones. The safe zone (i.e., $d > d_{wr}$) represents the condition where there is plenty of space between the two vessels; hence, the collision is not a concern. The TS could potentially be a thread if it is inside the warning zone (i.e., $d_{wr} \geq d > d_{cr}$), and the avoidance action may be triggered depending on the encounter condition. However, if the TS is inside the critical zone (i.e., $d_{cr} \geq d$), the OS would try its best to stay clear and maintain a safe distance between the two vessels. These zones can be defined as a fuzzy membership function as shown in Figure 2.

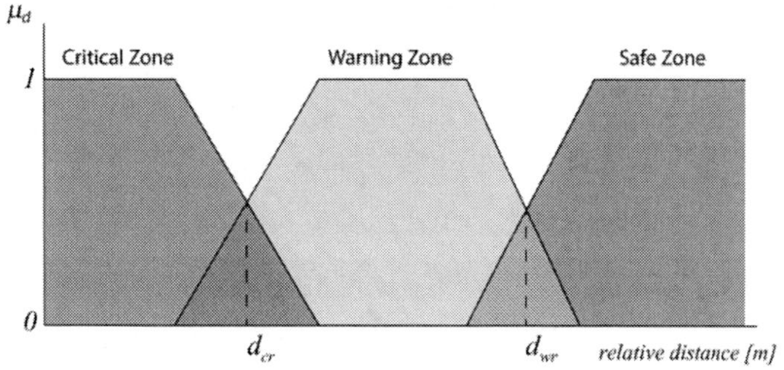

Figure 2. A fuzzy membership function to classify the collision risk in terms of the relative distance

3.2 Encounter angle membership function

The collision cone and the relative velocity vector are used for classifying the likelihood of the collision. In this regard, the cone is defined to have its apex at the TS and with the two sides of the cone tangential to the boundary of the critical zone of the OS [12], as illustrated in Figure 1. The half-apex angle of the collision cone is: $\alpha = \sin^{-1}(d_{cr}/d\,)$. For the relative velocity between the two vessels, it is denoted by $V_{TO} = V_{TS} - V_{OS}$, where V_{TS} and V_{OS} are the velocity vectors of the TS and the OS, respectively. The likelihood of the collision is determined according to the encounter angle, θ, which is the angle that V_{TO} makes to the centreline of the collision cone. The collision is identified as likely positive (LIKELY_P) when $\alpha \geq \theta \geq 0$. Likewise, it is said to be likely negative (LIKELY_N) when $0 > \theta \geq -\alpha$. These two likelihood sets help to decide which side the OS should deviate to quickly resolve the collision. Otherwise, the collision is recognised as unlikely for θ that is not contained inside the collision cone. The fuzzy membership function that represents the collision likelihood w.r.t. the collision angle is given in Figure 3.

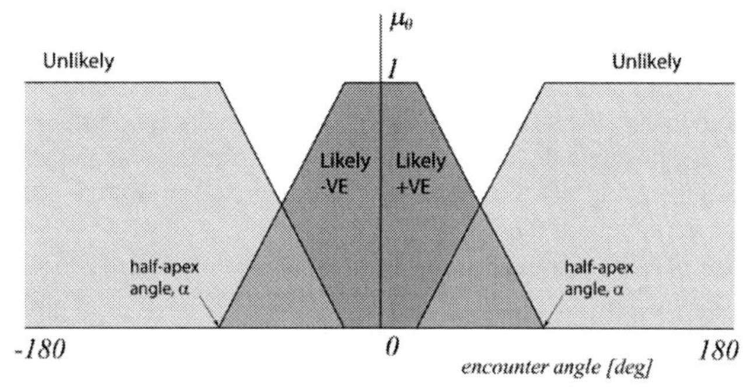

Figure 3. A fuzzy membership function to classify the collision risk in terms of the encounter angle

3.3 Determining avoidance action

The two fuzzy membership functions described above are used to identify the encounter condition. Then, the if-then rules are developed here to determine the appropriate avoidance actions accordingly. For example, if the distance is SAFE, then there is nothing to worry about, i.e., the ship would keep going on the desired track with the TRACK-KEEPING action. The same thing applies when the distance is WARNING but the collision is UNLIKELY. However, if the distance is WARNING and the collision is either LIKELY_P or LIKELY_N, then the appropriate VORTEX is triggered to steer the OS away on the side that quickly resolves the collision with the TS. Lastly, When the distance is CRITICAL, the rules trigger the STAY CLEAR action, or even discard the TRACK-KEEPING action in order to prioritise the safety over the progress on the path following. These if-then rules are summarised in Table 1. The total potential gradient that affects the OS is computed from the rules as follows:

$$\nabla U = K_a \nabla U_a + K_g \nabla U_g + K_{v,ccw} \nabla U_{v,ccw} + K_{v,cw} \nabla U_{v,cw} \tag{5}$$

where

$$K_a = R_1 + R_2 + R_3 + R_4 + R_5 \tag{6}$$

$$K_g = R_5 + R_6 + R_7 \tag{7}$$

$$K_{v,ccw} = R_3 + R_6 \tag{8}$$

$$K_{v,cw} = R_4 + R_7 \tag{9}$$

As can be noted, the vector ∇U_a is the gradient of the attractive potential that guides the OS to the desired heading and speed. This vector is assumed a constant during the simulation.

Table 1. If-then rules for the fuzzy inference system.

Rule	if	Distance is	and	Collision is	then do	Action 1	and	Action 2
R1	if	SAFE	-	-	then do	TRACK-KEEPING	-	-
R2	if	WARNING	and	UNLIKELY	then do	TRACK-KEEPING	-	-
R3	if	WARNING	and	LIKELY_P	then do	TRACK-KEEPING	and	VORTEX CCW
R4	if	WARNING	and	LIKELY_N	then do	TRACK-KEEPING	and	VORTEX CW
R5	if	CRITICAL	and	UNLIKELY	then do	TRACK-KEEPING	and	STAY CLEAR
R6	if	CRITICAL	and	LIKELY_P	then do	STAY CLEAR	and	VORTEX CCW
R7	if	CRITICAL	and	LIKELY_N	then do	STAY CLEAR	and	VORTEX CW

4. Results and discussion

Two cases of the simulation studies were implemented in MatLab to evaluate the resulting potential field that was generated by the proposed fuzzy-based collision avoidance framework. In the first case, the TS was set to travel north at the speed of 1 m/s whereas the OS travelled south at the speed of 2 m/s. Zone #1 in Figure 4 represents the situation where the two vessels encounter head-on, as well as where the OS encounters a static obstacle. When the TS was outside the critical zone, the vector field smoothly deviated the OS heading aside so the relative velocity vector pointed outside the collision cone and the OS smoothly circumvented the TS.

However, the vector field was stronger if the TS was inside the critical zone. In this case, the resulting vector field not only deviated the OS heading but also repelled it from the TS to ensure the safe distance between the two vessels. The OS did not need to worry about the collision if it was abaft of the TS because the two vessels were going apart (see zone #2 in Figure 4); in this regard, the vector field let the OS continue going south at the desired speed.

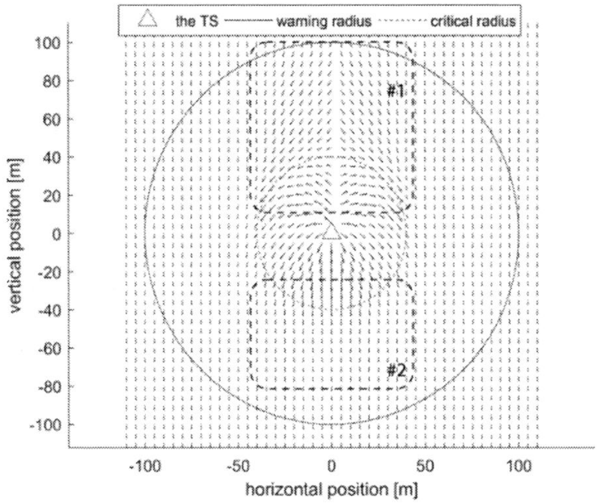

Figure 4. Vector field of the first scenario for a different state of the OS whereas the TS is fixed at the centre (the OS speed = 2 m/s, heading south; the TS speed = 1 m/s, heading north).

The second scenario represented the situation where the two vessels were crossing. Both vessels were set to travel at the speed of 2 m/s, but the OS was set to go east whereas the TS was set to go north. The diagonal dashed line in Figure 5 indicates the condition where the relative velocity vector directly led the OS to collide with the TS. Below the dashed line, zone #1 in Figure 5, the collision likelihood is identified as LIKELY_P. In this case, the CCW vortex potential was applied, inducing the OS to pass the TS from behind. Furthermore, if the OS got too close to the TS, the Gaussian repulsive potential component kicked in to also keep the clear distance between the two vehicles.

Zone #2 in Figure 5 was where the collision likelihood was recognised as LIKELY_N, so the OS was induced to speed up and pass in front of the TS because this is the quickest way to get the relative velocity vector outside the collision cone. However, one can argue that this may increase the risk of collision, especially when the OS cut in front of the TS at the close distance. More membership functions and fuzzy rules should be added to deal with this situation more delicately. Once the collision risk has relieved, the strength of the repulsive potential decayed so the OS can continue the original course and speed.

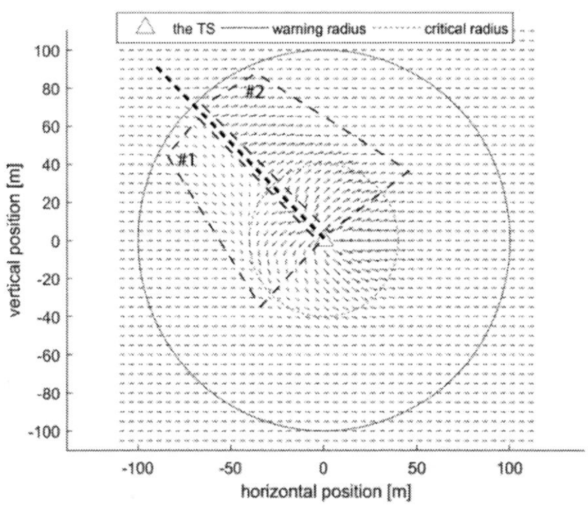

Figure 5. Vector field of the first scenario for a different state of the OS whereas the TS is fixed at the centre (the OS speed = 2 m/s, heading east; the TS speed = 2 m/s, heading north).

IOP Conf. Series: Materials Science and Engineering 1137 (2021) 012017 doi:10.1088/1757-899X/1137/1/012017

5. Conclusion

This work developed a fuzzy-based potential field collision avoidance framework for preventing an unmanned surface vehicle from crashing into another vessel or an obstacle. The basic idea is to use two fuzzy membership functions to classify the encounter condition between the two vessels. In addition, the fuzzy rules are developed to appropriately apply the potential functions in order to guide the own ship away from the collision path. The proposed framework has been demonstrated by the simulation to be effective for collision prevention. Once the collision has been resolved, the algorithm then lets the vehicle continue its original course and speed.

In the current state of this research, it is assumed that the state of the other vessel does not change, and the owner is the one that is entirely responsible for collision prevention. The future work will be focusing on developing membership functions and adding more fuzzy rules to make the algorithm COLLEGs compliance and be able to handle more complex encounter situations.

References

[1] Huang Y Chen L Chen P Negenborn R R and van Gelder P H 2019 *Safety Science* **121** 451-473
[2] Wang W Zhu M Wang X He S He J and Xu Z 2018 *International Journal of Advanced Robotic Systems* **15**(5) 1-13
[3] Niu H Lu Y Savvaris A and Tsourdos A 2016 *IFAC Papers OnLine* **49**(23) 121-126
[4] Tanakitkorn K Wilson P Turnock S and Phillips A 2014 *Proc. 2014 IEEE/OES Autonomous Underwater Vehicles* (Oxford, Mississippi) 1-8
[5] Rostami S M H Sangaiah A K Wang J and Liu X 2019 *Eurasip Journal on Wireless Communications and Networking* **2019:70** 1-19
[6] Xue Y Clelland D Lee B S and Han D 2011 *Ocean Engineering* **38**(17-18) 2290-2305
[7] Zhou L and Li W 2015 *Proc. 7th International Symposium on Computational Intelligence and Design* **2**(1) 429-432
[8] Hong M J and Arshad M R 2015 *Procedia Computer Science* **76**(Iris) 198-202
[9] Sfeir J Saad M and Saliah-Hassane H 2011 *Proc. 2011 IEEE International Symposium on Robotic and Sensors Environments* (IEEE) pp 208-213
[10] Kim D H and Shin S 2006 *Advanced Robotics* **20**(1) 115-135
[11] Lee M C Nieh C Y Kuo H C and Huang J C 2020 *Journal of Marine Science and Technology* **25** 925-942
[12] Oh H N H Tsourdos A and Savvaris A 2014 *IFAC Proceedings* **47**(3) 12174-12181
[13] Kufoalor D K Brekke E F and Johansen T A 2018 *Proc. 2018 IEEE/RSJ International Conference on Intelligent Robots and Systems* (IEEE) pp 2402-2409
[14] Song L Chen Z Dong Z Xiang Z Mao Y Su Y and Hu K 2019 *International Journal of Advanced Robotic Systems* **16**(3) 1-9
[15] Brcko T and švetak J 2013 *PROMET - Traffic & Transportation* **25**(6) 555-564
[16] Lee S M Kwon K Y and Joh J 2004 *International Journal of Control, Automation and Systems* **2**(2) 171-181
[17] Perera L P Carvalho J P and Soares C G 2010 *Advanced Ship Design for Pollution Prevention* (CRC Press) 205-216
[18] Perera L P Carvalho J P and Soares C G 2011 *Journal of Marine Science and Technology* **16**(1) 84-99
[19] United States Coast Guard 1999 *Navigation Rules: International - Inland* (Alexandria, Virginia: United States Department of Transportation)
[20] Rezaee H and Abdollahi F 2012 *Proc. 2012 IEEE/ASME International Conference on Advanced Intelligent Mechatronics* (Kaohsiung, Taiwan) pp 1-6

Parametric study of composite wing box for high altitude platform using finite element analysis

Nontawat Thaiprayoon[1], Kittigorn Chalernphon[2], Chompunut Somtua[3], Panya Aroonjarattham[1,*]

[1]Department of Mechanical Engineering, Faculty of Engineering, Mahidol University, Nakhonpathom Thailand 73170
[2]Geo-Informatics & Space Technology Development Agency (Public Organization), Chonburi Thailand 20230
[3]Department of Mechanical Engineering, Faculty of Engineering, Thonburi University, Bangkok Thailand 10160

* Corresponding Author: panya.aro@mahidol.ac.th

Abstract
This paper aims to evaluate the degree and the number of pile stack between five and six ribs wing box that created from the Eppler 397 Airfoil (e397-il) model. The model was varied three numbers of pile stack and six degrees of angle of pile stack to find the best result for the composite material to construct the High-Altitude Platform (HAP). The thickness of pile was fixed at 0.234 mm. The structure of wing box has the safety of factor 5.7 and receive load 3,052 N. The model was analysed by ABAQUS software to compute the vertical-static such as Tsai-Hill and skin defection that did not exceed 1 for Tsai-hill and 2 mm for skin defection. The results were shown that the wing box had six ribs model with the angle of pile stack as [+45/-45/90/90/0/90/90/-45/+45] was the best model in this research. It had Tsai-Hill 0.030 and skin defection 0.348 mm. The model had more ribs and number of pile stack increase the strength of wing box and the degree of pile stack between the outer and the adjacent layer should be a perpendicular direction.

Keywords: High Altitude Platform, Tsai-Hill, Skin defection.

1. Introduction
Geo-informatics and Space Technology Development Agency (GISTDA) has a project to create the High-Altitude Platform (HAP) using the composite materials that had a lightweight and stronger than aluminums. HAP can be used instead of the satellite because it can survey the resource on the earth, transmit and receive the radio signal at 20-50 kilometers over the ground with the lower operating and maintenances cost. In the maintenance process, HAP can be land and repair on the ground like a plane that easier maintain than the satellite. It uses the solar energy for work by install the solar cells on their wing. This research aim to evaluate the degree and the number of pile stack between five and six ribs wing box that created from the Eppler 397 Airfoil (e397-il) model. The results were computed the vertical-static such as Tsai-Hill and skin defection to find the best degree and numbers of pile stack that use to construct the HAP.

2. Materials and Methods

2.1 Three-dimensional model of wing box

Three-dimensional models of wing box were constructed using CATIA software. The spacing between ribs was 250 and 312.5 mm as shown in Figure 1.

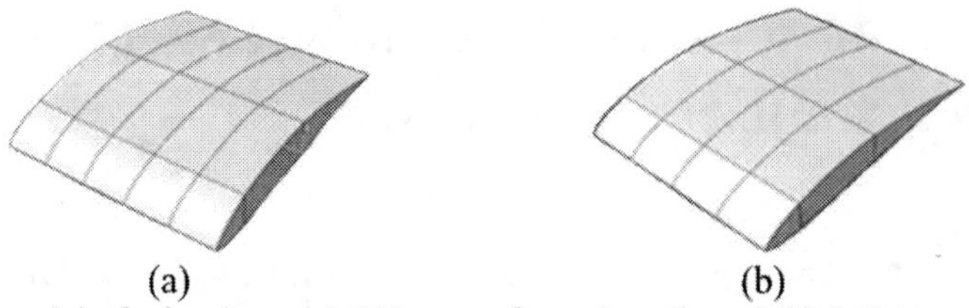

Figure 1. The model of wing box: (a) 250 mm of spacing rib and (b) 312.5 mm of spacing rib

Two spars of wing box as front and rear spar were used in this study. The front spar was installed at 15% from the anterior of wing box and the rear spar was installed at 60% from in front of wing box [1] as shown in Figure 2.

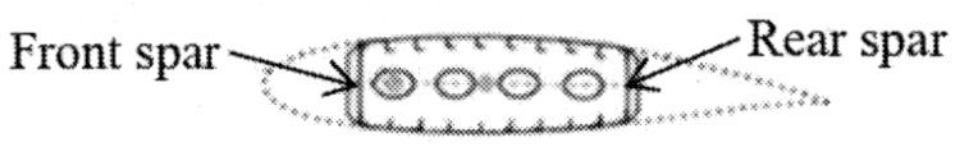

Figure 2. Position of front and rear spar of wing box [2].

2.2 Boundary condition of wing box

Wing box received the load from the wind action that analysed from CFD analysis. The angle of attack was set at 18 degree from Eppler 397 airfoil cl-alpha diagram. The wind's particle was act in the X-axis with the speed of 21 m/s. After that the maximum force in the X-axis was 550 N and Y-axis was 400 N as shown in Figure 3 [3].

Figure 3. The result of wind action on the X- and Y-axis [4].

The ultimate load act on the wing box was calculated by using the maximum force and load factor. The load factor can get from the V-n diagram that was shown the relationship between the load factor (n) and the velocity for flying (V) as shown in Figure 4.

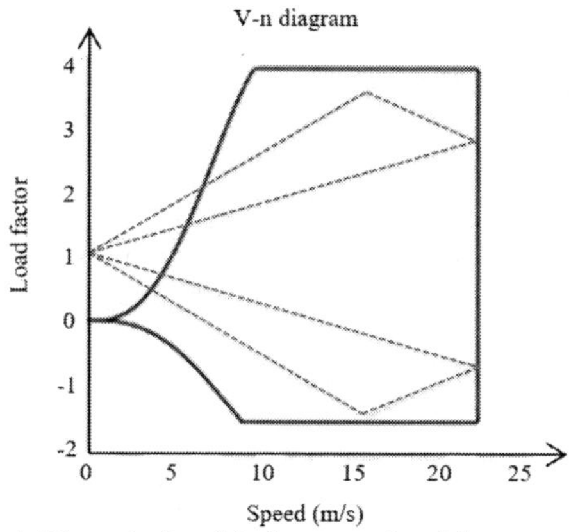

Figure 4. The relationship between load factor and velocity.

The operating flight strength of an airplane was presented in form of V-n diagram that shown the maximum performance of aircraft without breaking [5]. From the graph, the maximum load factor did not have a damage to the structure is 3.7 and the factor of safety is 1.5 so the ultimate load as shown in Figure 5 was the load factor x safety factor x maximum load

Ultimate load = 3.7 x 1.5 x 550
= 3,052 N

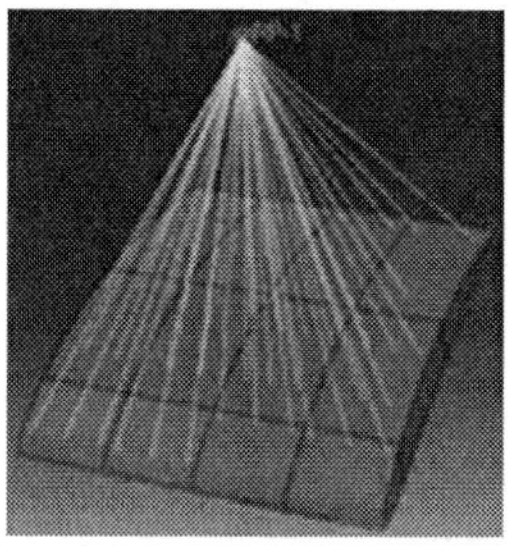

Figure 5. The ultimate load act on the wing box.

2.3 Skin-deflection

The force act on the surface of the wing box caused a point load to press the wing surface between the ribs that affect the bent of composite materials. The bendy skin of the wing box surface was called skin-deflection (δ) as shown in Figure 6. It should not bend more than 2 mm of the total skin thickness [6].

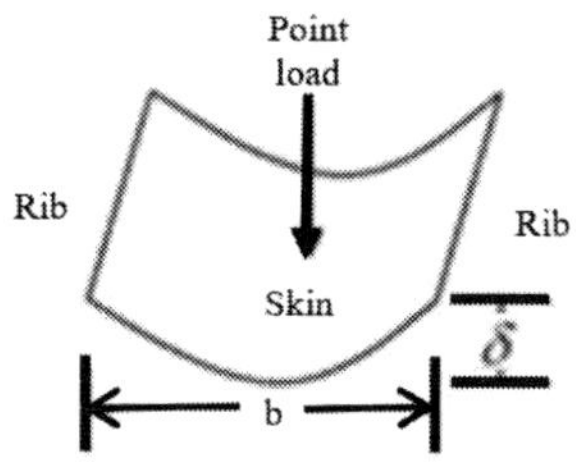

Figure 6. The skin-deflection of the wing box surface [6].

2.4 Tsai-Hill

Tsai-Hill was used to evaluate the failure condition of the laminated composite materials. It did not exceed than 1 for the safe condition [7].

IOP Conf. Series: Materials Science and Engineering 1137 (2021) 012018 doi:10.1088/1757-899X/1137/1/012018

$$\frac{\sigma_{11}^2}{X_{11}^2}-\frac{\sigma_{11}\sigma_{22}}{X_{11}^2}+\frac{\sigma_{22}^2}{X_{22}^2}+\frac{\sigma_{12}^2}{X_{12}^2}<1 \tag{1}$$

Where σ is normal stress
X_{11} is longitudinal tensile failure
X_{22} is transverse tensile failure
X_{12} is tensile failure between longitudinal and transverse

2.5 Basic stacking sequences rule

The composite material had three conditions for layup the pile stack as follow;

- The first was the symmetry condition. The degree of pile stack should be symmetry between upper layers and lower layers from the median layer as shown in Table 1 [8].

Table 1. The symmetry condition of the degree of pile stack.

Layer	Good layup pile stack		Bad layup pile stack	
	Degree of pile stack	Condition	Degree of pile stack	Condition
Outer	45	Symmetry	90	Symmetry
Upper 2	-45	Symmetry	-45	Asymmetry
Upper 1	45	Symmetry	45	Asymmetry
Median	0	None	0	None
Lower 1	45	Symmetry	-45	Asymmetry
Lower 2	-45	Symmetry	45	Asymmetry
Outer	45	Symmetry	90	Symmetry

- The second condition was specified the outer layer with 45 or -45 degree as shown in Table 2.

Table 2. The specify degree of pile stack at outer layer.

Layer	Good layup pile stack		Bad layup pile stack	
	Degree of pile stack	Condition	Degree of pile stack	Condition
Outer	45	Specify	90	Non-specify
Upper 2	-45	Symmetry	-45	Symmetry
Upper 1	45	Symmetry	45	Symmetry
Median	0	None	0	None
Lower 1	45	Symmetry	45	Symmetry
Lower 2	-45	Symmetry	-45	Symmetry
Outer	45	Specify	90	Non-specify

- The third condition was the close three layers should not layup with the same degree of pile stack as shown in Table 3.

Table 3. The symmetry condition of the degree of pile stack.

Layer	Good layup pile stack		Bad layup pile stack	
	Degree of pile stack	Condition	Degree of pile stack	Condition
Outer	45	Specify	-45	Specify
Upper 4	-45	Symmetry	90	Repeated degree

Upper 3	90	Symmetry	90	Repeated degree
Upper 2	-45	Symmetry	90	Repeated degree
Upper 1	45	Symmetry	45	Symmetry
Median	0	None	0	None
Lower 1	45	Symmetry	45	Symmetry
Lower 2	-45	Symmetry	90	Repeated degree
Lower 3	90	Symmetry	90	Repeated degree
Lower 4	-45	Symmetry	90	Repeated degree
Outer	45	Specify	-45	Specify

2.6 Case analysis
All models were analyzed Tsai-Hill and skin deflection on the five and six ribs wing box. The numbers of pile were varied with three values as shown in Table 4. The pile orientations were varied under three conditions for layup pile stack.

Table 4. List of case analysis.

Cases	Number of ribs	Number of piles	Pile orientation
1	5	7	[+45/0/0/-45/0/0/+45]
2	5	8	[+45/0/0/-45/-45/0/0/+45]
3	5	9	[+45/0/0/-45/90/-45/0/0/+45]
4	5	7	[-45/+45/0/90/0/+45/-45]
5	5	8	[-45/+45/0/90/90/0/+45/-45]
6	5	9	[+45/-45/90/90/0/90/90/-45/+45]
7	6	7	[+45/0/0/-45/0/0/+45]
8	6	8	[+45/0/0/-45/-45/0/0/+45]
9	6	9	[+45/0/0/-45/90/-45/0/0/+45]
10	6	7	[-45/+45/0/90/0/+45/-45]
11	6	8	[-45/+45/0/90/90/0/+45/-45]
12	6	9	[+45/-45/90/90/0/90/90/-45/+45]

2.7 Material properties
The material properties of unidirectional carbon fiber prepreg were used to calculate in this research was shown in Table 5.

Table 5. Material properties of carbon fiber T700/M21.

Variable	Values	Unit
E_1	131.100	GPa
$E_2 = E_3$	8.200	GPa
G_{12}	3.930	GPa
$G_{23} = G_{13}$	4.300	GPa
X_t	2.489	GPa
X_c	1.254	GPa
Y_t	53.000	MPa

Y_c	-202.000	MPa
S	63.300	MPa
$\nu_{12} = \nu_{23} = \nu_{13}$	0.350	-

2.8 Mesh generation

All mesh models were generated by ABAQUS software as shown in Figure 7. The mesh model had 126,554 nodes and 128,725 elements for five ribs wing box and 126,496 nodes and 128,769 elements for six ribs wing box.

Figure 7. Mesh model of six ribs wing box.

3. Result and Discussion

The results were evaluated Tsai-Hill and skin defection on the wing box of twelve cases analysis as shown in Table 6.

Table 6. The result of Tsai-Hill and skin-defection of 12 cases analysis.

Cases	Tsai-Hill	Skin-defection (mm)
1	0.149	1.434
2	0.111	1.072
3	0.084	0.826
4	0.097	1.160
5	0.083	0.886
6	0.042	0.616
7	0.095	0.869
8	0.074	0.661
9	0.056	0.518
10	0.065	0.702
11	0.053	0.533
12	0.030	0.348

The best wing box was a case no.12 that had the smallest Tsail-Hill and skin-defection. The contour of Tsai-Hill on the wing box was shown in Figure 8 and skin-defection was shown in Figure 9.

Figure 8. The contour of Tsai-Hill on mesh model.

Figure 9. The contour of skin-defection on mesh model.

All wing boxes constructed by composite material was safe in all cases analysis because all values of Tsai-Hill did not exceed 1 and skin-defection did not exceed 2 mm in all cases.

In case of five ribs, case no.6 was the best case because Tsail-Hail and skin-defection are lower than the other cases. It was shown more number of pile stack increase the strength of wing box. The case no.3 has a same number of piles as case no.6 but there are more Tsail-Hill and skin-defection than case no.6 that shown the degree of pile stack between upper 4 and outer layer and lower 4 and outer layer should be a perpendicular direction.

In case of six ribs, case no.12 was the best case because Tsail-Hail and skin-defection are lower than the other cases. It was shown more number of pile stack increase the strength of wing box same as five ribs case and the degree of pile stack between the outer and the adjacent layer should be a perpendicular direction.

4. Conclusion

This study aims to evaluate the parameters as number of ribs, number of pile stack and the degree of pile stack of wing box constructed by composite material by finite element analysis. The results were shown more ribs and number of pile stack help to increase the strength of wing box and the degree of pile stack between the outer and the adjacent layer should be a perpendicular direction. The next step is to validate the finite element result with the mechanical testing machine as shown in figure 10 to confirm the best model for construct the High-Altitude Platform (HAP).

Figure 10. The mechanical testing machine for wing box testing.

Acknowledgments

The author wishes to thank the Geo-informatics and Space Technology Development Agency (GISTDA) and Biomechanical Analysis and Orthopaedic Device Design Laboratory (B-AODD Lab), Faculty of Engineering, Mahidol University for their kind support of the facilities.

5. References

[1] Annabel R 2009 *M.S. Thesis* (Guildford: Department of Physics, University of Surrey)
[2] Niu M C Y 1992 *Composite airframe structures: practical design information and data, 3rd ed* (USA: Adaso/Adastra Engineering Center)
[3] Gregoire A Gabriel D 2016 *J Mech Phys Solids* **97** 168-196
[4] Jarukit S 2019 *B.E. Project*, (Nakhon Pathom: Department of Mechanical Engineering, Mahidol University)

[5] UAVNavigation In Depth: Flight Envelope 2019 https://www.uavnavigation.com/company/blog/uav-navigation-depth-flight-envelope, accessed 23 November 2019

[6] Edward A B Benjamin K S Keejoo L Curt S K Norman M W 2010 *J Intel Mat Syst Str* **21**(17) 1699-1717

[7] Engineering Sciences Data Unit 1995 *Elastic direct stresses and deflections for flat rectangular plates under uniformly distributed normal pressure, 71st ed* (England: Engineering Sciences Data Unit)

Investigation of power transfer efficiency: utilizing different coil designs in wireless charging of electric vehicles

M. E. Thein[1], J. Charoensuk[1], M. Masomtob[2], W. Onreabroy[3] and A. Kaewpradap[4]

[1]Department of Mechanical Engineering, School of Engineering, King Mongkut's Institute of Technology Ladkrabang, Chalongkrung Road, Ladkrabang, Bangkok 10520, Thailand
[2]National Energy Technology Center (ENTEC), National Science and Technology Development Agency (NSTDA), Pathum Thani 12120, Thailand
[3]Department of Physics, Faculty of Science, King Mongkut's University of Technology Thonburi, Pracha Uthit Road, Bangmod, Thungkhru, Bangkok 10140, Thailand
[4]Department of Mechanical Engineering, Faculty of Engineering, King Mongkut's University of Technology Thonburi, Pracha Uthit Road, Bangmod, Thungkhru, Bangkok 10140, Thailand

* Corresponding author: amornrat.kae@mail.kmutt.ac.th

Abstract
EV battery can be wirelessly charged by inductive coupling method between two electromagnetic couplers. In this study, the transmitter and receiver coils are designed. The effect of the number of turns for transmitter and receiver coils are studied. The results show that efficiency is decreased with the reduction of turns for both coils. Next, the spacing between each turn of coils is analyzed by 1mm of increasing for each simulation and the increased efficiency is only 2.6% for every 1mm spacing. Then, the power transfer efficiency is investigated for different coil designs. The efficiency can be improved with the proper consideration of air gap length, operation frequency, suitable coil designs and coil properties. Five coil designs such as circular, rectangular, double-D, double-D quadrature (DDQ) and bipolar designs are selected. The best design for maximum power transfer efficiency is DDQ design but the structure is complex since more wires are used in this design. The air-gap length between two coils is selected as 120mm and frequency is 85 kHz in this study.

Keywords: Inductive coupler, Power Transfer Efficiency, Wireless Charging, Electric Vehicles

1. Introduction

Electric Vehicles (EV) has been significantly popular in recent years since the required electric energy is obtained from various sources such as nuclear power and renewable energy sources [1]. However, one biggest problem is the time required for driving EV to a charging station and waiting for the battery to be fully charged. Wireless charging or Wireless Power Transfer (WPT) system brings an ultimate solution to that problem [2]. WPT is based on inductive coupling in which electrical energy is transmitted from a power source to an electrical load using two electromagnetic coils. The magnetic field is created between the coils and transferred over distance. Wireless charging for EV can be categorized into stationary, semi-dynamic and dynamic charging system. In dynamic charging, the transmitter/primary coil is installed under the ground connecting to the grid electricity and the receiver/secondary coil under the chassis of EV. The primary system consists of a grid network power supply, a cascade of AC-DC and DC-AC rectifier increasing the AC power line frequency (50Hz or 60Hz) to a high-frequency AC power (<100kHz) to energize the primary transmitter coils and a compensation circuit. The VA ratings of power electronics are reduced and the leakage inductance for magnetically coupled power transfer are neutralized to achieve the higher frequency by the compensation circuit. The secondary side includes AC-DC rectifier converting the induced high-frequency AC into DC and a secondary compensation circuit providing constant current, constant voltage, or step power to charge an onboard battery. The in-road transmitter must be delicately constructed to achieve the necessary power transfer efficiency and performance [3].

In order to design an efficient WPT system, the effect of significant operating parameters such as operating frequency, physical airgap spacing between transmitting and receiving charging coils, wire property used for coil and coil configuration must be primarily analyzed [4]. Moreover, the following parameters such as power level, maximum charging distance, charging tolerance, size and weight and efficiency are critical when WPT system for EV is designed [5]. Power level determines the required amount of time for full charging of the battery. The maximum charging distance is a distance between transmitter and receiver according to the ground clearance of EV to be charged. The charging tolerance is where a driver needs to park EV to adjust the horizontal and vertical position according to the charging equipment. The size and weight are for a receiver pad to be installed in EV. The receiver part was designed with the effect of shape, size of the wire, number of turns and size of the coil [6]. The efficiency of the WPT system could be compared with the plug-in system. In order to reduce the weight of the receiver pad and enhance the power transfer efficiency, this paper focuses on three parametric studies with the following criteria. The first parametric study is that the number of turns for transmitter and receiver coils were alternated in a way of 10 transmitter turns and 10 receiver turns, 10 transmitter turns and 5 receiver turns, and 5 transmitter turns and 5 receiver turns. The second one is the spacing of 1mm, 2mm and 3mm between consecutive turns that were compared for both coils simultaneously. Finally, the five coil designs for both transmitter and receiver coils were analyzed for maximum power transfer efficiency.

2. Methodology

During the design process of primary and secondary coils, the main parameters that should be maximized are quality factor Q and coupling coefficient k with the increasing air-gap lengths. Ferromagnetic materials (cores) can be used to guide magnetic flux to improve coupling [7]. Losses within the system contain core losses of the ferrite material and ohmic losses of the coils including proximity and skin effect losses. Litz wire can reduce skin effect losses while the flux density was lower than the saturation flux density of the material decreases core losses. However, limitations of power and space requirements are still challenging for the available design options. The efficiency of the power transfer can be enhanced by three design parameters coupling coefficient k or mutual inductance M, self-inductance of the coils L and frequency ω. An induced voltage in the secondary coil can be improved by increasing frequency but frequency-dependent losses will be also increased. Therefore, selecting the design frequency is also important. Increasing coil dimensions and the number of turns can be improved the self-inductance of the coils. However, there is also a limitation for maximum coil size which depends on the size of the underbody of EV. To obtain a higher coupling coefficient, decreasing the air-gap length between the coils, increasing the coil dimension, and having equally sized coils can be performed. The coupler size, coupling coefficient, quality factor of coils and maximum power transfer efficiency are calculated with the following equations.

$$d_{xo} = d_{xi} + 2 \cdot [N \cdot d_w + (N-1) \cdot \gamma] \quad (1)$$

where d_{xo} is the outer diameter of the coil, d_{xi} inner diameter of the coil, N number of turns of the coil, d_w diameter of the wire and γ the spacing between consecutive turns.

$$k = \frac{M}{L_1 \cdot L_2} \quad (2)$$

where k is coupling coefficient, M mutual inductance, L_1 self-inductance of primary coil and L_2 self-inductance of secondary coil.

$$Q = \frac{\omega \cdot L}{R} \quad (3)$$

where Q is the quality factor, ω frequency, L coil inductance and R coil resistance.

$$\eta_{max} = \frac{1}{1 + \frac{2}{k^2 Q_1 Q_2} + \frac{2\sqrt{k^2 Q_1 Q_2 + 1}}{k^2 Q_1 Q_2}} \quad (4)$$

where η_{max} maximum power transfer efficiency (%)
k coupling coefficient
Q_1 quality factor of the transmitter coil
Q_2 quality factor of the receiver coil

3. Conceptual model

This paper is only based on a simulation study using Ansys Electronics Suite software. Circular-circular coil design is selected as a base model for parametric studies. The outer coil diameter is chosen as 300 mm and the wire diameter is 7mm. Frequency of 85kHz is used and current is excited as 10A. An air gap between transmitter and receiver coils is chosen as 120mm. For the first parametric study which is the impact of the number of turns on efficiency, three different cases which are 10 turns for both transmitter and receiver coils, 10 turns for transmitter coil and 5 turns for receiver coil and 5 turns for both coils are analyzed. For the spacing between consecutive turns, 1mm, 2mm and 3mm spacings are examined for power transfer efficiency. For different coil designs, circular, rectangular, double-D, double-D quadrature (DDQ) and bipolar designs are observed.

4. Simulation results and discussion

To compare the power transfer efficiency of different coil designs, they are designed to be as analogous to each other as possible. This study ultimately aimed to investigate the efficiency of power transfer, so some received parameters such as coupling coefficient k, the quality factor of both transmitter Q_1 and receiver coils Q_2 are mainly compared. The software gives 6 decimal places for all results. The parameter that effects the efficiency is coupling coefficient k, so Q_1 and Q_2 are taken as only two decimal places. By comparing the power transfer efficiency given by different decimal places of k, it was discovered that the result with four decimal places of k is almost the same as six decimal places of k. Therefore, the important parameter coupling coefficient was taken as four decimal places and others are taken as only two decimals for every simulation result.

4.1 Number of Turns for transmitter and receiver coils

As a first step, the number of turns for transmitter and receiver coils are alternated as shown in Figure 1. A most general design (Circular-Circular) is chosen for simulation. Here, three cases are simulated and compared the power transfer efficiency. The spacing between consecutive turns is selected as 2mm for all three models to ensure a fair comparison. The magnetic coupling as well as efficiency increase dramatically with the increase in the number of turns for both coils as presented in Table 1. It can be

IOP Conf. Series: Materials Science and Engineering 1137 (2021) 012019 doi:10.1088/1757-899X/1137/1/012019

concluded that designing wireless charger coils with the highest possible number of coil turns is beneficial to achieve strong coupling between the coils. However, increasing the number of turns in a receiver coil will raise the weight of EV, and the resistance also increases. The effect of change in the number of turns affects significantly on the voltage in the secondary side since the ratio of the primary to secondary turn change. This effect is much more obvious than the effects due to parasitic capacitances or changes in the values of inductance and coupling.

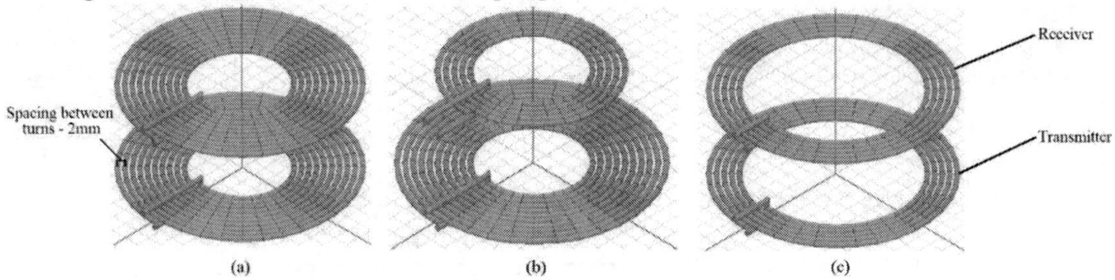

Figure 1. Models of circular coil design (a) 10-10 turns (b) 10-5 turns (c) 5-5 turns.

Table 1. Power transfer efficiency of three circular coil designs for a different number of turns.

Results	Transmitter turns – 10 Receiver turns – 10	Transmitter turns – 10 Receiver turns – 5	Transmitter turns – 5 Receiver turns – 5
Coupling coefficient k	0.0275	0.0230	0.0179
Quality factor Q_1	147.1	146.62	128.85
Quality factor Q_2	255.05	224.76	206.91
Power Transfer Efficiency η_{max}	68.85%	62.22%	51.09%

4.2 Spacing Between Consecutive Turns

The spacing between consecutive turns in coils is also crucial in wireless charging of EV. The power transfer efficiency depends on the spacing between turns as seen in Table 2. Here also, circular-circular coil design is simulated. For the second case, the spacing between turns is alternated while the number of turns for both transmitter and receiver coils are chosen as 5 turns. The reason for choosing 5 turns for both coils is because of less computational time. It is obviously seen that the efficiency can be improved by increasing the spacing between consecutive turns while maintaining the coil size constant. That makes the inner coil radius smaller. Inner coil radius has a significant influence on the magnetic coupling of the two coils [8]. If the coil size is not kept constant, increasing the spacing between turns of the coil makes the coil bigger. The coupling coefficient, as well as the mutual inductance also rise since the coupling coefficient is directly proportional to the radii of the coil.

Table 2. Power transfer efficiency of different spacing between consecutive turns.

Results	1mm Spacing	2mm Spacing	3mm Spacing
Coupling coefficient k	0.0183	0.0179	0.0205
Quality factor Q_1	115.44	128.85	119.58
Quality factor Q_2	188.12	206.91	199.87
Power Transfer Efficiency η_{max}	48.4%	51.09%	53.74%

4.3 Different Coil Designs

4.3.1 Circular – Circular Coil Designs (C-C)

In the early times, circular coil design was common due to simple design. Figure 2 shows the B-field of the circular coil design. For receiver coil which is mounted on vehicles, size and weight are preferred to be decreased. In addition, circular design can reduce the quantity of ferrite and maintain a critical coupling between the primary and the secondary coils [9]. Circular coil design has low non-directional misalignment tolerance. With a lateral offset, the magnetic flux comes into the coil from either side resulting in a null in power distribution. There is also a limitation in the achievable flux height. However, circular coils still display the highest magnetic coupling when compared to other equally sized coil configurations [8]. Hence, circular coils are best suitable for stationary wireless charging system where the coil performance can maximize using multi-objective optimization methods. In a single-coil design, circular pad shows the highest magnetic coupling and highest magnetic field for any number of turns comparing with other designs.

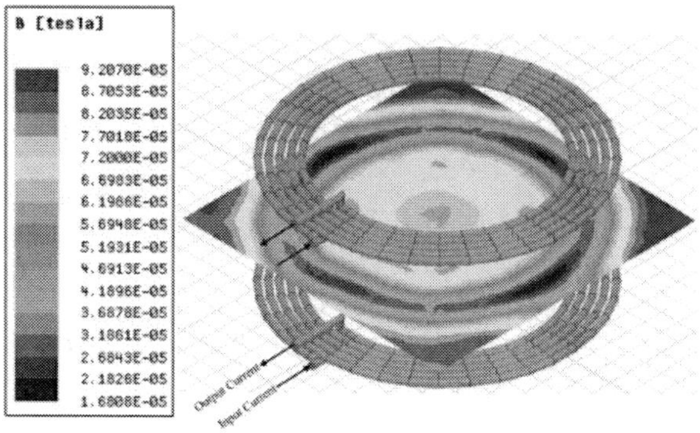

Figure 2. B-field of circular-circular coil design.

4.3.2 Rectangular – Rectangular Coil Designs

For dynamic wireless power transfer, rectangular design as presented in Figure 3 is the most common design for high longitudinal misalignment tolerance and efficient space using on the vehicle [10]. Rectangular coils are more preferable for a cost-effective option than circular and hexagonal designs [11]. Moreover, with a provided material allowance, rectangular pads transmit maximum power over a limited region. When the distances between coils increase, the coupling coefficient does not significantly change. However, the minimum power is shown in the center of the rectangular-rectangular coil and almost half of the leakage flux was lost in the bottom of the coil.

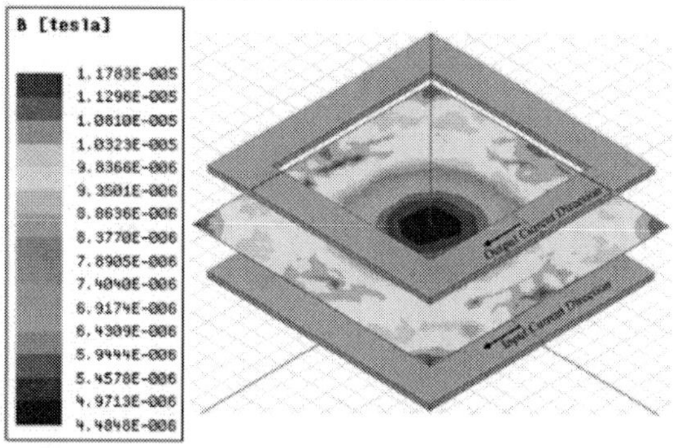

Figure 3. B-field of the rectangular-rectangular coil.

IOP Conf. Series: Materials Science and Engineering 1137 (2021) 012019 doi:10.1088/1757-899X/1137/1/012019

4.3.3 Double-D – Double-D Coil Designs (DD-DD)

DD is one of the most popular advanced coil structures producing a single-sided flux. The geometry is two rectangular coils on top of core material guiding magnetic flux and re-directing it to the front [12]. Thus, the magnetic flux is not produced from the backside of the coil. DD pad has a lower magnetic leakage field than rectangular pad [13] and higher coupling coefficient. DD coil only couples the horizontal flux while circular pad couples the vertical flux component when aligned centrally. The magnetic flux density plot is as shown in Figure 4. The input current directions are same with output current directions in the following figures. The magnetic coupling is significantly higher in multi-coils pad compared with circular and rectangular. Moreover, the coil resistances also increase since more wires are used. However, the efficiency is relatively high and the DD pad is assumed as the best suitable for primary coil.

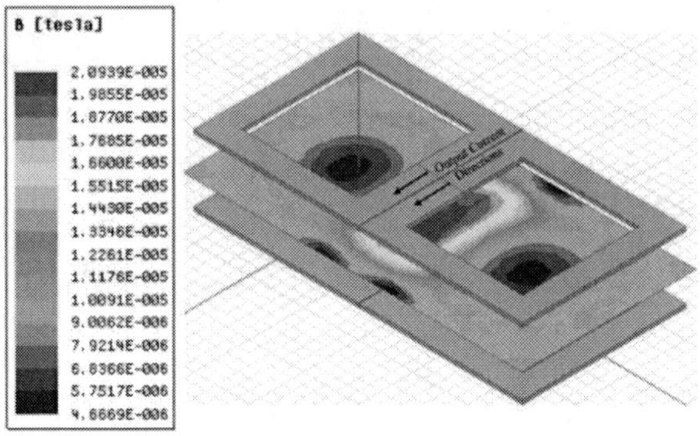

Figure 4. B-field of DD-DD coil design.

4.3.4 Double-D Quadrature – Double-D Quadrature Coil Designs (DDQ-DDQ)

DDQ is created by adding the quadrature coil to DD design as displayed in Figure 5 so not only the horizontal but also the vertical flux can be coupled [14]. DDQ pad combines two windings thus using more wires and having twice the flux height of a circular coil. DDQ has better performance than DD with different secondary topologies and is commonly used in secondary. DDQ pad has a higher tolerance to misalignment but a drawback is increased cost and space since the number of coils is increased. The coupling is highest from the simulation results and the efficiency is highest among all coil designs.

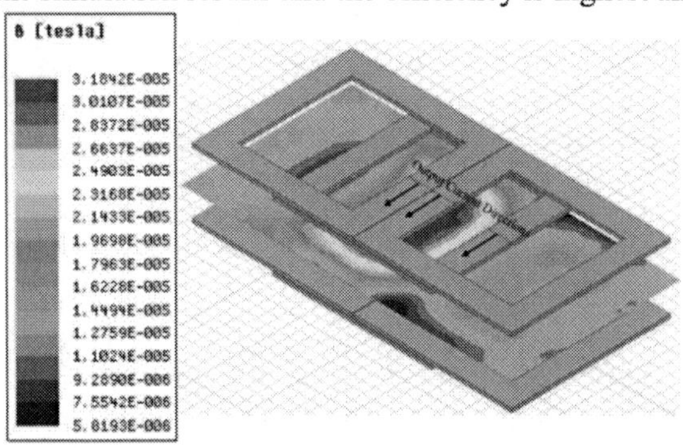

Figure 5. B-field of DDQ-DDQ coil design.

4.3.5 Bipolar – Bipolar Coil Designs (BP-BP)

Bipolar pad design has two windings like the DD design, but the coils are overlapped with each other as demonstrated in Figure 6. The two coils should be mutually decoupled as much as possible to obtain the appropriate overlapping distance. The using wire is 25% lower than DDQ pad but the power transfer abilities are quite similar [15]. DDQ pad has also a higher tolerance to misalignment. As the different current direction, the bipolar pad can act as a circular pad, a single coil or DD pad. Hence bipolar pad design is most suitable for multi-mode secondary pad design. Since the coils are overlapped, the available coil space can be reduced more than DD pad. All multi-coil design (DD, DDQ and Bipolar) require an independent power supply and synchronized converter [16].

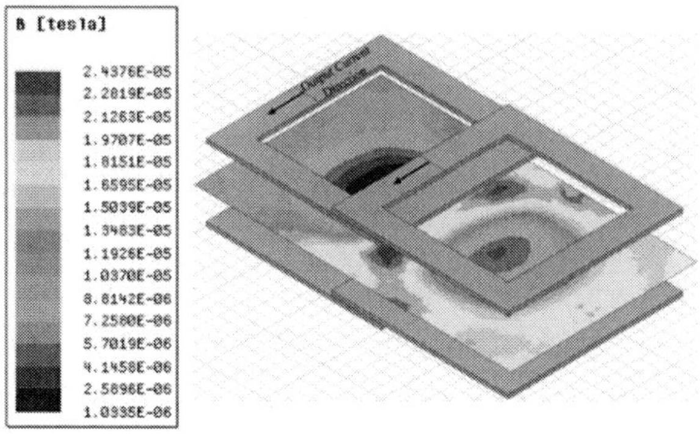

Figure 6. B-field of bipolar-bipolar coil design.

Table 3. Comparison of power transfer efficiency for different coil designs.

Results	C-C	R-R	DD-DD	DDQ-DDQ	BP-BP
Coupling coefficient k	0.0179	0.0135	0.0269	0.0345	0.0121
Quality factor Q_1	128.85	1674.05	2064.33	2485.72	1973.45
Quality factor Q_2	206.91	1882.77	1812.01	2073.97	1595.39
Power Transfer Efficiency η_{max}	51.09%	92%	96.23%	97.48%	91.11%

5. Conclusion

In this study, the impact of a number of turns for transmitter and receiver coils and spacing between each turn is investigated and power transfer efficiency of different lamped coil designs are simulated and analyzed. Increasing the number of turns for both coils and spacing between turns can enhance the power transfer efficiency. The increased efficiency is only 2.6% for rising 1mm spacing between turns. DDQ coil design has the highest power transfer efficiency, but it is not suitable for use as a receiver coil because of weight on the vehicle. The circular design is most simple, but the efficiency is not so high. Rectangular can transfer maximum power but the leakage of the magnetic field can occur from the other side of the pad. The bipolar design has the highest tolerance for misalignment, so it is appropriate for dynamic charging of EV along the road.

Acknowledgments

The author would like to appreciate for the research support from Office of Higher Education Commission Thailand, TAIST-Tokyo Tech AE Program and research facilities from National Science and Technology Development Agency, King Mongkut's Institute of Technology Ladkrabang and department of mechanical engineering, King Mongkut's University of Technology Thonburi.

IOP Conf. Series: Materials Science and Engineering 1137 (2021) 012019 doi:10.1088/1757-899X/1137/1/012019

References

[1] Ma H Balthasar F Tait N Riera-Palou X and Harrison A 2012 *Energy Policy* **44** 160-173
[2] Kurs A Karalis A Moffatt R Joannopoulos J D Fisher P and Soljačić M 2007 *Science (80-.)* **317** 83-86
[3] Throngnumchai K Hanamura A Naruse Y and Takeda K 2013 *World Electr. Veh. J.* **6** 848-857
[4] Sahany S Biswal S S Kar D P Sahoo P K and Bhuyan S 2019 *Prog. Electromagn. Res. M* **79** 187-197
[5] Lu M Junussov A and Bagheri M 2020 *Front. Energy* **14** 152-165
[6] Nalinnopphakhun P Onreabroy W and Kaewpradap A 2018 *Proc. 2018 IEEE Int. WIE Conf. Electr. Comput. Eng. WIECON-ECE 2018* 145-148
[7] Machura P and Li Q 2019 *Renew. Sustain. Energy Rev.* **104** 209-234
[8] Bosshard R Muhlethaler J Kolar J W and Stevanovic I 2013 *Conf. Proc. - IEEE Appl. Power Electron. Conf. Expo.* 1812-1819
[9] Budhia M Covic G A and Boys J T 2011 *IEEE Trans. Power Electron.* **26** 3096-3108
[10] Luo Z and Wei X 2018 *IEEE Trans. Ind. Electron.* **65** 331-341
[11] Chen W Liu C Lee C H T and Shan Z 2016 *Energies* **9** 906
[12] Budhia M Covic G A Boys J T and Huang C Y 2011 *Proc. IEEE Energy Convers. Congr. Expo. Energy Convers. Innov. a Clean Energy Futur.* 614-621
[13] Bosshard R Iruretagoyena U and Kolar J W 2016 *IEEE J. Emerg. Sel. Top. Power Electron.* **4** 1406-1415
[14] Elliott G A J Raabe S Covic G A and Boys J T 2010 *IEEE Trans. Ind. Electron.* **57** 1590-1598
[15] Zaheer A Kacprzak D and Covic G A 2012 *Proc. 2012 IEEE Energy Convers. Congr. Expo.* 283-290
[16] Zaheer A Hao H Covic G A and Kacprzak D 2015 *IEEE Trans. Power Electron.* **30** 1937-1955

Study of 3D printing for forming spoke of non-pneumatic tire using finite element method

Chakrit Suvanjumrat[1,2], **Ravivat Rugsaj**[1,2,*]

[1]Department of Mechanical Engineering, Faculty of Engineering, Mahidol University, Salaya, Nakhon Pathom, 73170, Thailand
[2]Laboratory of Computer Mechanics for Design (LCMD), Department of Mechanical Engineering, Faculty of Engineering, Mahidol University, Salaya, Nakhon Pathom, 73170, Thailand

* Corresponding Author: ravivat.rus@gmail.com

Abstract
Over the last decade, non-pneumatic tire (NPT) was developed by replacing air in traditional pneumatic tire with polyurethane (PU) spoke. However, there are some limitations in building complex shape of spokes. Recently, the 3D printing technique has been used to form complex part of real products. Consequently, it can use to build the complex spoke of NPT. This research aimed to study the possibility of using 3D printing technique to build the complex shape of spoke for NPT. This can be done by studying the mechanical properties of 3D printing material suitable for NPT's spoke. The PU filament was printed into tensile specimens to carry out the stress-strain relationship by using universal testing machine. The finite element (FE) model of NPT was then created using mechanical parameters of the tension result. The vertical stiffness testing on the NPT model was analyzed. The NPT performance to support the vertical load of real NPT and FE model with original material was carried out to compare with the 3D printing model. Furthermore, the advantage and disadvantage including limitation of using 3D printing technique to manufacture the NPT spoke were discussed.

Keywords: Non-pneumatic tire, 3D printing, Finite element method, Polyurethane

1. Introduction

Non-pneumatic tire (NPT) or airless tire was recently developed as an alternative to traditional pneumatic tire. The NPT utilizes an elastic spoke structure instead of air pressure to support vehicle load and absorb shock from road irregularities [1-3]. The spoke is made of polymer, which is polyurethane (PU) commonly used due to its high elastic properties and durability. Vertical stiffness is a tire property that can be used to indicate NPT's performance such as load carrying capacity and force-deformation profile [4, 5]. Finite element method (FEM) can be used to predict tire performance upon usage and testing [6-8]. The vertical stiffness testing of NPT can be simulated using FEM. The specimen was prepared from PU spoke of NPT using waterjet cutting technique. The material properties, which were required for finite element analysis (FEA), were derived from stress-strain relationship [9, 10]. The vertical stiffness testing was simulated and compared with the experiments using FEM. The method was very useful in designing NPT by optimizing the spoke number regarding vertical stiffness [11]. The geometry of spoke structure was found to have significant impact on NPT performance. The FEM can also be used to estimate effect of various spoke geometry to NPT's vertical stiffness. In addition, the novel NPT was designed and optimized for construction application using FEM [12]. While the FEM was proved to be useful tool for design and analysis of NPT, the proper material modeling corresponding to actual material is required to obtain accurate results.

Recently, 3D printing technique was used to build complex shape spoke for NPT. The acrylonitrile butadiene styrene (ABS) thermoplastic polyurethane (TPU) was used as materials to print honeycomb shear band. The effects of different shear band angles and loading conditions to riding comfort were then studied using the simulation [13]. The 3D printing technique can also be used to study the mechanical characteristic of complex structure. The NPT with anti-tetrachiral spoke structure was designed. The compressive test was carried out on the 3D printed structure, which the gradient type structure was shown good load carrying capacity [14]. As mentioned, the 3D printing technique was shown potential in creating complex part including NPT spoke structure. However, the in-depth study of mechanical properties of 3D printed material is still required. This research aimed to study the possibility to use the 3D printing technique to build the complex spoke structure for NPT. This can be done by studying the mechanical properties of 3D printing material suitable for NPT's spoke. The 3D printed specimen was prepared and tested to obtain required properties for FEM. Thus, the FEM of vertical stiffness testing of NPT was analyzed and compared with the actual NPT.

2. Specimens Preparation

2.1 3D printed specimens

The dumbbell shape specimen is prepared using the 3D printing technique according to ASTM D638 (Figure 1). The 3D printing machine, Sync Innovation C300, is developed by Sync Innovation corp., is used with printing parameters as shown in Table 1. As thermoplastic polyurethane (TPU) is suitable material for NPT spoke structure due to its high strength and flexibility. The Armadillo 3D printing filament for fuse deposition method (FDM), which was developed by Ninjatek corp., was used as printing material. The mechanical properties and some important printing properties as stated in its specification are shown in Table 1. The rectilinear printing pattern with 45% infill angle was selected as part of printing parameters.

2.2 Waterjet cut specimens

The NPT TWEEL 12N16.5 ALL TERRAIN, is developed by Michelin, is selected as subjected NPT in this research (Figure 2(a)). The spokes are extracted from the NPT using waterjet cutting technique (figure 2(b)). After that, the spoke is cut into dumbbell shape specimens using a die cutting machine according to ASTM D412 (Figure 2(c)). The spokes were sliced until a required thickness of 2 mm was obtained. In addition, the spokes were cutting with 2 directions, which is radial and width direction relating to the NPT, to study the effects of material's direction dependence.

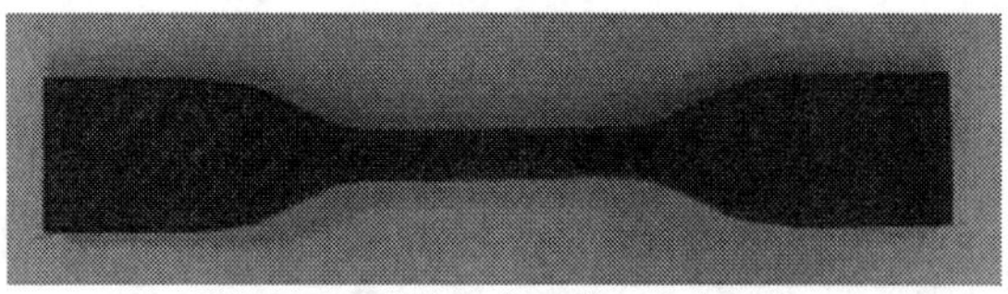

Figure 1. 3D printed specimen.

Table 1. 3D printing parameters and Mechanical properties of Armadillo 3D printing filament.

3D printing parameters		Mechanical properties	
Parameters	Value	Properties	Value
Nozzle temperature (°C)	235	Density	1.18
Bed temperature (°C)	50	Melting temperature (C)	212
Printing speed (mm/sec)	20	Yield stress (MPa)	27
Layer thickness (mm)	0.2	Ultimate stress (MPa)	48
Infill pattern	Rectilinear	Modulus of elasticity (MPa)	396
Infill angle	45°	Elongation at yield (%)	18
% infill	100 %	Elongation at break (%)	295

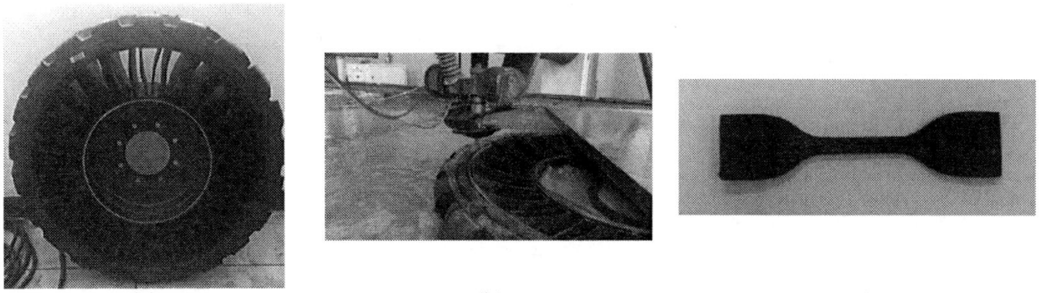

(a) (b) (c)

Figure 2. (a) The NPT, TWEEL 12N16.5 ALL TERRAIN, (b) the extraction of spokes using waterjet cutting technique, and (c) a waterjet cut specimen.

3. Mechanical Properties Testing

The 3D printed specimens and waterjet cut specimens are tested using universal testing machine (UTM), Instron 5969, as shown in Figure 3(a). The tensile tests are performed on both 3D printed (Figure 3(b)) and waterjet cut specimens (Figure 3(c)) according to ASTM D638 and ASTM D412, respectively. The tests were performed on 5 specimens for both preparation methods to ensure reliability and repeatability of the results. The stress-strain relationships obtain by the tension test on specimens can be plotted as shown in figure 4. The mechanical properties require for FE modelling can be summarized as shown in Table 2. It was found that waterjet specimens show nearly identical stress-strain profile for radial and width cutting direction. However, the ultimate stress and breaking stress were found to be significantly different.

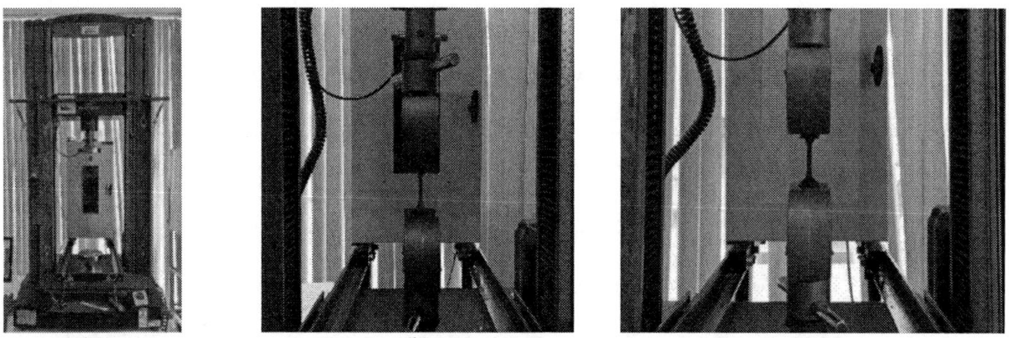

(a) (b) (c)

Figure 3. (a) The UTM, Instron 5969, (b) tensile testing of 3D printed specimen, and (c) tensile testing of waterjet cut specimen.

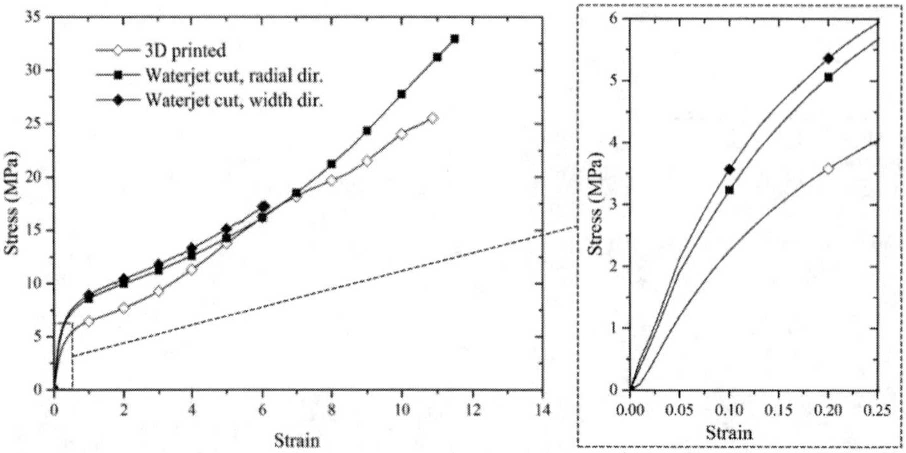

Figure 4. Stress-strain relationship of 3D printed specimens and waterjet cut specimens.

4. Finite Element Analysis of NPT

The FE model of NPT was created and the conditions were applied using FE software, MSC. Patran. The FE mesh of NPT along with its dimension are shown in Figure 5(a) The tread and shear band were modelled using continuum element, while the spoke was modelled using shell element. The linear elastic or Hookean material model was used to model the mechanical behaviour of NPT's components. The details of elements used in modelling the NPT are summarized as shown in table 3. Each component is assembled together using glued contact condition just like the actual NPT assemble process as shown in Figure 5(b). The belt layers were modelled using reinforce bar or rebar element. They were embed into shear band element using tying equation. The comparison of embed rebar element and actual belt's position is shown as schematic diagram in Figure 5(c). In addition, each rebar element composed of mathematically sublayers of belt. The outer layers, middle layers, and inner layers composed of 3, 1, and 2 sublayers, respectively. The bead wire diameter was estimated by measuring to be 1 mm, while the number of wires per unit length was estimated to 0.3582 mm^{-1}.

The simulation of vertical stiffness testing was performed using FE software, MSC. Marc. The FE mesh of NPT is combined with rigid plate, which represented tire testing machine's moving plate as shown in Figure 6(a). The contact algorithm with Coulomb's friction of 0.8 was used to calculate interactive force between NPT and rigid plate. The plate was assigned to press upward against the NPT in vertical direction with force of 20 kN or until the displacement of 26 mm was obtained. The NPT was assigned to be fixed at the inner spoke portion. This represented the locking adapter at the rim in the actual experiment. The analysis was then performed and the results were collected.

Table 2. The mechanical properties of 3D printed and waterjet cut specimen.

Specimens	Minimum breaking strain	Average breaking strain	Ultimate stress (MPa)	Modulus of elasticity (MPa)
3D printed	10.40	10.87	25.53	26.52
Waterjet cut, radial dir.	11.52	12.98	32.92	42.47
Waterjet cut, width dir.	6.06	9.26	17.24	38.29

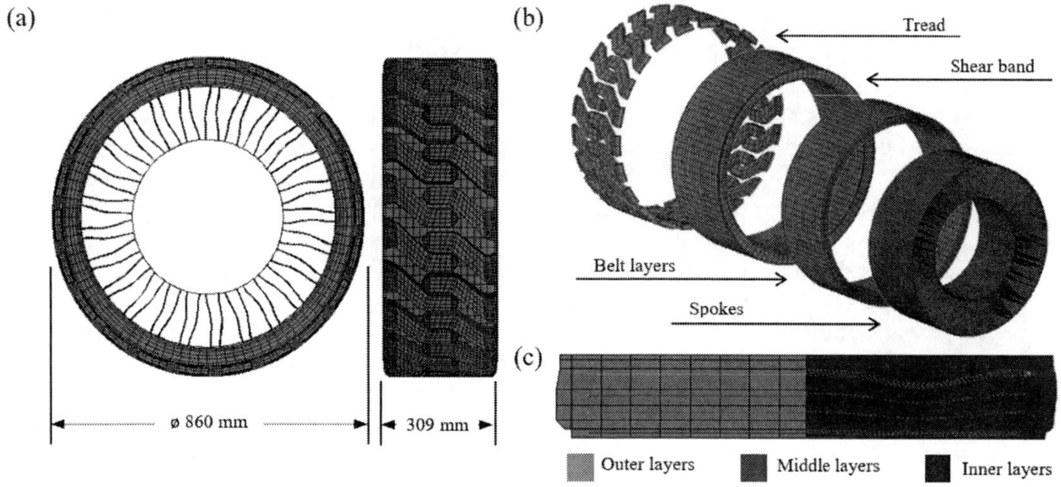

Figure 5. (a) FEM of NPT, (b) components of NPT, and (c) comparison of embed rebar element and actual belts' position.

Table 3. Details of FEM of NPT.

Parameters	NPT's components			
	Tread	Shear band	Spoke	Belts
Element type	Hexagonal	Hexagonal	Quadrilateral	Quadrilateral
Number of node per element	8-node	8-node	4-node	4-node
Number of element	2,288	11,904	35,500	6,144
Average element edge length (mm)	19.36	16.15	8.74	20.38
Average element thickness (mm)	-	-	5.8	-
Modulus of Elasticity	8 GPa	32 GPa	26.52/42.47 MPa	200 GPa

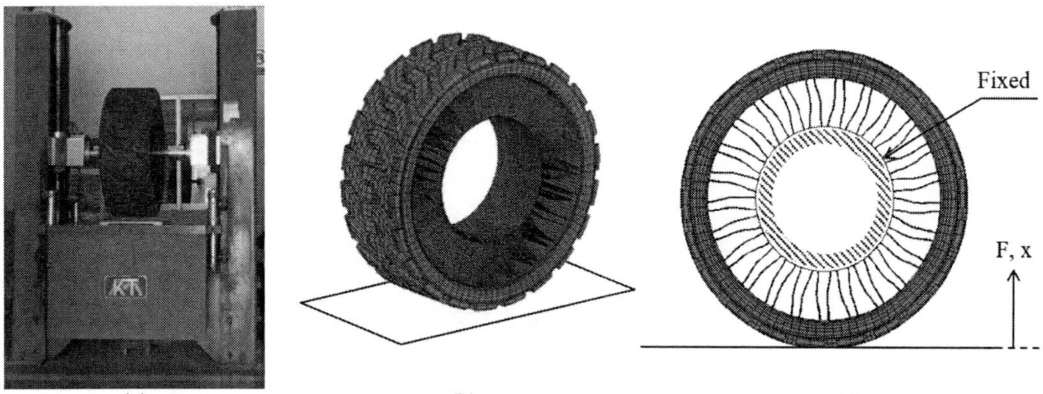

Figure 6. (a) Tire testing machine, Ektron PL-2003, (b) FE model of vertical stiffness testing of NPT, and (c) boundary conditions.

5. Results and Discussion

FEA of vertical stiffness testing of NPT was performed and the results were achieved. The stress and deformation results of FEA of both 3D printed and waterjet material are shown on Figure 7. The maximum stress value at spoke and deformation was found to be high on the 3D printed model due to its lower modulus of elasticity. The maximum stress value at spoke of both models are summarized in table 4. The maximum of stress that occurred on 3D printed NPT was estimated to be 12.21 % higher than the waterjet cut one. As the yield stress value was roughly estimated to be 6 MPa from the mechanical properties testing, the safety factor of 3D printed model and waterjet cut model was then estimated to be 4.08 and 4.58 times, respectively. It should be note that the tension was observed on the upper portion of spoke on both models. On the other hand, the bending and compression were observed on the lower portion of spoke on both models. This indicate the feature to uniformly distribute the load along all portion of spoke, not only on the lower portion which was directly contact the plate.

The vertical force and displacement were then collected. In addition, the force and displacement data were also collected from the vertical stiffness testing experiment of actual NPT. The experiment is performed by tire testing machine, Ektron PL-2003 (Figure 6). The vertical force and displacement relationship of the experiment and both FE model are plotted as shown in Figure 8. The vertical stiffness could be derived from ratio between vertical force and displacement. The estimated vertical stiffness values of FE models and experiment are shown in Table 5. The FEM's analysis error could be estimated by comparing between the waterjet FE model and the experiment because they were based on the same material and conditions. The comparison showed that the estimated vertical stiffness has error of 9.14%. There were proved the validity and accuracy of the FE model. According to the validated model, the vertical stiffness of NPT with 3D printed material is estimated to be 82.68% of original NPT made by traditional molding method. This indicate the mechanical performance and load carrying capacity of 3D printed NPT should be around 82.68% of original NPT as well. While the performances, which including maximum stress and vertical stiffness, of 3D printed NPT was found to be lower than original NPT, the 3D printing technique was proved to be superior in creating complex spoke shape for NPT. The difference in performance was found to be minor difference and negligible.

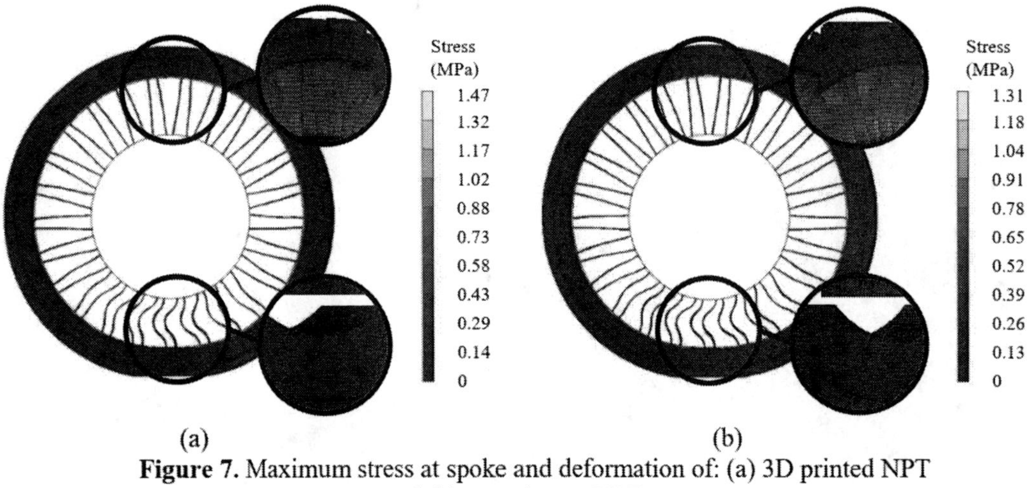

Figure 7. Maximum stress at spoke and deformation of: (a) 3D printed NPT and (b) waterjet cut NPT.

Table 4. Vertical stiffness and maximum stress of experiment and FEA results.

	Experiment	FEM, 3D printed	FEM, waterjet cut
Vertical stiffness (N/mm)	869.93	653.49	790.42
Maximum stress (MPa)	-	1.47	1.31
Safety factor	-	4.08	4.58

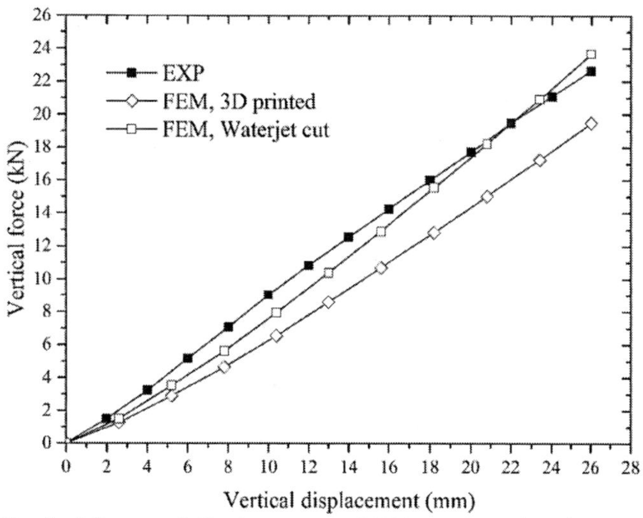

Figure 8. Vertical force and displacement of FEA compared to the experiment.

6. Results and Discussion

This research aimed to study the possibility to use 3D printing technique to build complex spoke structure for NPT. The conclusion can be summarized as follows:

1) The tensile specimens were prepared from 3D printing technique and waterjet cut from original NPT. The stress-strain relationships were collected by using UTM. The modulus of elasticity of the 3D printed specimen and waterjet cut specimen were found to be 26.52 and 42.47 MPa, respectively. The average breaking strains were found to be nearly identical at 10.87 and 12.98, respectively.

2) The FE model of NPT was then created using mechanical parameters of the tension result. The vertical stiffness testing on the NPT model was analyzed. The maximum stress of 3D printed NPT and waterjet cut NPT were found to be 1.47 and 1.31 MPa, respectively. This indicated that the maximum of stress that occurred on 3D printed NPT was estimated to be 12.21 % higher than the original NPT.

3) The vertical force and displacement were then collected from FEA and the experiment. The vertical stiffness was then calculated from vertical force and displacement relationship. The vertical stiffness of the experiment, FEM of 3D printed NPT, and FEM of waterjet cut NPT were found to be 869.93, 653.49, and 790.42 N/mm, respectively. The simulation error was found to be 9.14% compared to the experiment. The vertical stiffness of NPT with 3D printed material was estimated to be 82.68% of original NPT.

The results of this research indicate that the 3D printed NPT's performances, which including maximum stress and vertical stiffness, was found to be lower than original NPT. The performances of 3D printed NPT were found to be not significantly different to model NPT. However, the 3D printing technique was proved to be superior in creating complex spoke shape for NPT. Thus, the technique should be useful in development of NPT in the near future.

Acknowledgments

This work was financial supported by Rubber Technology Research Center and the Thailand Research Fund (TRF) under the TRF Research Grant No. RDG60T0140, and RDG62T0026.

References

[1] Gasmi A Joseph P F Rhyne T B and Cron S M 2012 *International Journal of Solids and Structures* **49**(13) 1723-1740
[2] Ju J Kim D-M and Kim K 2012 *Composite Structures* **94**(8) 2285-2295

[3] Phromjan J and Suvanjumrat C 2018 *Journal of Mechanical Science and Technology* **32**(4) 1539-1548
[4] Rugsaj R and Suvanjumrat C 2020 *IOP Conference Series: Materials Science and Engineering* **773** 1-4
[5] Jin X, Hou C Fan X Sun Y Lu J and Lu C 2018 *Composite Structures* **187** 27-35
[6] Phromjan J and Suvanjumrat C 2018 *Engineering Journal* **22** 141-155
[7] Phromjan J and Suvanjumrat C 2018 *Key Engineering Materials* **777** 416-420
[8] Phromjan J and Suvanjumrat C 2018 *Key Engineering Materials* **775** 560-564
[9] Rugsaj R and Suvanjumrat C 2018 *Key Engineering Materials* **775** 554-559
[10] Rugsaj R and Suvanjumrat C 2018 *Key Engineering Materials* **777** 411-415
[11] Rugsaj R and Suvanjumrat C 2019 *International Journal of Automotive Technology* **20**(4) 801-812
[12] Rugsaj R and Suvanjumrat C 2020 *Mechanics Based Design of Structures and Machines* In press
[13] Kannan P Shaik A Kumar Y and Bareddy N S 2019 *SAE Technical Paper 2019-28-0059* (United State: SAE International)
[14] Wu T Li M Zhu X and Lu X 2020 *Mechanics of Advanced Materials and Structures* In press

IOP Conf. Series: Materials Science and Engineering 1137 (2021) 012021 doi:10.1088/1757-899X/1137/1/012021

Effect of feed sprue size on porosity defects in Platinum 950 centrifugal investment casting via numerical modelling

K. Munpakdee[1], P. Ninpetch[1], S. Otarawanna[2], R. Canyook[3], and P. Kowitwarangkul[1*]

[1]The Sirindhorn International Thai-German Graduate School of Engineering, King Mongkut's University of Technology North Bangkok, Bangkok 10800, Thailand
[2]National Metal and Materials Technology Center (MTEC), National Science and Technology Development Agency (NSTDA), Pathum Thani 12120, Thailand
[3]Faculty of Engineering, King Mongkut's University of Technology North Bangkok, Bangkok 10800, Thailand

* Corresponding Author: pruet.k@tggs.kmutnb.ac.th

Abstract
The centrifugal investment casting process is typically used in the jewelry manufacturing industry. However, porosity defects are often found in jewelry castings made by this process. Porosity on the casting surface is a major concern for jewelry manufacturers as the defect leads to significant yield losses during the repair work. Improper feed sprue design is one of the major causes of porosity defects. The purpose of this work is to study the effect of feed sprue size on porosity defects in Platinum 950 jewelry made by the centrifugal investment casting process. Two feed sprue sizes were experimented in this study. FLOW-3D CAST software was employed for the casting process simulation. Simulation results were analysed together with experimental results quantified by an image analysis technique. The results show that the larger feed sprue provides castings with less porosity compared to the smaller feed sprue.

Keywords: Investment casting, Porosity, Platinum 950, Solidification, Computer simulation

1. Introduction

In the jewelry industry, centrifugal investment casting of precious metals such as platinum and gold, is a casting process providing the excellent finish surface and dimensions. However, only some castings are completely free of defects as this casting process has many variables that is difficult to control [1], e.g., casting temperature, flask temperature, centrifugal speed and casting atmosphere [2, 3]. Unsuitable process conditions lead to defects, such as porosity, in jewelry pieces. Porosity on the casting surface is one of major concerns for jewelry manufacturers. When it appears on the finish surface, it needs to be fixed. This results in the loss of productivity time and yield loss of raw materials.

Shrinkage porosity forms during solidification in the solidified part of specimen that lack of feeding to compensate the reduction in volume of the liquid-to-solid phase transition during the dendritic growth [4, 5]. In casting, the solidification can be divided into 2 types: directional solidification and progressive solidification [6]. During directional solidification, the molten metal starts to solidify from a smaller

IOP Conf. Series: Materials Science and Engineering 1137 (2021) 012021 doi:10.1088/1757-899X/1137/1/012021

modulus to a larger modulus [7]. For the progressive solidification, the molten metal solidifies from the casting surface to the casting center. When the progressive solidification dominates over the directional solidification, shrinkage porosity is likely to form [8].

In addition, the use of high superheat can enhance the fluidity during mold filling. However, it also causes the shrinkage porosity due to the volume expansion and contraction of the liquid as the consequence of the density change at high temperature [6]. To reduce the shrinkage porosity defects, proper feed sprues and component design which provides directional solidification and sufficient superheat of pouring temperature were commonly applied [6, 9-11].

Computer simulation technique is an effective method to solve the shrinkage porosity problem [12]. Over the past decade, the numerical modelling for centrifugal investment casting process has been developed to investigate and predict the effect of parameters in casting process [2, 13]. For instance, Fischer-Bühner J [14] studied casting defects of 18K gold alloys by using numerical simulation. They concluded that the different thermal properties of 18K gold alloys give rise to different sensitivity to shrinkage porosity. Chan et al. [15] investigated the influence of feed sprue design on the porosity of dental parts in investment casting process. It was found that the single main sprue design is less effective than double-sprue design to reduce internal porosity.

The aims of this research are to investigate the effect of feed sprue designs on porosity defects in jewelry casting by using numerical modelling. The material used in the study is Pt950Ru (95%Platinum - 5% Ruthenium), which is commonly used by jewelry manufacturers for high-end products [16]. The numerical simulation was performed by the commercial software "FLOW-3D CAST". This software is based on the Finite Volume Method (FVM) and Volume of fluid (VOF), which are commonly used to track the free surface of fluid flow during casting process [17]. Furthermore, image analysis technique was used to analyzed the simulation results and quantified experimental results

2. Methodology

The work in this study is divided into two parts: the first part is the numerical simulation which carried out by using commercial simulation software package FLOW-3D CAST Ver. 5.0 to perform 3D model simulation. In the second part, the experimental study was carried out to study the effect of pouring temperature and feed sprue design on porosity at Christy Gem Co., Ltd.

2.1 Process parameters and porosity characterization

The study was conducted by simulation and experiments. The casting experiments were carried out by using Yasui, Vacuum Casting Machine 2014. The investment mold with flask is rotated around the vertical axis, Z and is simultaneously driven by centrifugal force [18]. The main process parameters used in this study include pouring temperature of 1,980°C, flask temperature of 850°C and ladle temperature of 600°C obtained from Christy Gems Ltd. For porosity characteristic method, the specimens were ground and polished to analyze the cross-section area by image analysis in the experiment.

2.2 Mathematical Models of Numerical Simulation

In the simulation, the governing equations of fluid dynamic consist of three equations: (1) mass conservation, (2) momentum conservation and (3) energy conservation respectively [19, 20]. In addition, turbulent modeling (k-ω) and volume of fluid (VOF) are used to solve turbulence flow of molten metal in the casting mold [17, 21].

$$\nabla\cdot\vec{V}=0 \tag{1}$$

$$\frac{\partial\vec{V}}{\partial t}+(\vec{V}\cdot\nabla)\vec{V}=-\frac{1}{\rho}\nabla p+\nu\nabla^2\vec{V}+\vec{f} \tag{2}$$

$$\rho c_p\left[\frac{\partial T}{\partial t}+(\vec{V}\cdot\nabla)T\right]=\lambda\nabla^2 T+\emptyset \tag{3}$$

IOP Conf. Series: Materials Science and Engineering 1137 (2021) 012021 doi:10.1088/1757-899X/1137/1/012021

where ρ and $\vec{V}$ are density and velocity vector, p is the pressure, ν is the kinematic viscosity, $\vec{f}$ is the body force per unit mass, c_p is the specific heat at constant pressure, T is temperature, λ is thermal conductivity and ∅ is the dissipation function.

2.3 CFD simulation setup and thermal-physical properties

The 3D CAD model in this study comprises of ladle, mold, and casting tree. Four different feed sprue designs of the casting tree are shown in Figure 1(a). This paper primarily focuses on the results of the feed sprue design 1 (smaller feed sprue) and design 2 (larger feed sprue). The computational mesh of 9.6 million elements, with the mesh sizes ranging from 0.7 and 0.35 mm was used for the simulation. The 3D model is divided into 4 multi-mesh blocks with different mesh resolution as shown in Figure 1(b).

Furthermore, the probes were located at five positions of the ring to measure the temperature change of metal as illustrated in Figure 1(c). The gradient between probes P1 and P3 are used for the analysis of the shrinkage porosity in this study. Additionally, the material and thermophysical properties of Pt950Ru including solidus temperature (1,780°C), liquidus temperature (1,795°C), latent heat of fusion (122,000 J/kg), specific heat, density, thermal conductivity, and viscosity [2, 22, 23] were applied to the simulation setup.

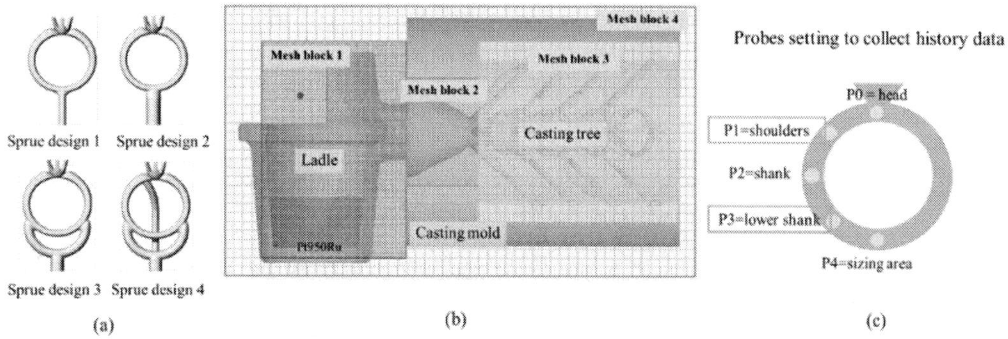

Figure 1. (a) The four different feeder sprue designs, (b) The assembling of 3D model casting tree with multi-mesh blocks with different mesh resolution, (c) schematic of probes setting in computer simulation.

3. Results and Discussion

The simulation results of casting at the pouring temperature of 1980°C from filling and solidification process were analyzed in this study. The molten metal filling process was done within 1 second and then the solidification started to simulate. The metal temperature was dropped from pouring temperatures to lower than liquidus temperature (1795°C) during the casting as shown in Figure 2. The color map showed the sequence of temperature variable in casting simulation process. The red and the blue color represented the initial pouring temperature and the solidus temperature at 1780°C. It could be seen that the larger feed sprue was filled sooner than small feed sprue because the diameter was bigger. The solidification results obtained from the simulation were further used to identify areas that could cause shrinkage porosity.

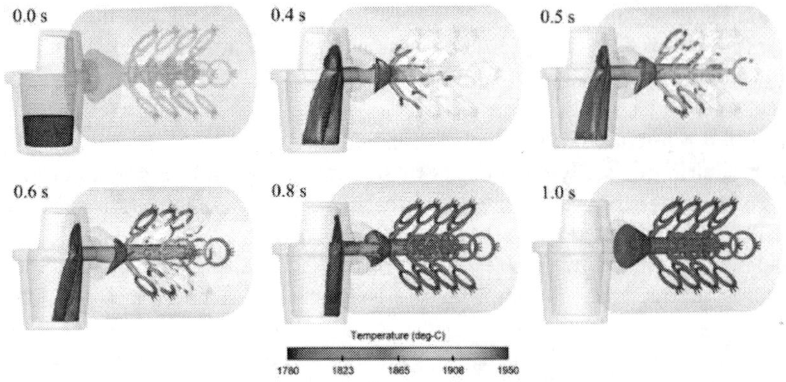

Figure 2. The temperature profile of filling and solidification process.

3.1 The effects of feed sprue designs on the directional solidification behaviour and thermal gradient

The shrinkage formation in solidification processes could be investigated and evaluated via the volume function through the color region. The 100% liquid metal and the solidified metal were indicated by red and grey colors. Figure 3 presented the fluid fraction of molten metal in the ring at the pouring temperature of 1980°C.

From the simulation results, it could be seen that when the casting time reaches at 1.8 and 1.9 seconds, the liquid metal in the ring with the smaller feed sprue started early in the stage of solidification. At 2.0 seconds, the liquid in the smaller feed sprue was completely solidified and blocks the feeding supply. Due to the lack of feeding supply for volume contraction during the growth of dendrite in solidification process, the smaller feed sprue with lower modulus volume could cause more shrinkage porosity formation. In contrast, the larger feed sprue with higher modulus volume that contained high temperature for a longer period of time could retard the heat transfer and cooling rate to allow more feeding supply for volume contraction. This process led to less formation of shrinkage porosity.

Moreover, the sequence of solidification of the ring with the larger feed sprue was more directional solidification which started from the head, shoulder, shank of ring compared to the ring with the smaller feed sprue. These results showed that the sprue designs had a significant effect on the directional solidification behavior and shrinkage porosity in jewelry casting.

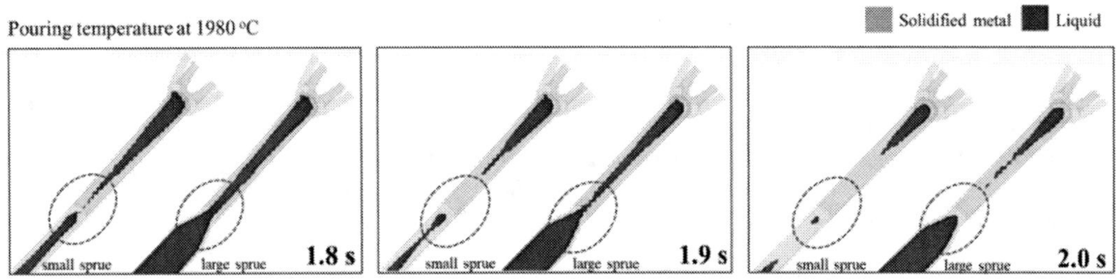

Figure 3. The fluid fraction of molten metal in the ring at pouring temperature at 1980°C

One of the quantitative indicators used in this study to indicate the directional solidification behavior was the solidification sequence that could be described by the thermal gradient between shoulder and shank of the ring. As described in the previous section, the probes P0-P4 were located on the ring to collect the change of temperature in casting process. Figure 4(a) and (b) illustrated the temperature change of probes P1 and P3 from the computer simulation with pouring temperature of 1980°C of the ring with smaller feed sprue and larger feed sprue. The dashed line represented the temperature change of P1 which was located in the shoulder area of the ring. The solid line represented the temperature change of P3 which was located in the lower shank area of the ring. In this study, the author observed the average of thermal gradient which was calculated from three temperature reference points of P3 including 1793°C, 1788°C and 1783°C. The higher average thermal gradient or more positive value of P1 minus P3 was, the more opportunity of shrinkage formation was.

The simulation results showed that the ring with the larger feed sprue had the average thermal gradient value of 1.9°C due to slow heat loss during the solidification in feed sprue compared to the other conditions. On the other hand, the ring with the smaller feed sprue had the average thermal gradient value of 2.8°C as illustrated in Figure 5(a). When the temperature of P1 was higher than P3, the progressive solidification at shank area tended to be more dominant compared to the directional solidification which led to lack of liquid supply and less directional solidification sequence from shoulder to shank of the ring. Figure 5(a). showed the thermal gradient of each ring which was calculated from variable temperature of P1 and P3. Moreover, Figure 5(b) showed the results of different of temperature change in the ring between P1 minus P3. The result of the smaller feed sprue presented higher difference of temperature between P1 minus P3 due to less modulus and faster temperature change in feed sprue.

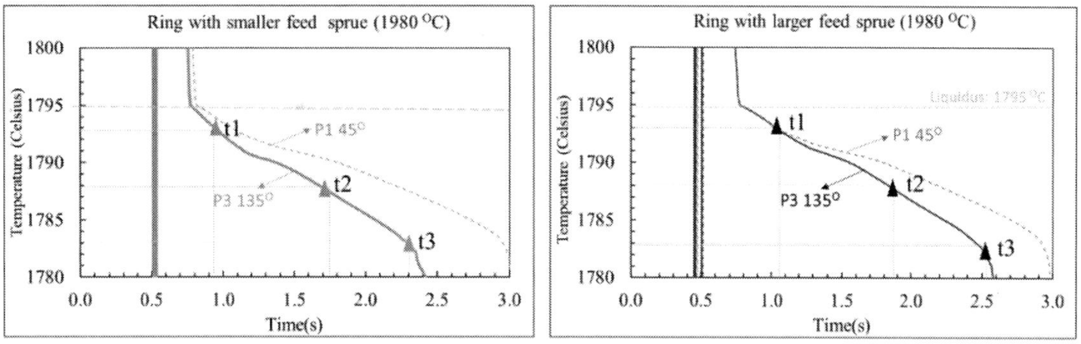

(a) (b)

Figure 4. Temperature change at probe P1 and P3 of the simulation at pouring temperature 1980 °C (a) smaller feed sprue, (b) larger feed sprue

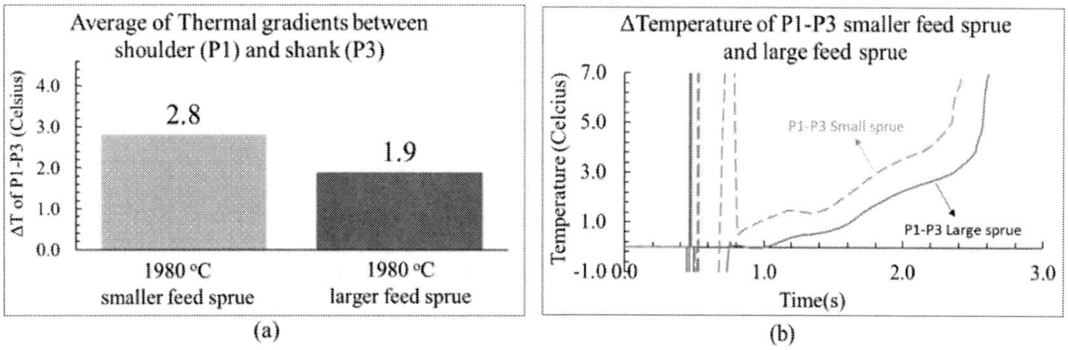

(a) (b)

Figure 5. (a) Average of thermal gradient of P1 minus P3 in the rings, (b) The temperature difference of P1 minus P3 in the rings.

3.2 Experiment results with image analysis

The cross section of the casting specimen from experiment were observed in this study by using the image analysis tools, optical microscope (OM) and "ImageJ" software. The total shrinkage porosity was indicated in percentage (%) compared to total area. Figure 6(a). shows the casting tree from the experiment with pouring temperature at of 1980°C. The porosity in cross section specimens can be found by the OM. To quantitatively analysis, ImageJ were used to count and calculate the total area of porosity on cross section area of specimen. As shown in Figure 6(b) the cross section of the ring with smaller sprue has porosity for approximately 0.84% compared to the total area and the cross section of the ring with larger sprue has porosity for approximately 0.0015%.

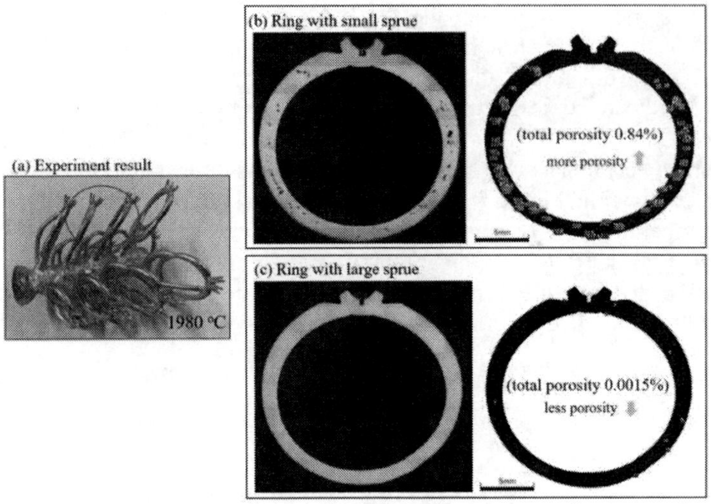

Figure 6. (a) The casting tree from experiment at pouring temperature of 1980°C, (b) Cross sectional ring with smaller sprue, (c) Cross sectional ring with larger sprue.

4. Conclusion

In this research, the numerical simulation and the experimental study were carried out to investigate the effect of feed sprue size on shrinkage porosity.

The simulation results show that the larger feed sprue provides a better result of the lowest average thermal gradient value of 1.9°C compared to the smaller feed sprue which has the highest average thermal gradient value of 2.8°C. The fluid fraction results reveal that feeding supply of molten metal in the larger feed sprue retards the temperature in the lower shank to prevent early solidification, while the molten metal solidified too early in the smaller feed sprue thus, it cannot compensate the volume contraction during the solidification.

In addition, the experimental results at the pouring temperature of 1980°C show that the total area of porosity on cross section of the ring with smaller feed sprue is 0.84%, while the porosity area of the ring with larger sprue is 0.0015%. The results indicate that the smaller feed sprue has the porosity more than the larger feed sprue.

Acknowledgments

The authors gratefully thank Christy Gem Co., Ltd. for corporation and support. This research was funded by the Thailand Science Research and Innovation (TSRI) and National Research Council of Thailand (NRCT) contract no. RDG62T0036.

References

[1] Dieter O 1998 *Handbook on Casting and Other Defects in Gold Jewelry Manufacture* (London: World Gold Council)

[2] Duncan M Tauriq K Penny P R Victoria H Ali B Irshad K and Candy L 2005 *Platinum Met.* Rev. **49**(4) 174-182

[3] Ulrich E K Tiziana H Dario T and Franz H 2015 *Jewelry Technology Forum* 2-10

[4] Silva F S 2010 *Proc. of 21st Santa Fe Symposium on Jewelry Manufacturing Technology* (New Mexico, USA)

[5] Pequet C Rappaz M and Gremaud M 2002 *Metal Mater Trans A.* **33**(7) 2095-2106

[6] Stefanescu D M 2008 *Science and Engineering of Casting Solidification, 2nd ed* (Berlin: Springer) 67
[7] Mark J 2005 *JOM.* **57**(5) 19-28
[8] Chastain S 2004 *Metal casting: a sand casting manual for the small foundry* (Jacksonville: Chastain Publishing)
[9] Dieter O 1997 *Gold Bulletin* **30**(1) 13-19
[10] Corti C and Holliday R 2010 Jewelry manufacturing technology Gold: Science and applications, (Boca Raton, FL: CRC Press)
[11] Eddie B 2002 *Gold Technol* **36** 3-11
[12] Mochamad A Poppy P Latief S and Yusuf A W 2020 *AIP Conf. Proc.* **2228** 030010-1- 030010-5
[13] Arblaster J W 2005 *Platinum Met. Rev.* **49**(3) 141-149
[14] Fischer-Bühner J 2007 *Proc. of 21st Santa Fe Symposium on Jewelry Manufacturing Technology* (New Mexico, USA)
[15] Daniel Ch Villa G Ronald B and Kwok-hung C 1997 *J. Prosthet. Dent.* **78**(4) 400-404
[16] Duncan M Tauriq K Penny P R Victoria H and Candy L 2005 *Platinum Met. Rev.* **49**(3) 110-117
[17] Tiziana H and Ulrich E K 2015 *Johnson Matthey Tech* **59**(2) 95-108
[18] Creeger G A 1994 Combination of Investment and Centrifugal Casting, *https://www.science.gov/topicpages/c/centrifugal+casting+technique.html*, accessed 1 May 2020
[19] Xuan-xuan Y Ling Ch and Yi-jie H 2012 *Energy Procedia* **17** 1864-1871
[20] Cüneyt S 2019 *Finite Element Analysis in Thermofluids* (Ankara: Middle East Technical University) 1-13
[21] Wilcox D C 2008 *AIAA Journal.* **46** 2823-2838
[22] Kent C and Christoph B 2008 *Proc. of 62nd SFSA Technical and Operating Conference* (Chicago, USA)
[23] Paul-François P 2014 *Johnson Matthey Tech* **58**(3) 124-136

Effect of Al_2O_3 nanoparticle additives on the performance characteristics of rough journal bearings

Waleephan Gunnuang[1], Chatchai Aiumbhornsin[2,*], Jittraporn Wongsa-Ngam[1], and Mongkol Mongkolwongrojn[3]

[1]Department of Mechanical Engineering, Faculty of Engineering,
King Mongkut's Institute of Technology Ladkrabang, Bangkok Thailand 10520
[2]Department of Maritime Engineering, Faculty of International Maritime Studies,
Kasetsart University, Chonburi Thailand 20230
[3]College of Advanced Manufacturing Innovation,
King Mongkut's Institute of Technology Ladkrabang, Bangkok Thailand 10520

* Corresponding Author: chataium@gmail.com

Abstract
This paper investigates the influence of Al_2O_3 nanoparticle additives on static and dynamic characteristics of water-lubricated rough journal bearings under thermo-hydrodynamic lubrication (THL). The time-dependent modified Reynolds equation and the approximated adiabatic energy equation have been formulated and solved numerically with boundary conditions and initial conditions. The results are presented for dimensionless maximum film pressure, dimensionless maximum film temperature, dimensionless minimum film thickness, eccentricity ratio, attitude angle, dimensionless friction force, spring constants, damping coefficients and mass parameters for both journal bearings with transverse roughness patterns and with longitudinal roughness patterns under a dimensionless load of 7. The results show that journal bearings with Al_2O_3 nanoparticle additives tend to significantly increase load carrying capacity and minimum film thickness by decreasing both the eccentricity ratio and temperature rise for longitudinal roughness patterns. For journal bearings with longitudinal roughness patterns and lubricated with Al_2O_3 nanoparticle additives, the stability region tends to increase.

Keywords: Journal bearing, Al_2O_3 nanoparticle additives, Longitudinal rough surface, Transverse rough surface

1. Introduction

In modern machines with minimized energy losses, ceramics are used as bearing materials for high-speed water-lubricated bearings. In ceramic bearings, the surface may be covered with the desired surface roughness patterns. Thermal effects in lubrication are important to the performance characteristics of bearings, see examples; Pinkus and Bupara [1], Smith and Tichy [2]. The performance characteristics of journal bearings have been studied by numerous researchers. Hashimoto and Mongkolwongrojn [3] studied the performance characteristics of turbulent partial journal bearings including surface roughness effects. They found that surface roughness significantly affected the bearing characteristics, especially for operating under turbulent film conditions. In 2004, Bouyer and Fillon [4] presented the incompressible thermoelastohydrodynamic film pressure has a significant effect on the performance characteristics of a plain journal bearing. The water film thickness can be very thin due to the thermohydrodynamic lubrication; therefore, the effect of surface roughness is very important. Majumdar and Ghosh [5], Tur Turaga, Sekhar [6] investigated the stability of a rough journal bearing using the perturbation technique. They found that stability can be improved significantly for rough surface journal bearings. Gururajan [7] studied the influence of surface roughness on bearing characteristics.

In recent years, many researches on nanoparticles show the enhancement of the heat transfer rate by increasing thermal conductivity and increasing thermal dispersion properties for cooling liquids (Nguyen, Desgranges [8], Xuan and Roetzel [9], Yu, Xie [10]). Recent research at the microscale is necessary to learn about the microstructure of nanofluids in order to understand the flow characteristics and heat transfer process of the nanofluid. Nanoparticle additives have attracted interest because of their enhanced thermal properties. Peng, Kang [11] showed that the use of diamond and SiO_2 nanoparticles as additives in liquid paraffin oil have better anti-wear and anti-friction properties than paraffin oil. The thermal conductivity and viscosity of nanofluids increase significantly with nanoparticle concentration (Murshed, Leong [12], Chandrasekar, Sures [13]). Shenoy, Binu [14] presented the effect of nanoparticle additives on hydrodynamic characteristics of an externally adjustable fluid film bearing. The bearing operated with API-SF engine oil blended with TiO_2 nanoparticles. The results showed that bearings with nanoparticle additives have better load-carrying capacity, reduced end leakage and increased friction when compared to API-SF engine oil and base oil without nanoparticle additives. Mongkolwongrojn and Rattapasakorn [15] studied a rough cylinder on a flat surface under TEHL with non-Newtonian lubricant blended with Al_2O_3 nanoparticle additives. The addition of Al_2O_3 nanoparticles significantly increased the film thickness both under normal loads and under sudden overloads due to the squeezing effect. The results have shown that nanoparticles significantly affected thermal elastohydrodynamic performance, especially under transient conditions. Mongkolwongrojn and Aiumpronsin [16] showed the effects of surface roughness in journal bearings under EHL; for longitudinal pattern, the increase in temperature was higher than both transverse and smooth patterns. As a result, the stability regions of the longitudinal pattern were smaller than a transverse pattern. Gunnuang, Aiumpornsin [17, 18] showed that Al_2O_3 nanoparticle additives can reduce the temperature.

The objective of this study is to investigate the performance characteristics of a journal bearing with minimize energy losses. In this analysis, the rough journal bearings with water blended with Al_2O_3 nanoparticle additives under THL have been formulated and solved using finite difference techniques.

2. Theory

2.1 Physical properties of nanofluids

The thermo-physical properties of nanofluids in this research assume that the nanoparticles are well dispersed in water and can be written as Equation (1) and equation (2) (Pak and Cho [19], Chandrasekar, Suresh [13])

$$\rho_{nf} = \varphi\rho_{par} + (1\text{-}\varphi)\rho_{water} \tag{1}$$

$$C_{P,nf} = \varphi C_{P,par} + (1\text{-}\varphi) C_{P,water} \tag{2}$$

Brinkman [20] suggested an equation for calculating the viscosity of the suspension, which is defined as follows:

$$\mu_{nf} = \frac{1}{(1\text{-}\varphi)^{2.5}}\mu_{water} \tag{3}$$

IOP Conf. Series: Materials Science and Engineering 1137 (2021) 012022 doi:10.1088/1757-899X/1137/1/012022

The lubricant density and viscosity depend on the temperature and can be written as follows:

$$\rho = \rho_{nf}\left(1-\beta(T-T_0)\right) \quad (4)$$

$$\mu = \mu_{nf}\exp\{-\gamma(T-T_0)\} \quad (5)$$

2.2 Linearization of bearing reaction

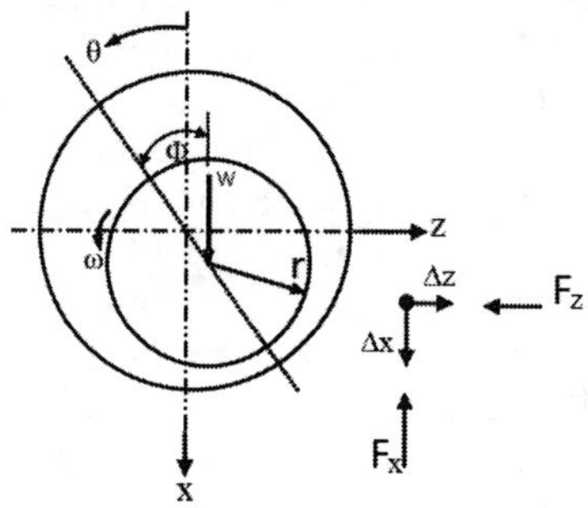

Figure 1. Geometry of the journal bearing

According to Lund [21], for the x-z coordinates in Figure 1, using the perturbation technique, five non-dimensional equations of bearing reaction can be obtained as follows:

$$\left[\frac{\partial}{\partial\theta}\left(\frac{\bar{\rho}\bar{h}_s^3}{12\bar{\mu}}\frac{\partial}{\partial\theta}\right)+\left(\frac{1}{\lambda^2}\right)\frac{\partial}{\partial y}\left(\frac{\bar{\rho}\bar{h}_s^3}{12\bar{\mu}}\frac{\partial}{\partial y}\right)\right]\begin{Bmatrix}\bar{p}_s\\ \bar{p}_x\\ \bar{p}_z\\ \bar{p}_{\dot{x}}\\ \bar{p}_{\dot{z}}\end{Bmatrix} = \begin{Bmatrix}\frac{1}{2}\frac{\partial}{\partial\theta}(\bar{\rho}\bar{h}_s)\\ \frac{1}{2}\frac{\partial}{\partial\theta}(\bar{\rho}\cos\theta)-\left[\frac{\partial}{\partial\theta}\left(\frac{\bar{\rho}\bar{h}_s^2}{4\bar{\mu}}\cos\theta\frac{\partial\bar{p}_s}{\partial\theta}\right)\right]-\left[\frac{1}{\lambda^2}\frac{\partial}{\partial y}\left(\frac{\bar{\rho}\bar{h}_s^2}{4\bar{\mu}}\cos\theta\frac{\partial\bar{p}_s}{\partial y}\right)\right]\\ \frac{1}{2}\frac{\partial}{\partial\theta}(\bar{\rho}\sin\theta)-\left[\frac{\partial}{\partial\theta}\left(\frac{\bar{\rho}\bar{h}_s^2}{4\bar{\mu}}\sin\theta\frac{\partial\bar{p}_s}{\partial\theta}\right)\right]-\left[\frac{1}{\lambda^2}\frac{\partial}{\partial y}\left(\frac{\bar{\rho}\bar{h}_s^2}{4\bar{\mu}}\sin\theta\frac{\partial\bar{p}_s}{\partial y}\right)\right]\\ \bar{\rho}\cos\theta\\ \bar{\rho}\sin\theta\end{Bmatrix} \quad (6)$$

where $\bar{p}_s = (\bar{p})_s, \bar{p}_x = \left(\frac{\partial\bar{p}}{\partial X}\right)_s, \bar{p}_z = \left(\frac{\partial\bar{p}}{\partial Z}\right)_s, \bar{p}_{\dot{x}} = \left(\frac{\partial\bar{p}}{\partial\dot{X}}\right)_s, \bar{p}_{\dot{z}} = \left(\frac{\partial\bar{p}}{\partial\dot{Z}}\right)_s, \bar{\rho} = \frac{\rho}{\rho_0}, \bar{\mu} = \frac{\mu}{\mu_0}$

The film thickness can be expressed as

$$\bar{h}_s = \left(1+\varepsilon\cos(\theta-\Phi)\right) - \delta\bar{h}_1 \quad (7)$$

where $\delta\bar{h}_1$ is the roughness height of the journal bearing and can be written as follows:

$$\delta\bar{h}_1 = \frac{A_h}{c}\sin(2\pi r\theta/\lambda_1) \quad \text{(for transverse roughness)} \quad (8)$$

$$\delta\bar{h}_1 = \frac{A_h}{c}\sin(2\pi r y/\lambda_1) \quad \text{(for longitudinal roughness)} \quad (9)$$

The boundary conditions for the film pressure at the start of cavitation and the bearing sides are

$$\bar{p}(\theta_c) = \left.\frac{\partial \bar{p}}{\partial \theta}\right|_{\theta_c} = 0,\ \bar{p}(\theta=0) = 0 \text{ and } \bar{p}(\bar{y}=1) = 0 \tag{10}$$

2.3 Energy equation

According to Majumdar [22], the dimensionless adiabatic energy equation can be written as shown in equation (11)

$$\left(6\bar{h} - \frac{\bar{h}^3}{\bar{\mu}}\frac{\partial \bar{p}}{\partial \theta}\right)\frac{\partial \bar{T}_m}{\partial \theta} - \left(\frac{1}{\lambda^2}\frac{\bar{h}^3}{\bar{\mu}}\frac{\partial \bar{p}}{\partial \bar{y}}\right)\frac{\partial \bar{T}_m}{\partial \bar{y}} = \frac{12\bar{\mu}}{\bar{h}} + \frac{\bar{h}^3}{\bar{\mu}}\left(\frac{\partial \bar{p}}{\partial \theta}\right)^2 + \frac{1}{\lambda^2}\frac{\bar{h}^3}{\bar{\mu}}\left(\frac{\partial \bar{p}}{\partial \bar{y}}\right)^2 \tag{11}$$

The boundary conditions for the film temperature at the film inlet and the gradient of the film temperature at the half of length are as follows:

$$\bar{T}(\theta=0)=0,\ \left.\frac{\partial \bar{T}}{\partial \bar{y}}\right|_{\bar{y}=\frac{1}{2}} = 0 \tag{12}$$

2.4 Load-carrying capacity and Friction force

The forces due to the hydrodynamic pressure on the journal in the x-z coordinate system are calculated as

$$\left(\bar{F}_x\right)_s = -\int_0^1\int_0^{2\pi} \bar{p}_s \cos\theta d\theta d\bar{y},\ \left(\bar{F}_z\right)_s = -\int_0^1\int_0^{2\pi} \bar{p}_s \sin\theta d\theta d\bar{y} \tag{13}$$

The friction force in the fluid film is obtained by integrating the shear stress on the journal surface. The non-dimensional friction force $\bar{F}_{fric}$ is given as follows:

$$\bar{F}_{fric} = \int_0^1\int_0^{2\pi} \frac{\bar{\mu}^*}{\bar{h}} d\theta d\bar{y} + \int_0^1\int_0^{2\pi} \frac{\bar{h}}{2}\frac{\partial \bar{p}}{\partial \theta} d\theta d\bar{y} \tag{14}$$

2.5 Journal bearing stability

The linearized equation of journal motion can be written as

$$\begin{bmatrix} M & 0 \\ 0 & M \end{bmatrix}\begin{bmatrix} \Delta\ddot{X} \\ \Delta\ddot{Z} \end{bmatrix} + \begin{bmatrix} B_{xx} & B_{xz} \\ B_{zx} & B_{zz} \end{bmatrix}\begin{bmatrix} \Delta\dot{X} \\ \Delta\dot{Z} \end{bmatrix} + \begin{bmatrix} K_{xx} & K_{xz} \\ K_{zx} & K_{zz} \end{bmatrix}\begin{bmatrix} \Delta X \\ \Delta Z \end{bmatrix} = 0 \tag{15}$$

where the non-dimensional spring and damping coefficients can be obtained as follows:

$$\begin{Bmatrix} K_{xx} & K_{xz} \\ K_{zx} & K_{zz} \end{Bmatrix} = \begin{Bmatrix} -\int_0^1\int_0^{2\pi} \bar{p}_x \cos\theta d\theta d\bar{y} & -\int_0^1\int_0^{2\pi} \bar{p}_z \cos\theta d\theta d\bar{y} \\ -\int_0^1\int_0^{2\pi} \bar{p}_x \sin\theta d\theta d\bar{y} & -\int_0^1\int_0^{2\pi} \bar{p}_z \sin\theta d\theta d\bar{y} \end{Bmatrix} \tag{16}$$

$$\begin{Bmatrix} B_{xx} & B_{xz} \\ B_{zx} & B_{zz} \end{Bmatrix} = \begin{Bmatrix} -\int_0^1\int_0^{2\pi} \bar{p}_{\dot{x}} \cos\theta d\theta d\bar{y} & -\int_0^1\int_0^{2\pi} \bar{p}_{\dot{z}} \cos\theta d\theta d\bar{y} \\ -\int_0^1\int_0^{2\pi} \bar{p}_{\dot{x}} \sin\theta d\theta d\bar{y} & -\int_0^1\int_0^{2\pi} \bar{p}_{\dot{z}} \sin\theta d\theta d\bar{y} \end{Bmatrix} \tag{17}$$

IOP Conf. Series: Materials Science and Engineering 1137 (2021) 012022 doi:10.1088/1757-899X/1137/1/012022

$$\text{where } \bar{p}_x = \left(\frac{\partial p}{\partial x}\right), \bar{p}_{\dot{x}} = \left(\frac{\partial p}{\partial \dot{x}}\right), \bar{p}_z = \left(\frac{\partial p}{\partial z}\right), \bar{p}_{\dot{z}} = \left(\frac{\partial p}{\partial \dot{z}}\right)$$

The solutions for dimensionless critical mass parameter $M_{CR}\bar{\Omega}^2$

$$M_{CR}\bar{\Omega}^2 = \frac{K_{xx}B_{zz}+K_{zz}B_{xx}-K_{xz}B_{zx}-K_{zx}B_{xz}}{B_{xx}+B_{zz}} \tag{18}$$

3. Numerical Solution

In the present study, the characteristics of full journal bearing with roughness surface were examined with the bearing parameters and lubricant properties as shown in Table 1. The diameter to width ratio of the bearing is 1, and the clearance to radius ratio is 0.002. The static pressure; in Equation (6) and temperature in Equation (11) are solved simultaneously. The second order central difference schemed described by Majumdar [22], were employed for the discretization of pressure equation and energy equation. The applied load is assumed to be constant in magnitude and implemented in the vertical direction. The pressure, temperature, eccentricity ratio, film thickness, viscosity, density and hydrodynamic force are also simultaneously calculated within the thermohydrodynamic regime. The relative accuracy in pressure, temperature and force components were less than or equal to 1×10^{-6}. Based on the analysis, a computer program was developed to solve the themohydrodynamic problems of finite journal bearing using 270 by 90 nodes.

Table 1. Physical properties of the journal bearing and lubricant properties

Parameter	Detail
Journal radius, r	0.015 m
Radius clearance, c	30 μm
Reference viscosity, μ_0	0.693 m Pa s
Viscosity-Temperature coefficient, γ	0.0153 K^{-1}
Bearing length, L	0.03 m
Inlet lubricant temperature, T_0	313 K
coefficient of thermal expansivity, β	0.000385 K^{-1}
Wavelength of roughness, λ_1	5 mm
Density of Al_2O_3, ρ_{par}	3890 kg m^{-3}
Specific heat of Al_2O_3, $C_{P,par}$	729 J $kg^{-1}K^{-1}$

4. Results

Figures 2-9 show the effects of the Al_2O_3 nanoparticle concentration on the steady-state characteristics of water-lubricated journal bearings under the dimensionless load W=7 for smooth bearings, transverse roughness pattern bearings and longitudinal roughness pattern bearings.

Figure 2 shows the effects of the variations in Al_2O_3 nanoparticle concentrations on the dimensionless maximum film pressure under dimensionless load W equal to 7. For the transverse roughness pattern, the dimensionless maximum film pressure of journal bearings lubricated with clean water and lubricated with water mixed with 2% Al_2O_3 nanoparticles are 13.56 and 13.16 respectively. The decrease in the maximum film pressure is 2.95%. For journal bearings with 2% Al_2O_3 nanoparticles, the dimensionless maximum film pressures of the longitudinal roughness pattern and smooth surface decrease 1.78% and 1.38% respectively, when compared with journal bearings lubricated without Al_2O_3 nanoparticles. The dimensionless maximum film pressure decreases slightly with an increase in Al_2O_3 nanoparticles added.

Figure 3 shows the effect of variations in the Al_2O_3 nanoparticle concentration on dimensionless maximum film temperature. The results show that addition of the Al_2O_3 nanoparticle additive decreases the dimensionless maximum film temperature, especially for the longitudinal roughness pattern. For smooth surface journal bearings, the dimensionless maximum film temperature of journal bearings lubricated with clean water and lubricated with water mixed with 2% Al_2O_3 nanoparticles are 66.92 and 63.08, which is a decrease of 5.74%. For transverse patterns, the addition of 0% to 2% Al_2O_3 nanoparticle additives can reduce dimensionless maximum film temperature from 82.33 to 76.18, a decrease of 7.47%. In the case of longitudinal patterns, the dimensionless maximum film temperature of journal bearings lubricated with clean water and lubricated with water mixed with 2% Al_2O_3 nanoparticles are 240.20 and 205.78, respectively, a decrease of 14.33%. The Al_2O_3 nanoparticle additive has a great effect on dimensionless maximum film temperature for journal bearings with longitudinal roughness patterns.

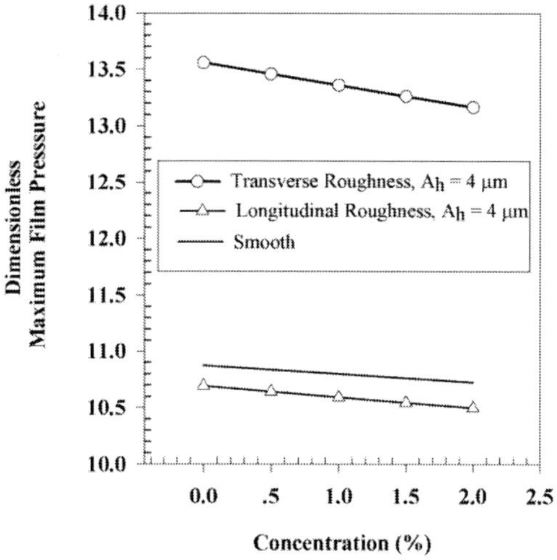

Figure 2. Variation of dimensionless maximum film pressure with Al_2O_3 nanoparticles concentration when journal bearing operated under dimensionless load W equal to 7

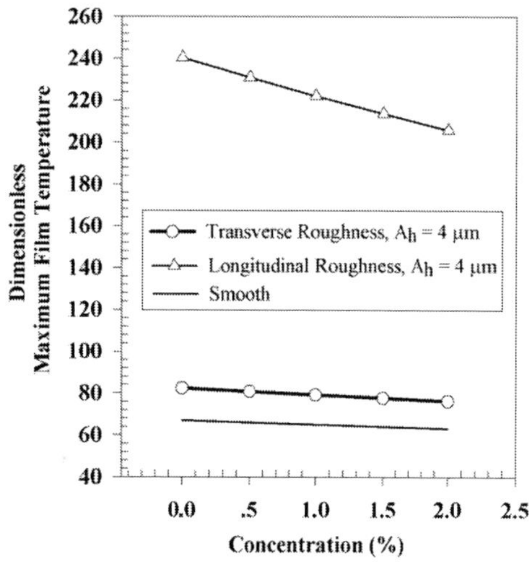

Figure 3. Variation of dimensionless maximum temperature with Al_2O_3 nano-particles concentration when journal bearing operated under dimensionless load W equal to 7

Figure 4 shows the variation of dimensionless maximum film temperature with surface roughness for longitudinal and transverse roughness patterns when the journal bearing operated under dimensionless load W equal to 7. The dimensionless maximum film temperature is nonlinearly increased with an increase of roughness amplitude. For journal bearings with 2% Al_2O_3 nanoparticle additive, the decrease of the dimensionless maximum film temperature for longitudinal roughness patterns with roughness amplitudes 5μm, 4μm, and 3μm are 23.3%, 14.33%, and 10.13%, respectively, when compared to journal bearings lubricated without nanoparticles added. The decreases in the dimensionless maximum film temperature for the transverse roughness pattern with roughness amplitudes 5 μm, 4 μm, and 3 μm are 7.66%, 7.47%, and 6.69%, respectively, when compared to journal bearings lubricated with clean water. The roughness surface has a significant effect on the increase of the film temperature, especially for the longitudinal roughness pattern. From the results in Figures 2-4, It is clear that for roughness amplitudes higher than 3 μm or larger than 10% of the clearance; $A_h/c > 0.1$, the addition of Al_2O_3 nanoparticles has a significant effect on the reduction of the temperature rise for longitudinal roughness patterns; the reduction of the temperature rise for transverse roughness patterns is small when compared with that for longitudinal roughness patterns. These results are useful in understanding the influence of Al_2O_3 nanoparticles on the longitudinal roughness patterns of journal bearings.

The effects of the concentration of Al_2O_3 nanoparticles on eccentricity ratio and attitude angle of the journal bearings are shown in Figure 5 and Figure 6. With the Al_2O_3 nanoparticle additive concentration ranging from 0% to 2%, the eccentricity ratio of journal bearings decreases from 0.793 to 0.786, a decrease of 0.88% for a smooth surface; the eccentricity ratio decreases from 0.747 to 0.739, a decrease of 1.07% for the transverse roughness pattern; and the eccentricity ratio decreases from 0.792 and 0.783, a decrease of 1.14% for the longitudinal roughness pattern, as shown in Figure 5. The attitude angle of the journal bearing increases from 36.69 to 37.36, an increase of 1.83% for a smooth surface; the attitude angle increases from 32.47 to 33.02, an increase of 1.69% for the transverse roughness pattern; and the attitude angle increases from 37.14 to 37.8, an increase of 1.78% for the longitudinal roughness pattern as shown in Figure 6.

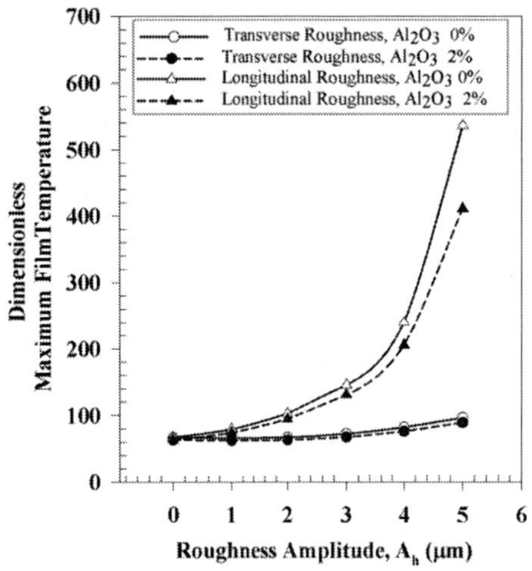

Figure 4. Variation of dimensionless maximum film temperature with roughness amplitude when the journal bearing operated under dimensionless load W equal to 7

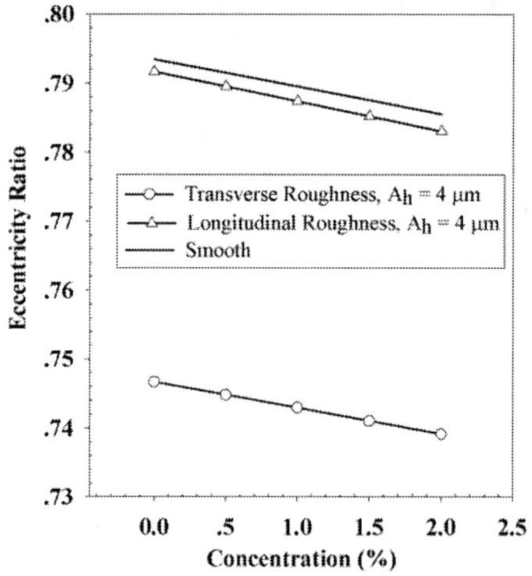

Figure 5. Variation of eccentricity ratio with Al_2O_3 nanoparticles concentration when journal bearing operated under dimensionless load W equal to 7

Figure 7 shows the variation of dimensionless minimum film thickness with roughness amplitude for longitudinal and transverse roughness patterns when the journal bearing is operated under dimensionless load W equal to 7. For a longitudinal roughness pattern at high roughness amplitude, the decrease of the dimensionless minimum film thickness is higher than for the transverse pattern; however, the addition of Al_2O_3 nanoparticles significantly affects the increase of the dimensionless minimum film thickness. For journal bearings with 2% Al_2O_3 nanoparticles, the increase of the dimensionless minimum film thickness for transverse roughness patterns with roughness amplitudes 5 μm, 4 μm and 3 μm are 6.67%, 6.35% and 6.12%, respectively, when compared to journal bearings lubricated without nanoparticles added. For journal bearings with 2% Al_2O_3 nanoparticles added, the increases of the dimensionless minimum film thickness are 21.4%, 10.53% and 8.33% for longitudinal roughness patterns with roughness amplitudes 5 μm, 4 μm and 3μm, respectively, when compared with bearings lubricated without nanoparticles added. The addition of Al_2O_3 nanoparticles has a significant effect on the increase of dimensionless minimum film thickness, as shown in Figure 8. The addition of Al_2O_3 nanoparticles increases the dimensionless friction force due to the increase of viscosity. Figure 9 shows that the

IOP Conf. Series: Materials Science and Engineering 1137 (2021) 012022 doi:10.1088/1757-899X/1137/1/012022

addition of Al_2O_3 nanoparticles from 0% to 2%, the dimensionless friction force of a longitudinal bearing with 5 μm in amplitude increases from 13.82 to 13.99, respectively, this is an increase of 1.23%. For transverse roughness patterns 5 μm in amplitude, the dimensionless friction force increases from 13.25 to 13.61, an increase of 2.72%.

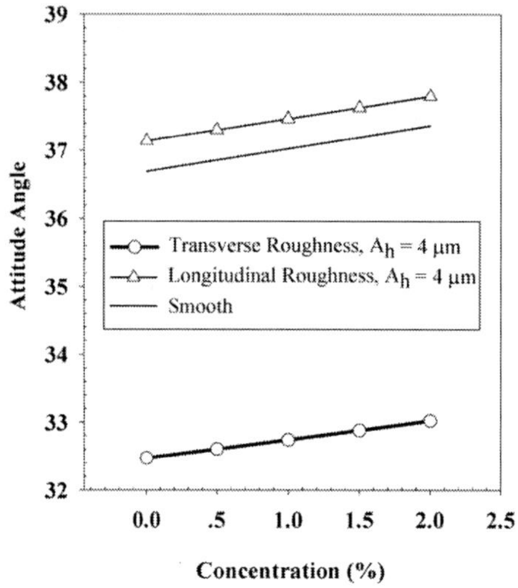

Figure 6. Variation of attitude angle with Al_2O_3 nanoparticles concentration when journal bearing operated under dimensionless load W equal to 7

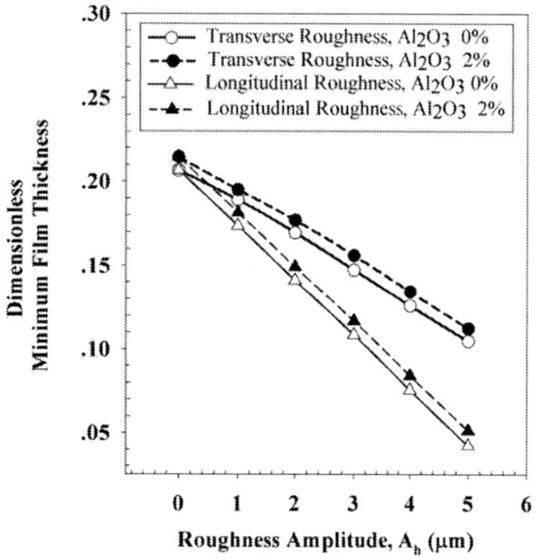

Figure 7. Variation of dimensionless minimum film thickness with roughness amplitude when journal bearing operated under dimensionless load W equal to 7

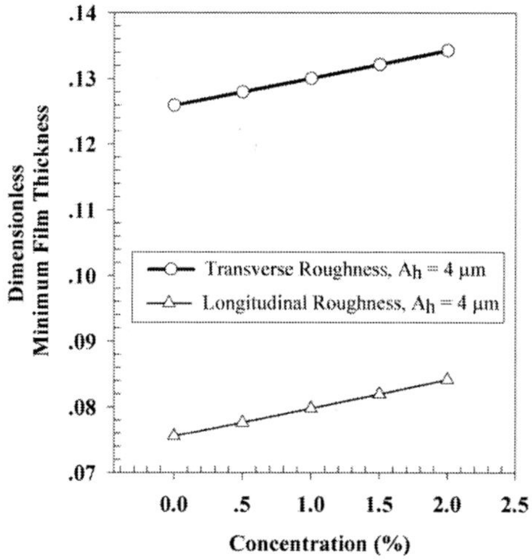

Figure 8. Variation of dimensionless minimum film thickness with Al_2O_3 nanoparticles concentration when journal bearing operated under dimensionless load W equal to 7

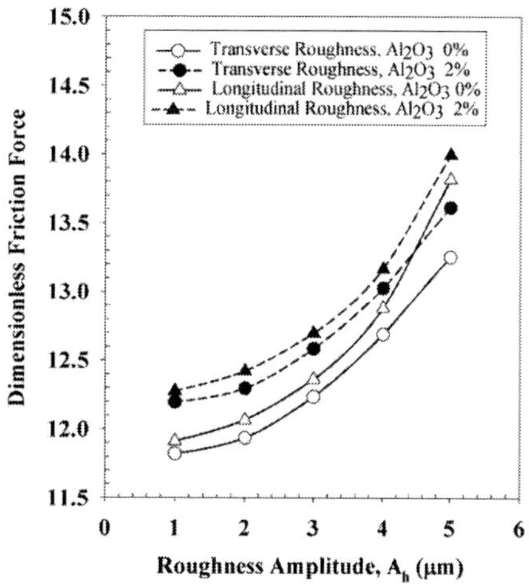

Figure 9. Variation of dimensionless friction force with roughness amplitude when journal operated under dimensionless load W equal to 7

The addition of Al_2O_3 nanoparticles decreases film temperature and increases minimum film thickness by reducing the eccentricity ratio or increasing the attitude angle. The Al_2O_3 nanoparticle additives have a significant effect on temperature when the roughness amplitude to clearance ratio, A_h/c, is higher than 0.1. The effects of Al_2O_3 nanoparticle additives on the dynamic characteristics of journal bearings under the dimensionless load W equal to 7 are shown in Figures 10-12.

Figure 10 shows the variation of dimensionless spring coefficients with the concentration of Al_2O_3 nanoparticles for different roughness patterns. The spring constants K_{xx} and K_{zx} for longitudinally rough surfaces increase significantly with Al_2O_3 concentration, but all the other dimensionless spring coefficients, K_{xx}, K_{xz}, K_{zx} and K_{zz}, for both transversely rough surfaces and smooth surfaces decrease with the concentration of Al_2O_3 nanoparticles. To compare bearings lubricated with water blended with 2% Al_2O_3 nanoparticles and lubricated with clean water for the longitudinally rough surface, the spring constant, K_{xx} increases from 0.65 to 0.93, an increase of 43%, and K_{zx} increases from -2.94 to -2.88, an increase of 2.04%. For longitudinally rough surfaces, spring constant K_{xz} decreases slightly from 4.33 to 4.31, a decrease of 0.46%. For both transversely rough surfaces and smooth surfaces, the spring constant K_{xz} decreases significantly by 1.45% and 2.09%, respectively, when the use of 2% Al_2O_3 nanoparticles is compared with the use of clean water. This means that the Al_2O_3 nanoparticle additive has great effect on increasing the spring constants K_{xx} and K_{zx} for journal bearings with longitudinal roughness patterns.

Figure 11 shows the variation of dimensionless damping coefficients with Al_2O_3 nanoparticle concentrations for different roughness patterns. The results clearly show that dimensionless damping coefficients B_{xx}, B_{xz} and B_{zx} decrease with the increase in the Al_2O_3 nanoparticle concentration, but B_{zz} increases with the increase in Al_2O_3 nanoparticle concentration for both transverse and longitudinal roughness patterns. For the transverse roughness patterns, damping coefficients are small when compared with the damping coefficients for smooth surfaces, but the damping coefficients for longitudinally rough surfaces are close to the damping coefficients for smooth surfaces. At 2% Al_2O_3 concentration, the damping coefficients B_{xx}, B_{xz}, B_{zx} and B_{zz} for a transversely rough surface are 14%, 29.3%, 29.3%, and 26% smaller, respectively, than those for smooth surfaces.

Figure 12 shows the variation of the critical mass parameter with Al_2O_3 nanoparticle concentration and roughness amplitude when journal bearings are operated under dimensionless load W equal to 7; the critical mass parameter indicates the stability region of the journal bearing. The stability region of the bearing with transverse roughness pattern increases with an increase in roughness amplitude, but the stability region of the longitudinal roughness bearing decreases with an increase in roughness amplitude, especially when the value of roughness amplitude to clearance ratio; A_h/c; is more than 0.1.

For the longitudinally rough surface, the mass parameter of water-blended Al_2O_3 is larger than that for clean water by 1.31%, 4.20% and 13.79% for roughness amplitudes of 3 μm, 4 μm and 5 μm, respectively. For the transversely rough surface, the mass parameter of water-mixed Al_2O_3 is smaller than that for clean water. The values of mass parameters decrease 0.62%, 0.69% and 0.78% for roughness amplitudes of 3 μm, 4 μm and 5 μm, respectively. The addition of Al_2O_3 nanoparticles tends to slightly decrease the stability region for transverse roughness patterns, but it is significant in increasing the stability region for longitudinal roughness patterns, especially when the roughness amplitude is higher than 0.3 μm. Comparing the mass parameter of water-blended 2% Al_2O_3 with the mass parameter of clean water, the transverse roughness pattern decreases 0.69% (from 2.464 to 2.447), the smooth surface decreases 0.52% (from 2.319 to 2.307) and the longitudinal roughness pattern significantly increases 4.20% (from 1.737 to 1.81). The results show that the concentration of Al_2O_3 nanoparticles tends to slightly decrease the stability region of both smooth surfaces and transverse roughness patterns, but for bearings with longitudinal roughness patterns, the stability region increases significantly with an increase in Al_2O_3 particle concentration.

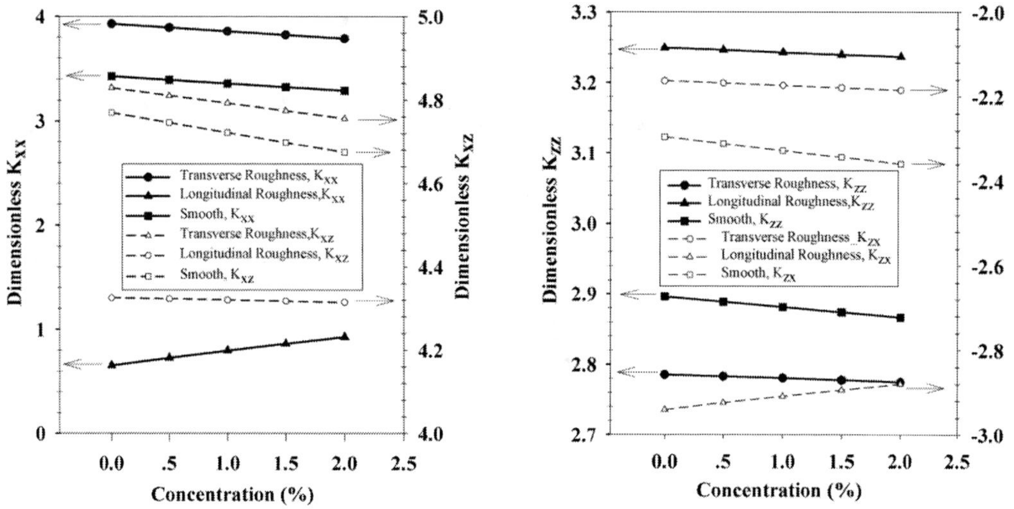

(a) (b)

Figure 10. Variation of dimensionless spring coefficients with Al_2O_3 nanoparticles concentration when journal bearing operated under dimensionless load W equal to 7 for (a) K_{xx}, K_{xz} and (b) K_{zz}, K_{zx}

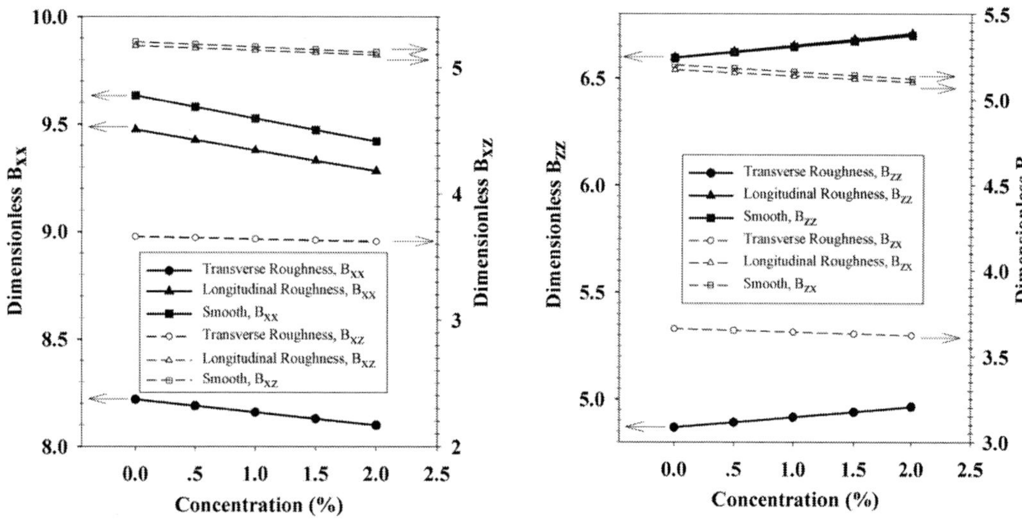

(a) (b)

Figure 11. Variation of dimensionless damping coefficients with Al_2O_3 nanoparticles concentration when journal bearing operated under dimensionless load W equal to 7 for (a) B_{xx}, B_{xz} and (b) B_{zz}, B_{zx}

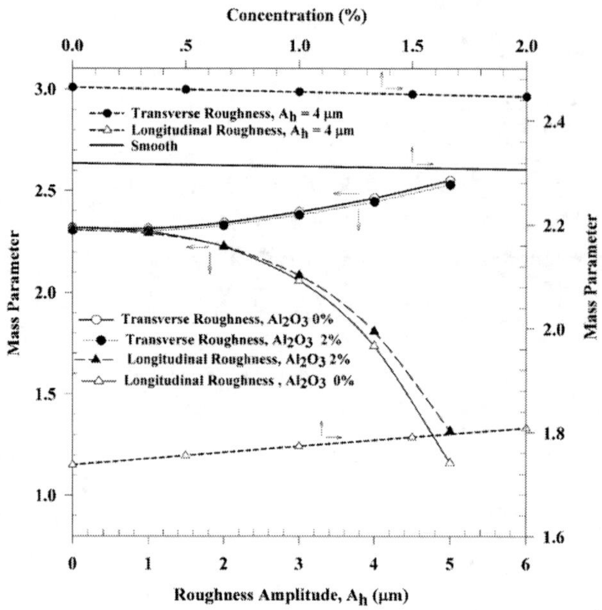

Figure 12. Variation of the mass parameter with Al_2O_3 nanoparticles concentration and roughness amplitude when journal bearing operated under dimensionless load W equal to 7

5. Conclusions

The static and dynamic characteristics of rough journal bearings with Al_2O_3 nanoparticle additives under thermohydrodynamic lubrication were theoretically examined. The time-dependent modified Reynolds equation and the approximated adiabatic energy equation were formulated and solved numerically. This research can be summarized as follows.

1) For journal bearings with smooth and rough surfaces with both transverse roughness patterns and longitudinal roughness patterns under static characteristics, the increase of Al_2O_3 nanoparticles significantly effect on the reduction of both the film temperature and the eccentricity ratio. Therefore, the load-carrying capacity and minimum film thickness are increased for journal bearings with Al_2O_3 nanoparticles.

2) For dynamics characteristics, journal bearings with longitudinally roughness and lubricated with Al_2O_3 nanoparticles tend significantly to increase the stability regions.

Nomenclature

A_h	Roughness amplitude, m
b	Damping coefficient, N s/m
$B_{xx} \dots B_{zz}$	Dimensionless damping coefficient, $B = c\,\omega b/w$
c	Radius clearance, m
$C_{P,nf}$	Specific heat capacity of nanofluids, J /kg K
$C_{P,par}$	Specific heat capacity of solid particles, J /kg K
$C_{P,water}$	Specific heat capacity of water, J /kg K
F_{fric}	Friction Force, N
$\bar{F}_{fric}$	Dimensionless friction force, $\bar{F}_{fric} = \left(c^2/\mu_0\,\omega r^3 L\right)F_{fric}$
h	Film thickness, m

$\bar{h}$	Dimensionless film thickness, $\bar{h}=h/c$
k	Stiffness coefficient, N/m
$K_{xx} \ldots K_{zz}$	Dimensionless stiffness coefficient, $K=ck/w$
L	Bearing length, m
M	Dimensionless mass, $M=\left(c\omega^2/w\right)m$
p	Fluid film pressure, Pa
$\bar{p}$	Dimensionless film pressure, $\bar{p}=\left(c^2/\mu_0\omega r^2\right)(p-p_{atm})$
r	Journal radius, m
T	Mean temperature of lubricant across the film, K
T_0	Inlet lubricant temperature, K
$\bar{T}$	Dimensionless film temperature, $\bar{T}=\left(\rho C_p c^2/\mu_0\omega r^2\right)(T-T_0)$
w	Load, N
W	Dimensionless load, $W=\left(c^2/\mu_0\omega r^3 L\right)w$
x	Coordinate in a vertical direction
y	Coordinate in an axial direction
z	Coordinate in a horizontal direction
Ω	Angular velocity of whirl, rad/s
$\bar{\Omega}$	Whirl ratio, $\bar{\Omega}=\Omega/\omega$
ω	Angular velocity, rad/s
ε	Eccentricity ratio of journal bearing, $\varepsilon=e/c$
θ	The circumferential angle measured from the vertical axis, rad
Φ	Attitude angle, rad
φ	Particle volume concentration
λ_1	The wave length of roughness, m
β	Coefficient of thermal expansivity, 1/K
γ	Viscosity-Temperature coefficient, 1/K
ρ	The apparent density of lubricant at the reference temperature, kg/m^3
ρ_{nf}	The density of nanofluids, kg/m^3
ρ_{par}	The density of solid particles, kg/m^3
ρ_{water}	The density of water, kg/m^3
μ	The apparent viscosity of the lubricant, Pa.s
μ_{nf}	The viscosity of nanofluids, Pa.s
μ_{water}	The viscosity of water, kg/m^3
ΔX	Dimensionless displacement in the x-direction, $\Delta X=\Delta x/c$
ΔZ	Dimensionless displacement in the z-direction, $\Delta Z=\Delta z/c$

References

[1] Pinkus O and Bupara S S 1979 *J. Lubr. Technol.* **101**(4) 492-496
[2] Smith R N and Tichy J A 1981 *J. Lubr. Technol.* **103**(3) 443-452
[3] Hashimoto H and Mongkolwongrojn M 1994 *ASME J. Tribol.* **116**(4) 672-680
[4] Bouyer J and Fillon M 2004 *ASME J. Tribol.* **126**(4) 819-822
[5] Majumdar B C and Ghosh M K 1990 *J. Tribol.* **112**(1) 73-77
[6] Turaga R Sekhar A S and Majumdar B C 1999 *Tribol. Int.* **32**(5) 231-236

[7] Gururajan K and Prakash J 1999 *J. Tribol.* **121**(1) 139-147
[8] Nguyen C T Desgranges F Galanis N Roy G Maré T, Boucher S and Angue M H 2008 *Int. J. Therm. Sci.* **47**(2) 103-111
[9] Xuan Y and Roetzel W 2000 *Int. J. Heat Mass Transfer.* **43**(19) 3701-3707
[10] Yu W Xie H Li Y Chen L and Wang Q 2011 *Colloids Surf.* A. **380**(1) 1-5
[11] Peng D X Kang Y Hwang R M Shyr S S and Chang Y P 2009 *Tribol. Int.* **42**(6) 911-917
[12] Murshed S M S Leong K C and Yang C 2008 *Int. J. Therm. Sci.* **47**(5) 560-568
[13] Chandrasekar M Suresh S and Chandra Bose A 2010 *Exp. Therm Fluid Sci.* **34**(2) 210-216
[14] Shenoy B Binu K Pai R Rao D and Pai R S 2012 *Tribol. Int.* **45**(1) 38-42
[15] Mongkolwongrojn M and Rattapasakorn S 2014 *J. Adv. Mech. Des. Syst.* **8**(3) *JAMDSM0042-JAMDSM0042*
[16] Mongkolwongrojn M and Aiumpronsin C 2010 *Tribol. Intl.* **43**(5) 1027-1034
[17] Gunnuang W Aiumpornsin C and Mongkolwongrojn M 2013 *Proc .Int .Conf. on Mechanical Engineering (Pattaya, Chonburi)* p 138
[18] Gunnuang W Aiumpornsin C and Mongkolwongrojn M 2015 *Appl. Mech. Mater.* **751** 137-142
[19] Pak B C and Cho Y I 1998 *Exp. Heat Transfer* **11**(2) 151-170
[20] Brinkman H 1952 *J. Chem Phys.* **20**(4) 571-571
[21] Lund J W 1987 *J. Tribol.* **109**(1) 37-41
[22] Majumdar B C 1975 *Wear.* **31**(2) 287-294

Comparison of two different surface for rolling airless tire by finite element method

Chakrit Suvanjumrat[1,2], Ravivat Rugsaj[1,2], Juthanee Phromjan[1,2,*]

[1]Department of Mechanical Engineering, Faculty of Engineering, Mahidol University, Nakhon Pathom Thailand 73170
[2]Laboratory of Computer Mechanics for Design (LCMD), Department of Mechanical Engineering, Faculty of Engineering, Mahidol University, Nakhon Pathom Thailand 73170

* Corresponding Author: juthaneephromjan@gmail.com

Abstract
Drum testing was carried out to determine rolling tire performance. The finite element method was used to model the drum testing instead. In order to yield accurate results, the generalized Maxwell's viscoelastic material model with hyperelastic material model was employed to simulate the drum testing on the airless tire. However, it consumed a lot of simulation time because of the contact surface between tire and drum was difficult to compute by using a finite element method. The finite element model of rolling an airless tire on a flat surface was developed. The contact patch and spoke deformation were carried out to investigate the surface effects. The spoke deformation of rolling an airless tire on a flat surface was found to be in accord with the curvature surface of the drum. The flat surface showed little difference in the contact area against the drum. Particularly, it was found that the analysis time of rolling an airless tire on a flat surface was less than the drum 10.13 times.

Keywords: Airless tire, Finite element method, Rolling tire, Tire testing

1. Introduction

Nowadays, airless tires or non-pneumatic tires (NPT) have been developed to overcome the disadvantages of pneumatic tires such as maintaining air pressure for inflation being and easy to damage. An airless tire can be developed to achieve various performance requirements by designing the spoke structure. Rugsaj and Suvanjumrat [1] have studied the effect of NPT spoke structures by using a finite element method (FEM). The spoke structures were grouped into 4 main types including (1) curved spokes, (2) hexagonal honeycomb spokes, (3) auxetic honeycomb spokes and (4) interconnected mesh spokes. The finite element analysis (FEA) showed the interconnected mesh spoke had the highest vertical stiffness. Furthermore, the optimum number of interconnected mesh spokes was 24. The cell expanding of honeycomb spokes was studied [2]. It was found that the smallest cell expanding angle provided the lowest rolling resistance. However, the mechanical behavior of tire has to be studied in many ways for the complete tire development such as dynamic characteristics and contact patches. The drum testing method which is a popular laboratory method for investigating tire rolling behavior which had been carried out to test on solid tires. The contact force between drum surface and tire was studied. It was found that the contact force which occurred by using a solid tire was higher than the contact force which occurred by using a pneumatic tire under the same conditions [3]. Liang et al. [4] have studied pneumatic tire footprint geometry for tread wear investigation by using a tire static loaded machine and Tekscan (tire pressure measurement system). In agriculture, the contact distributions on soil which were passed by tractor tire and rubber track were compared [5]. The bigger contact area provided lower soil compaction. Furthermore, traction force is one of important parameter to study the tire-soil interaction which is used in tractor tire development. A soilbin-wheel tester was built for investigating rolling tractor tire behavior in the laboratory [6, 7]. In addition, a FEM is an alternative to an empirical tire patch. Phromjan and Suvanjumrat [8] have investigated the most suitable constitutive model for solid tire compression by varying 6 famous material models by finite element analysis. The Ogden model was found to be the most suitable material model when compared with experimental results. The load capability of solid tire was also studied by FEA [9]. In addition, effects of tire compression on soil were also studied by FEM and had performance for parameter control and lower costs. The soil pressure distribution by a tire was studied [10, 11]. It was found low inflation pressure could reduce topsoil compaction but the highest pressure occurred at the tire edge which could damage a pneumatic tire.

This research aimed to study the contact patch of airless tire (TWEEL). Then, the simulation model of rotated TWEEL airless tire on different surface was investigated. The proper rolling airless tire model which is achieved by this study will be used to develop the novel airless tires in the future.

2. Tire Testing Methods

2.1 Tire stiffness testing

A TWEEL airless tire, which was developed by Michelin, was used to study a tire patch on a flat surface. Figure 1(a) presents the tire stiffness tester (Ektron PL-2003) which is used to investigate the vertical deformation of TWEEL airless tire under compression load. Vertical displacement and vertical force were recorded by a position transducer and a load cell, respectively. In this research, the airless tire was pressed by the tester with a compression force of 14 kN which was a usage load. The tire was fixed on a mounting axle and compressed by a measurement table which was driven by a hydraulic motor with the displacement error less than ±0.1 mm.

2.2 Drum testing

A drum testing machine (KAYTON) was used to investigate the rolling tire's behavior in the laboratory. The diameter of the drum is 1.70 meters. The advantages of this method are that it uses less of the testing area and safer than field testing while providing accurate control conditions. The TWEEL airless tire was fixed on the mounting arm and moved to press on the rolling drum which controlled the tire speed at 11 km/hr. The accuracy of velocity measurement on the drum is ±2 km/hr. The compression load of 14 kN had been applied in the horizontal direction, while the TWEEL airless tire was pressed. The spoke deformation while the TWEEL airless tire rolls on the drum is detected by a high-speed video camera as shown in Figure 1(b).

(a)

(b)

Figure 1. The tire testing methods comprising of: (a) tire stiffness testing and (b) drum testing.

IOP Conf. Series: Materials Science and Engineering 1137 (2021) 012023 doi:10.1088/1757-899X/1137/1/012023

3. Finite Element analysis

The TWEEL airless tire 12N16.5 ALL TERRAIN composes of 4 main components including rubber treads, shear bands, Polyurethane (PU) spoke and steel rims. The TWEEL airless tire was cut by using the water jet cutting technique to prepare the specimen for material properties testing. The rubber tread was cut into a cylindrical shape for compression testing according to ASTM D575 while the PU spoke of the airless tire was cut into dumbbell shape for a tensile test according to ASTM D412. Computer Aided Design (CAD) and Computer Aided Engineering (CAE) methods were used to create the TWEEL airless tire model. The TWEEL's diameter and width were 860 and 309 mm, respectively. The finite element (FE) model of the TWEEL was created and analyzed by using MSC.Patran and MSC.Marc software, respectively. These are the FE software. A hexagonal element was employed to create the tread and shear band model while belt layers and spoke models were created by quadrilateral elements. The steel belt layers were inserted into shear band by creating rebar element which were embedded using a tying equation. The parameters of the Ogden hyperelastic model, equation (1), and generalized Maxwell material model, Equation (2) and (3), which are suitable to describe material properties of airless tire components, are presented in Table 1 and Table 2, respectively [12]. To validate the FE model, the TWEEL airless tire model is compressed by using a rigid flat plate according to tire stiffness testing as shown in Figure 2. The Hermann element was selected to prevent the excessive stress due to volumetric locking.

$$ {}_0^t\tilde{W} = \sum_{n=1}^{3} \frac{\mu_n}{\alpha_n} (\lambda_1^{\alpha_n} + \lambda_2^{\alpha_n} + \lambda_3^{\alpha_n} - 3) \quad (1) $$

where ${}_0^t\tilde{W}$ is the strain energy density per unit volume, λ_i is the principal value of the stretch tensor while μ_i and α_i are material constants.

$$ G(t) = G_0 - \sum_{i=1}^{n} G_i (1 - e^{-t/\tau_i}) \quad (2) $$

$$ \tau_i = \eta_i / E_i \quad (3) $$

where $G(t)$ is shear relaxation modulus, G_0 is shear modulus at initial time (t=0), G_i is i^{th} term of shear modulus, τ_i is i^{th} term of relaxation time (s), E_i is modulus of elasticity and η_i is viscoelasticity.

Table 1. The constant of Ogden hyperelastic constitutive model of TWEEL.

Component	Ogden's constant			
	μ_1	μ_2	α_1	α_2
Tread and Shear Band	1.15673	1.06228	5.37146	-2.31827
Spoke	0.112983	-11.0664	3.1488	-1.75206

Table 2. The visco-elastic constants of TWEEL.

i^{th}	τ_i	G_i	
		Tread and Shear Band	Spoke
1	0.2	0.2	0.125
2	0.02	0.2	0.125
3	0.002	0.2	0.125

IOP Conf. Series: Materials Science and Engineering 1137 (2021) 012023 doi:10.1088/1757-899X/1137/1/012023

The boundary conditions of a rotated TWEEL airless tire on a drum surface according to drum testing methods are presented in figure 3(a). The diameter ratio between the drum and TWEEL airless tire was 2. Multipoint constraints (MPCs) were performed to mimic the rim which were tied together at tire center. Two steps to simulate the transient tire testing model were performed. The first step, the rigid drum was moved to press the tire model with a compression force of 14,000 N. The final step, the rigid drum was rotated at a speed of 11 km/hr. The tire model was induced to rotate at the desirable speed by using a contact algorithm. The friction coefficient between tire and surface was found to be 0.8 [13].

The boundary conditions of the rotated TWEEL airless tire on a flat surface are presented in Figure 3(b). The rigid rim was controlled by a FE node at the tire's center. In this case, a rigid flat surface was fixed while the airless tire model was moved to compress on the rigid flat plate with a compression force of 14 kN. After that, the airless tire model was rotated with a velocity of 11 km/hr. These boundaries were controlled on a rigid rim. The transient dynamic FEM was carried out by using the implicit method. The solution of transient FE implicit analysis is calculated by the single step Housbolt method as written by the following equations.

$$\dot{u}_{n+1} = \dot{u}_n + \gamma\Delta t\ddot{u}_n + \gamma_1\Delta t\ddot{u}_{n+1} \quad (4)$$

$$u_{n+1} = u_n + \Delta t\dot{u}_n + \beta\Delta t^2\dot{u}_n + \beta_1\Delta t^2\dot{u}_{n+1} \quad (5)$$

where u_i is the time-dependent displacement field, γ_i is a defined constant, t is time and β is a constant.

Figure 2. The FE model of the compression of the TWEEL airless tire on flat surface, (a) boundary conditions and (b) an isometric view.

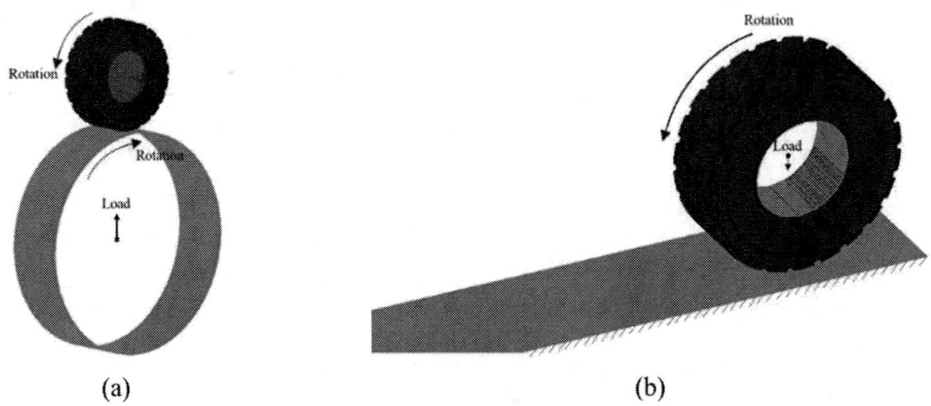

Figure 3. The boundary conditions of the rolling TWEEL airless tire on: (a) drum surface and (b) flat surface.

4. Results and Discussion

4.1 Validation

According to tire stiffness testing, the vertical force and deformation of the compressed TWEEL FE model were recorded. Figure 4(a) presents the comparison between the simulation and the experiment of the TWEEL airless tire deformation. The average tire stiffness of the experiment and the simulation were 951.01 and 846.42 N/mm, respectively. The simulation result had a good accord with the experimental result. It had an average error of less than 11%. The deformation of TWEEL airless tire model is presented in Figure 4(b). The maximum and minimum displacement were presented in yellow and blue, respectively. The result showed that the maximum deformation occurred on the compressed spokes. At the same time, the upper spokes which were not compressed, were stretched instead. That cooperation structure was one of the advantages of airless tires.

The spoke deformation at various rolling angle was detected by collecting nodal coordinates of spoke shown in Figure 5(a). Image processing using a high-speed video camera was performed to collected spoke positions from the experimental results. The comparison of spoke deformations at 30°, 60°, 90°, 120°, 150° and 180° is presented in Figure 5(b). The simulation results of the TWEEL airless tire rolling on a drum surface showed a good accord when compared to the experimental results. It had an average error of 3.51% with simulation time of 329.71 hours [13]. The simulation with the drum surface consumed a lot of simulation time due to complex contact and small-time step for analysis. Then, the FE model of the rotated TWEEL airless tire on flat surface was developed to reduce these complexities. The spoke deformation at each rolling angle between a rolling airless tire on a flat and drum surface was also found to be in a good accord. The deformation of spoke of TWEEL airless tire testing which was modelled on a flat surface was different on drum surface of 1.14%. The rolling airless tire on a flat surface model was found to have the advantage of using less analysis time than the drum surface model by 10.13 times.

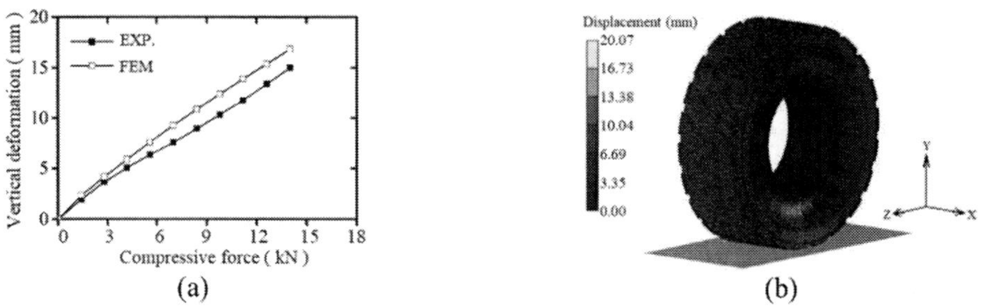

Figure 4. (a) The vertical displacement vs. compression force graphs and (b) the deformation of FE model under a compression force of 14 kN.

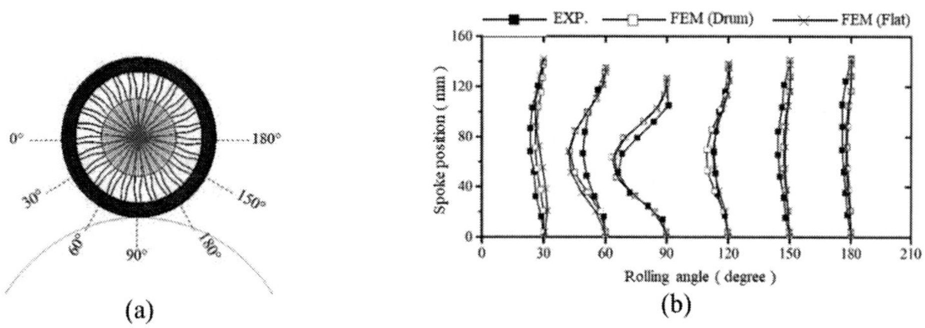

Figure 5. (a) The schematic diagram of collected rolling angle and (b) comparison of spoke deformation at various rolling angles.

4.2 Comparison of two different surface for rolling airless tires

Footprint analysis is an important criterion to indicate tire-road adhesion ability and tread wear. The footprint area (FA) is an overall contact area of a tire on the ground. On the other hand, the contact area (CA) is an actual contact area [14]. In this research, the contact area of rolling TWEEL airless tire model on flat surface was compared against the rotated model on drum. Figure 6 shows the comparison of contact patches on the different floor surface. At the initial state, the TWEEL airless tire was compressed by a compression force of 14,000 N. The contact area of TWEEL airless tire which was compressed by the drum was smaller than flat surface. This showed the effect of curvature on the surface of a drum. After the tire model was rotated at 90 degrees, the footprint still happened like in the initial position. It was found that the difference of contact patch at initial and 90 degrees could be neglected. The result indicated that contact area which occurred from the rolling tire model on a flat surface was a little bit bigger than the rolling on a drum surface. The contact area on a flat surface was different from the drum surface by just 56.78 cm^2.

In the stress analysis case, it was found that the stress distribution of the TWEEL airless tire rolling on flat surface was extensive with an analogous contour under a constant compression load of 14,000 N. The intense stress of TWEEL airless tire rolling on a drum surface is a little bit higher than the flat surface (Figure 7). The value of stress was separated as color bars which were placed beside each color contour plot.

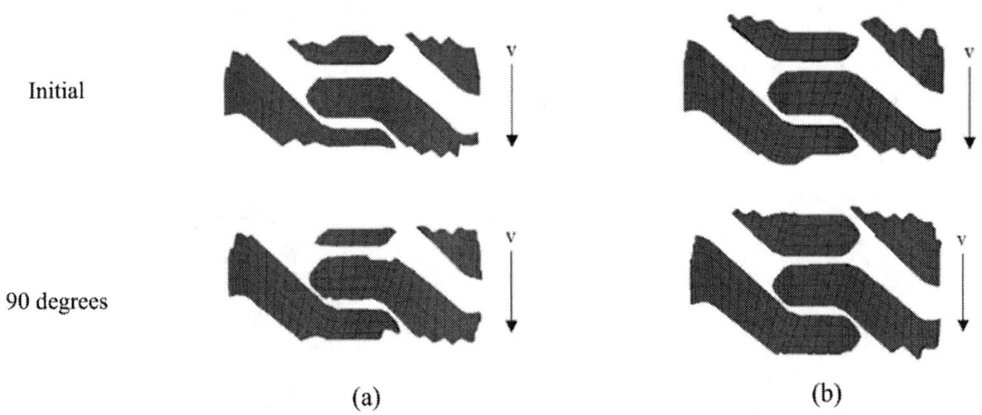

Figure 6. Footprint of the rolling TWEEL airless tire on: (a) drum surface and (b) flat surface.

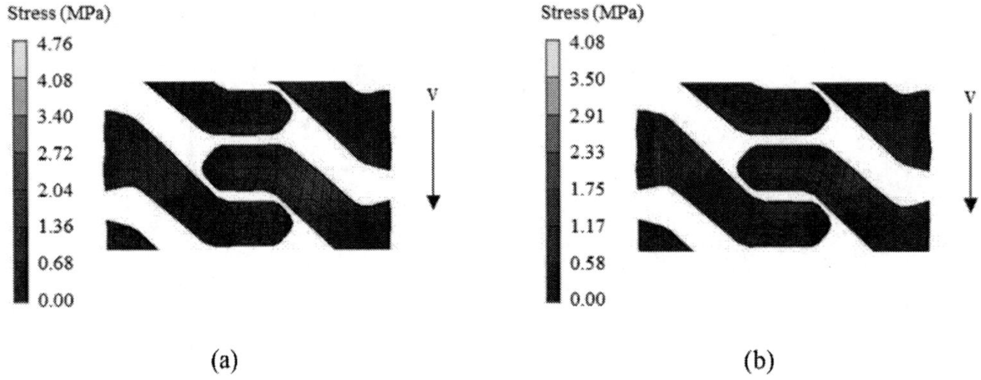

Figure 7. Contact stress of the rolling TWEEL airless tire on: (a) drum surface and (b) flat surface.

The effect of the curvature surface of the drum was found to be the reason for the different stress which could be observed at the center of contact patches. The arc of the rigid drum surface compressed the tread of the TWEEL airless tire model. The average difference of contact area and maximum stress of both models was only 26.97% and 14.22%, respectively. There are two criteria to indicate performance of these models. The overall results of the FE model of the rolling TWEEL airless tire on a flat surface were accepted to save computing time.

5. Conclusion

The TWEEL airless tire, which was created to overcome the disadvantages of a conventional pneumatic tire, was studied. Knowledge of the static and dynamic behavior of a tire is essential in complete tire development. To study tire static behavior, tire stiffness testing was performed and used to validate the TWEEL airless tire FE model. The vertical stiffness of the TWEEL tire model had a good accord with that of the experiment. Next, the TWEEL airless tire model was extended for the purpose of a dynamic behavior study. The FE model of rolling TWEEL airless tire on a flat surface was developed to provide high accuracy while having little simulation cost. A transient dynamic FE model with a visco-hyperelastic material model was developed. The spoke deformation and footprint analysis presented a good compatibility between the two different surfaces. The calculating time of the rolling tire on flat surface was 10.13 times less than the FE model of rolling a TWEEL airless tire on a drum surface. To achieve a desirable tire performance, the FE model of rolling a TWEEL airless tire on a flat surface in this research will be used to study various rolling airless tire parameters in the future.

Acknowledgments
This work was supported by Rubber Technology Research Center (RTEC), Mahidol University and the Thailand Research Fund (TRF) under the TRF Research Grant No. RDG60T0140.

References

[1] Rugsaj R and Suvanjumrat C 2020 *Mech. Based Des. Struct. Mach.* DOI:10.1080/15397734.2020.1777875
[2] Xiaochao J Cheng H Xueling F Yongle S Jinan L and Chunsheng L 2018 *Compos. Struct.* **187** 27-35
[3] Phromjan J and Suvanjumrat C 2018 *J. Mech. Sci. Technol.* **32**(4) 1539-1548
[4] Liang C Wang G An D and Ma Y 2013 *Chin. J. Mech. Eng.-En.* **26**(3) 506-511
[5] Ansorge D and Godwin R. J 2007 *Biosyst. Eng.* **98** 115-126
[6] Tiwari V K Pandey K P and Sharma A K 2009 *J. Terramechanics* **46** 293-298
[7] Mardahi A Shahidi K and Maslak H K 2010 *J. Food Agric. Environ.* **8**(2) 642-646
[8] Phromjan J and Suvanjumrat C 2018 *Engineering Journal* **22**(2) 141-155
[9] Phromjan J and Suvanjumrat C 2018 *Key Eng. Mater.* **777** 416-420
[10] Omar G C Ciro E I C Elvis L B Carlos A R M and Miguel H S 2016 *J. Terramechanics* **63** 61-67
[11] Farhadi P Golmohammadi A Malvajerdi A S and Shahgholi G 2019 *Comput. Electron. Agric.* **162** 793-806
[12] Rugsaj R and Suvanjumrat C 2019 *Int. J. Automot. Technol.* **20**(4) 801-812
[13] Rugsaj R and Suvanjumrat C 2020 *IOP Conf. Ser., Mater. Sci. Eng.* **886** 012056
[14] Phromjan J and Suvanjumrat C 2020 *IOP Conf. Ser., Mater. Sci. Eng.* **886** 012049

IOP Conf. Series: Materials Science and Engineering 1137 (2021) 012024 doi:10.1088/1757-899X/1137/1/012024

Development of a herbal cream mixer with a homogenizer-stirrer and circulating water cooling system

Suntorn Suttibak[1,*], Kamonrat Deesapa[2], Chayarnon Saengmanee[3], and Athika Chuntanapum[4]

[1*,2,3,4] Department of Mechanical, Udon Thani Rajabhat University, Udon Thani, Thailand 41000

* Corresponding Author: suntorn_su@hotmail.com

Abstract

This paper presents a study on the development of a herbal cream mixer with a homogenizer-stirrer and circulating water cooling system. The machine is designed to address problems in the production of herbal creams and lotions by the entrepreneurs of the Thangrak Company Limited, Chiang Yuen Sub district, Muang District, Udon Thani, Thailand. It has a capacity of around 100 liters and employs a circulating water-cooling system. The homogenizer uses a 3-horsepower electric motor to control the rotation speed of the stirring head, which in turn controls the mixer speed. The level of the machine is adjusted using an electric hoist. Cream was produced in 80 liter batches. Mixer speed was tested at four levels: 30 Hz (1,800 rpm), 35 Hz (2,100 rpm), 40 Hz (2,400 rpm) and 45 Hz (2,700 rpm). The total time of each run was approximately 30 minutes. The results showed that a temperature reduction of the ingredients from an initial temperature of around 70°C down to approximately 33°C. The process took around 1 hour and 30 minutes. From the experimental results, it was found that the product made at 1,800 rpm had a lower viscosity of around 78.4 Pa.s. The cream of this batch was not as homogenized as it should have been. However, tests at 2,100, 2,400, and 2,700 rpm produced a creamy material with increased viscosities of 94.3, 94.8 and 96.7 Pa.s, respectively. The herbal cream products made in the developed mixer with homogenizer-stirrer and a circulating water-cooling system were homogenous. They compared favorably with herbal cream products sold in the market. The designed mixer produced a cream product that is more homogenous, while the products now available on the market have separated cream and are not as well homogenized.

Keywords: Homogenizer-stirrer, Circulating water cooling system, Herbal cream

1. Introduction

Herbal creams are a group of cosmetic (skincare) products for nourishing, remedying dry skin, and addressing other skin problems. Most of these creams are in liquid form. The present community and small enterprises have a good reputation in cream production since these products rely on herbal extracts mixed in the cream. They increase the value of the cream and help to smooth skin. A herbal cream manufacturing process is needed that uses modern equipment in its manufacturing process to increase the competitiveness of these entrepreneurs. The current methods of stirring use high shear mixers, e.g., high shear reactor, rotor-stator mixers and high shear homogenizers [1, 2]. The stirrers of these units have high rotor tip speeds of about 10 – 50 m/s, very high shear rates of 20,000 – 100,000 s^{-1}. The energy distribution rate is high near the stirring head [3, 4]. High energy is required for the unit operations of homogenization, dispersion, emulsification, grinding and dissolving.

The entrepreneurs of Thangrak Company Limited, in Udon Thani Province, manufacture herbal cosmetic products, e.g., creams, lotions, shampoos, soaps, and other products. They are made for both retail and wholesale distribution. The average sales volume of these creams and lotions is about 4,000 bottles per month at an average price of 150 baht per bottle. Their monthly production is valued at more than 600,000 baht/month. Recently, these entrepreneurs recognized the need for innovation and improved technology for cosmetic product manufacture and processing, especially for the process used to mix the cream. The currently used equipment is antiquated, requires long processing times and has a low production capacity. Presently, the process of manufacturing herbal cosmetic cream products does not exceed 20 liters/batch. Manufacture requires five steps. The first is heating the cream mixture (coconut oil, volatile matter and herbal materials, among other ingredients), allowing them to dissolve at 45°C for 5 minutes. Then, distilled water is heated to 75°C. Next, the cream mixture is added to warm distilled water in the stirring tank. The batch is then cooled for about 30 minutes in an air-conditioned room. After that, cold stirring is done. The product cannot be stirred when it is warm because it can degrade. Stirring is done manually using a small stirrer. During the stirring process, a catalyst and enzyme are added. This step takes about 45 - 60 minutes. Finally, the cream product is placed in a manual filling machine to package the product in bottles. Then, the final product will be distributed. The production capacity of facility producing these cosmetic products is not more than 80 liters per day, i.e., four batches a day. Furthermore, it was found that the cream was not as homogeneous as other products currently available in the market. This is a result of sub-optimal mixing.

A review of recent research revealed a novel raw material mixing process using a homogenizer in the work of Bucher [5] who developed a homogenizer for processing chemical materials into a homogeneous form at a laboratory scale. André et al., [6] also used dimensional analysis considering the mixing time and energy consumption of a planetary mixer to develop a homogenization process for free-flowing powders. Ekpeni et al., [7] studied the parameters that influence the energy production of a high-pressure homogenizer (HPH) using yeast as a biomass source. This research produced a novel homogenizer stirrer.

The aim of the current research is to develop a herbal cream mixer employing a homogenizer stirrer with a circulating water cooling system for temperature control. A production capacity of about 100 liters/batch addresses the manufacturing problems in stirring cream mixtures and lotions faced by our entrepreneur clients. The time required for both the cooling and mixing steps was reduced. The product of this process is a homogeneous cream. Its production requires reduced labor in the stirring process and achieves a much higher production capacity.

2. Methodology

2.1 Design of a herbal cream mixer with homogenizer-stirrer employing a circulating water cooling system to control temperature

The developed mixer has a width × length × height of 100 × 145 × 230 cm, respectively. The capacity of this tank is approximately 100 liters. The storage tank employs a double layer construction. The outer layer cools the product by circulating a water coolant through a jacket using a ½ horsepower pump drawing a water coolant from an external tank. The homogenizer uses a 3-horsepower electric motor to control the rotation speed of the stirring head, which in turn controls the speed of the mixer. The maximum speed of the mixer is 3,000 rpm. Its level can be adjusted using an electric hoist. The components of this machine are shown in Figure 1.

Figure 2 illustrates the developed herbal cream mixer with a homogenizer-stirrer and circulating water cooling system designed and fabricated for entrepreneurs of the Thangrak Company Limited. The equipment has a modular design so it can be simply separated and moved. A controller is used to adjust

IOP Conf. Series: Materials Science and Engineering 1137 (2021) 012024 doi:10.1088/1757-899X/1137/1/012024

the stirring head speed. Its maximum speed is at 50 Hz producing 3,000 rpm at the exit channel of the cream product where a stainless valve is installed.

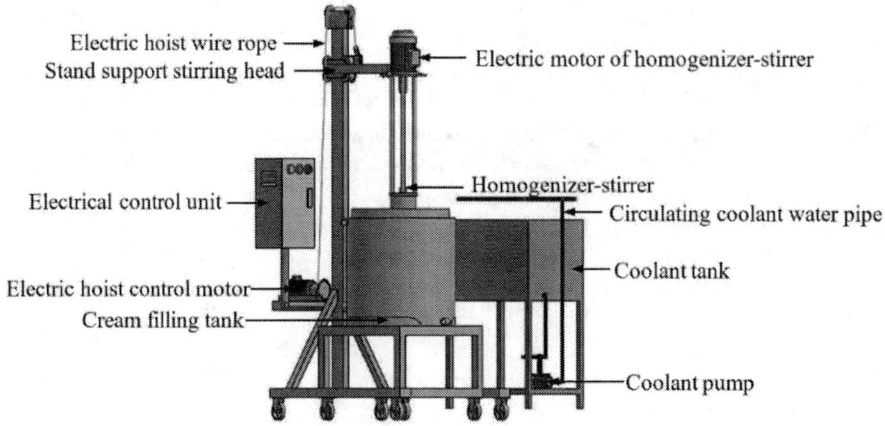

Figure 1. The components of the herbal cream mixer with homogenizer-stirrer with a circulating water-cooling system

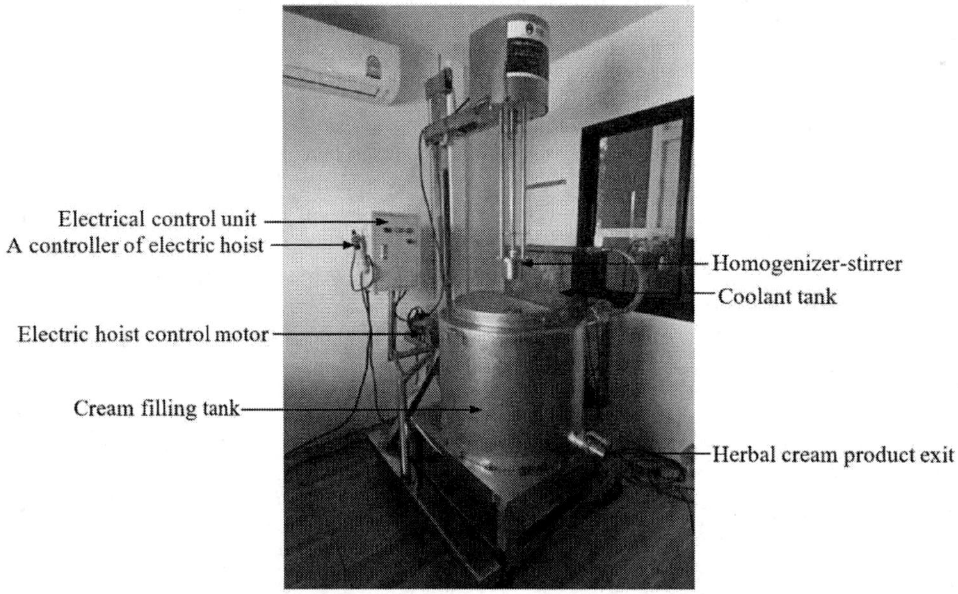

Figure 2. The herbal cream mixer with homogenizer-stirrer and circulating water cooling system

2.2 Equipment and Materials

1) The developed herbal cream mixer with a homogenizer-stirrer and circulating water cooling system, as shown in Figure 2.
2) A cream mixture (80 liters per batch)
3) Digital thermometer (Figure 4)
4) Rotor viscometer
5) 80 mL beakers
6) Cream products
7) A cream sample that is currently available in the market
8) Stopwatch

2.3 Test of the water-cooling system

Figure 3 depicts a result of the cooling time test of the cream mixture using a vortex system. Water in the tank was pumped through the outer jacket of the vessel containing the cream mixture. Then, a mixture consisting of distilled water heated to 75°C and herbal cream ingredients (coconut oil, volatile substances, and herbal fermentation enzymes, among others) dissolved at 45°C. The time required to cool the mixture to 30 – 35°C was determined using a digital thermometer and stopwatch, as shown in Figure 4. Then the mixture was stirred using a homogenizer pin in the next process.

Figure 3. Cooling reservoir tank with vortex cooling system

Figure 4. Measuring the temperature of the cream mixture using digital thermometer during cooling

2.4 Effect of mixing speed on the viscosity and characteristics of herbal cream products

Figure 5 presents the effects of mixing speed on the viscosity and characteristics of herbal cream products. The cream was produced 80-liter batches. The mixer speed was tested at four levels: 30 Hz (1,800 rpm), 35 Hz (2,100 rpm), 40 Hz (2,400 rpm), and 45 Hz (2,700 rpm). The total time of each run was approximately 30 minutes. Afterward, the product viscosities were determined using a rotor-viscometer and compared with those of commercially available creams and the original cream of the entrepreneur.

Figure 5. Mixing speed test of the cream mixture

2.5 Communication of research technology to target enterprises

The developed herbal cream mixer with a homogenizer-stirrer and circulating water cooling system was provided to the entrepreneurs of Thangrak Company Limited. They used it in the production of herbal creams and lotions to stir ingredient mixtures. This reduces manual labor and working time, as well to increase the production capacity and the quality of cream products.

3. Results and Discussion

3.1 The operability of the vortex cooling system

Figure 6 depicts/shows the results of cooling time tests of the cream mixture in a vortex cooling system. Eighty liters of the mixture were produced in each batch. After the test, the initial temperature of the mixture was around 70°C and it was reduced to 33°C, due to the cooling water circulation inside the external tank, the temperature of the cream mixture inside the internal tank was lowered further. Which is the temperature in which stirring starts. The developed process requires approximately 1 hour and 30 minutes, while the original method produces ten liters per batch, followed by 1 hour of cooling. The developed process requires a longer time, however, its production volume eight times greater. It also does not require an air-conditioned room for cooling, which can reduce the electrical expenses. The current process was operated using a small coolant tank. Increasing the volume of the coolant tank can further reduce cooling time.

3.2 The effect of mixing speed on the viscosity and characteristics of herbal cream products

Figure 7 shows the effect of stirring speed on the viscosity of the cream products manufactured in the developed herbal cream mixer. Stirring speed was examined at four levels, 30 Hz (1,800 rpm), 35 Hz (2,100 rpm), 40 Hz (2,400 rpm) and 45 Hz (2,700 rpm). In each test, stirring was done for 30 minutes. From the experimental results, the product made at 1,800 rpm showed a lower viscosity of 78.4 Pa.s. This product was not sufficiently homogenized. However, when produced at 2,100, 2,400, and 2,700 rpm, the cream products had increased viscosities of approximately 94.3, 94.8 and 96.7 Pa.s, respectively. In this study has chosen speed at 2,100 rpm. Due to the considered results of the viscosity which has no differences the speed used has no effect on stirrer vibration during processing. These creams were more homogenous as compared with herbal cream products sold in the market. The original product produced in this facility had separated cream and was not at all homogenous. Figure 8 demonstrates the physical appearance of products made by using the original and developed processes.

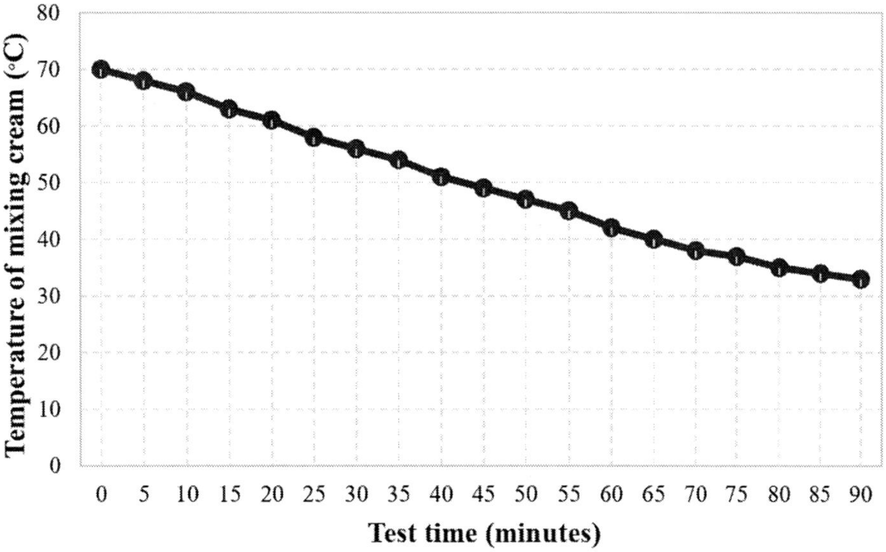

Figure 6. Time-temperature of the cream mixture in the developed herbal cream mixer with homogenizer-stirrer and vortex cooling system

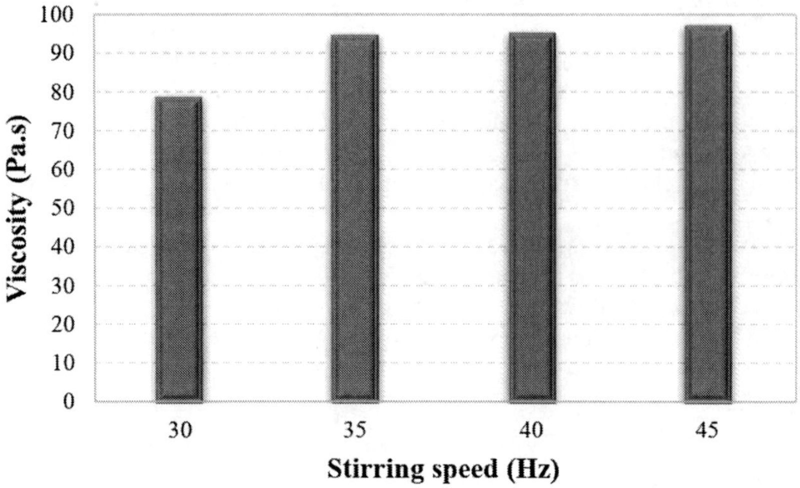

Figure 7. Effect of mixing speed on the viscosity of herbal cream products

Cream product obtained by stirring with a developed machine.

Cream product obtained by conventional stirring.

Figure 8. The characteristics of cream made by stirring using the developed machine compared with conventional stirring

Table 1 shows the viscosities and characteristics of cream products obtained from conventional stirring, stirring with the developed machine and Citra brand cream products. The characteristics of cream products indicate that the developed machine-stirred cream has a similar viscosity to that of the Citra cream with the same homogeneous texture. The viscosity of the original product is lower and it was not well homogenized.

Table 1. The viscosity of cream products

Cream products	Viscosity (Pa.s)	Characteristics of cream products
Cream products obtained from conventional stirring	75.0±2.65	Not well homogenized
Cream products obtained from stirring using the developed machine	94.3±1.15	Homogeneous
Citra brand cream products	93.7±0.58	Homogeneous

3.3 The communication of research technology to target enterprises

The developed herbal cream mixer with a homogenizer-stirrer was shared with the entrepreneurs of the Thangrak Company Limited. They will use it in the production of herbal creams and lotions. It can increase their daily production capacity to a level of not less than 100 liters (at least two times greater), and produce creams with higher quality. The saleable cream products and the working machinery are shown in Figures 9 and 10, respectively.

Figure 9. The saleable herbal cream products

Figure 10. The developed machinery in operation

4. Conclusions

Development of a herbal cream mixer with a homogenizer-stirrer and circulating water cooling system was done to solve problems in the production of herbal creams and lotions for the entrepreneurs of Thangrak Company Limited. The machine has dimensions of width × length × height of 100 × 145 × 230 cm, respectively. The capacity of the tank is around 100 liters. The design of the storage tank was double layer. The outer layer is a jacket employing a vortex cooling system with a ½ horsepower pump drawing coolant water from an external tank. The homogenizer uses a 3-horsepower electric motor to modulate the rotational speed of the stirring head, which controls the speed of the mixer. The cream was produced at a volume of 80 liters per batch. Stirring speed was tested at four levels, 30 Hz (1,800 rpm), 35 Hz (2,100 rpm), 40 Hz (2,400 rpm) and 45 Hz (2,700 rpm). The total time for each run was approximately 30 minutes from the beginning of stirring. The entire process requires around 1 hour and 30 minutes. However, the current design uses a small storage tank for coolant which leads to suboptimal cooling time. The size of this tank can be increased to further reduce the cooling time. From the experimental results, it was found that product made at 1,800 rpm had a low viscosity of around 78.4 Pa.s. This cream was not as homogenized as it should have been. However, when products were made at 2,100, 2,400, and 2,700 rpm, the cream had increased viscosities of 94.3, 94.8 and 96.7 Pa.s, respectively. These creams are more homogenous than the products sold in the market. The cream originally manufactured in this facility tended to separate and it was not well homogenized.

Acknowledgments

Financial support from the Innovation and Technology Assistance Program (ITAP), National Science and Technology Development Agency (NSTDA) and Mahasarakham University Network (the 2020 fiscal year), Thailand are gratefully acknowledged. Thanks, are also extended to Prof. Jeffrey Nash for his research assistance.

References

[1] Hert S C D and Rodgers T L 2017 *Chemical Engineering Science* **167**(6) 265-277
[2] Zhang J Xu S and Li W 2012 *Chemical Engineering and Processing: Process Intensification* **57-58**(4) 25-41
[3] Paul E L Atiemo-Obeng V A and Kresta S M 2003 *Handbook of Industrial Mixing* (New Jersey: Wiley) p 1483
[4] Atiemo-Obeng V A and Kresta S M 2004 *Handbook of Industrial Mixing: Science and Practice* (New Jersey: John Wiley & Sons) **2** 479-505
[5] Bucher F G 2007 *United States Patent* (US007165734B2) **1** 1-8
[6] André C Demeyre J F Gatumel C Berthiaux H and Delaplace G 2012 *Chemical Engineering Journal* **198-199**(5) 371-378.
[7] Ekpeni L E N Benyounis K Y Nkem-Ekpeni F F Stokes J and Olabi A G 2015 *Energy* **811**(7) 74-83

Numerical study on bending response of auxetic 2D-lattice plates

Pana Suttakul* and Thongchai Fongsamootr

Department of Mechanical Engineering, Faculty of Engineering,
Chiang Mai University, Chiang Mai Thailand 50200

* Corresponding Author: pana.s@cmu.ac.th

Abstract
In this study, the bending response of 2D-lattice plates having auxetic unit cells is determined by using the homogenization method based on equivalent strain energy. The auxetic unit cells in this study are different topologies of re-entrant hexagonal unit cells. In the homogenization method, the effective out-of-plane elastic properties of an auxetic 2D-lattice plate are obtained from the strain energy values of its unit cell under different curvature modes, determined by the finite element method. In the analysis, the auxetic unit cells are considered as frames whose struts are modeled as Euler beams. The effective elastic properties, i.e., the effective bending moduli, Poisson's ratios, and shear modulus, are described as the bending response of a 2D-lattice plate subjected to out-of-plane bending. In the validation, the effective elastic properties of some auxetic 2D-lattice plates obtained from their unit cells by the homogenization method are numerically compared with those obtained from direct structural analysis of the plates. Besides, the obtained results show how the bending response of the auxetic 2D-lattice plates can be adjusted by varying their unit-cell geometries, especially the internal cell angle.

Keywords: Auxetic structure, Re-entrant hexagon, Periodic plate, Bending modulus, Negative Poisson's ratio.

1. Introduction
The two-dimensional lattices are periodic structures that have a wide range of applications. The 2D lattices are characterized by having a unit-cell pattern repeated on a plane in two orthogonal directions. Since mechanical properties can be designed to yield the desired properties, directly depending on their unit-cell patterns, 2D lattices have been explored by many researchers [1-9]. Two-dimensional lattices can be modeled as frames and used as plane or plate structures. If there are enough unit cells in a 2D lattice, it can be considered as an equivalent homogeneous material. The material properties of the equivalent homogeneous material of a 2D lattice can be represented by effective elastic properties that can be determined by homogenization methods [3-6, 10-13]. One of the most widely used homogenization methods is the homogenization method based on equivalent strain energy. In the homogenization method, the strain energy in an equivalent homogeneous material, under uniform far-field boundary conditions that yield uniform stress and strain fields in the equivalent homogeneous material, is considered to be the same as the strain energy in its original inhomogeneous material, under

the same boundary conditions [14]. The effective material law of the inhomogeneous material can be obtained by considering the relationship between the average stresses and strains that occur in the material under these boundary conditions. Since 2D lattices are modeled as frames, their struts can be considered as beams. The effective material law of a 2D lattice can be determined from frame analysis of a selected unit cell of the lattice using the homogenization method.

One of the popular types of 2D lattices is honeycombs whose unit-cell patterns are typically in hexagonal geometries. Among various hexagonal unit-cell patterns, re-entrant hexagonal unit cells are featured because they can exhibit negative Poisson's ratio for 2D lattices or the so-called auxetic 2D lattices. An auxeticity of a material can be described by its specific geometries [15,16]. For an auxetic 2D lattice, the auxeticity can be designed by its unit-cell topologies. Mechanical properties of auxetic 2D lattices have been expansively investigated or designed by many researchers [1,4,10,15,17-20]. Most of the researchers focused on the elastic properties of auxetic 2D lattices subjected to in-plane loading. For example, auxetic cellular structures were suggested as 2D lattices with re-entrant unit cells by Gibson et al. [1], In their work, the auxetic effect of the 2D lattices was analytically determined. In the work by Scarpa et al. [19], the effective in-plane elastic properties, including the effective Young's moduli and Poisson's ratio, of 2D lattices with re-entrant hexagonal unit cells were investigated using finite element simulation and experiment. Effects of various geometries of the unit cells (aspect ratio, thickness, and internal cell angles) on the effective elastic properties were also investigated. Closed-form expressions of the effective in-plane elastic constants of conventional and re-entrant hexagonal lattices, considering variable dimensional parameters, e.g., aspect ratio and internal cell angle, were derived by Zhang et al. [10]. In the work by Alderson et al. [20], the finite element method was used to simulate the curvatures of various auxetic 2D lattices under out-of-plane bending.

A 2D lattice used as a plate structure can be termed as a 2D-lattice plate. When a common 2D-lattice plate is subjected to an out-of-plane bending moment, it is expected that the resulting curvature on the perpendicular edges occurs in the opposite direction of those on the edges that the moment applied. On the other hand, an auxetic 2D-lattice plate yields the resulting curvature in the same direction as those happening on the applied moment edges. This is a notable property resulting from negative Poisson's ratio of auxetic 2D-lattice plates. For this reason, in this study, the attractive properties of auxetic 2D-lattice plates are numerically investigated. The effective out-of-plane elastic properties of 2D-lattice plates with auxetic unit cells are determined using the homogenization method based on equivalent strain energy. The considered auxetic unit cells are different re-entrant hexagonal unit cells. These unit cells are obtained by adapting internal cell angles and the aspect ratio of hexagonal geometry. The effective elastic properties proposed in this study include the effective bending moduli, Poisson's ratios, and shear modulus. In the validation, the effective elastic properties of some auxetic 2D-lattice plates obtained from their unit cells by the homogenization method are compared with those obtained from direct structural analysis of the plates. All analyses are done by the finite element method. The obtained results show how the effective out-of-plane elastic properties of 2D-lattice plates with auxetic unit cells can be adjusted by varying the geometries of the unit cells.

2. Effective Out-of-Plane Properties of 2D-Lattice Plates

The effective out-of-plane properties in this study are used to represent the bending response of 2D-lattice plates and can be determined by using the homogenization method based on equivalent strain energy. In the homogenization method, the strain energy in an equivalent homogeneous plate, under periodic far-field boundary conditions, is equal to the strain energy in its original periodic plate, under the same boundary conditions [6,14]. Since under periodic far-field boundary conditions, the displacement, stress, and strain fields of a periodic plate are periodic, a unit cell of the plate under periodic boundary conditions is sufficient for creating these periodic response fields. The detailed information of the method can be found in [6]. Only important equations for the determination of the effective properties of 2D-lattice plates are briefly shown.

Consider a thin periodic plate, composed of a significantly large number of unit cells, in a coordinate system $x_1 - x_2 - x_3$ that its origin lies on a midplane A. The periodicity of the plate is only in the $x_1 -$

x_2 plane. A set of periodic kinematic boundary conditions is applied to the plate in a way that results in the following deflection w that is a displacement in x_3 direction, i.e.,

$$w(x_1,x_2) = w^o(x_1,x_2) + w^p(x_1,x_2) = \left[\frac{\kappa_{11}^o}{2}x_1^2 + \frac{\kappa_{22}^o}{2}x_2^2 + \kappa_{12}^o x_1 x_2\right] + w^p(x_1,x_2) \tag{1}$$

where κ_{ij}^o is a constant symmetric tensor, and w^p is a periodic function of x_1 and x_2. In addition, w^o represents the deflection of a homogeneous plate subjected to uniform kinematic boundary conditions that yield a constant curvature of κ_{ij}^o. The deflection w in equation (1) is extended everywhere in the midplane of the plate, including the inside of any voids, which can be considered as the limit cases of infinitely soft inclusions [14]. The rotation ϕ_i and curvature κ_{ij} of the periodic plate are respectively defined as $\phi_i = w_{,i}$ and $\kappa_{ij} = w_{,ij}$ [6].

The average of κ_{ij} over the midplane area A of the plate can be represented by κ_{ij}^o [6], i.e.,

$$\langle\kappa_{ij}\rangle = \frac{1}{A}\int_A \kappa_{ij}\,dA = \kappa_{ij}^o. \tag{2}$$

The equation of the effective stiffness tensor D_{ijkl}^* of the periodic plate is defined as

$$\langle M_{ij}\rangle = M_{ij}^o = \frac{1}{A}\int_A M_{ij}\,dA = -D_{ijkl}^*\langle\kappa_{kl}\rangle = -D_{ijkl}^*\kappa_{kl}^o \tag{3}$$

where $\langle M_{ij}\rangle = M_{ij}^o$ is the average of the bending moments M_{ij} over A.

The average strain energy of the plate $\bar{U}$ over A, and the strain energy of the unit cell U_C are respectively given by [6]

$$\bar{U} = -\frac{1}{2A}\int_A M_{ij}\kappa_{ij}\,dA = -\frac{1}{2}\langle M_{ij}\rangle\langle\kappa_{ij}\rangle = \frac{1}{2}D_{ijkl}^*\kappa_{kl}^o\kappa_{ij}^o; \tag{4}$$

$$U_C = \bar{U}A_C = \frac{1}{2}D_{ijkl}^*\kappa_{kl}^o\kappa_{ij}^o A_C \tag{5}$$

where, A_C denotes the midplane area of the unit cell. Equation (5) can be used to determine D_{ijkl}^* .In the determination, different modes of κ_{ij}^o are applied to the unit cell and the values of U_C are determined from structural analysis. Here, the curvature modes of κ_{ij}^o are created via periodic boundary conditions obtained from equation (1).

For an orthotropic periodic plate, equation (3) can be written in matrix form as

$$\begin{Bmatrix} M_{11}^o \\ M_{22}^o \\ M_{12}^o \end{Bmatrix} = -\begin{bmatrix} D_{11}^* & D_{12}^* & 0 \\ & D_{22}^* & 0 \\ Sym & & D_{33}^* \end{bmatrix}\begin{Bmatrix} \kappa_{11}^o \\ \kappa_{22}^o \\ 2\kappa_{12}^o \end{Bmatrix}. \tag{6}$$

Since there are four independent effective constants D_{ij}^* in the above equation, i.e., D_{11}^*, D_{22}^*, D_{33}^*, and D_{12}^*, four different modes of κ_{ij}^o in equation (5) are required [6]. Denote the strain energy of the unit cell to which $\kappa_{ij}^o = \kappa_{ij}^{o\,(i)}$, where i denotes the modes of κ_{ij}^o, is applied as $U_C^{(i)}$. The relationship between

$U_C^{(i)}$ and $\kappa_{ij}^{o\,(i)}$ are used to compute the corresponding values of D_{ij}^*. The effective out-of-plane properties of 2D-lattice plates can be written in terms of D_{ij}^* as

$$E_1^* = \frac{12D_{11}^*}{t^3}\left[1 - \frac{D_{12}^*}{D_{11}^* D_{22}^*}\right]; \quad E_2^* = \frac{D_{22}^*}{D_{11}^*}E_1^*;$$
$$\nu_{12}^* = \frac{D_{12}^*}{D_{22}^*}; \quad \nu_{21}^* = \frac{D_{12}^*}{D_{11}^*}; \quad G_{12}^* = \frac{12D_{33}^*}{t^3}. \tag{7}$$

Here, t is the thickness of the plate. Since D_{ij}^* is the function of $U_C^{(i)}$, the effective properties are also the function of $U_C^{(i)}$ and can be obtained directly in terms of $U_C^{(i)}$.

Besides, E_i^*, ν_{ij}^*, and G_{12}^* can be computed by direct structural analysis of a whole 2D-lattice plate. The effective stiffness equations for unidirectional bending of a thin periodic are given by [6]

$$E_1^* = \frac{12M_{11}^o}{t^3\kappa_{11}^o}; \quad E_2^* = \frac{12M_{22}^o}{t^3\kappa_{22}^o}; \quad \nu_{12}^* = -\frac{\kappa_{22}^o}{\kappa_{11}^o}; \quad \nu_{21}^* = -\frac{\kappa_{11}^o}{\kappa_{22}^o}; \quad G_{12}^* = \frac{6M_{12}^o}{t^3\kappa_{12}^o}. \tag{8}$$

3. Re-Entrant Hexagonal Unit-Cell Geometry

The auxetic unit cells considered in this study are re-entrant hexagonal unit cells as shown in figure 1. Two sets of unit-cell struts are defined as vertical struts of length h and inclined struts of length l. All struts are the same cross section area. A hexagonal unit-cell geometry is defined by some dimensionless parameters, e.g., internal cell angle θ and aspect ratio h/l of the unit cell. Different hexagonal unit-cell geometries can be obtained by varying θ and h/l. Figure 1, in which θ are all negative, shows different re-entrant hexagonal unit cells. For the unit cells with $h = l$, to keep an appropriate hexagonal shape, the negative internal cell angle should be limited when the angle is close to $-30°$. If the negative internal cell angle is required over the angle limit, the re-entrant hexagonal unit cells with $h > l$ can be used to avoid collision of vertical and inclined struts as shown in figure 1(b). The coordinate system in figure 1(a) is used for all unit cells. The boundaries of the unit cells are presented as dashed lines. The areas inside the boundary edges can be defined as the midplane areas A_C of the unit cells.

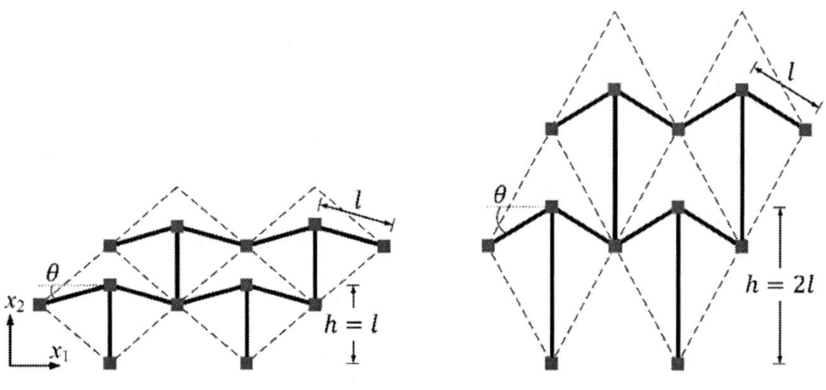

(a) Re-entrant Hexagon ($h = l$) (b) Re-entrant Hexagon ($h = 2l$)

Figure 1. Different geometries of re-entrant hexagonal unit cells

4. Results and Discussion

The effective out-of-plane elastic properties of 2D-lattice plates with re-entrant hexagonal unit cells shown in figure 1 are determined by using the homogenization method. In this investigation, all struts of the unit cells are modeled as Euler beams. The inclined strut's length l, defined in figure 1, is set to 1. The cross-sections of all struts in the 2D-lattice plates are set to be square sections of 0.1×0.1. The struts are placed such that the thickness of the 2D-lattice plates is also equal to 0.1. Young's modulus and Poisson's ratio of the base material that is isotropic are set to 1.0 and 0.2, respectively. Note that, to investigate the trend of bending response of the plates, mathematically, it is not necessary to use physically real values. In the homogenization method, numerical results of strain energy, $U_C^{(1)}$, $U_C^{(2)}$, $U_C^{(3)}$, and $U_C^{(4)}$, for the four modes of κ_{ij}^{o}, can be obtained from finite element analysis. After that, the effective elastic properties in equations (7) are numerically determined using the obtained strain energy values. Figure 2 shows plotting results of the effective elastic properties, E_i^*, ν_{ij}^*, and G_{12}^* against θ , plotted by every 5°, from 0° to -30°. It can be seen from the results that, when θ is different from 0° to negative, the obtained 2D-lattice plates are more auxetic. Reducing θ from 0° to -30° obviously increases the value of E_1^* but slightly reduces the value of E_2^*. Note that, E_1^* is always greater than E_2^* and, at $\theta = -30$°, E_1^* is significantly greater than E_2^*. For ν_{12}^* and ν_{21}^*, the negative values of both results increase when θ is reduced to more negative. The absolute values of ν_{12}^* are always greater than those of ν_{21}^*. Reducing θ continuously from 0° to -30° slightly increases the values of G_{12}^*.

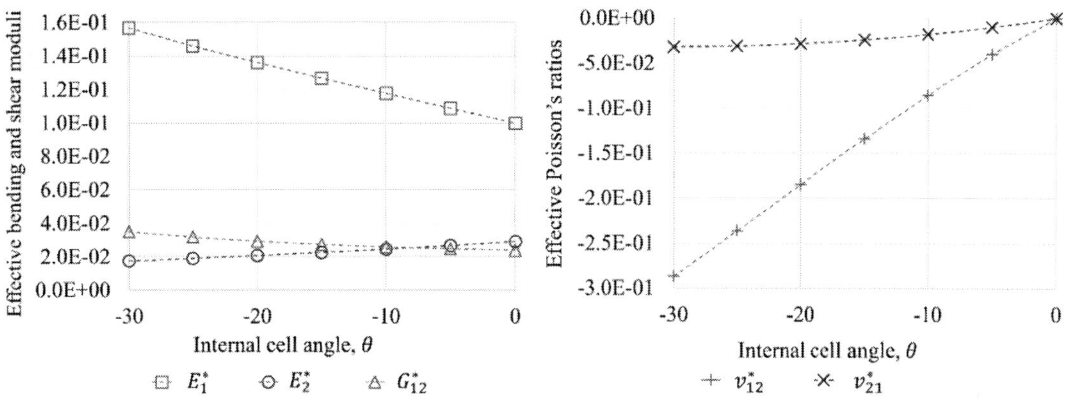

Figure 2. Plots of the effective elastic properties, E_i^*, ν_{ij}^*, and G_{12}^* against θ.

To validate the results obtained from the homogenization method, direct structural analysis of the 2D-lattice plates with auxetic unit cells is performed. Finite element models of 2D-lattice plates with the considered re-entrant hexagonal unit cells are created in a commercial finite element software, MSC Marc. In the direct structural analysis, the lattices, each of which consists of 64×64 unit cells, are subjected to pure bending and torsional loading for determining E_i^*, ν_{ij}^*, and G_{12}^* .For example, pure bending is applied by using the prescribed rotations on the two opposite edges, schematically shown in figure 3. The reactions at the non-free edges and the resulting rotations at the free edges are then used to compute the effective bending moduli and Poisson's ratios directly from equation (8).

Table 1 shows the effective bending moduli E_i^*, Poisson's ratios ν_{ij}^*, and shear modulus G_{12}^* of some auxetic 2D-lattice plates obtained from the homogenization method and the direct structural analysis. For the plates, four different topologies of re-entrant hexagonal unit cells, detailed in table 1, are selected to perform the validation in this study. Particularly good agreement is observed. For the values of G_{12}^*, there is a small fluctuation of the results between the homogenization method and the direct structural analysis since the effect of heterogeneity of the plate. The differences in table 1 are expected to reduce and converge to zero difference when there are enough unit cells in a 2D-lattice plate. Besides, when comparing the effective elastic properties of between the plates with re-entrant hexagonal unit cells having $h/l = 1$ and those having $h/l = 2$ at the same $\theta = -30°$, it is found that the values of E_1^* and G_{12}^* as well as the negative value of ν_{12}^* reduce while the values of E_2^* and the negative value of ν_{21}^* increase.

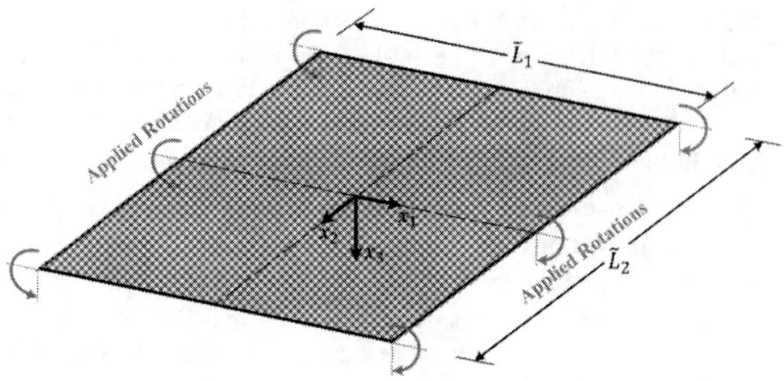

Figure 3. Example loading scheme of direct structural analysis of a lattice applied pure bending

Table 1. Comparison of the results from the homogenization method and direct structural analysis

Re-entrant hexagonal unit cell		Homogenization method, $\boldsymbol{a}$	Direct analysis of the lattices, $\boldsymbol{b}$	Difference, $[(\boldsymbol{b}-\boldsymbol{a})/\boldsymbol{a}] \times 100\%]$
Re-entrant hexagon with $\theta = -5°$ and $h = l$	E_1^*	1.088×10^{-1}	1.088×10^{-1}	0.000%
	E_2^*	2.680×10^{-2}	2.680×10^{-2}	0.000%
	ν_{12}^*	-3.992×10^{-2}	-3.992×10^{-2}	0.000%
	ν_{21}^*	-9.835×10^{-3}	-9.835×10^{-3}	0.000%
	G_{12}^*	2.484×10^{-2}	2.468×10^{-2}	-0.644%
Re-entrant hexagon with $\theta = -15°$ and $h = l$	E_1^*	1.267×10^{-1}	1.267×10^{-1}	0.000%
	E_2^*	2.261×10^{-2}	2.261×10^{-2}	0.000%
	ν_{12}^*	-1.339×10^{-1}	-1.339×10^{-1}	0.000%
	ν_{21}^*	-2.389×10^{-2}	-2.389×10^{-2}	0.000%
	G_{12}^*	2.740×10^{-2}	2.709×10^{-2}	-1.131%
Re-entrant hexagon with $\theta = -30°$ and $h = l$	E_1^*	1.567×10^{-1}	1.567×10^{-1}	0.000%
	E_2^*	1.741×10^{-2}	1.741×10^{-2}	0.000%
	ν_{12}^*	-2.867×10^{-1}	-2.867×10^{-1}	0.000%
	ν_{21}^*	-3.186×10^{-2}	-3.186×10^{-2}	0.000%
	G_{12}^*	3.500×10^{-2}	3.469×10^{-2}	-0.914%
Re-entrant hexagon with $\theta = -30°$ and $h = 2l$	E_1^*	5.222×10^{-2}	5.222×10^{-2}	0.000%
	E_2^*	3.258×10^{-2}	3.258×10^{-2}	0.000%
	ν_{12}^*	-9.558×10^{-2}	-9.558×10^{-2}	0.000%
	ν_{21}^*	-5.963×10^{-2}	-5.963×10^{-2}	0.000%
	G_{12}^*	1.637×10^{-2}	1.627×10^{-2}	-0.611%

5. Conclusion

In this study, the effective out-of-plane elastic properties of auxetic 2D-lattice plates are determined using the homogenization method based on equivalent strain energy. The considered unit cells, including different re-entrant hexagonal unit cells, are created by varying the internal cell angles and aspect ratios of hexagonal geometry. In the employed homogenization method, the strain energy values of unit cells under different curvature modes are computed by the finite element method and used to determine the effective elastic properties. The effective elastic properties of the considered 2D-lattice plates obtained from the homogenization method compared satisfactorily with those obtained from direct structural analysis of the plates. It is found that varying internal cell angles of hexagonal geometry can significantly change the values of the effective elastic properties of the resulting 2D-lattice plates, especially effective negative Poisson's ratio. The effective negative Poisson's ratio can significantly increase the degrees of auxeticity of these plates. Reducing the internal cell angles to negative also allows E_1^* of the plates under out-of-plane bending to be significantly increased. Moreover, increasing the aspect ratio of the re-entrant hexagonal unit cells causes reductions of E_1^* and G_{12}^*, while negative Poisson's ratios still appear whenever θ is negative.

Acknowledgment

The first author wishes to acknowledge the Department of Mechanical Engineering, Faculty of Engineering, Chiang Mai University and Computational Engineering Laboratory, School of Civil Engineering and Technology, Sirindhorn International Institute of Technology, Thammasat University.

References

[1] Gibson L J Ashby M F Schajer G and Robertson C 1982 *Proceedings of the Royal Society of London*. **382** 1782
[2] Gibson L J and Ashby M F 1999 *Cellular solids: Structure and properties* (Cambridge: Cambridge University Press)
[3] Gonella S and Ruzzene M 2008 *Int. J. Solids Struct.* **45** 10
[4] Dos Reis F and Ganghoffer J 2012 *Comput. Mater. Sci.* **51** 1
[5] Sam P Nanakorn P Theerakittayakorn K and Suttakul P 2017 *Int. J. Mech. Mater. Des.* **13** 3
[6] Suttakul P Nanakorn P and Vo D 2019 *Arch. Appl. Mech.* **89** 9
[7] Vasiliev V V Barynin V A and Rasin A F 2001 *Compos. Struct.* **54** 2–3
[8] Vigliotti A and Pasini D 2012 *Mech. Mater.* **46** 0
[9] Zhang Y H Qiu X M and Fang D N 2008 *Int. J. Solids Struct.* **45** 13
[10] Zhang K Deng Z-C Xu X-J Meng J-M and Jiang X-H 2017 *Compos. B. Eng.* **114**
[11] Zhang W Dai G Wang F Sun S and Bassir H 2007 *Acta Mech. Sin.* **23** 1
[12] Lebée A and Sab K 2013 *Comput. Struct.* **127**
[13] Hohe J and Becker W 1999 *Compos. Struct.* **45** 2
[14] Suquet P M 1987 Homogenization techniques for composite media *Lecture notes in physics* (Berlin: Springer-Verlag) pp 193-278.
[15] Gaspar N Ren X Smith C W Grima J and Evans K E 2005 *Acta Mater.* **53** 8
[16] Liu Y and Hu H 2010 *Sci. Res. Essays*. **5** 10
[17] Grima J N Gatt R Alderson A and Evans K 2005 *Mol. Simulat.* **31** 13
[18] Prall D and Lakes R 1997 *Int. J. Mech. Sci.* **39** 3
[19] Scarpa F Panayiotou P and Tomlinson G 2000 *J. Strain Anal. Eng. Des.* **35** 5
[20] Alderson A Alderson K Chirima G Ravirala N and Zied K 2010 *Compos. Sci. Technol.* **70** 7

Steel fiber/polyethylene terephtalate composites

Rossarin Duekhodthod [1], Usanee Kitkamthorn[1,*], Pusit Mitsomwang[1], Tawatchai. Intaing[1], Rattana Borrisutthekul [1]

[1]School of Metallurgical Engineering, Institution of Engineering,
Suranaree University of Technology, Nakhon Ratchasima, 30000, Thailand

* Corresponding Author: k_usanee@sut.ac.th

Abstract
Currently, lightweight materials with high performance have been strongly needed to meet the requirements of advanced structural materials. One of the promising lightweight materials is the metals fiber reinforced plastic composite, which usually gives good specific strength and high toughness. However, de-bonding between metal fiber and plastic is the main problem in the production and utilization of the composite. Recently, Katayama, et., al. studied the dissimilar materials joining between plastics and metals such as PET/SUS 304 and so on. The joint was succeeded with the reaction between the metal oxide and functional group in the molecule of the polymer. The possibility to join metal/plastic may provide the new approach to produce new design composites. Thus, the objective of the study is to investigate the feasibility to produce stainless steel fiber reinforced PET plastics composite (SFRP). In experiments, stainless steel mesh and PET sheet with 0.5 mm of thickness were used as materials to fabricate the SFRP composite. The hot pressing of sandwich stacks with inserted SUS304 wire mesh between two PET sheets was performed at 120°C. Results indicated that high strength composite without de-bonding could be obtained.

Keywords: Polyethylene terephthalate (PET), Stainless Steel grade 304 (SUS 304)

1. Introduction
Automotive part manufacturers have been challenged by new technologies concerned with manufacturing electric vehicles. In order to increase the distance of driving per charge, the application of lightweight materials to reduce the energy consumption rate of electric vehicles, is needed [1]. In the past 30 years ago, lightweight materials have been rapidly developed from high strength steel, to aluminum alloy, and then composites materials. Various composites materials have been developed based on using the difference of the raw materials used such as CFRP composites that are produced with carbon fiber and plastic. However, carbon fiber reinforced plastic provides low toughness, new fiber-reinforced plastic has been needed to be developed in order to improve toughness. The metal fiber-reinforced materials exhibit a 3 to 5 times higher elongation than typical continuous carbon and glass fiber composites and can therefore provide an answer to improve toughness [2-3]. However, the plastic matrix and metal reinforcement are difficult to bond. Therefore, after production, delamination between plastic matrix and metal fiber is easily found [4]. In order to solve the problem, using the compatibility

between steel fiber and plastic to make fiber reinforced plastic composites will provide the opportunity to obtain high quality materials. Katayama et al., [5] applied the laser bonding for stainless steel/PET composite fabrication. They succeeded in producing of good bonding between stainless steel and PET by the reaction between a functional group in the molecule of the PET and the chromium oxide film on the stainless-steel surface. Therefore, in this work, new composites materials developed by using PET as a matrix, and stainless-steel as fiber have been studied in order to study the feasibility of producing new composites materials.

2. Materials and Methods

2.1 Materials

Stainless steel fiber reinforced plastic composite (SFRP) was produced from stainless steel meshes and polyethylene terephthalate (PET) plastic. The stainless-steel wire meshes as reinforced materials were cut to a size of 70x20 mm. For the plastic matrix, the polyethylene terephthalate (PET) sheets with the size of 60x190x0.5 mm were used. The material characteristic and their mechanical properties of both materials are expressed in Table 1.

Table 1. Materials characteristics and their mechanical properties.

Materials	Wire Ø (mm)	Opening (mm)	Maximum Load at 15 mm of the width of the specimen (N)
PET	-	-	169
SUS304 mesh 60	0.15	271.0	485
SUS304 mesh 100	0.10	0.153	386
SUS304 mesh 200	0.055	0.432	314

2.2 Experimental procedures

SFRP composites were produced by using the compression molding technique with the MODEL PR2D-W3tech00L350 PM-WCL machine as shown in Figure 1. The sandwich stack of stainless-steel mesh in the middle between the PET sheets was performed as shown in Figure 2. The compressive molding process was performed at a mold temperature of 120 OC, the pressure of 60 psi, and a curing time of 5 minutes. After curing, the mold was opened and the sandwich stack was cooled by dipping into the water.

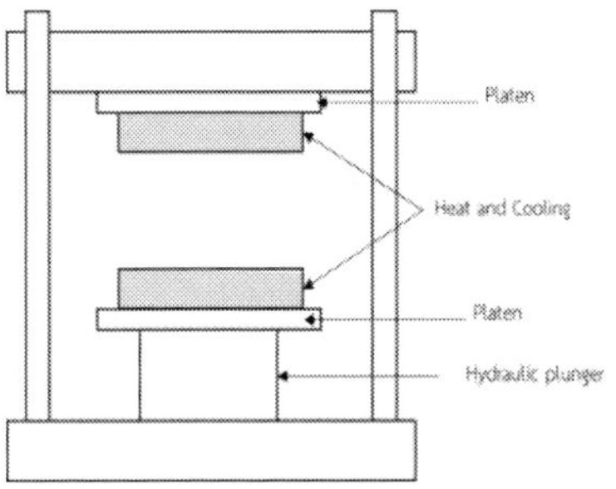

Figure 1. Compression Moulding machine

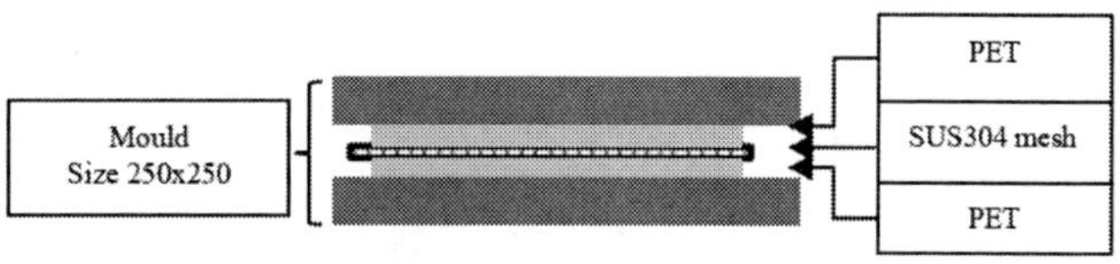

Figure 2. Schematic of SFRP composites stack layer. (PET/SUS304 mesh /PET)

The integrity of SFRP composites was then investigated by visual inspection, macrostructure observation. The macrostructure observation of SFRP composite cross-section was performed under the stereo microscope and scanning electron microscopes (SEM). Moreover, SFRP composites were cut to the tensile specimens as shown in Figure 3. The tensile testing was performed by Instron 4505 with a crosshead moving speed of 2 mm/min. After tensile testing, the failure samples were characterized under SEM observation.

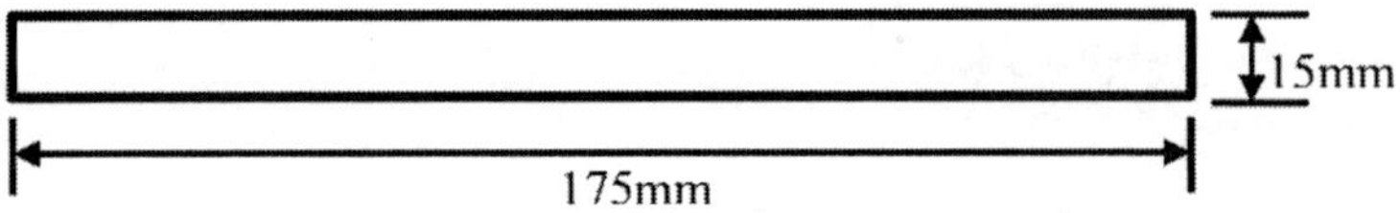

Figure 3. The tensile test specimen shape.

3. Results and discussion

3.1 Visual Inspection of SFRP composite

The visual inspection results of the SFRP composite samples with the different wire sizes of stainless-steel mesh were shown in Figure 4. Two layers of PET and stainless-steel mesh were perfectly welded into a single sheet of SFRP composites. The width at the end of all SFRP samples as indicated by the red dash line rectangular was wider than the width at the center of the SFRP sample. Moreover, in SFRP composites with #200 of stainless-steel wire mesh, the tearing of stainless-steel wire mesh was found at the edge of the sample as indicated in Figure 5. The tearing of stainless-steel wire at end of the SFRP sample was caused by that the pressing process was carried out with no constraint. Thus, in order to avoid such ending effect on the tensile test result, the tensile samples were cut only from the center of the SFRP composite sample.

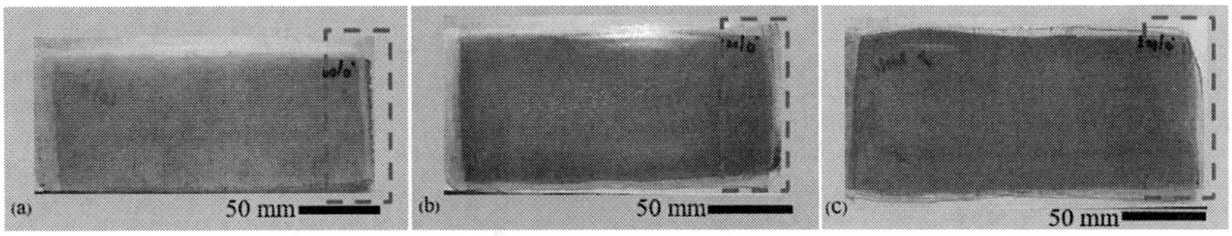

Figure 4. The SFRP composites formed with different stainless-steel wire mesh sizes (a) SFRP SUS304 mesh 60 (b) SFRP SUS304 mesh 100 (c) SFRP SUS304 mesh 200

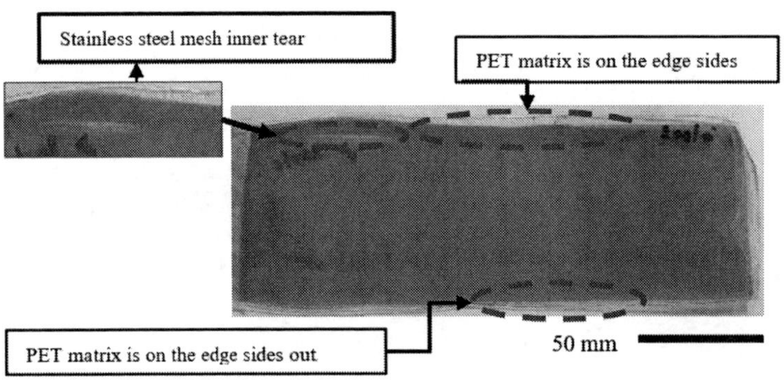

Figure 5. Defect on the sample SFRP with SUS304 mesh 200 composites.

3.2 Cross-sectional observations.

Figure 6 shows the cross-section observation of the SFRP composites reinforced with various stainless-steel mesh sizes. From Figure 6, it could be realized that the PET was fully filled into the gaps between the stainless-steel wire mesh. No porosity and gaps between the steel wire mesh and the PET matrix were found under optical microscope observation. However, at higher magnification observation of cross-section of SFRP composites, porosity, and PET unfilled in the hole between the stainless-steel wire mesh were observed as indicated in the dark spots in Figure 6. Although porosity and gaps were present, good bonding between PET stainless steel was achieved as realized in Figure 7(c), Figure 7(g), and Figure 7(j).

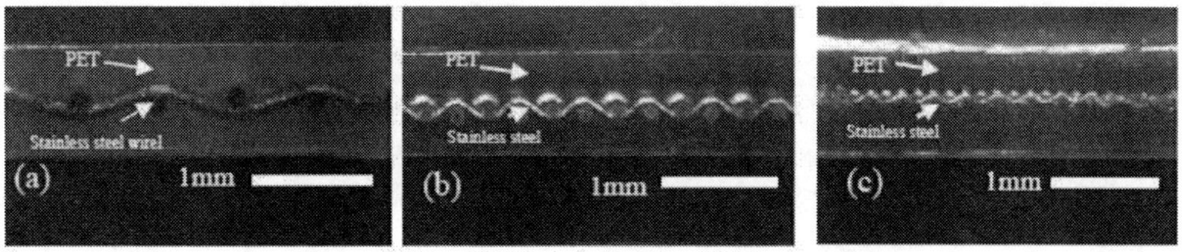

Figure 6. Macrostructure of a cross-section of the SFRP composites, difference stainless - steel wire mesh, (a) SFRP SUS304 mesh 60 (b) SFRP SUS304 mesh 100 (c) SFRP SUS304 mesh 200 composites.

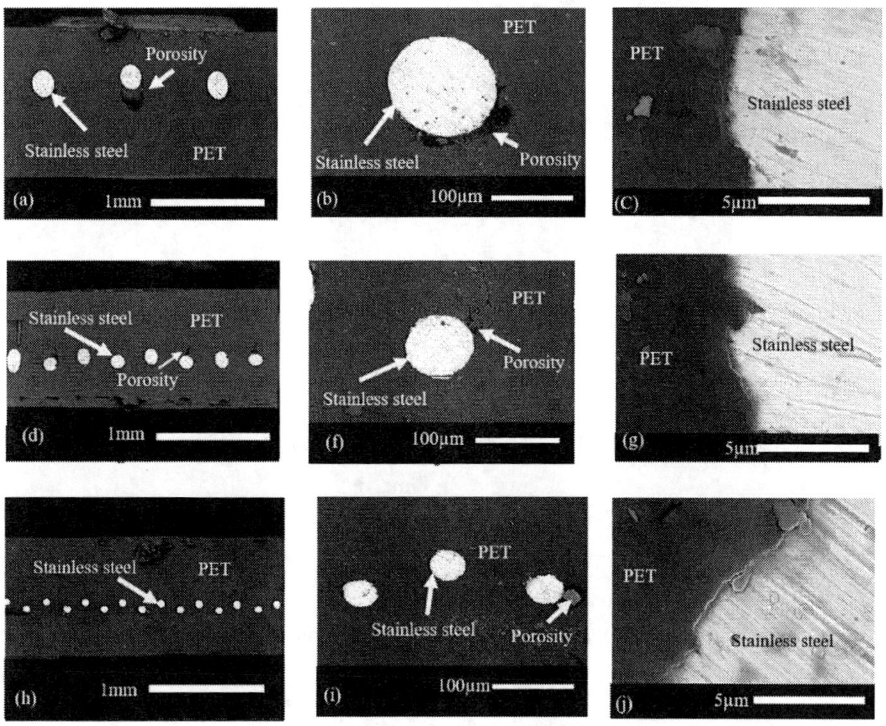

Figure 7. Shows the cross-section of the SFRP composites by SEM.

3.3 Tensile testing

Figure 8 shows the maximum tensile resistance loads of SFRP composites with different mesh sizes and the theoretical maximum tensile resistance load of SFRP composites. The theoretical maximum tensile resistance loads of SFRP composites were calculated from the summation of maximum load capacity of each material that were used to fabricate the SFRP composites. From Figure 8, the experimental maximum tensile resistance loads of the SFRP composites were better than that of the theoretical ones. Moreover, the maximum tensile load of SFRP composites decreased with the lower load capacity of stainless-steel wire mesh used as reinforcement. The elongation of SFRP with different stainless steel wire mesh was shown in Figure 9. It could be recognized that the SFRP composites were ductile materials due to high elongation. Using a larger stainless-steel wire mesh increased the elongation of the SFRP. Moreover, the overview of fracture specimens after the tensile test were shown in Figure 10. The stainless-steel wire was broken in the perpendicular direction to tensile loading. SEM micrographs as shown in Figure 11 reveal the ductile fracture of stainless-steel wire. Furthermore, the de-bonding of PET and stainless-steel wire after the tensile test was observed. Based on the above results, it could be inferred that new SFRP composites have good mechanical properties. Delamination of PET and stainless steel in the fracture samples occurred during tensile testing.

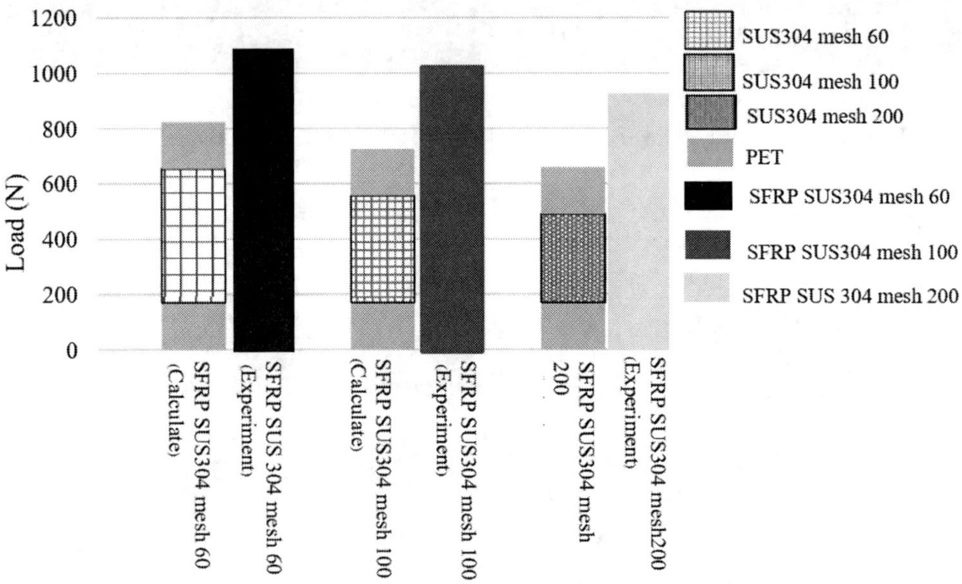

Figure 8. The Maximum tensile loading of SFRP in three different stainless-steel wire mesh sizes fiber in the PET matrix.

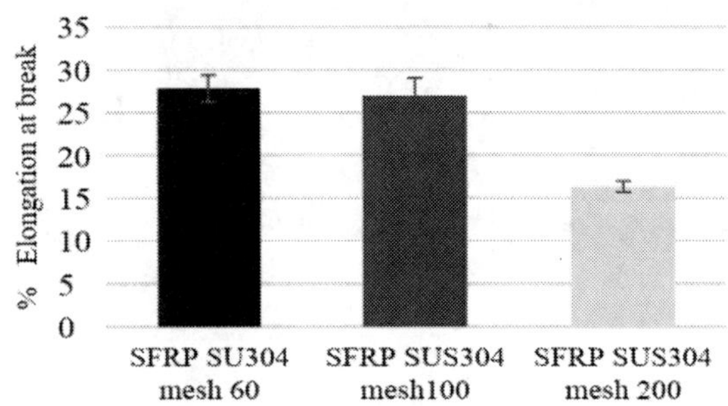

Figure 9. The % elongation at break of SFRP composites in three different stainless-steel wire mesh fiber size in the PET matrix

IOP Conf. Series: Materials Science and Engineering 1137 (2021) 012026 doi:10.1088/1757-899X/1137/1/012026

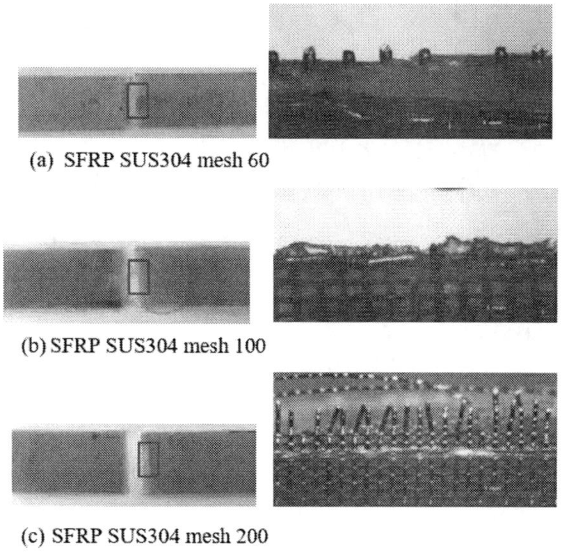

Figure 10. Fracture surface of the SFRP composites, three different stainless-steel wire mesh fiber size in the PET matrix.

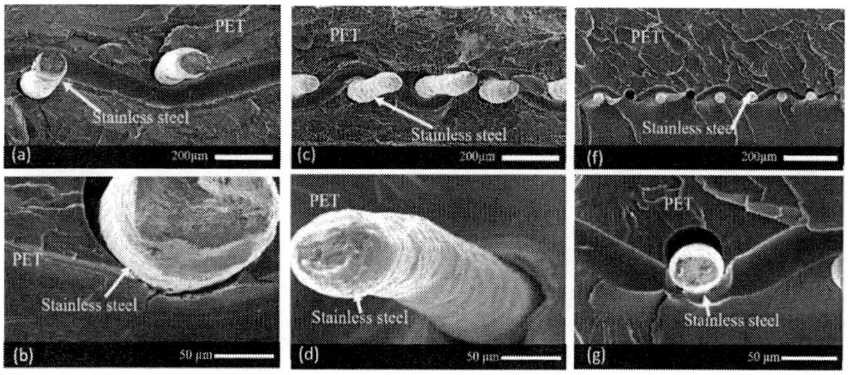

Figure 11. Fracture surfaces of the SFRP composites, (a, d) SFRP SUS304 mesh 60, (b, e) SFRP SUS 304 mesh 100, (c, f) SFRP SUS 304 mesh 200 composites by SEM.

4. Conclusion

According to the results, SFRP composites made by stainless steel wire mesh and PET were successfully produced by using a compressive molding process. Although some defects, such as porosities and no fill with PET, were found the mechanical properties of developed the SFRP composites were still better than that of raw materials. Moreover, good bonding between PET and stainless-steel wire mesh was achieved, after production. However, after the tensile test, de-bonding between the PET and the stainless-steel wire was still found. Ductile fracture of stainless-steel wire was also observed. The de-bonding between the PET and stainless steel after the tensile test will be possibly solved by refining the compressive molding process conditions in future work. Finally, it could be inferred that SFRP composites between PET and stainless steel are the potential candidate list as a new composite material used in lightweight structures.

Acknowledgments
The authors gratefully acknowledge the IMIARU Research Unit at the Suranaree University of Technology in support of the research work.

References
[1] Holmes M 2017 *Reinf. Plast.* **61**(5) 294-298
[2] Mosleh Y Clemens D Gorbatikh L Verpoest I and Vuure A 2015 *J. Reinf. Plast. Compos.* **34**(8) 624-635
[3] Callens M G Gorbatikh L and Larissa G 2014 *Compos. Part A Appl. Sci. Manuf.* **61** 235-244
[4] Yu Z and Zhou A 2013 *Mech. Adv. Mater. Struct.* **20**(5) 361-372
[5] Katayama S and Kawahito Y 2008 *Scr. Mater.* **59** 1247-1250

Sintered Fe-Cr-Mo-Si-C alloys produced from pre-alloyed Fe-Cr based powders admixed with 4% SiC for high performance applications

A.N. Soe [1], P. Wila [2], M. Morakotjinda [2], T. Yotkaew [2], P. Karin [1], N. Ohtake [3] and R. Tongsri [2,*]

[1]Department of Automotive Engineering, School of Engineering, King Mongkut's Institute of Technology Ladkrabang, Chalongkrung Road, Ladkrbang, Bangkok, 10520 Thailand
[2]Particulate Materials Processing Technology (PMPT) Laboratory, MTEC, 114 Thailand Science Park, Paholyothin Rd., Khlong Nueng, Khlong Luang, Pathum Thani, 12120 Thailand
[3]Department of Mechanical Engineering, Tokyo Institute of Technology, Chome -12-1 Ookayama, Meguro City, Tokyo 152-8550 Japan

* Corresponding Author: ruangdt@mtec.or.th

Abstract
The use of pre-alloyed Fe-Cr based powders has been steadily increasing in the automotive industry. Mechanical properties can be tailored due to phase transformations controlled by alloying elements and cooling rates. Employing silicon carbide (SiC) as a source of alloying elements has been proved to be possible. In this study, three different pre-alloyed Fe-Cr based powders, namely Astaloy CrA, Astaloy CrL, and Astaloy CrM, were alloyed with 4 wt. % SiC powder. The powders were mixed and pressed into tensile test bars. The green compacts were sintered at 1250 °C for 45 minutes under the cracked ammonia gas and slowly cooled (<0.1°C/s) to room temperature. All experimental sintered alloys manifested a common feature resembling those of full pearlitic ductile irons, i.e., the feature consisted of black nodular particles enveloped with hyper-eutectoid steel matrix (grain boundary or pro-eutectoid cementite plus pearlite). The tensile testing indicated that Alloy-3 (Astaloy CrM + 4%SiC) was observed to have better mechanical properties in comparison with Alloy-1 (Astaloy CrA + 4%SiC) and Alloy-2 (Astaloy CrL + 4%SiC), apart from slightly lower elongation than Alloy-2. Regardless of SiC addition, the presence of chromium (Cr) content was observed to substantially influence mechanical properties and microstructure. According to the investigation of fracture surfaces by SEM, mixtures of shallow dimples and cleavage fractures were examined in the sintered alloys.

Keyword: Sintered alloys, SiC addition, Fully pearlitic matrix, Interlamellar spacing

1. Introduction
Powder metallurgy (PM) has been recognized as a cost-effective as well as sustainable manufacturing process over decades due to low energy consumption and less material waste [1]. Traditionally, nickel

(Ni), copper (Cu), and molybdenum (Mo) have been used as essential alloying elements in PM steels. However, these elements are recently getting replaced by cost-effective alloying elements such as chromium (Cr) and manganese (Mn) due to fluctuation of price, the difficulty of recycling, and health issues in recent years. One issue concerning Cr and Mn is that they are oxidation-prone unless a proper sintering atmosphere is employed [2]. Despite the concern on oxidation vulnerability, the sintering of chromium-containing pre-alloyed-powder compacts has already been proved to be possible at high temperatures (1120 to 1250°C) under the proper sintering atmosphere (hydrogen/nitrogen mixture) [3]. Therefore, the utilization of pre-alloyed powder has been steadily increasing for high-performance applications since their properties can be tailored by sinter hardening, high-temperature sintering, warm compaction, and alloying addition [4]. As SiC exhibits a wide range of properties in terms of excellent thermal stability, thermal shock resistance, hardness and wear resistance, there has been a lot of on-going studies related to the sinterability of SiC with Fe and Fe based powders to produce metal matrix composites (MMC) [5-9]. Tensile strength and hardness of sintered Fe-SiC were observed to improve with increasing sintering temperature and smaller SiC particle size [7]. Likewise, the addition of SiC led to the improvement of strain at break in the case of sintered 316L+SiC composite, except for the reduction in sintered density [8]. It was previously reported that the interaction between SiC and iron (Fe) or Fe based powders above 1100 °C resulted in decomposition of SiC, and both Si and C atoms diffused into the Fe or Fe based powders, contributing to the formation of fully pearlite or ferrite/pearlitic microstructure [9]. Similar behavior was detected to exist when SiC was admixed into Fe-Mo-Si-C alloys [10-11]. In addition to that, sintering of Fe-Cr-Mo + SiC compacts followed by a slow cooling rate (0.1°C/s) was found to achieve a microstructure similar to that of fully pearlitic ductile iron [12].

In the present work, the authors aim to explore the effect of Cr content in pre-alloyed Fe-Cr based powders on microstructure and mechanical properties of sintered alloys produced from pre-alloyed Fe-Cr based + SiC powder compacts.

2. Experimental procedure

2.1 Material Preparation

The starting powders used in this study were water atomized pre-alloyed powder grades, namely Astaloy CrA, Astaloy CrL, and Astaloy CrM supplied by Höganäs AB, Sweden. As-received powder sizes were in the range of 45-150 μm. The pre-alloyed powders were admixed with 1% zinc stearate as a lubricant and 4% SiC as an alloying source of C and Si. Powder mixtures were blended in a mixer for 1 hr and pressed into tensile test bars using a uniaxial compaction machine at room temperature, according to MPIF-10 in Figure 1a. In the sintering process, the samples were delubricated by holding at 600 °C for 1 hour and 700°C for 30 minutes under the argon (Ar) atmosphere, and the samples were heated with a heating rate of $4.8°C/min$ up to isothermal sintering temperature. The isothermal sintering was conducted at 1250°C for 45 minutes under the cracked ammonia gas (75% Hydrogen and 25% Nitrogen). The sintered parts were cooled down to room temperature with a rate of $2.89\ °C/min$ in the sintering furnace. The nominal composition of investigated materials is summarized in Table 1.

2.2 Material characterization

The tensile test was performed on a Universal Testing Machine (Instron 8801) at a cross speed rate of 0.1 mm/min in Figure 1b. The hardness value of sintered materials was evaluated by conducting on Rockwell B macro hardness tester. To observe the microstructure, the cross-section of tensile test bars was cut and hot mounted using thermoplastic resin for 15 minutes. The microstructure was analyzed by optical microscopy (OM) and scanning electron microscopy (SEM). The area fraction of microstructural components was evaluated by an image analysis software (Image-J) utilizing several OM micrographs taken on polished samples. In pearlitic steel, the interlamellar spacing and pearlite colony size are very important parameters that control the mechanical properties. Therefore, interlamellar spacing was measured by the circular grid test method [14], and pearlite colony size was determined by the linear intercept method following the procedure of ASTM-E112 (2010). In the case of fractography analysis, the tensile fracture surface of each sintered alloy was studied by SEM.

(a)

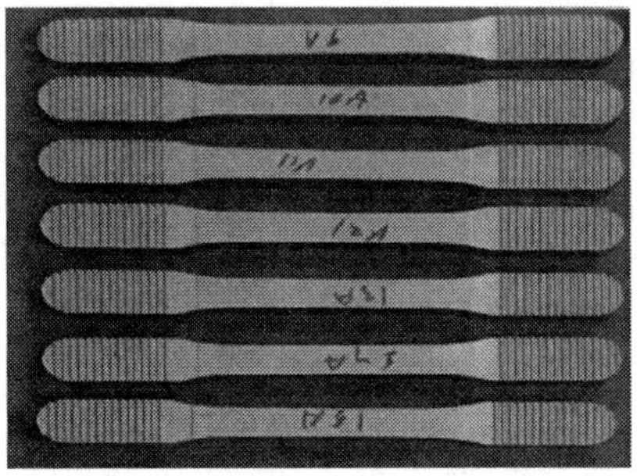

(b)

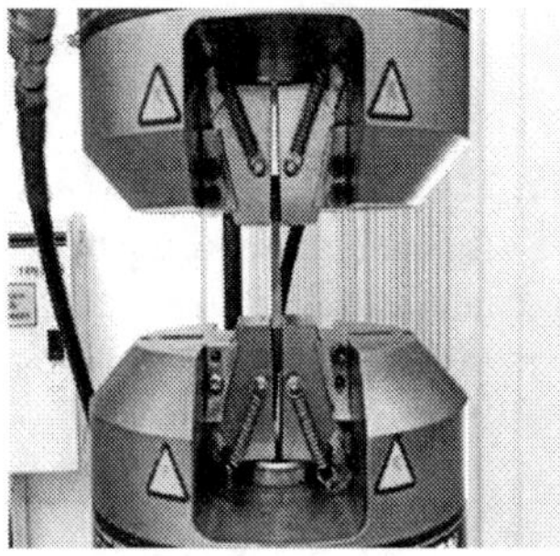

Figure 1. MPIF-10 standard tensile test bar (a) [13] and an image of tensile testing (b).

Table 1. The nominal composition of powder mixes (wt.%).

Remark	Base powder	Admixed alloying element	Fe	Cr	Mo	Si	C
Alloy-1	Astaloy CrA	SiC	Bal.	1.8	-	2.8	1.2
Alloy-2	Astaloy CrL	SiC	Bal.	1.5	0.2	2.8	1.2
Alloy-3	Astaloy CrM	SiC	Bal.	3.0	0.5	2.8	1.2

3. Results and discussion

3.1 Microstructure

Even though the composition differed to some extent, the microstructure of all sintered alloys somehow unveiled a common feature closed to those of fully pearlitic ductile irons. The feature generally consisted of black nodular particles enveloped with pearlitic matrix in which pro-eutectoid cementite network (PCN) is precipitated along prior-austenite grain boundaries (PAGBs) in Figure 2a-2c and a small fraction of porosity. The area percentage of microstructure components is displayed in Figure 3. Measurements of pearlite structure parameters including interlamellar spacing and pearlite colony size were conducted and given in Table 2. The measured pearlite interlamellar spacing decreases with increasing Cr content in sintered alloys (Table 2).

The sintered alloys with nominal compositions corresponding to those of Alloy-2 and Alloy-3, produced under sintering temperature of 1250°C and vacuum atmosphere also generated a fully pearlitic ductile iron-like microstructure [15], but with smaller PCN size and amount. The formation of PCN at PAGBs prior to pearlite transformation is a common phenomenon occurring in hypereutectoid Fe-C steels. In a previous study, austenitization temperature and transformation temperature affected PCN size and amount [15]. Large PCN was favored by large prior austenite grain sizes obtained at high austenitization temperatures. In our work, the sintering temperature, which is assumed to be equivalent to austenitization temperature, is 1250°C. Thus, it can be expected to have large prior austenite grain sizes in sintered alloys. The post-sintering cooling rate employed in our work was 2.89°C /min or 0.05°C/s, which was slow enough for C diffusion to form large carbide particles at high temperatures (in austenite (γ) + cementite (θ) filed). The slow cooling rate is unable to make the austenite phase bypass cementite precipitation. The disappearance of ferrite halo from all sintered Cr-containing alloys in this work, compared to sintered Fe-Mo-Si-C alloys [10-11], may imply that Cr addition suppresses the formation of ferrite halo around the black nodular particle in Figure 2d. In other words, Cr promotes pearlite transformation.

The reduction of measured pearlite interlamellar spacing with increasing Cr content in sintered alloys, as given in Table 2, is in good agreement with the analysis results using the neural network model

[16]. The effect of alloying element concentration on interlamellar spacing (S_0), measured in μm, is shown in Equation (1):

$$\log S_0 = -2.212 + 0.0514 \times [\mathrm{Mn}] - 0.0396 \times [\mathrm{Cr}] + 0.0967 \times [\mathrm{Ni}] - 0.002 \times [\mathrm{Si}] - 0.4812 \times [\mathrm{Mo}] - \log\left(\frac{T_E - T}{T_E}\right) \quad (1)$$

where [Mn], [Cr], [Ni], [Si] and [Mo] are alloying contents in wt. %, T_E is eutectoid temperature and $\Delta T = T_E - T$ is the undercooling below the eutectoid temperature. The addition of Mo and Cr decreases the pearlite interlamellar spacing whereas that of Si slightly influences interlamellar spacing. The sintered alloy with higher Cr content has fine interlamellar spacing.

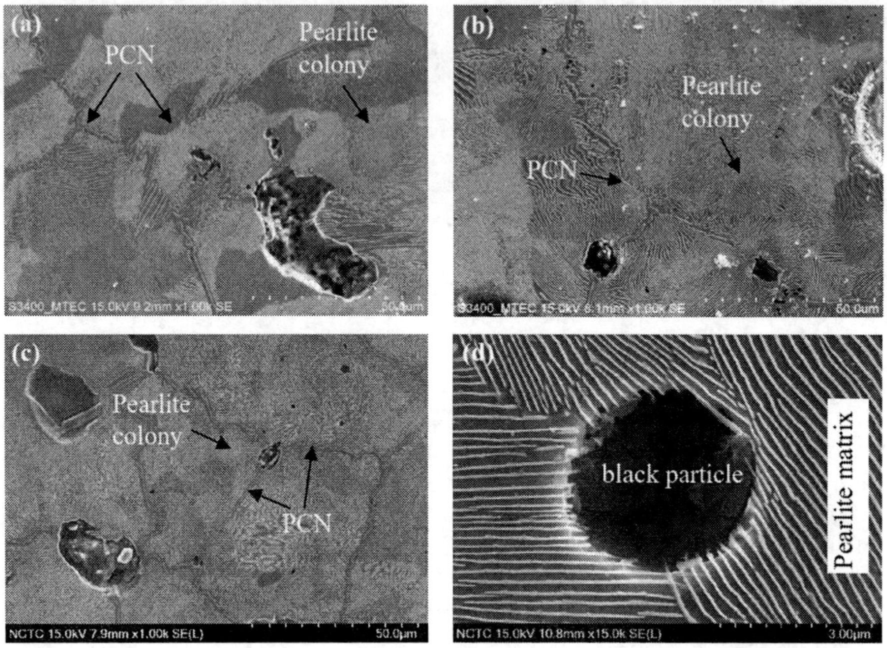

Figure 2. SEM micrographs of sintered Alloy-1 (a, d), Alloy-2 (b), and Alloy-3 (c).

Table 2. Evaluated interlamellar spacing and pearlite colony size.

Remark	Interlamellar Spacing (nm)	Pearlite colony size (μm)
Alloy-1	194.93	9.42
Alloy-2	278.68	9.65
Alloy-3	152.13	8.48

3.2 Mechanical properties

All experimental sintered alloys showed high ultimate tensile strengths (UTS) in the range of 700-800 MPa with yield strengths (YS) in the range of 300-400 MPa (Figure 4a). Both strengths varied slightly with Cr content in sintered alloys. The hardness value and elongation varied with Cr content (Figure 4b). Since the sintered alloys exhibited composite microstructure consisting of black particles, pores, and PCN-containing pearlite matrix (Figures 2 and 3), it was supposed that all microstructural components should have contributions to tensile strength. Taking the area fractions of microstructural components (Figure 3) into account, the pearlite matrices may be considered as the most important contributor to the mechanical properties of sintered alloys. As given in the literature [17-20], the strength and hardness of pearlite structure relied on refinements of interlamellar spacing and pearlite colony size, as given by the Hall-Petch relationship (Equation 2):

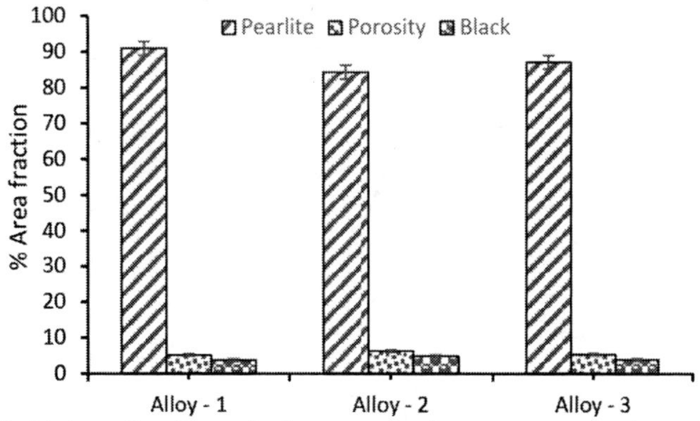

Figure 3. % Area fractions of microstructural components in sintered alloys.

$$\sigma = \sigma_0 + KS_0^{-1/2} \tag{2}$$

where σ is the stress, σ_0 is the friction stress, K is the Hall-Petch constant. With a comparison between tensile strength and hardness, given in Figure 4, it is noticed that the hardness is quite sensible to the Hall-Petch relationship. For the elongation value, it is also sensible to interlamellar spacing values of the pearlite structure (Figure 4b). The relative elongation of the steels with coarse lamellar pearlite is higher than that of the steels with fine lamellar pearlite [21]. However, microstructural parameters, such as interlamellar spacing, austenite grain size, and pearlite colony size are not sufficient to fully describe the structure and the mechanical strength of pearlite [22]. Further studies on the effects of strengthening mechanisms in pearlite will be needed.

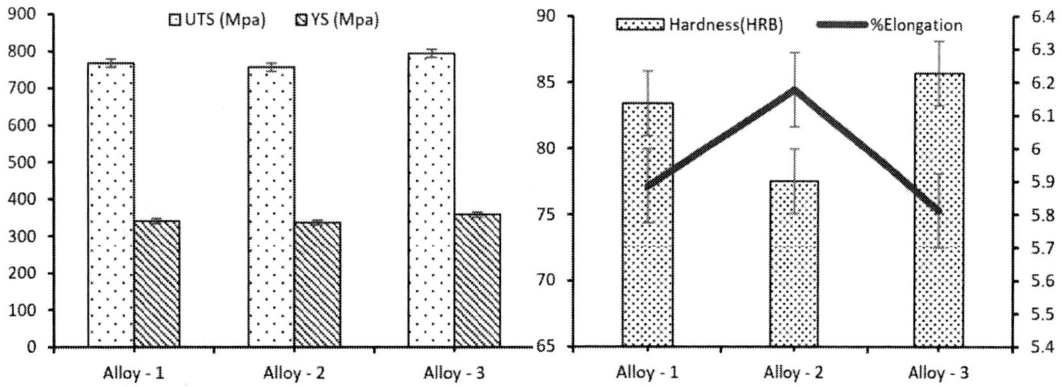

(a) **(b)**

Figure 4. Mechanical properties of sintered alloys (a) strengths and (b) hardness and %elongation.

3.3 Fractography Analysis

The fracture surface of the experimental sintered alloys is shown in Figure 5. All sintered alloys exhibited a large portion of cleavage characterized by a river pattern, small fractions of quasi-cleavage, and micro-dimples with sizes comparable to interlamellar spacing. The sites of black particles can be seen as empty holes in Figure 5. The fracture surface characters are related to tensile properties. As all sintered alloys have low elongation values but high tensile strengths, the fracture surfaces are dominated by cleavage, and smaller areas with dimples were observed. Some areas with quasi-cleavage are observed in Alloy-2 (Figure 5c) due to its highest elongation values among all experimental sintered alloys.

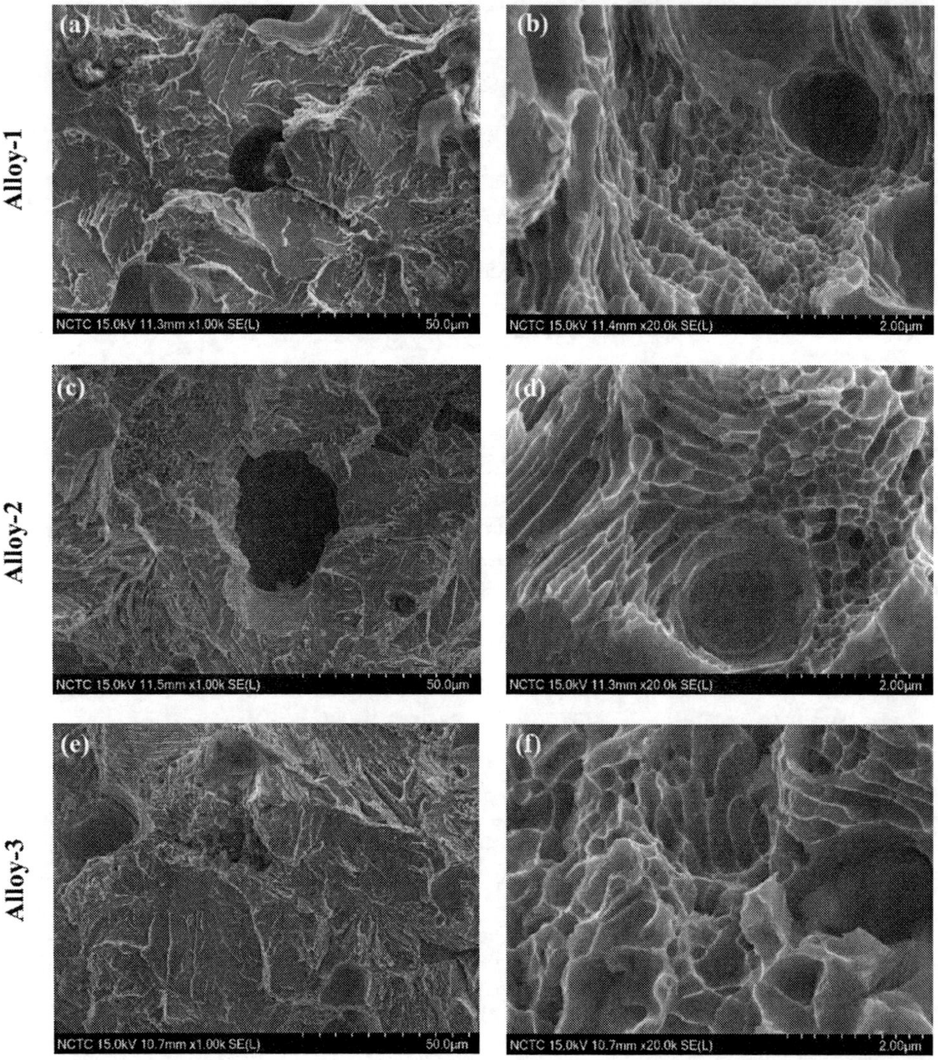

Figure 5. Fractography of sintered Alloy-1 (a-b), Alloy-2 (c-d), and Alloy-3 (e-f).

4. Conclusions

The microstructure of all sintered alloys shows a common feature similar to that of fully pearlitic ductile iron, i.e., the feature consists of black nodular particles enveloped with hyper-eutectoid steel matrix (grain boundary or pro-eutectoid cementite plus pearlite). Even though the ferrite hallo around black particles (core-shell structure) was discovered to exist in sintered Fe-Mo-Si-C composites [10-11], this phenomenon was not found in this study. Therefore, it can be concluded that the existence of a small portion of molybdenum (Mo) in this work might not substantially influence the final microstructure. Instead, the presence of chromium suppressed the ferrite halo formation around black nodular particles and might promote pearlite formation in some ways.

Surprisingly, although the fabrication and sintering were conducted under the same condition, refining of the interlamellar spacing by increasing chromium (Cr) content was observed in this work. Moreover, sintered Alloy-3 achieved higher strength and hardness values in comparison with Alloy-1 and Alloy-2. All experimental sintered alloys have a microstructure-property relationship explained by the Hall-Petch relationship. Fractography analysis confirmed that sintered alloys exhibited a large portion of cleavage fracture characterized by a river pattern, small fractions of quasi-cleavage, and micro-dimples with sizes comparable to the interlamellar spacing.

Acknowledgment

This work is funded by the project P1951261 supported by National Science and Technology Development Agency (NSTDA). The first author is also financially supported by TAIST-Tokyo Tech scholarship. The authors are grateful for supports from Thailand National Metal and Materials Technology Centre (MTEC) and King Mongkut's Institute of Technology Ladkrabung (KMITL).

References

[1] Han Y Engström U Zhang Q and Cao G 2018 *WorldPM* **2018** 32-37
[2] Hryha E Cajkova L Dudrova E and Nyborg L 2008 *Proc. Euro Int. Powder Metall. Congr. Exhib. Euro PM 2008* **1** 109-114
[3] Arvidsson J and Tryggmo A 1998 *Proc. PM World Congress & exhibition* (Shrewsbury, UK) **2** 253
[4] Ferraro C and Tunon E G 2015 *Material Today* **18** 117
[5] Frykholm R and Bergman O 2005 *Proc. 2005 Int. Conf. Powder Metall. Part. Mater. PowderMet* 211-226
[6] Chakthin S Poolthong N and Tongsri R 2008 *Adv. Mater. Res.* **55-57** 357-360
[7] Chakthin S Morakotjinda M Yodkaew T Torsangtum N Krataithong R Siriphol P Coovattanachai O Vetayanugul B Thavarungkul N and Poolthong N 2008 *J. Met. Mater. Min.* **18** 67-70
[8] Coovattanachai O Mima S Yodkaew T Krataitong R Morkotjinda M Daraphan A Tosangthum N Vetayanugul B Panumas A Poolthong N and Tongsri R 2006 *Proc. 2006 Int. Conf. Powder Metall. Part. Mater. PowderMet* 7161-7171
[9] Binder C Bendo T Pereira R V Hammes G de Mello J D B and Klein A N 2016 *Powder Metall.* **59** 384-393
[10] Ruangchai K Wiengmoon A Morakotjinda M Krataitong R Tanprayoon D Yotkaew T Tosangthum N Patakham U and Tongsri R 2017 *Key Eng. Mater.* **751** 47-52
[11] Ruangchai K Wiengmoon A Morakotjinda M Tosangthum Y and Tongsri R 2018 *Journal of Physics: Conf. Series* **1144** 012099
[12] Ruangchai K Wiengmoon A Krataitong M Yotkaew T Tosangthum N and Tongsri R 2018 *Journal of* Physics: Conf. *Series* **1144** 012147
[13] Čiripová L Hryha E Dudrová E and Výrostková A 2012 *Mater. Des.* **35** 619-625
[14] Vander Voort G F 2015 *Pract. Metallogr.* **52** 419-436
[15] Elwazri A M Wanjara P and Yue S 2005 *Metall. Mater. Trans. A Phys. Metall. Mater. Sci.* **36** 2297-2305

[16] Capdevila C Caballero F G and Andrés C G D 2005 *ISIJ Int.* **45** 229-237
[17] Modi O P Deshmukh N Mondal D P Jha A K Yegneswaran AH and Khaira H K 2001 *Mater. Charact.* **46** 347
[18] Taleff E M Lewandowski J J and Pourladian B 2002 *JOM.* **54** 25-30
[19] Toribio J González B Matos J C and Ayaso F J 2016 *Metals* (Basel) **6** 318
[20] Elwazri A M Wanjara P and Yue S 2005 *Mater. Sci. Eng. A* **404** 91-98
[21] Izotov V I Pozdnyakov V A Luk'yanenko E V Usanova O Y and Filippov G A 2007 *Phys. Met. Metallogr.* **103** 519-529
[22] Takahashi T Ponge D and Raabe D 2007 *Steel Res. Int.* **78** 38-44

IOP Conf. Series: Materials Science and Engineering 1137 (2021) 012028 doi:10.1088/1757-899X/1137/1/012028

The effect of molybdenum on the microstructure and mechanical behaviour of the sintered Fe-Mo-Mn-Si-C composite

Thunyamon Nithimethakul[1], Preechar Karin[1], Naoto Ohtake[2], Pongsak Wila[3], Thanyaporn Yodkaew[3], Bhanu Vetayanugul[3,*], Monnapas Morakotjinda[3], and Ruangdaj Tongsri [3]

[1]School of Engineering, King mongkut's Institute of Technology Ladkrabang, Chalongkrung Road, Ladkrabang, Bangkok 10520, Thailand.
[2]Department of Mechanical Engineering, Major in Engineering Sciences and Design, Tokyo Institute of Technology, Tokyo, 152-8550, Japan.
[3]Particulate Materials Processing Technology (PMPT), National Metal and Materials Technology Center, 114 Paholyothin Road, Khlong Nueng, Khlong Luang, Pathumthani 12120, Thailand.

* Corresponding Author: bhanuv@mtec.or.th

Abstract
Experimental sintered Fe-Mo-Mn-Si-C composites were prepared from 3 different pre-alloyed Fe-Mo-Mn powders, namely Fe-0.50Mo-0.15Mn, Fe-0.85Mo-0.15Mn and Fe-1.50Mo-0.15Mn, mixed with fixed 4 wt.% silicon carbide powder. Sintered Fe-SiC composite was also prepared, as a reference material, from pure Fe powder mixed with fixed 4 wt.% silicon carbide powder. All specimens were processed by using the 'press and sinter' method. Sintering was performed in a vacuum furnace at 1250°C for 45 minutes and slow cooling in the furnace. The microstructures of most sintered alloys showed a common feature consisting of a black particle enveloped with ferrite and pearlite. Sintered composites produced from high-molybdenum powders showed different matrices, in which some pearlite regions were replaced by BF/M- A structures. The Tensile and hardness test showed that the sintered alloys exhibited an increase of tensile strength, yield strength, and hardness with increasing molybdenum content. The pearlite + BF/M-A fractions of sintered composites played important roles in material strengthening. In the sintered composites produced from high-molybdenum powders, the carbide-free BF/M-A structure showed a promising strengthening effect.

Keywords: Sintering, Fe-Mo-Mn-Si-C, Pearlite, Carbide-free bainite, Mechanical property.

1. Introduction

The traditional process of 'press and sinter' is commonly used for producing particulate-reinforcement composites. The early attempts to produce silicon carbide (SiC)-reinforced metal matrix composites (MMCs) revealed that interfacial reaction between SiC and iron-base matrix occurred during heating and sintering [1-3]. In 316L/SiC system [1], interfacial reaction resulted in intergranular liquid phase formation. In Fe/SiC system [2, 3], SiC residual particle was enveloped with ferrite-pearlite matrix. Later, pre-alloyed powders, such as Fe-Mo [4] and Fe-Cr-Mo [5] powders, and diffusion bonded powders [6], were employed as matrices. The final microstructures of sintered composites strongly depended on pre-alloyed powder composition. In the Fe-Mo/SiC systems [4], ductile iron-like microstructure consisting of black (graphite-containing) particle surrounded with ferrite-pearlite matrix was a common microstructural feature. In the Fe-Cr-Mo/SiC systems [5], pearlitic ductile iron-like microstructure was obtained. In the diffusion bonded powders added with SiC systems [6], the matrices surrounding black particles contained bainite instead of pearlite.

These sintered composites based on pre-alloyed powders [4-5] showed promising tensile properties and had graphite-containing particles embedded in their matrices. The formation of graphite-containing particles is attributed to SiC decomposition, interfacial reaction, melting and re-solidification in SiC site vicinities. The interactions between SiC and iron-base powder particles was given by experimental evidence [7]. These interactions were also considered by thermodynamic models as given in [8, 9].

The materials consisting of graphite-containing particles embedded in iron base matrices are classified as self-lubricating composites due to lubricity of graphite [10-12]. The self-lubricating composites containing graphite particles are commonly applied as train brake linings [13,14]. Definitively, tribological properties of self-lubricating composites or brakes pads depend on lubricant characteristics [15, 16]. However, according to the recent study [17], the matrix of train brake linings plays an important role on braking performance, such as stable coefficient of friction (COF) and wear rate. To have such good braking performance, the brake matrix should have mechanical properties as good as tribological properties.

Thus, the sintered composites, having graphite-containing particles embedded in iron-based matrices, have potential for applications as train brake linings. This present work focuses on the influence of molybdenum content in pre-alloyed powders, used for making composite matrices, on mechanical properties of sintered composites. Pure iron powder was also used for producing a sintered composite as the reference material.

2. Materials and Methods

2.1 Specimen Preparation

Sintered composites were prepared from 4 different metal powders, including pure iron (Fe) and 3 pre-alloyed Fe-Mo powders, mixed with fixed 4 wt.% silicon carbide powder and 1 wt.% zinc stearate as a lubricant. The nominal compositions of experimental sintered composites are given in Table 1. Specimen preparation procedure included cold pressing and the sintering of powder mixture compacts. Green specimens were produced by compacting powder mixtures into standard tensile test bars (MPIF standard 10) with green density of 6.50 ± 0.01 g/cm^3. Green compacts were sintered at 1250°C for 45 minutes in a vacuum atmosphere. After sintering, the specimens were slowly cooled in the sintering furnace.

2.2 Hardness and tensile tests

Rockwell scale B (HRB) hardness test, with ball diameter of 1/16 inch and 100 kgf load, was used to measure the hardness of each specimen. This test was carried out on un-etched surface of a specimen. Five indentations on each sintered composite were pressed for obtaining a mean hardness value. For tensile test, sintered composites were tensioned at room temperature by using Instron 8801 Universal Instrument with load cell of 100kN and speed of 1 mm/min. Three specimens of sintered composite were tested for obtaining average the tensile properties.

2.3 Microstructure Observation

Microstructures of sintered composites were observed by optical microscopy (OM) and scanning electron microscopy (SEM) equipped with energy-dispersive X-ray spectroscopy (EDS). A surface of a specimen was abraded on silicon carbide grit papers to number 4ooo and then polished with 6, 3, and 1 μm diamond pastes. The polished samples were etched with 2% Nital and cleaned by ethanol. The SEM was operated at 20 kV to observe the details of microstructure. Phase identification was carried out by using X-ray diffraction (XRD) technique.

Table 1. Nominal composition of experimental sintered composites

Composite	Metal powder	SiC (wt. %)	Composite nominal composition (wt. %)				
			C	Si	Mo	Mn	Fe
0.00Mo	Fe	4.0	1.20	2.80	0.00	0.00	Bal.
0.50Mo	Fe-0.5Mo-0.15Mn	4.0	1.20	2.80	0.48	0.14	Bal.
0.85Mo	Fe-0.85Mo-0.15Mn	4.0	1.20	2.80	0.82	0.14	Bal.
1.50Mo	Fe-1.5Mo-0.15Mn	4.0	1.20	2.80	1.44	0.14	Bal.

3. Results and Discussion

3.1 Microstructure

All experimental sintered composites exhibited common microstructural features consisting of black particle(s) surrounded with white and dark areas (Figure 1). The black particles had different shapes from irregular to spherical ones. They were identified as graphite-containing particles. The white area contained polygonal ferrite grains the forms of halos around black particles. The dark area was revealed by SEM (Figure 2) to contain a pearlite structure and bainitic ferrite (BF) and martensite-austenite (M-A) constituent. The BF/M-A structure with no carbide precipitation was observed in sintered composites with high molybdenum contents (sintered 0.85Mo and 1.50Mo composites). Area fractions of microstructural components in sintered composites were measured and presented in Table 2.

The microstructural features as shown in Figure 1 resemble those of sintered composites [3,4, 10-12] and of ferritic-pearlitic ductile irons [17, 18]. The formation of ductile iron-like microstructure was briefly explained in [4]. The conditions, under which ferrite and pearlite transformations occur in ductile irons, are also explained in [19]. Discussion on such issues is not repeated here but the attention is focused on the molybdenum effect on the sintered composite matrix feature.

Pearlite structures are found in all experimental sintered composites produced under slowl continuous cooling. The existence of a pearlite structure indicates that cementite (Fe_3C) precipitation does form at high temperatures, under which substitutional alloying elements (silicon and molybdenum) are possible to undergo partitioning. XRD patterns given in Figure 3 show strong peaks corresponding to α-iron or ferrite (labelled) and weak peaks corresponding to cementite (not labelled). Other alloy carbides are not detected by XRD. This suggests that the partitioning of silicon from cementite to ferrite occurs to favor cementite precipitation and hence pearlite transformation. It is well-known that silicon retards the precipitation of cementite from different parent materials or phases, such as cast ions, martensite and austenite [20]. This is due to the fact that silicon has low solubility in cementite (Fe_3C), i.e., up to 4.5 at. % Si [21]. Theoretical study suggests that the formation of $(FeSi)_3C$ carbide is associated with high formation energy [22], which is not thermodynamically favored.

In the sintered composites with high molybdenum contents, such as sintered 0.85Mo and 1.50Mo composites, the BF/M-A structure forms at low temperatures. The BF/M-A formation is strongly under the influence of molybdenum. Under continuous cooling, molybdenum additionally retards ferrite and pearlite transformations by shifting the corresponding transformation fields to lower cooling rates [23, 24]. The delay of high-temperature transformations opens up the window for low-temperature transformations even under a slow cooling. The BF/M-A constituent contains no carbide particles. This indicates that carbide precipitation retardation occurs at low temperatures in such sintered composites.

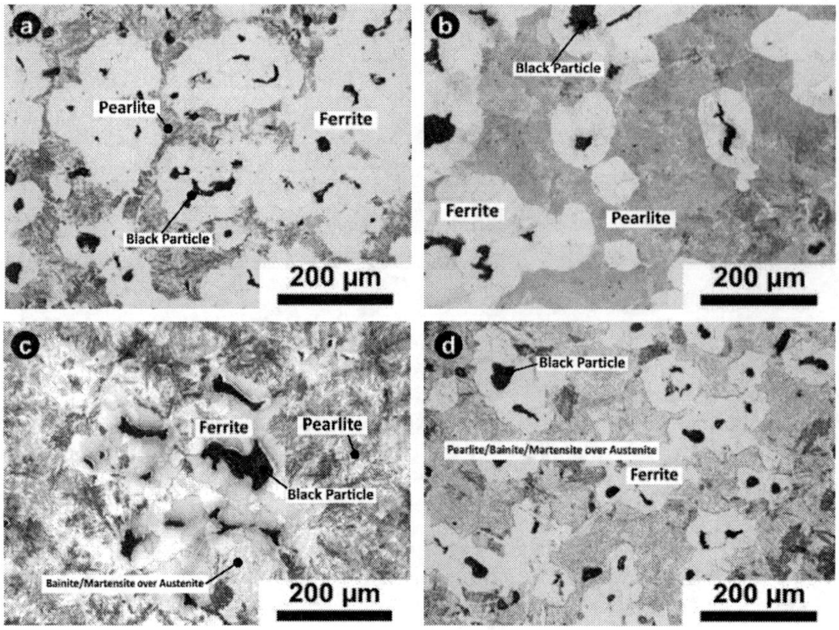

Figure 1. Optical micrograph of sintered composites (a) 0.00Mo composite, (b) 0.50Mo composite, (c) 0.85Mo composite, and (d) 1.50Mo composite.

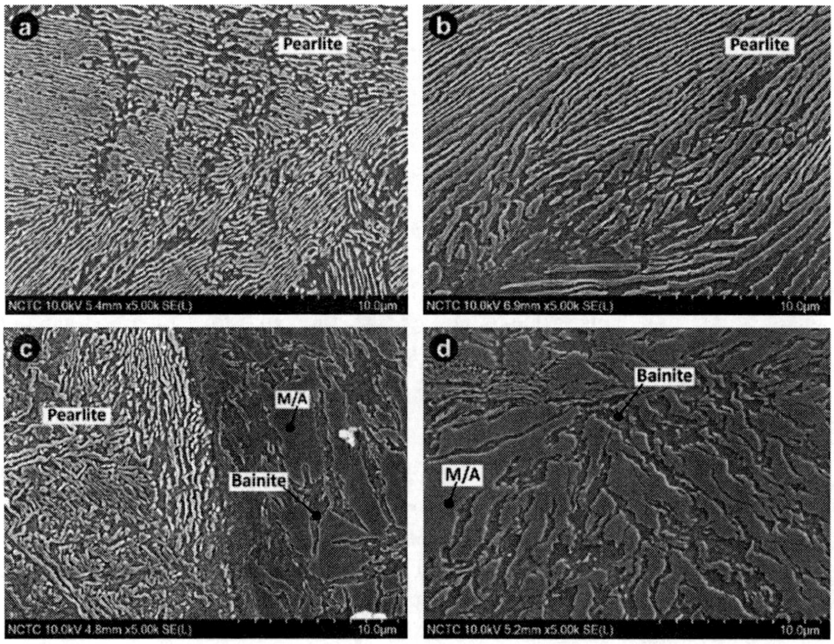

Figure 2. SEM micrographs of sintered composites (a) 0.00Mo composite, (b) 0.50Mo composite, (c) 0.85Mo composite, and (d) 1.50Mo composite.

IOP Conf. Series: Materials Science and Engineering 1137 (2021) 012028 doi:10.1088/1757-899X/1137/1/012028

Table 2. Total area of each phase of sintered materials.

Composite	Area fraction (%)		
	Black particle	Ferrite	Pearlite + BF/M-A
0.00Mo	3.8	66.2	30.0
0.50Mo	8.5	36.1	55.4
0.85Mo	19.0	23.2	57.8
1.50Mo	19.0	8.5	72.5

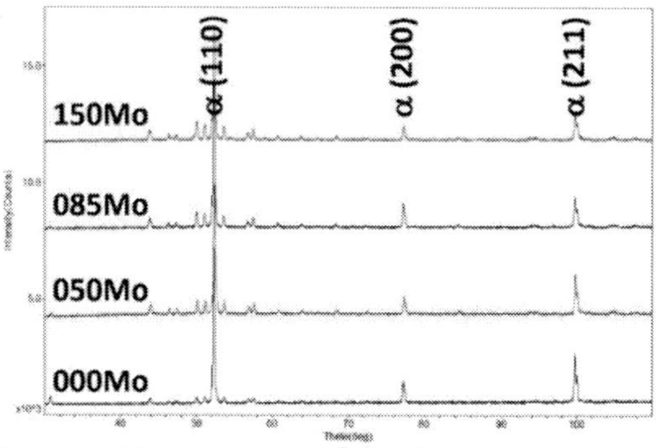

Figure 3. XRD patterns of the sintered composites.

3.2 Mechanical property

The Tensile strength and hardness of sintered composites increased with increasing molybdenum content while the values of elongation at break were in the range of 2.5-4.0 % (Figure 4). The strength and hardness of sintered composites depends strongly on pearlite + BF/M-A fractions, which are promoted by high molybdenum content. The fraction of soft phase or ferrite is supposed to dictate ductility of sintered composites, but experimental results do not give any evidences to support such a concept.

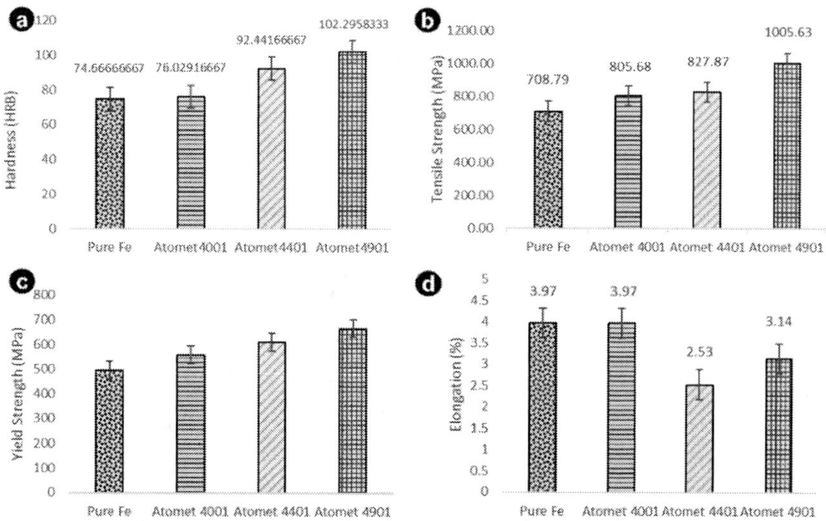

Figure 4. Mechanical properties of the sintered composites; (a)hardness, (b) ultimate tensile strength, (c) yield strength, (d) elongation at break

4. Conclusion

The influences of molybdenum on microstructure and mechanical property of sintered composites were investigated and the conclusions were drawn as given below.

1) General microstructures of sintered composites resembled those of ferrite-pearlite ductile irons.
2) Molybdenum retards ferrite and pearlite transformations and promotes BF/M-A formation.
3) Strength and hardness of sintered composites are influenced by pearlite + BF/M-A fraction.

Acknowledgments

This work was funded by the project P1951261 supported by National Science and Technology Development Agency (NSTDA). The first author is also financially supported by TAIST-Tokyo Tech scholarship. The authors are grateful for supports from Thailand National Metal and Materials Technology Center (MTEC) and King Mongkut's Institute of Technology Ladkrabung (KMITL). The authors are also grateful for the material for this research, courtesy from Rio Tinto Metal Powders of Canada (via P.S. Steel Co., Ltd., of Thailand and Rio Tinto Iron & Titanium (Suzhou) Co., Ltd., of China).

References

[1] Coovattanachai O Mima S Yodkaew T Krataitong R Morkotjinda M Daraphan A Tosangthum N Vetayanugul B Panumas A Poolthong N and Tongsri R 2006 *Proc. 2006 Int. Conf. Powder Metall. Part. Mater. PowderMet* 7161-7171
[2] Yodkaew T Morakotjinda M Tosangthum N Coovattanachai O Krataitong R Siriphol P Vetayanugul B Chakthin S and Poolthong N 2008 *J. Met. Mater. Miner.* **18** 57-61
[3] Chakthin S Poolthong N and Tongsri R 2008 *Adv. Mater. Res.* **55-57** 357-360
[4] Ruangchai K Wiengmoon A Morakotjinda M Tosangthum N and Tongsri R 2018 *J. Phys. Conf. Ser.* **1144** 012099
[5] Ruangchai K Wiengmoon A Krataitong R Yotkaew T Tosangthum N and Tongsri R 2018 *J. Phys. Conf. Ser.* **1144** 012147
[6] Szewczyk-nykiel A 2017 *Czas. Tech.* **6** 179-190
[7] Tongsri R and Vetayanugul B 2010 *J. Met. Mater. Miner.* **20** 45-49
[8] Klein A N Furlan K P Schroeder R M Hammes G Binder C Rodrigues Neto J B Probst S H and de Mello J D B 2015 *Powder Technol.* **271** 193-203
[9] Araya N Neves G O Filho A I R Aguilar C Biasoli de Mello J D Binder C Klein A N and Hammes G 2020 *Mater. Chem. Phys.* **253** 123442
[10] Araya N I Binder C Klein A N Hammes G De Mello J D B and Aguilar C 2018 *Mater. Res.* **21** 1-10
[11] Binder C Bendo T Hammes G Neves G O Binder R de Mello J D B and Klein A N 2017 *Carbon N. Y.* **124** 685-692
[12] De Mello J D B Binder C Hammes G Binder R and Klein A N 2017 *Friction* **5** 285-307
[13] Günay M Korkmaz M E and Özmen R 2020 *Eng. Sci. Technol. an Int. J.* **23** 421-431
[14] Wang X and Ru H 2019 *Materials* (Basel) **12**(2) 313
[15] Chen B Bi Q Yang J Xia Y and Hao J 2008 *Tribol. Int.* **41** 1145-1152
[16] Peng T Yan Q Li G, Zhang X Wen Z and Jin X 2017 *Tribol. Lett.* **65** 1-13
[17] Peng T Yan Q and Zhang X 2018 *Tribol. Lett.* **66** 1-13
[18] Iacoviello F Di Bartolomeo O Di Cocco V and Piacente V 2008 *Mater. Sci. Eng. A* **478** 181-186
[19] Lacaze J Sertucha J and Magnusson Åberg L 2016 *ISIJ Int.* **56** 1606-1615
[20] Kozeschnik E and Bhadeshia H K D H 2008 *Mater. Sci. Technol.* **24** 343-347
[21] Filonenko N and Babachenko A 2019 *East Eur. J. Phys.* **51** 46-51
[22] Jang J H Kim I G and Bhadeshia H K D H 2009 *Comput. Mater. Sci.* **44** 1319-1326
[23] Ackermann M Resiak B Buessler P Michaut B and Bleck W 2020 *Steel Res. Int.* **1900663** 1-9
[24] Hannula J Porter D Kaijalainen A Somani M and Kömi J 2019 *Metals* (Basel) **9**(3) 350

IOP Conf. Series: Materials Science and Engineering 1137 (2021) 012029 doi:10.1088/1757-899X/1137/1/012029

Effect of carbon content on microstructure and mechanical properties of sintered Fe-Mo-Mn-C alloys

A. Wanalerkngam[1], S. Boonmee[1], R. Krataitong[2], M. Morakotjinda[2], W. Pongsak[2], N. Tosangthum[2,*], and R. Tongsri[2]

[1]School of Metallurgical Engineering, Institute of Engineering, Suranaree University of Technology 111 University Avenue, Suranaree Sub-District, Muang Nakhon Ratchasima District, Nakhon Ratchasima 30000 Thailand
[2]Particulate Materials Processing Technology (PMPT), National Metal and Materials Technology Center, 114 Paholyothin Road, Khlong Nueng, Khlong Luang, Pathum Thani 12120 Thailand

* Corresponding Author: nattayt@mtec.or.th

Abstract
Sintered Fe-Mo-Mn-C alloys were produced by sintering of mixtures of pre-alloyed Fe-0.5Mo-0.15Mn powder and varied carbon amounts (0.30-1.20 wt.% with 0.15% increments) followed by slow cooling in furnace. Microstructures and mechanical properties of the sintered alloys varied with added carbon content. With up to 0.75 wt.% carbon additions, the sintered alloys exhibited polygonal ferrite plus non-cooperative eutectoid decomposition products. With 0.90 wt.% carbon addition, the whole microstructure of the sintered alloy mainly consisted of non-cooperative eutectoid decomposition products. With 1.05 and 1.20 wt.% carbon additions, the microstructures of the sintered alloys consisted of large grain boundary carbides and mixed non-cooperative and cooperative eutectoid decomposition products within grains. Tensile strength showed the maximum value in the sintered alloy with 1.05 wt.% carbon addition. Elongation values decreased sharply with increasing carbon contents of up to 0.60 wt.%, beyond which the values were constant.

1. Introduction

One advantage of powder metallurgy is the ability to combine sintering and heat treatment into a one-step process, widely known as sinter hardening. It can tailor the microstructures to obtain a variety of mechanical properties by using alloy design and a controlled cooling rate [1]. Carbon steel is widely used in many applications according to its mechanical properties, which strongly depend on carbon content and manufacturing process. Generally, the higher carbon content results in better tensile properties, with less ductility and fabrication difficulty. As a carbon steel is slowly cooled, primary ferrite (hypoeutectoid steel) or primary cementite (hypereutectoid steel) starts to nucleate at austenite grain boundaries until below eutectoid temperature, the rest austenite decomposes into a lamellar structure of two stable phases, by the cooperative growth of cementite (Fe_3C) and ferrite at a common transformation front with the parent austenite by a eutectoid reaction [2]. Although the microstructure becomes lamellar of fully pearlitic as carbon content approaches 0.8 wt.%C, Lu et al. [3] reported that the precipitation of small cementite particles has been verified in unalloyed steels containing carbon 0.16, 0.60, 0.8, and 1.24 wt.% by using thermal cycling treatment.

The addition of alloying elements can alter the phase transformation scenarios in steels. So far as we know, all alloying elements retard pearlite transformation kinetics with the exception of Cobolt (Co) [4]. Several works have been done about the kinetics of pearlite transformation caused by alloying design [5-8]. Bain [5] proposed the reduction of proeutectoid ferrite fraction because Manganese (Mn) changed the position of the critical point of steels to lower temperature and lower carbon regions. Whereas Ochiai et al. [6] found that the Chromium (Cr) and Molybdenum (Mo) addition increased the eutectoid temperature, which promoted the high strength of pearlite due to the refinement of the inter-lamellar spacing. Moreover, it has been reported that the Silicon (Si) addition could suppress the spheroidization of the cementite lamellae [7]. Previous work reported that the addition of $\geq$ 1 wt.% Ni could promote bainitic transformation at the expense of proeutectoid ferrite in the Fe-Cr-Mo-C sintered steels [8].

Pandit et al. [2] claimed that when the isothermal heat treatment was employed, the divorced eutectoid transformation (DET) product was clearly shown as spheroidal cementite particles in a ferrite matrix. DET occurred by a non-cooperative growth mode resulted from the parent austenitic phase contains cementite particles with a few microns of spacing. DET occurred due to the incorporation of partitioned carbon into the existing cementite particles. Andersson [9] reported that by using a slow cooling rate, the microstructure of a high-carbon steel alloyed with molybdenum consisted of coarse degenerated pearlite, nanometer-sized molybdenum carbides and ferrite. Upper and lower bainite structures also formed by non-cooperative eutectoid transformations [10]. Taleff et al., [11] found that very fine cementite particles were clearly observed in high-carbon steels around 1.5 wt.%. The transformation front, having a cellular-shaped austenite/ferrite boundary, advanced into an array of austenite plus fine cementite particles in a ferrite matrix. It has been reported by Howell et al. [12] that pearlite with lamellae of ferrite and $M_{23}C_6$ carbide instead of M_3C plate has been found in Fe-C-Cr alloys. The partitioning of Mn, Al, and C solutes was clearly observed in the $M_{23}C_6$ pearlite of the Fe-C-Mn-Al alloy [13].

Numerous studies have been reported that softening processes, of bearing steels, to create final structure as coarse cementite particles dispersed in a matrix of ferrite [14, 15, 16]. It also known as a divorced pearlite, since the product phases no longer grew cooperatively. Moreover, Verhoeven et al. [17] found that addition of Cr is most effective in minimizing carbide size. Because the morphology and dispersity of cementite and alloy carbide are important for the strength and ductility of alloyed steels therefore the balance between the carbon content and other alloying in order to control phase transformation, is the key to obtain high performance sintered steel. Mo with amounts of 0.5-1.0 wt.%, is commonly used in high strength steels. It has high affinity for C and trends to transform into carbides during sintering. In addition, Mo also partitions to other carbides and forms its own series of carbides as M_3C, $M_{23}C_6$, M_2C and M_6C [18].

Nonetheless, there are a few researches work about the effect of high-level carbon addition on phase transformation behaviour's in quaternary alloy sintered steel. Therefore, the present work aims to investigate the effect of medium-to high-carbon contents on the final microstructures and mechanical properties of pre-alloyed Fe-0.5Mo-0.15Mn sintered steel using slow cooling process.

2. Experimental procedure

As received water atomised steel powder pre-alloyed with 0.5 wt.% Mo and 0.15 wt.% Mn (Atomet 4001) was used as a base material in this investigation. It is produced by Rio Tinto Metal Powders of Canada with average particle size of less than 250 μm. Eight sets of specimens were prepared by varying amounts of added carbon from 0.30 to 1.20 wt.% with 0.15 increment. The set of specimens with no carbon addition was also produced. Zinc stearate of 1.0 wt.% was also added as a lubricant before blending, by using a barrel mixer rotating with the speed of 30 rpm for 1 hour. After that, each set of mixed powders was cold compacted into standard tensile test bars, according to MPIF standard 10 by

using a uniaxial pressing machine. Densities of the green compacts were 6.50 ± 0.05 g/cm^3. The densities of the specimens were measured by the Archimedes method according to the MPIF Standard 42
The green compacts were sintered at temperatures of 1,280°C under vacuum atmosphere for 45 min. After that, the sintered specimens were cooled down in the furnace with the rate of 0.1°C/s. For metallographic examination the specimen cross sections were polished as mirror-like surfaces and etched with a solution of 2% nital for characterization using optical microscopy (OM) using Olympus STM7 and also scanning electron microscopy (SEM) using FESEM-HITACHI/SU8230. Macrohardness of the sintered specimens was carried out using a hardness tester (Rockwell scale B). Tensile properties were tested by using the universal testing machine (Instron 8801).

3. Results and Discussion

3.1 Microstructures

The microstructures of Fe-0.5Mo-0.15Mn sintered at 1280°C under vacuum atmosphere and cooled at the rate of 0.1°C/s were observed by OM and SEM. The OM and SEM images indicate that sintered alloy microstructure has polygonal ferrite (PF) grains as shown in Figure 1(a) and Figure 1(b) respectively.

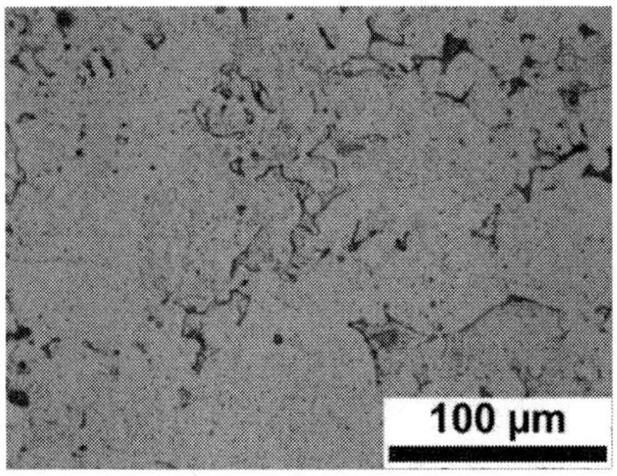

(a)

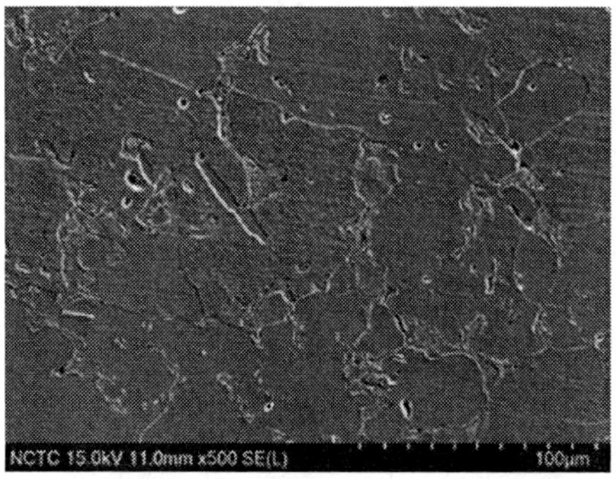

(b)

Figure 1. OM (a) and SEM (b) micrographs of Fe-0.5Mo-0.15Mn sintered alloys.

The microstructures of sintered alloys containing carbon contents in the range of 0.30-0.75 wt.% (with increment of 0.15%) changed according to carbon content as shown in Figure 2. The microstructures had dual-phase structures, which composed of PF grains and zones of non-cooperative eutectoid decomposition products. The latter zones contained particles with different sizes and morphologies, which depended on the degree of undercooling, embedded in a soft ferritic matrix. The non-cooperative eutectoid decomposition products formed in these sintered alloys are attributed mainly to the effects of alloying Mo and Mn elements, which can retard the formation of pearlite [19]. Besides, the fraction of these products also increases with increasing carbon content.

The formation of the divorced eutectoid transformation (DET) product occurred under conditions that low undercooling and small inter-particle spacing or fine array of pre-existing cementite due to the advancing transformation front begins to pull-away from cementite [20, 21]. Austenite and pearlite coexist in equilibrium were presented in the three-phase regions, austenite (γ) + ferrite (α) + cementite (θ) of Fe-Fe_3C phase diagram due to Mo and Mn moved the eutectoid temperature and also changed the eutectoid composition to lower carbon content [2].

The whole microstructure of the sintered alloy added with 0.90 wt.% carbon consisted mainly of non-cooperative eutectoid decomposition products. These products appeared as fine needle-shaped particles dispersed in the ferrite matrix similar to upper bainite as shown in Figure 3. The microstructures of sintered alloys containing 1.05 and 1.20 wt.% carbon consisted of large grain boundary carbides and mixtures of non-cooperative and cooperative eutectoid decomposition products. Comparison between these two sintered alloys, the more carbon content, the larger grain boundary carbide was observed.

IOP Conf. Series: Materials Science and Engineering 1137 (2021) 012029 doi:10.1088/1757-899X/1137/1/012029

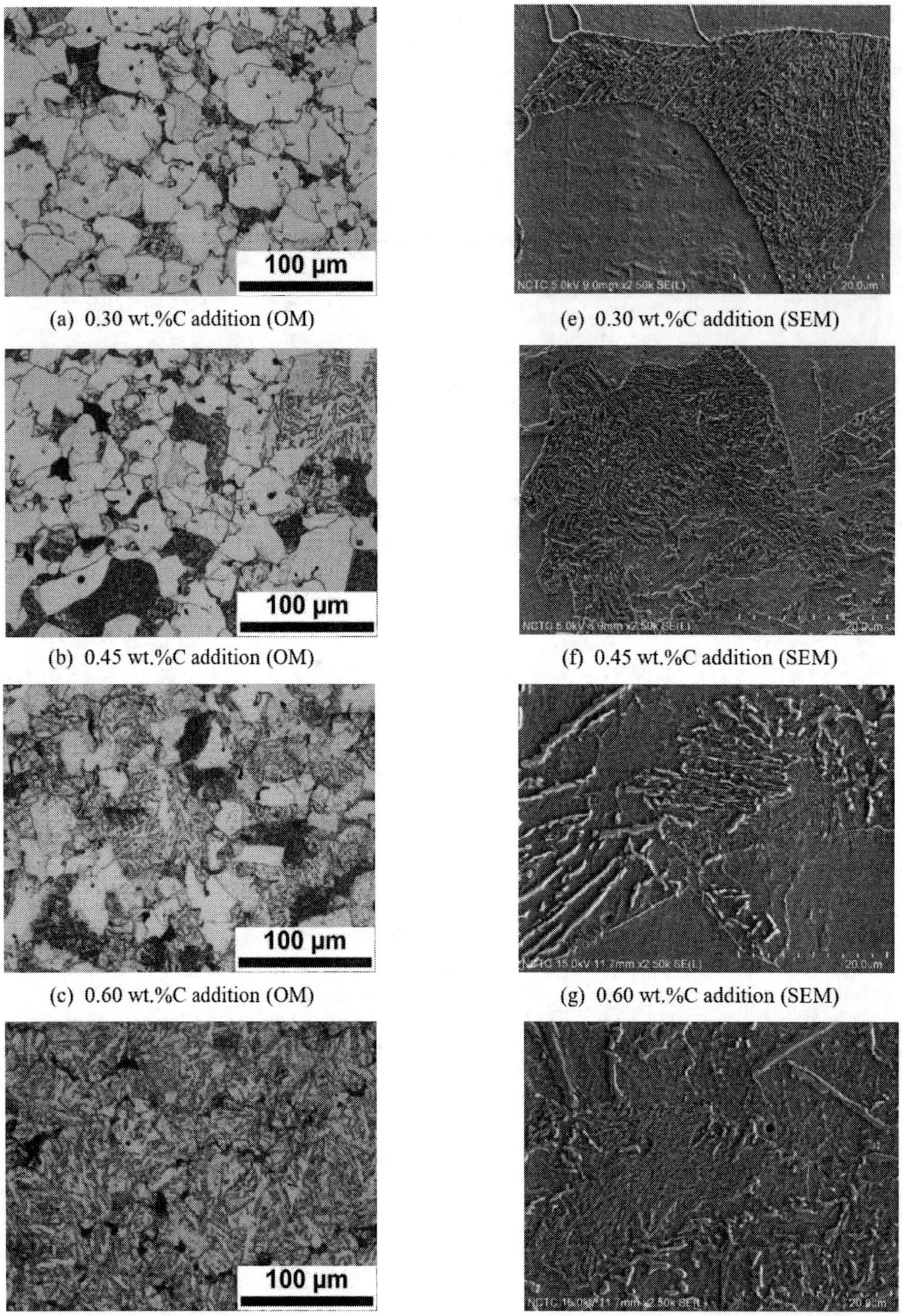

(a) 0.30 wt.%C addition (OM)

(b) 0.45 wt.%C addition (OM)

(c) 0.60 wt.%C addition (OM)

(d) 0.75 wt.%C addition (OM)

(e) 0.30 wt.%C addition (SEM)

(f) 0.45 wt.%C addition (SEM)

(g) 0.60 wt.%C addition (SEM)

(h) 0.75 wt.%C addition (SEM)

Figure 2. Optical (a-d) and SEM (e-h) micrographs of Fe-Mo-Mn sintered steels. Variation of carbon contents are specified.

IOP Conf. Series: Materials Science and Engineering 1137 (2021) 012029 doi:10.1088/1757-899X/1137/1/012029

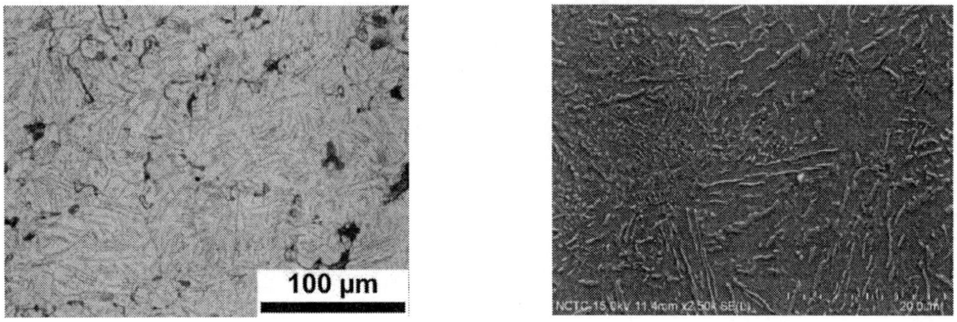

(a) (b)

Figure 3. OM (a) and SEM (b) micrographs of Fe-0.5Mo-0.15Mn with 0.9 wt.%C addition sintered steels.

The micrographs of these sintered alloys are shown in Figure 4. It can be explained that the transformation started by proeutectoid carbide precipitation at grain boundary. After that, the austenite transformed by eutectoid reaction into mixtures of non-cooperative and cooperative of ferrite and carbide. With increasing undercooling and carbon content, the transformation products became smaller and more densely dispersed. The transformation sequence can be explained by partitioning of the substitutional and interstitial element C coupled with a change in the mechanism of transformation with undercooling [22]. The results given above are in line with Andersson's work [9], in which under a slow

(a) 1.05 wt.%C addition (OM)

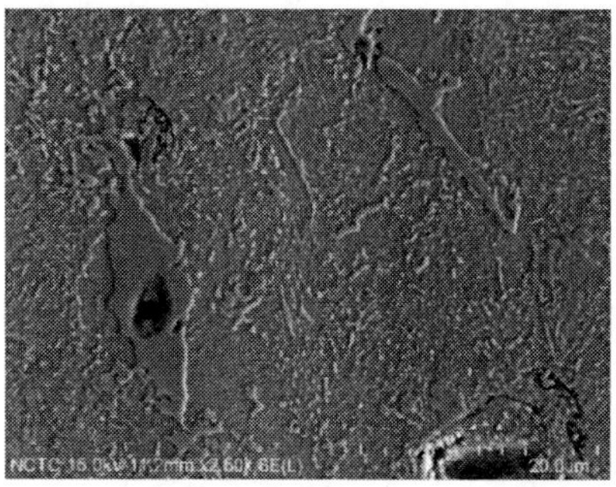

(c) 1.05 wt.%C addition (SEM)

(b) 1.20 wt.%C addition (OM)

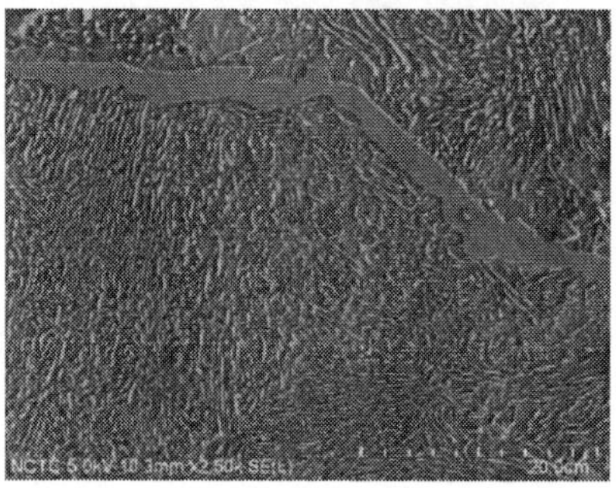

(d) 1.20 wt.%C addition (SEM)

Figure 4. OM (a-b) and SEM (c-d) micrographs of Fe-0.5Mo-0.15Mn Variation of carbon contents are specified.

cooling rate, the microstructure of a high-carbon steel alloyed with molybdenum consisted of a coarse degenerated pearlite, nanometre-sized of molybdenum carbides and ferrite and the work of Taleff et al. [11] also found that very fine cementite particles were clearly observed in high-carbon steels around 1.5 wt.%C. In such work, a transformation front, in which a cellular-shaped austenite/ferrite boundary, advanced into an array of austenite plus fine cementite particles in a ferrite matrix.

3.2 Mechanical properties

The mechanical properties of sintered Fe-0.5Mo-0.15Mn-xC alloys (x = 0.0, 0.30, 0.45, 0.60, 0.75, 0.90, 1.05 and 1.20 wt.% carbon), such as tensile strength, yield strength and hardness, strongly depended on carbon content. They showed a similar trend, i.e., they notably increased with increasing carbon content up to 1.05 wt.% but beyond that carbon content the values decreased as shown in Figure 5. Due to the increase of precipitate volume fraction with increasing carbon content, the precipitation strengthening phenomenon contributes its effect to mechanical properties of sintered alloys. The decrease of tensile properties and hardness in specimen containing 1.2 wt.% C, may be caused by large continuous networks of carbides along the grain boundaries.

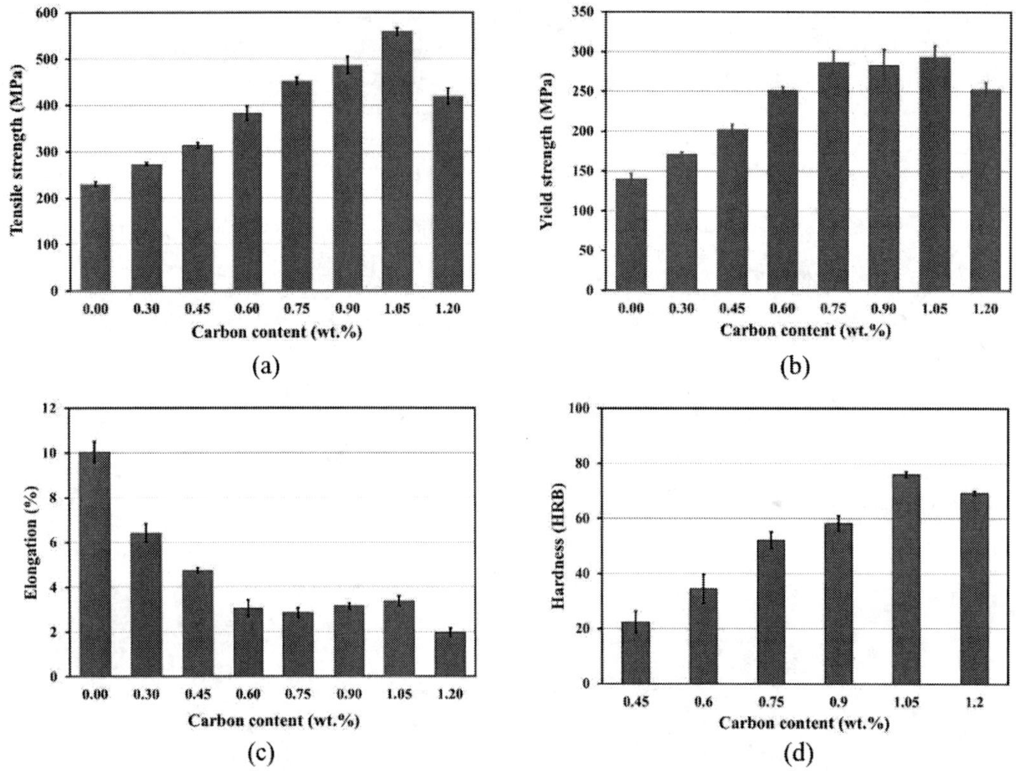

Figure 5. Mechanical properties, (a) ultimate tensile strength, (b) yield strength (c) elongation and (d) hardness of Fe-Mo-Mn sintered steels with variation in carbon content.

Elongation values decreased sharply with increasing carbon contents of up to 0.60 wt.%, beyond which the values were constant. The decrease in soft ferrite phase fraction results in a decrease in the ductility of sintered alloys. In sintered alloys with C contents higher than 0.60 wt.%, there are no polygonal ferrite grains. The constant ductility value may be due to the similar microstructural feature of ferrite and carbide mixtures in such alloys. Perhaps, the fine scale of ferrite and carbide mixtures has something to control alloy ductility. According to the work conducted by using thermomechanical processing on the two steels with 0.46% C and 0.76% C, finer pearlite interlamellar spacing and spheroidized pearlite showed increase of strength without loss of toughness and ductility [23]. In our work, there are no hardness values reported for the specimens containing less than 0.45 wt.% C because they were too soft to be measured by using the HRB scale.

4. Conclusion

The influence of varied carbon contents (0.30-1.20 wt.% with 0.15% increments) on microstructures and the mechanical property of sintered alloys were investigated. By using a sintering temperature of 1280°C and cooled in the furnace with 0.1°C/s, the experimental results indicated that carbon played important roles in microstructural development by changing the microstructure from dual-phase structure composed of polygonal ferrite grains and non-cooperative eutectoid decomposition products to fully non-cooperative eutectoid decomposition products with different sizes and shapes depending on the degree of undercooling. Increase of carbon addition tended to increase tensile properties and macro hardness. The formation of thick continuous carbide networks along the grain boundaries in sintered alloy containing 1.2 wt.% carbon resulted in the decrease of mechanical properties.

Acknowledgments

This work is funded by the project P1951261 supported by National Science and Technology Development Agency (NSTDA). The authors are grateful for supports from Thailand National Metal and Materials Technology Center (MTEC). The authors are also grateful for the material for this research (ATOMET4001), courtesy from Rio Tinto Metal Powders of Canada (via P.S. Steel Co., Ltd., of Thailand and Rio Tinto Iron & Titanium (Suzhou) Co., Ltd., of China).

References

[1] Srijumpan W Wiengmoon A Morakotjinda M Krataitong R Yotkaew T Tosangthum N and Tongsri R 2015 *Mater. Des.* **88** 693-701
[2] Pandit A S and Bhadeshia H K D S 2012 *Proc. R. Soc.* **A468** 2767–2778
[3] Lu Z Wang B Wang Z Sun S and Fu W 2013 *Mater. Sci. Eng.* **A547** 143-148
[4] Smallman RE 2013 *Modern Physical Metallurgy* (Amsterdam, Netherlands: Elsevier)
[5] Bain E C 1939 *Functions of the Alloying Elements in Steel* (Cleveland, Ohio: ASM)
[6] Ochiai I Nishida S and Tashiro H 1993 *Wire J. Int.* **26** 50-61
[7] Guo K Fang T T Wang J Y Wu A A Wang Y Z and Jiang J 2014 *Mater. Lett.* **134** 290-294
[8] Tosangthum N Morakotjinda M Krataitong R Wila P Yotkaew T Vetayanugul B Boontatim W and Tongsri R 2018 *Mater Today Proceeding* vol **5** 9351-9358
[9] Andersson D I and Lund T 1993 *Creative Use of Bearing Steels ASTM International* 171-184
[10] Lee H Spanos G J Shiflet J and Aaronson H I 1998 *Acta. Metal.l* **36** 1129-1140
[11] Taleff E M Syn C Lesuer D and Sherby O 1996 *Metall. Mater. Trans.* **27A** 111-118
[12] Howell P R and Bee J V 1979 *Metall. Trans.* **10**A 1213-1222
[13] Cheng W and Li Y 2012 *Metall. Mater. Trans.* **43A** 1817-1825
[14] Honda K and Saito S 1920 *J. Iron Steel Inst.* **102** 261-267
[15] Whitley J 1922 *J. Iron Steel Inst.* **105** 339-357
[16] Luzginova N Zhao L and Sietsma J 2008 *Metall. Mater. Trans.* **A39** 513-521
[17] Verhoeven J and Gibson E 1998 *Metall. Mater. Trans.* **A29** 1181-1191
[18] Hackenberg R E and Shiflet G 2003 *Acta. Mater.* **51** 2131
[19] Wang J and Van Der Zwaag S 2001 *Metall. Mater. Trans.* **32A** 1527-1539
[20] Martin J W 2007 *Encyclopedia of the Structure of Materials* (Amsterdam, Netherlands: Elsevier)
[21] Verhoeven J 2000 *Metall. Mater. Trans.* **31A**(10) 2431-2438
[22] Aaronson H I and Domian H 1966 *Trans. Metall. Soc. AIME* **236** 781-796
[23] Radko K Hans G and Karl H 1991 *Mater. Technol.* **62** 405–441

FEM model of protection bullet helmet made from composite material.

R. Promma[1], E. Phongphinittana[1] and P. Nimdum[1*]

[1]Department of Mechanical and Aerospace Engineering, Faculty of Engineering, King Mongkut's University of Technology North Bangkok, 1518 Pracharat 1 Road, Wongsawang, Bangsue, Bangkok 10800, Thailand

* Corresponding Author: pongsak.n@eng.kmutnb.ac.th,
Telephone Number 0-2555-2000 ext 8315, Fax Number 0-2586-9541

Abstract
Damage evolution, mechanical behaviour, energy absorbtion, back-face deformation (BFD) and the perforation performance of Kevlar epoxy woven composite with and without foam (PU foam) were examined for various impact energies based on the drop weight test. The Kevlar woven composite induces different damage and deformation depending on its hybrid patterns. The Kevlar composite with PU foam decreases significantly, and the BFD of shell composite and transform locally deformation become a global nature. In addition, damage type - as fibres break - is primary dominant for Kevlar composite without PU foam. However, the role major damage as delamination, significantly appear for Kevlar composite with PU foam. In order to investigate the damage mechanism, energy absorbtion and BFD, FEM simulation were also performed. The FEM simulation is in very good agreement with the drop weight test. Finally, the FEM study showed that PU foam is able to decrease significantly, the BFD of composite shell, and the contact force on the skull, during ballistic impact.

Keywords: Helmet, Ballistic Impact, Perforation Impact, Back-Face Deformation, Polyurethane Foam

1. Introduction

Currently, the protection bullet helmet was developed for use in battlefields. The first model of the helmet made of steel (Steel M1) was good against perforation from bullets, but it weighed a lot. Consequently, helmet weight and perforation performance needed to be considered. The Personal Armor System for Ground Troops (PASGT) model and Advance Combat Helmet (ACH) model made from Kevlar fiber and resin/epoxy were developed for use in the U.S. Army and the U.S. Marines [1-3].

The previous study found that the helmets worked against ballistic impact based on two standards. The first is the MIL-H-44099A standard which used 1.1g of 0.22 Caliber FSP with an impact velocity of 610 m/s and impact energy of 204.66 J [1]. The second is NIJ-STD-0106.01 Type II standard which used the 9 mm FMJ bullet with an impact velocity of 358 ± 15 m/s, mass of 8 g, and impact energy of 512.66 ± 15 J [1-4]. Both of them are considered only with ballistic impact performance. However, head injuries are not only due to perforation impact but also due to BFD affecting the BHBT. Then, BFD is another one parameter that should be considered for ballistic impact analysis on the helmet shell composite. The previous study [2] found that BFD values is dependent on the helmet size.

Nowadays, the helmet is considered to increase the comfort effect and decrease the energy imparted to the human head. The PU foam is a good candidate inserted between the helmet shell composite and the skull. Its density is 63 kg/m³ for hard foam and 61 kg/m³ for soft foam [2]. The simulation result shown that the foam pad is capable of absorbing impact energy and reducing the energy imparted to the human head [3]. In addition, the effect of foam pad hardness results in the decrease BFD values about 25% after ballistic impact on the combat helmet [3].

The aim of this study is to understand the absorbtion energy due to the drop weight test of Kevlar shell composite, and to simulate for ballistic impact by using FEM simulation based on NIJ-STD-0106.01 Type II. A simple skull model with Kevlar shell composite and PU foam was created to study its performance, and BFD effect.

2. Material model & identification of mechanical properties

2.1 Kevlar Composite

In this study, the woven Kevlar fabric as an orthotropic material with resin/epoxy are used. In order to identify composite behavior and damage evolution for FEM simulation, the tensile test was performed based on ASTM D638-14 standard [5]. The identification parameters from the tensile test are shown as Figure 1 and its mechanical properties are listed in Table 1.

Table 1. The elastic constants, Hashin Damage and Damage Evolution of the Kevlar composite.

Elastic Constants (Mechanical Properties)	
Young's Modulus	E_{11} = 14158 MPa E_{22} = 14158 MPa E_{33} = 1948 MPa
Poisson's Ratio	v_{12} = 0.08 v_{13} = 0.7 v_{23} = 0.08
Shear Modulus	G_{12} = 1152.16 MPa G_{13} = 1100 MPa G_{23} = 1100 MPa
Hashin Damage (MPa)	
Longitudinal Tensile Strength	360
Longitudinal Compressive Strength	360
Transverse Tensile Strength	360
Transverse Compressive Strength	360
Longitudinal Shear Strength	105
Transverse Shear Strength	105
Damage Evolution (mJ)	
Longitudinal Tensile Fracture Energy	8.42
Longitudinal Compressive Fracture Energy	8.42
Transverse Tensile Fracture Energy	8.42
Transverse Compressive Fracture Energy	8.42

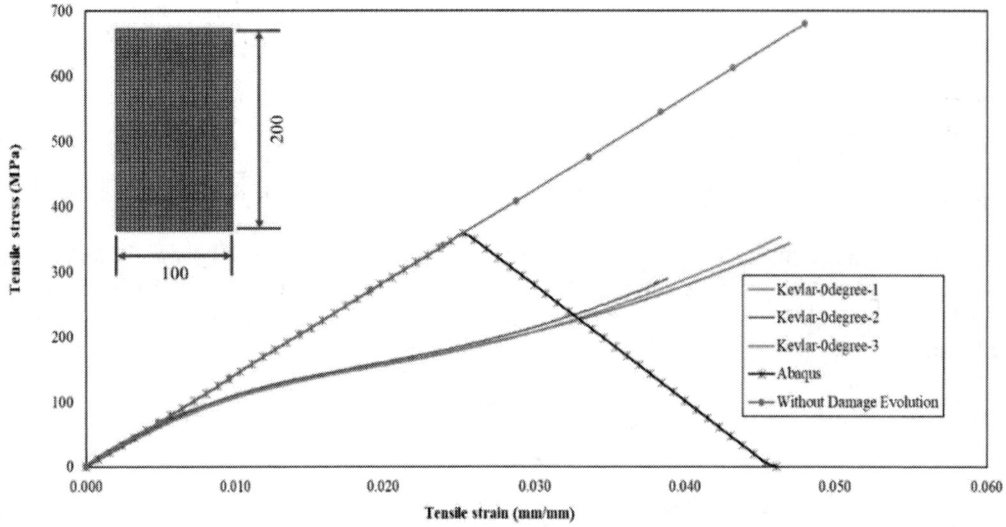

Figure 1. Identification parameters of Kevlar composite.

2.2 Foam pad model and material identify

In order to investigate the comfort effect, PU foam was inserted between the helmet shell and skull [3]. From a previous study [6-8], Ogden's material model was used to simulate the mechanical behavior of PU foam as in Equation (1).

$$W = \sum_{i=1}^{N} \frac{2\mu_i}{\alpha_i^2} [\hat{\lambda}_1^{\alpha_i} + \hat{\lambda}_2^{\alpha_i} + \hat{\lambda}_3^{\alpha_i} - 3] + \sum_{i=1}^{N} \frac{1}{D_i} (J^{el} - 1)^{2i}] \quad (1)$$

Where $\hat{\lambda}_i$, μ_i, $\alpha_i$and D_i are material parameters identified by compressive test is used [9, 10]. Dimension of specimen is 18 mm x 36 mm x 36 mm (Figure 2). The head speed in the compressive test base on ASTM D 1621 standard is 0.3, 3, 15, 30 and 45 mm/min, respectively. The experimental set-up is shown as Figure 3.

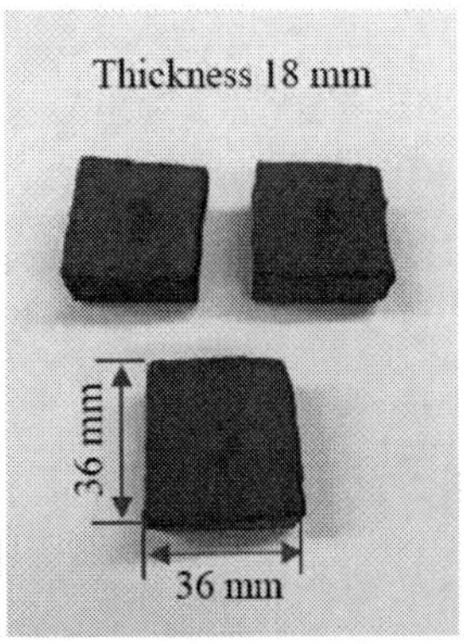

Figure 2. PU foam specimen Test

Figure 3. Instron Universal Testing Machine Model 5567 with specimen test

The experimental result shows that the mechanical properties of PU foam are independent with strain rate (Figure 4). Identification of Parameter based on Ogden's material model order2nd are listed Table 2.

IOP Conf. Series: Materials Science and Engineering 1137 (2021) 012030 doi:10.1088/1757-899X/1137/1/012030

Table 2. Material parameters of Ogden's material model

μ_1	μ_2	α_1	α_2	D_1	D_2
1.78×10^{-5}	0.02	16.4	12.2	0	0

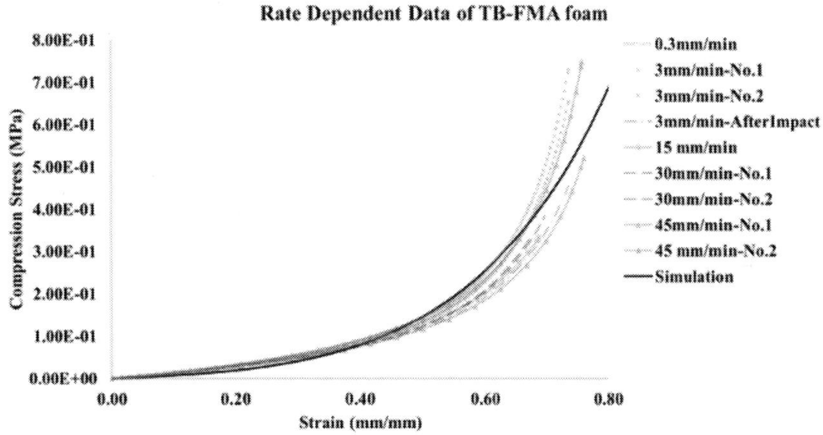

Figure 4. Relation graph between Compression Stress and Strain of PU foams.

2.3 Bullet and skull

The bullet is the 9 mm FMJ base on NIJ-STD-0106.0 Type II. The ABAQUS geometry model of 8 g is shown as Figure 5. Johnson-Cook Model was used as in Equation (2) [2, 3]. The parameter constants are listed in Table 3.

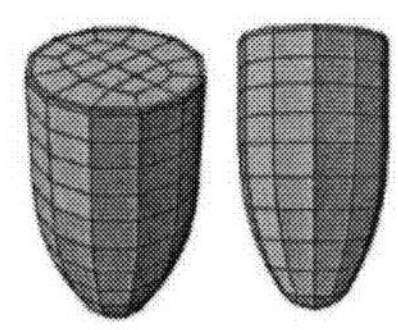

Figure 5. Geometry of 9 mm FMJ Bullet.

$$\bar{\sigma} = (A + B\bar{\varepsilon}^n)\left(1 + Cln\frac{\dot{\bar{\varepsilon}}}{\dot{\bar{\varepsilon}}_0}\right)[1 - (T^*)^m] \quad (2)$$

Where A, B, C and n are Johnson-Cook constants, $\bar{\sigma}$ is the von Mises equivalent stress, $\bar{\varepsilon}$ is the equivalent plastic strain, $\dot{\bar{\varepsilon}}$ is the plastic strain rate, $\bar{\varepsilon}_0$ are reference strain rate and $T^*$is the homologous temperature defined by $T^* = (T - T_r)/(T_m - T_r)$ when T_r is room temperature, T_m is melting temperature and T is specimen temperature.

Table 3. Material constants of Johnson-Cook Model [3]

Material	Yield Stress	Strain Hardening		Strain Rate		Temperature Softening		
	A (MPa)	B (MPa)	n	$\dot{\varepsilon}_0(s^{-1})$	C	T_r(K)	T_m (K)	m
Lead Core	24	300	1.00	$5x10^{-4}$	0.1	293	760	1.00
Brass Jacket	206	505	0.42	$5x10^{-4}$	0.01	293	1189	1.68

As mentioed above, the skull injuries will be analyzed during the ballistic impact. The mechanical properties of the skull are listed in Table 4.

Table 4. Mechanical properties of skull [11, 12].

Young modulus (MPa)	Density (Kg/m^3)	Tensile strength (MPa)	Compressive strength (MPa)
15000	2000	132	90

3. Experimental set-up

3.1 Drop Weight Test

In order to validate of the FEM simulation, drop weight tests were used. The experimental set-up of the drop weight test is shown as Figure 6. The impact event is recorded by a high-speed camera. The impact energy varying with the height is 0.9, 1.0 and 1.4 m with the impact velocity (V_{in}) is 3665.43, 3952.01 and 4709.91 mm/s, respectively. The mechanical behavior of PU foam for impact absorbed will be studied. Figure 7 presents the specimen for drop the weight test.

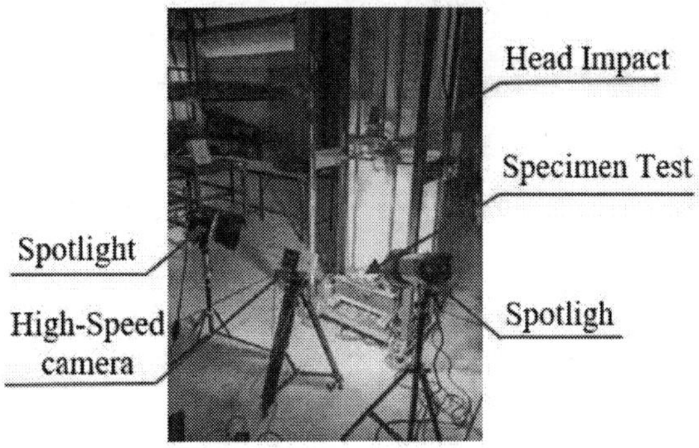

Figure 6. The experimental set up of the Drop weight test.

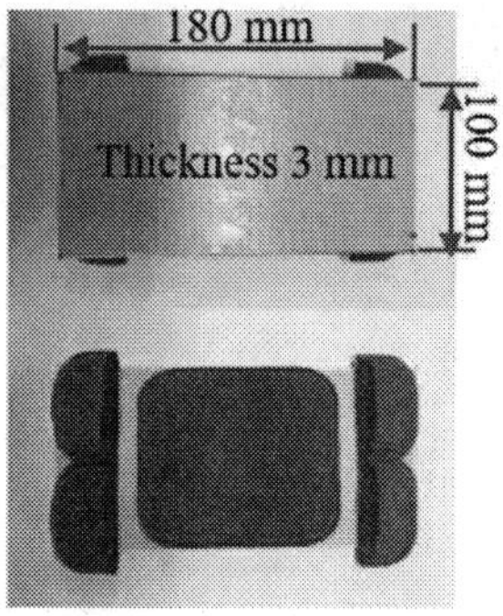

Figure 7. Specimen of the Drop weight test.

4. FEM simulation

4.1 Drop Weight

In order to understand and validate the experimental result, the drop weight simulation is performed. Figure 8 presents the FEM simulation set-up which consist of head impact, Kevlar composite, PU foam and upper/lower support.

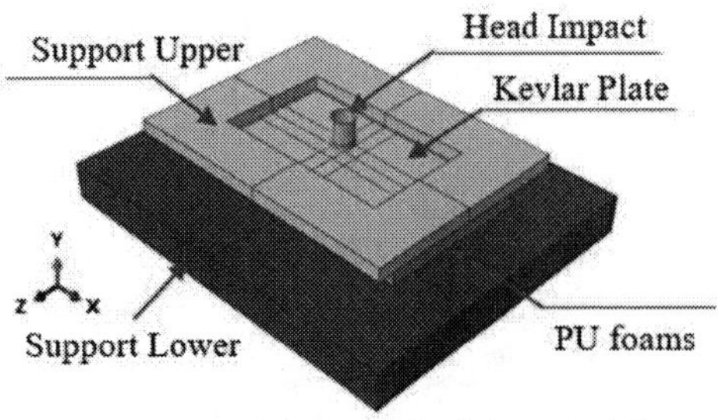

Figure 8. Assembly model of drop weight test simulation.

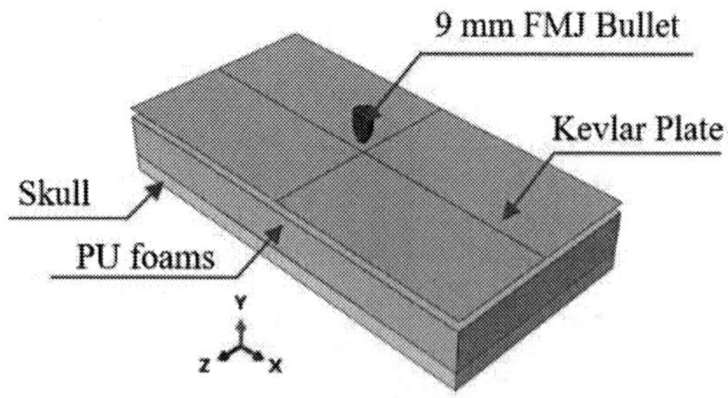

Figure 9. Assembly model of 9mm FMJ, Kevlar Shell, PU foam and Skull.

4.2 Ballistic Impact

In order to analyze the skull injuries, the FEM simulations were performed for the ballistic impact based on NIJ-STD-0106.01 Type II. Figure 9 presents the FEM model for the ballistic impact model which consists of the bullet, Kevlar shell composite, PU foam and the skull. The bullet with velocity of 358 ± 15 m/s and the impact energy of 512.66 ± 15 J is used [2, 3]. The Kevlar composite, PU foam, skull and brass jacket with lead core are simulated by using a hexahedral element. Mesh size of 4 mm for the Kevlar shell composite, PU foam and the skull, 1.8 mm for the brass jacket and 1.6 mm for the lead core were used [13, 14]. To simulate, the skull are fixed at x, y, z directions and the bullet is striking in y-direction and to be impacted on the Kevlar shell composite. The PU foam and skull appeared to be perfect bonding with the Kevlar shell composite.

5. Result

5.1 The experimental result

The experimental result showed that all energy impact levels, of the Kevlar shell composite with PU foam, is not perforation. Figure 10 presents the high-speed images of the drop weight test at 0.9 m. Figure 11 shows the BFD results with and without PU foam. Figure 12 presents the small damage area occurring after impact for Kevlar shell composite with PU foam. The presence of PU foam allows to increasing time response induce global deformation on Kevlar shell composite. As a result, delamination can be observed primarily after impact (Figure 13). While Figure 14 shows local damage occurring on the impact zone for the Kevlar shell composite without PU foam. The fiber break had primary damage.

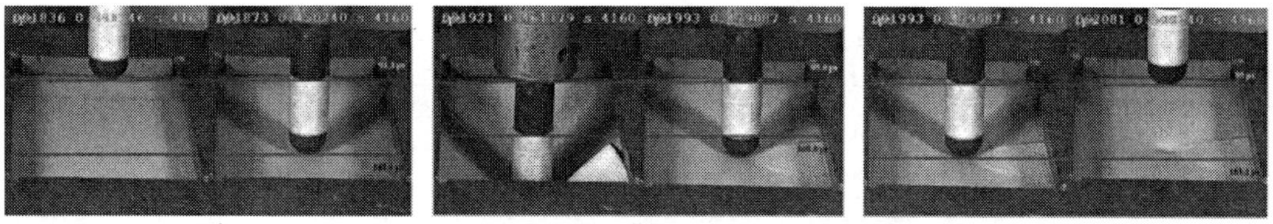

(a) (b) (c)

Figure 10. The high-speed images of Drop weight test at 0.9 m: (a) Before (b) During and (c) After the impact

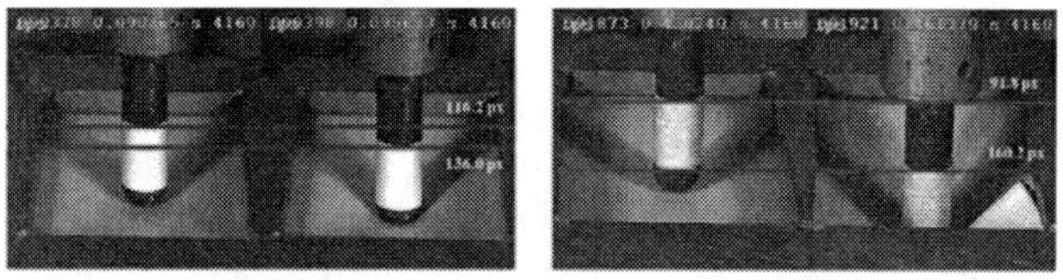

(a) (b)

Figure 11. The compare of high-speed images at 0.9 m: (a) Kevlar plate without PU foams, (b) Kevlar plate with PU foams.

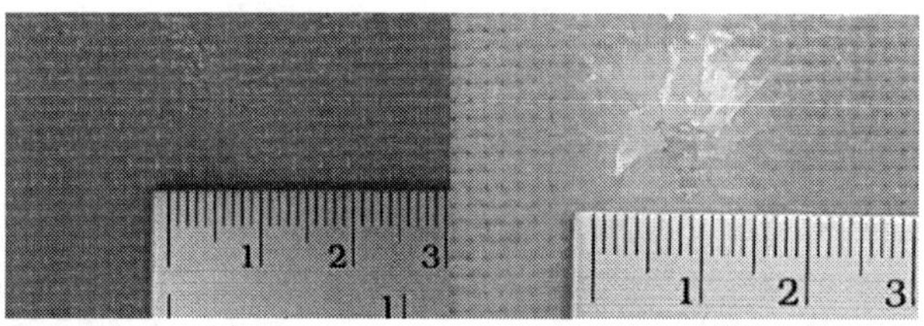

(a) (b)

Figure 12. Kevlar shell composite with PU foams at 0.9 m: (a) Upper side and (b) Lower side.

(a) (b) (c)

Figure 13. The delamination damage of Kevlar shell composite with PU foams with different height (different energy impact levels): (a) 0.9, (b) 1.0 and (c) 1.4 m

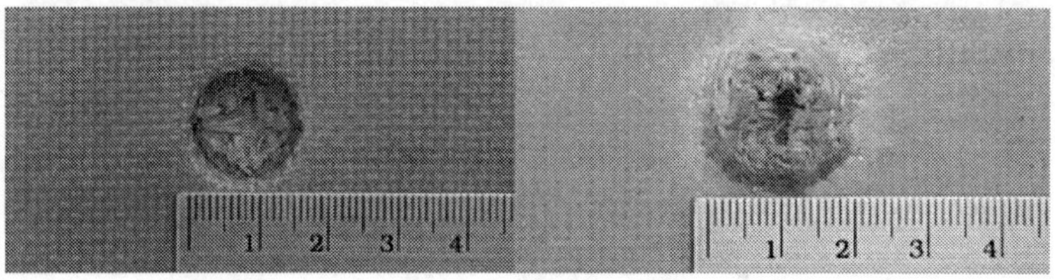

(a) (b)

Figure 14. Kevlar shell composite without PU foams at 0.9 m: (a) Upper and (b) Lower side.

Furthermore, the presence of PU foam in the Kevlar shell composite aid to increase the perforation performance but to increase significantly the BFD (Table 5). However, its increased is necessary to analyze head injury by considering the pressure stress on skull that we will discuss in section 5.3.

Table 5. Comparison the BFD of with and without PU foam case of experimental

Kevlar shell composite without PU foam			Kevlar shell composite with foam		
Height (m)	BFD (mm)	Remark	Height (m)	BFD (mm)	Remark
0.9	6.15	Not-Perforation	0.9	24.62	Not-Perforation
1.0	10.18	Perforation	1.0	30.89	Not-Perforation
1.4	-	-	1.4	31.61	Not-Perforation

5.2 The comparison between the simulation and the experimental result

In order to validate the drop weight test, the FEM simulation was performed. In the case of - without foam, FEM shows that BFD induces locally the deformation response. On the other hand, the BFD response of Kevlar shell composite with PU foam occurs globally the deformation as shown in Figure 15. To consider the simulation result, we found that the Kevlar shell composite is not perforate. The FEM model with damage evolution allow more production of material deformation due to its damage such as fiber break during impact. As a result, a good correlate between the BFD simulation and experiments can be obtained. Then, we can summarize that when the Kevlar shell composite become a large deformation due to its damage, the FEM simulation with damage evolution models are necessaries. The comparison results between the experiment and FEM simulation is listed in Table 6.

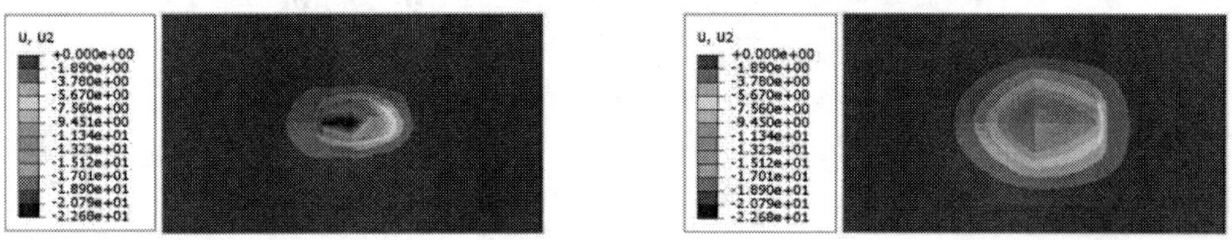

(a) (b)

Figure 15. The BFD on Kevlar plate at 0.9 m height: (a) Without foam and (b) With foam.

Table 6. Comparison Result of experimental and simulation.

Values	Experimental	Simulation (Without Damage Evolution)	Simulation (With Damage Evolution)
Height 0.7 m	Without foams		
V_{in} (mm/s)	3178.00	3178.00	3178.00
V_{out} (mm/s)	1734.50	3173.83	1433.94
BFD (mm)	12.35	9.55	14.78
Perforation	No	No	No
Values	Experimental	Simulation (Without Damage Evolution)	Simulation (With Damage Evolution)
Height 0.9 m	With foams		
V_{in} (mm/s)	3665.43	3665.43	3665.43
V_{out} (mm/s)	1425.85	3451.47	1973.21
BFD (mm)	24.62	9.23	16.16
Perforation	No	No	No
Values	Experimental	Simulation (Without Damage Evolution)	Simulation (With Damage Evolution)
Height 1.0 m	With foams		
V_{in} (mm/s)	3952.01	3952.01	3952.01
V_{out} (mm/s)	1581.86	3717.79	1944.06
BFD (mm)	30.89	9.63	21.42
Perforation	No	No	No
Values	Experimental	Simulation (Without Damage Evolution)	Simulation (With Damage Evolution)
Height 1.4 m	With foams		
V_{in} (mm/s)	4709.91	4709.91	4709.91
V_{out} (mm/s)	1524.13	4425.00	703.92
BFD (mm)	31.61	10.90	30.07
Perforation	No	No	No

In addition, the fiber breaks on the Kevlar shell composite are also considered. We found that the fiber breaks respond more with height (Figure 16). The presence of PU foam on the Kevlar shell composite aids to significantly reduce the fiber breakage. Figure 17 presents the fiber breakage on the Kevlar shell composite with and without foam cases at 0.9 m. However, Hashin's damage model with shell element type cannot predict the delamination onset due to the absent of out-of-plan stress. As results, the BFD prediction is not precision since the damage evolution due to delamination does not take into account.

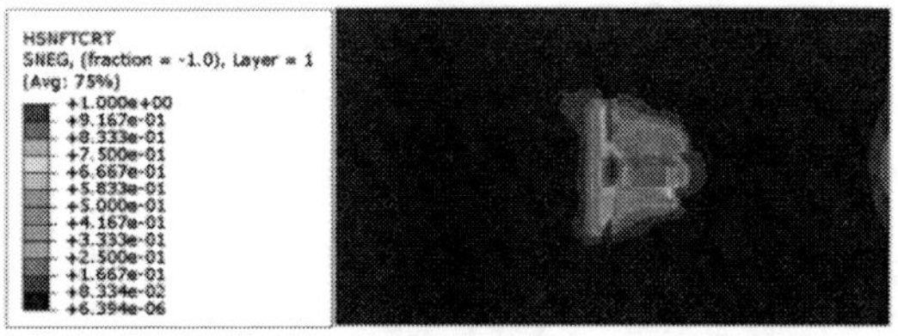

(a)

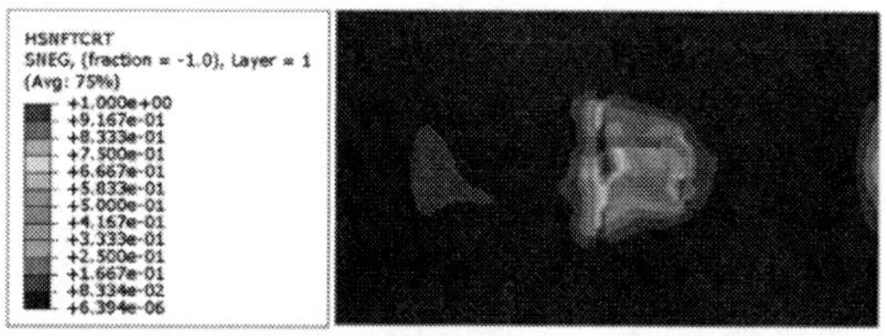

(b)

Figure 16. Hashin's tensile fiber initiation criterion failure on Kevlar plate at the height of drop weight test of with foam case: (a) 1.0 m and (b) 1.4 m.

IOP Conf. Series: Materials Science and Engineering 1137 (2021) 012030 doi:10.1088/1757-899X/1137/1/012030

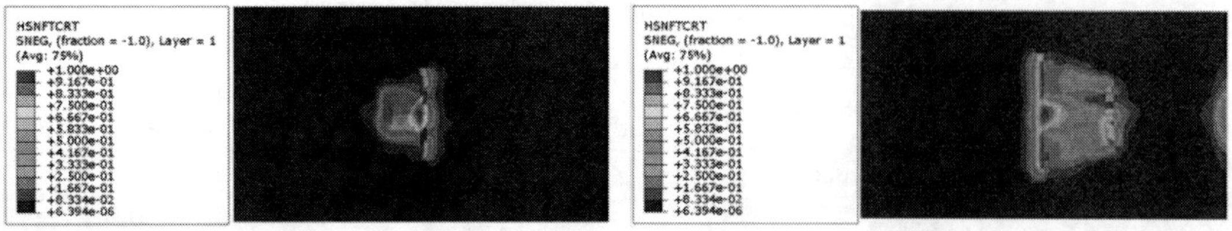

(a) (b)

Figure 17. Hashin's tensile fiber initiation criterion failure on Kevlar plate at the height of drop weight test at 0.9 m: (a) without foam case and (b) with foam case

5.3 The FEM simulation result for the ballistic impact

In order to consider the skull injuries under ballistic impact, the FEM simulation on simplified models were studied. The FEM simulation showed that the von Mises stress on the skull with and without PU foam is significantly different (Figure 18). The von Mises stress decrease considerably for the Kevlar shell composite with PU foam. While, the bullet for both cases is deformed similarly after impact with the mushroom shape.

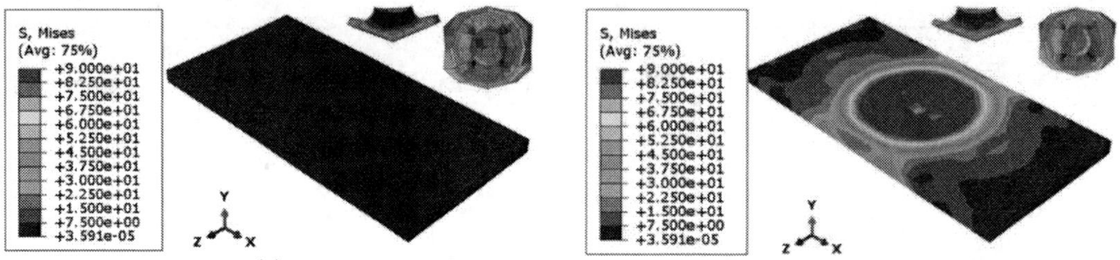

(a) (b)

Figure 18. The maximum von Mises stress of the skull: (a) With foam and (b) Without foam case.

Figure 18 represents the maximum von Mises stress of the case (a) and (b) are 0.03 and 90 MPa, respectively. Figure 19 presents the maximum pressure on the skull that the maximum pressure of case (a) and (b) are 0.03 and 79.93 MPa, respectively. We can summarize that the presence of PU foam can significantly increase the BFD. However, this effect does not induce head injuries and its comfort since the von Mises stress and contact pressure can considerably decrease, respectively. This result is in agreement with the study of Li, et al [3].

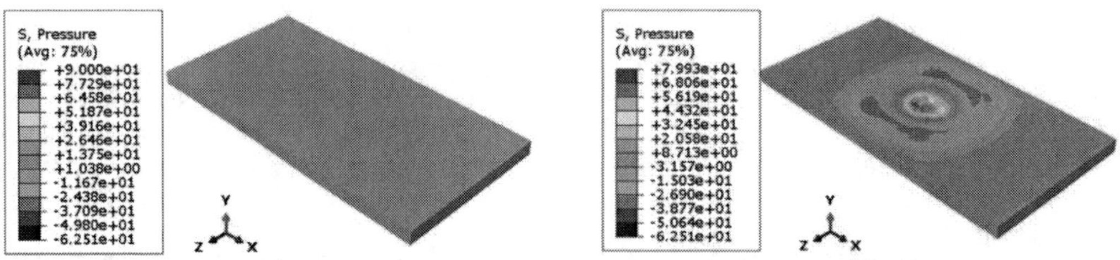

(a) (b)

Figure 19. The maximum pressure of the skull: (a) With foam and (b) Without foam case.

6. Conclusions

In this paper, the experimental investigations and FEM simulation have been presented. This study can be summarized as follows:

- The drop weight test is able to predict the energy absorbed by the Kevlar shell composite; this procedure can be used to preliminary assess the perforation performance of ballistic impact.
- The effect of PU foam is able to significantly increase the perforation performance. They against to damage initiation and allow to increasing time response induce global deformation on Kevlar shell composite.
- The FEM simulation is a good prediction for the fiber break damage. However, Hashin's criterion model with shell element type cannot predict the delamination onset due to the absent of out-of-plan stress.
- The ballistic simulation was carried out and shown that PU foam is able to decrease the maximum von Mises stress and the maximum pressure of the skull. As a result, they can reduce the head injuries effect and increase comfort.

Acknowledgments

This work was financially supported by the Faculty of Engineering at King Mongkut's University of Technology North Bangkok (ENG-63-85).

References

[1] Tham C Y Tan V B C and Lee H P 2008 *International Journal of Impact Engineering* **35** 304-318
[2] Li Y Q Li X G and Gao X L 2015 *Journal of Applied Mechanics* **82**111004-1-111004-9
[3] Li X G Gao X L and Kleiven S 2016 *International Journal of Impact Engineering* 91 56-67
[4] National Institute of Justice 1981 *NIJ Standard for Ballistic Helmets* (Washington, District of Columbia: U.S. Department of Justice) Standard No. 0106.01
[5] ASTM International 2015 *Standard test Method for Tensile Properties of Plastics, Designation: D 638-14* (West Conshohocken, Pensylvania)
[6] Crocker L E Duncan B C Hughes R G and Urquhart J M 1999 *NPL Report CMMT(A)183* (United Kingdom)
[7] Briody C Duignan B and Jerrams S 2011 *Proc. the International conference on Materials, Tribology and Recycling (MATRIB)* (Vela Luka, Croatia)
[8] Thanakhuna K and Puttapitukpornb T 2019 *Engineering Journal* **23**(6) 381-398
[9] ASTM International 2000 *Standard test Method for Compressive Properties of Rigid Cellular Plastic, Designation: D 1621-00* (West Conshohocken, Pensylvania)
[10] Park K B Kim H T Her N Y and Lee J M 2019 *Materials* **12**(17):2672
[11] Svein Kleiven 2007 Predicrors for Traumatic brain injuries evaluated through accident reconstructions *Stapp Car Crash Journal* **51** 81-114
[12] Sahoo D Deck C Yoganandan N and Willinger R 2016 *Journal of The Mechanical Behavior of Biomedical Materials* **57** 24-41
[13] Rodríguez-Millán M Ito T Loya J A Olmedo A and Miguélez M H 2016 *Materials and Design* **110** 391-403
[14] Palomar M Lozano-Mínguez E Rodríguez-Millán M Miguélez M H and Giner E 2018 *Composite Structures* **201** 49-61

The study on the characterization of sintered hot forging and austenite stainless steels

Sittipong Eiumtadanai[1], Sithipong Mahathanabodee[1], Ruangdaj Tongsri[2], Bhanu Vetayanugul[2], Sumate, Poomiapiradee[3,*]

[1]Department of Production Engineering, Faculty of Engineering, King Mongkut's University of Technology North Bangkok (KMUTNB), 1518 Pracharat 1, Bang-Sue, Wongsawang, Bangkok, 10800, Thailand
[2]Particulate Materials Processing Technology Laboratory (PMPT), National Metal and Materials Technology Center, 114 Paholyothin Road, Khlong Nueng, Khlong Luang, Pathum Thani, 12120, Thailand
[3]Thailand Institute of Scientific and Technological Research, 35 Mu 3 Technopolis, Tambon Khlong Ha, Amphoe Khlong Luang, Pathum Thani 12120, Thailand

* Corresponding Author: 20sittipongeiu@gmail.com

Abstract
The powder metallurgical (PM) process, consisting of compaction and sintering, is commonly used to produce sintered structural parts applied in various industrial fields. Sintered parts always have porosity due to limited materials transport by solid-state diffusion from powder particles to voids between them. Sintered specimens with low porosity show superior mechanical properties. Hot forging is one of processes for enhancing density of sintered parts. This process was applied to sintered austenitic stainless steels (303L, 304L, 310L and 316L). The hot-forged specimens showed increased density. This led to remarkable increase of hardness (up to 5 times). Due to the hot working nature of the process, microscopic changes in the hot-forged specimens during the process were examined. It was found that grain size and volume fraction of twins were significantly deceased.

Keywords: Powder metallurgy, Stainless steel, Hot forging, Mechanical properties

1. Introduction

Powder metallurgy (PM) involves metal and alloy parts manufacturing from metal and alloy powders [1, 2]. The use of PM for producing sintered stainless-steel parts minimizes the need for machining and deformation operations and thus allows economical production of near net-shaped products [3]. Sintered stainless steels possess promising properties such as corrosion and oxidation resistance, appreciable mechanical strength as well as good formability [4]. Due to such promising properties, they have been applied in several fields including automotive [5], biomedical [6, 7] and others [8].

Typical sintered stainless steels contain pores, which show adverse effects on mechanical [7] and chemical [9] properties. Thus, some research works [10-12] have been conducted with attempt to get rid of porosity in sintered metals and alloys by taking the benefit of liquid phase sintering [13]. The increase of sintered density due to porosity reduction by liquid phase sintering is traded off by elongation reduction due to the presence of intergranular phases.

Hot forging is one of hot forming processes used to form shaped parts. It applies loads on a heated specimen to cause metal deformation, transformation and shape formation. In addition, hot forging is also used to refine the microstructure and to redistribute the alloying element producing high quality steel [14]. The microstructural refinement in hot-deformed stainless steels (bulk alloys) is due to dynamic recrystallization (DRX) [15, 16].

In this work, it is expected that hot forging would enhance density of sintered austenitic stainless steels (303L, 304L, 310L and 316L) due to due to metal flow or deformation during hot working. Since the presence of porosity in sintered austenitic stainless steels may affect the driving force for some dynamic softening processes it is thus worth investigating microstructural change after hot forging as shown in Figure 1.

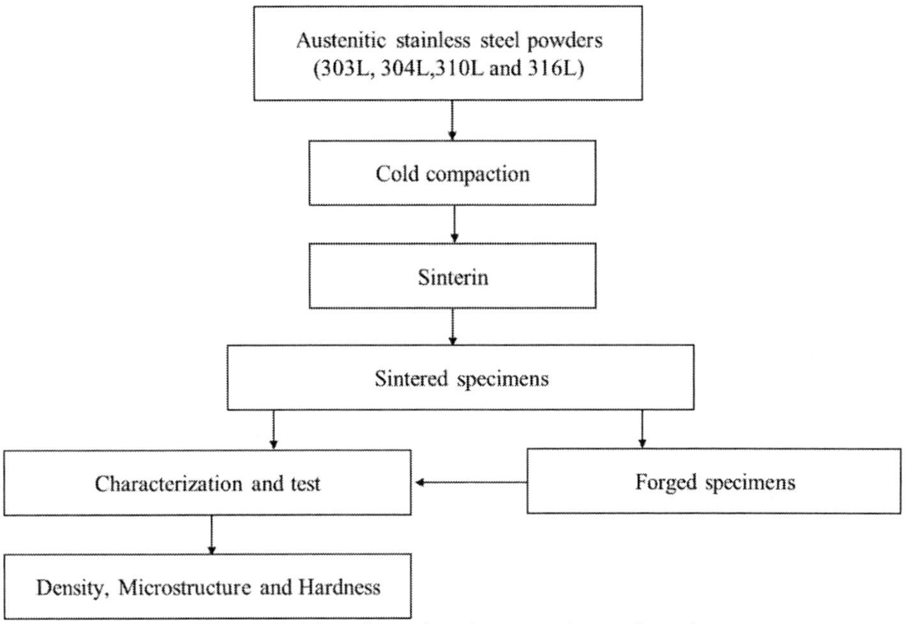

Figure 1. Flowchart showing experimental works

2. Materials and Methods

2.1 Specimen preparation

The flowchart showing experimental works is given in Figure 1. Four austenitic stainless steel powders (303L, 304L, 310L and 316L) with composition given in Table 1 were used for this investigation. Each powder was mixed with 1 wt% lubricant and cold compacted into a cylinder-shape specimen with dimensions of 14.80 mm diameter and 10.0 mm height under a constant pressure of 58.5 bars. After compaction, green specimens were sintered at 1280°C for 45 minutes under pure hydrogen.

For hot forging, sintered specimens were reheated to 1100°C for 25 minutes. The heated specimens were then hot-forged immediately to reduce the specimen height to 2.0 mm (80 % height reduction). The table must be arranged as shown in Table 1.

Table 1. Compositions (wt. %) of austenitic stainless steel powders.

Grade	Composition (wt. %)							
	C	Si	Mn	N	O	Ni	Cr	Mo
303L	0.016	0.85	0.10	-	0.24	11.40	17.70	-
304L	0.02	0.90	0.10	0.05	0.22	11.20	18.50	-
310L	0.29	1.03	0.11	0.047	0.205	20.0	24.70	-
316L	0.02	0.85	0.10	0.04	0.20	13.0	16.80	2.20

2.2 Characterization and testing

Due to simple specimen geometry, the density of the as-sintered and as-forged specimens was measured by using mass and volume of the specimen. The as-sintered and as-forged specimens were prepared for metallography observation following standard methods. Microstructures were observed by using scanning electron microscopy (SEM) equipped with electron backscatter diffraction (EBSD) to measure grain size and twin boundary fraction.

3. Results and Discussion

3.1 Densities of sintered and forged specimens

Under the same PM process (compaction and sintering), each sintered stainless steel showed different sintered densities the table must be arranged as shown in Table 2. The sintered 316L exhibited the highest sintered density whereas the sintered 303L showed the lowest sintered density. Sintered density results from shrinkage or swelling phenomena during sintering state. In general, sintered austenitic stainless steels gain increased density after sintering [15, 16]. The increased density after sintering relates directly to shrinkage, which is in turn directly proportional to sintered neck size ratio [17]. The sintered neck size or volume depends on contacting areas in powder compacts after cold compaction. Thus, different densification levels in sintered austenitic stainless steels are due to the nature of powder response on compaction action

Table 2. Properties of sintered and forged austenitic stainless steels.

Material	Density (g/cm^3)	Porosity (%)	Grain size (μm.)	Twin fraction (%)	Hardness (HV1)
Sintered 303L	6.92	8.98	43.36	15.19	66.31
Forged 303L	7.6	0.27	12.13	3.05	308.87
Sintered 304L	7	11.96	49.27	16.12	62.31
Forged 304l	7.58	2.13	12.88	3.07	351.29
Sintered 310L	7.1	8.64	40.77	11.34	61.18
Forged 310L	7.43	1.17	10.83	4.18	323.33
Sintered 316L	6.89	11.41	44.87	13.05	69.01
Forged 316L	7.7	1.03	11.47	3.16	350.57

All forged specimens gained increased densities but with different increments the table must be arranged as shown in Table 2. The most density increment was observed in forged 316L. The porosity images in sintered and forged austenitic stainless steels, given as shown in Figure 2. Confirm sintered and forged densities given the table must be arranged as shown in Table 2. The density increment obtained from hot forging is due to metal deformation and flow at high temperatures [18, 19]. Other actions of hot forging on density enhancement are pore closure and bonding [20]. The experimental results implement the hypothesis,

IOP Conf. Series: Materials Science and Engineering 1137 (2021) 012031 doi:10.1088/1757-899X/1137/1/012031

in which density would be increased after hot forging. However, different austenitic stainless steels have different responses to density increase by hot forging.

3.2 Microstructures of sintered and forged specimens

Sintered austenitic stainless steels exhibited equiaxed grains with sizes in the range of 40 .0 -49 .0 μm. There were twins within grains. The wins and pores also existed in forged specimens, except that the forged grains with sizes in the range of 10 .0 -12 .0 μm. The examples of sintered and hot-forged grains as shown in Figure 3.

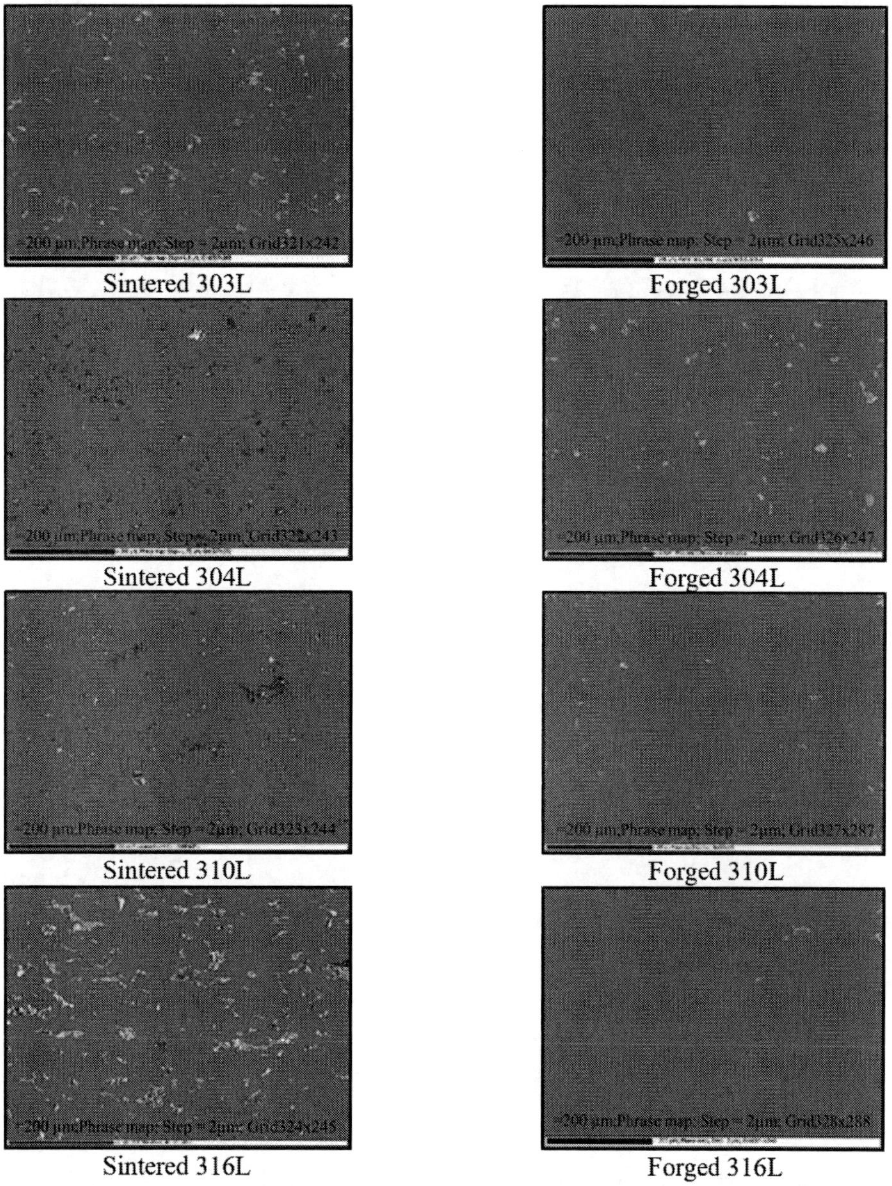

Figure 2. EBSD Phrase map images showing porosity natures of sintered and forged austenitic stainless steels.

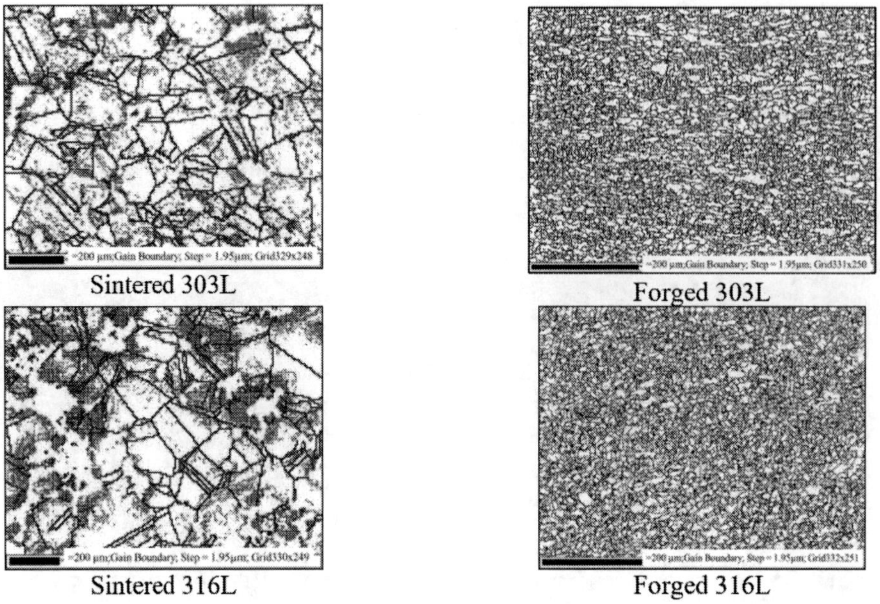

Figure 3. EBSD Grain Boundary images showing grain size of sintered and forged austenitic stain less steels.

3.3 Hardness values of sintered and forged specimens

The hardness values of forged specimens increased appreciably from those of sintered specimens as shown in Table 2 and Figure 4. From the experimental evidences, it is apparently that the significant improvement of hardness is resulted from the increased density and grain refinement.

To find other causes of hardness improvement, the microstructural change regarding twin fraction was examined using EBSD. Quantification of twin fraction was conducted and it found that twin fractions in all forged specimens were lower than those in the respective sintered specimens as shown in Table 2. From this experimental result, the twin fraction has inversely relationship with hardness.

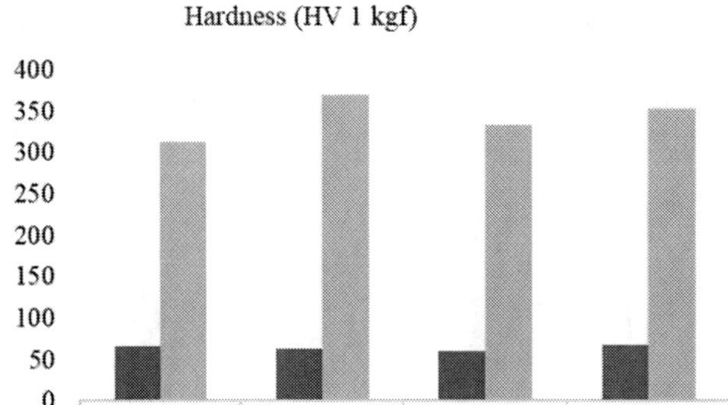

	PM303L	PM304L	PM310L	PM316L
■ Sintering	65.62	62.79	60.84	68.29
■ Forging 80%	312.69	369.27	334.14	352.92

Figure 4. Hardness values of sintered and forged austenitic stain less steels.

4. Conclusion

Under the same PM processing conditions (compaction and sintering), different sintered austenitic stainless steels showed different sintered densities. It is assumed that the different sintered densities are to differences in campressibility and sinterability of austenitic stainless steel powders. Further improved densities of austenitic stainless steels were obtained by hot forging. The improvement of density led to hardness increase. On the other hand, the grain size and volume fraction of twin were significantly deceased.

Acknowledgments
This research was financially supported by Thailand Institute of Scientific and Technological Research (TISTR) and Thailand National Metal and Materials Technology Center (MTEC).

References

[1] German R M 1998 *Powder Metallurgy of Iron and Steel* (New York: John Wiley)
[2] Al-Mangour B 2015 *Powder metallurgy of stainless steel: State-of-the art, challenges, and development Stainless Steel* Eds A Pramanik and A K Basak (New York: Nova Science Publishers)
[3] Cheng L H and Hwang K S 2012 *Mater. Trans.* **53** 179
[4] Marnier G C Noudem J and Hug E 2014 *Mater. Des.* **63** 633
[5] Hale T 1998 *Met. Powder Rep.* **53** 22
[6] Dudek A and Włodarczyk R 2013 *Mater. Sci. Eng. C.* **33** 434
[7] Kurgan N 2014 *Mater. Des.* **55** 235
[8] Klar E and Samal P K 2007 *Powder Metallurgy Stainless Steels: Processing, Microstructures, and Properties* (Ohio: ASM International)
[9] Chaiyarat N Daopiset S Mahathanabodee S and Tongsri R 2018 *Mater. Today Proc.* **5** 9384
[10] Tosangthum N Muangtong P Coovatanachai O Morakotjinda M Yodkaew T Wila P Krataitong R Vetayanugul B and Tongsri R 2008 *J. Met. Mater. Miner.* **18** 47
[11] Coovattanachai O Tosangthum N Morakotjinda M Yotkaew T Daraphan A Krataitong R Vetayanugul B and Tongsri R 2007 *Mater. Sci. Eng. A.* **445-446** 440
[12] Ninpetch P Luechaisirikul A Morakotjinda M Yotkaew T Krataitong R Tosangthum N Mahathanabodee S and Tongsri R 2018 *Mater. Today Proc.* **5** 9409
[13] German R M Suri P and Park S J 2009 *J. Mater. Sci.* **44** 1
[14] Farahat A I Z Hamed O El-Sisi A and Hawash M 2011 *Mater. Sci. Eng. A.* **530** 98
[15] Coovattanachai O Tosangthum N Morakotjinda M Yotkaew T Krataitong R Vetayanugul B and Tongsri R 2010 *Songklanakarin J. Sci. Technol.* **32** 163
[16] Panda S S Singh V Upadhyaya A and Agrawal D 2006 *Scripta Mater.* **54** 2179
[17] Xu X Lu P and German R M 2002 *J. Mater. Sci.* **37** 117
[18] Lippmann H and Iankov R 1997 *J. Mech. Sci.* **39** 585
[19] Rajeshkannan A Pandey K S Shanmugam S and Narayanasamy R 2008 *Mater. Des.* **29** 1862
[20] Bai Q Lin J Jiang J Dean T A Zou J and Tian G 2015 *Mater. Sci. Eng. A.* **621** 68

Comparative study on the wood-based PLA fabricated by compression molding and additive manufacturing

Kanjanaphorn Chansoda[1], Chakrit Suvanjumrat[2], and Watcharapong Chookaew[1,*]

[1]Material and Manufacturing Innovation Research Group, Department of Mechanical Engineering, Faculty of Engineering, Mahidol University, Nakhon Pathom Thailand 73170
[2]Department of Mechanical Engineering, Faculty of Engineering, Mahidol University, Nakhon Pathom Thailand 73170

* Corresponding Author: watcharapong.chk@mahidol.ac.th

Abstract
Additive manufacturing namely 3-Dimensional (3D) printing and it's feeding materials are increasingly used for producing customized products. Fused Deposition Modelling (FDM) in conventional 3D printers is a cost-effective solution. This is due to the low cost of the machine and various size selections. Although the machine cost is reasonable, the filament material is a higher cost. In general, the current filament is priced about 4 times more than granulated plastic. Moreover, the special filament containing wood powder shows a high cost of approximately 20 times that of pellets. This work proposes a comparative study of the components obtained from different processes. Compression molding and no-mold techniques were used. A previous design, nozzle from extrusion concept, was also used for fabricating the specimens. PLA and wood-based PLA were carried out to produce the testing samples. The results showed that parts obtained from the compressive mold showed low tensile strength. It was indicated that the residue voids were significantly affected by its strength. Taking into account the 3D printed parts, the articles from the designed extruder present a higher strength than the results derived from the conventional 3D printer and compression machine. On the other side, it was found that the moving direction of the nozzle (raster angle) affected directly the maximum tensile strength. Furthermore, the neat PLA had higher stress than the wood-composite materials used for all fabrication methods. However, in the case of using an extrusion device, it represented the enhancing of mechanical strength, which was caused by the homogeneous texture of polymers melting in a hot barrel. In terms of fabrication cost, the molding technique showed the highest cost even though the feeding material has the lowest cost. A similar cost was found for providing parts from the conventional printer and our design nozzle. However, a pellet-based extruder would be printed the sample filled 15%wt, conventional but the 3D printer is unavailable. Furthermore, recycled materials or waste of 3D printed parts can be utilized by using a conceptual extruder.

Keywords: Additive manufacturing, Fused deposition modelling, Wood composite, Pellet-based extruder

1. Introduction

Currently, Additive Manufacturing (AM) and 3-Dimensional (3D) printing are rapidly growing. Many varieties of materials such as plastics, metals, alloys and bio-materials are applied for supporting AM. Thermoplastics have been widely used in Fused Deposition Modelling (FDM), which is an important type of AM. Other forms of thermoplastics, in particular, the molding process, is often comparatively studied. The difference of investment costs and mechanical strength is an interesting point. The molding process, especially injection molding, will always give an advantage of mechanical property over 3D printing [1-7].

Acrylonitrile butadiene styrene (ABS) was experimentally investigated, which was prepared by injection moulding and 3D printing. It was found that in almost all cases injected parts exhibited enhanced mechanical strength which resulted from the higher material compaction [8]. Another result confirmed that the tensile strength of a printed part was lower than those of an injected part by approximately 50% and the other mechanical properties were reduced as well [9]. Processing parameters of 3D printing, infill patterns or part density, was one of reducing parameter which destroys 3D printed parts [10].

However, the cost of parts obtained from an injection molding process was more than several times higher than that of 3D printed parts. The investment cost depended on a variety of parameters e.g., machine, mold design and manufacturing, the number of products and energy consumption etc. The cost was greatly increased in the case of parts having a complex geometry. Normally, the supplied materials for a molded article were pellets or granulate, which were low cost in respect of the filaments of a 3D printer. A secondary process of filament extruding was necessary. Thus, the number of plastic selections was limited. Also, the special filaments, the material filling wood particles, had a very high cost. Previous work achieved the preparation of a wood composite which had a lower-cost material [11]. However, the obtained material always presents as brittle in nature and it may give in the granulate form of a high-wood content.

The above mentioned indicated that a compression molding technique was an alternative method to be considered in terms of investment cost and the low-cost of materials. Because compression molding was a low-cost process and the plastic materials were supplied in the form of small granulates. A further approach, from current literature, combining a 3D printer combining with an extrusion concept was another interesting development [12-15]. This way is usually called the pellet-based 3D printer. The machine was introduced to conveniently use pellets for solving unsuitable forms of feeding material. A 3D printer based on an extrusion machine was proposed to perform the high viscosity of polyether-ether-ketone (PEEK) [12]. A pellet-based printer was also applied for making large scale printed parts [13].

This work was designing and constructing a pellet-based extruder to eliminate the high cost of 3D printed materials. Our machine can use granulates as material to be used in compression molding. A further advantage was that it was also applied to a wood composite which is a special type of printing material. A comparison of plastic formed by compression molding, traditional 3D printing and pellet-based extrusion was investigated in this work. The wood contents and printing orientations were also characterized in detail.

2. Materials and methods

For reducing variations of material properties, polylactic acid (PLA) from the same source was used throughout this work. It was mixed by loading it into a twin-screw extruder. The barrel and die temperature of 200°C and 180°C were controlled for melting material and maintaining it in a molten state. The melting polymer was cooled in a water bath and collected by the rolling machine as depicted in Figure 1. All wood-powder material was kindly supported from the Thai Hevea Wood Association. The varying of particle sizes was determined by screening with a sieve shaker to ensure the particle sizes were in the range of 106-425 μm. Parawood powder and PLA were compounded by twin screw extruder machine in wood content ranges of 5%wt-15%wt. The wood filament was divided into 2 categories.

The first, filament form, was directly used for the traditional 3D printer as in Figure 2. The second one was chopped into granulate approximately 3-5 mm and carried out for compression molding and pellet-based 3D printer.

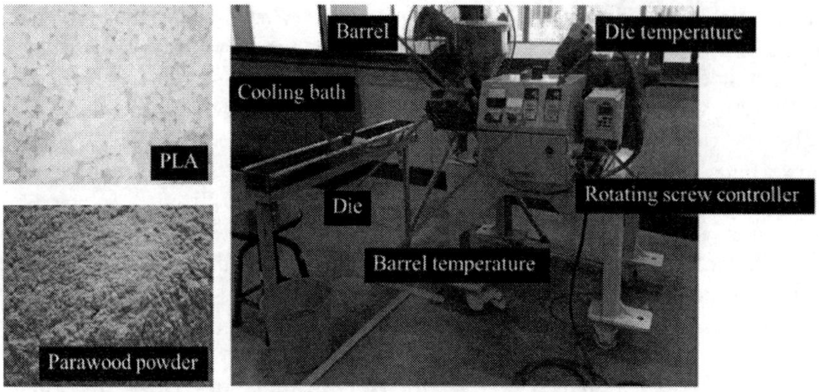

Figure 1. Twin screw extruder when rolling filament.

(a) (b)

Figure 2. Shredding composite materials (a) chopped machine (b) composite material.

In the compression molding process, aluminum foil was laid on the bottom and the top of the mold cavity. Each composite was weighed and then placed between aluminum foils. All samples were compressed under a pressure of 15 MPa and a mold temperature of 190oC for 15 minutes. The polymer sheets were punched as the geometry of ASTM D638 specification which is shown in Figure 3. In the printing process, the dumbbell shape of ASTM D638 was also built up in 3 directions as in Figure 4. The obtained samples were used for evaluating the mechanical properties by using Universal Testing Machine (UTM, COMETECH Model LRK-20). According to ASTM D638 procedure, a dumbbell-shaped testing sample was installed between the grips with a gauge length of 50 mm and 5 mm/min of crosshead speed was also set. The raw data from five specimens of each condition was collected to find the average value and the standard deviation.

IOP Conf. Series: Materials Science and Engineering 1137 (2021) 012032 doi:10.1088/1757-899X/1137/1/012032

Figure 3. Cut the workpiece from the extrusion sheet.

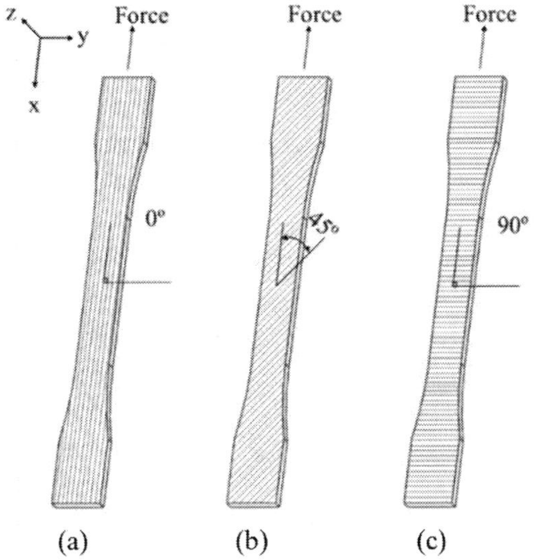

Figure 4. Different orientation directions (a) 0 degree (parallel to force) (b) ±45 degree (c) 90 degree (Perpendicular to force).

3. Results and discussions

3.1 Effect of forming processes

The tensile properties of different types of plastics fabricated using FDM and compression molding are presented in Figure 5. There was shown a significant difference between compression molding and additive manufacturing. Molded articles presented lower tensile strength and gave better elongations. Generally, samples from the molding process are compressed using high pressure which promotes a strong and stiff polymer material. However, the short-compressed stroke of the open flash mold caused insufficient pressure to release air in the raw material. Hence, the air bubbles which occurred inside the test sample [9,10] are presented in Figure 6(a). This reason led to the lower failure strength. The short-cycle time of polymer melts in the molding process also resulted in the stretchability of materials. In additive manufacturing, heated positions promoted an excellent diffusion of polymer chains leading to a stronger bonding strength. A slight difference between a traditional 3D printer and a pellet extruder was observed. This pellet-based extruder was stronger than traditional 3D printer due to better cross-layer diffusion as shown in Figure 6(b) and (c). However, the part of pellet extruder had a lower flexibility caused by the chain scission of polymer melt in a barrel.

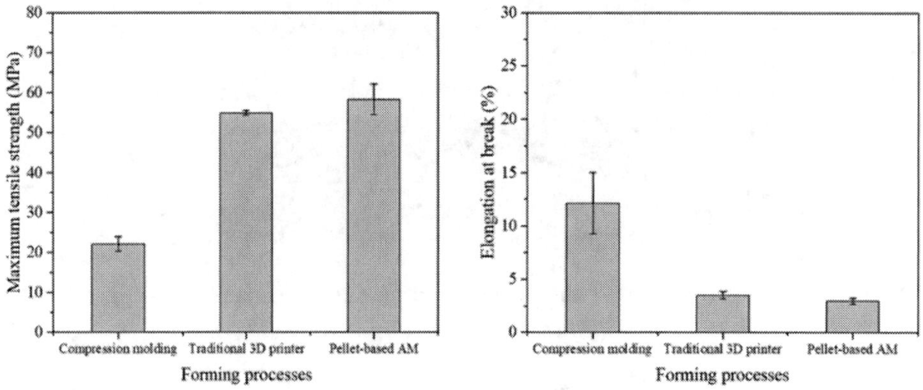

(a) (b)

Figure 5. Mechanical property comparison between compression molding and additive manufacturing processes (a) maximum tensile strength (b) elongation at break.

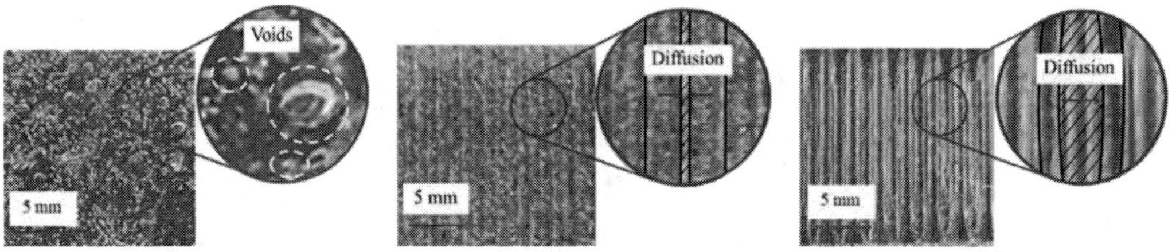

(a) (b) (c)

Figure 6. Composite surface fabricating in different processes (a) compression molding (b) traditional 3D printer (c) pellet-based AM.

3.2 Effect of parawood contents

Figure 7(a) and (b) show the results of wood composites which composed of parawood flour and PLA. The maximum tensile strength decreased as wood content was added. In contrast, material elongation showed t better effects (except for the traditional 3D printer). This observation can be explained by the fact that an increase of wooden power gradually decreased the polymer area which supported the applied force. However, the nature soft of wood attributed to the stretchability of the composite materials. Surprisingly, in the case of a pellet-based extruder, stronger strength than those of a conventional printer was found. It can be interpreted that the long-heated barrel induced the cumulative heat of melting polymer. This led to an excellent molecular diffusion higher than the results obtained from the traditional 3D printer. The unexpected result was found that the maximum strength and elongation became more pronounced in case wood filling 15%wt. As considered the pellet-based mechanism, the slow-rotating screw in the heated barrel contributed to the good dispersion and distribution of wood particle and plastic matrix.

IOP Conf. Series: Materials Science and Engineering 1137 (2021) 012032 doi:10.1088/1757-899X/1137/1/012032

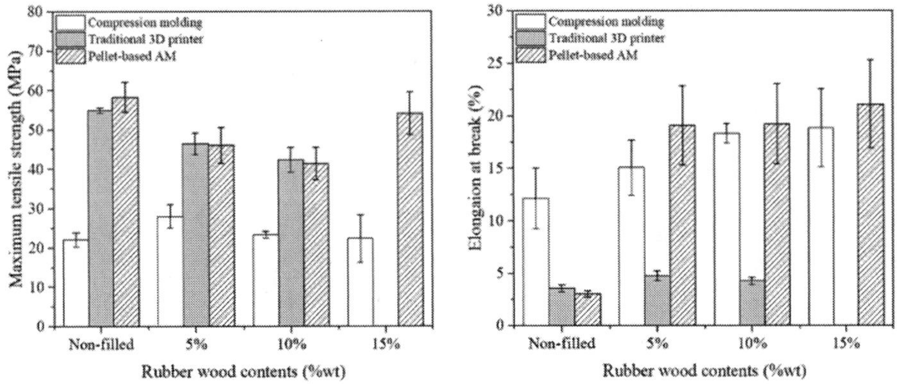

(a) (b)

Figure 7. Effect of parawood content on the mechanical properties for varying forming processes (a) maximum tensile strength (b) elongation at break.

3.3 Effect of filled orientation

The three filling paths, which were compared to compression molding, are presented in Figure 8. The corresponding results presented the higher strength of FDM techniques as mention earlier. This difference in tensile strength of the orientations can be explained by the arrangement of the filling path in the specimens. The 0 degrees, parallel to the applied force, was presented the highest strength. As the raster angle increased, it showed a reduction in mechanical failure, especially for a 90-degree orientation. This was due to the influence of molecular orientation and bonding strength. As considered the AM techniques, it can be explained by two parameters i.e., polymer diffusion and surface roughness. Normally, the diffusion depth of the polymer molecular promoted the strength of the material, while the surface roughness destroyed it properties. In the case of 0 angles (parallel to force direction), a pellet-based extruder showed an improvement in tensile strength. This observation was due to the high-temperature pellet-based AM inducing better diffusion. It meant that the diffusion effect (pellet-based extruder) governed the roughness effect (traditional printer). For a 90-degree arrangement, a traditional 3D printer displayed good maximum tensile strength. This result indicated that the surface roughness of apart from a pellet-based extruder was more affected by the reduction of tensile properties. In the middle angle of 45 degrees, the two effects from roughness and diffusion were similarly found.

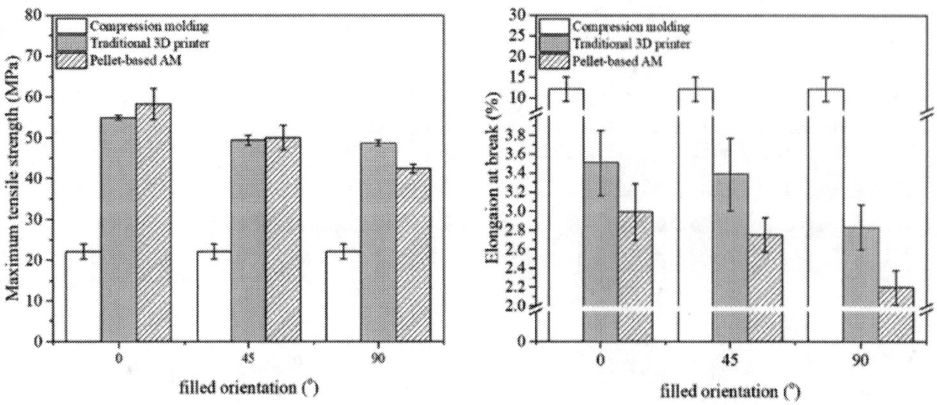

(a) (b)

Figure 8. Effect of filling orientation of additive manufacturing (a) maximum tensile strength (b) elongation at break.

4. Conclusion

This study presents the comparative performances between a compression molded article and two types of AM parts. It can be seen that AM parts showed good strength above the compression molding technique. This was from the homogenous texture of the fabricating parts. The obtained results indicated that an increase of rubber-wood powder tends to decrease the maximum tensile strength of all introduced processes. Surprisingly, the stretchability of parts obtained from a pellet-based extruder was close to the molded part. However, the effect of filling, orientation was evidently presented both for the traditional 3D printer and pellet-based extruder. Our design, the nozzle motivated from the extrusion process, could supply low-costing material in the form of neat pellet and wood-plastic composites. While the mechanical strength was higher than that of a compression molded part. A further advantage of this extrusion-based concept was that it could provide the same complex shapes as a traditional 3D printer. Moreover, it could use recycled materials.

Acknowledgments

This research was funded by National Research Council of Thailand (NRCT) and Thailand Science Research and Innovation (TSRI) under contract number RDG62T0009.

References

[1] Dawoud M Taha I and Ebeid S J 2016 *J. Manuf. Process.* **21** 39-45
[2] Franchetti M and Kress C 2017 *Int. J. Adv. Manuf. Technol.* **88** 2573-2579
[3] Krishna L S R and Gundeti A 2017 *IJMET* **5** 12
[4] Kaynak C and Varsavas S D 2019 *J. Thermoplast. Compos. Mater* **32** 501-520
[5] Cisneros-López E O Pal A K Rodriguez AU Wu F Misra M Mielewski D F Kiziltas A and Mohanty A K 2020 *Mater. Today* **7** 100027
[6] Toro E V D Sobrino J C Martínez A M Eguía V M and Pérez J A 2020 *Materials* **13** 672
[7] Zandi M D Jerez-Mesa R Llumà J Jorba-Peiro J and Travieso-Rodriguez J A 2020 *Int. J. Adv. Manuf. Technol.* **108** 1725-1735
[8] Dawoud M Taha I and Ebeid S J 2016 *J. Manuf. Process* **21** 39-45

[9] Lay M Thajudin N L N Hamid Z A A Rusli A Abdullah M K and Shuib R K 2019 *Compos. B. Eng* **176** 107341

[10] Priya M S Naresh K Jayaganthan R and Velmurugan R 2019 *Mater. Res. Express* **6**(8) 085345

[11] Chansoda K Suwanjamrat C and Chookaew W 2020 *Proc. Int. Conf. on Materials Research and Innovation* vol **773** (Bangkok, Thailand) 012053

[12] Tseng J W Liu C Y Yen Y K Belkner J Bremicker T Liu B H Sun T J and Wang A B 2018 *Mater. Des* **140** 209-221

[13] Nieto D M López V C and Molina S I 2018 *Addit. Manuf* **23** 79-85

[14] Whyman S Arif K M and Potgieter J 2018 *Int. J. Adv. Manuf. Tech* **96**(9-12) 3417-3428

[15] Liu S Zhao P Wu S Zhang C Fu J and Chen Z 2019 *Adv. Polym. Technol.* **2019** 5075327

Influences of cooling rate and carbon content microstructure and mechanical properties of sintered Fe-1.5Mo-xC alloys

Arisara Wanalerkngam[1], Sarum Boonmee[1], Thapanee Srichumpong[2], Monnapas Morakotjinda[2], Nattaya Tosangthum [2,*], Ruangdaj Tongsri[2]

[1]School of Metallurgical Engineering, Institute of Engineering, Suranaree University of Technology 111 University Avenue, Suranaree Sub-District, Muang Nakhon Ratchasima District, Nakhon Ratchasima Thailand 30000
[2]Particulate Materials Processing Technology (PMPT), National Metal and Materials Technology Center, 114 Paholyothin Road, Khlong Nueng, Khlong Luang, Pathum Thani Thailand 12120

* Corresponding Author: nattayt@mtec.or.th

Abstract
Sintered Fe-1.50Mo-xC alloys were produced by the sintering of powder compacts made from mixtures of pre-alloyed Fe-1.5Mo powder and varied carbon amounts (0.30-1.20 wt.% with 0.15 increment) followed by slow and fast cooling rates. The slowly cooled sintered Fe-1.50Mo-xC alloys (for carbon contents of up to 0.45 wt.%) showed microstructures consisting of polygonal ferrite grains and eutectoid transformation products. When carbon contents were higher than 0.45 wt.%, eutectoid transformation products were dominant. The fast-cooled sintered Fe-1.50Mo-xC alloys (for carbon contents of up to 0.75 wt.%) showed microstructures consisting of upper bainite. When carbon contents were higher than 0.75 wt.%, upper bainite and inverse bainite were dominant. Tensile strength and hardness values of sintered Fe-1.50Mo-xC alloys increased with increasing carbon content. In addition, fast cooling further enhanced mechanical properties of the sintered alloys. It was found that values of ultimate tensile strength (UTM) and hardness on slow and fast cooling rates were 385-565 MPa, 564-743 MPa, 43-78 HRB and 60-82 HRB, respectively.

Keywords: Sintering, Carbon content, Cooling rate, Microstructural development, Mechanical property

1. Introduction

Powder metallurgy (PM) ferrous alloys are increasingly employed in the automotive industry. The automotive industry has accepted PM technology, with the result that over 70% of its production worldwide is aimed at the automobile manufacturing sector [1]. The Fe-1.50Mo (AstaloyMo) is a ferrous based powder pre alloyed with 1.5 wt.% molybdenum (Mo) [2], which is one of alloying elements used for enhancing the mechanical properties of sintered steels [3]. Improvements of mechanical properties in sintered Mo-containing ferrous alloys is due to their sinter hardening ability [4]. Sinter hardening refers to a process where the cooling rate experienced in the cooling zone of the sintering furnace is fast enough that a significant portion of the material matrix transforms to martensite [5]. Since the cooling rate obtained in a commercial conveyor belt sintering furnace is quite low, the pre-alloyed ferrous powders are designed to have a sinter hardening ability due to the influence of their alloying elements. It is widely known that molybdenum suppresses ferrite and pearlite formations during continuous cooling by shifting both phase transformation fields, to the slow cooling side of a continuous cooling transformation (CCT) curve [6-8]. This means that the Mo effect opens the window for low temperature phase transformations (bainite and martensite transformations) to occur. In previous studies, carbon (C) content and post-sintering cooling rate showed strong influences on microstructures and mechanical properties of sintered ferrous alloys [9-11].

This work was aimed to understand the sinter hardening behavior of sintered Fe-Mo-C alloys. To achieve such goal, different C contents and cooling rates were employed to produce experimental sintered alloys. The influences of these two parameters on mechanical properties of sintered alloys were, as such, investigated.

2. Material and methods

2.1 Materials preparations

The base material employed in this work was pre-alloyed Fe-1.5Mo (AstaloyMo) powder. This powder was mixed with 1.0 wt.% of zinc stearate and varied graphite amounts to obtain 0.30, 0.45, 0.60, 0.75, 0.90, 1.05 and 1.20 wt.% contents in sintered alloys. After mixing, the dog-bone specimens were produced by pressing at 400 MPa to MPIF standard 10, ASTM B783 with green density of 6.58 ± 0.05 g/cm3 (Figure 1) [12]. All specimens were sintered at 1280°C in a vacuum furnace for 45 min. Two different post-sintering cooling rates (0.1 and 5.4°C/s) were employed. The sintering cycle profile is shown in Figure 2.

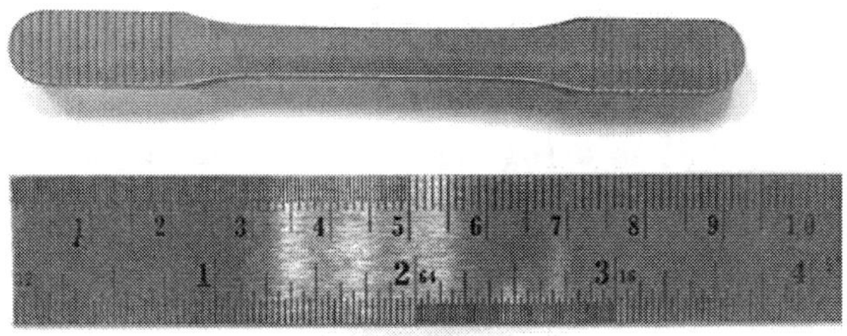

Figure 1. The tensile test specimen follows by MPIF standard 10, ASTM B783

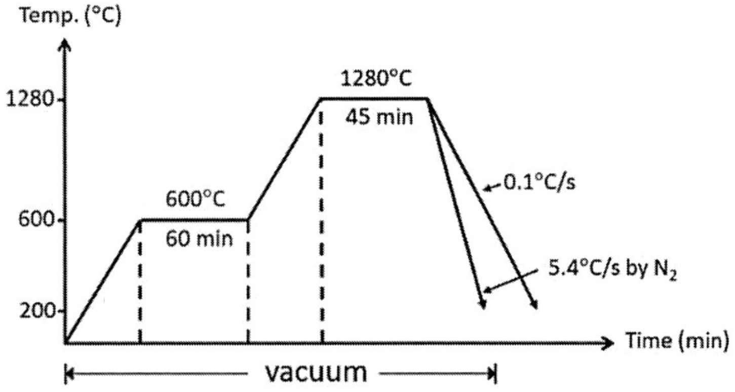

Figure 2. Temperature versus time in setting up the sintering process.

2.2 Materials characterizations

For metallographic examination, all specimens were grinded by 60, 120, 240, 400, 600, 800, 1200-grit silicon carbide abrasive papers, respectively, then polished by 1, 3 and 6μm diamond pastes. Etching was carried out by using 2% Nital acid solution for microstructural analysis. The microstructures were observed under optical microscopy (OM) by using Olympus STM7, Tokyo Japan and under scanning electron microscopy (SEM) by using HITACHI SU-8030, Tokyo, Japan.

Tensile tests were conducted using the Instron Universal testing machine employing a strain rate of 5 mm/min. Hardness studies were tested by using Rockwell scale B hardness testing 20, in which specimens were indented with the load of 0.3 kg and loading time of 15 s.

3. Results and discussions

3.1 Microstructures of slowly cooled sintered alloys

Microstructures of slowly cooled sintered Fe-1.50Mo-xC alloys are shown in Figure 3 marked as (a) to (d) to represent C contents of 0.30 to 1.20 wt.%, respectively. With 0.30 wt.% C addition (figure 3 (a), the sintered alloy showed proeutectoid ferrite grains in a polygonal shape and eutectoid transformation products in the forms of ferrite/carbide (F/C) aggregates. The formation of polygonal ferrite grains and F/C aggregates in sintered Fe-1.50Mo-0.30C alloy (figure 3 (a)) is similar to that of sintered low-carbon Fe-Cr-Mo-C steels [10], and hypoeutectoid steels [13]. However, the F/C aggregates in sintered Fe-1.50Mo-0.30C alloy was not identified as pearlite due to non-cooperative growth of ferrite and carbide components. With 0.45 wt.% C addition (Figure 3 (b)), proeutectoid ferrite grains changed shape from polygonal to elongated ones and the eutectoid transformation zones distributed more homogeneously. SEM image in Figure 4 (a) clearly showed the shape of proeutectoid ferrite grains and F/C aggregates. The sintered alloys with 0.60 and 0.75 wt.% C contents (Figure 3 (c), (d) and Figure 4 (b)) had F/C aggregates as dominant microstructural feature whereas polygonal ferrite grains were hardly observed. Pro-eutectoid carbide particles started to form, in addition to F/C aggregates, in sintered Fe-1.50Mo-0.90C alloy (Figure 3 (e)). This type of carbide particle become larger and prominent in sintered alloys with 1.05 and 1.20 wt.% C contents (Figure 3 (f) and (g)). A new F/C aggregate feature with nodular carbide particles formed next to a large pro-eutectoid carbide particle shown in Figure 4 (c). The disappearance of polygonal ferrite and conventional lamellar pearlite in most of slowly cooled sintered Fe-1.50Mo-xC alloys confirms the influence of Mo on suppression of ferrite and pearlite transformations [6-8]. The Mo influence becomes stronger when C contents are higher. However, the synergy of Mo and C is insufficient to keep austenite stable down below martensite transformation start temperature. Thus, only F/C aggregates with features like that of upper bainite can be obtained in slowly cooled sintered Fe-1.50Mo-xC alloys.

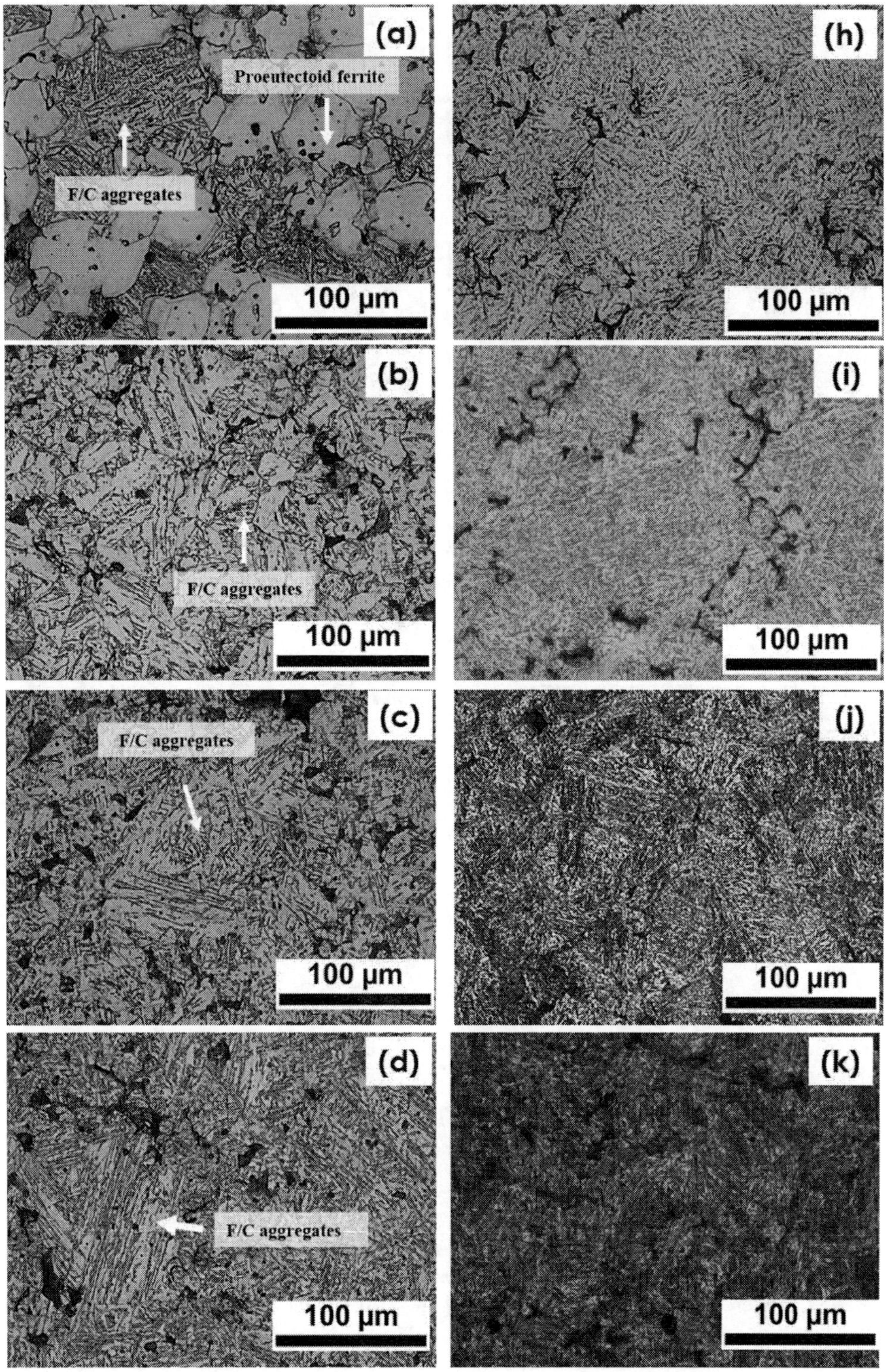
(a)
Proeutectoid ferrite
F/C aggregates
100 µm
(h)
100 µm
(b)
F/C aggregates
100 µm
(i)
100 µm
(c)
F/C aggregates
100 µm
(j)
100 µm
(d)
F/C aggregates
100 µm
(k)
100 µm

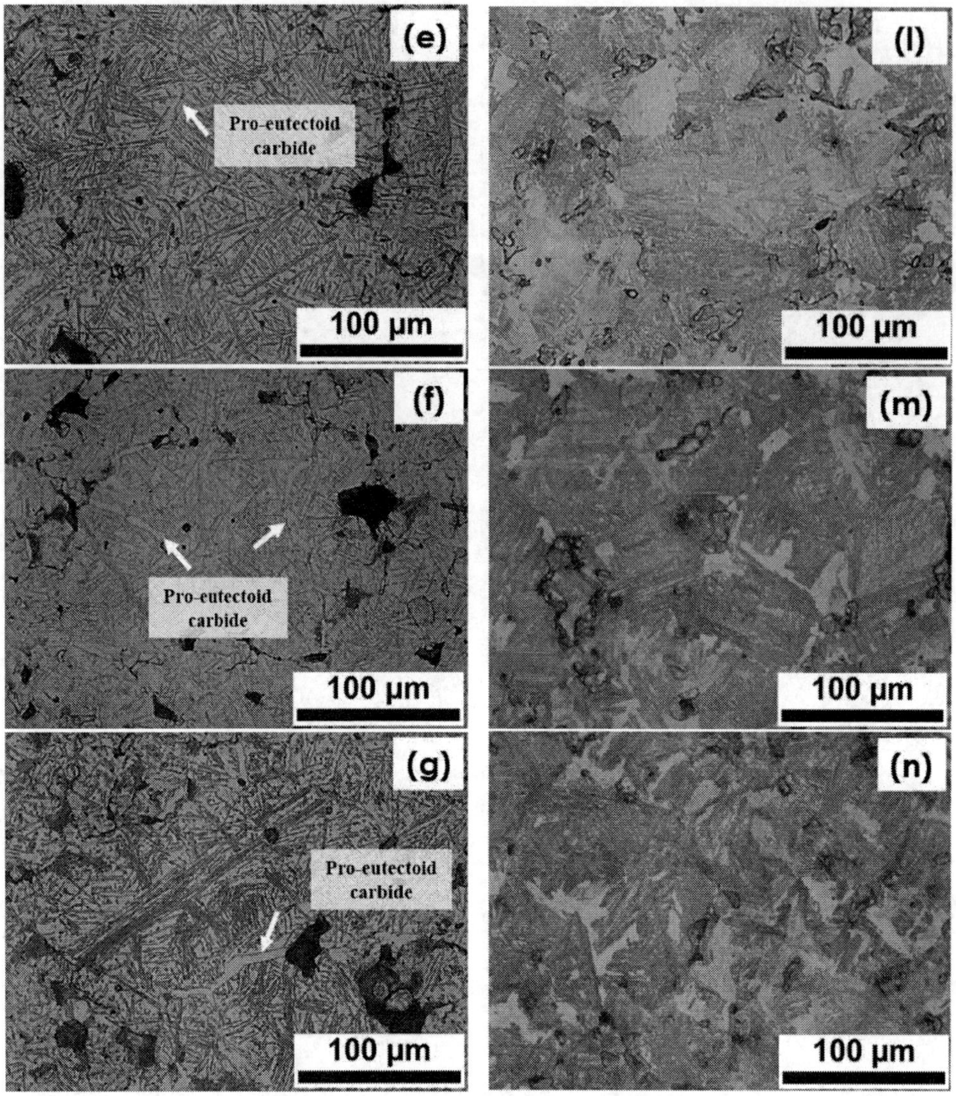

Figure 3. OM images showing microstructures of sintered Fe-1.50Mo-xC alloys: (a) to (g) represent slowly-cooled sintered alloys with C contents from 0.30 to 1.20 wt.% and (h) to (m) represent fast-cooled sintered alloys with C contents from 0.30 to 1.20 wt.%.

3.2 Microstructures of fast-cooled sintered alloys

Microstructures of fast cooled sintered Fe-1.50Mo-xC alloys are shown in Figure 3 marked as (h) to (n) to represent C contents of 0.30 to 1.20 wt.%, respectively. There were no polygonal ferrite grains observed in all sintered alloys. Clearly shown in Figure 4 (d), (e) and (f), only F/C aggregates dominated microstructures of all fast-cooled sintered Fe-1.50Mo-xC alloys. With close observation on the microstructure of the sintered Fe-1.50Mo-1.05C alloy (Figure 4 (f)), there were two types of F/C aggregates, i.e., the first was upper bainite and the second contained carbide particles showing nature of splines and branches, which are characteristics of inverse bainite [14]. Due to a similar response to etching, the OM images in Figure 3 (k), (l) and (m) suggest that inverse bainite starts to form in addition to upper bainite in sintered alloys with C contents of ≥ 0.90 wt. %. Again, the synergy of Mo, C content and fast cooling rate (5.4°C/s) is insufficient to keep austenite stable down below martensite transformation start temperature. Thus, only finer F/C aggregates can be obtained in fast-cooled sintered Fe-1.50Mo-xC alloys.

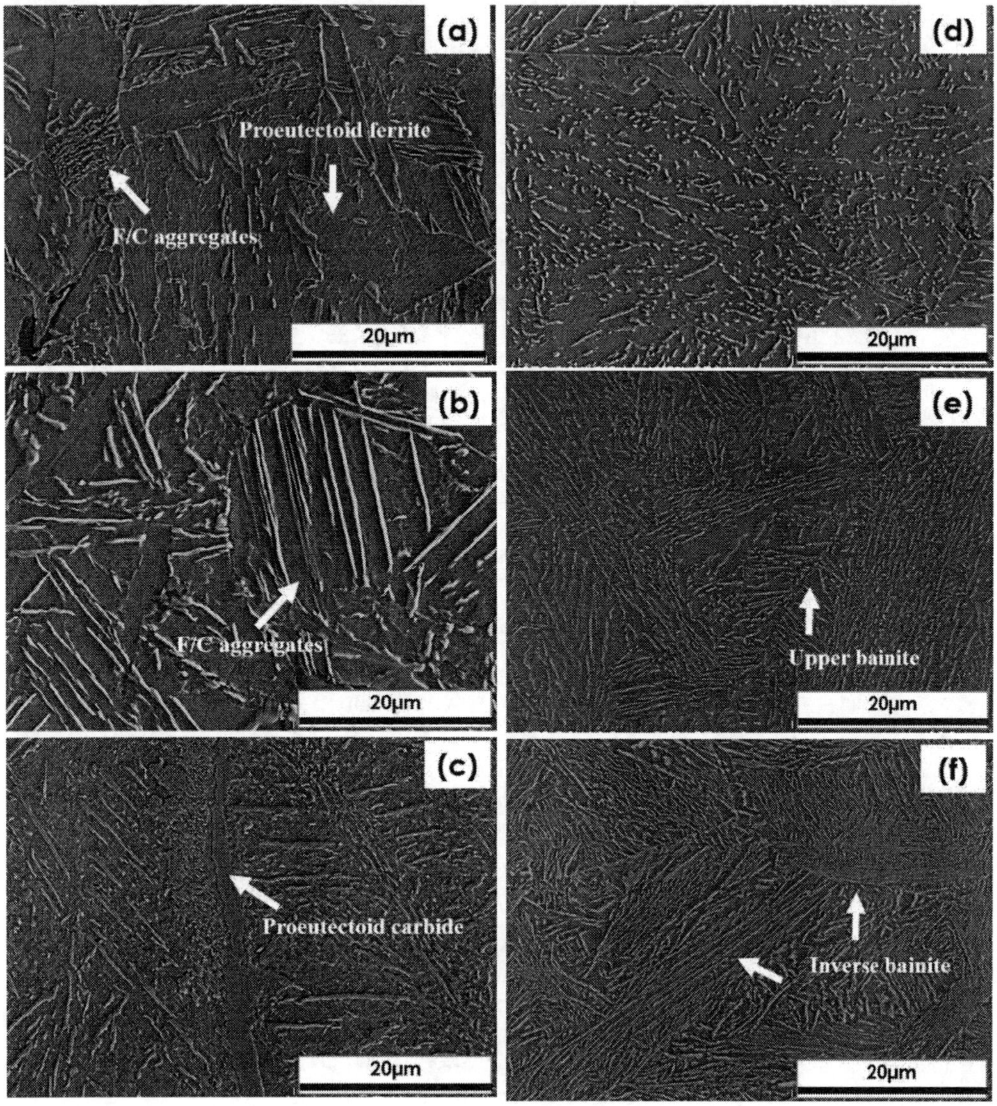

Figure 4. SEM images showing microstructures of sintered Fe-1.50Mo-*x*C alloys: (a), (b) and (c) represent slowly-cooled sintered alloys with C contents of 0.45, 0.75 and 1.05 wt. %, respectively and (d), (e) and (f) represent fast-cooled sintered alloys with C contents of 0.45, 0.75 and 1.05 wt. %, respectively.

3.3 Mechanical property

Tensile properties (ultimate tensile strength (UTM) yield strength (YS and elongation) and hardness of all experimental sintered alloys are shown in Figure 5. All mechanical properties of the fast-cooled sintered alloys were better than those of the slowly cooled sintered ones. This is attributed to finer scale

IOP Conf. Series: Materials Science and Engineering 1137 (2021) 012033 doi:10.1088/1757-899X/1137/1/012033

of F/C aggregates in the former group of alloys. The inferior mechanical properties of the slowly cooled sintered alloys with high C contents are due to microstructural heterogeneity and the presence of large proeutectoid carbide particles. The proeutectoid carbide particle is long and large at high carbon content which affects the strength properties decrease, but the hardness properties increase. Especially, carbon contents higher than 0.75 wt.%, pro-eutectoid carbide is clearly visible (Figure 3 (e), (f) (g) and Figure 4 (c)).

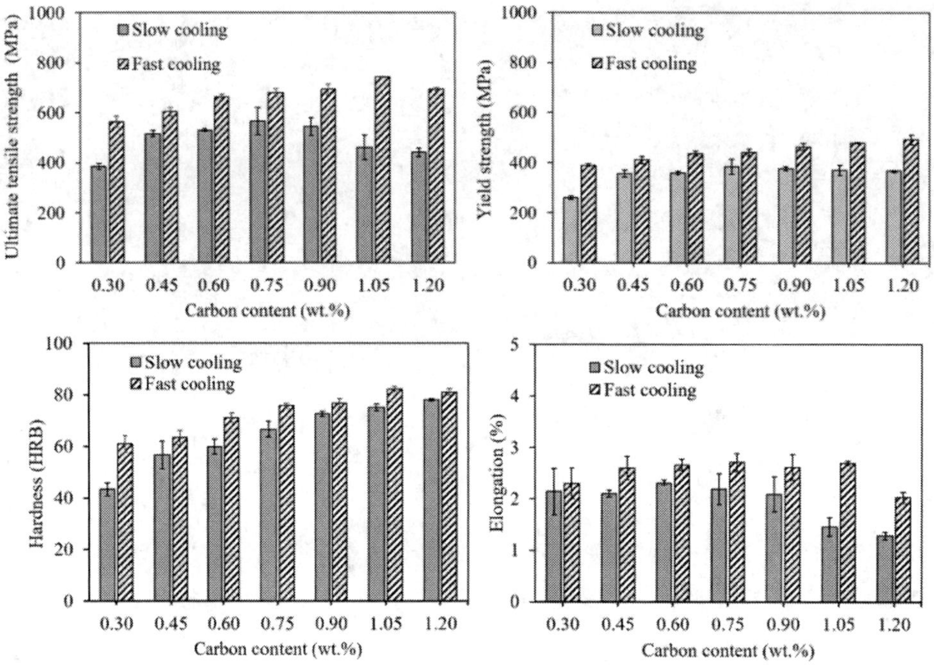

Figure 5. Effects of carbon content and cooling rate on mechanical properties of sintered alloys.

4. Conclusions

The effects of C content and cooling rate on mechanical properties were studied. Under slow cooling, microstructures of sintered Fe-1.50Mo-xC alloys become finer due to morphological changes of ferrite grains and F/C aggregates. When C contents were $\geq$ 0.90 wt.%, large proeutectoid carbide particles formed in addition to F/C aggregates. Large carbide particles deteriorated mechanical properties. Under fast cooling, only F/C aggregates (upper bainite and inverse bainite) dominated microstructures. When C contents were $\geq$ 0.90 wt.%, inverse bainite formed in addition to upper bainite. Mechanical properties of the fast-cooled sintered alloys were superior to those of the slowly-cooled sintered alloys due to microstructural refinement and homogeneity of the former alloys.

Acknowledgments

The authors are grateful for financial support from National Science and Technology Development Agency (NSTDA) under the project P1951261. The authors are also grateful for support from National Metal and Materials Technology Center.

References

[1] Ramakrishnan P 2013 *Advances in Powder Metallurgy,* ed I Chang and Y Zhao (USA: Woodhead Publishing) pp 493-519

[2] Höganäs A B 2017 *Iron and steel powders for sintered components* (Sweden)

[3] Thakur S N Newkirk J Fillari G B Murphy T and Narasimhan K 2004 *Int J Powder Metall* **40**(3) 45-54
[4] Moghaddam K S Ghambari M Farhangi H and Solimanjad N 2012 *J. Iron Steel Res. Int.* **19**(10) 43-46
[5] Rutz H Graham A H and Davala A 1997 *International Conference on Powder Metallurgy and Particulate Materials,* (Chicago, Illinois)
[6] Ackermann M Resiak B Buessler P Michaut B and Bleck W 2020 *Steel Res. Int.* **91** 1-9
[7] Capdevila C Ferrer J P Garcica-Mateo C Caballero F G Victor L and Andres C G 2006 *ISIJ International* **46**(7) 1093-1100
[8] Hannula J Porter D Kaijalainen A Somani M and Kömi J 2019 *Metals* **9**(3) 350
[9] Koetniyom W Chantawet P Tosangthum N Morakotjinda M Thanyaporn Y Wila P and Tongsri R 2019 *J. Met., Mater. Miner.* **29**(1) 22-30
[10] Srijampan W Morakotjinda M Krataitong R Thanyaporn Y Tosangthum N Wiengmoon A and Tongsri R 2016 *Chiang Mai J. Sci.* **43**(2) 358-364
[11] Srijampan W Wiengmoon A Morakotjinda M Krataitong R Yotkaew T Tosangthum N and Tongsri R 2015 *Mater. Des.* **88** 693-701
[12] ASTM B783-19 2019 *ASTM International,* (West Conshohocken, Pennsylvania)
[13] Choi S 2003 *Mater. Sci. Eng. A* **363**(1) 72-80
[14] Borgenstam A Hedström P Hillert M Kolmskog P Stormvinter A and Ågren J 2011 *Metall. Mater. Trans. A* **42**(6) 1558-1574

Sintered Fe-Mo-Mn-Si-C alloys

P. Nakornkaew[1], T. Patcharawit[1], P. Wila[2], M. Morakotjinda[2], T. Yotkaew[2, *], N. Tosangthum[2], and R. Tongsri[2]

[1]School of Metallurgical Engineering, Institute of Engineering, Suranaree University of Technology 111 University Avenue, Suranaree Sub-District, Muang Nakhon Ratchasima District, Nakhon Ratchasima 30000 Thailand
[2]Particulate Materials Processing Technology)PMPT(, National Metal and Materials Technology Center, 114 Paholyothin Road, Khlong Nueng, Khlong Luang, Pathum Thani 12120 Thailand

* Corresponding Author: thanyy@mtec.or.th

Abstract
Sintered Fe-Mo-Mn-Si-C alloys were produced by alloying silicon and carbon elements to pre-alloyed Fe-0.5Mo-0.15Mn powder via sintering of mixtures of pre-alloyed powder and varied silicon carbide amounts (0.5-4.0 wt.% with 0.5 increment). The sintered alloys were cooled in a sintering furnace. Microstructures and mechanical properties of the sintered alloys were altered depending on added silicon carbide content. With up to 2.0 wt.% silicon carbide additions, the sintered alloys exhibited polygonal ferrite plus eutectoid decomposition products, showing non-cooperative ferrite and carbide growth mode. With silicon carbide contents of $\geq$ 2.5 wt.%, the sintered alloys exhibited microstructural feature similar to that of a ductile iron, i.e., a common feature consisted of a black particle enveloped with ferrite and pearlite. Tensile strengths of the sintered alloys increased but elongation values decreased with increasing added silicon carbide content. The sintered alloys produced by 3.5 and 4.0 wt.% silicon carbide showed high strengths and considerably good elongation values.

Keywords: Sintering, Silicon carbide, Eutectoid decomposition, Pearlite, Mechanical property.

1. Introduction

The continuing and growing demands for high-performance and cost-effective parts in automotive industry are the motivation driving the innovation in sintered steel technology. Sintered steel parts with superior performance are produced from different iron (Fe)-based powder grades, such as pre-alloyed and diffusion-bonded powders [1, 2]. These powders contain low alloying element concentrations in the forms of solid solution in pre-alloyed powders or elemental particles bonded on pure-Fe or pre-alloyed powders. Important alloying elements include chromium (Cr), molybdenum (Mo), Nickel (Ni), and manganese (Mn). They act as substitutional alloying elements, which are able to influence phase transformations resulting in transformation product morphologies and types desirable for better properties. Both pre-alloyed and diffusion-alloyed powder grades are designed for flexible carbon (C) additions in order that varieties of mechanical properties can be manipulated by sintered steel parts manufacturers. Different added C contents dictate microstructural feature formation in sintered steels

IOP Conf. Series: Materials Science and Engineering 1137 (2021) 012034 doi:10.1088/1757-899X/1137/1/012034

and hence affect mechanical properties [3, 4]. In addition to C content, post-sintering cooling rate is also the important parameter controlling microstructure and mechanical property of sintered steels [5].

Recently, it has been discovered that not only graphite but non-oxide silicon carbide (SiC) ceramic can be used as a carbon source for C alloying in sintered steel fabrication [6, 7]. Lately, high content (4.0 wt.%) SiC was admixed to pre-alloyed Fe-Mo and Fe-Cr-Mo powders to produce sintered composites having microstructures similar to those of cast ductile irons [8-10]. The experimental results in the literatures given above suggest that several phenomena, including SiC decomposition, dissolution of Si and C atoms, and phase transformations of Si and C-containing solid solutions, resulting from the interactions between Fe-based and SiC powders during sintering cycle, are complicated but important for sintered steel technology development. They also indicate that the amount of added SiC powder controls microstructural development. In this work, alloying behaviour of different Si and C contents from different SiC amounts in sintered Fe-based powder with lean alloying element contents (pre-alloyed Fe-0.50Mo-0.15Mn powder) has been explored. This work was focused on phase transformations during post-sintering cooling with a slow rate.

2. Experimental procedure

2.1 Sample preparation

The metal powder used in this study was pre-alloyed Fe-0.50Mo-0.15Mn powder, courtesy from Rio Tinto Metal Powders of Canada (via P.S. Steel Co., Ltd., of Thailand and Rio Tinto Iron & Titanium (Suzhou) Co., Ltd., of China). The powder mixtures were made by blending of the pre-alloyed powder with varied SiC amounts of 0.5-4.0 wt.% (with increment of 0.5wt.%) show in Table 1. The powder mixtures were compacted into tensile test bars according to MPIF standard 10 with green density 6.50 ± 0,05 g/cm^3. The green specimens were sintered at 1280°C for 45 minutes in a vacuum furnace at pressure around 1.0 x 10^{-5} mbar. After sintering, specimens were continuously cooled in a vacuum furnace with cooling rate of about 0.1°C/s. The densities of sintered steels were determined using the Archimedes method according to MPIF Standard 42.

Table 1. The composition of the powder mixtures.

Material	Composition (wt.%)								
SiC	0	0.5	1.0	1.5	2.0	2.5	3.0	3.5	4.0
Pre-alloyed Fe-0.5Mo-0.15Mn	Bal.	Bal.	Bal.	Bal.	Bal.	Bal.	Bal.	Bal.	Bal.

2.2 Microstructural investigation

Microstructure of sintered steel specimens were observed by using optical microscopy (OM) and scanning electron microscopy (SEM). Specimens were prepared according to standard metallography practice and etched with 2% Nital. Phase identification was conducted by using X-ray diffraction (XRD) technique using Co K_{α} radiation (wavelength = 1.7902 Å). The diffracting angle (2θ) was scanned from 30 to 110, with scan speed of 0.04° /s and step size of 0.02° in PANalytical model X'Pert PRO.

2.3 Hardness and tensile tested Microhardness measurement was conducted by using Rockwell hardness tester (Wilson, Rockwell 574) in F scale using 60 kgf load for 6 indentations per a sample. Tensile test was carried out by using Instron Universal Instrument. The samples were tested according to the ASTM E8M standard.

3. Results and discussion

3.1 Microstructure and phase analysis

Microstructures of sintered Fe-Mo-Mn-Si-C alloys varied with added SiC contents as shown in Figure 1. Without SiC addition, the sintered alloy had only polygonal ferrite (PF) grains (Figure a1, b1 and c1). With 0.5 wt.% (Figure 1 (a2, b2 and c2)) and 1.0 wt.% (Figure 1 (a3, b3 and c3)) SiC additions, the sintered alloys had PF grains and eutectoid transformation zones (ETZs). The ETZ showed Widmanstätten ferrite (WF) plates and ferrite/discrete carbide mixture, termed as bainite (B). With 1.5 wt. % (Figure 1 (a4, b4 and c4)) and 2.0 wt.% (Figure 1 (a5, b5 and c5)) SiC additions, the fraction of PF grains was reduced whereas that of ETZs was increased. The carbide particles in some ferrite/carbide mixtures changed from discrete to lamellar shape. The ferrite/carbide lamellae dominated and PF grains returned to the microstructure of the sintered Fe-Mo-Mn-Si-C alloy with 2.5 wt.% SiC addition (Figure 1 (a6, b6 and c6)).

The microstructural feature consisting of residuals of SiC decomposition, appeared as black particles, surrounded with PF grain halo and ferrite/carbide lamellae were observed in sintered alloyed added with 3.0 wt.% (Figure 1 (a7, b7 and c7)), 3.5 wt.% (Figure 1 (a8, b8 and c8)) and 4.0 wt. % (Figure 1 (a9, b9 and c9)).

XRD patterns of some sintered alloys are shown in Figure 2. Strong peaks were corresponding to body-centered cubic (bcc) crystal structure of α-ferrite. Weak peaks were corresponding to face-centered cubic (fcc) crystal structure of $M_{23}C_6$ carbide in sintered alloys with low SiC additions. In the sintered alloy with 4 wt.% SiC addition, XRD pattern showed weak peaks corresponding to M_3C carbide.

The microstructures given in Figure 1 suggest that added SiC powder particles undergo decomposition, which results in alloying of Fe-based matrix with Si and C atoms. Similar alloying phenomena were previously reported in [6-10]. With low SiC additions, no SiC decomposition residuals were remained in microstructures [6, 7]. In contrast with 4.0 wt.% [8-10] and 5 wt.% or more [11, 12] SiC additions, SiC decomposition residuals were reported to remain as black particles.

The Si and C atoms, alloyed in Fe-Mo-Mn matrices, influence phase transformations during post-sintering cooling. In low Si and C matrices, phase transformations start with ferrite transformation and followed by eutectoid decomposition. In the latter, the decomposition proceeds by WF plate formation followed by decomposition of austenite strips between WF plates. The decomposition of austenite strips depends strongly on the added SiC content, indicating Si and C content dependence. The decomposition resulting in ferrite/discrete carbide mixture occurs in sintered alloys with low added SiC content whereas that resulting in ferrite/carbide lamellae occurs in sintered alloys with low added SiC content. The austenite decomposition into WF plates and ferrite/carbide mixtures in strips between WF plates was previously investigated and the mechanisms involved were explained clearly [13].

Precipitation of carbides in sintered Si-containing alloy matrices is an interesting subject as it is generally accepted that cementite is inhibited in steels containing judicious Si contents [14-17]. The reason is due to the fact that the dissolution of Si in cementite is thermodynamically unflavored condition [18]. In our works, cementite precipitation is truly inhibited but is replaced by $M_{23}C_6$ carbide precipitation in low Si-containing matrices. Moreover, in high Si-containing matrices M_3C carbide precipitation is not prevented.

IOP Conf. Series: Materials Science and Engineering 1137 (2021) 012034 doi:10.1088/1757-899X/1137/1/012034

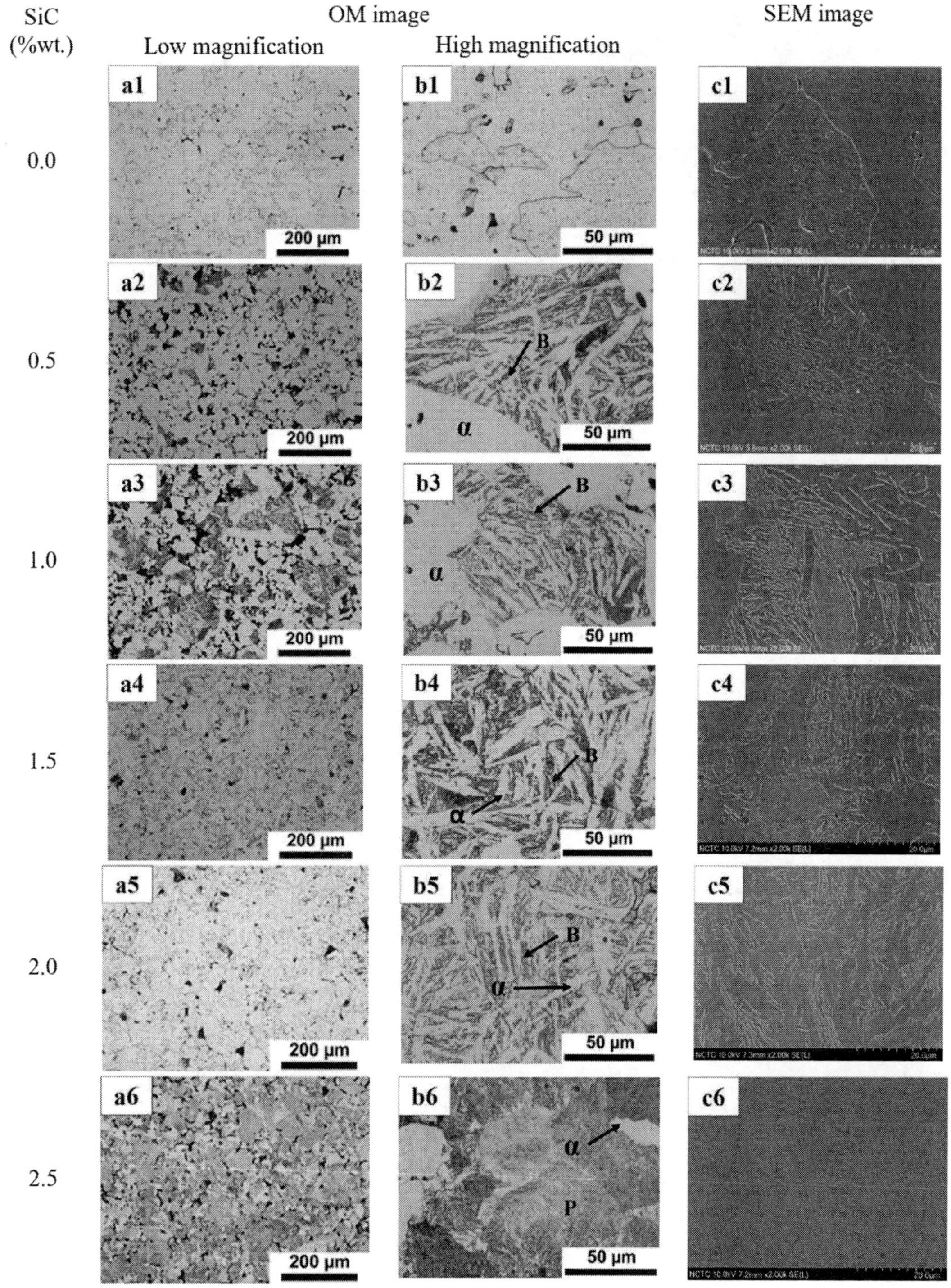

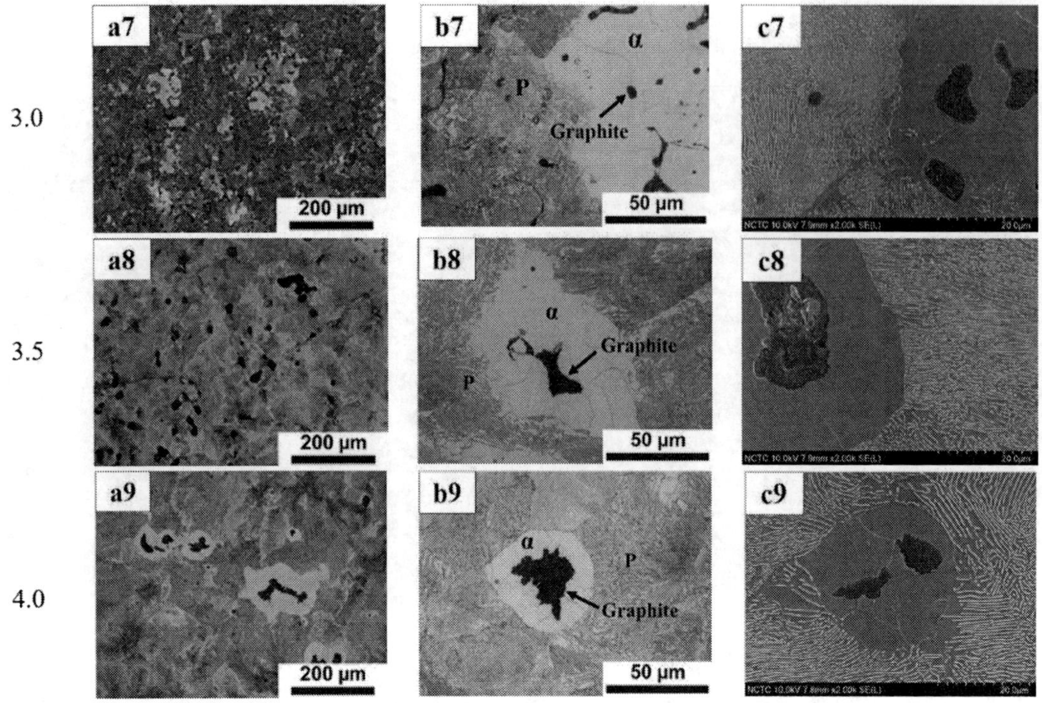

Figure 1. OM and SEM images of experimental sintered alloys.

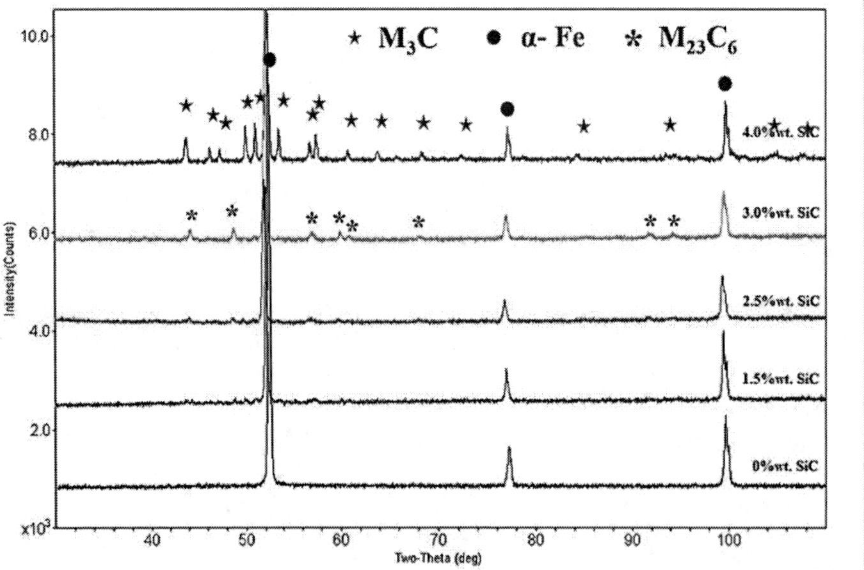

Figure 2. XRD patterns of experimental sintered alloys.

3.2 Mechanical property

Tensile strength and hardness increased whereas elongation decreased with added SiC contents (Figure 3). The mechanical properties of sintered alloys are influenced by microstructures. The increase of strength and hardness is sharp when microstructures contain both PF grains and ETZs. The ETZs with ferrite/discrete carbide mixtures contribute strongly on strength and hardness. When the microstructures have only ETZs but with ferrite/ carbide lamellae nature, their contribution on strength and hardness is observed to be low. The contribution of ferrite/ carbide lamellae may be compensated by another unknown factors. This issue will be further studied.

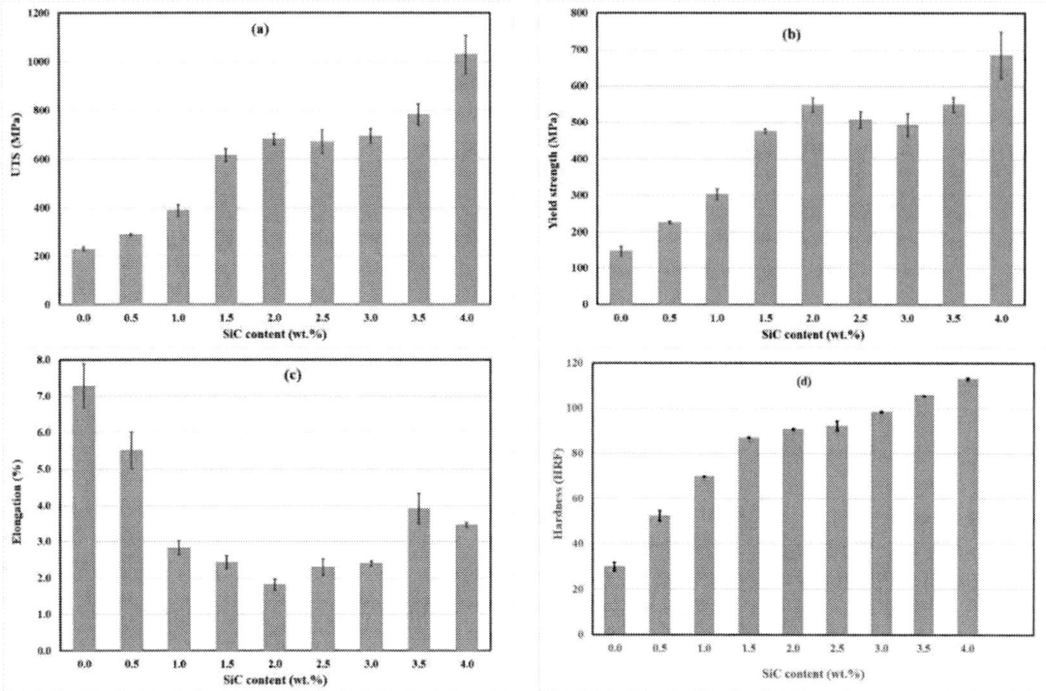

Figure 3. Mechanical properties of experimental sintered alloys.

4. Conclusions

Experimental sintered alloys, produced from pre-alloyed Fe-0.50Mo-0.15Mn powder mixed with varied SiC contents, were studied .It was found that with up to 2.0 wt %.SiC additions, the sintered alloys exhibited PF grains and ETZs, showing ferrite/discrete carbide mixtures .With SiC contents in the range 2.0 to 2.5 wt%., the fraction of PF grains decreased and that of ETZs increased .Some ETZs showed ferrite/carbide lamellae .With SiC contents ≤2.5 wt%., the sintered alloys exhibited microstructural feature similar to that of a ductile iron, i.e., a common feature consisted of a black particle enveloped with ferrite and pearlite .Tensile strengths of the sintered alloys increased but elongation values decreased with increasing added silicon carbide content .The sintered alloys produced by 3.5 and 4.0 wt %.silicon carbide showed high strengths and considerably good elongation values.

Acknowledgements

This work is funded by the project P1951261 supported by National Science and Technology Development Agency)NSTDA .(The authors are grateful for supports from Thailand National Metal and Materials Technology Center)MTEC .(The authors are also grateful for the material for this research)ATOMET4001(, courtesy from Rio Tinto Metal Powders of Canada)via P.S .Steel Co., Ltd., of Thailand and Rio Tinto Iron & Titanium)Suzhou (Co., Ltd., of China(.

References

[1] Brian J W 1991 *Metal Powder Report* **46** 26-32
[2] Dizdar S 2012 *Gear Technology* (A critical review) 60-65

[3] Srijampan W Morakotjinda M Krataitong R Yotkaew T Tosangthum N Wiengmoon A and Tongsri R 2016 *Chiang Mai J Sci* **43** 358-64
[4] Koetniyom W Chantawet P Tosangthum N Morakotjinda M Yotkaew T Wila P and Tongsri R 2019
JMMM **29** 22-30
[5] Srijampan W Wiengmoon A Morakotjinda M Krataitong R Yotkaew T Tosangthum N and Tongsri R 2015 *Materials and* **88** 693-701
[6] Kiatdherarat W Mungsantisuk P Tongsri R Mahathanabodee S Sirivedin K Krataitong R Morakotjinda M Yotkaew T and Tosangthum N 2015 *Proceedings of the Burapha University International* Conference 454-462
[7] Hebda M Dębecka H and Dilatometric K J 2016 *J Therm Anal Calorim* **125** 1319-1326
[8] Ruangchai K Wiengmoon A Morakotjinda Krataitong R Tanprayoon D Yotkaew T Tosangthum N Patakham U and Tongsri R 2017 *Key Engineering Materials* **751** 47-52
[9] Ruangchai K Wiengmoon A Morakotjinda M Tosangthum N and Tongsri R 2018 *Journal of Physics :Conf.Series* **1144** 12099
[10] Ruangchai K Wiengmoon A Krataitong R Yotkaew T Tosangthum N and Tongsri R 2018 *Journal of Physics :Conference Series* **1144** 012147
[11] Yodkaew T Morakotjinda M Tosangthum N Coovattanachai O Krataithong R Siriphol P Vetayanugul B Chakthin S Poolthong N and Tongsri R 2008 *JMMM* **18** 57-61
[12] Chakthin S Poolthong N and Tongsri R 2008 *Advanced Materials Research* **55-57** 357-360
[13] Yin J Hillert M, Borgenstam A 2017 *Metallurgical and Materials Transactions* **A 48** 1444-1458
[14] Bhadeshia H K D H and Edmonds D V 1979 *Metallurgical Transactions* **A 10** 895-907
[15] Jacques P Girault E Catlin T Geerlofs N Kop T Zwaag van der S and Delannay F 1999 *Materials Science and* **A 273-275** 475-79
[16] Kozeschnik E and Bhadeshia H K D H 2008 *Materials Science and Technology* **24** 343-47
[17] Suzuki T Ono Y Miyamoto G and Furuhara T 2010 *ISIJ International* **50** 1476-1482
[18] Jang J H Kim I G and Bhadeshia H K D H 2009 *Computational Materials* **44** 1319-1326

Sintered Fe-Ni-Si-C alloys

Tiwat Kaewkam[1], Panya Kansuwan[1], Naoto Ohtake[2], Pongsak Wila[3], Rungtip Krataithong[3], Nattaya Tosangthum[3], Thanyaporn Yotkaew[3,*], and Ruangdaj Tongsri[3]

[1]Department of Mechanical Engineering, Faculty of Engineering, King Mongkut's Institute of Technology Ladkrabang (KMITL), Bangkok Thailand 10520
[2]Department of Mechanical Engineering, Major in Engineering Sciences and Design, Tokyo Institute of Technology, Tokyo Japan 152-8550
[3]Particulate Materials Processing Technology (PMPT), Metal and Manufacturing Precess Research Group, Natianal Metal and Materials Technilogy Center, 114 Paholyothin Road, Khlong Nueng, Klong Luang, Pathum Thani Thailand 12120

* Corresponding Author: thanyy@mtec.or.th

Abstract
The aim of this research is to explore the effect of nickel content on microstructural development and mechanical properties of sintered Fe-Ni-Si-C alloys. The sintered alloys were prepared from powder mixtures of pre-alloyed Fe-Ni-based powders (nickel contents varied as 0.45, 0.90, 1.80, and 4.00 wt.%) and fixed 4 wt.% silicon carbide powder by using a powder metallurgical process. Sintering was performed in a vacuum furnace at 1250°C for 45 minutes and slow cooling in the furnace. Microstructures of sintered alloys varied with nickel contents. Nickel showed a strong influence on promoting bainitic ferrite/martensite-austenite (BF/M-A) structure formation. Retained austenite in BF/M-A structure was found to increase with increasing nickel content. Tensile properties (strength and elongation value) of sintered alloys increased with increasing nickel content.

Keywords: Sintering, Fe-Ni-Si-C alloys, Ferrite, Martensite-austenite, Mechanical property.

1. Introduction

Thailand has a geographic location suitable to become a primary regional economic hub. Not only in trade, investment, and tourism but also in communication and transportation networks, that connect to other regions around the world. The Thai government has, in recent years, made investment in domestic infrastructure, a top priority. Railway transportation is increasingly important as a mass transit, to link destinations with manageable travelling time. With the project of high-speed railway development, sintered friction materials (SFMs) are expected to be widely used. These materials are used in train brake system and have many advantages, such as high performance, wear resistance, improved fade resistance, enhanced wet-effectiveness, and extended life SFMs all important components of rail brakes to ensure safety and comfort expectation [1]. SFMs are composites commonly consisting of graphite particles embedded in metallic matrix [2], [3]. The graphite particles act as solid lubricant for anti-wear purpose. According to the recent study [4], the SFM matrix has important role on braking performance such as stable coefficient of friction (COF) and wear rate. This means the composite matrix integrity is important for high-speed train SFM brake pads. Previous works have shown that sintered composites having graphite particles embedded in iron-base matrices could be produced by using the traditional 'press and sinter process' [5–8]. These sintered composites have different matrices depending on the compositions of iron-base powders employed. The matrix microstructure has prime contribution to mechanical properties of a sintered composites.

Research works have shown that nickel addition to steels is able to modify microstructure hence improve mechanical property. The increase in Ni contents (0-6%) causes microstructural change from dominant granular bainite to lath bainite and/or lath martensite [9]. The roles of nickel addition include (i) lowering the pearlite start (Ps) and the bainite start (Bs) temperatures and shifting both pearlites, (ii) and bainite transformation curves to the right-hand side of continuous cooling transformation (CCT) diagram and (iii) retardation of isothermal bainite transformation kinetics by shift the bainite transformation curve to the right-hand side of time-temperature-transformation (TTT) diagram [10]. The decrease of Bs and martensite start (Ms) temperature with nickel addition [11] or increasing nickel content [12] were previously reported.

The shifting of the pearlite transformation curve to the right-hand side of CCT diagram opens the opportunity for a steel produced under a considerably slow cooling rate to avoid pearlite transformation at a high temperature thus undergoes other lower temperature phase transformations. The additions of 1-3 wt.% nickel powder to sintered Fe-Cr-Mo base steels, processed under a slow or furnace cooling, seem to prove that concept [13]. The retardation of isothermal bainite transformation kinetics due to nickel addition is reported in [10, 14].

Due to the important roles of nickel addition as given above, it was hypothesized that by using pre-alloyed nickel-containing powders, the matrices of sintered SFMs would be developed under the influence of nickel content. Due to the delay of pearlite transformation, other transformations occurring below the pearlite field are expected. One of low temperature transformation products of ductile cast ions known as ausferrite consisting of ferrite and carbon-rich austenite is expected to form by the nickel influence in sintered nickel-containing SFMs. The ausferrite is one type of microstructural features important for mechanical property improvement in cast ductile irons [15–19]. The effects of microstructural components, varied with nickel content, on mechanical properties of sintered nickel-containing SFMs had been investigated in this work.

2. Materials and Methods

2.1 Materials preparation

The sintered alloys were prepared from powder mixtures of pre-alloyed Fe-Ni based powders (nickel contents varied as 0.45, 0.90, 1.80, and 4.00 wt.%) and fixed 4 wt.% silicon carbide powder (shown in Table 1) by using a powder metallurgical process. The powder mixtures were compacted into tensile test bars with a green density of 6.50 ± 0.05 g/cm3. For each steel alloys, sintering of green compacts were performed in a vacuum furnace at 1250°C for 45 minutes and slowly cooled in the furnace. The density of green and sintered specimens was determined in accordance with MPIF Standard 10, ASTM B783).

Table 1. Nominal compositions of experimental Fe-Ni-Si-C alloys.

Alloy	Pre-alloyed powder	SiC (wt.%)	Nominal composition (wt. %)						
			C	Si	Cr	Ni	Mo	Mn	Fe
0.45Ni	Fe-0.45Ni-0.60Mo-0.15Mn	4.0	1.2	2.8	0.0	0.45	0.85	0.15	Bal.
0.90Ni	Fe- 0.90Ni- 1.00Mo- 0.45Cr-0.45Mn	4.0	1.2	2.8	0.45	0.90	1.00	0.45	Bal.
1.80Ni	Fe-1.80Ni-0.55Mo-0.20Mn	4.0	1.2	2.8	0.0	1.80	0.55	0.20	Bal.
4.00Ni	Fe-4.00Ni-0.50Mo-0.20Mn	4.0	1.2	2.8	0.0	4.00	0.50	0.20	Bal.

2.2 Microstructure investigation

Specimens for metallographic examination were ground on silicon carbide papers from 80 to 4000 grits and then progressively polished with 6, 3, and 1μm diamond pastes, respectively. The polished samples were etched with 2% Nital. Microstructures of the sintered specimens were observed by using optical microscopy (OM) and scanning electron microscopy (SEM) equipped with an EDS.

2.3 Hardness and tensile test

Rockwell scale B hardness test (HRB) was carried out on un-etched specimens using 100 kg-f loads with 15s indentation time (Standard ISO 6508-1). The mean values were based on ten different areas. The tensile property of the sintered materials was carried out at room temperature using Instron Universal Instrument with speed 1 mm/min. Three samples were tested for each sintered steel and the average value was reported.

3. Results and discussion

3.1 Microstructure and phase characterization

Sintered Fe-Ni-Si-C alloys showed microstructures resembling those of ductile cast irons (Figure 1). The sintered 0.45Ni (Figure 1a) and 1.80Ni (Figure 1c) alloys showed common microstructural feature consisting of black particle surrounded with ferrite and striated structure, whereas sintered 0.90Ni (Figure 1b) and 4.00Ni (Figure 1d) had black particle surrounded with only striated structure. Microstructural observation using SEM revealed that the striated structure in the sintered 0.45Ni alloy (Figure 2a) consisted of pearlite and bainitic ferrite/martensite-austenite (BF/M-A) with higher proportion of pearlite. The striated structure in the sintered 0.90Ni alloy (Figure. 2b) consisted of BF/M-A structure. The striated structure in the sintered 1.80Ni alloy (Figure 2c) consisted of BF/M-A and pearlite with lower proportion of pearlite. The matrix of the sintered 4.00Ni alloy (Fig. 2d) alloy consisted of BF/M-A and martensite with lower proportion of martensite. The XRD patterns of the sintered Fe-Ni-Si-C alloys as given in Figure 3 show peaks corresponding to two different crystal structures, i.e., body-centered cubic (bcc) crystal structure belonging to ferrite (α) phase and face-centered cubic (fcc) crystal structure belonging to austenite (γ) phase.

The formation of microstructure consisting of black particle surrounded by ferrite and striated structure in sintered Fe-Ni-Si-C alloys is already explained and given in [6]. However, the striated patterns in the sintered nickel-containing alloys are quite different from those observed in sintered Fe-Ni-Si-C alloys. Thus, additional discussion in this work is focused on the formation of striated structure. When the striated structures in sintered 0.45Ni alloy (Figure 2a), 1.80Ni alloy (Figure 2c) and 4.00Ni alloy (Figure 2d) are compared, the effect of nickel content on phase transformations is remarkable. The proportion of pearlite decreases but that of BF/M-A structure increases with increasing nickel content. Taking the sintered Fe-Mo-Si-C alloys [6,7]as references, all sintered nickel-containing alloys (in this work) have higher proportions of BF/M-A structure. The reduction or disappearance of ferrite and pearlite in sintered nickel-containing alloys confirms the effect of nickel on shifting of high-temperature transformation curves to the right-hand side of CCT diagram [10]. Experimental results suggest that the higher nickel content the farther high-temperature transformation curves are shifted and the wider window is opened for the low-temperature transformation products (BF/M-A structure and martensite) to form.

The formation of low-temperature transformation products dictates the whole matrices of the sintered 4.00Ni alloy (Figure 2b) and 0.90Ni alloy (Figure 2d). In the former alloy, the low-temperature transformations are clearly influenced by nickel. In the latter one, nickel concentration alone cannot totally suppress high-temperature transformations. In this alloy, the synergistic effects on high-temperature transformation suppression are supposed to obtain from chromium (1.0 wt.%) and molybdenum (1.0 wt.%). The strong and medium effects of molybdenum and chromium, respectively, in suppressing pearlite formation were previously reported [20]. Taking the overall microstructures of these two alloys into consideration, their microstructures are similar to those of austempered ductile irons. The microstructure of sintered 0.90Ni and 4.00Ni alloys consist of black particle (graphite-containing particle according to [6–8]) surrounded with the BF/M-A structure matrix while that of austempered ductile irons consist of graphite particle surrounded with auferrite (ferrite and carbon-rich austenite) matrix [5–19].

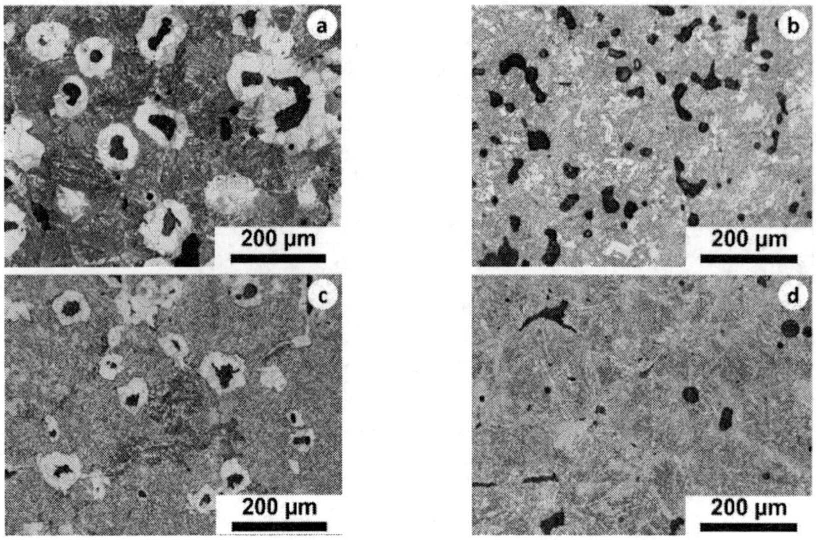

Figure 1. Optical microstructures of sintered Fe-Ni-Si-C alloys.

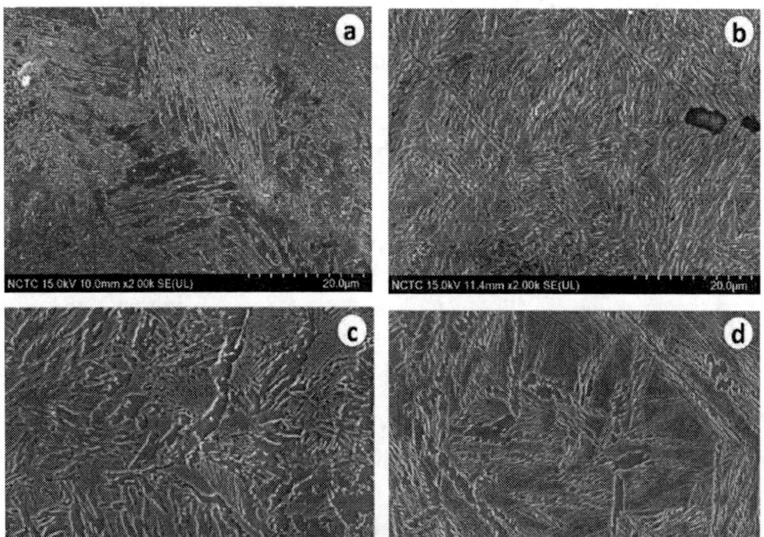

Figure 2. SEM microstructures of sintered Fe-Ni-Si-C alloys.

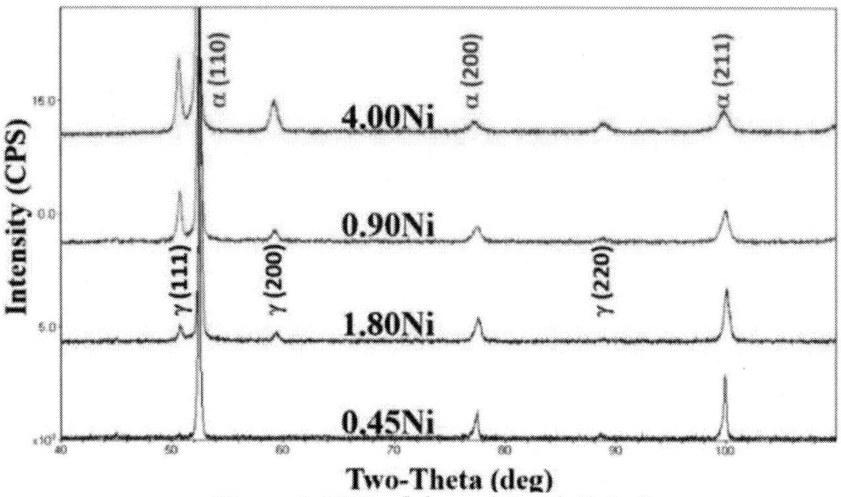

Figure 3. XRD of sintered Fe-Ni-Si-C alloys.

3.2 Mechanical property

The sintered Fe-Ni-Si-C alloys showed appreciably high tensile strengths (ultimate tensile and yield strengths) as given in Figure 4. Ductility values were quite low, no higher than 3.0%. Both strength and elongation values increased with increasing nickel content. Rockwell scale B hardness value showed relationship with nickel content for most sintered alloys, except the value drop in the sintered 1.80Ni alloy. Note that the prominent strengths and hardness of the sintered 0.90Ni alloy is also attributed to high Cr and Mo contents.

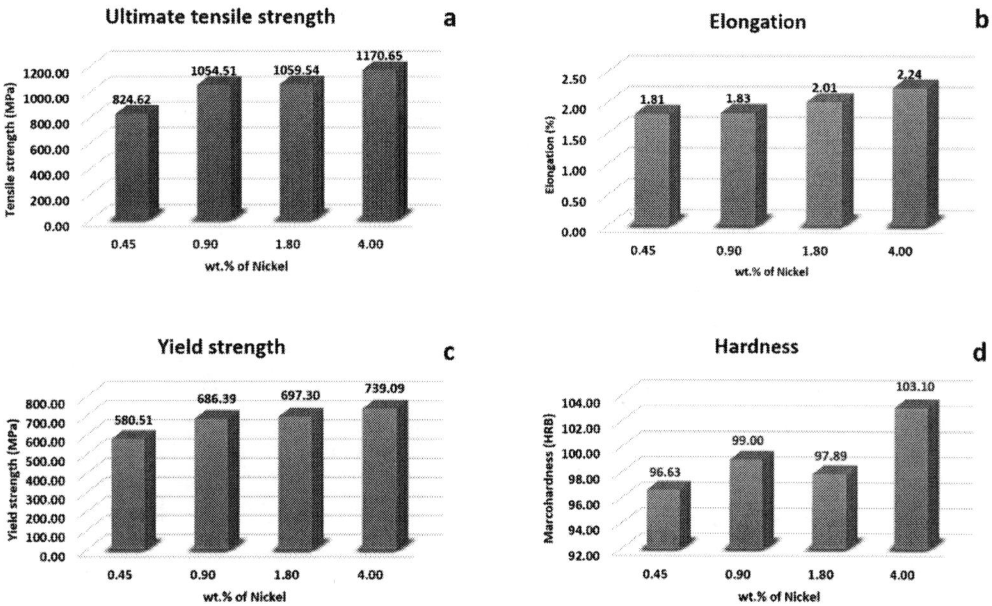

Figure 4. Mechanical properties of sintered Fe-Ni-Si-C alloys, according to nickel content.

The increase of tensile strength is attributed to microstructural components. Experimental results (Figure 2) provide the evidences of low-temperature transformation products, such as BF/M-A and martensite, as the root causes of alloy strengthening. Particularly, the sintered 4.00Ni alloy having tensile strength over 1 GPa has matrix consisting mostly of BF/M-A and martensite. The elongation value increases with increasing nickel content (Figure 4c) but is not directly related to retained austenite volume indicated by the γ (111) peak intensities shown by XRD patterns in Figure 3. The ductility of the sintered alloys is partly influenced by retained austenite. The hardness (HRB) value has no direct relationship with nickel content (Figure 4d). The hardness value drops in the sintered 1.80Ni alloy is attributed to pearlite structures distributed in BF/M-A matrix.

3.3 Fracture surface

Fracture surface of the sintered 0.45Ni alloy with the lowest elongation value showed cleavage with river pattern (Figure 5a) whereas that of the sintered 4.00Ni alloy with higher elongation value showed mixture of cleavage (on the right) and quasi-cleavage (on the left) (Figure 5b). Cleavage, as a typical sign of brittle fracture, found in pearlite-containing matrix of sintered 0.45Ni alloy, may not be uncommon as it was found on a fracture surface of a fully pearlitic eutectoid steel [21,22]. Quasi-cleavage, as one form of cleavage, exhibits both cleavage and plastic deformation [23]. This fracture mode found in the sintered 4.00Ni alloy reflects small plastic deformation as this alloy has high strength but small elongation value.

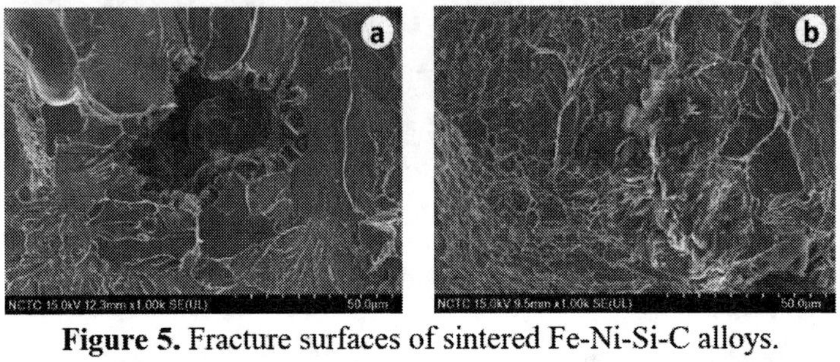

Figure 5. Fracture surfaces of sintered Fe-Ni-Si-C alloys.

4. Conclusion

Sintered Fe-Ni-Si-C alloys, designed and fabricated for application as SFMs, could be produced from mixtures of pre-alloyed Fe-Ni and fixed 4 wt.% silicon carbide powders by using traditional press and sinter process. Nickel content in pre-alloyed showed influence on microstructures and mechanical properties of sintered alloys. High nickel content could suppress ferrite and pearlite transformations and open a window for low-temperature transformations of BF/M-A and martensite. The low-temperature transformation products showed strong influence on mechanical properties of sintered alloys.

Acknowledgments

This work is funded by the project P1951261 under National Science and Technology Development Agency (NSTDA). The first author is also financially supported by TAIST-Tokyo Tech scholarship. The authors are grateful for supports from Thailand National Metal and Materials Technology Center (MTEC) and King Mongkut's Institute of Technology Ladkrabung (KMITL). The authors are also grateful for the material for this research, courtesy from Rio Tinto Metal Powders of Canada (via P.S. Steel Co., Ltd., of Thailand and Rio Tinto Iron & Titanium (Suzhou) Co., Ltd., of China).

References

[1] Günay M Korkmaz M E and Özmen R 2020 *Eng. Sci. Technol. an Int. J.* **23** 421-431
[2] Xiao J K Xiao S X Chen J and Zhang C 2020 *Tribol. Int.* **150** 106357
[3] Pérez B and Echeberria J 2019 Influence of abrasives and graphite on processing and properties of sintered metallic friction materials *Heliyon* **5**(8) E02311
[4] Peng T Yan Q and Zhang X 2018 *Tribol. Lett.* **66** 1-13
[5] Chakthin S Poolthong N and Tongsri R 2008 *Adv. Mater. Res.* **55-57** 357-360

[6] Ruangchai K Wiengmoon A Morakotjinda M Krataitong R Tanprayoon D Yotkaew T Tosangthum N Patakham U and Tongsri R 2017 *Key Eng. Mater.* **751** 47-52
[7] Ruangchai K Wiengmoon A Morakotjinda M Tosangthum N and Tongsri R 2018 *J. Phys . Conf. Ser.* **1144** 012099
[8] Ruangchai K Wiengmoon A, Krataitong R Yotkaew T Tosangthum N and Tongsri R 2018 *J. Phys. Conf. Ser.* **1144** 012147
[9] Mao G Cao R Yang J Jiang Y Wang S Guo X Yuan J Zhang X and Chen J 2017 *J. Mater . Eng. Perform.* **26** 2057-2071
[10] Tian J Xu G Jiang Z Hu H and Zhou M 2018 *Met. Mater. Int.* **24** 1202-1212
[11] Kong J and Xie C 2006 *Mater. Des.* **27** 1169-1173
[12] Yang H S and Bhadeshia H K D H 2008 *Mater. Sci. Technol.* **24** 335-342
[13] Tosangthum N Morakotjinda M Krataitong R Wila P Yodkaew T Vetayanugul B Boontetim W and Tongsri R 2018 *Mater. Today Proc.* **5** 9351-9358
[14] Yao Z Xu G Hu H Yuan Q Tian J and Zhou M 2019 *Trans. Indian Inst. Met.* **72** 1167-1174
[15] Zimba J Simbi D J and Navara E 2003 *Cem. Concr. Compos.* **25** 643-649
[16] Eric O Sidjanin L Miskovic Z Zec S and Jovanovic M T 2004 *Mater. Lett.* **58** 2707-2711
[17] Erić O Rajnović D Zec S Sidjanin L and Jovanović M T 2006 *Mater. Charact.* **57** 211-217
[18] Ovali I Kilicli V and Erdogan M 2013 *ISIJ Int.* **53** 375-381
[19] Kaczorowski M and Krzyńska A 2007 *Arch. Foundry Eng.* **7** 161-166
[20] Park Y J and Fletcher F B 1986 *J. Heat Treat.* **4** 247-252
[21] Bernstein I M and Alexander D J 1989 *Metall. Trans. a* **20** 2321-2335
[22] Lewandowski J J and Thompson A W 1986 *Metall. Mater. Trans. A* **17** 1769-1786
[23] Tartaglia J M Lazzari K A Hui G P and Hayrynen K L 2008 *Metall. Mater. Trans. A Phys. Metall. Mater. Sci.* **39 A** 559-576

Sintered 316L/Cu/h-BN composites

E. Chusong[1], P. Kansuwan[1], R. Khosakul[2], N. Ohtake[3], P. Wila[4], N. Tosangthum[4], B. Vetayanugul[4,*], and R. Tongsri[4]

[1]Department of Mechanical engineering, Faculty of Engineering, King Mongkut's Institute of Technology Ladkrabang (KMITL), Bangkok, 10520, Thailand
[2]Department of Materials Science and Technology, Faculty of Science, Prince of Songkla University, Hat Yai, Songkhla 90112 Thailand
[3]Department of Mechanical engineering, Major in Engineering Sciences and Design, Tokyo Institute of Technology, Tokyo, 152-8550, Japan
[4]Particulate Materials Processing Technology (PMPT), National Metal and Materials Technology Center, 114 Paholyothin Road, Khlong Nueng, Khlong Luang, Pathum Thani 12120 Thailand

* Corresponding Author: bhanuv@mtec.or.th

Abstract
Sintered 316L/h-BN composites show high potential as self-lubricating materials, which can be applied as moving parts where lube oils and greases are not applicable and inaccessible for maintenance. The past production of such composites by sintering, faced two problems. Firstly - interaction between 316L matrix and h-BN resulting in loss of h-BN content and its lubricity. And secondly - poor 316L matrix integrity. This work employed two approaches to solve such problems. The first approach was the use of nitrogen-containing atmosphere for sintering, to retard 316L matrix and h-BN interaction. And the second approach, was copper addition to promote sintering of 316L powder particles. Sintered self-lubricating 316L/Cu/h-BN composites were produced from mixtures of 3 different 316L/Cu matrices (made by additions of 2.0, 4.0 and 6.0 wt.% copper powder to 316L powder) and 15 vol.% of h-BN powder. Green compacts of powder mixtures were sintered under cracked ammonia (75 vol.% hydrogen + 25 vol. % nitrogen) at 1300°C for 60 minutes. The sintered composites were cooled in a Linn high temperature sintering furnace. It was found that copper additions led to the decrease of densities of sintered 316L/Cu/h-BN composites, compared to that of sintered 316L/h-BN composite, due to porosity left behind by copper powder melting and dissolution. Tensile strengths of sintered 316L/Cu/h-BN composites were improved when copper contents were 4 and 6 wt.%. There was no intergranular phase, the evidence of 316L matrix and h-BN interaction in all experimental sintered composites. Microstructural observation by scanning electron microscopy (SEM) also revealed that h-BN flakes still existed in pores of sintered composites. The unreacted h-BN flakes are expected to provide lubricity of sintered composites.

Keywords: Sintering, self-lubricating composites, 316L/Cu/h-BN, mechanical properties

1. Introduction

Self-lubricating components are widely used with many applications such as bearings, gears, seals and connectors [1]. These components are able to extend service life, and reduce the maintenance operation of moving components, by releasing and forming a lubricant film on a contact surface [2]. The lubricant film prevents direct contact between two rubbing surfaces, even under high loads, resulting in reducing wear and heat in relative motion.

Among various types of self-lubricating components, a self-lubricating composite is suitable for such an application. Sintering a mixture of metal powder and solid lubricant is a traditional method for processing a self-lubricating composite. Sintered materials are usually porosity, by nature. The pore formation is due to the presence of cusp-shape voids, generated at powder particle corners in a powder compact [3-6]. This porosity is considered as an ideal site for impregnating a solid lubricant, to form a self-lubricating composite.

Various types of solid lubricants including molybdenum disulfide, graphite, dichalcogenide and boron nitride [7]. Molybdenum disulfide [8] and graphite [9] are widely used in many self-lubricating composites, generally produced by sintering. However, these materials experience one major problem, which involves their decomposition under high sintering temperatures. Hexagonal boron nitride (h-BN), which has high thermal and chemical stability, is considered for this research because it has good lubricating properties due to its crystal structure similar to that of graphite. Due to the lubricating potential of h-BN, it was introduced to sintered stainless steel 316L matrix by 'the press and sinter' process [10].

A previous study of sintered 316L/h-BN composites revealed that there was an intergranular boride phase formed in the materials. The intergranular boride phase resulted from a reaction between boron from h-BN decomposition and iron-based powder particles. The interaction between 316L matrix and h-BN resulted in a loss of h-BN content and its lubricity.

However, it has been found that the intergranular boride phase, formed with very small amounts under a pure hydrogen atmosphere [11, 12]. With the presence of nitrogen in cracked ammonia (75 % hydrogen and 25 % nitrogen) atmosphere, it is expected that the interaction between h-BN and stainless steel 316L matrix to form borides, will be minimized.

A recent discovery found that the sintering atmosphere of cracked ammonia could retard h-BN decomposition and subsequent boride formation, but the sintered composite materials still had inferior mechanical properties due to poor 316L matrix integrity [13]. This work employed two approaches to solve such problems. The first approach was to use a cracked ammonia atmosphere for sintering, to retard 316L matrix and h-BN interaction. The second approach was to add copper, which was expected to promote the liquid phase sintering of 316L powder particles.

2. Experimental procedure

2.1 Materials preparation

The materials employed for this investigation included 316L stainless steel powder, solid lubricant powder (15 Vol.% h-BN) and copper powder (2, 4, 6 wt.%) were prepared by using a tumbling mixer with a speed of 40 rpm. Powder mixtures were then compacted into green parts with a green density of 6.5 g/cm^3. Green compacts were sintered under cracked ammonia at 1300°C for 60 minutes and cooled in a Linn high temperature sintering furnace with the rate of 0.1°C/s.

2.2 Materials characterization

Sintered density was determined by using the Archimedes method. Microstructural observation was performed by following a standard metallography method. The sintered specimens were cut and polished down to 1 μm and etched with 2% Nital. To observe the general microstructure, an optical microscopy (OM). Scanning electron microscopy (SEM) technique was performed on the specimens. For observing the decomposition of h-BN energy, a dispersive spectroscopy (EDS) mapping technique was used.

Mechanical properties of the sintered specimens were tested by using a universal testing machine and hardness tester

3. Results and Discussion

3.1 Microstructure

The experimental sintered Cu-bearing composites showed porosity higher than that of the sintered Cu-free 316L specimen (Figure 1). Microstructures of sintered 316L/Cu/h-BN composites with various Cu contents)0-6 wt(%. showed in Figure 1 .The experimental sintered Cu-bearing composites showed porosity higher than that of the sintered Cu-free 316L specimen .It was found that copper additions led to the decrease of densities of sintered 316L/Cu/h-BN composites) Figure 4(, due to porosity left behind by copper powder melting and dissolution. It was difficult to identify solid lubricant h-BN particles by observation under OM technique and SEM SE mode (in the top row of Figure 2). Differentiation of matrix, pore and h-BN was available by using SEM SE mode (in the bottom row of Figure 2). No free copper was detected in experimental sintered composites. EDS mapping analysis was performed on a sintered composite added with h-BN (Figure 3). The results clearly showed the coexistence of nitrogen (N) and boron (B) atoms.

High porosity in the experimental sintered composites (Figure 4) is due to copper (Cu) dissolution into 316L matrix. The pores left behind by Cu dissolution are not remedied or refilled. Cu is supposed to become liquid during heating and sintering. During cooling the liquid Cu is expected to solidify and fill empty pores in a sintered composite. However, no free copper is detected in any sintered composites. There are two possible ways related to undetectable free Cu. The first is that solubility of Cu in 316L matrix is higher than expectation. The second is that the dissolved Cu precipitate as nano Cu particles in 316L matrix. Since these nano Cu particles cannot be detected by the equipment used in this work and the formation of nano Cu particles in 316L matrix was already reported in some research works [14, 15], the nano Cu precipitation is highly possible for undetectable free Cu in the sintered composites.

The h-BN retention in the experimental sintered composites is strongly supported by the evidences given in the bottom row of Figure 2 and the EDS analysis result given in Figure 3. In previous works [10, 16, 17], the interaction between h-BN and 316L matrix results in formation of intergranular boride phase, which affects tribological property of sintered 316L/h-BN composites. The formation of intergranular boride phase retarded by employing cracked ammonia sintering atmosphere has been recently reported in [13]. Thus, the h-BN retention in the experimental sintered composites strongly confirms the hypothesis that cracked ammonia sintering atmosphere can prevent h-BN decomposition, even it is in contact with Fe-based materials at high temperatures. The mechanism responsible for h-BN decomposition is proposed here as the reversible reaction.

IOP Conf. Series: Materials Science and Engineering 1137 (2021) 012036 doi:10.1088/1757-899X/1137/1/012036

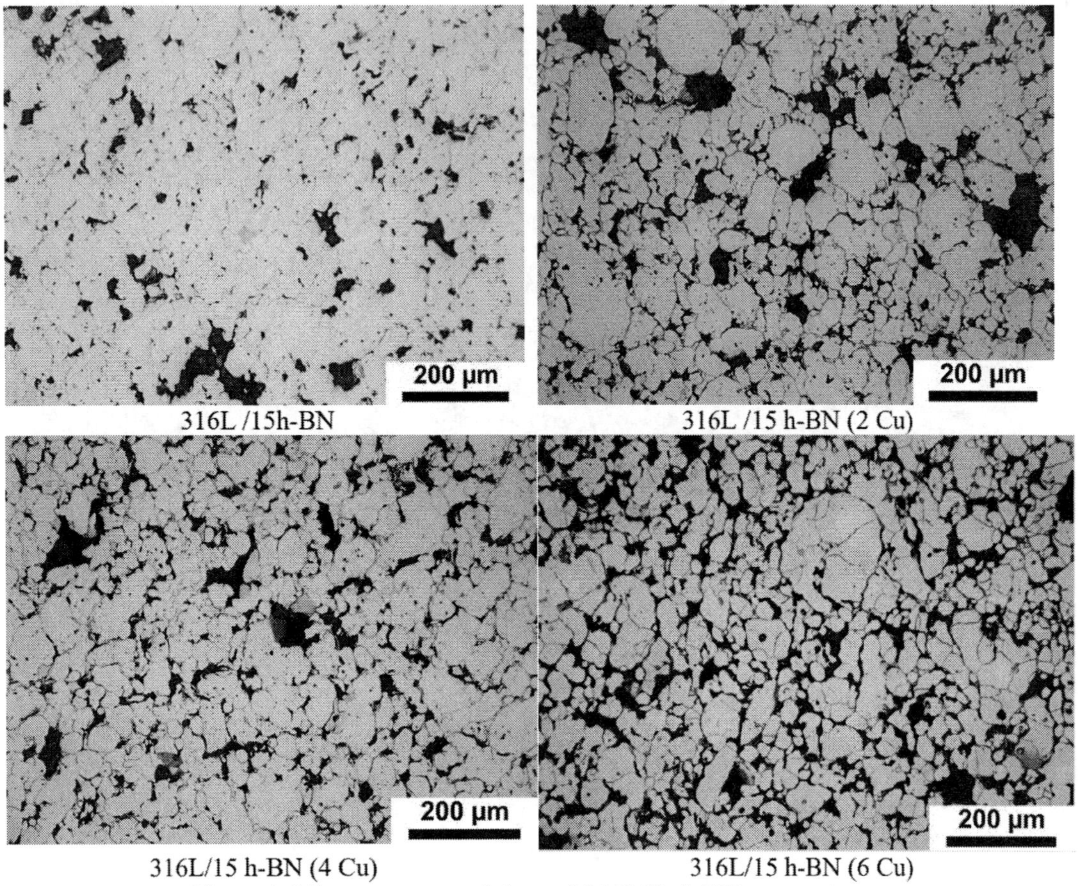

Figure 1. Microstructures of sintered 316L/Cu/h-BN composites.

$$2BN \leftrightarrow 2B + N_2 \quad (1)$$

During heating and sintering, if BN is allowed to decompose the reaction products are free B and nitrogen gas. The high concentration of nitrogen gas due to cracked ammonia atmosphere (75 % H_2 and 25 % N_2) will cause reversible reaction, which yields BN again. Due to such reversible reaction, the B amount dissolved in 316L is very small. Consequently, the eutectic melting according to Equation (2) hardly occurs.

$$Fe + Fe_2B \rightarrow Liquid \quad (2)$$

In case that very small amount or no liquid is formed, solidification to form the intergranular boride phase will be limited or not occur. Microstructures of the sintered composites hardly show the intergranular boride. Moreover, close observation by SEM also reveals h-BN particle flake shape (Figure 5), which confirms the h-BN retention in the experimental sintered composites.

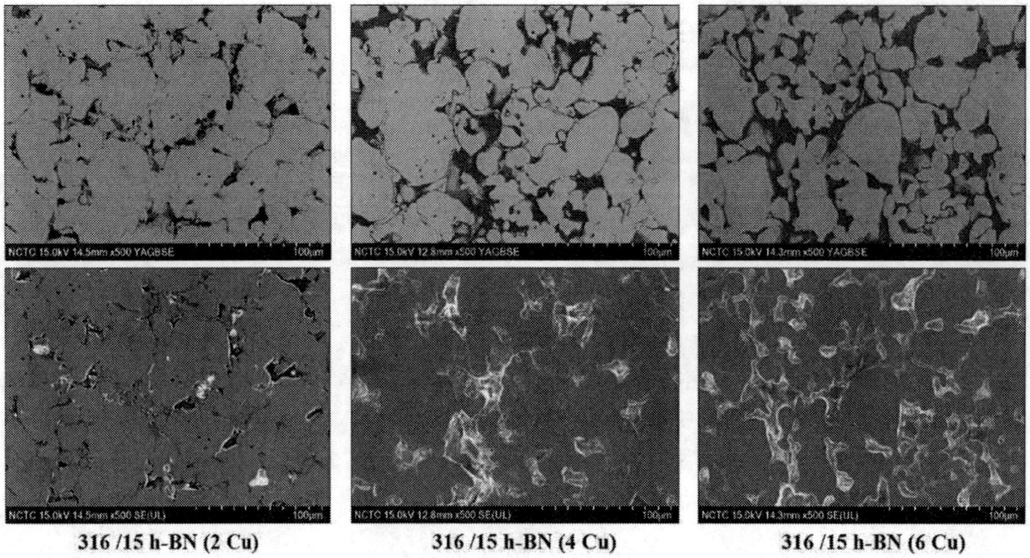

Figure 2.. SEM micrograph (SE in the top row and BSE in the bottom row) of sintered 316L/h-BN composites.

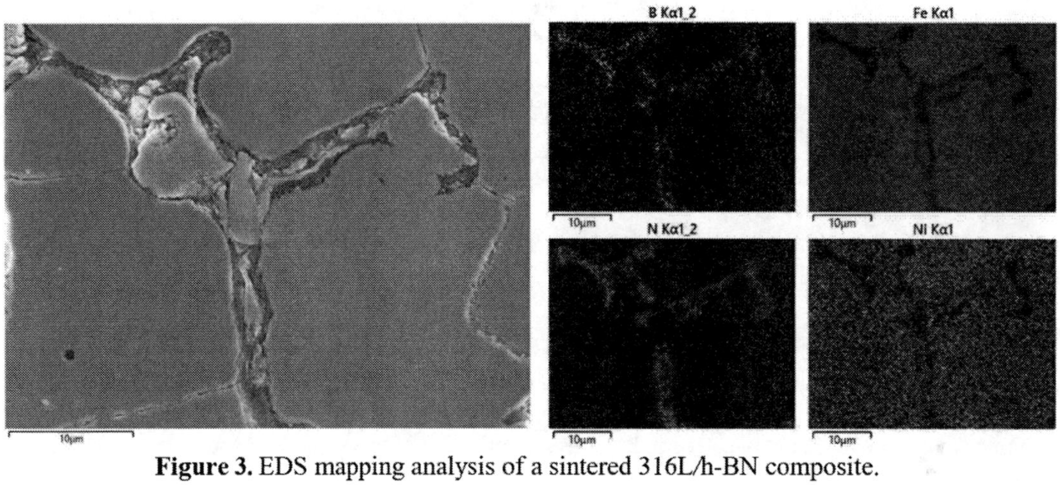

Figure 3. EDS mapping analysis of a sintered 316L/h-BN composite.

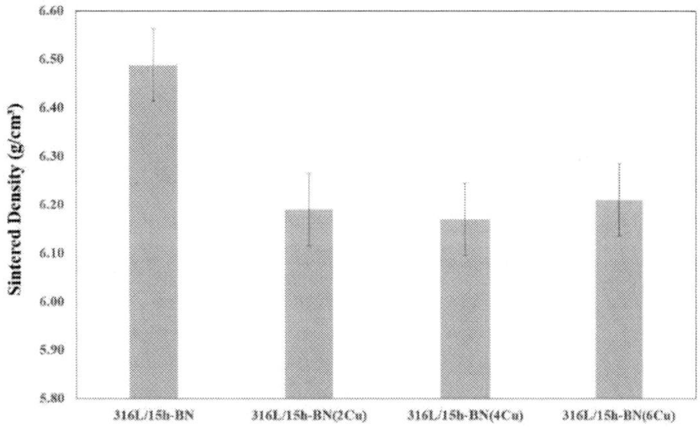

Figure 4. Sintered density of sintered 316L/Cu/h-BN composite.

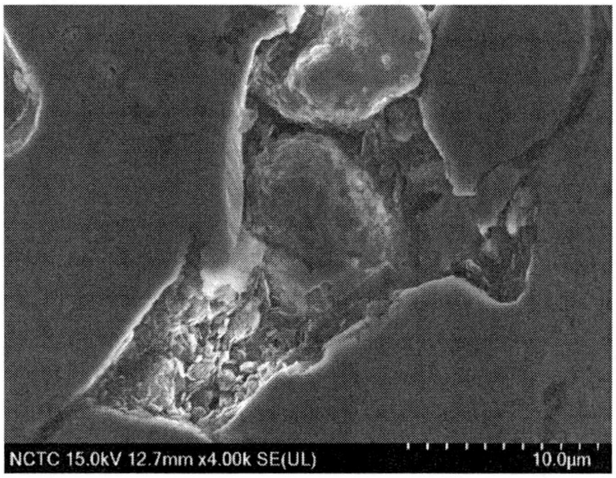

Figure 5. Flaky h-BN particles in a sintered composite.

3.2 Mechanical Properties

Tensile strength and elongation of sintered 316L/Cu/h-BN composites improved with increasing copper content whereas the drop of hardness was found in 2 and 4 wt. % Cu additions (Figure 6). The overall mechanical properties of the sintered composites are still low. The main reason for poor mechanical properties is high porosity left behind by Cu dissolution. It was expected in this work that Cu addition would help improving sintered density by liquid phase sintering, by which liquid Cu would fill empty pores and wet grain boundaries between 316L grains. However, there is no evidence for free Cu in all sintered composites. It is possible that the added Cu contents may be too low to form free Cu, which improves sintered density and mechanical properties of sintered composites. Further work is necessary, to obtain liquid phase sintering in this kind of sintered composites.

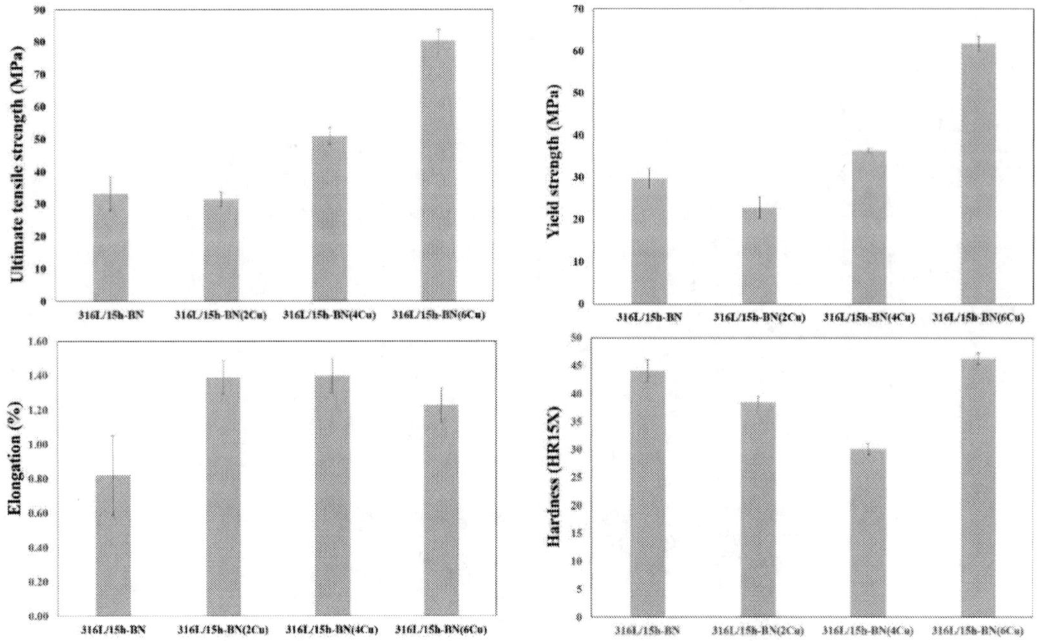

Figure 6. Mechanical properties of sintered composites.

4. Conclusions

1) Under a cracked ammonia atmosphere and at sintering temperatures of 1300°C, h-BN was still encapsulated in pores of sintered composites .The remaining h-BN, indicated by the absence of intergranular boride phase, was confirmed by EDS analysis. The hypothesis, in which nitrogen-containing sintering atmosphere retards h-BN decomposition and boride formation, was proved. The unreacted h-BN flakes were expected to provide lubricity of sintered composites.
2) Copper additions led to the decrease of densities of sintered 316L/Cu/h-BN composites, compared to that of sintered 316L/h-BN composite, due to porosity left behind by Cu powder melting and dissolution.
3) Tensile strengths of sintered 316L/Cu/h-BN composites were improved when Cu contents were 4 and 6 wt.%. It was possible that the Cu addition improved mechanical properties of sintered composites due to the solution strengthening by copper.

Acknowledgments

This work is funded by the project P1951261 supported by National Science and Technology Development Agency (NSTDA). The first author is also financially supported by TAIST-Tokyo Tech scholarship. The authors are grateful for supports from Thailand National Metal and Materials Technology Center (MTEC) and King Mongkut's Institute of Technology Ladkrabung (KMITL).

References

[1] Ahmed N S and Nassar A M 2013 *Lubrication and lubricants, in Tribology fundamentals and advancements* 55
[2] Gunda R and Narala S 2017 *Surf. Coat. Technol.* **332** 341
[3] Bocchini G F 1986 *Powder Rep* **41** 829
[4] Coovattanachai O Tosangthum N Morakotjinda M Yotkaew T Krataitong R Vetayanugul B and Tongsri R 2010 *Songklanakarin J. Sci. Technol* **32** 163
[5] Xu Z Hodgson M A Chang K Chen G Yuan X and Cao P 2017 *Metals* **7** 81
[6] Zaręmbski K Putyra P and Mierzwiński D 2019 *Metals* 9 264

[7] Clauss F J 1972 *Solid lubricants and self-lubricating solids* (Massachusetts: Academic Press)
[8] Vazirisereshk M Martini A Strubbe D and Baykara M 2019 *Lubricants* **7** 57
[9] Ravindran P Manisekar K Narayanasamy R and Narayanasamy P 2013 *Ceram. Int.* **39** 1169
[10] Mahathanabodee S Palathai T Raadnui S Tongsri R and Sombatsompop N 2013 *Mater. Des.* **46** 588
[11] Koetniyom W Chantawet P Tosangthum N Morakotjinda M Yotkaew T Wila P and Tongsri R 2019 *J. Met. Mater. Miner.* **29** 22
[12] Tosangthum N Kunnam P Morakotjinda M Koetniyom W Krataitong R Wila P and Tongsri R 2019 *Key Eng. Mater.* **798** 9
[13] Chusong E Kansuwan P Ohtake N Wila P Tosangthum N and Tongsri R 2020 *J. Met. Mater. Miner.* **30** 117
[14] Xi T Shahzad M B Xu D Zhao J Yang C Qi M and Yan K 2016 *Mater. Sci. Eng. A.* **675** 243
[15] Xi T Shahzad M B Xu D Sun Z Zhao J Yang C Qi M and Yang K 2017 *Mater. Sci. Eng. C.* **71** 1079
[16] Mahathanabodee S Palathai T Raadnui S Tongsri R and Sombatsompop N 2013 *Adv. Mater. Res.* **747** 307
[17] Mahathanabodee S Palathai T Raadnui S Tongsri R and Sombatsompop N 2014 *Wear* **316** 37
[18] Xu J Jian Z Dang B Zhang D and Liu F 2017 *Metall. Mater. Trans. A.* **48** 1817

Phase transformation and mechanical properties of sintered Fe-Mo-Si-C-(Cu) alloys

Monnapas Morakotjinda[1,*], Kittikhun Ruangchai[2], Bhanu Vetayanugul[1], Rungtip Krataitong[1], Nattaya Tosangthum[1], Amporn Wiengmoon[2], and Ruangdaj Tongsri[1]

[1]Particulate Materials Processing Technology (PMPT), National Metal and Materials Technology Center, 114 Paholyothin Road, Khlong Nueng, Khlong Luang, Pathum Thani 12120, Thailand
[2]Department of Physics, Faculty of Science, Naresuan University, Pitsanulok-Nakhon Sawan Road, Tha Pho, Muang, Pitsanulok, 65000, Thailand

* Corresponding Author: monnapm@mtec.or.th

Abstract
Sintered Fe-Mo-Si-C alloys, prepared from pre-alloyed Fe-0.85Mo powder, fixed 4wt. % silicon carbide powder and varied graphite powder contents, showed microstructures similar to those of ductile cast irons, i.e., the microstructural feature consisted of a black particle enveloped with matrix consisting of ferrite halo and pearlite. The varied added-graphite content caused morphological change from black nodular to black vermicular particles. With increasing added-graphite content resulted in the decrease of black nodular/vermicular particle fraction, the increase of pearlite fraction and the slight change of ferrite fraction. The black nodular particles were either graphite or Fe-Mo-Si-C/graphite core-shell particles whereas black vermicular particles were totally composed of carbon. With 2 wt.% copper addition, the morphologies of black nodular and vermicular particles in sintered Fe-Mo-Si-C-Cu alloys were not affected, but one component of the matrix changed from pearlite to duplex structure consisting of bainitic ferrite (BF) and martensite-austenite (M-A) constituent. The change of matrix component from pearlite to duplex BF/M-A structure led to drops of ultimate tensile strengths and elongation values but small effects on yield strengths and hardness values. No benefits of tensile properties were gained from the duplex structure due to coarse BF/M-A size.

Keywords: Sintered Fe-Mo-Si-C alloys, Copper addition, Phase transformation, Mechanical properties

1. Introduction

Silicon carbide (SiC) powder can react with iron (Fe) or Fe-base powders at temperatures close to normal sintering temperatures for such Fe or Fe-base powder compacts [1, 2]. Sintering of Fe + 5wt.% SiC powder compacts at 1200°C resulted in the microstructure consisting of nodular graphite-like particles surrounded with ferrite and a pearlite structure. This microstructural feature was designated as ductile cast iron-like microstructure [1]. Similar microstructures were also reported in other systems [3, 4].

Copper (Cu) is a common additive powder mixed with Fe powder to produce sintered Fe-base materials. The addition of Cu in Fe-0.85Mo + 0.6 wt.% graphite powder mixtures resulted in mechanical property modification. The Cu additions with contents of 0.07 to 0.85 wt.% to 0.90 carbon rail steels resulted in small pearlite transformation delay but appreciable pearlite interlamellar spacing refinement [5]. The copper addition to 0.15C-1.5Mn-1.5Si TRIP-aided multiphase cold-rolled steel sheets resulted in the increase of retained austenite volume fraction, although it did not affect the stability [6]. Similarly, the study on Cu addition in a hot rolled ferrite-based lightweight steel revealed that Cu addition delayed bainitic transformation, resulting in higher volume fraction of retained austenite with lower stability in the steel [7]. In ductile cast irons added with 0.2 to 2.0 wt.% Cu, the graphite nodules are not influenced by Cu but the pearlite fraction increases with increasing added Cu contents in the range of 0.2-0.8 wt. %, and it maintains at roughly constant with further increase of Cu [8].

Due to the effects of Cu on phase transformation behaviours in sintered and wrought steels given above, it is interesting to explore how Cu addition affect the matrices of sintered Fe-Mo-Si-C alloys produced under slow cooling rate.

2. Materials and methods

Sintered Fe-Mo-Si-C alloys without and with Cu addition were prepared by sintering powder mixture compacts of pre-alloyed Fe-0.85Mo powder, fixed 4wt. % silicon carbide powder and varied graphite powder contents. The nominal compositions of sintered alloys are given in Table 1. The powder mixtures were compacted into tensile test bars with green density of 6.5 g/cm^3. The green compacts were sintered at 1,250 °C for 45 min in a vacuum furnace. After sintering, sintered specimens were cooled in the sintering furnace with cooling rate of 0.1 °C/s. Microstructures were investigated by using optical microscopy (OM) and scanning electron microscopy (SEM). Phase identification was conducted by using X-ray diffraction (XRD) technique. Mechanical properties of the sintered materials, such as tensile and hardness, were tested by a tensile testing machine (Instron model 8801) and hardness tester (Rockwell scale B), respectively.

Table 1. Composition of the powder mixtures

Alloy	SiC	Graphite	Copper	Nominal composition (wt. %)				
	(wt. %)	(wt.%)	(wt. %)	C	Si	Mo	Cu	Fe
0Cu00G	4.0	0.0	0.0	1.2	2.8	0.82	0.0	Bal.
0Cu01G	4.0	0.1	0.0	1.3	2.8	0.82	0.0	Bal.
0Cu02G	4.0	0.2	0.0	1.4	2.8	0.81	0.0	Bal.
0Cu03G	4.0	0.3	0.0	1.5	2.8	0.81	0.0	Bal.
0Cu04G	4.0	0.4	0.0	1.6	2.8	0.81	0.0	Bal.
2Cu00G	4.0	0.0	2.0	1.2	2.8	0.80	2.0	Bal.
2Cu01G	4.0	0.1	2.0	1.3	2.8	0.80	2.0	Bal.
2Cu02G	4.0	0.2	2.0	1.4	2.8	0.80	2.0	Bal.
2Cu03G	4.0	0.3	2.0	1.5	2.8	0.80	2.0	Bal.
2Cu04G	4.0	0.4	2.0	1.6	2.8	0.80	2.0	Bal.

3. Results and discussion

3.1 Microstructure

The sintered Cu-free alloys (Figure 1a-1e) showed common microstructural features consisting of a black particle surrounded with ferrite halo and pearlite. The SEM image showing the common microstructural feature is shown in Figure 2a. Pearlite structures consisting of ferrite and carbide lamellae are shown in Figure 2b. Black particle morphology changed from nodular to a vermicular shape when the added graphite content was increased. XRD patterns of the sintered Cu-free alloys (Figure 3a) show only body-centered cubic (bcc) crystal structure of α-ferrite. No peaks corresponding to face-centered cubic (fcc) crystal structure of α-ferrite indicate that there is a very small amount or no austenite remaining in the the sintered Cu-free alloys. The characteristics and formation mechanism of black particles were already given in [9]. The formation mechanisms for ferrite and pearlite in ductile cast irons given in [10] can be used for explaining ferrite and pearlite transformations in the sintered Cu-free alloys of this work. Thus, the explanation for ferrite and pearlite formations is not repeated here. The sintered Cu-containing alloys (Figure 1f-1j) also showed common microstructural feature similar to those of the sintered Cu-free alloys. The black particle and ferrite halo in the sintered Cu-containing alloys were not affected by Cu addition. However, the pearlite structure disappeared and was replaced by duplex structure in the sintered Cu-containing alloys. The SEM image showing the common microstructural feature in the sintered Cu-containing alloys is given in Figure 2c. The SEM image showing the duplex structure is given in Figure 2d. XRD patterns of the sintered Cu-containing alloys (Figure 3b) show both bcc and fcc crystal structures. This indicates that there are both ferrite and austenite in the sintered Cu-containing alloys. According to the SEM image given in Figure. 2d and XRD patterns given in Figure 3b, the duplex structure is characterized by alternating bainitic ferrite (BF) plates and martensite-austenite (M-A) islands. The size and shape of both BF plates with M-A islands are roughly similar and equal. There are all microstructural features classified according to [11], but no carbide particles are observed in the duplex BF/M-A structure.

The duplex BF/M-A structure given in Figure 2d is similar to microstructures of steels transformed at low temperatures [11-13]. This implies that the duplex BF/M-A structure is the result of phase transformation from parent austenite phase to duplex structure at a low temperature. Since the cooling rate employed in this work is quite low (0.1°C/s), the parent austenite must have stability sufficient to reach a low temperature phase transformation field. Due to no carbide precipitation in the duplex BF/M-A structure, this microstructure type may be classified as one of carbide-free bainite structures.

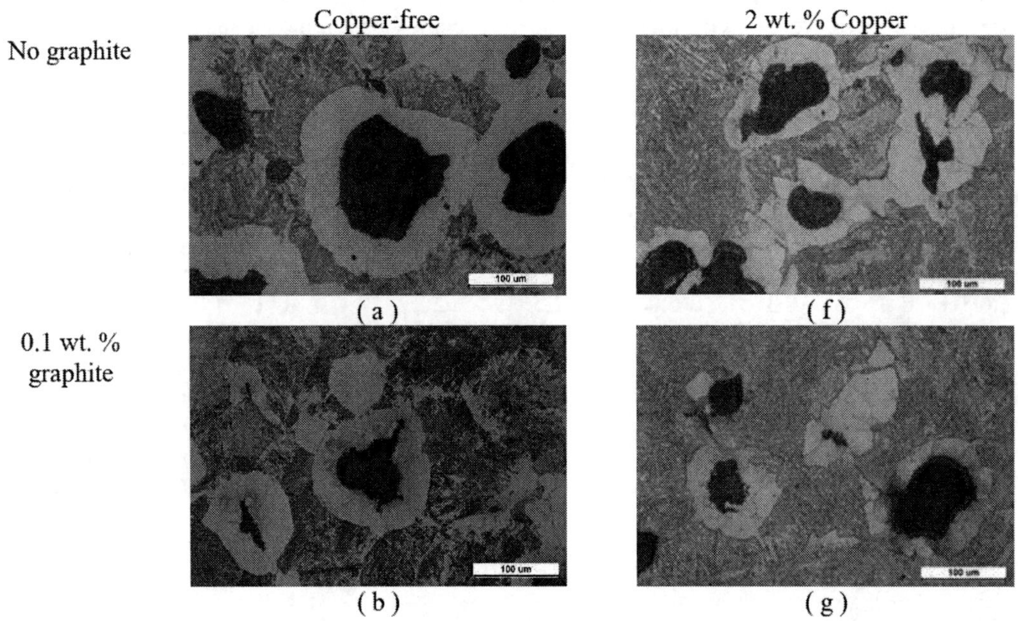

IOP Conf. Series: Materials Science and Engineering 1137 (2021) 012037 doi:10.1088/1757-899X/1137/1/012037

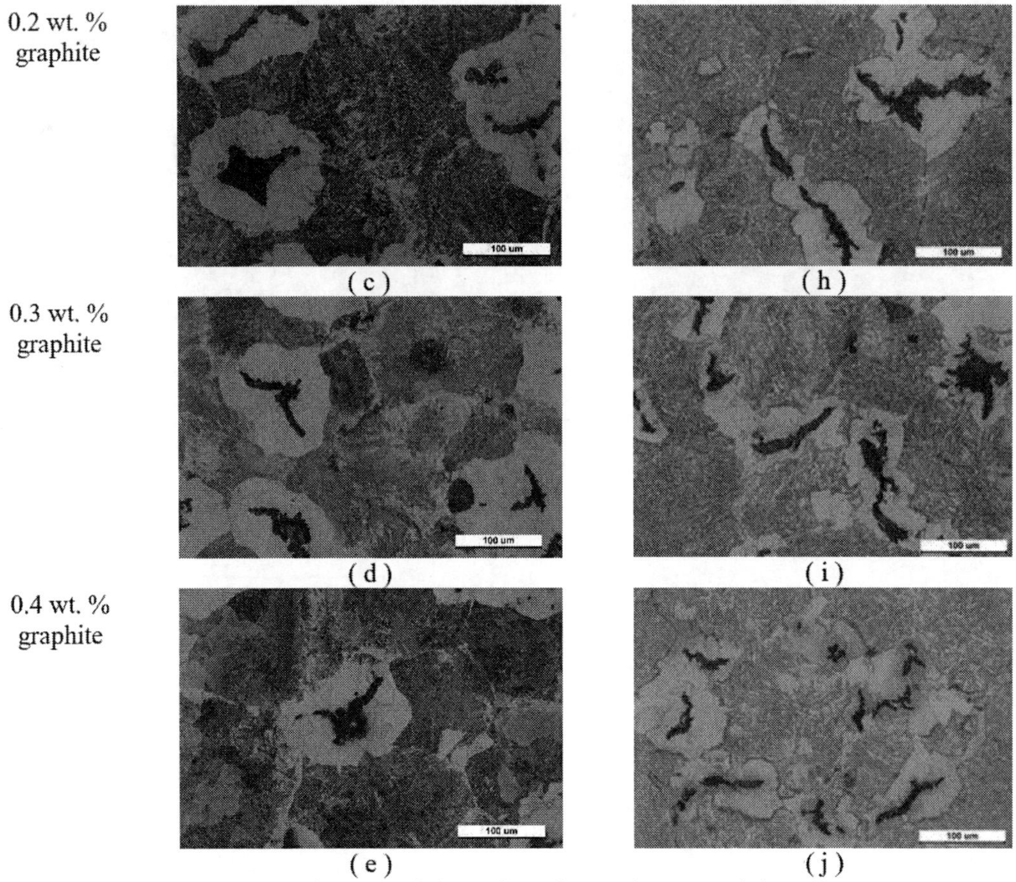

Figure 1. OM images of sintered Cu-free and Cu-containing alloys.

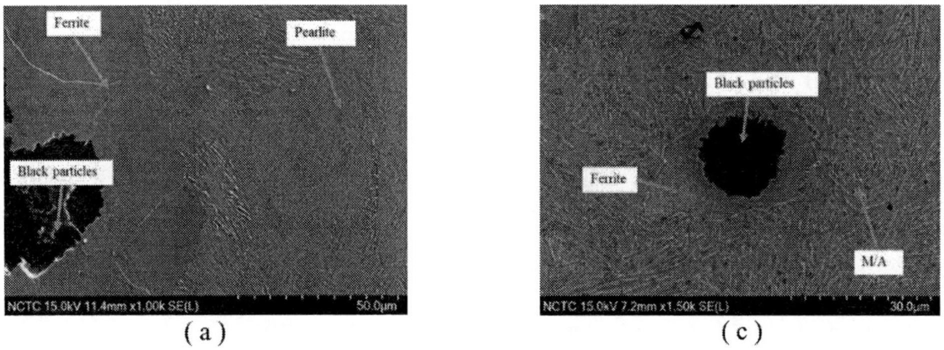

IOP Conf. Series: Materials Science and Engineering 1137 (2021) 012037 doi:10.1088/1757-899X/1137/1/012037

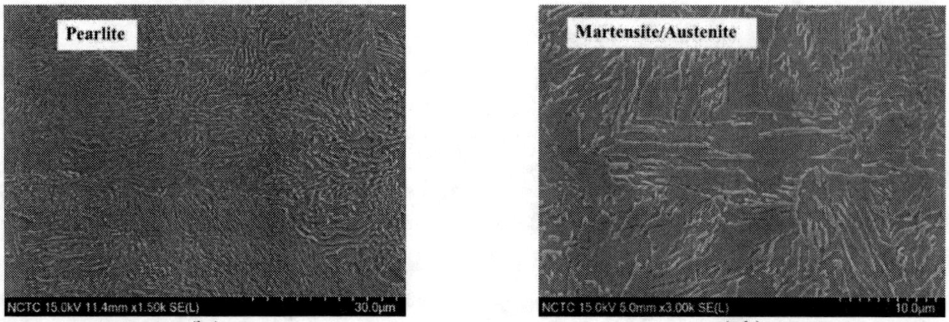

(b) (d)

Figure 2. SEM micrographs of sintered Fe-Mo-Si-C-(Cu) alloys: (a) overall microstructure of sintered Fe-Mo-Si-0.3C alloy, (b) Pearlite structure of sintered Fe-Mo-Si-0.3C alloy, (c) overall microstructure of sintered Fe-Mo-Si-0.3C-2Cu alloy, and (d) duplex BF/M-A structure of sintered Fe-Mo-Si-0.3C-2Cu alloy.

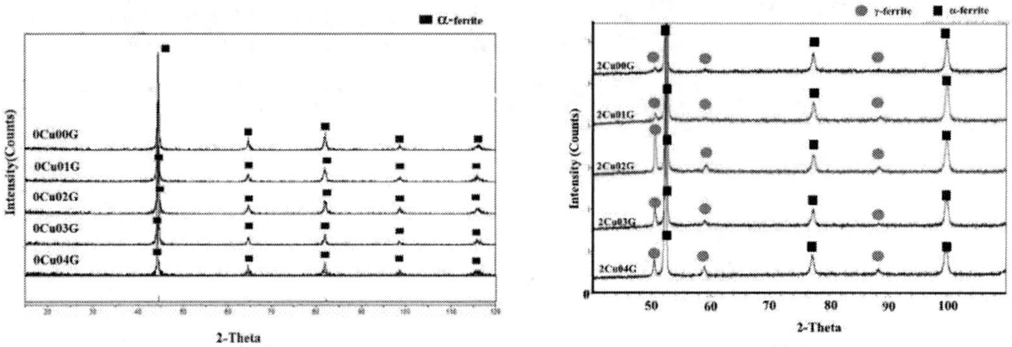

(a) (b)

Figure 3. XRD patterns of sintered Fe-Mo-Si-C alloys with varied graphite contents; (a) without Cu addition, and (b) with 2 wt. % Cu addition.

3.2 Mechanical properties

With increasing carbon content, tensile strengths (ultimate tensile strength (UTS) and yield strength or (YS)) of experimental sintered alloys increased to optimum values and then slightly dropped as shown in Figure 4. Elongation followed similar trend of the tensile strength. Both strength and ductility of the sintered Cu-containing alloys were inferior to the sintered Cu-free ones. The existence of strength and elongation optimal values is attributed to composite microstructure consisting of black particle, ferrite and pearlite or duplex structure. The optimum strength or elongation value is resulted from the contributions of these microstructural components. The inferior strength in the sintered Cu-containing alloys is attributed to coarser scales of duplex BF/M-A structure, compared to finer scales of ferrite and cementite lamellae in pearlite of the sintered Cu-free alloys. The superior strength of superbainitic steels depends strongly on very fine slender microstructural components [14, 15].

IOP Conf. Series: Materials Science and Engineering 1137 (2021) 012037 doi:10.1088/1757-899X/1137/1/012037

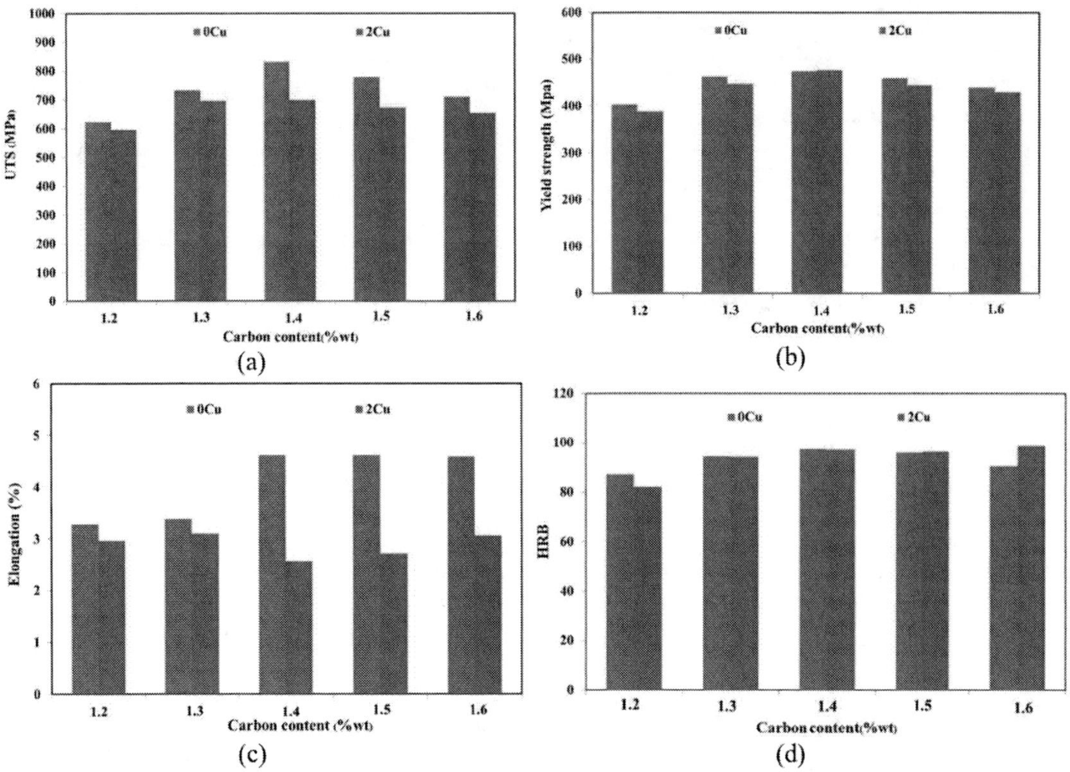

Figure 4. Mechanical properties of sintered Fe-Mo-Si-C alloys without and with Cu modification; (a) UTS, (b) Yield strength, (c) Elongation and (d) Hardness.

Although, the sintered Cu-containing alloys have a detectable amount of retained austenite (Figure 3b), they do not gain the benefit of transformation-induced plasticity (TRIP) effect, which is depending on the transformation of austenite to martensite during deformation [16, 17]. The inferior ductility of the sintered Cu-containing alloys may be related to retained austenite characters. Optimum retained austenite characters (volume fraction, morphology, and stability) are required for ductility enhancement [18-21]. In general, compared to block retained austenite, the thin film one is more stable and has higher C concentration. According to SEM image given in Figure 2d, the inferior ductility of the sintered Cu-containing alloys is related to retained austenite with a blocky shape. Regarding hardness, Figure 4d indicates that Cu addition does not significantly affect hardness values.

4. Conclusion

The Cu addition to sintered Fe-Mo-Si-C alloys led to changes of microstructures and tensile properties as summarized below.

(1) The Cu addition led to change of pearlite to duplex BF/M-A structure but little changes of black particle and ferrite halo.

(2) The coarser scales of BF/M-A components led to inferior tensile strength in the sintered Cu-containing alloys.

(3) The inferior ductility in the sintered Cu-containing alloys may be attributed to retained austenite characters in duplex BF/M-A structure.

Acknowledgments
The authors are grateful for financial support from National Science and Technology Development Agency (NSTDA) under the project P1951261. The authors are also grateful for support from National Metal and Materials Technology Center.

References
[1] Chakthin S Poolthongt N and Tongsri R 2008 *Advanced Materials Research* **55-57** 357-360
[2] Tongsri R and Vetayanugul B 2010 *Journal of Metals Materials and Minerals* **20** 45-49
[3] Binder C Hammes G Schroeder R Klein A N de Mello J D B Binder R and Junior W R 2010 *Met.Powder Rep.* **65** 29-37
[4] Klein A N Furlan K P Schroeder R M Hammes G Binder C Neto J B R Probst S H and de Mello J D B 2015 *Powder Technology* **271** 193-203
[5] Eavenson G T 2025 Master of Science Thesis (USA: Colorado School of Mines)
[6] Kim S J Lee C G Lee T H and Oh C S 2002 *ISIJ International* **42** 1452-1456
[7] Wang J Wang Z Wang X Yang Q Jin X Wang L Wang J Wang Z Wang X Yang Q Jin X and Wang L 2017 *Scripta Materialia* **129** 25-29
[8] Yi P Guo E Wang L Feng Y and Wang C 2019 *China Foundry* **16** 168-176
[9] Ruangchai K Wiengmoon A Morakotjinda M Krataitong R Tanprayoon D Yotkaew T Tosangthum N Patakham U and Tongsri R 2017 *Key Engineering Materials* **751** 47-52
[10] Lacaze J Sertucha J and Åberg L M 2016 *ISIJ International* **56** 1606-1615
[11] López A N Hidalgo J Sietsma J and Santofimia M J 2017 *Materials Characterization* **128** 248-256
[12] Tian J Xu G Jiang Z Wan X Hu H and Yuan Q 2019 *Steel Research International* **90** 1800474
[13] Liu M Wang J Zhang Q Hu H and Xu G 2020 Metals and Materials International https://doi.org/10.1007/s12540-020-00726-5
[14] Bhadeshia H K D H 2010 *Proceedings of the Royal Society A* **466** 3-18
[15] Hu F Hodgson P D and Wu K M 2014 *Materials Letters* **122** 240-243
[16] Jacques P J 2004 *Current Opinion in Solid State and Materials Science* **8** 259-265
[17] Soleimani M Kalhor A and Mirzadeh H 2020 *Materials Science and Engineering A* **795** 140023
[18] Zhang M X Kelly P M Bekessy L K and J D Gates 2000 *Materials Characterization* **45** 39-49
[19] Sugimoto K Usui N Kobayashi M and Hashimoto S 1992 *ISIJ International* **32** 1311-1318
[20] Saha A Podder I Lonardelli Molinari A and Bhadeshia H K D H 2011 *Proceedings of the Society A* **467** 3141-3156
[21] Mateo C G and Caballero F G 2005 *Materials Transactions* **46** 1839-1846

Evaluation of conventional invasive measurements and examination of non-invasive measurement technique on human body core temperature

Atsumasa Yoshida[1*], Rinzo Kamon[1], Takuma Naka[1], Narihisa Chigusa[1], Shinichi Kinoshita[1], and Takashi Kawabata[2]

[1]Department of Mechanical Engineering, Osaka Prefecture University, 1-1 Gakuen-cho, Naka-ku, Sakai, 599-8531, Japan
[2]Department of Health and Well-being, Kansai University, 1-11-1 Kaorigaoka-cho, Sakai-ku, Sakai, 590-8515, Japan

* Corresponding Author: ayoshida@me.osakafu-u.ac.jp

Abstract
It is important in many cases to measure and monitor, human body core temperatures, to prevent the likes of heat stroke and hypothermia. Measuring core body temperatures is also important for example, in relation to the basal body temperatures of women, and for improving the quality of life for vulnerable people such as infants, and those with cervical spine injuries. In doing so, thermal environmental changes can be monitored. However, todays conventional measuring method is apparently invasive because a temperature sensor has to be inserted into the body, from the outside. Characteristics of the body parts measured were considered by the subject experiments. There was a difference in core temperatures, depending on the measurement body parts, and it was found that the temperature decreased in the order of rectum, sublingual, and tympanic during normal times. It was confirmed that the tympanic temperature showed the most significant increase in core temperature, with running and the sublingual temperature the best in responsiveness. As a non-invasive core temperature measurement method, the basic characteristics were examined by the heat transfer experiment for the dual-heat-flux method. Thus, it was clarified that adequate temperature accuracy can be guaranteed if appropriate materials, thicknesses, and sizes are selected and the adiabatic condition for the peripheral part is fulfilled.

Keywords: Core temperature, Invasive and non-invasive measurements, Subject measurement, Body part, Heat transfer

1. Introduction

The human body core temperature is a physiologically important numerical value in clinical management of patients, modeling of thermoregulatory mechanism, and evaluation of thermal load on the human body. It is necessary to monitor the core temperature for preventing heat stroke and hypothermia [1-2]. Conventional core temperature measurement is invasive and exhibits different behavior depending on the body part measured [3-6]. In this study, the measurements were made in the tympanic, sublingual, and rectum by subject experiments. We compared the core temperatures of different body parts and considered their characteristics. On the other hand, the practical application of a non-invasive core temperature measurement method with a few measurement restrictions is desired. The zero-heat-flow [7-10] and the dual-heat-flux methods [11-14] have been proposed as non-invasive core temperature measurement methods. In the zero-heat-flow method, the core temperature is obtained by achieving thermal equilibrium on the skin surface using a thermal insulating material and a heater. It is already commercialized and applied to manage the body temperature during surgery [8], however, its application under the non-restricted conditions of daily life is difficult. The dual-heat-flux method is one of the deflection measurement methods not requiring a heater, thus, may have a wider application range. Although the basic principle is explained, there are multiple issues for practical use, and quantitative evaluation cannot be adequate. In this study, in consideration of future developments in daily life, a basic study was conducted by the heat transfer method [11], which simulated human skin.

2. Subject experiment of invasive human body core temperature

2.1 Measurement items and subjects

Measured environmental factors were air temperature, relative humidity, and solar radiation (outdoor only), and physiological factors measured were skin temperature at four sites (chest, upper arm, thigh, and lower leg), core temperature at three sites (sublingual, ear cavity (tympanic), and rectum), blood pressure, heart rate, expiratory metabolism, and perspiration. The core temperature was measured every 10 s using an infrared sensor inside the ear and thermistors for other parts. The subjects were six healthy male students aged 22-24 years. The subjects initiated the experiment after resting for 30 min in a room at 26°C. They performed the exercise by running on a treadmill.

2.2 Experimental method

2.2.1 Outdoor exercise load experiment

The subject experiment was carried out on the wood deck of a building courtyard at the Nakamozu Campus of Osaka Prefecture University. To examine the effect of solar radiation on the core temperature, the experiments were conducted using the protocol shown in Figure 1, and the exercise was performed in the sunny area.

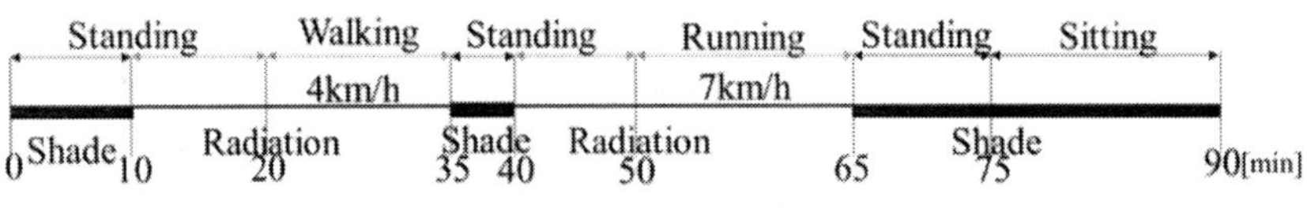

Figure 1. Protocol of subject experiment 1.

2.2.2 Indoor exercise load experiment

The subject experiment was carried out at a room (air conditioning and cooling 26°C setting) in a building at the Nakamozu Campus of Osaka Prefecture University. To examine the influence of body posture and exercise on core temperature, the experiments were conducted using the protocols shown in Figures 2 and 3.

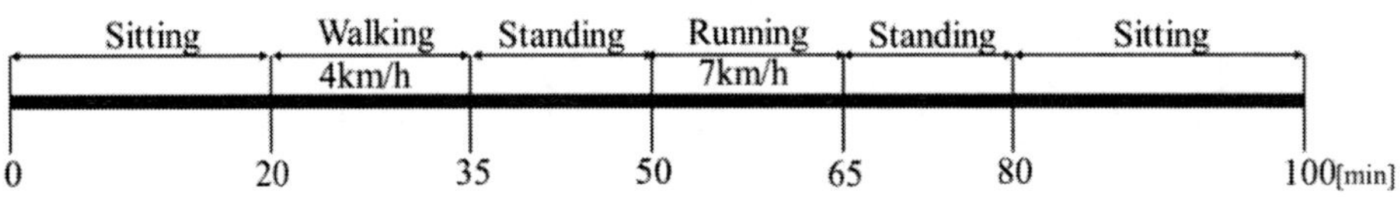

Figure 2. Protocol of subject experiment 2.

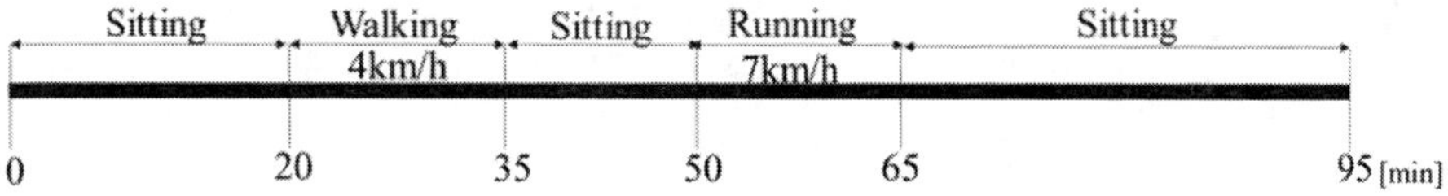

Figure 3. Protocol of subject experiment 3.

2.3 Experimental results and discussion

Figure 4 shows a comparison of the indoor and outdoor core temperatures (rectal and sublingual) similar for the same subject and at 23°C. The moving average value for 1 minute is displayed. The same was applied below. Although, the outdoor experiment was expected to have a higher core temperature than the indoor experiment due to the effects of solar radiation, the core temperature of the indoor experiment was slightly higher. Therefore, the effect of solar radiation on the core temperature was lesser than the effects of posture, exercise, and physical condition of the subject.

Figures 5 and 6 show the measurement results of experiments 2 and 3 (rectal, sublingual, and tympanic). In all cases, the number of subjects with no missing measurement data was five, the curve represents the average value, and the error bar represents the standard deviation at the switching time of posture and exercise. The mouth was sometimes opened during the exercise, and a decrease in sublingual temperature was observed. Variations in tympanic temperature were observed due to changes in posture and body movements during exercise.

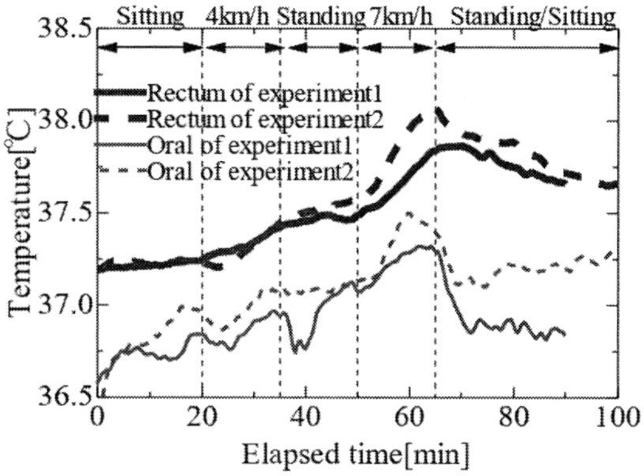

Figure 4. Core temperature comparison with and without solar radiation.

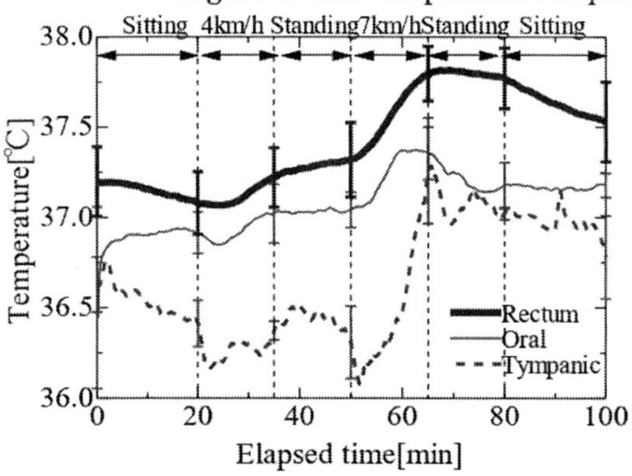

Figure 5. Temporal change of core temperature in experiment 2.

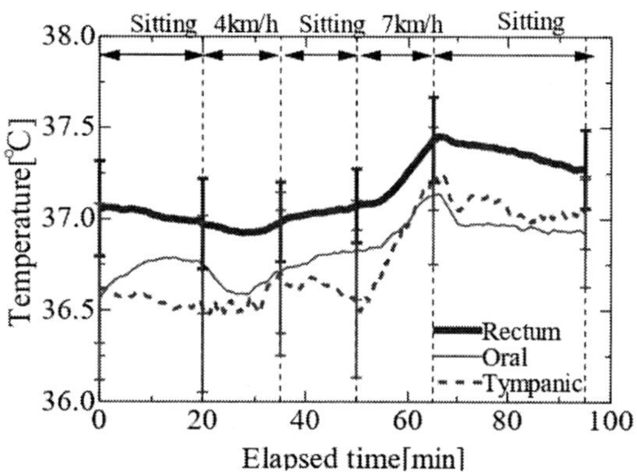

Figure 6. Temporal change of core temperature in experiment 3.

Figure 7 shows a comparison of each core temperature in the sitting position before exercise. Each core temperature varied depending on the measured body part, and the average value decreased in the order of rectum, sublingual, and tympanic. The rectal temperature was highest [3-5]. Similar results were obtained over the entire experiment time in Figures 4-6. Significant differences in measured temperature were confirmed between the rectum and sublingual, and rectum and tympanic.

Figure 8 shows the changes in each core temperature during the exercise period. Although there was no significant difference between the regions at a running speed of 4 km/h, a significant difference was observed at a traveling speed of 7 km/h. The tympanic was significantly changed compared with the rectum and sublingual, with a greater change in the rectum than in the sublingual. The tympanic is closely associated with the behavior of the brain temperature in the hypothalamus, which controls the amount of sweating [4].

Regarding the responsiveness of the core temperature, Figure 9 shows the time from the start of exercise (running speed 4 km/h) until the rise of the core temperature. There was a significant difference between the sublingual and the other responses, with a shorter sublingual response time than that of the rectum and tympanic. The responsiveness of pulmonary arterial blood temperature, significantly affecting changes in body temperature, is close to sublingual temperature [4]. As shown in Figures 5 and 6, the decrease in rectal temperature was lesser than that in the tympanic and sublingual after the exercise.

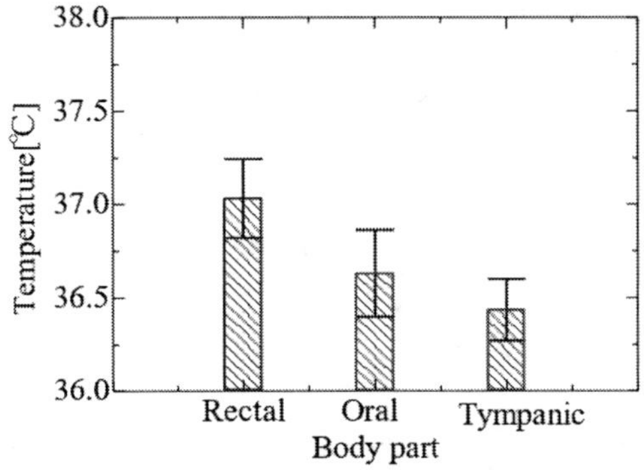

Figure 7. Comparison of core temperature before exercise.

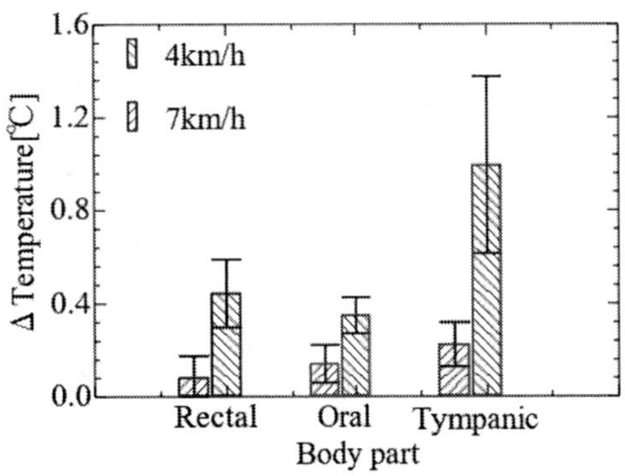

Figure 8. Core temperature change in walking and running periods.

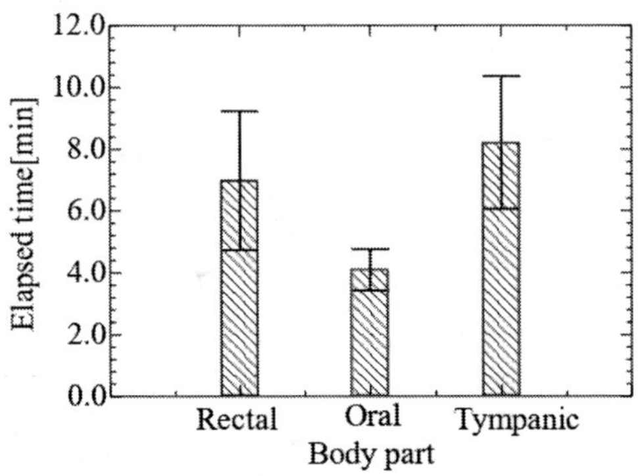

Figure 9. Response time of core temperature rise after walking exercise.

3. Examination of measurement method of non-invasive core temperature

3.1 Measurement principle

The dual-heat-flux method [11] was adopted for measuring the non-invasive human body core temperature. It is assumed that one-dimensional heat is transferred from the core part of the body to the skin and two types of flat plate materials (insulation materials) with different thicknesses cover the skin, and thermal equilibrium. This is a method to calculate the heat flux passing through each thermal insulating material from the temperature at both ends of the thermal insulating material and obtain the core temperature T_{core} without using the thermal resistance value R_{sk} of the skin tissue. Figure 10 shows the concept related to the measurement principle. The temperatures at both ends of the insulation are T_1 to T_4, and the thermal resistance of the insulation is R_1 and R_2. The core temperature obtained from each measured temperature is shown in Equations (1) and (2).

$$T_{core} = T_1 + \frac{(T_1 - T_3)R_{sk}}{R_1} \tag{1}$$

$$T_{core} = T_2 + \frac{(T_2 - T_4)R_{sk}}{R_2} \tag{2}$$

The thermal resistance value R_{sk} of the skin tissue is deleted from both equations, and the deep temperature T_{core} is obtained.

$$T_{core} = T_1 + \frac{(T_1 - T_2)(T_1 - T_3)}{K(T_2 - T_4) - (T_1 - T_3)} \tag{3}$$

Where, $K=R_1/R_2$. When the contact thermal resistance value R_t between the skin and the thermal insulating material is considered, the deep temperature T_{core} is shown in Equation (4).

$$T_{core} = T_1 + \frac{(T_1 - T_2)(T_1 - T_3)}{S(T_2 - T_4) - (T_1 - T_3)} \tag{4}$$

Where, $S=R_1(R_{t2}+R_{sk})/R_2(R_{t1}+R_{sk})$. When R_{t1}, R_{t2} $<<R_{sk}$, S=const, and the core temperature can be calculated considering the contact thermal resistance. It is desirable to minimize the contact thermal resistance.

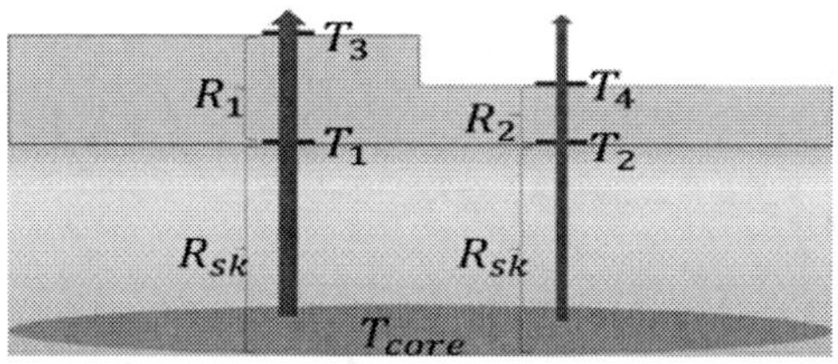

Figure 10. Schematic diagram of duel-heat-flux method.

3.2 Heat transfer experiment simulating biological tissue

A simulation experiment was performed controlling the core temperature. The temperatures were measured based on the dual-heat-flux method, and the measurement of core temperature accurately from the viewpoint of accuracy and responsiveness was verified. Figure 11 shows the measurement setup. The measurement items were air temperature (5 mm directly above the thermal insulating material and room temperature), temperatures at the top and bottom edges of the thermal insulating material (T_1, T_2, T_3, T_4), side temperature of the thermal insulating material, core (agar bottom) temperature, and water temperature. Agar simulating biological tissue (agar powder: water = 7 g: 100 mL, thickness 15 mm) was used to make a petri dish with an outer diameter of 125 mm, and the temperature was controlled in a constant temperature bath. The air temperatures were measured with a thermocouple and the others were measured with a thermistor every 10 s. To reduce the contact thermal resistance, an aluminum sheet of 0.3 mm thickness was inserted between the heat insulating material and agar, and the heat insulating material (50 x 50 mm) was fixed with tape. The material of the heat insulating material was polystyrene foam which was easy to process.

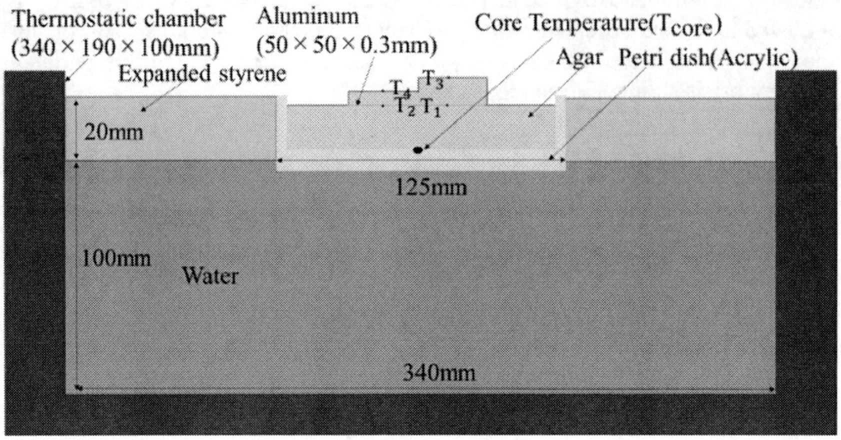

Figure 11. Experimental apparatus.

Figure 12 shows the time variations of the upper and lower end temperatures of the thermal insulating material, the core part (agar bottom part) temperature, and the estimated core part temperature by thedual-heat-flux method. The air temperature was 31°C. About 30 min after the start of the experiment, a temperature close to the measured data of the core temperature was obtained by the dual-heat-flux method. Table 1 shows the measurement accuracy of the core temperature when the thickness of the thermal insulating material was different. The measurement accuracy was the worst when the thickness of the insulation was 5 mm and 9 mm. This was related to the surface roughness when the thermal insulating material was cut. Sufficient measurement accuracy was confirmed for the other two cases.

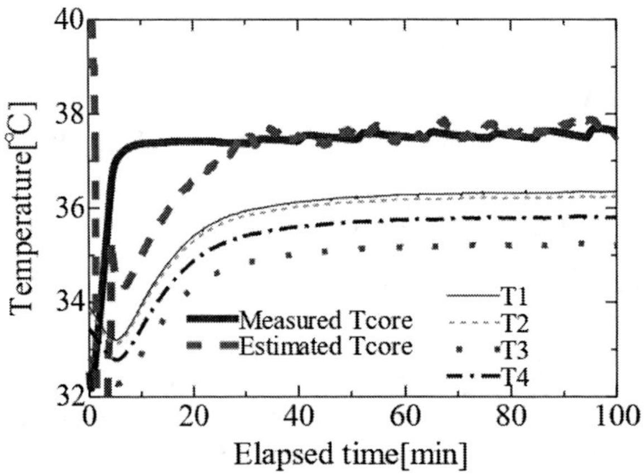

Figure 12. Steady-state experiment result for 5 mm and 9 mm in thermal insulation thickness.

An experiment was conducted in which the temperature was kept constant for 60 min from the initiation of measurement, and then the set temperature of the thermostatic chamber was increased by 0.3°C. The temperature was 32-33°C. Table 2 shows the response time when the thickness of the thermal insulating material was different, and the measurement accuracy of the core temperature after changing the set temperature. The response time was about 20 min. In case of a combination of 6 mm and 12 mm thick insulation, the temperature difference was large and the accuracy was not good. The smaller the thickness, the smaller the temperature difference, and the better the measurement accuracy. This is due to the small heat loss from the side surface of the thermal insulating material, which meets the one-dimensional heat conduction assumption more strictly.

Table 1. Comparison of measured and estimated core temperature for steady-state experiment.

Thermal insulation	6 mm/12 mm	5 mm/10 mm	5 mm/9 mm
Measured T_{core} [°C]	37.37±0.05	37.47±0.03	37.52±0.07
Estimated T_{core} [°C]	37.34±0.07	37.49±0.05	37.58±0.13
ΔT_{core} [°C]	0.03	-0.02	-0.06

Table 2. Comparison of measured and estimated core temperature after setting temperature change of water bath.

Thermal insulation	6 mm/12 mm	5 mm/10 mm	5 mm/9 mm
Response time [min]	20.8	24.8	19.2
Measured T_{core} [°C]	37.77±0.02	37.77±0.02	37.81±0.04
Estimated T_{core} [°C]	38.03±0.04	37.84±0.03	37.78±0.04
ΔT_{core} [°C]	-0.26±0.03	-0.08±0.03	0.03±0.06

4. Conclusion

(1) There was a difference in core temperatures depending on the measured body part, the rectum being the highest, followed by the sublingual, and the tympanic.
(2) It was confirmed that the increase in core temperature with running motion was the largest in the tympanic with the best response by the sublingual.
(3) A non-invasive measurement of core temperature based on the dual-heat-flux method was examined by heat transfer experiments, and sufficient temperature accuracy was guaranteed.

Acknowledgments

This study was partially funded by the Grant-in-Aid for Scientific Research (17H00802, Principal Investigator, Atsumasa Yoshida) under the Japanese Society for Promotion of Science. The authors would like to thank the society for this grant.

References

[1] Shin J Kim J Song K and Kwak Y 2013 *Resuscitation* **84** 810
[2] Zhao J Wang H Li Y Xiao F and Deng Q 2020 *Building and Environment* **173** 106752
[3] Sound-Levander M Forsberg C and Wahren L K 2002 *Scandinavian Journal of Caring Sciences* **16** 122
[4] Matsumoto T Kosaka M Yamauchi M Ohwatari N Tsuchiya K Lee J M Yang G J Tsuruta M Yokoyama N Izumi M and Nagataki S 1992 *Japanese Journal of Biometeorology* **29** 119
[5] Tayler N A S Tipton M J and Kenny G P 2014 *Journal of Thermal Biology* **46** 72

[6] Opersteny E Anderson H Bates J Davenport K Husby J and Myking K 2017 *Journal of Pediatric Nursing* **35** 36
[7] Fox R H Solman A J Isaacs R Fry A J and MacDonald C I 1973 *Clinical Science* **44** 81
[8] Demach K Yoshida T and Tsuneoka H 2011 *Japanese Journal of Biometeorology* **48** 119
[9] Atallah L Bongers E Lamichhane B and Bambang-Oetomo S 2016 *IEEE of Biomedical and Health Informatics* **20** 100
[10] West N Cooke E Morse D Merchant R N and Gorges M 2020 *Journal of Clinical Monitoring and Computing* **34** 1121
[11] Kitamura K Zhu X Chen W and Nemoto T 2010 *Medical Engineering & Physics* **32** 1
[12] Huang M Tamura T Chen W and Kanaya S 2015 *Journal of Thermal Biology* **47** 26
[13] Huang M Tamura T Tang Z Chen W and Kanaya S 2016 *Journal of Sensors* **2016** 4828093
[14] Fang J Zhou C and Ye X 2019 *IOP Conference Series: Material Science and Engineering* **677** 032006

Numerical simulation of tooth extraction socket: alveolar bone regeneration

Rintaro Furukawa[1], Katsuya Nagayama[1*], and Masato Matsuo[2,*]

[1]Kyushu Institute of Technology, 680-4 Kawazu, Iizuka, Fukuoka, 820-8502, Japan
[2]Kanagawa Dental University, 82 Inaokacho, Yokosuka, Kanagawa, 238-8580, Japan

[*] Corresponding Author: nagayama@mse.kyutech.ac.jp, Tel&Fax 81-948-29-7778

Abstract
Recently, regenerative medicine has received attention owing to the progress of an aging society. Regenerative dentistry, such as bone regenerative therapy and implant treatment to regain bone loss as a result of active periodontal disease, is actively performed. The alveolar bone is a functional bone that supports teeth and buffers occlusal pressure. In the implant treatment, the implant is inserted into the tooth after it has been extracted. After the implant is inserted, the alveolar bone will regenerate over about three months and the implant will be fixed in place. During this process of alveolar bone regeneration, there have been problems with the implant and alveolar bone not bonding. This work was contributed to dental regenerative medicine by building a simulation model of alveolar bone regeneration. After tooth extraction is performed, the alveolar bone regenerates by chemically converting calcium (Ca) secreted from the newly formed blood vessels into hydroxyapatite, which is the main component of bone. Thus, its internal structure and external shape are mechanically affected around the bone. The main purpose of this study is to establish a mathematical model, introduced as the Ca transport model that is secreted from the newly formed blood vessels, and mechanical factor model expressed as bone remodeling for analyzing and predicting the interaction of angiogenesis and bone regeneration. As a result, we were able to obtain a good agreement between the qualitative features of the experimental images and the simulation results.

Keywords: Alveolar bone, Angiogenesis,

1. Research background

In recent years, regenerative medicine has been attracting attention in Japan due to the aging population and the discovery of iPS cells (Induced pluripotent stem cells). Regenerative medicine is the most advanced medical technology to regenerate tissues and organs that have been damaged by disease or injury using one's own cells. In this study, we focused on dental regenerative medicine, which is performed on periodontal tissues such as gums and bones that support teeth. A typical example of dental regenerative medicine is implant treatment. Implant therapy is a treatment in which an artificial tooth root made of titanium is implanted in the alveolar bone (Fig. 1) [1], which is the bone that supports the teeth, and is bonded to the alveolar bone, which is then covered with artificial bone. In the process of alveolar bone regeneration, the implant does not bond with the alveolar bone, leading to treatment errors.

Therefore, alveolar bone regeneration is important in dental regenerative medicine, and we aim to contribute to dental regenerative medicine by elucidating this phenomenon.

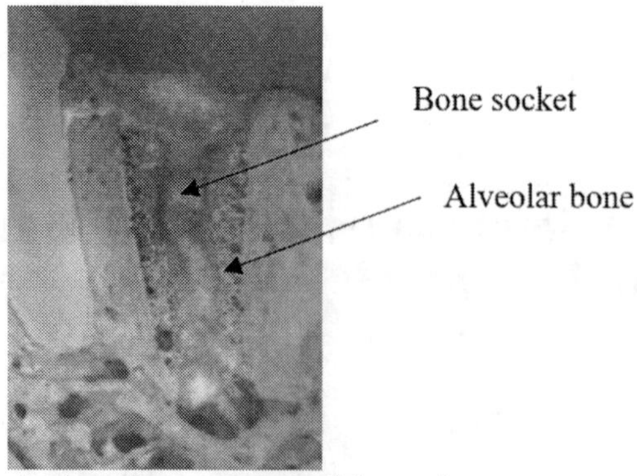

Figure 1. Socket after extraction of the tooth

2. Object of analysis

After dentification, the holes (Figure 1) require 90 days to regenerate under the influence of gravity and other mechanical loads and calcium (Ca) secretion from blood vessels [2]. At intervals of 14, 30, and 90 days, the regenerated bones are referred to as new bone, juvenile bone, and osteoblastic structures, respectively (Figure 2(a), (b), (c)) [3].

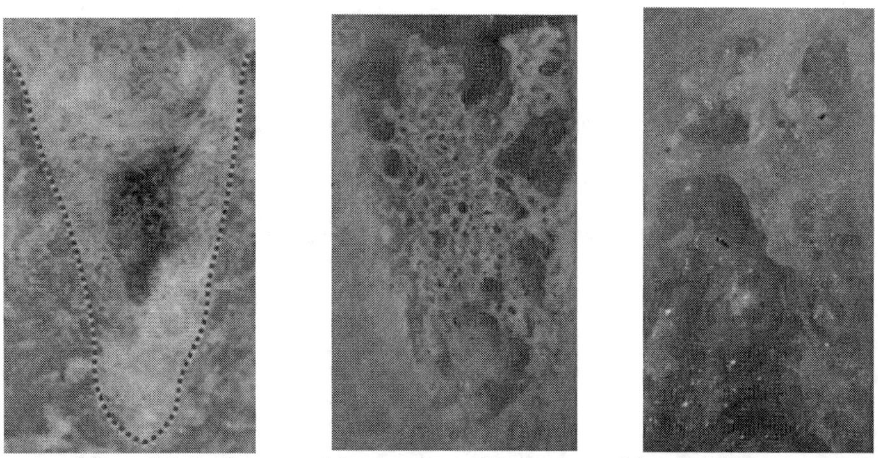

(a) neonatal bone (14 days) (b) young bone (30 days) (c) trabecular structure (90 days)

Figure 2. Experimental image of alveolar bone regeneration

In this study, we analyzed these phenomena using particles, which are very close to the real world. Herein, the term "particle" is used to describe a wide range of particles, ranging from individual cells to groups of cells or tissue fragments, and it is distinguished from vascular and bone particles.

3. Calculation model

3.1 Angiogenesis

Initially, vascular triggers (blood clots) are placed at random numbers. Next, a search is performed around the calculation point (particle) that represents the tip of the blood vessel, and a calculation point suitable for new blood vessel formation is determined based on the value of the inducing substance and other factors, and the blood vessel is renewed toward that calculation point. The bifurcation angle was set to 60-80° based on fractal nature [4].

3.2 Ca transportation

The main component of bone is hydroxyapatite, a hard crystalline form of basic phosphate Ca^{2+}, which is formed from Ca^{2+} secreted from neovascular vessels, as shown in equation (1) [5].

$$10Ca(OH)_2 + 6H_3PO_4 \rightarrow Ca_{10}(PO_4)_6(OH)_2 + 18H_2O \quad (1)$$

Using Reynolds' transport theorem, which considers Ca^{2+} secreted from the neovascular system in migration phenomenology, equation (2) can be expressed as

$$D\nabla^2 A = \frac{1}{\beta}\Delta\rho. \quad (2)$$

However, the convection term was neglected under the assumption that the effect of convection by the secretion of Ca^{2+} from neovascular tissue is small. Moreover, the nonstationary term was ignored under the assumption of steady-state secretion and diffusion of Ca^{2+}. Because the local bone formation and Ca^{2+}consumptions are conservative, the generation term is defined as the local bone formation rate $\Delta\rho$.
In equation (2), A denotes the concentration of Ca^{2+}, ρ denotes the bone density, D represents the diffusion coefficient, and β denotes the mass ratio of Ca^{2+} to the bone in Equation (1). Hence, the transport equation is represented by the parallel equation for Ca^{2+} diffusion and consumption.

3.3 Bone remodeling

The phenomenon of bone remodeling refers to a change in the bone structure due to the surrounding mechanical load. For example, the bones of a sick, bedridden person or an astronaut become brittle because of a lack of mechanical load.

The experiments conducted by Goodship et al. showed a correlation between local strain and remodeling, and this correlation is based on the mechanical stimulus [6]. When the mechanical stimulus is locally strong, a microscopic level of remodeling formation occurs. Conversely, when the mechanical stimulus is locally weak, remodeling absorption occurs.

The mathematical model proposed by Huiskes et al. uses strain energy density U to represent mechanical stimuli and expresses bone growth in terms of Young's modulus E [7].
It is modeled as in equation (3), assuming that there is a threshold of strain energy density toward bone growth and absorption.

$$\frac{dE}{dt} = C(U - U_h), \quad (3)$$

where C_U denotes the rate constant and U_h represents the reference strain energy density.
The relationship of equation (4) is also experimentally demonstrated in the report by Carter et al. [8]

$$E = 3790 \cdot \dot{\varepsilon}^{0.06} \cdot \rho^3, \quad (4)$$

where $\dot{\varepsilon}$ is the strain rate.
From equations (3) and (4), equation (5) is realized.

$$\frac{d\rho}{dt} = \frac{C_U}{3\rho^2}(U - U_h) \quad (5)$$

3.4 Bone Formation Equation

From terms 2-1 and 2-2, the bone structure formation equation is modeled using Equation (6)[9]

$$\frac{d\rho}{dt} = C_A(A - A_h) + \frac{C_U}{3\rho^2}(U - U_h) + d_\rho \nabla^2\rho \quad (6)$$

where A represents the concentration of Ca^{2+} secreted by neovascularization and A_h denotes the concentration of a reference Ca^{2+} in the blood.
We introduced $\nabla^2\rho$, which is responsible for locally flattening the formation and absorption effects of modeling. d_ρ is the diffusion constant.
The various parameters are set as shown in C_A=20, Cu=3.0× 10^{-3}, dp=8.64× 10^{-3}.

IOP Conf. Series: Materials Science and Engineering 1137 (2021) 012039 doi:10.1088/1757-899X/1137/1/012039

4. Analysis conditions

4.1 Initial conditions

The extraction socket was modeled using particles (Fig. 3). Each particle was considered as a cell, and an analytical model close to the real phenomenon was constructed.

For simplicity, a three-dimensional cubic region of 12 mm per side was used as the analysis area, and a total of 729,000 computational nodes (particles), 90 of which were placed in each direction, were placed. The extraction sockets (Figure 2) were 8 mm in diameter and 12 mm in height. The time tick width was set to 0.1 day/step. The initial bone density ρ immediately after tooth extraction was randomly assigned in the range of $10^{-6} \leq \rho \leq 10^{-3}$. The initial Ca concentration was set at 4.95 [mg/dl].

4.2 Boundary conditions

The upper surface of the extraction socket region was set as a free particle as a boundary condition, and the forced displacement as shown in Figure 3 (b) was applied as a mechanical stimulus for analysis. The sides were set as particles with a calcium concentration and bone density of 0.

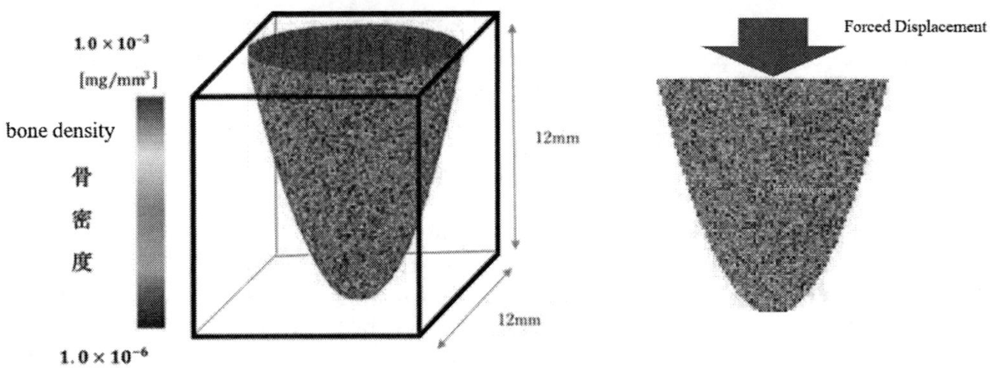

(a) Extraction socket analysis area (b) Cross-sectional view of the analysis area

Figure 3. Extraction socket analysis area

5. Analysis result

5.1 Angiogenesis and vascular regression

In the analysis, the vessels in the region of tooth extraction fossa actively regenerated up to the 15th day after tooth extraction, buried in the bone by the 30th day, and retracted by the 90th day. Thus, the results of the analysis are shown in Figure 3, which represents the above characteristics.

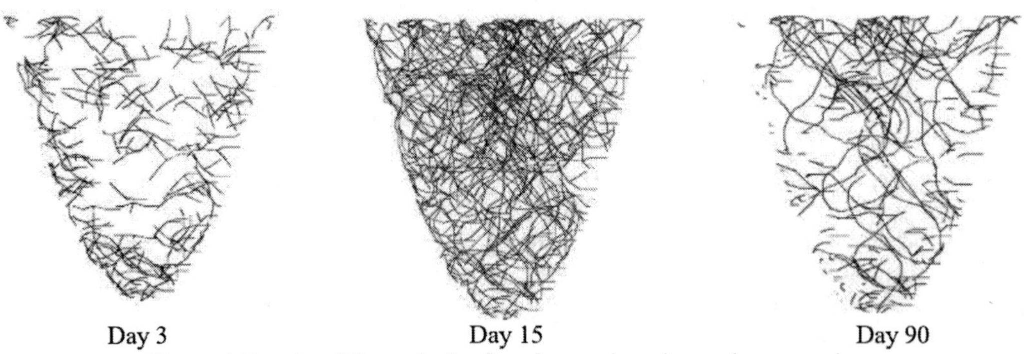

Day 3 Day 15 Day 90

Figure 4. Results of the analysis of angiogenesis and vascular regression

5.2 Ca^{2+} transport analysis

The results of the Ca^{2+} transport analysis (Figure 5) shows that Ca^{2+} was secreted from the vessels and evenly diffused throughout the extraction fossa as time progressed from day 30 (Figure 4 (a)) to day 90 (Figure 4 (b)).

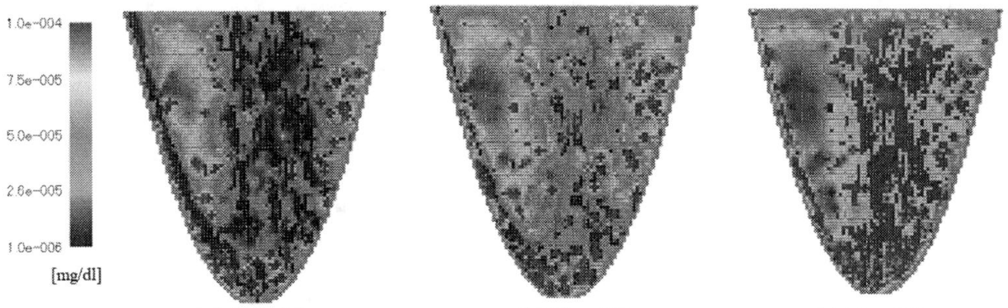

(a) Day 15 (b) Day 30 (c) Day 90

Figure 5. Results of the Ca^{2+} transport analysis

5.3 Bone formation analysis

The results of the bone formation analysis (Figure 6) showed bone formation overtime at 14, 30, and 90 days. These results (Figure 6) corresponded to the experimental results of (Figure 2) and showed agreement in qualitative features.

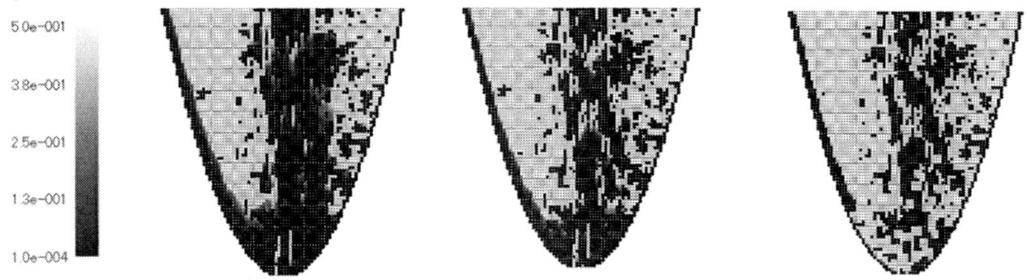

[mg/mm^3] (a) Day 15 (b) Day 30 (c) Day 90

Figure 6. Results of the bone formation analysis

6. Conclusions

1) Alveolar bone regeneration after tooth extraction was reproduced by simulation using particles.
2) In order to simulate the real world, we have formulated a bone formation equation that takes into account calcium transport from blood vessels and mechanical stimulation, which are major factors in bone regeneration. agreement with the realized life.
3) The analysis results of bone formation analysis (Figure 6) were compared with those of the experimental data (Figure 2) and were found to be qualitatively consistent.
4) In the future, we would like to conduct a quantitative evaluation of the simulation results and change the conditions by gender and age.

References

[1] Okudera T Masato M and Iwamiya M 2010 *Journal of Japanese Society of Oral Implantology* **23** 18-25

[2] Masato M Okudera T and Iwamiya M 2011 *Microvascular Reviews and Communications IV* **4** 12-17

[3] Matsuo M Iwamiya M Saito M Todoki K and Kishi Y 2007 *Bulletin Kanagawa Dental College* **35** 25-33

[4] Matsushita A 1992 *Fractals in Medicine and Biology, 1st ed* (Tokyo: Asakura Bookstore) pp 53-64

[5] Hitomi T Takeda N and Iriya K 2009 *Calculation of Calcium Diffusion Coefficient of Cement Hardenings Using Minute Pore Data* (Japan: A house magazine of Obayasi Co., Ltd) 73

[6] Ruimerman R 2005 *Modeling and remodeling in bone tissue* (Eindhoven, Netherlands: Technische Universiteit Eindhoven)

[7] Yoshiki H Adachi T Tezuka K and Tomita Y 2004 *The Japan Society of Mechanical Engineers* **04-40** 327-328

[8] Carter D R and Hayes W C 1997 *The Journal of Bone and Joint Surgery* **59** 954-962

[9] Kitamoto A and Nagayama K 2015 *Proc. the 6th TSME Int. Conf. on Mechanical Engineering* (Hua Hin, Thailand) BME0009

IOP Conf. Series: Materials Science and Engineering 1137 (2021) 012040 doi:10.1088/1757-899X/1137/1/012040

Numerical simulation of skin formation: epidermal stem cells and undulation of the basal layer

Shu Oba[1] and Katsuya Nagayama[1,*]

[1]Graduate School of Computer Science and Systems Engineering,
Kyushu Institute of Technology, 680-4 Kawazu, Iizuka, Fukuoka, 820-8502, Japan

* Corresponding Author: nagayama.katsuya304@mail.kyutech.jp, +81-948-29-7778

Abstract

The basal layer at the base of the epidermis produces new epidermal cells once a week and plays an important role in skin formation. Interestingly, the basal layer undulates, and the amplitude of this undulation is dependent on age or disease. However, the mechanisms driving this phenomenon have not been well studied. This work simulated the skin formation process using an original method to understand the mechanism of the undulation formation in the basal layer. The basal layer has cells with both short and long cell cycles, but only the former has been examined in previous studies. Accordingly, we more accurately simulated the skin formation process by including cells with a long cell cycle and observed undulation of the basal layer. Based on our findings, we propose that the basal layer undulates due to the interaction between two types of basal layer cells.

Keywords: Basal Layer, Undulation, Epidermal Stem Cell, Skin

1. Introduction

The skin is the largest organ in the body and plays roles in water retention and protection from external threats. However, it is extremely difficult to observe the interior layers of the epidermis, as epidermal cells spend about a month before they reach the skin surface. This work simulated the inside of the skin to understand the mechanism of its formation and is expected to contribute to product development in the beauty industry.

This study focuses on the basal layer, which is located at the base of the epidermis. The basal layer consists of progenitor epidermal cells and plays an important role in skin formation. Interestingly, the basal layer undulates, and the amplitude of this undulation is dependent on age or disease [1]. The undulation amplitude in our simulation results does not correspond with that described in the literature because the phenomenon has not previously been studied [2–8]. A goal of this work is to understand the mechanism of the basal layer undulation so that this characteristic may be applied to simulations in the future.

The basal layer contains cells with short and long cell cycles, but only the former has been considered in previous studies [2–8]. We hypothesized that differences in the behavior of these two types of cells could influence undulation formation in the basal layer. Accordingly, we chose to include cells with a long cell cycle in our simulation of the skin formation process.

2. Analysis target

As shown in Figure 1, the skin is mainly composed of three layers: the epidermis, dermis, and subcutaneous tissue. The topmost of these layers, the epidermis, is itself composed of four layers, the bottom most of which is the basal layer. The basal layer plays a role in producing new epidermal cells.

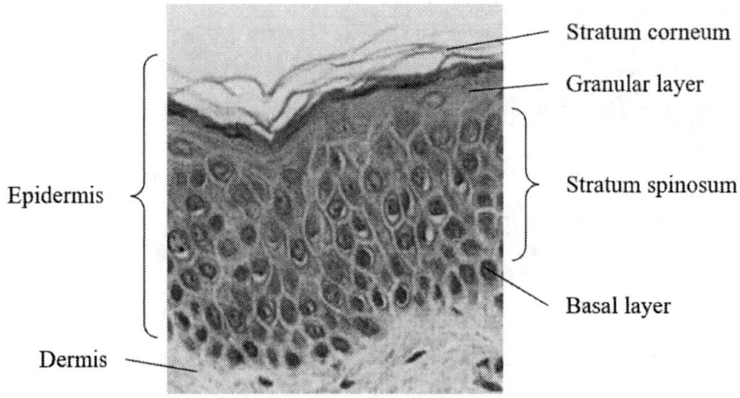

Figure 1. Cross section of the skin [9].

3. Analysis method

3.1 Particle method

The particle method represents a continuum using a finite number of particles and calculates the behavior of the continuum through the movement of the particles. Each moving particle has a characteristic set of variables associated with it. This method does not use the mesh required by the finite difference method or the finite-element method. For this study, the particle method was used because each particle can be regarded as a cell, and the particles can move freely.

3.2 Cell dynamics

Skin cells move by exerting forces on each other. In the case of epidermal cells, the division of basal cells pushes neighboring cells upward. Therefore, for this study, cell-to-cell interactions were represented using a body force and a spring force so that the particles could move by the equilibrium of the forces [2, 3]. Equation (1) shows the body force, and Equation (2) shows the spring force. Since the body force mainly works to push the cells up, the repulsive force is larger than the attractive force. The spring force works to strengthen the binding force. Each cell moves according to Equation (3),

$$F = k\left(1 - \frac{r}{1.106 \times r_1}\right)\left(1 - \frac{r}{1.4 \times r_1}\right) \quad (1)$$

$$f = k'\left(1 - \frac{r}{1.106 \times r_1}\right) \quad (2)$$

$$\Delta x = \alpha\left(\sum F + \sum f\right) \quad (3)$$

where r is the inter-particle distance; r_1 is the cell diameter; k and k' are inter-particle coefficients; α is the movement coefficient.

3.3 Maintenance of basal layer

The basal layer is composed of one cell layer. However, it is difficult to maintain this layer in a simulation by relying only on the force between cells. Therefore, for this study, a method for maintaining the layer is proposed. As shown in Figure 2, it is possible to maintain the layer by placing virtual particles above and below the basal cell particles and exerting a spring force on them.

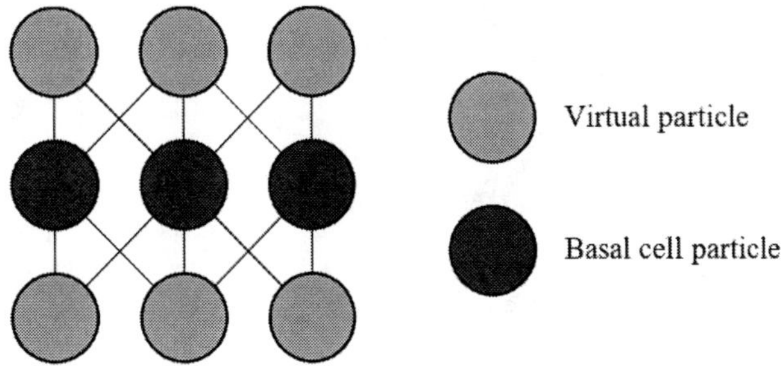

Figure 2. Particle placement for the maintenance of the basal layer.

3.4 Cell division

The basal layer has two types of cells: epidermal stem cells and transit-amplifying (TA) cells. Epidermal stem cells divide about 4–6 times a year [10]. On the other hand, TA cells divide about once a week [10]. As shown in Figure 3, both epidermal stem cells and TA cells have three division patterns. There is about a 10% chance of duplication, an 80% chance of asymmetric division, and a 10% chance of differentiation [10]. For this study, these profiles are applied to two types of cells.

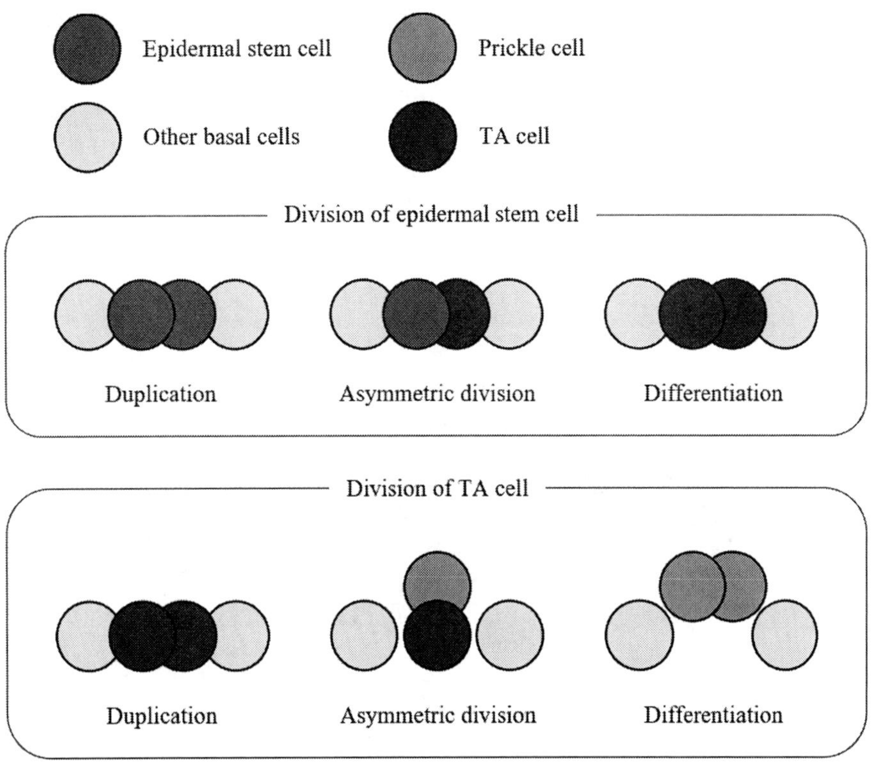

Figure 3. Division of two types of cells.

IOP Conf. Series: Materials Science and Engineering 1137 (2021) 012040 doi:10.1088/1757-899X/1137/1/012040

4. Analysis flow

The analysis flow is shown in Figure 4. The analysis frequency is 2500 times a day, and the analysis period is 300 days.

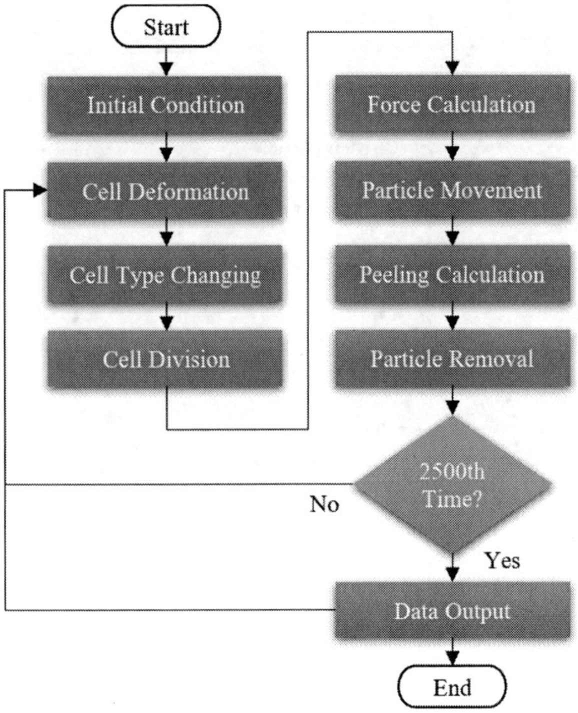

Figure 4. Analysis flow. For cell deformation, the epidermal cells change to thinner as they grow into prickle cells, granule cells, and corneum cells. For cell type changing, after a certain period, the epidermal cells change from prickle cells to granule cells or from granule cells to corneum cells. For peeling calculation, dead corneum cells peel off [4, 5].

5. Analysis conditions

The overall initial placement is shown in Figure 5. Each particle diameter is 10 μm. The analysis area is 300 × 300 × 300 μm^3. A periodic boundary condition was adopted in the horizontal direction. In the vertical direction, particles outside the analysis area were removed.

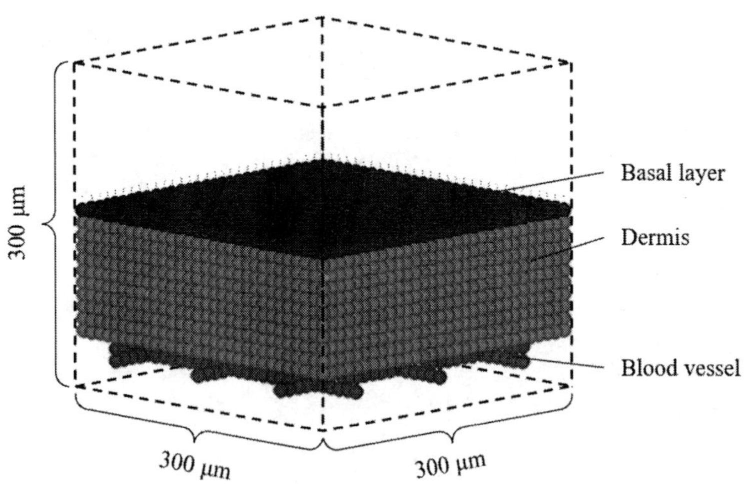

Figure 5. Overall initial placement.

IOP Conf. Series: Materials Science and Engineering 1137 (2021) 012040 doi:10.1088/1757-899X/1137/1/012040

The placement of basal cells is shown in Figure 6. For 100 days after the start of the analysis, the basal layer was simulated using only TA cells and was restrained, which allowed the simulation to reach equilibrium. Then, as shown in Figure 6(a)–(c), three patterns of cell placement were simulated from 100 days to 300 days. In addition, only the basal layer is displayed.

(a) Pattern 1 (b) Pattern 2 (c) Pattern 3

Figure 6. Placement of basal cells. Epidermal stem cells are red, whereas TA cells are blue.

6. Analysis results

The analysis results are shown in Figure 7(a)–(c). The basal layer for each pattern undulated over time. In addition, we noted that epidermal stem cells tended to be distributed more in the convex regions of the basal layers.

The distribution ratio for epidermal stem cells in the convex regions is shown in Figure 8. Over 90% of epidermal stem cells were distributed in the convex regions from 100 days to 150 days. After this point, the distribution ratio for epidermal stem cells in the convex regions generally showed a downward trend, but rose again at around 300 days. From these results, it was shown that epidermal stem cells tended to be distributed in the convex regions of the undulating basal layer, which corresponded with the literature [1].

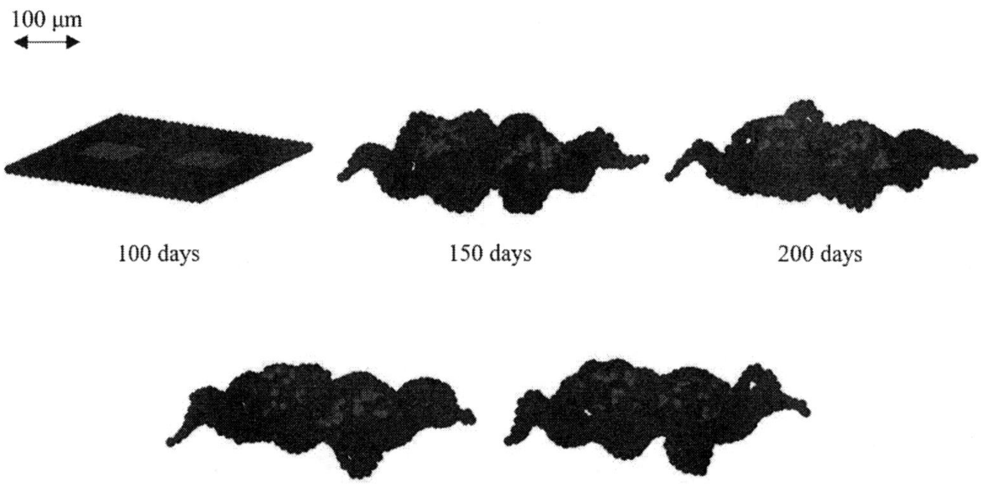

(a) Pattern 1

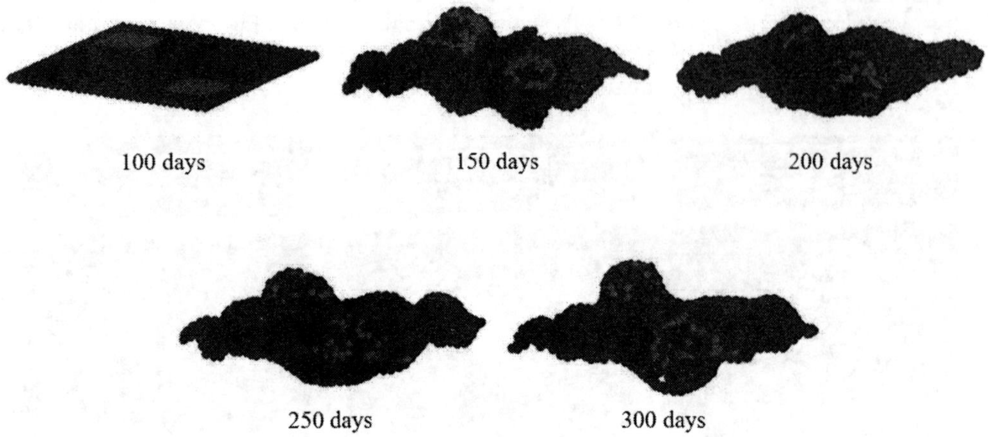

(b) Pattern 2

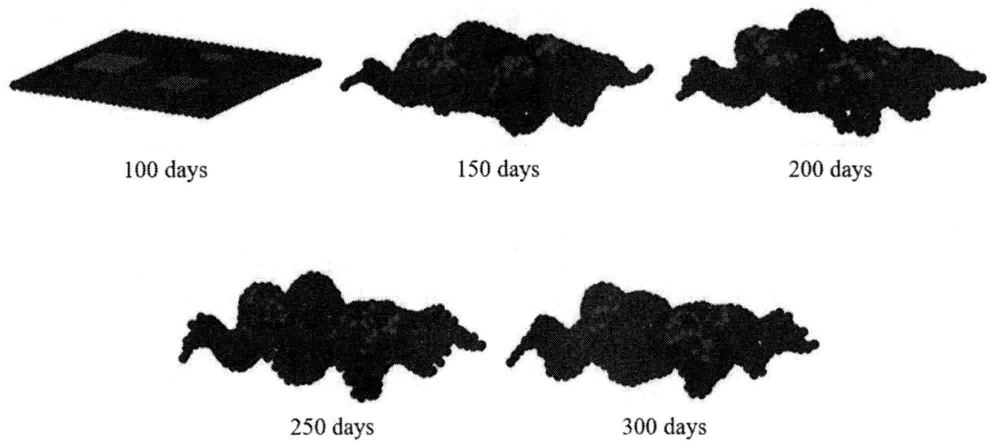

(c) Pattern 3

Figure 7. Analysis results. Epidermal stem cells are red, whereas TA cells are blue.

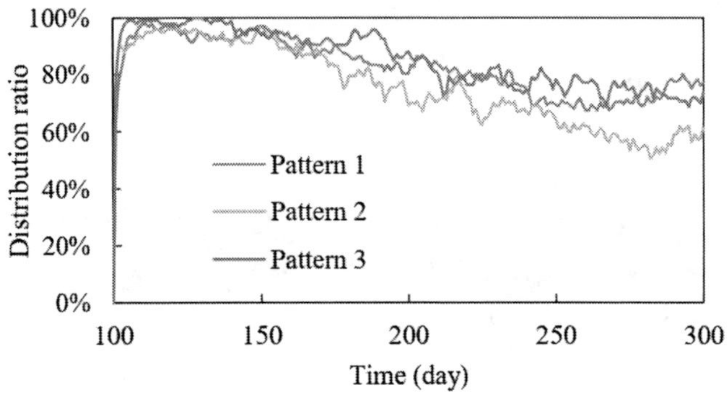

Figure 8. Distribution ratio for epidermal stem cells in convex regions.

7. Discussion

We hypothesized that the reason the basal layer undulates is related to the interaction of two types of basal cell divisions. As shown in Figure 3, epidermal stem cells only divide horizontally. Therefore, epidermal stem cell numbers increase horizontally, which could cause buckling of the basal layer. On the other hand, TA cells divide either horizontally or vertically. Therefore, TA cells may temporarily be pushed down by the generation of prickle cells. In addition, TA cells could frequently be pushed down because these cells divide more often than epidermal stem cells. From these findings, we suggest that basal cells caused the buckling of its layer and that epidermal stem cells and TA cells moved to the convex and the concave regions of the basal layer, respectively, leading to the undulation formation.

8. Conclusions

In this study, we succeeded in simulating the basal layer using epidermal stem cells. We demonstrated that the differences in the cell cycle and the division direction between epidermal stem cells and TA cells contribute to the undulation formation in the basal layer. In addition, we also showed that epidermal stem cells tend to be distributed in the convex regions of the basal layer.

In the future, we will simulate how the undulation amplitude of the basal layer changes with age or disease.

References

[1] Yamada T Hasegawa S Miyachi K Date Y Inoue Y Yagami A Arima M Iwata Y Yamamoto N Nakata S Matsunaga K Sugiura K and Akamatsu H 2018 *Mech. Ageing dev.* **171** 37-46
[2] Kurihara T and Nagayama K 2015 *Proc. 6th TSME Int. Conf. Mech. Eng.* (Petchburi: Mahidol University)
[3] Kurihara T and Nagayama K 2016 *Proc. 7th TSME Int. Conf. Mech. Eng.* (Chiang Mai: Chulalongkorn University)
[4] Shobuda K and Nagayama K 2019 *IOP Conf. Ser.: Mater. Sci. Eng.* **501** 012032
[5] Oba S and Nagayama K 2020 *IOP Conf. Ser.: Mater. Sci. Eng.* **886** 012019
[6] Nagayama K Uehara T Amano Y and Tanahashi M 2015 3D *J. Biosci. Med.* **3** 45-49
[7] Nagayama K Kurihara T Amano Y and Tanahashi M 2016 *J. Biosci. Med.* **4** 33-77
[8] Nagayama K and Kurihara T 2018 *J. Appl. Math. Phy.* **6** 1757-1762
[9] Saita T 2009 *Introduction to Dermatopathology Diagnosis 2nd ed* (Tokyo: Nankodo) pp 2-13
[10] Mascré G Dekoninck S Drogat B Youssef K K Brohée S Sotiropoulou P A Simons B D and Blanpain C 2012 *Nature* **489** 257-262

Influence of axial force and characteristics of axial force affecting on total knee replacement constraint test

Tanapat Musikul[1], Usanee Kitkamthorn[1], Pusit Mitsomwang[1], Bura Sindhupakorn[2], Rattana Borrisutthekul[1,*]

[1]School of Metallurgical Engineering, Institution of Engineering, Suranaree University of Technology, Nakhon Ratchasima, 30000, Thailand
[2]Institute of Medicine, Suranaree University of Technology, Nakhon Ratchasima, 30000, Thailand

* Corresponding Author: rattana@g.sut.ac.th

Abstract
The result of testing for determination of Total Knee Replacement (TKR) Constraint (ASTM F1223) is necessary for surgeons for matching the knee prosthesis with the different degrees of severities of knee osteoarthritis patients. In the knee replacement constraint test, the axial compressive load and translation displacement are the control and the adjusting variables. The magnitude of axial compressive load during testing is affected by the slope of knee bearing surface. According to ASTM 1223 standard, the axial compressive load is not clarifyingly mentioned. Namely, it should be equal 710 N during testing or only in state of initiation. It may lead to testers to perform the test into different two modes which consequently in possibly different constraint result. Therefore, the objective of this research is to clarify the differences in test results when applying two axial compressive load characteristics during testing. The six-axis testing machine (Prosim Universal Joint Simulator) was used to determine the anterior posterior constraint (AP) of fix bearing posterior sacrifice (FB PS) system. Bovine serum was directly applied to the bearing surfaces as a lubricant. An initial axial compressive load 710 N was used. Two axial compressive load characteristics during testing, 1) keeping the axial force at given value as mentioned during testing (load-controlled testing) and 2) with allowing the change in axial compressive load during testing (displacement-controlled test), were used. From the results, the AP load resistance hysteresis loop patterns obtained by both testing modes were similar. The magnitude in forward direction load resistance is higher, while the backward AP load resistance showed the lower, in case of displacement control mode. It resulted in span between AP load resistance was larger. Moreover, the higher constraint in case of displacement control was indicated. Thus, if the test modes are not mentioned in test result, the surgeons will be possibly choosing the wrong TKR for their patient.

Keywords: Total knee replacement (TKR), TKR constraint, TKR testing

IOP Conf. Series: Materials Science and Engineering 1137 (2021) 012041 doi:10.1088/1757-899X/1137/1/012041

1. Introduction

Before knee replacement surgery, patients are required to undergo an assessment of their knee osteoarthritis condition, in order to match the appropriate total knee prostheses. The available bone stock and soft tissue capacities for stabilization, are the parameters surgeons use to assess knee osteoarthritis conditions [1]. For example, patients with good soft tissue restraints, may require a lesser constraint prosthesis. Whereas patients with major bone loss or destroyed ligamentous structures, may require a prosthesis with a higher degree of constraint. Measurements for the constraint of total knee prosthesis are defined according to the Standard Test Method for Determination of Total Knee Replacement Constraint (ASTM F1223). According to ASTM F1223-14, the axial compressive load of 710 N is mentioned to apply during the test. Also, both bovine serum and deionized water is allowed to use as lubricant between the knee insert and the femoral component. In constraints testing, the femoral component and the tibial component will be moved relatively in only the testing direction, while other directions movements will not be allowed. From the year 2005 until present, there are many studies aiming to evaluate the disadvantage of the current TKR constraint test. Firstly, Moran et al. (2008) [2] assessed the constraint of the TKR device experimentally in order to validate their computer simulation technique developed based on the ASTM guideline.

In their study, the axial compressive load of 200 N was applied initially by using the MTS 858 Bionix test machine. However, after testing, they found that the measured axial compressive load found fluctuated. Secondary, Haider and Walker (2005) [3] used the test methods base on ASTM F1223 of 2005 version to assess the constraint of three commercial knee designs. However, their test was conducted by application of the constant axial compressive load. Under Haider and Walker's works, the secondary motions were discovered. Thirdly, Halewood et al. (2018) [4] conducted the AP constraint test with Instron 5565 that also allowed to control the constant axial compressive load from 700 N and 2000 N during testing. From Halewood et al. study, it was found that a secondary motion was also occurred. Moreover, the different axial compressive load resulted in the different constraints result. From the above studies, we could realize that the axial compressive load characteristics should have the effects on the constraint results. However, no research works have been performed to understand the effect of axial load characteristics on the constraint result. Moreover, the ASTM standard itself does not include guidance on the axial load characteristics. It mentions only using the axial compressive load at the set point of 710 N. Thus, the current research work aims to clarify the differences in constraint test results when applying two axial compressive load characteristics during testing.

2. Materials and Methods

The tests will be done in similarity with the ASTM-F1223 (2014) standard. Any changes of standard have been highlighted in the sections that follow.

2.1 Test equipment setup

The six-axis testing machine (Prosim Universal Joint Simulator) was used as the testing machine. Fixtures were designed and were produced to be suitable with the femoral and the tibial components used in the study. After that, the femoral component and tibial component were mounted to the fixture using polymethylmethacrylate (PMMA) bone cement as shown in Figure 1. Then, the tibial insert was locked up on the tibial tray as shown in Figure 1.

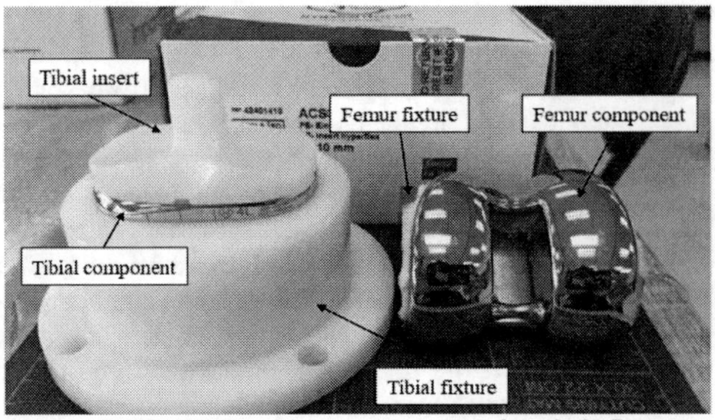

Figure 1. The TKR components mounted to their fixtures.

Later, fixture mounted with the femoral component was installed to the top part of the testing machine, which the flexion angle and varus-valgus (VV) rotation can be adjusted. Also, the tibial component mounted on fixture was installed on the lower part of the testing machine that anterior-posterior (AP) translation, medial-lateral (ML) translation, and internal-external (IE) rotation can be controlled. All the setting of the parts on testing machine is shown in Figure 2.

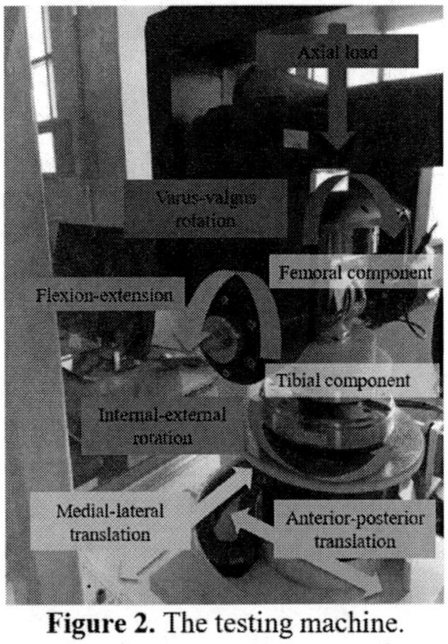

Figure 2. The testing machine.

In the next step, the determination of neutral point of the tibial component relative with femoral component was set up. In neutral point setup, the 100 N axial compressive load was applied to the TKR. Then, tibial component was moved in AP direction. The lowest point of the tibial insert compared to femoral component was noticed. The lowest point is assigned to be the neutral point. Then, the pressure film was inserted between the femoral component and the tibial insert to identify the contact point where femoral component touch to tibial insert at difference flexion angles. The contact point measurement technique with pressure film was presented in TMETC12 [5].

For determining AP displacement limit, which was used to program in controlling the test, manual AP movement in both AP forward and backward direction of the tibial component withholds others movement direction was done. The AP movement was continuously done until the conditions of 1) slightly increasing or decreasing AP load with increasing displacement and 2) dramatically AP load

IOP Conf. Series: Materials Science and Engineering 1137 (2021) 012041 doi:10.1088/1757-899X/1137/1/012041

change. The maximum movement in forward and backward AP direction is assigned to constraint limit of AP movement. The determining AP limit of the TKR with 0°, 30°, 60° and 90° of knee flexion angles were shown in Table 1.

Table 1. The AP displacement limit

Knee flexion angles	Anterior translation (mm)	Posterior translation (mm)	Total displacement (mm)
0°	+1	-3	4
30°	+3.6	-7	10.6
60°	+9	0	9
90°	+3	0	3

2.2 Implant & Lubrication
The constraint test was conducted on fix bearing posterior substituted (FB PS) system of TKR. The TKR used had the symmetrical condylar geometries. The lubrication used was the bovine serum.

2.3 Test method
Before testing, bovine serum was applied onto the tibial insert surface [6]. In the constraint test, two experiments; 1) experiment to examine the hysteresis loop of AP load resistance and 2) experiment to determine AP constraint were carried out. Two experimental conditions 1) Control axial load to be at 710 N during the initial setting and during the test (load-controlled test), and 2) fix axial displacement after setting axial load at 710 N during the initial setting (displacement-controlled test) were studied.

For examine the hysteresis loop of AP resistance load, the AP displacement limit was applied and was used to program the machine. The AP motion rate during the test is 1 mm/sec and other movement direction is prohibited during the test. The AP resistance load correlated with movement distance was obtained and was used to construct the hysteresis loop of AP resistance load. After that, determining AP constraint was measured from the hysteresis loop base on the slope of the curve became constant or change dramatically

3. Results and Discussion

3.1 The contact point of TKR at various flexion angles
The contact point of various flexion angles was defined with the red area on the film as shown in Figure 3.

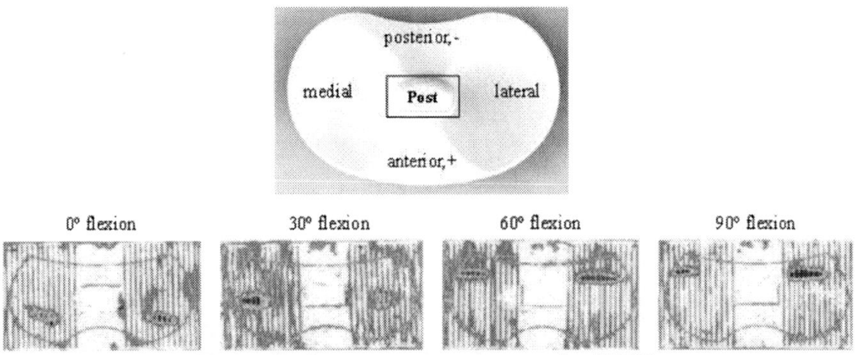

Figure 3. the contact pressure result of TKR.

From Figure 3, the tibial insert shape was also mapped to tested film in order to recognize the position of contact area on TKR. At 0° flexion angle, the contact point is near to the anterior side of tibial insert, while increasing the flexion angle, the contact point was moved closer to the posterior side.

3.2 The hysteresis loop of AP resistance load

Before going to the AP resistance load result, the axial compressive load in both testing conditions with different test angles during the test is revealed as shown in Figure 4.

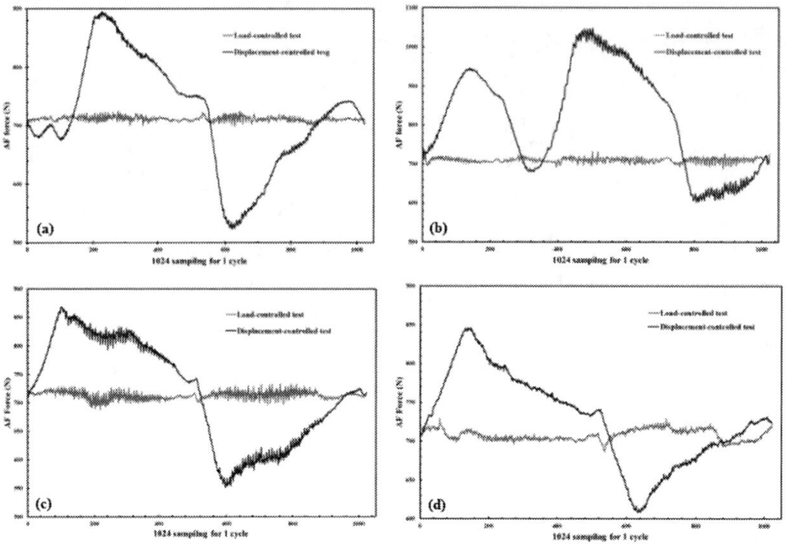

Figure 4. The axial compressive load during the test; (a) 0° flexion angle, (b) 30° flexion angle, (c) 60° flexion angle and (d) 90° flexion angle

From Figure 4, the characteristics of axial compressive load in two test conditions are not similarity. In load-controlled test (Orange line), the constant load characteristic throughout the test was observed, while displacement-controlled test (Blue line) gives the fluctuation of axial load characteristic. From axial load characteristic result, we could refer that two test conditions were successes controlled. Moreover, in the displacement-controlled testing, the fluctuation of axial load was occurred because the femoral component was fixed in the axial position during test, and when it was moved on the curve of tibial insert the axial load was changed. On the other hand, the load-controlled testing provides the constant axial load compressive because automatically adjusted by machine

The hysteresis loop of AP resistance load was the plot between AP displacement and AP resistance. The hysteresis loop of AP resistance load for the tests as shown in Figure 5. From Figure 5, the hysteresis loop of load-controlled test (Orange line) was a significantly narrower span than that of the displacement-controlled test (Blue line). Moreover, it can be seen that, in the displacement-controlled test mode, the AP resistance load is higher magnitude than that of the load-controlled test mode in both forward and backward movement. It is due to that the higher axial compressive load acting on the TKRs results in increasing the friction between femoral component and tibial insert and finally affects to increase AP load resistance force

3.3 The AP constraint

Determination of the AP constraint was performed with using the hysteresis loop of AP constraint. According to the ASTM stipulates, "dislocation of the components is imminent. . . or if a dangerous or unrealistic situation is about to occur was mention". Then, the AP constraint was defined as the point at which the force-displacement graph started to plateau or overshoot (the slope of the curve became constant or dramatically change) as also shown in Figure 5. The constraint evaluation was manually chosen.

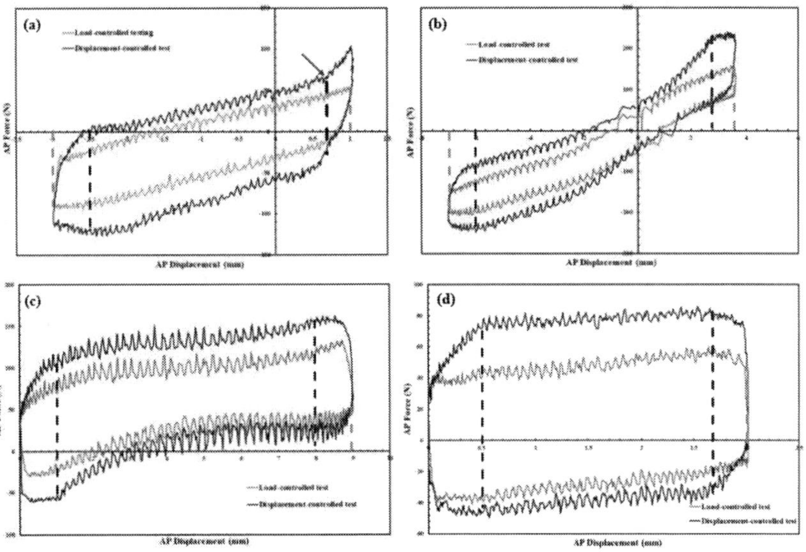

Figure 5. Determination of AP constraint; (a) 0° flexion angle, (b) 30° flexion angle, (c) 60° flexion angle and (d) 90° flexion angle

For example, in Figure 5(a), the blue dash line with red arrow was assigned by the dramatically change of slope of graph. The blue line indicates the constraint in the moving direction. The AP constraint results are summarized in Table 2. From Table 2, at 0 flexion angle, the total movable length in AP direction of TKR is 4 mm for load-controlled test and 3.2 mm for displacement-controlled test. When increasing flexion angle, the total movable length in AP direction was increasing until flexion angle of 30°. Then, when higher than 30° flexion angle was used to study, the total movable length in AP direction was reducing. Moreover, the trend of total movable length in AP direction various with flexion angle from both test condition is similar.

Table 2. The AP constraint results

Knee flexion angles	Anterior translation (mm)		Posterior translation (mm)		Total translation length (mm)	
	Load controlled	Displacement controlled	Load controlled	Displacement controlled	Load controlled	Displacement controlled
0°	+1	0.7	-3	-2.5	4	3.2
30°	+3.6	2.8	-7	-6	10.6	8.8
60°	+9	8	0	1	9	7
90°	+3	2.7	0	0.5	3	2.2

However, it found that the AP constraint in case of the load-controlled test is narrower than the displacement-controlled test in all test flexion angles. It is to that the fluctuation of axial compressive load in case of displacement-controlled test effect on easier change in AP load resistance.

4. Conclusion

According to the results, the tests with different axial load characteristics gave different constraint results of total knee replacement. Mostly, the displacement-controlled test gives a higher constraint than the load-controlled test mode, even though the same TKR was used. Thus, in order to correctly understand constraint results, the remarks of axial load type (constant/non-constant) should be included in constraint test reports, for the correct consideration of surgeons.

Acknowledgment
The authors gratefully acknowledge the Health Systems Research Institute (HSRI) In support of the research work.

5. References

[1] Kakarlapudi T K and Bickerstaff D R 2000 *Br. J. Sports Med.* **34** 395-400.
[2] Moran M F Bhimji S Racanelli J and Piazza S J 2008 *J. Biomech.* **41**(9) 2013-2020
[3] Haider H and Walker P S 2005 *J. Biomech.* **38**(2) 341–348.
[4] Halewood C Athwal K K and Amis A A 2018 *J. Biomech.* **73** 153-160
[5] Musikula T Mitsomwanga P Sindhupakornb B and Borrisutthekula R 2019 *TMETC12*
[6] Emely L B Baptiste C and Roman H 2015 *Lubricants* **3** 664-686

Experimental investigation on durian thorns

Bundit Phungsara, Ekkarin Phongphinittana, and Petch Jearanaisilawong*

Department of Mechanical and Aerospace Engineering,
King Mongkut's University of Technology North Bangkok,
1518 Pracharat 1 Rd Wongsawang Bangsue, Bangkok, 10800, Thailand

* Corresponding Author: petch.j@eng.kmutnb.ac.th

Abstract
Durian is a fruit that has a round shape with thorny shell. When a durian is ripe and falling down from a tall branch to the ground, its thorn help absorb shock from an impact with the ground and lessen the damage to its soft internal flesh. This research aims to investigate the structure of durian thorn and its mechanical properties. Among varieties of durian cultivars, Monthong durian is selected as a representative variety because of its popularity and availability. Thorns and components of durian peels are harvested from a ripened durian and stored in vacuum refrigerated bags. Modulus of Elasticity of each component is measured by performing uniaxial tensile tests on coupon samples. The results show the gradients of modulus of elasticity at various points on the durian peels. The experimental results can be used as inputs for finite element simulations of durian thorns for further analysis.

1. Introduction

Durian is a fruit native to Indonesia. It is a thorny oblong fruit with aromatic soft yellow flesh, popular among Thai and foreign consumers. While numerous species of durian are found in Thailand, six varieties are grown commercially: Gan-yao, Monthong, Cha-nee, Kra-dum, Long-Lub-Lae, Puangmanee. Among them, Monthong is the most popular variety and commercially available.

Aside from the distinct aroma, durian thorns are the most prominent feature of this fruit. The thorns function as a protective mechanism and a deterrent to natural predators. In addition, when a durian is ripe and falling from the tree, the thorny part of durian sinks into the ground to reduce the effects of impact on the inside flesh. Thus, the structure of its thorns is naturally designed to withstand an impact load.

Durian thorns begin developing from its flower. Once pollinated, the ovary within a flower transforms into a bud of greenish skin with pimples on the surface - as shown in Figure 1a. As the bud grows, the pimples elongate into hardened thorns (Figure 1b). The thorn base is enlarged as the fruit expands in size. The ripened durian is shown in Figure 1c.

A durian thorn has a pyramid-shaped structure with a polygonal base. Thorn sizes depend on the size of the fruit as well as locations on the shell. These thorns are joined at the base. Durian thorns can be divided into several types such as pyramid or cone, with polygonal bases. The tips of these thorns can be hardened and pointy, while others are crooked.

Figure 1. Growth of Durian (a) Durian bulb at 1 month (b) Durian fruit at 2 months (c) Durian fruit at 4 months.

Previous studies on durian thorn structure as impact absorbers are limited due to its exotic nature. Ha et al., [2] studied the structure of durian peels by partitioning into three layers: exocarp, mesocarp and endocarp corresponding to external surface of the thorn, soft flesh inside the thorn, and a thin layer under the thorn, respectively. Compressive tests on individual thorns were performed to measure their strengths. Bulk properties of the thorns were also investigated by compressing rectangular samples of 4, 9, 16 and 25 thorns. Results of the stress tests showed that the stress curves were divided into three phases: the first phase was when the stress-strain curve increased linearly, the second phase when the samples deformed at relatively constant stress, and the third phase when stress increased significantly due to compaction of the thorns. These results show that the deformation energy increases as the number of thorns increases but tapers off after 16 thorns.

A related study was the characterization of jujube branches by Wang et al. [8]. Nanoindentation tests and uniaxial tension tests were performed to measure modulus of elasticity and tensile strength along longitudinal and transverse directions of the branches.

Studies of material structures resembling a thorn of durian under impact loads were performed by Ahmad et al., [3-4]. In these works, thin-walled pyramid- or cone-shaped tubes and those with internal foam inserts were subjected to compressive and impact loads. Effects of taper angles on the strength of tubular cones were investigated. They showed that a tube with a taper angle of 15 degrees had the highest strength. Also, thin-walled cones with foam inserts can absorb impact energy more than hallow thin-walled cones. These studies demonstrate that a composite structure with a hard outer layer and a soft core can be used for absorbing impact energy.

Inspired by previous studies that use structures similar to durian thorns as impact absorbers, this research focuses on characterizing mechanical properties of durian thorns and components of durian peels. The results can lead to an insight into underlying structural design and arrangement of durian thorns that are responsible for the impact absorption capability.

2. Characterization of Durian Peels Structure

Durian is an oblong ellipsoidal fruit covered by thorns on the outer. Inside a durian is divided into several carpels that house arils (flesh) and seeds. A central spine connects all the carpels. For clarification, longitudinal direction of the fruit is defined as the direction along the central spine and transverse direction is perpendicular to the central spine as shown in Figure 2a. A carpel is shown in Figure 2b.

This research focuses on the structure of durian peels. An inspection on micrographs of the peels shows two distinct layers divided by a thin boundary: an outer layer that consists of hardened thorns and their bases, and an inner layer that represents the soft white locule that houses the durian flesh. The layers of durian peels are shown in Figure 3.

Figure 2. Durian fruit (a) Direction of durian (b) A durian carpel.

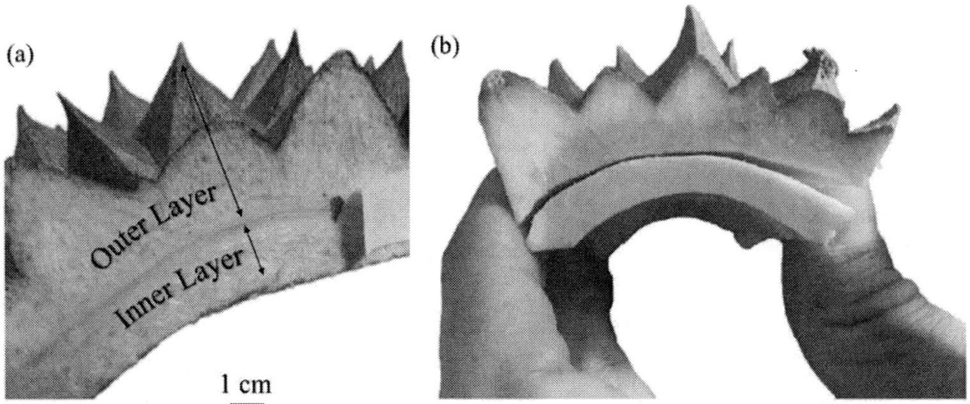

Figure 3. (a) Layer of durian peels (b) Cut section of outer and inner layers.

The outer layer is categorized by hardness into three substructures: Thorn tip - a dry hard apex, Thorn skin – its greenish exterior shell, and Thorn core – a white soft material under the skin. In comparison, the inner layer is divided by the texture into two substructures: a soft locule and a stretchy thin membrane attached to the flesh of the durian. Figure 4 shows the substructures of durian peels.

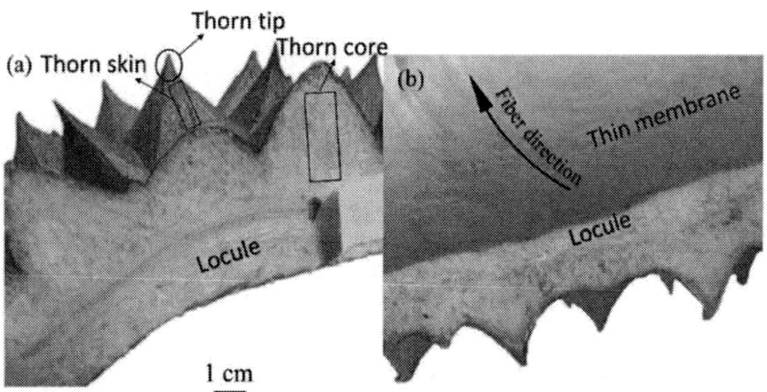

Figure 4. (a) Structure of outer layer (b) Structure of inner layer

3. Mechanical Testing of Durian Peels

Mechanical properties of durian peels' substructures are investigated. Due to size limitation, the mechanical properties of thorn tip are not covered in this investigation. The properties of four other substructures were tested using uniaxial tensile tests. A Monthong durian was selected as a test sample because of its availability. This durian had a total mass of 7.5 kg and a shape as close to an oval. Durian peels were cut by a thin knife to rectangular pieces with 60 mm width and 60 mm length. After cutting, the flesh was removed. The durian peels were further cut into outer and inner layers as shown in Figure 3b. The durian thorns on the outer layer were dissected into thorn skin and thorn core.

The thorn skin samples had fibers directed along its length, and they were dryer than other structures. The thorn core had similar structure to the thorn skin but exhibited more porosity. All of the fibers in the thorn core align lengthwise from the thorn tip to the base. Samples of the substructures were harvested from these pieces as shown in Figure 5. The arrows in Figure 5 indicate the directions of fibers within the substructures. The cut samples reveal that a thorn is a composite structure reinforced by fibers along its length. The inner layer is characterized by a composite material with long fibers in the transverse direction as shown in Figure 4b. Samples of the inner layer along both longitudinal and transverse directions were prepared to investigate the effects of anisotropy. Figure 5c-5f show the samples of locule and membrane layers along both longitudinal and transverse directions. Due to the uneven nature of thorns and substructures, the samples were cut into rectangular coupons with varying dimensions as listed in Table 1.

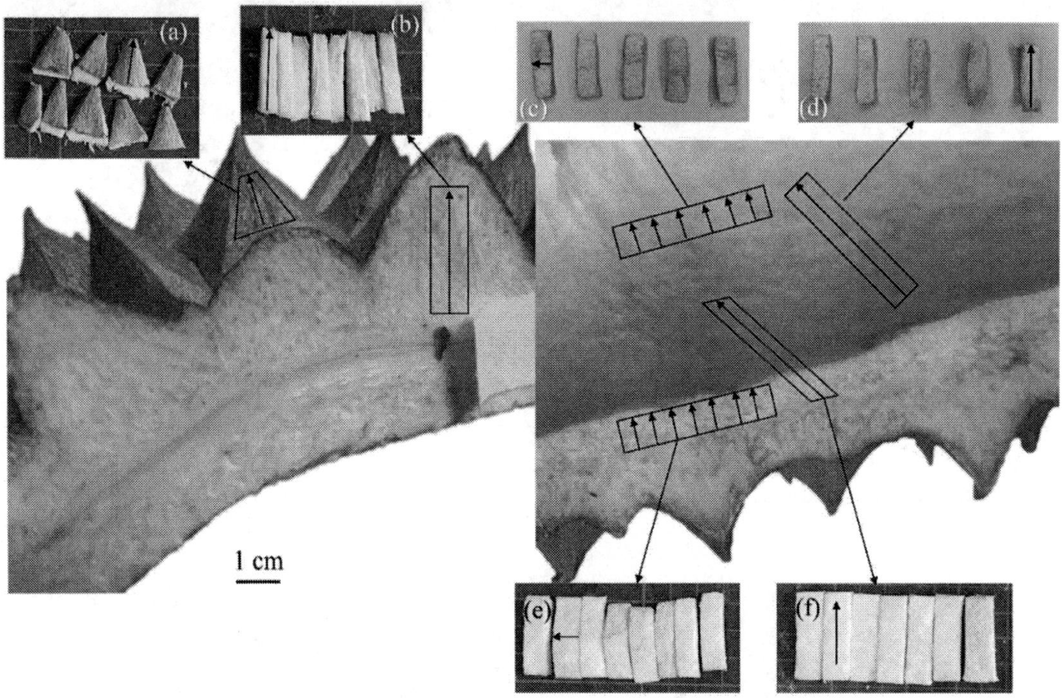

Figure 5. Samples harvested from a durian peel (a) Thorn skin (b) Thorn core (c) Locule layer along longitudinal direction (d) Locule layer along transverse direction (e) Membrane layer along longitudinal direction and (f) Membrane layer along transverse direction

After preparation, the samples were stored in a controlled environment at 70% humidity and 25°C for 24 hours before being tested. Prior to testing a sample was bonded to a pair of wood panels by epoxy resin to increase gripping area and avoid premature failure at the grips. Figure 6 shows representative samples for tensile test.

Table 1. Dimensions of all samples (unit in millimeter)

	Sample	W	t	Lo		Sample	W	t	Lo
Thorn skin	No.1	3	1	8.5	Thorn core	No.1	1.9	1.3	11
	No.2	2.8	1.3	8.5		No.2	1.9	1.2	11.2
	No.3	3	0.7	8.5		No.3	1.9	1.5	11.3
	No.4	3.1	1	8.5		No.4	2.2	1.4	11.3
	No.5	2.6	0.8	8.5		No.5	2	1.7	11.3
Locule longitudinal	No.1	6.6	1.3	15	Locule transverse	No.1	9.9	1.5	20
	No.2	6.3	1.5	14.7		No.2	9.3	1.2	19.8
	No.3	6.7	1.4	15		No.3	10.3	1.4	20
	No.4	6.7	1.4	14.7		No.4	9.7	1.8	20
	No.5	7.3	1.5	14.6		No.5	10	1.2	20
Thin skin membrane longitudinal	No.1	7.1	0.5	20.5	Thin skin membrane longitudinal	No.1	8.4	0.6	20.5
	No.2	8	0.6	20.5		No.2	8.9	0.5	19.8
	No.3	8.5	0.6	20		No.3	9.2	0.6	20.5
	No.4	7.6	0.5	20.2		No.4	10	0.5	20
	No.5	7.9	0.6	20.7		No.5	9.4	0.5	20.5

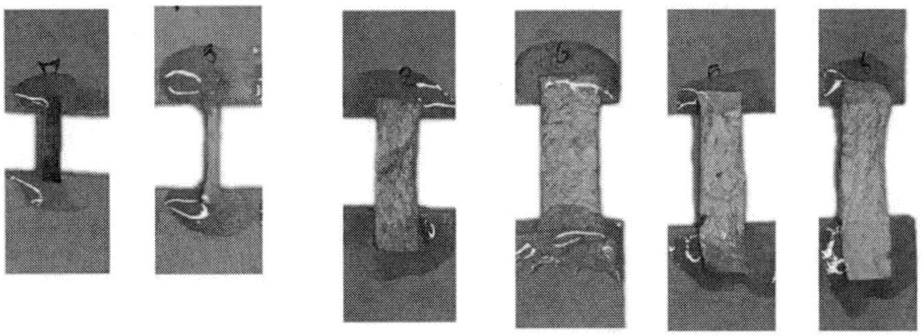

(a) (b) (c) (d) (e) (f)
1 cm

Figure 6. (a) Thorn skin sample (b) Thorn core sample (c) Longitudinal locule layer sample (d) Transverse locule layer sample (e) Longitudinal membrane layer sample and (f) Transverse membrane layer sample

The tensile tests were performed on a standard universal testing machine at a rate of deformation of 1 mm/min. Strain was directly calculated from the crosshead displacement. Tests were conducted until the samples failed.

4. Results and Discussions

Stress-strain plots of thorn skin, thorn core, locule along longitudinal direction, locule along transverse direction, membrane along longitudinal direction and membrane along transverse direction are shown in Figure 7 to 12, respectively. Modulus of elasticity, ultimate tensile strength and elongation at break of each substructure are listed in Table 2. The thorn skins show a linear stress-strain response to failure. In comparison, the thorn cores exhibit an initially linear followed by a softening response. The thorn cores have lower modulus but higher elongation at break compared to the thorn skin. This is because the thorn cores have a porous structure whereas the thorn skins are homogeneous.

IOP Conf. Series: Materials Science and Engineering 1137 (2021) 012042 doi:10.1088/1757-899X/1137/1/012042

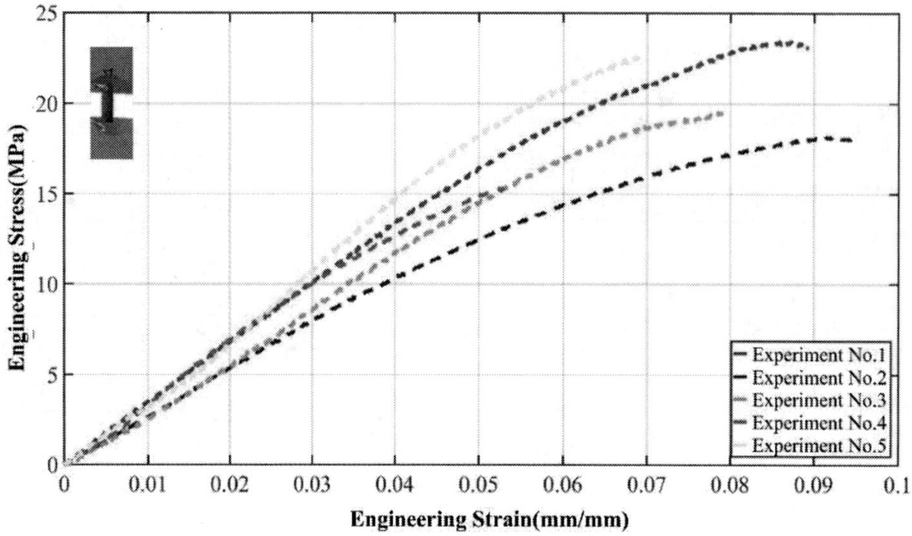

Figure 7. Stress strain curve of thorn skins

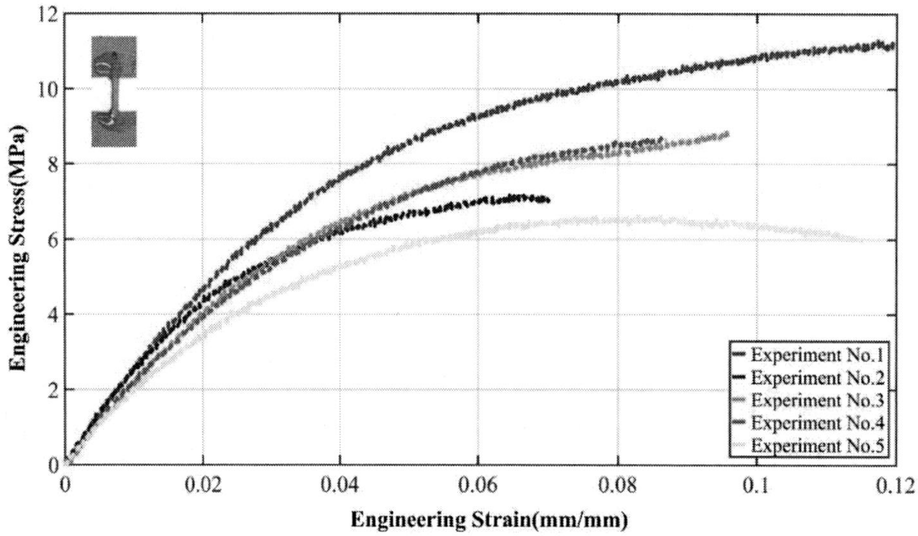

Figure 8. Stress strain curve of thorn cores

The stress-strain response of durian locule along both directions exhibit a linear behavior similar to the thorn skin. Comparisons among the responses of the two directions show that the transverse direction is stiffer than then longitudinal direction because the reinforcing fibers align along the transverse direction. This confirms material anisotropy of the locule.

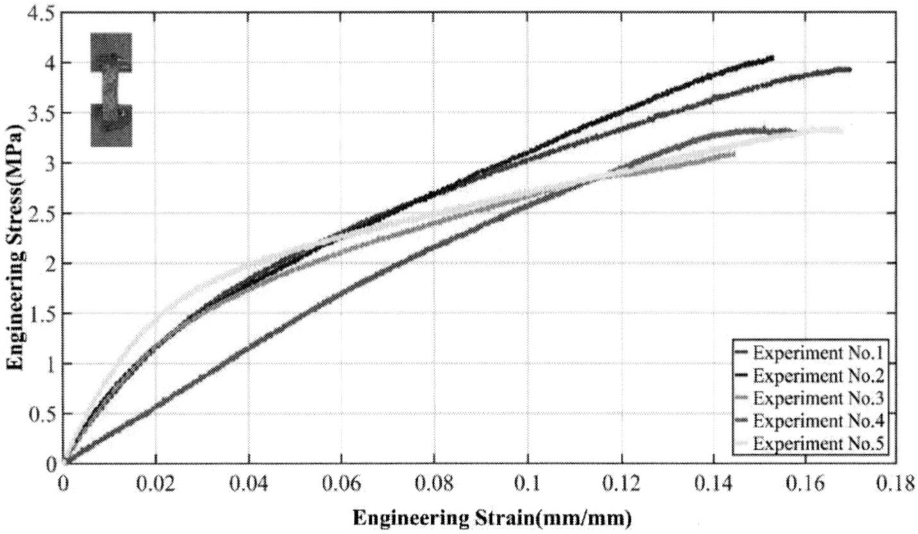

Figure 9. Stress strain curve of locules along longitudinal direction

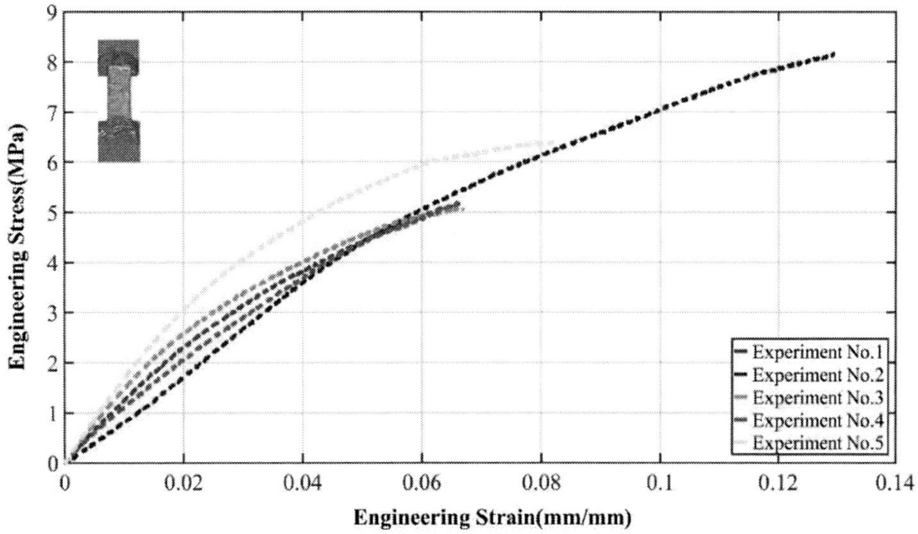

Figure 10. Stress strain curve of locules along transverse direction

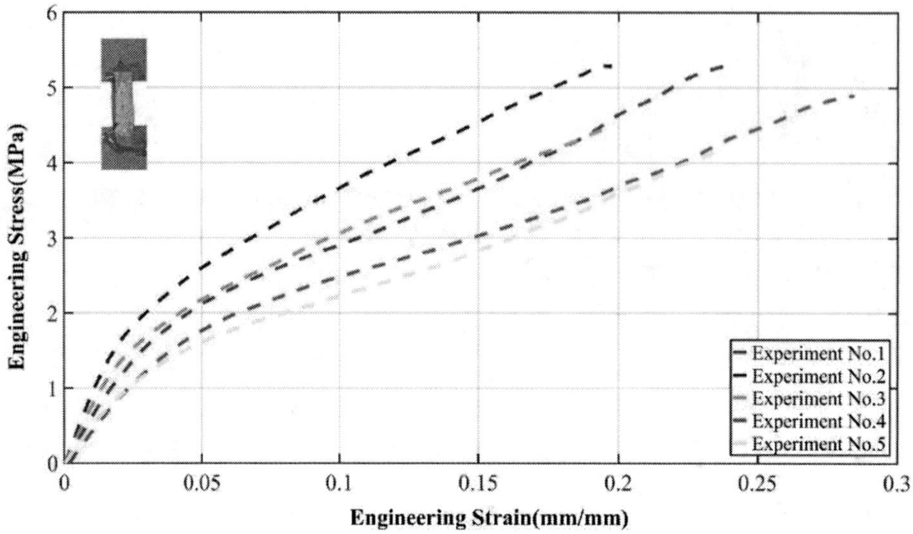

Figure 11. Stress strain curve of membrane along longitudinal direction

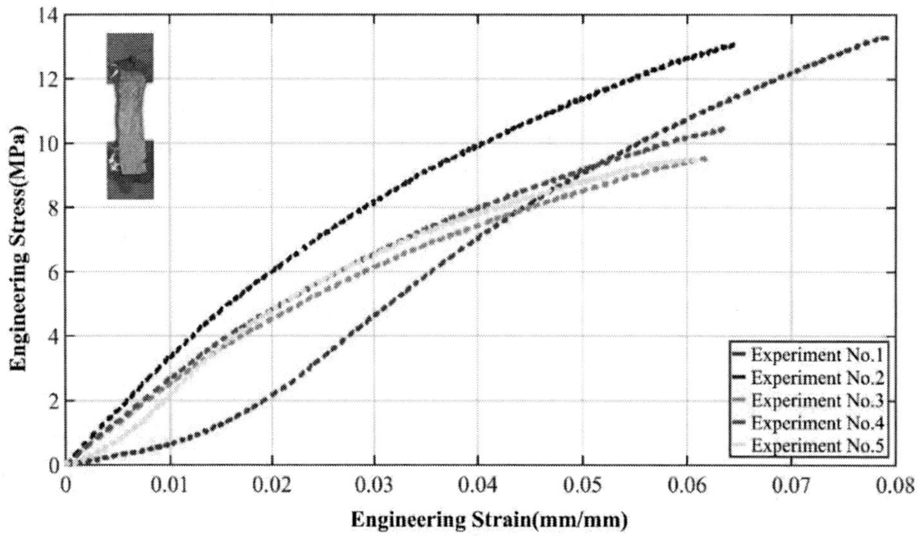

Figure 12. Stress strain curve of membrane along transverse direction

The responses of durian membranes are also similar to the locules. However, the modulus and tensile strength of the membranes are higher despite being thinner layers than those of the locules. The average ultimate tensile strength of the membrane along the longitudinal direction is lower than that along the transverse direction because the reinforcing fibers in the membrane layers appear to be along the transverse direction as well.

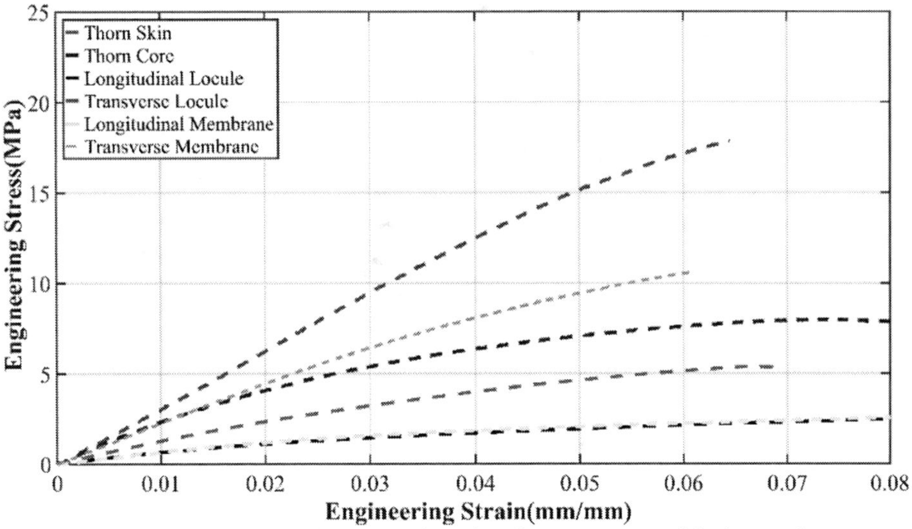

Figure 13. Averaged stress and strain of components of durian peels

Table 2. Mechanical properties of components of durian peels

Sample	Modulus (MPa)	Tensile Strength (MPa)	Maximum elongation (mm/mm)
Thorn skin	307.188±39.5	20.070±3.109	0.085±0.010
Thorn core	216.714±23.611	8.527±1.818	0.098±0.020
Locule Longitudinal	58.864±18.693	3.610±0.426	0.193±0.013
Locule Transverse	121.521±29.304	6.019±1.323	0.100±0.030
Membrane Longitudinal	62.133±17.037	4.876±0.534	0.232±0.034
Membrane Transverse	221.908±70.523	11.249±1.876	0.070±0.011

Figure 13 shows the averaged stress-strain relation of all six substructures of a durian peel. Among these components, the thorn has the highest modulus and the locule along longitudinal direction has the lowest because thorn skin is the least porous structure whereas the inner parts of the peels are less dense as shown in Figure 14. Fiber directions also influence the mechanical responses of the locule and the membrane layers. The average modulus of the thin membrane is higher than the locule because the locule layer appears to be less dense than the membrane.

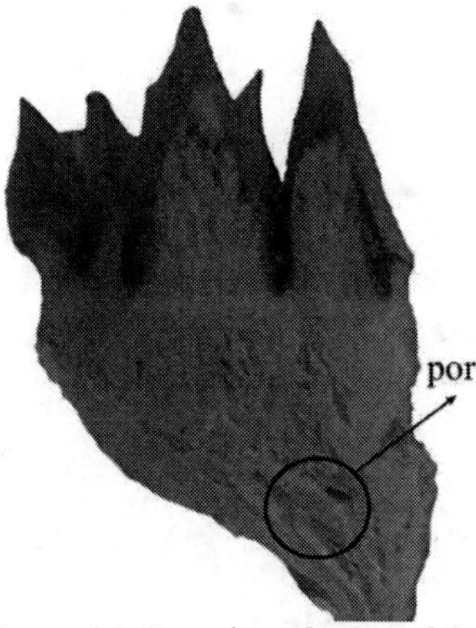

Figure 14. Porosity of a cut of durian peel

5. Conclusion

In this study, the mechanical properties of components of a durian peel were investigated through tensile tests. The modulus of components of a durian peel ranks, from the highest to the lowest, and from the outer to inner layers. Factors that affect the mechanical properties of durian peel are alignment of underlying fibers and porosity. The locule and the membrane layers exhibit stiffer responses along the transverse direction of durian fruit. The substructure with a higher porosity has less modulus than a denser one.

References

[1] Manshor M R Anuar H Aimi M N N Fitrie M I A Nazri W B W Sapuan S M El-Shekeil Y A and Wahit M U 2014 *J. Materials and Design* **59** 279-286

[2] Ha N S Lu G Shu D W and Yu T X 2020 *Journal of the mechanical behavior of biomedical materials* **104** 103603

[3] Ahmad Z 2009 *Impact and energy absorption of empty and foam-filled conical tubes* (Queensland University)

[4] Ahmad Z Thambiratnam D P and Tan A C C 2010 *International Journal of Impact Engineering* **37** 475-488

[5] Li Z Chen W and Hao H 2019 *International Journal of Impact Engineering* **133** 103341

[6] Wisutiamonkul A Dwamena C A Allan A C and Ketsa S 2017 *Scientia Horticulturae* **220** 233-242

[7] Zhao X Zhu G Zhou C and Yu Q 2019 *Composite Structure* **222** 110920

[8] Wang B He M Li C Wang L and Meng H 2020 *Biosystem*

Effects of the environment on thermo physiological responses via 3D human eye model with porous media heat transfer theory under inconstant solar irradiation

P. Chawengwanicha[1], P. Rattanadecho[2] and P. Keangin[1,*]

[1]Department of Mechanical Engineering, Faculty of Engineering, Mahidol University, 25/25 Phutthamonthon 4 Road, Salaya, Nakhon Pathom 73170, Thailand.
[2]Department of Mechanical Engineering, Faculty of Engineering, Thammasat University (Rangsit Campus), 99 mu 18, Paholyothin Road, Klong Nueng, Klong Luang, Pathumthani, 12120, Thailand.

* Corresponding Author: pornthip.kea@mahidol.ac.th

Abstract
The human eye is one of the most sensitive parts of our entire body. The lens is an important component of the eye, and it also plays an important role in vision. A lens has a normal temperature below 38.5°C. Thermal effects with solar irradiation result in the temperature of the lens to increase to more than 38.5 °C and may significantly influence the thermal physiologic response which, in turn, will cause the of deterioration of the lens. Therefore, investigation to get knowledge and guidelines to avoid thermal effects from the environment with solar irradiation, will help prevent the eyes from deterioration. However, the study of invasion, within the human body is impracticable. The study of environmental effects on thermo physiological responses within the human eye, from numerical analysis, is another alternative that is gaining attention worldwide. The purpose of this research is to learn about thermal physiologic response on human eyes exposed to solar irradiation, using the three-dimensional (3D) model by finite element method (FEM) via a computer program. This study uses the heat transfer equation and the Navier-Stroke equation to describe heat transfer phenomena in porous mediums and fluid flow phenomena within the 3D human eye model, based on considering the natural convection heat transfer of aqueous humor and vitreous humor, under inconstant solar irradiation. The effects of the environmental parameters such as ambient temperature and air convection heat transfer coefficient are investigated. The results show that the case study which makes the best cooling mechanism is the effect of ambient temperature with ambient temperature at 25°C and effect of air convection heat transfer coefficient at 40 $W/m^2 \cdot K$ that the eye can absorb maximum solar irradiance at 575 W/m^2 at the lens reaches 38.5°C and the worst cooling mechanism is the effect of ambient temperature with ambient temperature at 35°C that the eye can absorb maximum solar irradiance at 360 W/m^2 at the lens reaches 38.5°C. The results from this study provide the essential aspects for a fundamental understanding of thermal physiologic response within human eye subjected to solar irradiation.

Keywords: Human eye, Porous media, Finite element method, Environmental effects

1. Introduction

The human eye is an important organ used for seeing. An important part of the eye, sensitive to heat, is the lens. The lens is a medium, that controls light going to the retina, and then the sensory system generates the image [1]. Lenses deteriorate faster as heat increases their temperature. The International Commission on Non-Ionizing Radiation Protection (ICNIRP) Guidelines specified the lens's normal temperature should be below 38.5°C [2]. The three types of solar radiation are ultraviolet (UV), visible light and infrared (A and B) [3]. ICNIRP Guidelines indicated that the iris absorbs visible light and infrared-A region which are the main part of solar radiation [3-4]. Iris heat generation is performed on the lens which results in a higher lens temperature [5-6]. However, the invasive study of human bodies to measure some parameters such as the lens's temperature, it is limited. Therefore, to gain insight into the heat transfer phenomena and thermo physiological responses in the human body, especially within the human eye subject to solar irradiation, investigations are carried out via numerical simulation. Furthermore, numerical simulation under various conditions can be utilized to determine and identify essential parameters, as well as providing guidance for different applications.

Previously, many researchers have studied heat transfer phenomenons in the eye. In 1982 Lagendijk J. J. W. [7] set an invasive experiment to measure the thermal properties of rabbit eyes. Some thermal properties were used in the rabbits' eye model to analyze heat transfer using the finite-difference method (FDM). The results showed that the temperature of rabbit eyes, between the experimental and the numerical results, were similar in each component. Scott, J. A. [8] studied temperature distribution on the two-dimensional (2D) of human eye model based on the bio-heat transfer equation and solved the governing equations with boundary conditions by the finite element method (FEM). The human eye model was constructed consisting of four homogeneous domains, and thermal properties of the human eye component from previous research was used. The effects of the ambient temperature and the body-core temperature on the temperature distribution within the human eye was considered. After that, the human eye model was developed by Ng, E. Y. K. and Ooi, E. H. in 2006 [9]. The three-dimensional (3D) symmetry of human eye model with six components were presented. The bio-heat transfer equation to study the temperature distribution of human eye model were solved using FEM via computer simulation. Ooi, E. H. and Ng, E. Y. K. in 2008 [10] improved their human eye model from the previous human eye model. The flow phenomenon of aqueous humor (AH), which remains behind the cornea, with buoyancy force due to the temperature difference of AH (Boussinesq approximation) was considered. The lens deterioration from high ambient temperature was investigated by Sharon N. et al. [11]. The computational fluid dynamics (CFD) was used to solve the problem and to determine the temperature in the eye lens when exposed to environmental temperature fluctuations. It was found that the damage appeared earlier in the 6 h exposure and proceed from the lens anterior suture to its center. Optical damage was recovered in lenses exposed 1 h to 39.5°C, but the damage remained in the lens epithelial cells. Heat transfer in porous media was used in the human eye model to represent the actual eye because the actual eye components consisted of 2 phases in each component i.e., tissue and blood phases. Wessapan T. and Rattanadecho P. [12] applied the porous media heat transfer theory with modified Pennes bio-heat transfer to describe the temperature distribution on the 2D human eye model which iris and sclera were treated as the porous medium domains. The temperature distribution under the electromagnetic field at 900 MHz with the influences of ambient temperature were studied. The porous media heat transfer theory with local thermal equilibrium (LTE) model was widely used for investigating temperature changes in various tissues during thermal ablation [13-14]. There are many treatment methods for eye such as laser treatment using numerical simulation was useful for investigating the thermal effect and thermal damage with the appropriate power, time and processing position [15]. Garcia O. P. et al. [16] investigated the effects of vitreous humor (VH) flow phenomenon under the heating by laser-induced on the 2D human eye model with various size tumors at the retina under the various power of the laser. Although there have been studies on heat transfer phenomenon in human eye model based on LTE assumption, very few works have studied thermo physiological responses via 3D human eye model under inconstant solar irradiation with porous media heat transfer theory. Furthermore, the human eye models from the literature review are mentioned before, there are still differences from the actual human eye because of the complexity of the model. The model that is representative of the real human eye will contribute to a greater clarity concerning attributes more closely to actual behavior of heat transfer phenomenon in human eye subjected to solar irradiation.

In this study, improves the 3D human eye model base on porous media heat transfer theory by considering the natural convection heat transfer from aqueous humor and vitreous humor to represent the actual human eye. The heating load of this research focuses on inconstant solar irradiation with increasing function depend on the time is absorbed by iris only then convert to the heat generation. The effects of the environment such as the effect of ambient temperature and the effect of air convection heat transfer coefficient are considered. This research purpose to investigate the intensity of solar irradiance that makes the lens's temperature reach the maximum normal temperature of the lens (38.5 °C). The analysis from this study serves as essential fundament for the understanding of thermal physiologic response within human eye exposed to the solar irradiation.

2. Methodology

This section is divided into 3 sections to explain the details of this research which consist of the modeling and materials, the governing equation and boundary condition and the numerical method.

2.1 Modeling and materials

This research constructs the 3D human eye model with the dimension of human eye elements as following the human eye model from the literature reviews. The human eye model has 25 mm diameter, cornea thickness 0.5 mm and 7.7 mm radius, anterior chamber length 3.4 mm, iris thickness 0.5 mm, ciliary body thickness 2 mm, lens thickness 4.2 mm and height 9.6 mm, pupil radius 2.5 mm, sclera thickness 0.99 mm, choroid thickness 0.2 mm and retina thickness 0.25 mm [17] as shown in Figure 1. The essential thermal properties for simulation as shown in Table 1. The heat load, which is an increased function of solar irradiation used in the analysis, as shown in Figure 2. The solar irradiance was measured during a clear sky day in Bangkok, Thailand [18]. This research considers only 2 regions of the electromagnetic wave in solar irradiation, there are visible light region and near-infrared or infrared-A region.

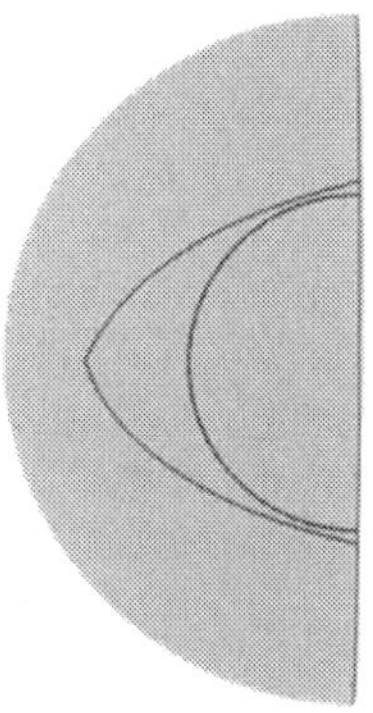

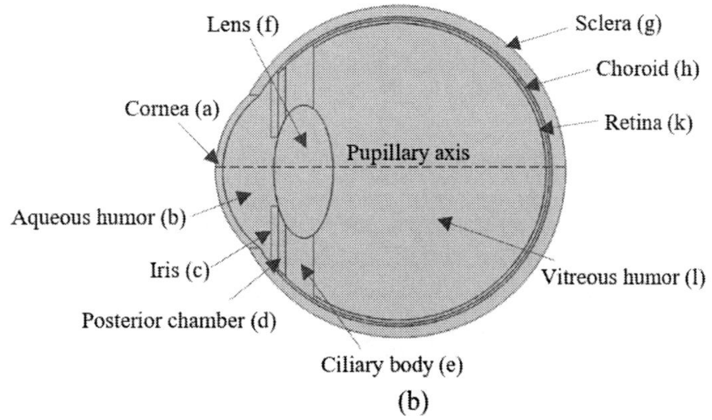

Figure 1. The 3D human eye model in (a) front view and (b) side view

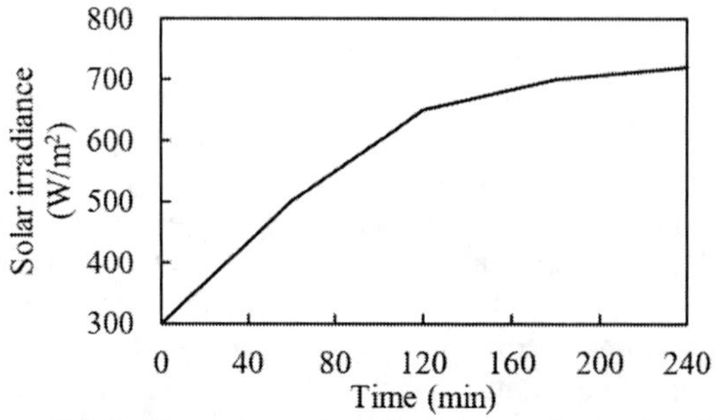

Figure 2. The time-dependent functions of solar irradiance

Table 1. Thermal properties in each element [17]

Human eye model's element	Density ρ (kg/m³)	Thermal conductivity k (W/m·K)	Specific heat capacity c_p (J/kg·K)
Cornea (a)	1050	0.58	4178
Aqueous humor (b)	996	0.58	3997
Iris (c)	1100	1.0042	3180
Posterior chamber (d)	996	0.58	3997
Ciliary body (e)	1040	0.498	3430
Lens (f)	1050	0.4	3000
Sclera (g)	1100	1.0042	3180
Choroid (h)	1060	0.53	3840
Retina (k)	1039	0.565	3680
Vitreous humor (l)	1000	0.603	4178

2.2 Governing equation and boundary condition

An analysis of heat transfer and fluid flow within the 3D human eye model subjected to the solar irradiation will be illustrated. The governing equations as well as initial and boundary conditions are solved numerically using the FEM to determine the temperature distribution and the velocity profile of the 3D human eye model. Heat transfer analysis of the human eye model has the assumptions to simplify the problem as follows: 1. There is no phase change in each domain. 2. Thermal properties in each domain are constant. 3. Porous mediums analyze with the local thermal equilibrium between tissue and blood. 4. There is no chemical reaction in each domain. 5. The thermoregulation mechanism is neglected. The heat transfer equation which implements to the porous medium domains (c, h) based on the modified Pennes bio-heat transfer as given in equation (1), the homogeneous solid domains (a, d, e, f, g, k) based on the classical conduction heat transfer equation as given in Equation (2) and the homogeneous fluid domains (b, l) based on convection heat transfer equation as given in Equation (3).

$$(1-\phi)\rho_i c_{p,i}\frac{\partial T_i}{\partial t} = \nabla\cdot((1-\phi)k_i\nabla T_i) + \rho_{bl}c_{p,bl}\omega_{bl}(T_{bl}-T_i) + Q_{ext,j};\ Q_{ext} = \alpha I(t) \quad (1)$$

where T is the temperature (°C), ρ is the density (kg/m³), c_p is the specific heat capacity (J/kg·K), k is the thermal conductivity (W/m·K), $\phi = 0.85$ is the porosity [19], $\rho_{bl} = 1060$ kg/m³ is the density of blood [17], $\omega_{bl} = 0.022$ 1/s is the blood perfusion rate [20], $c_{p,bl} = 3594$ J/kg·K is the specific heat capacity of blood [17], $T_{bl} = 37$°C is the blood temperature [12], Q_{ext} is the external heat source generates by solar irradiation (W/m³), $\alpha = 2250$ 1/m is the average absorption coefficient at the wavelength between 400 nm to 1400 nm [21] and $I(t)$ is the solar irradiance (W/m²). When i and j denote the domains, which use Equation (1), thus i is the iris and choroid and j is only the iris.

$$\rho_i c_{p,i} \frac{\partial T_i}{\partial t} = \nabla \cdot (k_i \nabla T_i) \tag{2}$$

when i denote the domains, which use Equation (2), thus i is the cornea, posterior chamber, ciliary body, lens, retina and sclera.

$$\rho_i c_{p,i} \frac{\partial T_i}{\partial t} = \nabla \cdot (k_i \nabla T_i) - \rho_i c_{p,i} \mathbf{u}_i \cdot \nabla T_i \tag{3}$$

where **u** is the velocity of a fluid (m/s). When i denote the domains, which use Equation (3), thus i is the AH and VH.

Fluid flow analysis of the human eye model has the assumptions to simplify the problem as follows: 1. AH and VH are incompressible Newtonian fluid with laminar flow. 2. The buoyancy force of fluid is defined by the Boussinesq approximation when the states that the density of fluid changes slightly with temperature but negligibly with pressure. The Navier-strokes equations include 2 equations, the continuity equation and the momentum equation as in Equation (4) and Equation (5).

$$\nabla \cdot \mathbf{u}_i = 0 \tag{4}$$

$$\rho_i \frac{\partial \mathbf{u}_i}{\partial t} + \rho_i \mathbf{u}_i \nabla \cdot \mathbf{u}_i = -\nabla p_i + \nabla \cdot \left[\mu (\nabla \mathbf{u}_i + \nabla \mathbf{u}_i^{\mathrm{T}}) \right] + \rho_i \mathbf{g} \beta_i (T_i - T_{ref}) \tag{5}$$

where p is the pressure (N/m^2), $\mu = 0.00074$ and 0.7 N·s/m^2 is the viscosity of AH and VH [10,22], $\mathbf{g}$ = 9.81 m/s^2 is the gravitational acceleration, β = 0.000337 1/K is the volume thermal expansion coefficient of AH and VH [10] and T_{ref} = 37°C is the reference temperature [12]. When i denote the domains which use these equations, thus i is the AH and VH.

The boundary conditions of the human eye model are applied to each region as shown in Figure 3. Heat transfer boundary conditions include 4 equations. There is the convection boundary condition by air, radiative boundary condition and evaporation boundary condition as given in equation (6) is applied to boundary 1 ($B1$) as shown in Figure 3. The convection boundary condition which represents the convection heat transfer coefficient between sclera and fat as given in equation (7) is applied to boundary 2 ($B2$) as shown in Figure 3. The symmetric boundary condition as given in equation (8) is applied to boundary 3 ($B3$) as shown in Figure 3. and the continuity boundary condition as given in equation (9) is applied to contact surface between domains.

$$-\mathbf{n} \cdot (-k \nabla T) = h_c (T - T_{amb}) + \varepsilon \sigma (T^4 - T_{amb}^4) + E \quad at\, B1 \tag{6}$$

where E = 40 W/m^2 is the heat loss due to tear evaporation [17], ε = 0.975 is the emissivity [12], σ = 5.67x10^{-8} W/m^2·K^4 is the Stefan-Boltzmann constant, h_c is the ambient convection heat transfer coefficient (W/m^2·K) and T_{amb} is the ambient temperature (°C).

$$-\mathbf{n} \cdot (-k \nabla T) = h_s (T - T_b) \qquad at\, B2 \tag{7}$$

where h_s = 20 W/m^2·K is the convection heat transfer coefficient of fat [17].

$$-\mathbf{n} \cdot (k \nabla T) = 0 \quad at\, B3 \tag{8}$$

$$-\mathbf{n}_{dst} \cdot (k_{dst} \nabla T_{dst}) = \mathbf{n}_{src} \cdot (k_{src} \nabla T_{src}),\ T_{dst} = T_{src} \tag{9}$$

where subscript dst and src denote destination and source pair of the contact surface.

Fluid flow boundary conditions include 2 equations. There is no-slip boundary condition as equation (10) is applied to boundary 4 ($B4$) and symmetric boundary condition as Equation (11) is applied to boundary 5 ($B5$) as shown in Figure 3.

$$\mathbf{u} = 0 \quad at\, B4 \tag{10}$$

$$\mathbf{u} \cdot \mathbf{n} = 0 \ at\, B5 \tag{11}$$

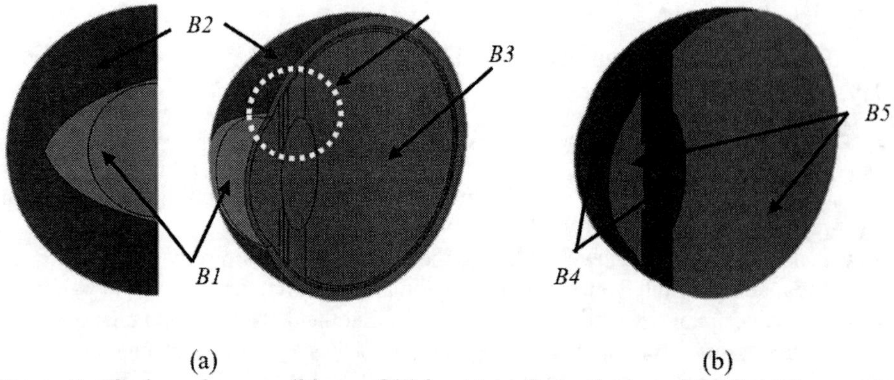

Figure 3. The boundary conditions of (a) heat transfer analysis and (b) fluid flow analysis

2.3 Numerical method

The system of governing equations to predict the temperature distribution and the velocity profile of the 3D human eye model is in the form of nonlinear partial differential equation. The governing equations are solved numerically using the FEM via the commercial program COMSOL Multiphysics. This program accommodates all steps for FEM such as meshing, computational process and post-processing. The human eye model provides mesh independent analysis for the minimum number of the elements which the result slightly changes as shown in Figure 4. This convergence test leads to the 3D human eye model is meshed by 195,000 linear tetrahedral elements as shown in Figure 5. It is reasonable to assume that, at this element number, the accuracy of the simulation results is independent of the number of elements. The numerical model presented is verified by Lagendijk's experiment results in 1982 which specimen is the rabbit eye and installed the thermometer in the rabbit eye with 3 positions along the pupillary axis. In the validation, the parameters are used with the same Lagendijk's experiment such as ambient temperature is 23°C, convection heat transfer coefficient of air is 20 W/m^{2}K and body core temperature 38.8 °C (rabbit body). The comparison between temperature distribution obtained from this numerical study and obtained by Lagendijk's experiment as shown in Figure 6. The figure indicates that the slight temperature difference between Lagendijk's experiment and present study at 3 positions. The estimated error in 3 positions is 0.85 %, 0.99 % and 0.26 % that the average error is 0.7 %. These estimate errors are satisfied for investigating the environmental effects of thermo physiological responses on the 3D human eye model.

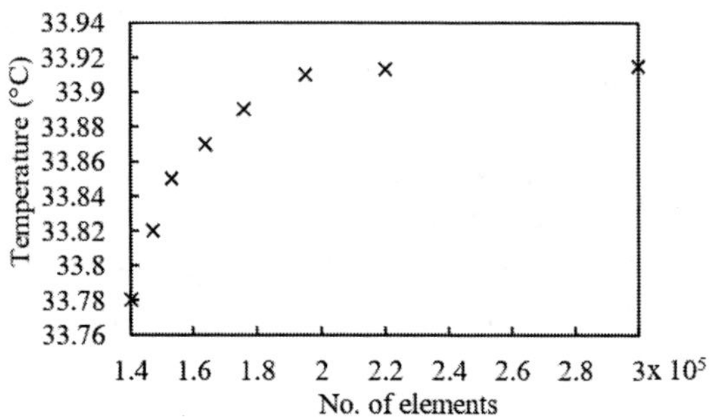

Figure 4. Mesh independent of the human eye model

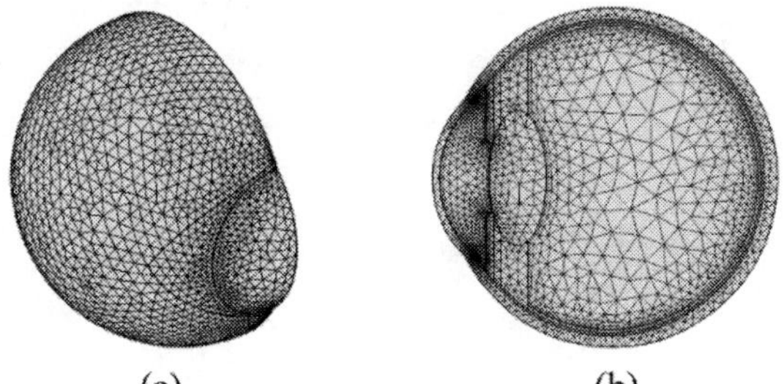

Figure 5. The human eye model with meshed (a) isometric view and (b) side view

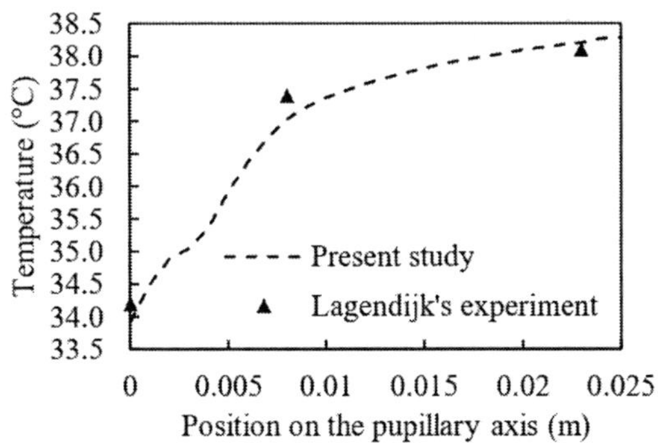

Figure 6. The comparison of temperature results between present study and Lagendijk's experiment

3. Results and discussion

The results of environmental effects are investigated in this section and provide the intensity of solar irradiance that the lens's temperature reaches 38.5°C in any case study while exposed to inconstant solar irradiation.

3.1 Effect of ambient temperature

In this research, the effect of ambient temperature includes 25°C, 30°C and 35°C based on the other parameter is constant (the air convection heat transfer coefficient at 20 $W/m^2{\cdot}K$) in each case. The temperature distribution and velocity profile on the symmetry plane of the 3D human eye model at various time exposed to inconstant solar irradiation as shown in Figures 7 and Figures 8, respectively. From the Figures 7 and Figures 8 show the ambient temperature affects directly to cooling mechanism because the different temperatures between air and anterior ocular make the buoyancy force of Boussinesq approximate at AH which flows with higher velocity when the different temperature is higher and convection heat transfer between cornea and ambient is higher too. The flow direction of the cold fluid and hot fluid is downward and upward respectively. Thus, the lowest ambient temperature makes the anterior ocular's temperature lower when compared to the results of other ambient temperatures at the same time in Figure 7. In Figure 8 displays the velocity profile of VH is stationary even the different temperatures between anterior ocular and VH are high due to the viscosity of VH is high. The lens's maximum temperature along the time that appears at the top lens near the iris as shown in Figure 9. The ambient temperature is the significance of the cooling mechanism while the eye exposed to inconstant solar irradiation. Figure 9 shows the effect of ambient temperature on the maximum temperature in various exposure time. In Figure 9 shows, the ambient temperature at 25°C that makes the lens's maximum temperature reach 38.5°C at 90 min and solar irradiance equal to 575 W/m^2, at 30 °C that makes the lens's maximum temperature reach 38.5°C at 48 min and solar irradiance equal to 460 W/m^2 and at 35°C that makes the lens's maximum temperature reach 38.5°C at 18 min with solar irradiance equal to 360 W/m^2 which the solar irradiation corresponds to Figure 2.

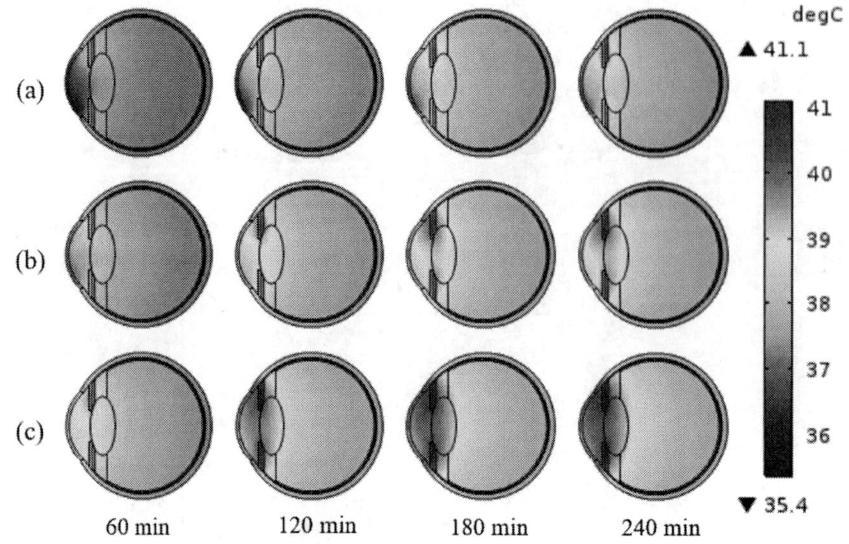

Figure 7. Temperature distribution on symmetry plane of human eye model with ambient temperature at (a) 25°C, (b) 30°C and (c) 35°C

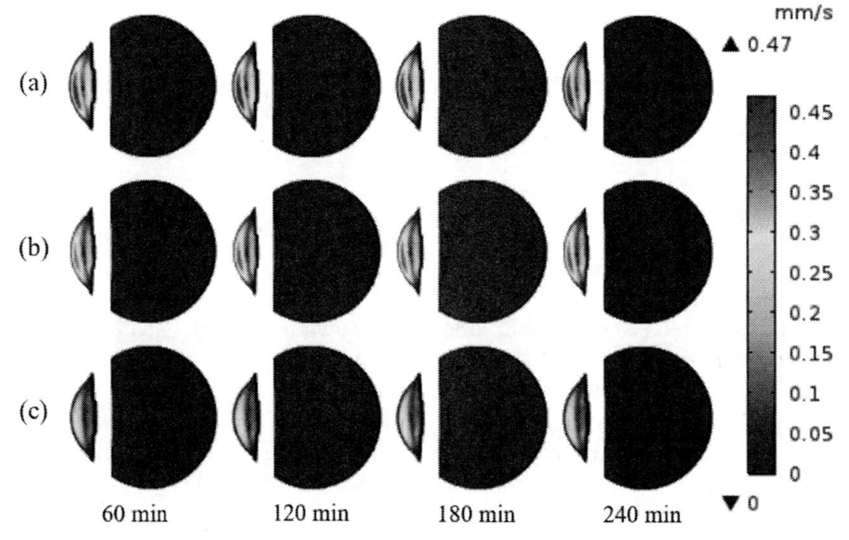

Figure 8. Velocity profile on symmetry plane of aqueous humor and vitreous humor with ambient temperature (a) 25°C, (b) 30°C and (c) 35°C

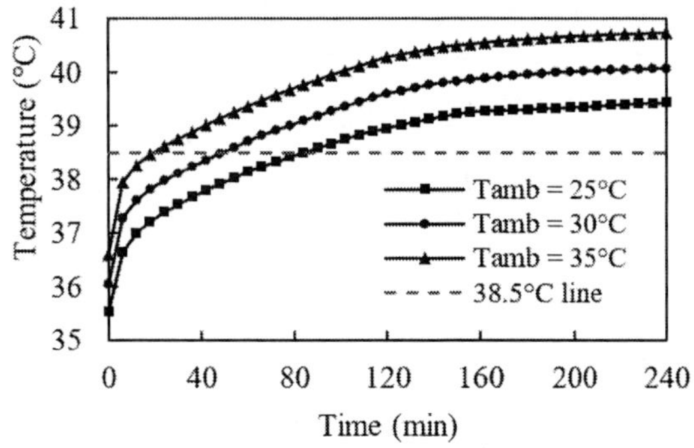

Figure 9. Maximum temperature of the lens along the time with the effect of ambient temperature

3.2 Effect of air convection heat transfer coefficient

The effect of air convection heat transfer coefficient includes 10 $W/m^2{\cdot}K$, 20 $W/m^2{\cdot}K$ and 40 $W/m^2{\cdot}K$ based on the other parameter is constant (the ambient temperature at 30°C) in each case. The temperature distribution and velocity profile on the symmetry plane of the 3D human eye model at various time exposed to inconstant solar irradiation as shown in Figure 10 and Figure 11, respectively. These figures show the air convection heat transfer coefficients directly affect the cooling mechanism on anterior ocular when air convection heat transfer coefficient is high that represents the velocity of ambient air is faster (windy) than the lower air convection heat transfer coefficient (lightly windy). The velocity profile of VH is also high when air convection heat transfer coefficients are higher which makes the high-temperature differential between the cornea and AH. Thus, the velocity of AH is high from the buoyancy force with high-temperature differences. The flow direction of the cold fluid and hot fluid is downward and upward respectively. In the part of VH, the velocity profile in each case are the same results as the effect of ambient temperature, the VH is stationary. The lens's maximum temperature time that appears at the top lens near the iris as shown in Figure 12. In Figure 12 shows, the air convection heat transfer coefficients at 10 $W/m^2{\cdot}K$ that makes the lens's maximum temperature reach 38.5°C at 30 min with solar irradiance equal to 400 W/m^2, the air convection heat transfer coefficients at 20 $W/m^2{\cdot}K$ that makes the lens's maximum temperature reach 38.5°C at 48 min with solar irradiance equal to 460 W/m^2 and the air convection heat transfer coefficients at 40 $W/m^2{\cdot}K$ that makes the lens's maximum temperature reach 38.5°C at 90 min with solar irradiance equal to 575 W/m^2 which the solar irradiation corresponds to Figure 2.

4. Conclusion

The environmental effects on the thermo physiological via the 3D human eye model under inconstant solar irradiation are investigated. The heat transfer model is developed based on porous media heat transfer by considering the natural convection of AH and VH. The main findings of this study are:

1. Both the effect of ambient temperatures and the effect of air convection heat transfer coefficients, are significant to the cooling mechanism of the human eye model, when the human eye model is exposed to inconstant solar irradiation.

2. The best and worst cases for the cooling mechanism of the human eye model, in the effect of ambient temperatures, are the ambient temperatures of 25°C and 35°C that can absorb the solar irradiance at 575 W/m^2 and 360 W/m^2, respectively when the lens's maximum temperature reaches 38.5°C.

3. The best and worst cases for the cooling mechanism of the human eye model in the effect of air convection heat transfer coefficients of 40 $W/m^2{\cdot}K$ and 10 $W/m^2{\cdot}K$ that absorbs the solar irradiance at 575 W/m^2 and 400 W/m^2, respectively when the lens's maximum temperature reaches 38.5°C.

4. To reduce the computational times and computational resources, due to the results of the velocity profile of VH, are slightly changed based on the condition in this study. This research recommends treating only the AH as a fluid and VH can treat as a solid because the velocity of VH is stationary from the result. However, the VH is considered to be fluid when the VH's viscosity is low as AH (Eye floaters: the VH gel becomes liquid phase) that the flow of VH has occurred.

In further studies, the human eye model will be developed for representing the actual human eye by considering some mechanisms of the eye such as the eye-blinking effect, which is an important mechanism of the human eye, that may affect temperature distribution on the human eye model.

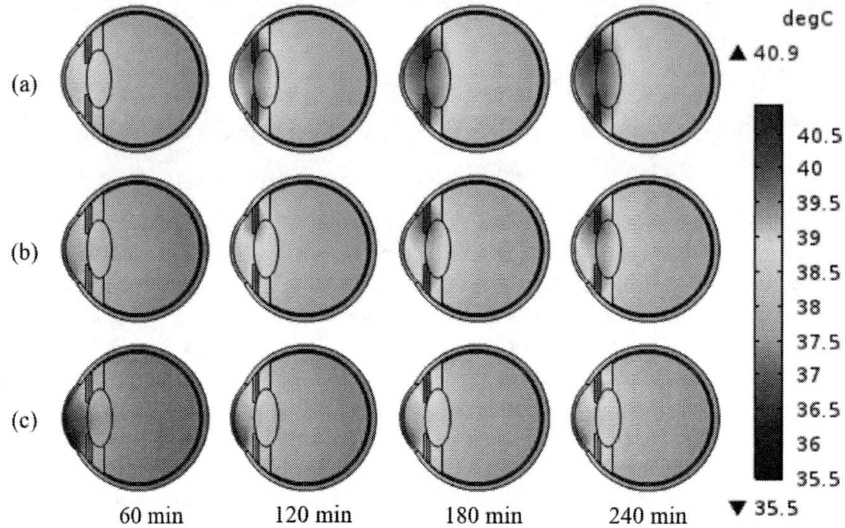

Figure 10. Temperature distribution on symmetry plane of human eye model with air convection heat transfer coefficient (a) 10 $W/m^2 \cdot K$ (b) 20 $W/m^2 \cdot K$ and (c) 40 $W/m^2 \cdot K$

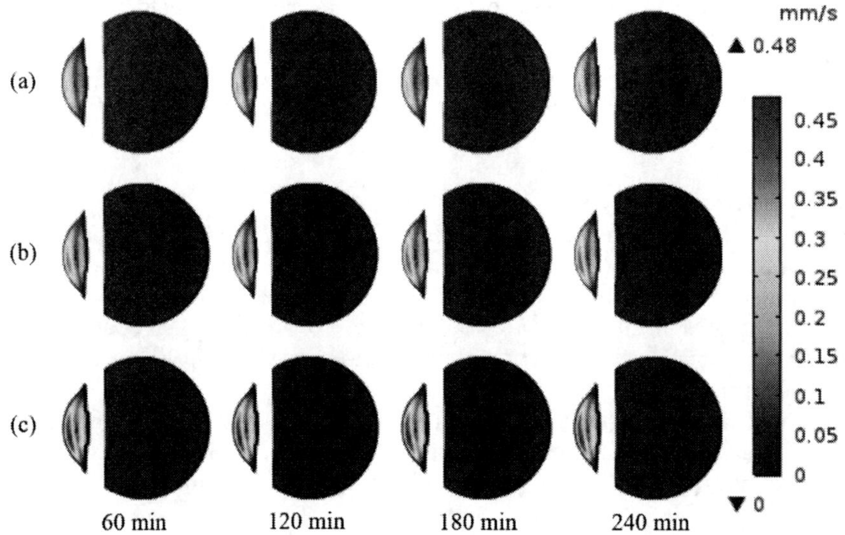

Figure 11. Velocity profile on symmetry plane of aqueous humor and vitreous humor with air convection heat transfer coefficient (a) 10 $W/m^2 \cdot K$ (b) 20 $W/m^2 \cdot K$ and (c) 40 $W/m^2 \cdot K$

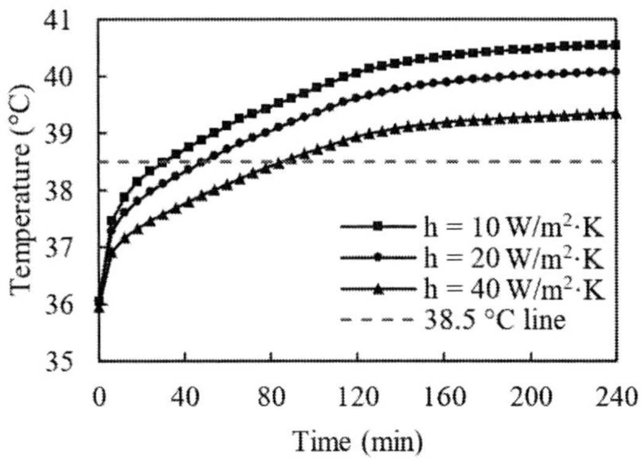

Figure 12. Maximum temperature of the lens along the time with the effect of air convection heat transfer coefficient

Acknowledgments
This research has been supported by the 60th Year Supreme Reign of His Majesty King Bhumibol Adulyadej Scholarship academic year 2019, Faculty of graduate studies, Mahidol University.

References
[1] Cassel G H Billig M D and Randall H G 2004 *The eye book: A complete guide to eye disorders and health* (Australia: Vision Australia Information and Library Service)
[2] International Commission on Non-Ionizing Radiation Protection 2020 *Health Phys.* **118** 483-524
[3] Petty G W 2006 *A first course in atmospheric radiation* (Madison: Sundog Publishing)
[4] International Commission on Non-Ionizing Radiation Protection 2013 *Health Phys.* **105** 74-96
[5] Okuno T 1991 *Ann. Occup. Hyg.* **35** 1-12
[6] Tsutomu O 1994 *Ann. Occup. Hyg.* **38** 351-359
[7] Lagendijk J J W 1982 *Phys. Med. Biol.* **27** 1301-1311
[8] Scott J A 1988 *Phys. Med. Biol.* **33** 227-41
[9] Ng E Y K and Ooi E H 2006 *Comput. Meth. Prog. Bio.* **82** 268-276
[10] Ooi E H and Ng E Y K 2008 *Comput. Meth. Prog. Bio.* **38** 252-262
[11] Sharon N Bar Y P Z Bormusov E and Dovrat A 2008 *Exp. Eye Res.* **87** 49-55
[12] Wessapan T and Rattanadecho P 2014 *Int. J. Heat Mass Transf.* **70** 378-388
[13] Keangin P and Rattanadecho P 2018 *Adv. Mech. Eng.* **10** 1-13
[14] Keangin P and Rattanadecho P 2013 *Int. J. Heat Mass Transf.* **67** 46-60
[15] Semenyuk V 2017 *Int. J. Heat Mass Transf.* **112** 480-8
[16] Garcia O P Lyra P R M, Fernandes A and de Lima R D C F 2019 *Int. J. Therm. Sci.* **136** 444-456
[17] Heussner N Holl L Nowak T Beuth T Spitzer M S and Stork W 2014 *Comput. Biol. Med.* **51** 35-43
[18] Sansomboon A Luewarasirikul N Ittipongse A Phae-Ngam W and Pattarapanitchai S 2013 *Adv. Mat. Res.* **770** 229-232
[19] Ehrlich R Harris A Wentz S M Moore N A and Siesky B A 2016 *The Curated Reference Collection in Neuroscience and Biobehavioral Psychology* 73-82
[20] Flyckt V M M Raaymakers B W and Lagendijk J J W 2006 *Phys. Med. Biol.* **51** 5007-5021
[21] Regal S O' Connor D Brige P Delattre R Djenizian T and Ramuz M 2019 *J. Biophotonics.* **12** e201800398
[22] Repetto R Siggers J H and Stocchino A 2010 *Biomech. Model. Mechanobiol.* **9** 65-76
[23] Mauro A and Mohamed S 2020 *Int. J. Heat Mass Transf.* **158** 119994

IOP Conf. Series: Materials Science and Engineering 1137 (2021) 012044 doi:10.1088/1757-899X/1137/1/012044

Assessment of turbulence models for low turbulent natural convection heat transfer in rectangular enclosed cavity using OpenFOAM

Kittipos Loksupapaiboon[1,2] and Chakrit Suvanjumrat[1,2,*]

[1]Department of Mechanical Engineering, Faculty of Engineering, Mahidol University, Salaya, Nakhon Pathom, 73170, Thailand
[2]Laboratory of Computer Mechanics for Design (LCMD), Department of Mechanical Engineering, Faculty of Engineering, Mahidol University, Salaya, Nakhon Pathom, 73170, Thailand

* Corresponding Author: chakrit.suv@mahidol.ac.th

Abstract
This research aimed to study two-equation turbulence models, for the simulation of low turbulent natural convection heat transfer, which happened in a rectangular enclosure. These models include $k-\varepsilon$, $k-\omega$, $RNG\ k-\varepsilon$, $k-\omega\ SST$ and $Realizable\ k-\varepsilon$ which are two-equation turbulence models in OpenFOAM. The simulation was carried out to compare against experiment in literatures which gave a Rayleigh number (Ra) of 1.58×10^9. The 3D cell structure by the $50\times50\times100$ cells for the rectangular enclosure dimensions of 0.75 m (width)×0.75 m (height)×1.50 m (length) was modeled. It gave the large y+ of 30 but was still in the initial sub-layer. It was found that the $k-\omega$ model provided the best prediction of turbulent natural convection heat transfer. The standard k and ω wall functions were carried out to simulate well with the $k-\omega$ model. It was good modeling technique which the fine cells were not necessary to define nearly enclosure walls.

Keywords: Turbulence model, Natural convection, Heat transfer, Cavity

1. Introduction

The natural convection heat transfer in enclosures was widely carried out to study by many researchers. They mentioned both experimental and simulating approaches [1-4]. The experiment of convection motion was generated by the buoyancy force on fluid in rectangular enclosures which held two sides at different temperatures. This phenomenon was investigated in various engineering systems such as electronic cooling equipment, building, air conditioning application, solar energy collectors, etc. The numerical approach was carried out by the computational fluid dynamics (CFD). To govern the turbulent flow, the $k-\varepsilon$ turbulence model was the most simulation model which used for this phenomenon. Mao and Zhang [5] had evaluated turbulence models for the simulation of natural convection in a tall cavity. These models comprised $k-\varepsilon$, $RNG\ k-\varepsilon$ turbulence model and large eddy simulation (LES). They found that the $k-\varepsilon$ model had a high accuracy of velocity prediction while the LES model had good performance for temperature distribution. Rathore and Das [6] had studied the buoyancy-driven flow or natural convection in a tall cavity, with an aspect ratio of 5. They applied the $k-\varepsilon$, $k-\omega$ and $k-\omega\ SST$ turbulence model for simulating turbulent flow. The $k-\varepsilon$ model based on the SIMPLEC algorithm for pressure-velocity coupling and y^+ less than 1 performed better.

The natural convection in a cavity, with or without an internal body, has been widely investigated. Square, circular and elliptic cylinders were the internal bodies [7]. The study of natural convection in a rectangular enclosure was extremely enhanced, whilst coupled with other effects. Cintolesi et al. [8] simulated natural convection with water evaporation and condensation in a cold enclosed cubic cavity, containing a hot rectangular plate. The internal surface of the cavity was covered by a thin water film. The LES was employed to simulate this phenomenon. Chen et al. [9] studied the natural convection in cavities with a horizontal fin on a hot sidewall. The zero-equation turbulence model was suitable for this study. The vortices at the upper two corners of the cavity, and near the upper surface of the fin edge, were varied by the fin positions. Dehbi et al. [10] studied the thermal radiation influence on natural convection in enclosures. The LES was performed with and without wall-to-wall radiation modeling. Simulation results showed that the radiation was significant enough to increase the natural convection heat transfer. Iyi and Hasan [11] investigated moist air and heat transfer inside a rectangular cavity with different vertical heating walls by using CFD. The $k-\varepsilon$ turbulence model was used with the moisture transport model for simulation. They found that the natural convection was sensitive to moisture content variation. Other shapes of enclosure - not square - were also investigated, in relation to natural convection heat transfer. The proper design of enclosures to achieve the highest heat transfer rate, was the objective for these studies [12]. Enayati et al. [13] studied natural convection heat transfer in cylindrical enclosures, shaped as a reactor. It contained a rack and 80 seeds mounted inside. The LES was used to simulate the heat transfer phenomenon. They found the important number of seeds, in relation to turbulence intensity. Previous literatures presented the importance of investigation for turbulent natural convection heat transfer in enclosures. The challenge being the numerical models for the prediction of turbulent buoyancy flow. Most researchers used commercial software having an expensive license cost for simulation. Consequently, the researchers without supporting funds for the software license, were obstructed. This research proposes using open source software, Open FOAM, to investigate the natural convection heat transfer in enclosed cavities. The RANS turbulence models, which were specifically the two-equation transport models, demonstrated accuracy for the simulation. The suitable model which is detected in this research can be used for applying natural convection in various engineering systems.

2. Natural Convection Flow Models

2.1 Governing equation

The turbulent natural convection flow of compressible fluid in a square enclosure is governed by the conservative equations as follows:

$$\frac{\partial(\rho U_i)}{\partial t}+\frac{\partial(\rho U_i)}{\partial x_i}=0 \tag{1}$$

$$\frac{\partial(\rho U_i)}{\partial t}+\frac{\partial(\rho U_i U_j)}{\partial x_j}=-\frac{\partial P}{\partial x_j}+\rho g_i+\frac{\partial}{\partial x_j}\left((\mu+\mu_t)\left(\left(\frac{\partial U_i}{\partial x_j}+\frac{\partial U_j}{\partial x_i}\right)-\frac{2}{3}\frac{\partial U_i}{\partial x_i}\right)\right) \tag{2}$$

$$\frac{\partial(\rho h)}{\partial t}+\frac{\partial(\rho U_i h)}{\partial x_j}+\frac{\partial(\rho K)}{\partial t}+\frac{\partial}{\partial x_j}(\rho U_i K)-\frac{\partial P}{\partial t}=\frac{\partial}{\partial x_j}\left(\left(\frac{\mu_t}{P_{rt}}+\frac{k_c}{C_p}\right)\frac{\partial h}{\partial x_j}\right)+\rho U_i g_i \tag{3}$$

$$\frac{\partial(\rho e)}{\partial t}+\frac{\partial(\rho U_i e)}{\partial x_j}+\frac{\partial(\rho K)}{\partial t}+\frac{\partial}{\partial x_j}(\rho U_i K)+\frac{\partial U_i P}{\partial x_j}=\frac{\partial}{\partial x_j}\left(\left(\frac{\mu_t}{P_{rt}}+\frac{k_c}{C_p}\right)\frac{\partial e}{\partial x_j}\right)+\rho U_i g_i \tag{4}$$

2.2 The two equation turbulence model

The number of equations of the RANS turbulence is counted by using the transport equation number of turbulence parameters for estimating the Reynolds stress term. Consequently, the two-equation

IOP Conf. Series: Materials Science and Engineering 1137 (2021) 012044 doi:10.1088/1757-899X/1137/1/012044

turbulence models in OpenFOAM consist of $k-\varepsilon$, $k-\omega$, $RNG\ k-\varepsilon$, $k-\omega\ SST$ and $Realizable\ k-\varepsilon$.

2.2.1 The $k-\varepsilon$ model

The $k-\varepsilon$ model composes the turbulent kinetic energy (k) and rate of viscous dissipation (ε) transport equations [14, 15]. These equations can be written as follows:

$$\frac{\partial(\rho k)}{\partial t}+\frac{\partial(\rho k U_i)}{\partial x_i}=\frac{\partial}{\partial x_i}\left(\frac{\mu_t}{\sigma_k}\frac{\partial k}{\partial x_i}\right)+2\mu_t S_{ij}\cdot S_{ij}-\rho\varepsilon \tag{5}$$

$$\frac{\partial(\rho\varepsilon)}{\partial t}+\frac{\partial(\rho\varepsilon U_i)}{\partial x_i}=\frac{\partial}{\partial x_i}\left(\frac{\mu_t}{\sigma_\varepsilon}\frac{\partial \varepsilon}{\partial x_i}\right)+C_{1\varepsilon}\frac{\varepsilon}{k}2\mu_t S_{ij}\cdot S_{ij}-C_{2\varepsilon}\rho\frac{\varepsilon^2}{k} \tag{6}$$

$$\mu_t=\rho C_\mu\frac{k^2}{\varepsilon}, S_{ij}=\frac{1}{2}\left(\frac{\partial U_i}{\partial x_j}+\frac{\partial U_j}{\partial x_i}\right) \tag{7}$$

where $C_\mu=0.09$, $\sigma_k=1.00$, $\sigma_\varepsilon=1.30$, $C_{1\varepsilon}=1.44$, $C_{2\varepsilon}=1.92$.

2.2.2 The $k-\omega$ model

The turbulence frequency (ω) had been developed for the length scale determining variable of two-equation turbulence models. The $k-\omega$ model composes the turbulent kinetic energy (k) and turbulence frequency (ω) transport equations [16]. These equations can be written as follows:

$$\frac{\partial(\rho k)}{\partial t}+\frac{\partial(\rho k U_i)}{\partial x_i}=\frac{\partial}{\partial x_i}\left(\left(\mu+\frac{\mu_t}{\sigma_k}\right)\frac{\partial k}{\partial x_i}\right)+2\mu_t S_{ij}\cdot S_{ij}-\frac{2}{3}\rho k\frac{\partial U_i}{\partial x_j}\delta_{ij}-\beta^*\rho k\omega \tag{8}$$

$$\frac{\partial(\rho\omega)}{\partial t}+\frac{\partial(\rho\omega U_i)}{\partial x_i}=\frac{\partial}{\partial x_i}\left(\left(\mu+\frac{\mu_t}{\sigma_k}\right)\frac{\partial \omega}{\partial x_i}\right)+\gamma_1\left(2\rho S_{ij}\cdot S_{ij}-\frac{2}{3}\rho\omega\frac{\partial U_i}{\partial x_j}\delta_{ij}\right)-\beta_1\rho\omega^2 \tag{9}$$

where $\mu_t=\rho\frac{k}{\omega}$, $\sigma_k=2.00$, $\sigma_\omega=2.00$, $\gamma_1=0.553$, $\beta^*=0.09$, $\beta_1=0.075$.

2.2.3 The $RNG\ k-\varepsilon$ model

The $RNG\ k-\varepsilon$ model expresses effect of small scale removing by the large scale motion term and modification of viscosity. This equation can be written as follows:

$$\frac{\partial(\rho k)}{\partial t}+\frac{\partial(\rho k U_i)}{\partial x_i}=\frac{\partial}{\partial x_i}\left(\alpha_k\left(\mu+\rho C_\mu\frac{k^2}{\varepsilon}\right)\frac{\partial k}{\partial x_i}\right)+\tau_{ij}\cdot S_{ij}-\rho\varepsilon \tag{10}$$

$$\frac{\partial(\rho\varepsilon)}{\partial t}+\frac{\partial(\rho\varepsilon U_i)}{\partial x_i}=\frac{\partial}{\partial x_i}\left(\alpha_\varepsilon\left(\mu+\rho C_\mu\frac{k^2}{\varepsilon}\right)\frac{\partial \varepsilon}{\partial x_i}\right)+C_{1\varepsilon}\frac{\varepsilon}{k}\tau_{ij}\cdot S_{ij}-C_{2\varepsilon}\rho\frac{\varepsilon^2}{k} \tag{11}$$

where $C_\mu=0.0845$, $\alpha_k=1.39$, $\alpha_\varepsilon=1.39$, $C_{1\varepsilon}=1.42$, $C_{2\varepsilon}=1.68$.

2.2.4 The $k-\omega\ SST$ model

The rate of viscous dissipation in $k-\varepsilon$ model is equal to $k\omega$. Consequently, the $k-\omega\ SST$ model can be written by the following equations.

$$\frac{\partial(\rho k)}{\partial t}+\frac{\partial(\rho k U_i)}{\partial x_i}=\frac{\partial}{\partial x_i}\left(\left(\mu+\frac{\mu_t}{\sigma_k}\right)\frac{\partial k}{\partial x_i}\right)+2\mu_t S_{ij}\cdot S_{ij}-\frac{2}{3}\rho k\frac{\partial U_i}{\partial x_j}\delta_{ij}-\beta^*\rho k\omega \tag{12}$$

$$\frac{\partial(\rho\omega)}{\partial t}+\frac{\partial(\rho\omega U_i)}{\partial x_i}=\frac{\partial}{\partial x_i}\left(\left(\mu+\frac{\mu_t}{\sigma_k}\right)\frac{\partial \omega}{\partial x_i}\right)+\gamma_1\left(2\rho S_{ij}\cdot S_{ij}-\frac{2}{3}\rho\omega\frac{\partial U_i}{\partial x_j}\delta_{ij}\right)-\beta_1\rho\omega^2+\frac{2\rho}{\sigma_{\omega 2}\omega}\frac{\partial k}{\partial x_j}\frac{\partial \omega}{\partial x_j} \tag{13}$$

$$\mu_t=\rho\frac{a_1 k}{max\left(a_1\omega_1\sqrt{2S_{ij}S_{ij}F_2}\right)} \tag{14}$$

where $\sigma_k=1.00$, $\sigma_{\omega 1}=2.00$, $\sigma_{\omega 2}=1.17$, $\gamma_1=0.44$, $\beta^*=0.09$, $\beta_1=0.083$, F_2 is the blending function.

2.2.5 The $Realizable\ k-\varepsilon$ model

The rate of viscous dissipation transport equations and realizable eddy viscosity formulation are proposed to improve the $k-\varepsilon$ model. The $Realizable\ k-\varepsilon$ model composes the following equations.

$$\frac{\partial(\rho k)}{\partial t}+\frac{\partial(\rho k U_i)}{\partial x_i}=\frac{\partial}{\partial x_i}\left(\frac{\mu_t}{\sigma_k}\frac{\partial k}{\partial x_i}\right)+2\mu_t S_{ij}\cdot S_{ij}-\rho\varepsilon \tag{15}$$

$$\frac{\partial(\rho\varepsilon)}{\partial t}+\frac{\partial(\rho\varepsilon U_i)}{\partial x_i}=\frac{\partial}{\partial x_i}\left(\frac{\mu_t}{\sigma_\varepsilon}\frac{\partial \varepsilon}{\partial x_i}\right)+C_1\rho\varepsilon\sqrt{2S_{ij}\cdot S_{ij}}-C_2\rho\frac{\varepsilon^2}{k+\sqrt{\nu\varepsilon}} \tag{16}$$

$$C_1=max\left\{0.43,\frac{\eta}{5+\eta}\right\},\eta=\frac{Sk}{\varepsilon},C_\mu=\frac{1}{A_0+A_sU^*\frac{k}{\varepsilon}},U^*=\sqrt{S_{ij}\cdot S_{ij}+\tilde{\Omega}_{ij}\cdot\tilde{\Omega}_{ij}},A_s=\sqrt{6}cos\phi \tag{17}$$

where $\sigma_k = 1.00$, $\sigma_\varepsilon = 1.20$, $C_2 = 1.9$, $A_0 = 4.0$.

2.3 Boundary conditions

The number of equations of the RANS turbulence is counted by using the transport equation number. The wall function has been developed to satisfy the physic of flow in near wall region. It bridges the inner region between the wall and the turbulence fully developed region. In the viscous sub-layer ($y^+ < 5$), the k, ε and ω wall functions [17] are written as follows:

$$k=\frac{\tau_w C_\mu^{-1/2}}{\rho},\varepsilon=C_\mu^{3/4}\frac{k^{3/2}}{\kappa y},\omega=\frac{6\tau_w}{\beta_1\mu(y^+)^2} \tag{18}$$

In the logarithmic area ($30<y^+< 200$), the turbulent variations are dominated by the following equations.

$$k=\frac{u_\tau^2}{\sqrt{C_\mu}},\varepsilon=\frac{u_\tau^3}{\kappa y},\omega=\frac{u_\tau^2}{\nu\kappa\sqrt{C_\mu}y^+} \tag{19}$$

where u_τ is the friction velocity and ν is the kinematic viscosity.

2.4 The wall heat flux

The wall heat transfer coefficient is evaluated using the Nusselt number as follows:

$$Nu=\frac{QL}{k(T_h-T_c)} \tag{20}$$

where Q is the wall heat flux, L is the width of the cavity, k is the thermal conductivity, T_h is the hot wall temperature and T_c is the cold wall temperature.

3. Computational Fluid Dynamics Model

The experiment of low turbulent natural convection flow of air in a rectangular enclosure had been performed by Tian and Karayiannis [18, 19]. The dimensions of enclosure were 0.75 m × 0.75 m × 1.50 m. The hot and cold walls of the enclosure were isothermal at 50°C and 10°C, respectively. The experiment was studied in the 2D plane at the mid-length of the enclosure. In order to achieve CFD domain as same as the experimental enclosure, the rectangular cavity has been created as shown in figure 1. The wall function was controlled in the inertial sub-layer then the y^+ was chosen at 30. Subsequently, the cell structure was assembled by the hexahedron cells with the only one dimension that was 0.015×

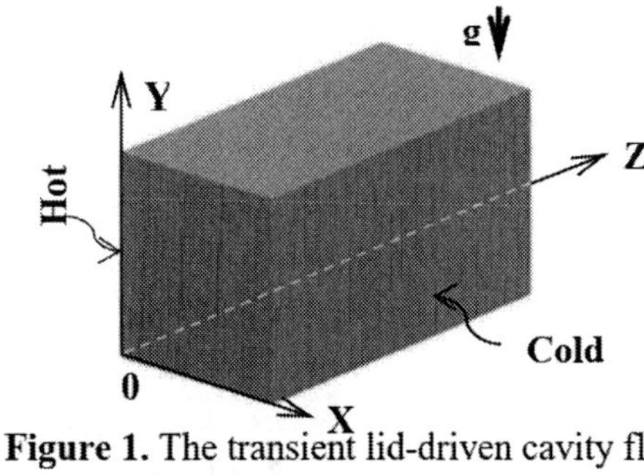

Figure 1. The transient lid-driven cavity flow schematic.

0.015× 0.015 m. This cell structure was built by the total cells of 250,000 cells. The wall of domain at the left hand (X = 0 m) is the hot wall. It had been defined the constant temperature of 50°C. The adversative wall was the cold wall (X = 0.75 m) with the constant temperature of 10°C. The other walls of enclosure were defined the adiabatic wall. The air inside the CFD domain was assigned to be the compressible fluid. At the initial time, the air temperature inside the domain was 30 °C. The initial pressure inside domain was 1.0×10^5 Pa. The pressure gradient was the pressure boundary on enclosure walls. To activate the buoyant flow, the acceleration gravity was specified carefully in the –Y direction. The no slip condition was assigned on the enclosure walls for the air velocity. The two-equation turbulence models composed of $k-\varepsilon$, $k-\omega$, $RNG\ k-\varepsilon$, $k-\omega\ SST$ and $Realizable\ k-\varepsilon$ were assigned to govern the low turbulent flow; therefore, the k, ε and ω wall functions were defined on the enclosure walls following the turbulence models. The wall functions would activate the turbulent flow nearly the enclosure walls; therefore, it was not concerned for the grid independence of enclosure domain.

The solutions had been obtained by solving the transport equations. The upwind and central-differencing scheme were used to discretize the convective and diffusive terms, respectively. The velocity-pressure coupling was performed using PIMPLE algorithm [17]. All of the low turbulent natural convection flow cases, the personal computer with the i7-8700K processor and 8 GB DDR RAMmemory was intended to calculate the effect of wall functions together with the two-equation turbulence models of OpenFOAM.

4. Results and Discussion

4.1. The temperature distribution

Figure 2a shows an example of the temperature distribution in the mid-length plane (Z = 0.75 m) of enclosure by using $RNG\ k-\varepsilon$ model. The color contour was indicated the air temperature. The hot air was red while the cold air was blue. The red color started buoyance up at the bottom edge of hot wall. The hot air buoyant to impinge with the top wall of enclosure. The cold air was displaced by the hot air; therefore, the boundary layer started at the top edge of the cold wall. It flow down to the bottom of enclosure nearly the cold wall. The flow region near the walls was turbulent flow caused of the viscosity of air on these walls. The temperature distribution by the two-equation turbulence models at the height of 0.225 m is compared with the experiment as shown in figure 2b. The simulation points inside CFD domain along the X-axis accorded the experimental data. The temperature changed across the enclosure was steep nearly both isothermal walls. It shot at X = 0.0263 and 0.7238 m, respectively. The comparison

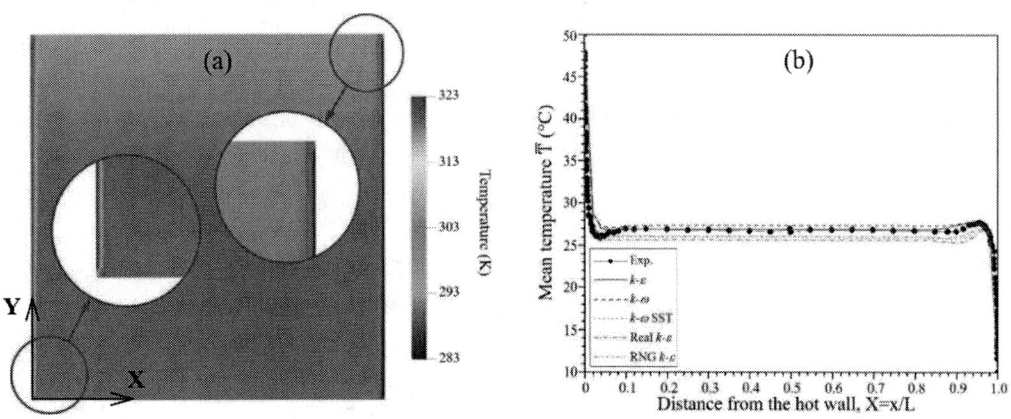

Figure 2. (a) Temperature distribution on the mid-length plan of enclosure by $RNG\ k-\varepsilon$ model and (b) Temperature distribution across the mid-length plan of enclosure at the height of 0.255 m.

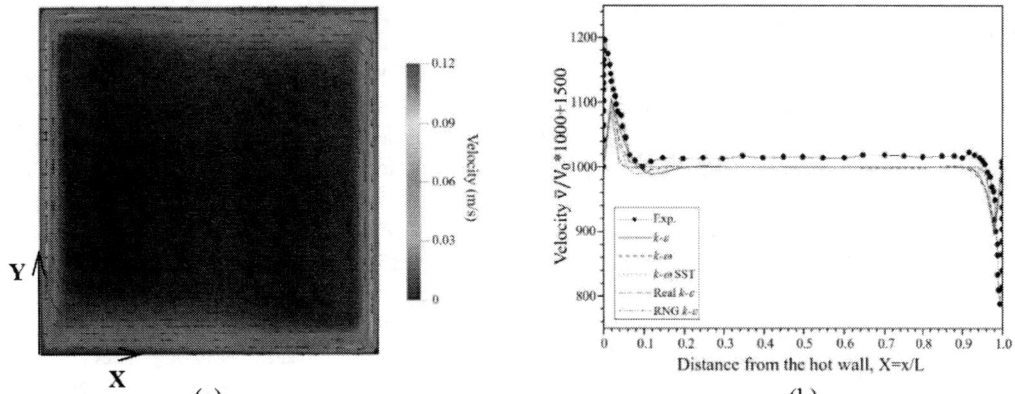

(a) (b)

Figure 3. (a) Velocity distribution on the mid-length plan of enclosure by $k-\varepsilon$ model and (b) The v-velocity distribution across the mid-length plane of enclosure at the height of 0.525 m.

had been performed every 10% of the enclosure height. It was found that the $RNG\ k-\varepsilon$ showed the temperature distribution in the most agreement with the experiment. The $k-\varepsilon$ and $k-\omega$ turbulence models were the second and third accuracy of results, respectively. The average error of the $RNG\ k-$ $-\ \varepsilon$, $k-\varepsilon$ and $k-\omega$ turbulence models was 19.48%, 19.58% and 19.60%, respectively. In addition, the temperature profile was linear nearly the hot and cold walls.

4.2. The velocity distribution

The air velocity was investigated in X and Y directions. The color contour also used to interpret the air velocity of the simulation results and the arrows were added to investigate the flow direction. The high velocity was red while low velocity was blue. Figure 3a shows color contour of the velocity distribution on the mid-length plane of enclosure by $k-\varepsilon$ model. The velocity values in front of the central region of the hot and cold walls were higher than the other regions. Nearly the hot wall, the arrows of air velocity pointed upward. It confirmed the buoyant flow of air happened nearly the hot wall. On the other hand, the air flow down nearly the cold wall. The vortex of air flow happened distinctly in front of the hot and cold walls in the clockwise direction. It confirmed that the turbulent flow phenomenon was happened. The circulation of air flow happened distinctly near the top and bottom walls. It was caused by the flow past upper right angle between hot and top walls and lower right angle between cold and bottom walls. The great difference of air velocity between nearly isothermal wall and inside enclosure

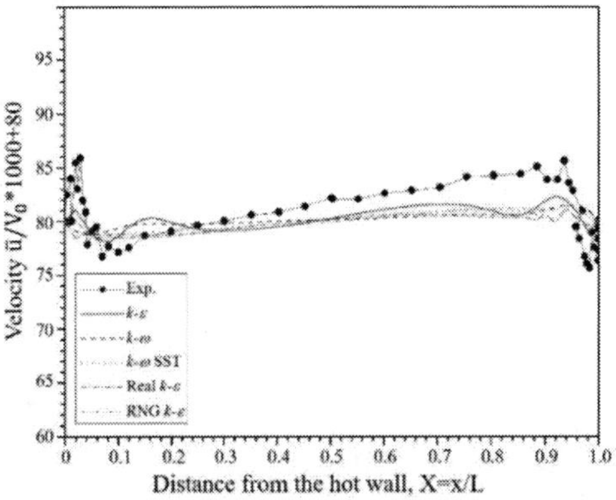

Figure 4. The u-velocity distribution across the enclosure at the height of 0.375 m.

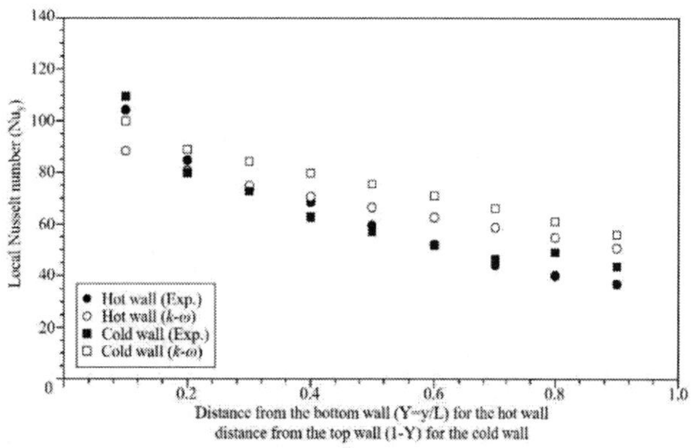

Figure 5. Comparison of the local Nusselt number.

buoyance velocity was equal to 1. The comparison also performed every 10% of the enclosure height. It was found that the $k-\varepsilon$ showed the vertical velocity distributed in the most agreement with the was the other cause that the turbulent flow happened. The v-velocities of simulation on the mid-length plane at the height of 0.525 m is compared with the experiment as shown in figure 3b. V_0 was the experimental data. The $RNG\ k-\varepsilon$ and $k-\omega$ turbulence models were the second and third accuracy, respectively. The average error of the $k-\varepsilon$, $RNG\ k-\varepsilon$, and $k-\omega$ turbulence models was 12.55%, 13.12% and 13.20%, respectively.

Figure 4 shows the comparison graphs of u-velocity between simulation and experiment at the height of 0.375 m. The $k-\omega$ showed the horizontal velocity distributed in the most agreement with the experimental data. The $RNG\ k-\varepsilon$ and $k-\omega\ SST$ turbulence models were the second and third accuracy, respectively. The average error of the $k-\omega$, $RNG\ k-\varepsilon$, and $k-\omega\ SST$ turbulence models was 14.57%, 19.53% and 24.50%, respectively. The horizontal velocity value was one order lower than the vertical velocity. It was significantly distributed at different height; therefore, the error of CFD happened distinctly. Figure 5 shows the comparison of Nusselt number between experiment and simulation. The $k-\omega$ model was good agreement with experiment. The average error was 21.84% showed the accuracy of this model. The Ra in this study was 1.58×10^9 which depended on the different temperature between hot and cold wall and then the Nusselt number was increased following the Ra.

5. Conclusion

The temperature and velocity changed sharply near the isothermal walls, and conducted the turbulence natural convection in an air-filled square cavity. A 3D cell structure was built to study the temperature and velocity distribution inside the enclosure. The cell size was large and uniform to control the y^+ as 30. The standard wall functions include the k, ε and ω wall functions were applied on the enclosure walls. The two-equation turbulence models of Open FOAM including $k-\varepsilon$, $k-\omega$, $RNG\ k-\varepsilon$, $k-\omega\ SST$ and $Realizable\ k-\varepsilon$ were employed to simulate the turbulent buoyance flow. It was found that the $k-\omega$ model was the most accurate turbulence model, to represent the low turbulent natural convection heat transfer, in rectangular enclosed cavities. Particularly, the k and ω wall functions suitably matched with this turbulence model. Instead, fine cells could be used to control the small y^+ less than 1. Consequently, simulation time was not consumed. The average error of this model was less than 15.79% when the temperature, v-velocity and u-velocity across the mid-length plane of enclosure every 10% of the height were investigated. This study depicts a good choice of turbulence model in Open FOAM software, for applying natural convection in various engineering systems.

Acknowledgments

This research was funded by The Thailand Research Fund (TRF) and W.A. Rubbermate Co., Ltd under Research and Researchers for Industry Program, Grant No. PHD61I0003.

References

[1] Miroshnichenko I V and Sheremet M A 2018 *Renewable and Sustainable Energy Reviews* **82** 40-59
[2] Chaiworapuek W and Kittichaikarn C 2016 *Exp. Therm. Fluid Sci.* **77** 179-187
[3] Inworn N and Chaiworapuek W 2018 *Int. J. Heat Mass Tran.* **122** 1153-1161
[4] Rakpakdee W Kittichaikarn C and Chaiworapuek W 2019 *Int. J. Heat Mass Tran.* **135** 491-510
[5] Mao Y and Zhang Y 2013 *Numerical Heat Transfer Part B – Fundamentals* **64** 351-364
[6] Rathore S and Das M 2016 *Heat Mass Transfer* **52**(3), 437-457
[7] Pandey S Park Y G and Ha M Y 2019 *Int. J. Heat Mass Tran.* **138** 762-795
[8] Cintolesi C Petronio A and Armenio V 2016 *Int. J. Heat Mass Tran.* **101** 1123-1137
[9] Chen H T Lin M C and Chang J R 2018 *Int. J. Heat Mass Tran.* **124** 1217-1229
[10] Dehbi A Kelm S Kalilainen J and Mueller H 2019 *Nuclear Engineering and Design* **341** 176-185
[11] Iyi D and Hasan R 2019 *Int. J. Heat Mass Tran.* **136** 543-554
[12] Das D Roy M and Basak T 2017 *Int. J. Heat Mass Tran.* **106** 356-406
[13] Enayati H Chandy A J and Braun M J 2018 *International Journal of Thermal Sciences* **123** 42-57

[14] Suvanjumrat C 2017 *Engineering Journal* **21**(5) 225-241
[15] Suvanjumrat C Loksupapaiboon K 2020 *Engineering Journal* **24**(2) 109-120
[16] Suvanjumrat C 2017 *Engineering Journal* **21**(3) 207-221
[17] Robertson E Choudhury V Bhushan S Walters D K 2015 *Computers & Fluids* **123** 122-145
[18] Tian Y S Karayiannis T G 2000 *Int. J. Heat Mass Tran.* **43**(6) 849-866
[19] Tian Y S Karayiannis T G 2000 *Int. J. Heat Mass Tran.* **43**(6) 867-884

Evaluation of wall functions for RANS turbulence models by unsteady lid-driven cavity flow

Chakrit Suvanjumrat[1,2], Kittipos Loksupapaiboon[1,2,*]

[1]Department of Mechanical Engineering, Faculty of Engineering, Mahidol University, Nakhon Pathom, 73170, Thailand
[2]Laboratory of Computer Mechanics for Design (LCMD), Department of Mechanical Engineering, Faculty of Engineering, Mahidol University, Nakhon Pathom, 73170, Thailand

* Corresponding Author: kittipos.lok@gmail.com

Abstract
Turbulence models do not perform well near walls because the Reynolds number in this interesting area is low. The wall functions and the advantage method can be efficiently used to model fluid flow at the near wall region. The Reynolds averaged Navier-Stokes turbulence models which had the developed wall functions in Open FOAM, the open source CFD software, were evaluated by the turbulent flow. The 3-D lid-driven cavity flow at the turbulence Reynolds number of 10,000 with various spanwise aspect ratio had been referred to validate these turbulence models. The comparison shown that the realizable $k-\varepsilon$ turbulence model with the standard k and ε wall functions of OpenFOAM was the most powerful model to represent the velocity profile of water flow in the 3-D lid-driven cavities.

Keywords: Wall function, RANS, Turbulence model, Cavity

1. Introduction

The Reynolds averaged Navier-Stokes (RANS) turbulence model obtained by the evaluation of Reynolds stress in the RANS equation [1]. To calculate the Reynolds stress, the transport equations of the turbulent energy were carried out for the computational fluid dynamics (CFD) modeling. Many transport equations models had been used to estimate the Reynolds stress. The $k-\varepsilon$ and $k-\omega$ models are two-equation transport models widely used to simulate turbulence flow, based on flow over immersed bodies and duct flow [2, 3]. The $k-k_L-\omega$ turbulence model is the three-equation transport model included in the RANS turbulence model which is also used to simulate turbulent flow [4]. The $\overline{v^2}-f$ model otherwise the RANS turbulence model is composed of four transport equations. It met with great success to simulate some kinds of the open channel flow [5-8]. The turbulence models which were reviewed usually simulated the complex flow domains. The fine grids at the rigid wall of complex domain were required to process these turbulence models efficiently. Consequently, the computational costs were consumed importantly to capture the turbulent flow at the near wall region.

The advantage approach which can reduce the computational costs of the turbulent flow simulation wall functions. In this way, the near wall region of complex flow domain will yield coarse grids. Many wall functions were improved and developed for the remedy of this problem. For example, Parente et al. [9] had been improved $k - \varepsilon$ model and wall function to simulate the atmospheric boundary layer flow. The source term was added to the dissipation rate equation while the rough surface was improved for wall function. The flow around bluff-body was carried out to demonstrate the improved $k - \varepsilon$ model and wall function. Nazif et al. [10] derived an analytic wall function which significantly reduced the computational costs. The Robin-type wall function had been implemented by using two benchmark cases, near-equilibrium channel flow and flow over a backward-facing step. This function obtained computational costs slightly higher than the standard wall function but better kept the DNS results. Chedevergne [11] had developed wall function by including the roughness corrections in the analytical wall function model to investigate the roughness effect on velocity profile. However, the improved wall function adapted only to the $k - \omega\ SST$ model. Beaugendre and Morency [12] proposed a wall function which was compatible with the penalization method and vortex formulation to simulate fluid flow around complex geometries.The velocity profiles of the Spalart-Allmaras model, RANS turbulence model, with a wall function agreed with the analytic law of the wall solution. In addition to the turbulent flow, Cao et al. [13] improved heat transfer model based on an enhanced thermal wall function. The $RNG\ k - \varepsilon$ turbulence model of gas flow was utilized in the heat transfer model. The stable and accuracy predictions were achieved by the improved model. Consequently, the wall functions advantage to reduce simulating time by increase of grid size at near wall but still keep an accurate result.

The flow simulation with turbulence model and wall function which were reviewed were generally performed by the commercial CFD software. It was the great barrier to the researchers whom had not had funding for the license software support. This research aimed to carry out turbulent flow simulation using the open source CFD software, Open FOAM. The RANS turbulence model and wall functions were implemented and valid by the experimental data of reference papers [14-16]. The capable RANS turbulence model and wall functions were proposed to advance the flow simulation in the further works.

2. Wall functions derivations

In the process of derivation of wall function, turbulence parameters are given in non-dimensional form as follows:

$$U^+ = \frac{\bar{U}}{u_\tau}, k^+ = \frac{k}{u_\tau^2}, \varepsilon^+ = \frac{\varepsilon v}{u_\tau^4}, \omega^+ = \frac{\omega v}{u_\tau^2}, \bar{v}^{2+} = \frac{\overline{v^2}}{u_\tau^2}, f^+ = \frac{f v}{u_\tau^2}, u_\tau = \sqrt{\frac{\tau_w}{\rho}} \tag{1}$$

where τ_w is the local wall shear stress, $\bar{U}$ is averaged velocity, u_τ is the friction velocity, k is the turbulent kinetic energy, ε is the rate of viscous dissipation, v is the kinematic viscosity, ω is the turbulence frequency, $\overline{v^2}$ is the wall-normal stress, and f is the wall-normal fluctuation.

2.1 The $k - \varepsilon$ turbulence model

The standard $k - \varepsilon$ turbulence model composes two transport equations to solve two turbulence quantities, the turbulence kinetic energy and rate of viscous dissipation [17]. In the viscous sub-layer, the wall shear stress is equal to the shear stress of fluid flow which is written as follows:

$$\tau = \tau_w = \mu \frac{\partial \bar{U}}{\partial y} \tag{2}$$

The velocity on the solid wall is assessed by the no-slip condition, therefore the solutions of two transport equations are written as follows:

$$k = u_\tau^2 C_\mu{}^{-1/2}, \varepsilon = C_\mu^{3/4} \frac{k^{3/2}}{\kappa y} \tag{3}$$

where κ is the Karman coefficient which is equal to 0.41, and C_μ is equal to 0.09.

The fluid flow inside the inertial sub-layer, therefore the fluid velocity can be written by following equation.

$$U^+ = \frac{\bar{U}}{u_\tau} = \frac{1}{\kappa} ln(Ey^+), y^+ = \frac{y}{v}\sqrt{\frac{\tau_w}{\rho}} \tag{4}$$

where E roughness coefficient.

The turbulence kinetic energy and rate of viscous dissipation are written by following equations.

$$k = u_\tau^2 C_\mu^{-1/2}, \varepsilon = \frac{u_\tau^3}{\kappa y} \tag{5}$$

2.2 The $k-\omega$ model

The $k-\omega$ turbulence model is composed the turbulence kinetic energy and turbulence frequency transport equations [18]. In the viscous sub-layer, these transport equations have solution as written by:

$$\omega^+ = \frac{6}{\beta_1 (y^+)^2}, k^+ = C_\mu (y^+)^{3.23} \tag{6}$$

In the inertial sub-layer, the diffusion term is assumed to be small when compared to the other two terms of the frequency transport equation. Using $k^+ = \kappa y^+ \omega^+$, the turbulence kinetic energy equation can balance to provide the ω^+ and give k^+ by the following equations.

$$\omega^+ = \frac{1}{\kappa \sqrt{C_\mu} y^+}, k^+ = \frac{1}{\sqrt{C_\mu}} \tag{7}$$

2.3 The $\overline{v^2} - f$ model

The $\overline{v^2} - f$ is four transport equations in RANS turbulence model [1]. In the viscous sub-layer, the ε^+ and k^+ can be analytic, and written as follows:

$$\varepsilon^+ = \frac{2400}{C_{\varepsilon 2}^2} \frac{1}{(y^+ + C)^4}, k^+ = \frac{2400}{C_{\varepsilon 2}^2} \left[\frac{1}{(y^+ + C)^4} + \frac{2y^+}{C^3} - \frac{1}{C^2} \right] \tag{8}$$

The boundary condition of f^+ is derived by assuming that $\left(\overline{v^2}\right)^+$ is equal to $C_{v2}(y^+)^4$, therefore the solution of $\overline{v^2} - f$ equations is written as follows:

$$f^+ = \frac{-4(6-N)\left(\overline{v^2}\right)^+}{\varepsilon^+ (y^+)^2} \tag{9}$$

In the inertial sub-layer, the diffusion term is small, therefore the diffusive flux is constant as follows:

$$\kappa y^+ \frac{d^2 k^+}{dy^+} = C_k \tag{10}$$

Integrating, equation (10) yields the following equation.

$$k^+ = \frac{C_k}{\kappa} log(y^+) + B_k \tag{11}$$

In the same way, the $\left(\overline{v^2}\right)^+$, ε^+ and f^+ are obtained as follows:

$$\left(\overline{v^2}\right)^+ = \frac{C_{v2}}{\kappa} log(y^+) + B_{v2}, \varepsilon^+ = \frac{1}{\kappa y^+}, f^+ = N \frac{\left(\overline{v^2}\right)^+}{(k^+)^2} \varepsilon^+ \tag{12}$$

To fit the numerical solution of $\overline{v^2} - f$ equation, the following values for the constants are given as follows:

$$C_k = -0.416, B_k = 8.366, C_{v2} = 0.193, B_{v2} = -0.940 \tag{13}$$

3. Verifications

The lid-driven cavity flow models was developed following the dimension of three rectangular cavities. The cross section of each cavity was equal to the square size of 150×150 mm^2 (B×H) while the length was various to investigate the side wall effect. The spanwise aspect ratio (SAR) composes, the ration between length (L) and width (B), 0.5/1, 1/1, and 3/1, respectively. Figure 1 shows the schematic of the rectangular cavity flow. The fluid of cavity flow was water which had the constant density and kinematic viscosity of 997.05 kg/m^3 and 0.89×10^{-6} m^2/s, respectively. The water flow over the cavity along the width, characteristic length, with the constant velocity of 0.05953 m/s, therefore it sturdily occupied the Reynolds number of lid-driven cavity flow at 10,000. The velocity on five sides of the cavity wall was fixed by the no-slip boundary condition. The pressure boundary condition was the zero gradient on every

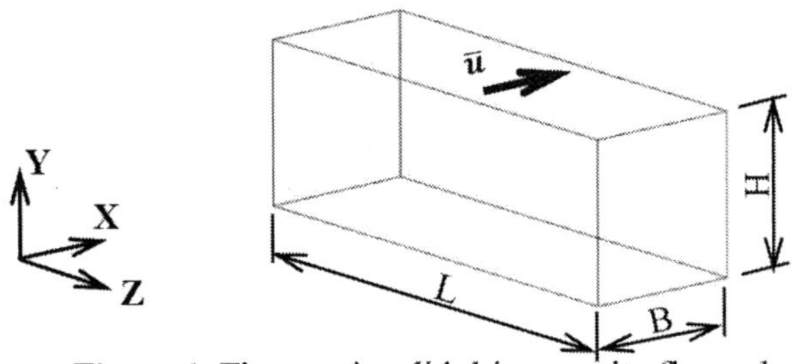

Figure 1. The transient lid-driven cavity flow schematic.

side of cavity flow model. The dimensionless wall distance (y^+) at the wall region was five which were the maximum size that be still in the viscous sub-layer. The grid structure of the lid-driven cavity flow model was possessed by y^+; therefore, the uniform grids which were divided the rectangular cavities have the cubical size with the dimension of 2.5×2.5×2.5 mm³. The RANS turbulence model which was used to simulate the transient lid-driven cavity flow composed the $k-\varepsilon$, $k-\omega$, $k-\omega\ SST$, $RNG\ k-\varepsilon$, $Realizable\ k-\varepsilon$ and $\overline{v^2}-f$ turbulence model. The turbulence parameters on the rigid wall of the lid-driven cavity flow model including k, ε, ω, $\overline{v^2}$ and f were defined according to the wall functions based on the viscous sub-layer. The transient flow was estimate by Euler implicit method with the constant time step of 0.01 seconds. Effects of velocity of water flow inside the cavity was assessed by the limited linear differencing scheme which was class of the total variation diminishing (TVD) schemes equation. The iteration for the pressure-velocity coupling problem of flow had been solved by PISO method. All of lid-driven cavity flow cases, the personal computer with the i7-8700K processor and 8 GB DDR RAM memory was intended to calculate the effect of wall functions together with the RANS turbulence model of OpenFOAM.

4. Results and discussion

The simulation results were illustrated on the cross-section plan at the mid length of cavities. To perceive the phenomenon of lid-driven cavity flow, the stream tracer was used to present the eddy flow in cavities. Figure 2 shows the stream tracer in the cross-section plan at the mid length of lid-driven cavity with a SAR of 3/1. The transient flow inside lid-driven cavity had scenario of the secondary eddy forming. It consisted of one main circulation flow and three secondary eddies. The first secondary eddy established at the right bottom corner of cavity, the downstream secondary eddy, by the near boundary flow past a right-side wall and right angle. The upstream secondary eddy occurred at the left bottom corner of cavity by the near boundary flow past bottom wall and right angle, which was the second order secondary eddy. The final secondary eddy was the upper secondary eddy, which formed at the left top of cavity cause of the near boundary flow past just left side wall of cavity. The shear stress and flow direction were the main cause of the separated flow that the secondary eddies were formed.

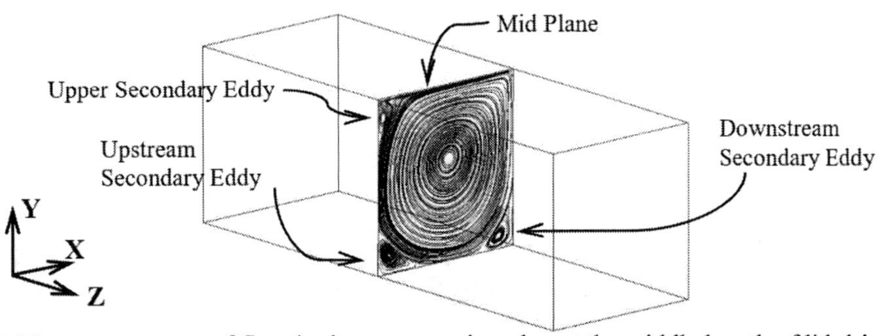

Figure 2. The stream tracer of flow in the cross-section plan at the middle length of lid-driven cavity for SAR 3/1.

The transient lid-driven cavity flow was investigated in the time period of 150 seconds. The RANS turbulence models could calculate to form the secondary eddies in different time. In the lid- driven cavity with a SAR of 3/1, the downstream eddy by the $k-\varepsilon$, $k-\omega$, $k-\omega\ SST$ and $RNG\ k-\varepsilon$ turbulence model was formed after 35, 35, 35 and 35 seconds, respectively. Subsequently, the upstream eddy by the $k-\varepsilon$, $k-\omega$, $k-\omega\ SST$ and $RNG\ k-\varepsilon$ turbulence model was formed after 45, 45, 50 and 50 seconds, respectively. The downstream and upstream secondary eddies of the $Realizable\ k-\varepsilon$ and $\overline{v^2}-f$ turbulence model occurred almost in the same time (20 seconds). The upper secondary eddy was formed by every RANS turbulence model. The $k-\varepsilon$, $k-\omega$, $k-\omega\ SST$, $RNG\ k-\varepsilon$, $Realizable\ k-\varepsilon$ and $\overline{v^2}-f$ turbulence model could form the upper secondary eddy after 125, 70, 70, 90, 100 and 90 seconds, respectively. Cause of the proper wall functions, three secondary eddies could calculate by RANS turbulence models and the sluggish forming of upper secondary was observed by the $k-\varepsilon$ turbulence model. In the near-wall region, the $k-\omega$ model had good to implement boundary layer. The $k-\omega$ and $k-\omega\ SST$ were incapable preserving the conformation of secondary eddies completely to 150 seconds due to the end-wall effect arrived to the mid plane of cavity faster than other models. Figure 3 shows the stream tracer in the lid-driven cavity with the SAR of 3/1 by the $Realizable\ k-\varepsilon$ turbulence model. The secondary eddy forming happened at three regions along the transient flow simulation [18]. Consequently, entire RANS turbulence models which used wall functions except the $Realizable\ k-\varepsilon$ and $\overline{v^2}-f$ turbulence models could calculate secondary eddies in order to experiment.

In the lid-driven cavity with a SAR of 1/1, the upper secondary eddy generation was affected by its end walls. The end walls drag the main circulation of water flow. The upper secondary eddy could not be generated by each RANS turbulence model. The $k-\varepsilon$, $k-\omega\ SST$, $Realizable\ k-\varepsilon$ and $\overline{v^2}-f$ turbulence models could not form the upper secondary eddy while it was happened by the $k-\omega$ and $RNG\ k-\varepsilon$ turbulence models after 70 and 85 seconds, respectively. Even though the $k-\omega$ and $RNG\ k-\varepsilon$ turbulence model could calculate the upper secondary eddy successfully, it was not the complete pattern of the secondary eddies forming. The end walls were close to the mid plane of lid-driven cavity with a SAR of 0.5/1 were more influent than other larger cavity. The last secondary eddy could ever be formed by any RANS turbulence model. It was eliminated caused of the drag effect grew stronger at the middle plane.

All RANS turbulence model which governed all size of lid-driven cavity mostly formed the secondary eddies corresponding the experiment after 60 seconds. Consequently, the normalized mean velocity profiles, $\overline{U}/U_B$ and $\overline{V}/U_B$, by the RANS turbulence model in the lid-driven cavity for three SARs at the corresponding time are plotted by graphs as shown in left of figure 4. In the lid-driven cavity with SAR of 3/1, the $Realizable\ k-\varepsilon$ turbulence model was in the superb agreement with the experimental data and then the $k-\varepsilon$ turbulence model was the worst agreement. The different value of $\overline{U}/U_B$ and $\overline{V}/U_B$ from the experiment just were 0.0356 and 0.0344. The $Realizable\ k-\varepsilon$ turbulence model was still superb agreement with the experiment in the lid-driven cavity for the SAR of 1/1 while the worst agreement was change to be the $k-\omega$ turbulence model. It had the different value of $\overline{U}/U_B$

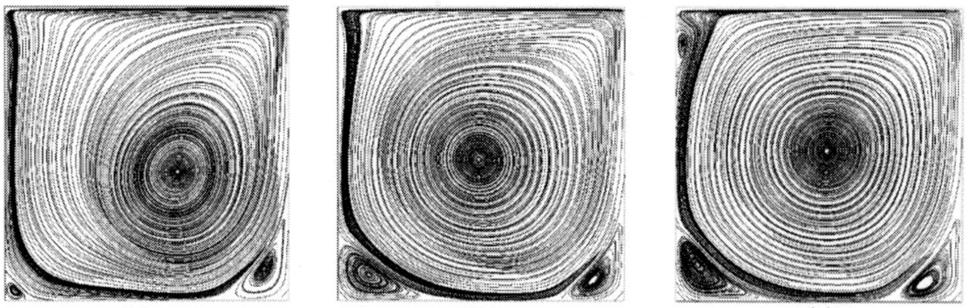

(a) (b) (c)

Figure 3. The transient flow inside lid-driven cavity for SAR of 3/1 by the $Realizable\ k-\varepsilon$ turbulence model after (a) 40, (b) 60 and (c) 100 seconds.

IOP Conf. Series: Materials Science and Engineering 1137 (2021) 012045 doi:10.1088/1757-899X/1137/1/012045

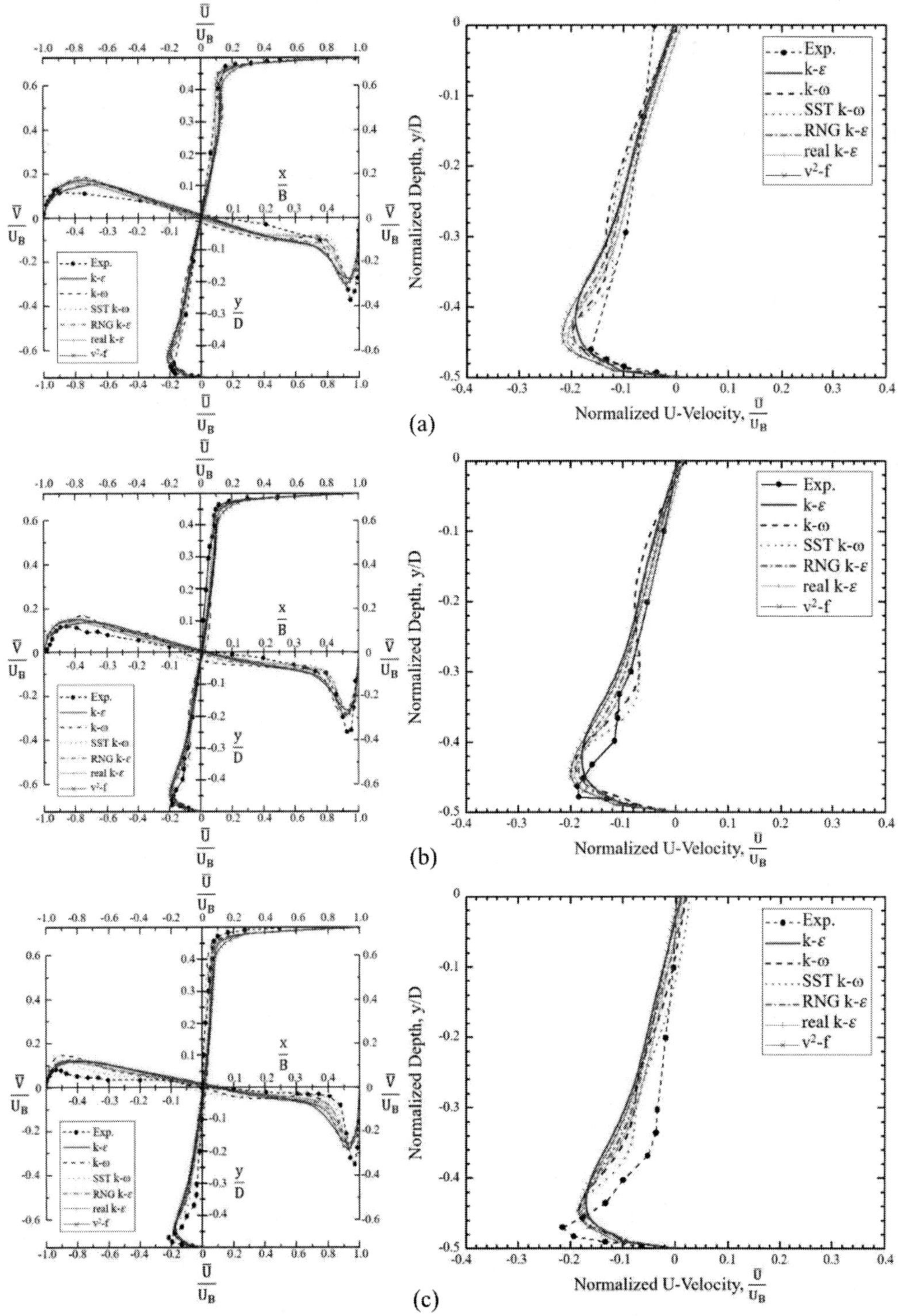

Figure 4. The comparison of normalized mean *U*- and *V*-velocity profiles inside, and normalized mean *U*-velocity profiles over the bottom wall of lid-driven cavity for SAR of (a) the 3/1, (b) 1/1 and (c) 0.5/1.

and $\bar{V}/U_B$ from the experiment about 0.0180 and 0.0308. In the most end-wall effect as the lid-driven cavity with the SAR of 0.5/1, the $Realizable\ k-\varepsilon$ turbulence model was the last turbulence model standing for the good agreement with experiment with the different value of $\bar{U}/U_B$ and $\bar{V}/U_B$ from the experiment were 0.0627 and 0.0401, respectively.

In the right of figure 4 shows the comparison graph of lower part of normalized *U*-velocity profiles between experiment and simulation of the lid-driven cavity with the SARs of 0.5/1, 1/1 and 3/1. The comparison showed that the $Realizable\ k-\varepsilon$ turbulence model had the normalized mean *U*-velocity profile in the good agreement with experiment more than the other models. It was a good end-wall effect sensitivity. This study had been verified the wall functions. The k and ε wall functions had been used efficiently with the $Realizable\ k-\varepsilon$ turbulence model even though both wall functions were derived by the $k-\varepsilon$ turbulence model. Therefore, the simulation result will be good agreement with experiment rather than wall functions of this study if it is derived by pertinent model. Further investigation, the mean *U*-velocity profile could be attributed to the effect of the end wall. As the SAR of cavity decreased, the end-wall effect forced a higher peak *U*-velocity at the near-wall region of bottom cavity.

5. Conclusion

The transient lid-driven cavity flow was modelled by the RANS turbulence models of Open FOAM in this study. According to the simulation results, the RANS turbulence models with the viscous sub-layer wall functions at the maximum y^+ of 5 could illustrate sequence of secondary eddy forming in different time periods. It was found that four turbulence models concluding the $k-\varepsilon$, $k-\omega$, $k-\omega\ SST$ and $RNG\ k-\varepsilon$ turbulence models could form the secondary eddies respectively in accordance with the experiment. Decrease of wall distance from the mid plane increased effect to generation of the upper secondary eddy, and conserved the complete pattern of secondary eddies. Particularly, the end-wall effect increased the peak of normalized mean *U*-velocity at the near-wall region over the bottom wall of cavity. It was found that only the $Realizable\ k-\varepsilon$ turbulence model with the k and ε wall functions could show very good agreement with the experiment. These wall functions which were implemented in Open FOAM could represent the normalized mean *U*-velocity profile in the near-wall region efficiently even though they were derivative by the $k-\varepsilon$ turbulence model. The velocity profile at the near-wall region by the $Realizable\ k-\varepsilon$ turbulence model may be very accurate if the k and ε wall functions are derived from the pertinent turbulence model in the future work. However, this explored model advantageously obtained accurate results when simulated near wall flow problems.

Acknowledgments

This research was funded by The Thailand Research Fund (TRF) and W.A. Rubbermate Co., Ltd under Research and Researchers for Industry Program, Grant No. PHD61I0003.

References

[1] Suvanjumrat C 2020 *The Finite Volume Method with OpenFOAM* (Bangkok: CU-Press)
[2] Suvanjumrat C 2017 *Engineering Journal* **21**(3) 207-221
[3] Suvanjumrat C and Loksupapaiboon K 2020 *Engineering Journal* **24**(2) 109-120
[4] Salimipour E 2019 *Computers and Fluids* **181** 67-76
[5] Priyadumkol J and Kittichaikarn C 2013 *Applied Mechanics and Materials* **249-250** 126-131
[6] Chaiworapuek W and Kittichaikarn C 2016 *Exp. Therm. Fluid Sci.* **77** 179-187
[7] Inworn N and Chaiworapuek W 2018 *Int. J. Heat Mass Tran.* **122** 1153-1161
[8] Rakpakdee W Kittichaikarn C and Chaiworapuek W 2019 *Int. J. Heat Mass Tran.* **135** 491-510
[9] Parente A Gorle C Beeck J and Benocci C 2011 *J. Wind Eng. Ind. Aerod.* **99**(4) 267-278
[10] Nazif H R Tabrizi H B and Ferhadpour F A 2014 *European Journal of Mechanics B/Fluids* **44** 22-31
[11] Chedevergne F 2018 *Int. J. Heat Fluid Fl.* **73** 258-269
[12] Beaugendre H and Morency F 2018 *Computers and Fluids* **170** 313-323
[13] Cao J Jia M Niu B Chang Y Xu Z and Liu H 2019 *Energy Conservation and Management* **195** 748-759
[14] Koseff J R and Street R L 1984 *J. Fluid Eng-T ASME* **160**(4) 390-398

[15] Rhee H S Koseff J R and Street RL 1984 *Exp. Fluids* **2** 57-64
[16] Prasad A K and Koseff J R 1989 *Physics of Fluids A: Fluid Dynamics* **1**(2) 208-218
[17] Suvanjumrat C 2017 *Engineering Journal* **21**(5) 225-241
[18] Isaev S A Baranov P A Sudakov A G and Mordynsky N A 2008 *Thermodynamics and Aeromechanics* **15**(3), 463-476

IOP Conf. Series: Materials Science and Engineering 1137 (2021) 012046 doi:10.1088/1757-899X/1137/1/012046

Numerical simulation of flow over a passive disturbance and backward-facing step

Kittipos Loksupapaiboon[1,2], **Chakrit Suvanjumrat**[1,2,*]

[1]Department of Mechanical Engineering, Faculty of Engineering, Mahidol University, Nakhon Pathom, 73170, Thailand
[2]Laboratory of Computer Mechanics for Design (LCMD), Department of Mechanical Engineering, Faculty of Engineering, Mahidol University, Nakhon Pathom, 73170, Thailand

* Corresponding Author: chakrit.suv@mahidol.ac.th

Abstract
This research aimed to simulate the flow past a rib and backward-facing step (BFS). The Reynolds and sub-grid-scale (SGS) stresses were modelled. These models consisted of $k-\varepsilon$, Smagorinksy-Lilly, SGS turbulence kinetic energy (k_{SGS}), dynamic SGS and wall-adapting local eddy-viscosity (WALE) which implemented in Open FOAM. The simulation was carried out to compare against the experiment in literatures which gave a high Reynolds number of 15,500. The simulation results showed that the $k-\varepsilon$ model provided the best prediction of the turbulent flow over a passive disturbance and BFS. The absence of rib was carried out to compare with the presence of the rib. It was found that the rib positions had affected to flow reattachment behind the BFS.

Keywords: Simulation, Flow, Passive disturbance, Backward-facing step

1. Introduction

Separated flows are investigated in a wide range of devices or structures contacting air flow [1]. Generally, they are caused by a change of local structure, such as a channel with sudden expansion. The reattached flow location behind a backward-facing step (BFS) is an important phenomenon that attracts many researchers to manage [2]. The possibility of separated flow control is active and passive methods. The periodic disturbances inserted into the main flow or near wall of channel by a speaker [3] or a jet actuator [4-6] were the active method. These methods were effective but difficult to implement. The passive method is preferred. There are the additional elements with smaller sizes than the main separated flows or various step angles. The plate was additional on the step to study the passive control method. It changed the mean profiles of downstream reattached flows [7]. The rib was located on the channel entrance. An effect was found to happen to the downstream reattachment flow, when the distance between the rib and the BFS was proper [8, 9]. The numerical method is usually used for the simulation of turbulent flow over BFS. The turbulence modeling method is commonly used to simulate the large eddy flow in a high range of Reynolds number. The turbulence models are carried out to solve the Reynolds stresses in the Reynolds averaged Navier-Stokes (RANS) equation which governs the eddy flow [10]. Otherwise, simulation method, the spatial filtering operations are used to develop the Navier-Stokes equations. The sub-grid-scale (SGS) stresses are terms which happen by the filtering operation.

Many models were developed to calculate the SGS stresses. There were included Smagorinksy-Lilly, SGS turbulence kinetic energy (k_{SGS}), dynamic SGS and wall-adapting local eddy-viscosity (WALE) models [11, 12]. The selection of the SGS model play an important role in the large eddy simulation (LES) method. The flow over BFS with the expansion ratio of 2.00 and Reynolds number of 9,000 was investigated by $k-\varepsilon$ model and LES method. The Smagorinsky-Lilly model was used to calculate SGS stresses and successful to simulate the instantaneous large-scale vortices in BFS flow [13]. The dynamic SGS was selected to model the BFS with the expansion ratio of 2.25 and Reynolds number of 6,400. It was compared with the direct numerical simulation (DNS) and linear $k-\varepsilon$ model. The LES results were in good agreement with the DNS results in spite of the simulation domain was 1.7 times shorter than in DNS [14]. To investigate effects of a passive disturbance to the reattachment flow, the Smagorinksy-Lilly and one-equation models were used for the SGS stresses. The one-equation model described the velocity field characteristics on the step edge and the development dynamics of the separation region behind the step better than the Smagorinksy-Lilly model [15].

In this research, the Reynolds and SGS stresses models for the flow over the passive disturbance and BFS had been compared. There were $k-\varepsilon$, Smagorinksy-Lilly, k_{SGS}, dynamic SGS and WALE models in Open FOAM, open-source software. Particularly, effects of a location of the passive disturbance to the reattachment point behind the BFS were investigated. The best model found in this study will be used to design the passive disturbance of the BFS geometry, of the devices or structure which are in contact with air in future work.

2. Simulation methods

2.1 Reynolds averaged Navier-Stokes

The mass conservative and Navier-Stokes equation for incompressible fluid have been averaged by time can be written by following equations.

$$\frac{\partial\bar{\rho}}{\partial t}+\nabla\cdot\left(\bar{\rho}\bar{\boldsymbol{U}}+\overline{\rho'\boldsymbol{U}'}\right)=0 \tag{1}$$

$$\frac{\partial\bar{\boldsymbol{U}}}{\partial t}+div(\bar{\boldsymbol{U}}\bar{\boldsymbol{U}})=-\frac{1}{\rho}grad(p)+vdiv(\nabla\bar{\boldsymbol{U}})-div(\overline{\boldsymbol{U}'\boldsymbol{U}'}) \tag{2}$$

2.1.1 The $k-\varepsilon$ model

The Reynolds stresses ($-(\overline{\boldsymbol{U}'\boldsymbol{U}'})$) as shown in equation (2) has been estimated by turbulence model, $k-\varepsilon$ model. The $k-\varepsilon$ model composes the turbulent kinetic energy (k) and rate of viscous dissipation (ε) transport equations [16]. These equations can be written as follows:

$$\frac{\partial(\rho k)}{\partial t}+\frac{\partial(\rho kU_i)}{\partial x_i}=\frac{\partial}{\partial x_i}\left(\frac{\mu_t}{\sigma_k}\frac{\partial k}{\partial x_i}\right)+2\mu_t S_{ij}\cdot S_{ij}-\rho\varepsilon \tag{3}$$

$$\frac{\partial(\rho\varepsilon)}{\partial t}+\frac{\partial(\rho\varepsilon U_i)}{\partial x_i}=\frac{\partial}{\partial x_i}\left(\frac{\mu_t}{\sigma_\varepsilon}\frac{\partial\varepsilon}{\partial x_i}\right)+C_{1\varepsilon}\frac{\varepsilon}{k}2\mu_t S_{ij}\cdot S_{ij}-C_{2\varepsilon}\rho\frac{\varepsilon^2}{k} \tag{4}$$

$$\mu_t=\rho C_\mu\frac{k^2}{\varepsilon},S_{ij}=\frac{1}{2}\left(\frac{\partial U_i}{\partial x_j}+\frac{\partial U_j}{\partial x_i}\right) \tag{5}$$

where $C_\mu = 0.09$, $\sigma_k = 1.00$, $\sigma_\varepsilon = 1.30$, $C_{1\varepsilon} = 1.44$, $C_{2\varepsilon} = 1.92$

2.2 Large eddy simulation method

The LES method uses a spatial filtering operation to separate the larger and smaller eddies. The filtering function performs on the transient flow equation and gives rise to SGS stresses. The unsteady Navier-Stokes equations for a fluid which are performed by the filtering operation are given by:

$$\frac{\partial\rho}{\partial t}+\nabla\cdot(\rho\tilde{\boldsymbol{U}})=0 \tag{6}$$

$$\frac{\partial(\rho\tilde{\boldsymbol{U}})}{\partial t}+div(\rho\tilde{\boldsymbol{U}}\tilde{\boldsymbol{U}})=-grad(\tilde{p})+\mu\,div(\nabla\tilde{\boldsymbol{U}})-\left\{div(\rho\widetilde{\boldsymbol{U}\boldsymbol{U}})+div(\rho\tilde{\boldsymbol{U}}\tilde{\boldsymbol{U}})\right\} \tag{7}$$

The last term of the LES momentum can be written as follows:

$$div(\rho\widetilde{u_i}\boldsymbol{U} - \rho\tilde{u}_i\widetilde{\boldsymbol{U}}) = div(\rho\widetilde{u_i}\tilde{u}_j - \rho\tilde{u}_i\tilde{u}_j) = \frac{\partial \tau_{ij}}{\partial x_j} \tag{8}$$

$$\tau_{ij} = \rho\widetilde{u_i}\tilde{u}_j - \rho\tilde{u}_i\tilde{u}_j = L_{ij} + C_{ij} + R_{ij} \tag{9}$$

$$L_{ij} = \rho\widetilde{\tilde{u}_i\tilde{u}_j} - \rho\tilde{u}_i\tilde{u}_j \tag{10}$$

$$C_{ij} = \rho\widetilde{\tilde{u}_i u'_j} + \rho\widetilde{u'_i\tilde{u}_j} \tag{11}$$

$$R_{ij} = \rho\widetilde{u'_i u'_j} \tag{12}$$

where τ_{ij} is the SGS stresses which contain three groups of stress comprising of 1) Leonard stresses (L_{ij}), 2) cross-stresses (C_{ij}), and 3) LES Reynolds stresses (R_{ij}).

2.2.1 Smagorinsky-Lilly model

The SGS stresses relate to the rate of strain ($\tilde{S}_{ij}$) which can be written by following equations [17].

$$\tau_{ij} = -2\mu_{SGS}\tilde{S}_{ij} + \frac{1}{3}\tau_{ii}\delta_{ij} = -\mu_{SGS}\left(\frac{\partial \tilde{u}_i}{\partial x_j} + \frac{\partial \tilde{u}_j}{\partial x_i}\right) + \frac{1}{3}\tau_{ii}\delta_{ij} \tag{13}$$

$$\mu_{SGS} = \rho(C_{SGS}\Delta)^2|\bar{S}| = \rho(C_{SGS}\Delta)^2\sqrt{2\tilde{S}_{ij}\tilde{S}_{ij}}, \tilde{S}_{ij} = \frac{1}{2}\left(\frac{\partial \tilde{u}_i}{\partial x_j} + \frac{\partial \tilde{u}_j}{\partial x_i}\right) \tag{14}$$

where μ_{SGS} is SGS dynamic viscosity, $|\bar{S}|$ is the average strain rate, C_{SGS} is constant and Δ is the filter cutoff width.

2.2.2 SGS turbulence kinetic energy model

The average strain rate is replaced by the SGS turbulent kinetic energy, therefore the SGS dynamic viscosity is written by:

$$\mu_{SGS} = \rho C'_{SGS}\Delta\sqrt{k_{SGS}} \tag{15}$$

where C'_{SGS} is 0.094.

The distribution of the SGS turbulent kinetic energy can be solved with the transport equation as follows:

$$\frac{\partial \rho k_{SGS}}{\partial t} + div(\rho k_{SGS}\bar{\boldsymbol{U}}) = div\left(\frac{\mu_{SGS}}{\sigma_k}\cdot\nabla k_{SGS}\right) + 2\mu_{SGS}S_{ij}\cdot S_{ij} - \rho C_\varepsilon\frac{(k_{SGS})^{\frac{3}{2}}}{\Delta} \tag{16}$$

where C_ε is constant and σ_k is the turbulent kinetic energy stress.

2.2.3 Dynamic SGS model

Two different SGS stresses by different cutoff widths can be evaluated from following equations.

$$\tau_{ij}^2 - \tau_{ij}^1 = \rho L_{ij} = \rho\widetilde{\tilde{u}_i\tilde{u}_j} - \rho\tilde{u}_i\tilde{u}_j \tag{17}$$

$$L_{ij} - \frac{1}{3}L_{kk}\delta_{ij} = -C_{SGS}^2\left(2\Delta_2^2|\bar{S}|\tilde{S}_{ij} - 2\Delta_1^2\widetilde{|\bar{S}|\tilde{S}_{ij}}\right), C_{SGS}^2 = \frac{\langle L_{ij}M_{ij}\rangle}{\langle M_{ij}M_{ij}\rangle}, M_{ij} = -2\Delta_2^2|\bar{S}|\tilde{S}_{ij} + 2\Delta_1^2\widetilde{|\bar{S}|\tilde{S}_{ij}} \tag{18}$$

where C_{SGS}^2 is constant, Δ is the cutoff width.

2.2.4 WALE

The dynamic viscosity has been written by:

$$\mu_{SGS} = \rho(C_w^2\Delta)^2\frac{(S_{ij}S_{ij})^{3/2}}{(\tilde{S}_{ij}\tilde{S}_{ij})^{5/2} + (S_{ij}S_{ij})^{5/4}}, S_{ij} = \frac{1}{2}\left(\frac{\partial \tilde{u}_k}{\partial x_i}\frac{\partial \tilde{u}_j}{\partial x_k} + \frac{\partial \tilde{u}_k}{\partial x_j}\frac{\partial \tilde{u}_i}{\partial x_k}\right) + \frac{1}{3}\delta_{ij}\frac{\partial \tilde{u}_k}{\partial x_l}\frac{\partial \tilde{u}_l}{\partial x_k} \tag{19}$$

where C_w is 0.325.

3. Backward-facing step with and without a disturbance model

The flow investigation had been performed by modeling the 2D velocity field of the narrow air duct with a BFS before a rib was placed upstream of the BFS, whose geometry was described in [9]. The air duct had a length of 1,000 mm and the BFS had a constant height of 9 mm located at a distance of 600 mm from the air duct entrance. The expansion ratio of the air duct, the ratio between air duct outlet and inlet height, was 1.43. The rib high and width were 3 and 3 mm, respectively. The locations for installation of the ribs were varied between 0 to 60 m upstream of the BFS. The schematic diagram of the channel is shown in figure 1.

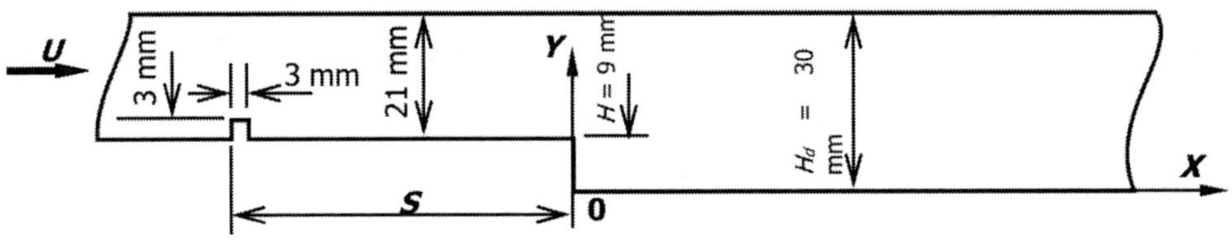

Figure 1. The schematic diagram of the air duct section

To model the narrow air duct without a rib, 221,400 hexagonal cells were carried out by the cell independent method to discrete this domain. In the same manner, the cell number of the air duct with a rib was 221,319. The inlet flow was constant and uniform with air velocity of 25 m/s. Consequently, the Reynolds number of flows over the BFS was 15,500. The outlet boundary condition was the zero gradient of pressure. The wall of air duct was set up with the no-slip boundary condition. The $k-\varepsilon$, Smagorinksy-Lilly, k_{SGS}, dynamic SGS and WALE models which implemented in OpenFOAM were carried out to simulate the velocity field in the air duct domain. The Euler implicit was used for time discretization. The upwind method was used to protect the high-speed effect of air flow for the pressure-velocity coupling problem. The PISO (Pressure Implicit with Split Operator) iteration was used for the accuracy solution. All simulation cases were computed by the personal computer with the computing property on Intel Core-i5 3.2 GHz CPU and 8 GB DDR2 SDRAM memory.

4. Results and discussion

The numerical results of the BFS flow without a rib show the velocity field by a color contour (figure 2). The air velocity was high at the mid height of air duct after past the BFS. The reattachment flow was investigated behind the step. There was the large recirculation flow. When the SGS models were investigated, a smaller one happened at the top wall and deep step corner of the narrow air duct behind the BFS. It was distinctly different flow behavior from the $k-\varepsilon$ model. However, the experimental result could not represent this phenomenon. The comparison of velocity field of simulation is carried out by non-dimensional longitudinal velocity profiles as shown in figure 3. The R-squared value was carried out to decide the accuracy of each model. It was found that the $k-\varepsilon$ model gave the best accuracy in agreement with the experiment. Its R-square was 0.9150. However, the middle range of velocity profile was incorrect slightly from the experimental results. The reattachment point could indicate by negative direction of velocity or reverse flow on bottom wall behind the BFS. It was showed at $X/H = 2\text{-}5$.

Figure 4 represents the numerical results of air flow past the BFS with a rib at $S = 60$. The $k-\varepsilon$ represented the velocity field like the air duct without BFS. The length of reattachment flow behind the BFS was reduced. There was effect by a rib. The air flow was disturbed by the rib; therefore, the vortex happened a lot after past the BFS for the SGS models. The air flow phenomenon by these SGS models could not measure by experiment. Figure 5 shows the comparison of velocity field of these simulations by non-dimensional longitudinal velocity profiles. There was found that the $k-\varepsilon$ model was a good agreement with the experiment because of it performed particularly well in confined flows [11]. Its R-square was 0.8716. The WALE model gave the R-square of 0.3964 which more accurate than other SGS

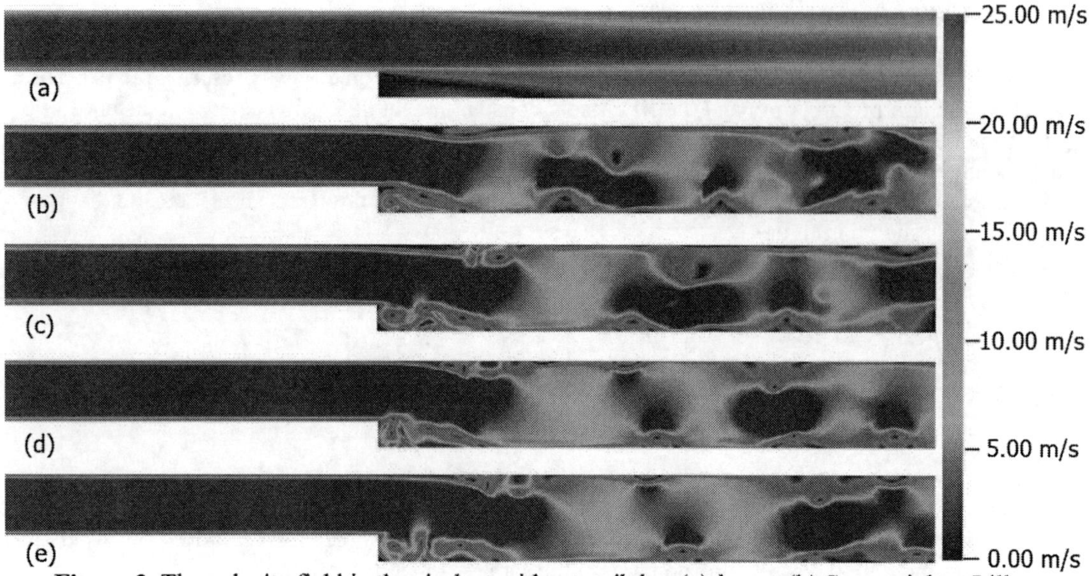

Figure 2. The velocity field in the air duct without a rib by: (a) $k-\varepsilon$, (b) Smagorinksy-Lilly, (c) k_{SGS}, (d) dynamic SGS, and (e) WALE model.

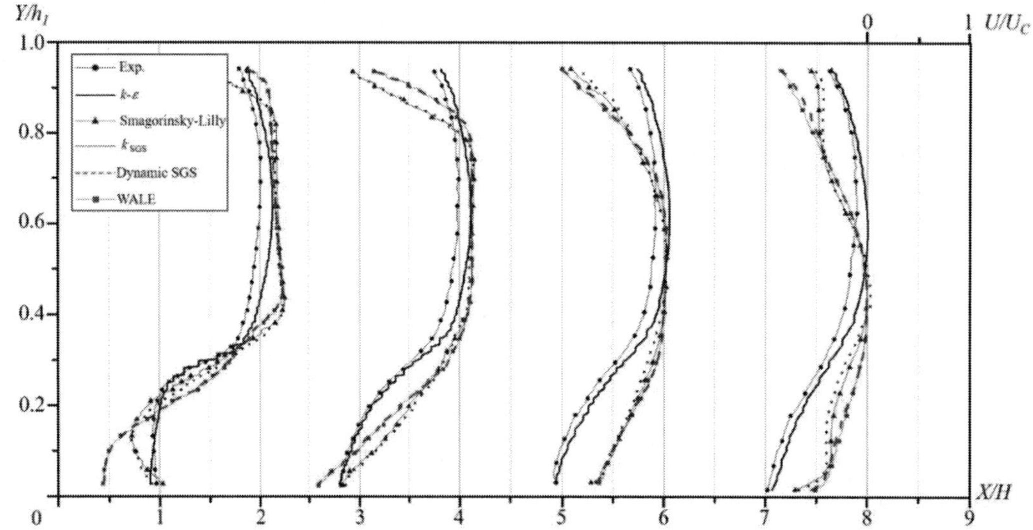

Figure 3. Profiles of streamwise velocity behind the BFS.

models. The rib position had affected to the length between the step and reattachment point extremely, therefore the recirculation length reduced distinctly when compared to the air duct without a rib. The reattachment point happened at X/H = 1-3.

The rib was removed and installed at 3 mm upstream of the step. Figure 6 shows the comparison of velocity field of simulation by non-dimensional longitudinal velocity profiles. The R-square of the $k-\varepsilon$ model was 0.9176. The SGS models gave the minus of R-square value. The reattachment length increased distinctly when compared to the flow past BFS with a rib at S = 60 and without a rib. The reattachment point happened at X/H = 2-7. These simulations rendered the rib was installed at the location which added the height of the BFS. There was found that the height of the BFS increased the recirculation length.

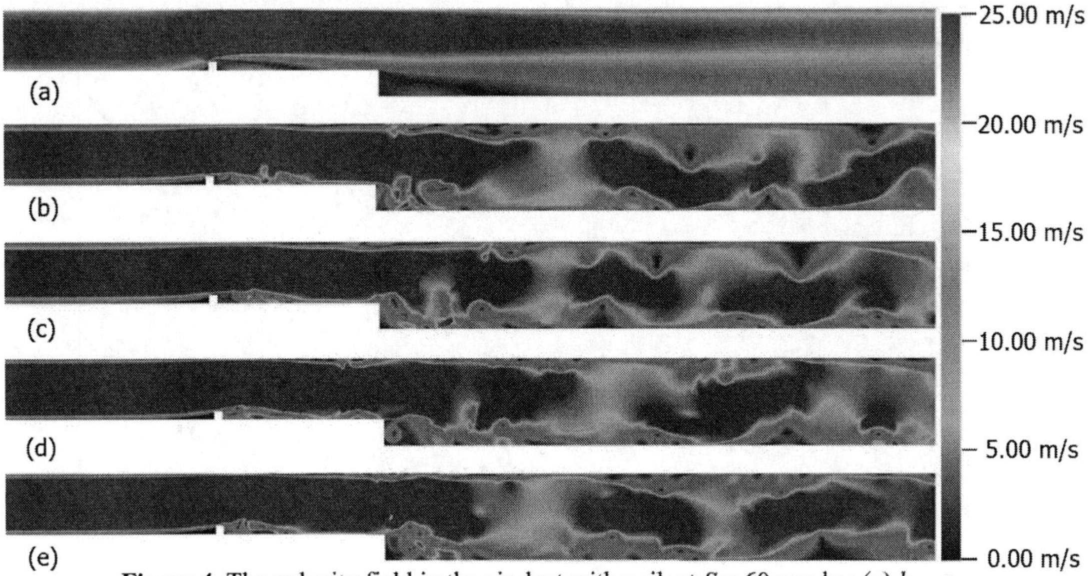

Figure 4. The velocity field in the air duct with a rib at $S = 60$ mm by: (a) $k - \varepsilon$, (b) Smagorinksy-Lilly, (c) k_{SGS}, (d) dynamic SGS, and (e) WALE model.

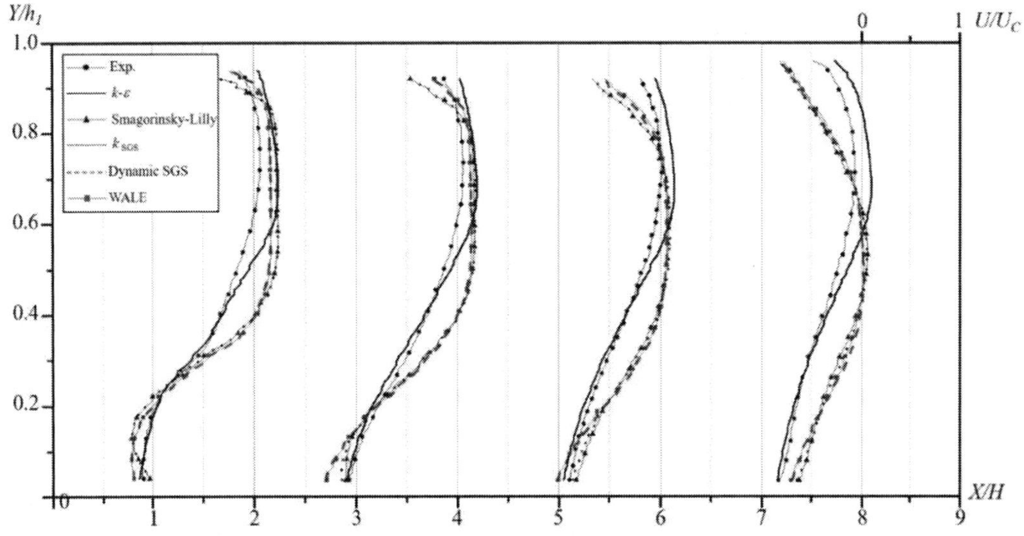

Figure 5. Profiles of streamwise velocity behind the BFS in the presence of the rib at the distance between the rib and the BFS of 60 mm.

5. Conclusion

The velocity field of air flow inside a narrow air duct, with and without a rib upstream of a step, was modelled and investigated. The simulation models included the $k - \varepsilon$, Smagorinksy-Lilly, k_{SGS}, dynamic SGS and WALE models. These models were implemented in Open FOAM. The R-square was carried out to compare with the experiment result for the accuracy of each model. It was found that the best model for simulating these cases was the $k - \varepsilon$ model by the average R-square of 0.9014. The 2D models of SGS stresses, the LES, in Open FOAM did not act properly, for the simulation of flow inside the narrow air duct with BFS, and also with a step and rib. The position for installing the rib had affected

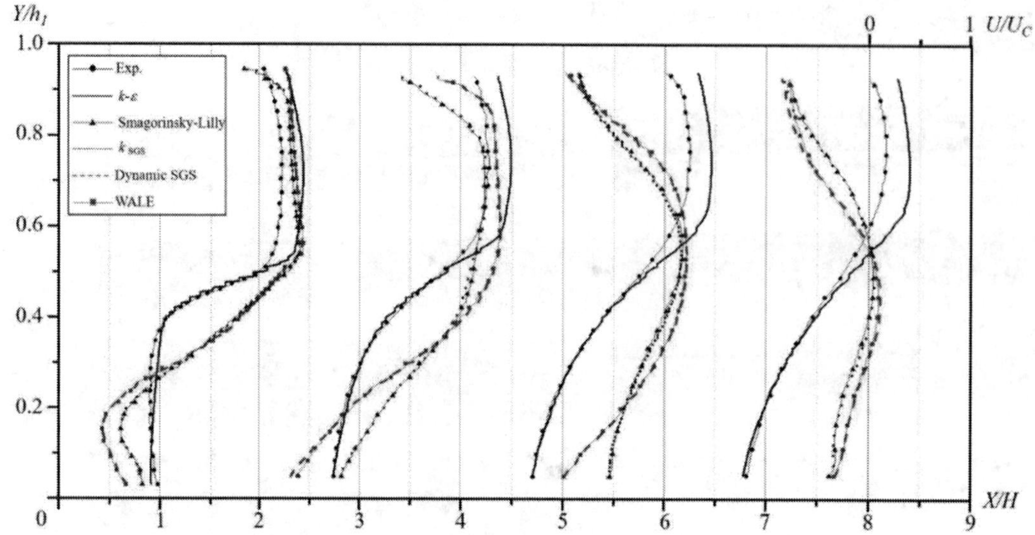

Figure 6. Profiles of streamwise velocity behind the BFS in the presence of the rib at the distance between the rib and the BFS of 3 mm.

to the reattachment point of the recirculation flow behind the BFS. The length between the rib and BFS increased would reduce the recirculation length. Moreover, the height of the BFS also affected to the recirculation length. Therefore, the BFS height increased, would increase the reattachment point behind the BFS. The 2D model which was the $k-\varepsilon$ was found in this study will be used to design the passive disturbance of the BFS geometry for devices or structures in the future work.

Acknowledgments
This research was funded by The Thailand Research Fund (TRF) and W.A. Rubbermate Co., Ltd under Research and Researchers for Industry Program, Grant No. PHD61I0003.

References
[1] Suvanjumrat C 2017 *Engineering Journal* **21**(3) 207-221
[2] Chen L Asai K Nonomura T Xi G and Liu T 2018 *Thermal Science and Engineering Progress* **6** 194-216
[3] Li Z Y Guo S Bai H L and Gao N 2019 *Int. J. Heat Mass Tran.* **130** 240-251
[4] Chaiworapuek W and Kittichaikarn C 2016 *Exp. Therm. Fluid Sci.* **77** 179-187
[5] Inworn N and Chaiworapuek W 2018 *Int. J. Heat Mass Tran.* **122** 1153-1161
[6] Rakpakdee W Kittichaikarn C and Chaiworapuek W 2019 *Int. J. Heat Mass Tran.* **135** 491-510
[7] Priyadumkol J Khaothong K and Chaiworapuek W 2019 *IOP Conference Series: Materials Science and Engineering* **501**(1) 012054
[8] Terekhov V I Smul'skii Y I and Sharov K A 2012 *Tech. Phys. Lett.* **38**(2) 125-128
[9] Terekhov V I Smul'skii Y I and Sharov K A 2016 *J. Appl. Mech. Tech. Phy.* **57**(1) 180-187
[10] Suvanjumrat C and Loksupapaiboon K 2020 *Engineering Journal* **24**(2) 109-120
[11] Versteeg H K and Malalasekera W 2007 *An Introduction to Computational Fluid Dynamics* (Malaysia: Pearson)
[12] Suvanjumrat C 2020 *The Finite Volume Method with OpenFOAM* (Bangkok: CU-Press)
[13] Wang F F Wu S Q and Zhu S L 2019 *Water Science and Engineering* **12**(2) 145
[14] Oder J Tiselj I Jager W Schaub T Hering W Otic I and Shams A 2020 *Nucl. Eng. Des.* **359** 110460
[15] Zhdanov V L Ivanov D A Smul'skii Y I and Terekhov V I 2018 *Journal of Engineering Physics and Thermophysics* **91**(3) 628-640
[16] Suvanjumrat C 2017 *Engineering Journal* **21**(5) 225-241
[17] Smagorinksy J 1963 *Mon. Weather Rev.* **91**(3) 99-164

Design of opener shape for improving gimbal tongue shift on y-axis

Warut Klayaubol[1] and Monsak Pimsarn[1,*]

[1,1,*]Department of Mechanical Engineering, Faculty of Engineering,
King Mongkut's Institute of Technology Ladkrabang, Bangkok 10520, Thailand.

* Corresponding Author: warut.klayaubol@seagate.com

Abstract
There are four necessary components for testing read-write efficiency of Slider, which are [1] Slider electrical testing machine [2] Fixture, a tool for holding Slider while testing [3] Slider to read and write data in hard disk drive and [4] Opener, a mechanism to open a fixture for loading and unloading Slider. The opener should open a fixture only in x direction but it appears that the opener drives a fixture in both x and y directions which results in misalignment of Slider in a fixture. A poor Slider alignment definitely affects a read-write efficiency of Slider. Therefore, this article will show a study of opener shape that relates to fixture behaviour and also a comparison of four designs of opener-arm angle (15 degrees, -15 degrees, -30 degrees and -40 degrees) by using Finite Element Analysis (FEA). Result shows -40 degrees opener is the best one that can open the fixture only in y direction and meets the design requirement.

Keywords: Opener, Slider

1. Introduction

Nowadays, the competition in data storage technology is very intensive. Quality and price are main factors for customers to select the products. Therefore, Seagate Technology (Thailand) Limited, as the leader of data storage business, needs to research and develop Slider testing processes to deliver products with the best quality for clients. There are four components necessary for testing the read-write efficiency of Slider.

1) Slider electrical testing machine
2) Fixture (a tool for holding Slider while testing)
3) Slider (an element to read & write data in hard disk drive)
4) Opener (a mechanism to open a fixture for loading and unloading Slider)

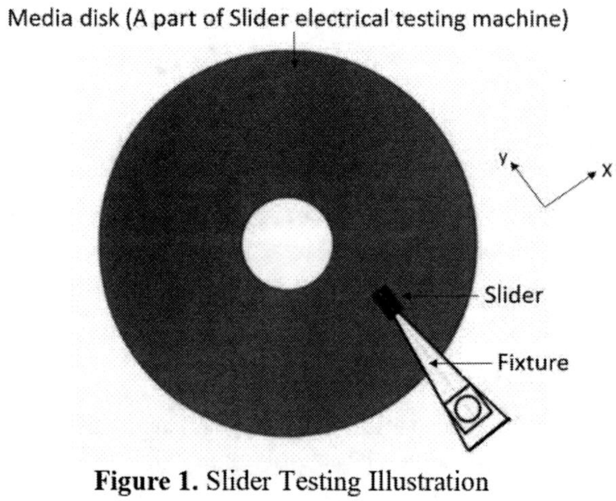

Figure 1. Slider Testing Illustration

A result of loading Slider into a fixture showed Slider misalignment about 30 microns. After investigation, it found that there is a movement of fixture in y direction causing Slider shift from a loading target. This 30-micron misalignments definitely impacts Slider testing performance.

Research and development team had investigated deeply this problem and found that the tip of opener has motion in both x and y directions instead of only x direction as desire. So, when a fixture is opened, it will be moved together with the opener in positive y direction. And it will be moved back in negative y direction when it is closed. Regarding Slider loading sequence, Slider is aligned perfectly with a fixture while it is opened. But while fixture is closing, Slider shifts out from target and results in Slider misalignment. So, the real root cause is a motion of opener tip in y direction.

This study will show a relationship between opener shape and fixture behaviour and also a comparison of four new designs of opener-arm (15 degrees, -15 degrees, -30 degrees and -40 degrees) by using Finite Element Analysis. However, the same failures of this kind of mechanism had been investigated by many researchers, and reported in many literatures. [1] Finite Element Analysis with ANSYS program for optimization by considering stress, weight comparison & reducing of size of connecting rod. [2] Re-design of a failed clutch fork using topology and shape optimization by the response surface method. [3] Finite Element Modeling of structural stainless-steel cross-sections. [4] An automatic Finite Element Modeling for deformation analysis of composite structures. [5] A general mesh smoothing method for Finite Element. [6] Investigation of finite element mesh independence in rate dependent materials. [7] Finite Element Analysis of Stress-Strain Response at the Tool Pin during Friction Stir Process

Although there are many researches in mechanism failures and shape optimization, but a study in micro motion as the opener is rarely considered. In this research, four new conceptual designs of opener are determined a deformation shape to consider the best design of opener that has the minimum y motion.

2. Method

Finite Element Method (FEM), a numerical method, is used for solving engineering problems. Typically, many engineering problems (structural analysis, heat transfer, fluid flow, mass transport, complicated geometries, loadings and material properties) can be solved by using Finite Element Method.

This paper examined the opener shape by using Finite Element Analysis to study the opener characteristics after opening.

Opener mechanism (micro mechanism) is shown in Figure 2. There is a dowel pin, which does not show in a picture, to constraint the opener at the center hole. So, the opener can rotate around the z-axis as red arrow. The external forces, 4.5 N each, are applied to opener arms to activate the opener tip. Then opener tip moves apart in x direction as green arrows.2. Materials and Methods

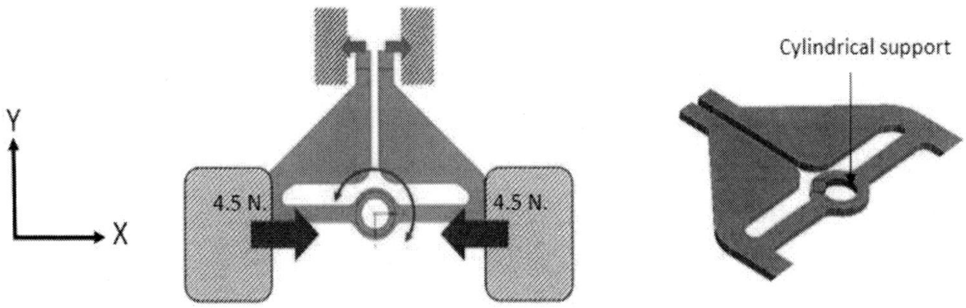

Figure 2. Opener Mechanism

3. Modeling

In this study, software Solidworks 2019 SP3 is used to create opener models and ANSYS Workbench 2020 R1 is used to analysis the opener behaviour with boundary conditions and meshing element as following.

3.1 Boundary condition

In order to make all models be ready for the experiment. It is necessary to determine boundary conditions. For example: material properties, constraints, loads, and etc., need to be represent the actual condition.

Table 1. The material properties of the materials

Materials	Property 1 (Unit)	Property 2
Stainless Steel 440C	HRC	59

Mechanical properties of stainless steel 440C are determined as below:

1. Tempering Temperature (°C) = 204
2. Yield Strength (MPa) = 1900
3. Elongation (% in 50 mm) = 4

3.2 Mesh Independence

It is necessary to consider a mesh convergence at preprocessing process in order to get the accurate result from Finite Element Analysis.

Figure 3. x-axis shows numbers of opener meshing elements and y-axis shows opener tip deformation. Therefore, it can be determined that y deformation will converge to the exact value if number of elements is equal to or higher than 15,000 elements onward. Therefore, the minimum element

IOP Conf. Series: Materials Science and Engineering 1137 (2021) 012047 doi:10.1088/1757-899X/1137/1/012047

that can be used for the opener analysis is 15,000 elements in order to reduce the analytical time of ANSYS program and still deliver a reliable result.

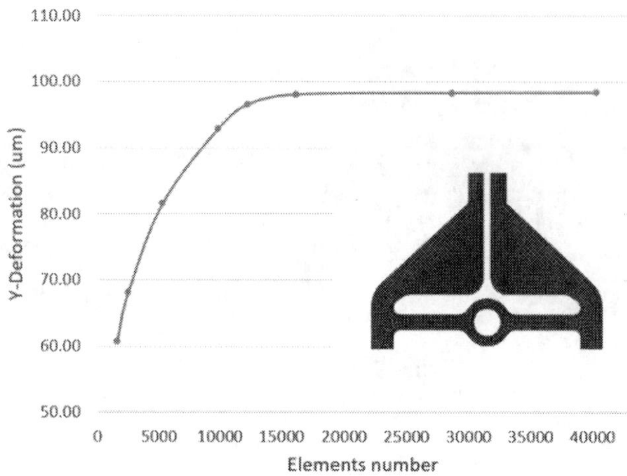

Figure 3. Mesh Independence Study

4. Result

4.1 Model Validation

The commercial Finite Element software (ANSYS) was employed to simulate the response of opener with or without geometry defects. The mesh was discretized using the hex dominant element type. A mesh Independence study was also applied until the simulation results showed negligible change with increasing the mesh elements.

Y deformation of visual inspection and Finite Element Analysis shows in Figure 4. Result between visual inspection data and Finite Element Analysis. There is offset value approximately 6 microns.

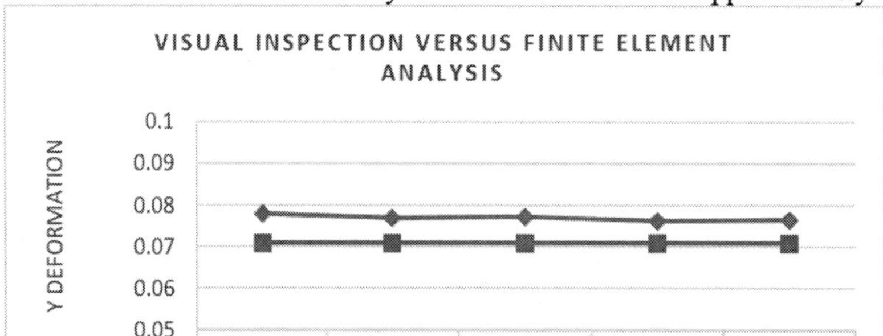

	Round 1	Round 2	Round 3	Round 4	Round 5
VISION	0.078	0.077	0.0772	0.0762	0.0765
FEM	0.0709	0.0709	0.0709	0.0709	0.0709

Figure 4. Visual inspection versus Finite Element Analysis result

Figure 5. Shows result from Finite Element Analysis. The opener tip moves 0.070 mm. in y-axis and stress occurs in both arms, 974.65 MPa, which is still under yield strength of 1,900 MPa. Thus, the deformation is still under elastic zone.

IOP Conf. Series: Materials Science and Engineering 1137 (2021) 012047 doi:10.1088/1757-899X/1137/1/012047

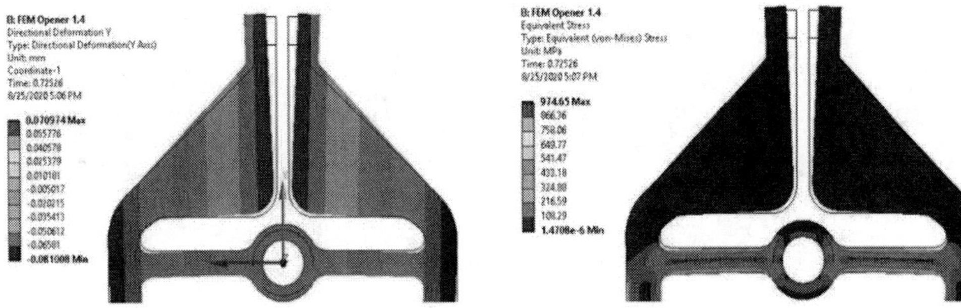

Figure 5. Deformation and stress of the opener

4.2 Design Improvement.

A study is designed to consider parameters as shown in Table 1. Due to a limitation of design packaging, only the opener-arm can be modified. Therefore, the angle of opener-arm is varied from 15 degrees to -40 degrees. Thickness and material (steel sheet grade SST440C) of new designs are maintained the same as existing opener.

Table 2. Parameter for study

Parameter	Nominal	Min.	Max.
Alpha angle	-	15 degrees	-40 degrees
Thickness of Opener	0.200 mm.	-	-
Material of Opener	SST 440C	-	-

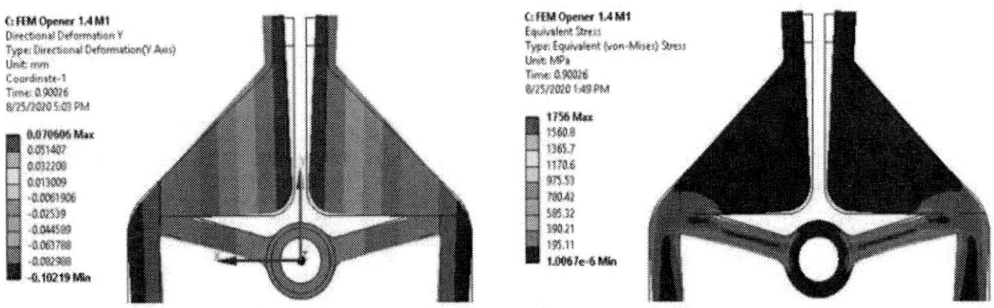

Figure 6. An improved version A (15 degrees)

IOP Conf. Series: Materials Science and Engineering 1137 (2021) 012047 doi:10.1088/1757-899X/1137/1/012047

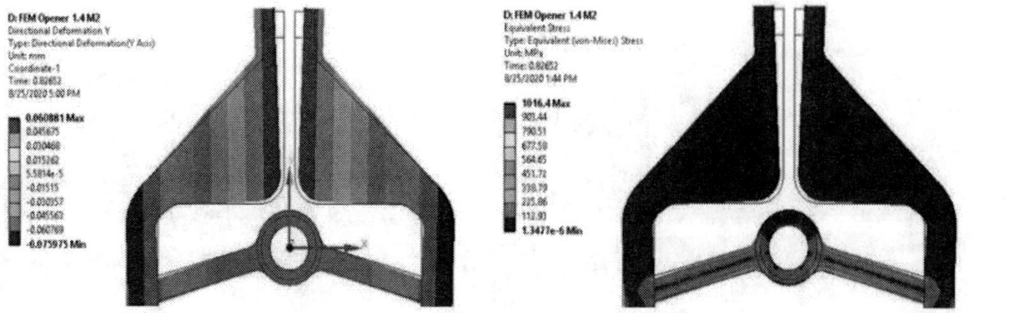

Figure 7. An improved version B (-15 degrees)

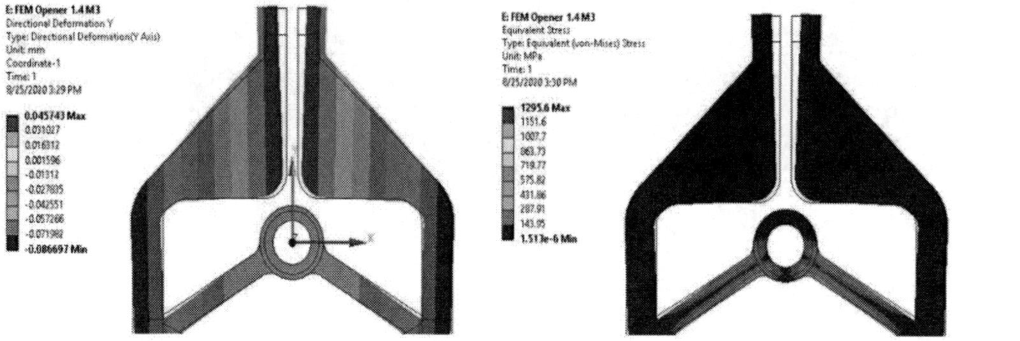

Figure 8. An improved version B (-30 degrees)

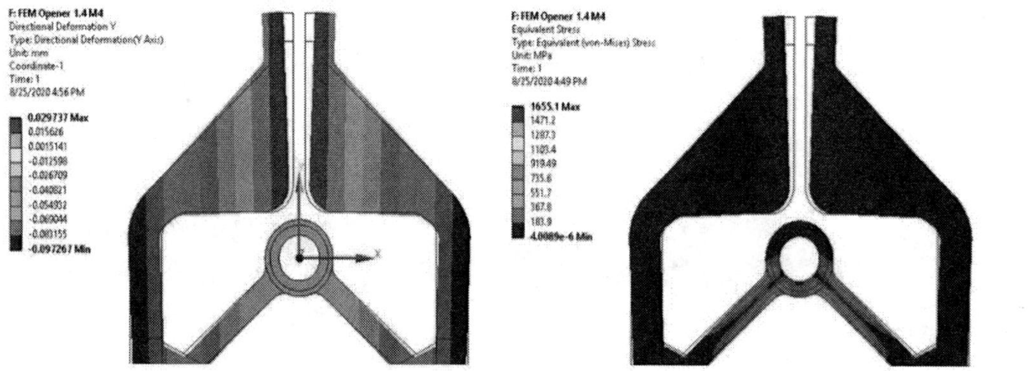

Figure 9. An improved version B (-40 degrees)

Table 3. Result for FEM Study

Opener shape	Y-axis deformation	Von Mises stresses
Improved version A 15 degrees	0.0706 mm.	1756 Mpa.
Improved version B -15 degrees	0.0608 mm.	1016.4 Mpa.
Improved version B-30 degrees	00457 mm.	1295.6 Mpa.
Improved version B -40 degrees	0.0297 mm.	1655.1 Mpa.

Table 3. It shows that the more changing of the angle in negative way, the less y deformation of opener tip. An improved version B-40 degree has minimum y deformation of 0.0297 mm. versus 0.0706 mm. (existing opener). However, design B-40 degrees may need to be observed more in term of life time, closing time, and etc.

5. Conclusion

The parameters that are considered by FEM in this study are: 1) the value of y deformation of opener tip, and 2) equivalent (von-Mises) stress of opener body. The maximum y deformation can be improved by revising the angle of opener arm. Therefore, this research can provide the following conclusions.

- Regarding four new conceptual designs of opener, it can be determined that the more changing of the angle in a negative way, the less y deformation of the opener tip.
- Mesh independence, 15,000 elements for mesh quantity, and hex-dominant mesh is enough to make the result convert to an exact value.
- Opener improved version B (-40 degrees) has minimum y deformation of 0.0297 mm. versus 0.0706 mm. (existing opener).
- The maximum stress at the fillet of opener arms should be minimized in order to improve lifetime.

Acknowledgments

This work is a part of current project funded by Seagate Technology (Thailand), Equipment Development Engineering Department.

References

[1] Bharti Y K Singh V Hussain A Singh D Lal S B and Dwivedi S K 2013 *International Journal of Scientific and Engineering Research* **4**(6) 1796-1803
[2] Kaya N Karen I and Öztürk F 2010 *Materials and Design* **31** 3008-3014
[3] Ashraf M Gardner L and Nethercot D A 2006 *Thin-Walled Structures* **44**(10) 1048-1062
[4] Yanga H Xub X and Neumann I 2019 *Composite Structures* **212** 434-438
[5] Durand R Pantoja B G and Oliveira V 2019 *Finite Elements in Analysis and Design* **158** 17-30
[6] Harewood F J and McHugh P E 2006 *Computational Materials Science* **37** 442-453
[7] Jaffarullah M S Low C Y Saedon J B Shaari M S B and Jaffar A 2015 *Procedia Computer Science* **76** 522-527

Crash analysis of sandwich composite electric microbus

P. Jongpradist[1,*], S. Limpanawisut[1], A. Siritana[1], P. Saardying[1], N. Saingam[1], P. Tangthamsathit[1], P. Chanpaibool[1]

[1]Department of Mechanical Engineering, Faculty of Engineering,
King Mongkut's University of Technology Thonburi, 126 Pracha Uthit Rd.,
Bang Mod, Thung Khru, Bangkok, Thailand 10140

* Corresponding Author: pattaramon.jon@mail.kmutt.ac.th

Abstract
Sandwich composite is ideal for lightweight monocoque structure owing to not only its superiority in strength-to-weight ratio and specific flexural properties but also its manufacturability to any desired complex shape. In this work, an optimum lightweight design of monocoque composite sandwich-structure microbus subjected to bending and torsion stiffness criteria and natural frequency constraints is further investigated under crash conditions through finite element analysis. A composite sandwich panel made of 5.4-mm woven glass fabric-epoxy face sheets and 50-mm rigid polyurethane foam core is tested under low-velocity impact with a hemisphere headed impactor following ASTM D7136 to characterize the failure characteristics. An enhanced composite damage material model so-called MAT54 in LS-DYNA with damage progression in each lamina utilizing strength-based Chang-Chang criteria is implemented with various formulations to validate and compare for their efficiency and accuracy under impact test. The reduced-integration Balytschko-Tsay shell element with 5 through-thickness integration points is applied to the monocoque composite microbus model. The structural crashworthiness under the frontal crash of the bus at 20, 40 and 50 km/h to a rigid wall barrier is examined. The front structure absorbs the most frontal impact energy up to 60% and shows severe damage when the crash velocity is 40 km/h or higher. For the bus impact velocity of 50 km/h, the maximum deformation of the bus front is 809 mm resulting in the intrusion of the structure into the specified driver survival space. Design improvement of the sandwich structure should be further explored to enhance its energy absorption and safety of the bus under crash events.

Keywords: Crash simulation, Sandwich structure, Composite material, Microbus

1. Introduction

An electric vehicle is a green alternative to traditional internal combustion engine vehicle that benefits in reducing emissions of greenhouse gas and air pollution contributing to climate change and smog. One of the key challenges in designing electric vehicles lies in lightweight design and energy efficiency. However, the safety of the vehicle structures must not be compromised. Sandwich composite is ideal for lightweight monocoque structure owing to its superiority in strength-to-weight ratio and specific flexural properties as well as its manufacturability to any desired complex shape. Many researchers have been focused on developing composite automotive parts to enhance vehicle strength and crashworthiness. Kang et al. [1] appended carbon composite roll bar components to the bus structure to improve safety under rollover in which the maximum deformation of the bus structure could be reduced by 40% at the end of the collision. Ko et al. [2] designed a composite bus structure made of glass fabric/epoxy with stainless steel inserts as the bus frame. Safety of the bus under frontal crash to a rigid wall at a speed of 60 km/h and rollover according to ECE-R66 standard were verified via finite element analysis using LS-DYNA. Several studies [3-5] designed lightweight superstructure for the bus roof strength based on ECE-R66 rollover test. It was found that pillar strength highly affects the bus crashworthiness under rollover accident [6-7].

Recently, Kunakorn-ong et al. [8] designed a lightweight monocoque composite sandwich-structure microbus based on bending and torsion stiffness criteria and natural frequency constraints. The monocoque design proves to considerably reduce the body weight while the bus stiffness under working conditions are reassured. The current study further investigates the performance of the bus structure under full frontal crash conditions by using finite element analysis using LS-DYNA. Recommendations on safety improvement are also provided.

2. Modelling and Validation

2.1 Validation of material models through an impact drop test

The material models defined in LS-DYNA are verified following ASTM D7136 Measuring the Damage Resistance of a Fiber-Reinforced Polymer Matrix Composite to a Drop-Weight Impact Event [10]. This test method determines the damage resistance by using a drop weight machine with a hemispherical striker tip as displayed in Figure 1a. A sandwich structure composes of 4 plies of 0/90 G600 glass fiber fabric and 5 plies of ±45 G600 glass fiber fabric for each skin and 50-mm H100 foam core. Four composite specimens made by vacuum infusion process are tested under impact energy of 100 J, 150 J, and 200 J. The impactor mass is 10 kg and the impact velocity ranges are between 3.1 and 5.4 m/s. Damages of the specimens for different impact energy are shown in Figure 1b. It can be seen that the damage area increases as the impact energy increases. Fiber fracture and delamination are observed in all specimens and permanent foam deformations occur without perforation of the composite panels.

The impact test scenarios are simulated in LS-DYNA by using a mesh size of 2 mm. For explicit dynamic simulation using plate and shell elements in LS-DYNA, the critical time step is automatically determined by the wave propagation velocity for membrane deformations governed by the size of the smallest element. The MAT024 Piecewise linear plasticity is assigned to H100 foam core while the composite material model is assigned as MAT054 Enhanced composite damage with Chang-Chang strength-based damage criteria. The 2-way fiber behavior option is triggered to account for the properties of woven fabric. Thus, the failure criteria e in material coordinates a- and b-directions (corresponding to the fiber direction of 0 and 90 degrees) are given as follows:

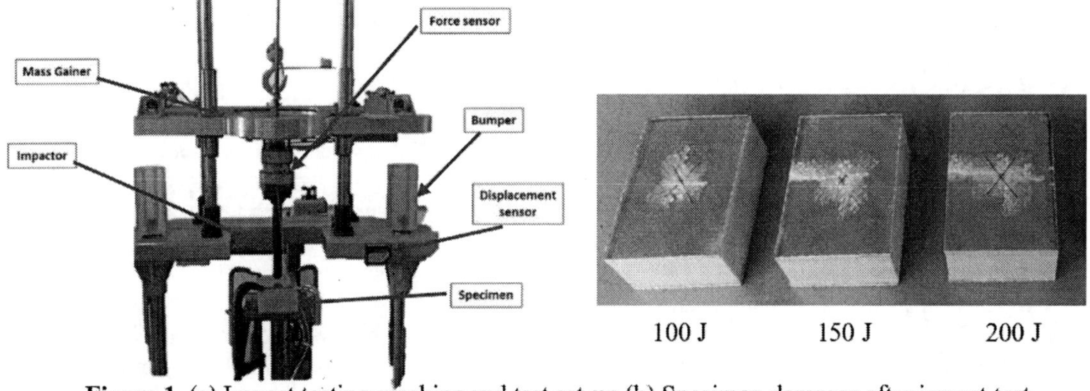

Figure 1. (a) Impact testing machine and test set-up (b) Specimen damages after impact test

for the tensile fiber mode in both directions,

$$\sigma_{aa} > 0, e_f^2 = \left(\frac{\sigma_{aa}}{X_t}\right)^2 + \beta\left(\frac{\sigma_{ab}}{S_c}\right) - 1\begin{cases}\geq 0 & \text{failed}\\ < 0 & \text{elastic}\end{cases}$$

and

$$\sigma_{bb} > 0, e_f^2 = \left(\frac{\sigma_{bb}}{Y_t}\right)^2 + \beta\left(\frac{\sigma_{ab}}{S_c}\right) - 1\begin{cases}\geq 0 & \text{failed}\\ < 0 & \text{elastic,}\end{cases} \quad (1)$$

for the compressive fiber mode,

$$\sigma_{aa} < 0, e_c^2 = \left(\frac{\sigma_{aa}}{X_c}\right)^2 - 1\begin{cases}\geq 0 & \text{failed}\\ < 0 & \text{elastic}\end{cases} \text{ and } \sigma_{bb} < 0, e_c^2 = \left(\frac{\sigma_{bb}}{Y_c}\right)^2 - 1\begin{cases}\geq 0 & \text{failed}\\ < 0 & \text{elastic,}\end{cases} \quad (2)$$

for matrix failure criterion,

$$e_m^2 = \left(\frac{\sigma_{ab}}{S_c}\right)^2 - 1\begin{cases}\geq 0 & \text{failed}\\ < 0 & \text{elastic}\end{cases} \quad (3)$$

After the material failure criterion is satisfied, the material properties are perfectly plastic until the defined DFAIL. Information on the damage status at each integration point and each element can be plotted using additional integration point history variables written to the LS-DYNA database by the *DATABASE_EXTENT_BINARY definition as variable NEIPS. The history variables 1, 2, 3 are failure flags for tensile fiber mode e_f, compressive failure mode e_c, and matrix mode e_m, respectively.

Sandwich structure is modelled through PART_COMPOSITE. Laminated Shell Theory is activated using parameter LAMSHT = 1 to corrects the default assumption of uniform transverse shear strain through thickness when the composite shell has layers of dissimilar materials where large stiffness changes across thickness such as sandwich structures. Results from different shell element formulations ELFORM=2 and 16 are investigated for model accuracy and robustness. Both formulations are based on Reissner-Mindlin kinematic assumption. The shell deformation is described by the mid-surface displacements and rotations with 5 DOF in a local coordinate system which globally yields 6 DOF. The shell cross-section remains straight and unstretched while shear deformations are possible. The ELFORM=2 is Balytschko-Tsay shell which uses one-point integration and bi-linear nodal interpolation. Thus, the element is highly efficient, but warpage may be an issue in coarse models. The ELFORM=16 corresponds to 2x2 full-integration shell formulation and uses Bathe-Dvorkin transverse shear treatment to eliminate w-mode hourglassing. This formulation generally provides more accurate results than using ELFORM=2

The number of through-thickness integration points can be defined in *PART_COMPOSITE. The BETA angles contribute to the initial orientation of the material coordinate system for each integration point. In this study, the number of through-thickness integration points of 5 points (one point for each fiber orientation for skin and one point for foam core), 19 points (one point for each ply and one point for the core), 29 points (one point for each fiber direction and 25 points for the core), and 38 points (one point for each ply and 25 points for the core) are studied. By default, the output results in the composite shell are reported at 3 integration points, i.e., top, middle and bottom points. More results at the through-thickness integration points can be requested by defining MAXINT in *DATABASE_EXTENT_BINARY.

Figure 2a demonstrates comparison of the force-displacement curves of impact test obtained from different models. It is found that simulation results when ELFORM=2 generally returns a slightly higher maximum force and displacement before the impactor rebound as well as a larger permanent deformation. Impact modelling of sandwich structure using ELFORM=16 is accurate despite the number of through-thickness integration points. However, models using ELFORM=16 requires the computational cost of approximately 3-4 times larger than the similar model with ELFORM=2 (Figure 3b). In views of accuracy, robustness and efficiency of a simple impact problem the use of 5 integration points and ELFORM=16 is recommended. For the current application to the microbus model consisting of a large number of nodes and elements, the ELFORM2 element with one integration point for each fiber direction in the sandwich faces and one integration point in the core which delivers moderately stiffer results is therefore chosen for implementation to achieve a cost-effective model.

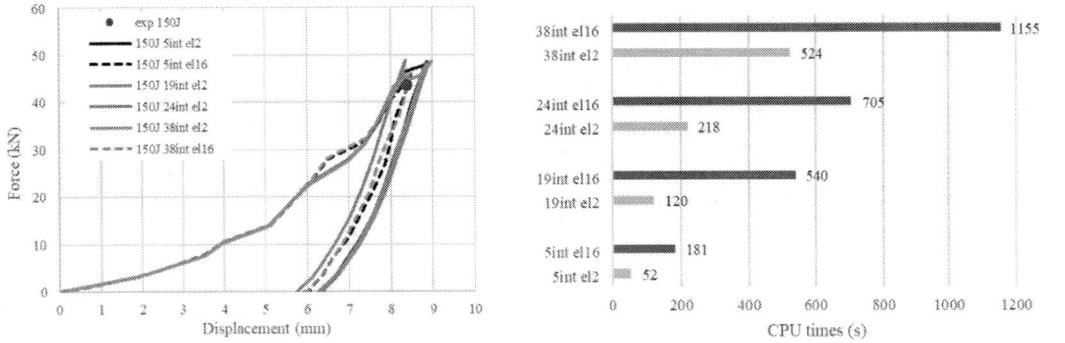

Figure 2. Simulated results under impact drop and CPU times using various model formulations

Table 1 illustrates the force-displacement curves obtained from numerical simulation are compared with the maximum forces recorded from the experiments. For all impact energy of interest, the sandwich structures are damaged with permanent deformations after the impactor rebound. The higher the impact energy, the higher the impact force and permanent deformation. It is also noticed that the maximum forces from impact test are somewhat less than that obtained from FEA whereas the predicted displacements at the maximum force are comparable. Results discrepancies are more apparent when the impact velocity is higher. The maximum force obtained from the experiment of 200-J impact seems to be rather lower than the expected value. This could be owing to inhomogeneous material properties of the composite laminate which cannot be captured in the simulation results. The results can be further investigated via reliability analysis to account for this contribution.

Table 1. Experimental and simulation result for forces and displacement under an impact drop test.

Test no.	Impact energy (J)	Maximum force			Maximum displacement		
		Exp (kN)	Sim (kN)	Error	Exp (mm)	Sim (mm)	Error
1	100	31.1	35.6	14.4%	6.8	7.5	9.8%
2	150	43.6	47.0	7.8%	8.4	8.5	1.6%
3	200	44.3	55.2	24.4%	8.9	9.4	6.4%

2.2 Baseline design

The sandwich composite microbus used in the current study is adopted from Kunakorn-ong [8] as illustrated in Figure 3. The bus consists of nine components distinguished by their functions including the bottom, floor, roof, beam, ceiling, pillar, battery tray, subfloor, and vent. The bus is 2.4 meters wide, 7.8 meters long and 3.2 meters height with 24 passenger seats. To correctly compute the dynamic behavior of the bus under crash events, the mass of other important components is uniformly distributed in the FE model at their corresponding locations. The mass elements for axle systems, windows and doors of 1892 kg are assigned to the bottom component. The driver and 24-passenger load and seats of 2282 kg are given to the floor. The 1536-kg battery is applied to the battery tray. The total body weight of the baseline bus 7,610 kg.

In this paper, the FE simulation is performed by using LS-DYNA. The shell elements of size 30 mm^2 were implemented to all components of the bus model. The total numbers of elements in the baseline model are 104,829 elements. The sandwich structure is composed of H100 foam core and E-glass/epoxy faces with their thicknesses and ply orientations as listed in Table 2 with the color code explanation in Table 3. The glass fiber fabric with a surface density of 400 g/m^2 (G400) and 600 g/m^2 (G600) was mainly selected for the design of structure due to its lightweight and high specific stiffness, strength, and toughness except that the plain-weave carbon fiber fabric with a surface density of 200 g/m^2 (C200) was implemented at the battery tray where much higher bending stiffness of the component was required

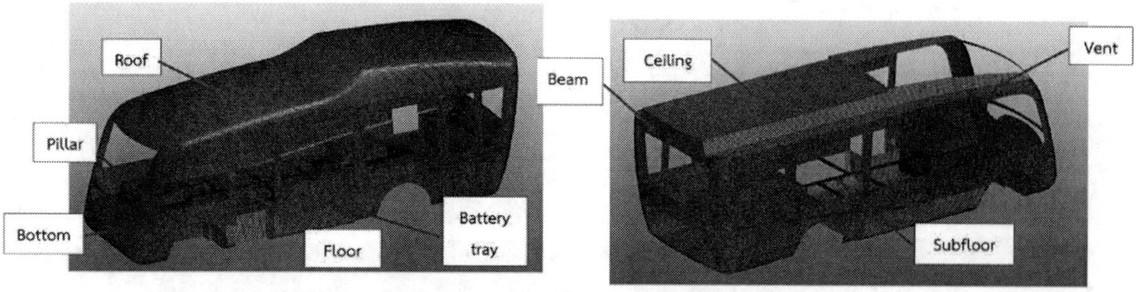

Figure 3. Components of composite microbus.

Table 2. Material layup for composite microbus components.

Component	Laminate layup	Thickness (mm)	weight (kg)
Bottom		61.2	524
Vent		38.8	130
Battery Tray		69.2	270
Pillar		55.2	204
Floor		47.2	232
Roof		4	125
Ceiling		70	264
Beam		51.2	44
Subfloor		6.6	5

Table 3. Material, ply thickness and fiber orientation for color codes used in Table 1.

Material	Ply thickness (mm)	Fiber direction (degree)	Color code	Fiber direction (degree)	Color code
G400	0.4	0/90		±45	
G600	0.4	0/90		±45	
C200	0.4	0/90		±45	
H100 (core)	5	-			

to support the weight of the battery packs. Each layer of fabric has a nominal thickness of 0.4 mm. The zero-degree fiber directions for all components were assigned in the orientation along the horizontal axis parallel to the floor in the plane of each the component panel. Mechanical material properties for the composite faces and foam core are obtained from [8].

3. Frontal crash simulation

The bus structure under full frontal crash to a rigid wall barrier is simulated. The bus is constrained in z-axis at the wheel locations and the initial velocities v at 20, 40 and 50 km/h are prescribed in +x-direction (see Figure 4a). Automatic single surface contact interaction is assigned to all parts. The driver survival space is measured according to the survival space in ECE-R29 [11] corresponding to the reference point of the driver's center of gravity. Therefore, the maximum deformation of the bus frontal structure should not be greater than 600 mm after full-frontal crash. An example of bus deformation behavior for the frontal crash with an initial velocity of 50 km/h is represented in Figure 4b. The bus impacts the wall at the time = 29 ms and structural damage start occurring at 60 ms until the crash finished at 160 ms.

The energy absorption of each component under the initial impact velocity of 20, 40 and 50 km/h are displayed and compared in Figure 5. The values indicated in the plot is the internal absorbed energy by each component after the crash while the y-axis shows the percentage of energy absorption for each part. For all crash velocities, most of the kinetic energy is transferred to the bottom and floor of the microbus by which approximately 90% of crash energy are absorbed. The bottom component especially the front panel absorbs more percentage of impact energy when the crash velocity increases.

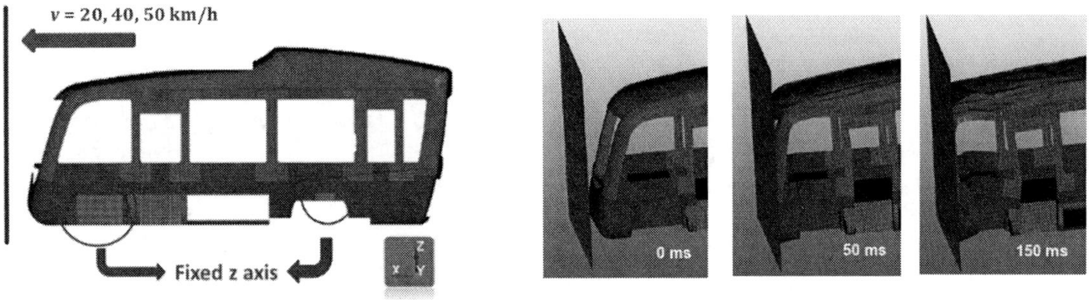

Figure 4. (a) Boundary and initial conditions (b) Deformation behavior for a frontal crash at 50 km/h.

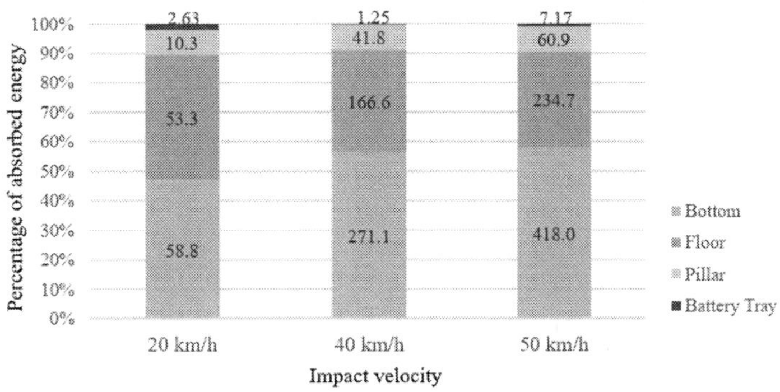

Figure 5. Percentage energy absorbed by each component under the frontal crash.

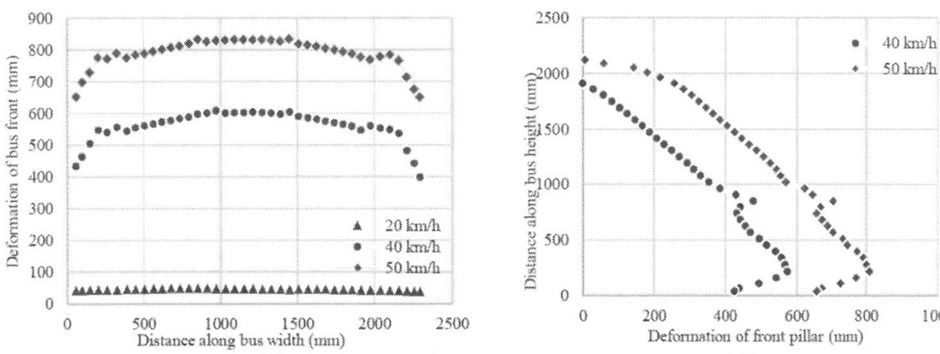

(a) (b)

Figure 6. Deformations of (a) front panel and (b) right front pillar.

IOP Conf. Series: Materials Science and Engineering 1137 (2021) 012048 doi:10.1088/1757-899X/1137/1/012048

Figure 6 provides the deformations of the front panel at 400-mm height from the lower floor of the bus and the deformations along with the right pillar's height for different impact velocity. The deformation of the frontal part at a slow impact velocity of 20 km/h is found to be less than 50 mm. Nonetheless, damages in the compressive fiber mode are already detected at the outer ply of the sandwich face in the front panel of the bus. At an impact velocity of 40 km/h, the front panel has damaged in all failure modes with the largest damage in compressive fiber mode. However, the deformations of the front panel and pillar are still less than the criteria for survival space. When the crash speed reaches 50 km/h, the maximum deformation is found to be 809 mm in which the driver's survival space is intruded by 59 mm. This implies that the driver is not safe during the crash accident and the front panel should be re-design to improve the bus crashworthiness under the frontal crash.

4. Conclusions

This study set out to present the method to simulate an impact drop test to a sandwich structure with a different number of integration points and shell element formulation. The use of highly computing-efficient Balytschko-Tsay element with the minimum through-thickness integration points has proved to provide comparable and reliable results with much more detailed elements. The simple model is thus applied to crash analysis of a complex sandwich composite microbus. Most of the impact energy from the frontal crash are found to dissipate via the bottom components through front panel's deformation. Consequently, the baseline bus model is not safe under the frontal crash. Overall, this study strengthens the idea that the design of these two parts should be modified to increase the energy absorption of the part to improve the crashworthiness of the structure under full-front collision.

References

[1] Kang K Chun H Na W Park J Lee J Hwang I and Hong H 2011 *Proc. the 18th International conference on composite materials* (Jeju, Korea)
[2] Ko H Y Shin K B Jeon K W and Cho S H 2009 *J. Mech. Sci. and Technol.* **23**(10) 2686-2693
[3] Gursel K T and Gursesli S 2010 *Gazi Univ. J. Sci.* **23**(1) 71-79
[4] Liang C C and Le G N 2009 *Int. J. Crashworthiness* **14**(6) 623-639
[5] Karlinski J Ptak M Dzialak P and Rusinski E 2014 *Arch Civ. Mech. Eng.* **14**(1) 342-343
[6] Kongwat S Jongpradist P and Hasegawa H 2020 *Int. J. Automot. Technol.* **21**(4) 981-991
[7] Kunakron-ong P and Jongpradist P 2020 *Int. J. Automot. Technol.* **21**(1) 215-25
[8] Kunakorn-ong P Ruangjirakit K Jongpradist P Aimmanee S and Laoonual Y 2020 *Proc. Inst. Mech. Eng., Part C* 3-1
[9] Sirisobrana P Ruangjirakit K Jongpradist P and Aimmanee S 2019 *Proc. the 22nd International Conference on Composites Materials* (Melbourne, Australia)
[10] ASTM D7136 *Standard Test Method for Measuring the Damage Resistance of a Fiber-Reinforced Polymer Matrix Composite to a Drop-Weight Impact Event*
[11] United Nations 2011 *Economic Commission for Europe agreement regulation No.29: Uniform Provisions Concerning the Approval of Vehicles with regard to the Protection of the Occupants of the Cab of a Commercial Vehicle*

Computational study of the flow field characteristics of the radial-swirl burners for a small gas turbine

Tosaporn Soontornpasatch

Department of Mechanical and Aerospace Engineering, Faculty of Engineering, King Mongkut's University of Technology North Bangkok, Bangkok 10800, Thailand

* Corresponding Author:tosaporn.s@eng.kmutnb.ac.th

Abstract
The aim of this study is to investigate the flow field characteristics of the radial-swirl burner which is developed for a small gas turbine. To create a stable flame during the combustion process, the burner has to create a reversal flow area or a recirculation flow. Without this zone, the combustion process cannot be created. Therefore, the characteristics of the flow field in the burner is very critical. The current study applied two types of the radial-swirl burner. One has a straight-type vane and the other has a nozzle-type vane. The flow angle of both types are varied from 40°, 50° and 60°. The mass flow rate of air is varied from 1.0 to 1.8 kg/min. The commercial computational fluid dynamic program, ANSYS FLUENT is also used. The results of the simulation are discussed in this paper.

Keywords: Radial-swirl burner, Computational Fluid Dynamics, Gas turbine.

1. Introduction

The burner is one of the most important parts in the combustor of a gas turbine. The primary function of the burner is to mix air with fuel and create a continuous and stable combustion process. To create a stable flame, the burner has to create a reversal flow area or recirculation flow. In this flow area, a high-temperature combustion gas will flow back to the burner and mix with the incoming air and fuel which create a continuous combustion process. Without the recirculation flow, the flame blowout will occur. One of the most effective methods to create recirculation flow is to use a radial-swirl burner [6]. The air flows in the radius direction into the burner and changes the inward flow angle by the flow passages. This will create the toroidal flow, which has a recirculation zone inside. In the past, there were many researchers who studied the flow characteristics in the swirl burners. A much-debated question is whether the complexity of this type of flow cannot be predicted by using this kind of analytical method. Existing research recognises the critical role played by using experiment approach. Cheng and Simone [3] studied the flow field of the swirl burner of a model gas turbine by experiment approach by using PIV method. Mohammad et al. [7] recognised the relationship between discharge coefficient of various radial air swirler at each Reynolds number by experimenting on the test rig. A considerable amount of literature has been published on the experimental work conducted by David [8]. However, the development of high-speed computer technology allows the researchers to perform the study by using a numerical model instead. There were many studies that performed by using this approach. Hasem et al.

IOP Conf. Series: Materials Science and Engineering 1137 (2021) 012049 doi:10.1088/1757-899X/1137/1/012049

[1] used CFD to investigate the combustion process in 100kW in swirl burner which has various inlet angle conditions. Mohd et al., [2,4] focused on the fluid dynamic and the emission performance of the double radial swirlers and a combustor aerodynamic with radial swirler by simulation approach. Vondal and Hajek [5] carried CFD to predict the flow through the swirl generator and validate the result with the measured data. The results from these works indicated that the numerical model can be used as a primary tool in the study and can give a reasonable result. Therefore, the numerical approach was also used in this study.

The specific object of this study was to study the characteristics of the flow field inside the two types of the radial-swirl burner. One has a straight-type flow passages and another has a nozzle-type. The angle of the passage varied from 40°, 50° and 60°. The mass flow rate of air was varied from 1.0 to 1.8 kg/min which is in the range that the burners are expected to be used. The size of the recirculation zone, the maximum reversal velocity and the location that maximum reversal velocity occurs are the primary variables. The commercial computational fluid dynamics code, ANSYS FLUENT 17.2, was used as a tool for this study. This study emphasized on the relation between characteristics of air flow and shape parameters of the flow passage only. The injection and combustion process are included from this study. The result from this study will be used as a basis for designing a burner for a small gas turbine. This study provides new insights into gas turbine to be used as a tool for the aerospace engineering laboratory subject in the department.

2. Radial-swirl burner

There are two types of radial-swirl burner chosen in the current study. One has a straight-type and another has nozzle-type passages and the main dimension of these burners share some similarities (Figure 1). The air flows into the burner in the radial direction and the fuel is injected from the gas port at the center of the burner. This study considers only air flow characteristics, the gas port was set to be a solid wall instead. There are seven flow passages which have 7 mm-width and 13 mm-depth at the inside surface in the burner. In the case of straight-type, the width of the passage is constant. For a nozzle-type, the width of the passage is 10.2 mm at the outside surface.

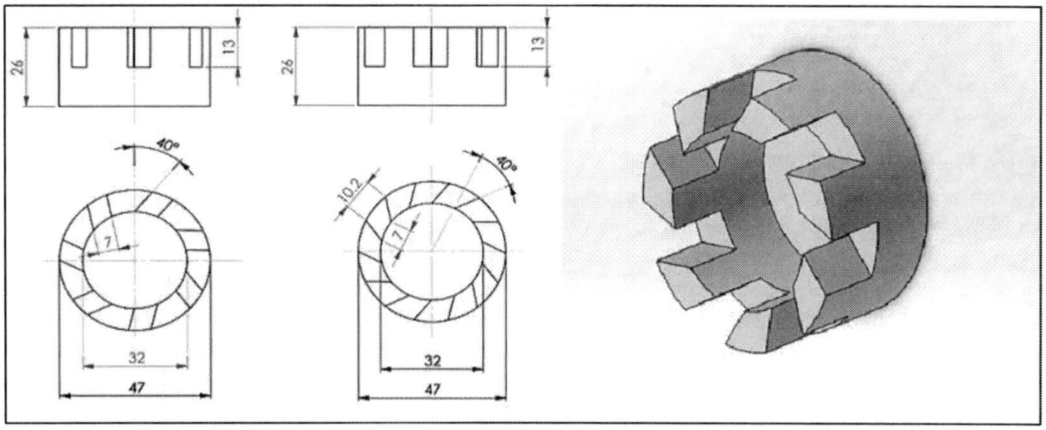

Figure 1. Shows the main dimension of the burner

3. Computational method

The flow field was calculated by using the commercial computational fluid dynamics code, ANSYS FLUENT 17.2. To make sure that the boundary condition will not allow a false solution in the computational domain, the diameter of the control volume was set at 7.5 times. In particular, the length was set at 40 times of the burner inner diameter. As can be seen from Figure 2, the air inlet was set as a mass flow inlet and the outer surfaces of the control volume were set as outflow as shown in Figure 2.

The surface that divides the burner and the main control volume was set as a wall. The unstructured tetrahedron mesh scheme was used to create meshes in the control volume. For time-saving during the computational process, the fine meshes were applied in the area around the burner outlet and the coarse meshes were used at the area that far from the burner outlet (see Figure 3). The pressure-based solver and SIMPLE algorithm were used in the calculation and the second-order upwind scheme is applied to calculate all flow variables. The program was set to stop calculation when the residual values of all flow parameters are lower than 10^{-4}. The temperature of the air at the burner inlet was set at 300 K. The turbulence model $k - \varepsilon$ realizable was also chosen in the current study. This turbulence model has some benefits to the standard $k - \varepsilon$ model and RNG $k - \varepsilon$ model. As such, it is suitable for the swirling flow which has a recirculation zone inside. It has been also used by many researchers regarding the study of the flow of the swirl-burners [1,4,5].

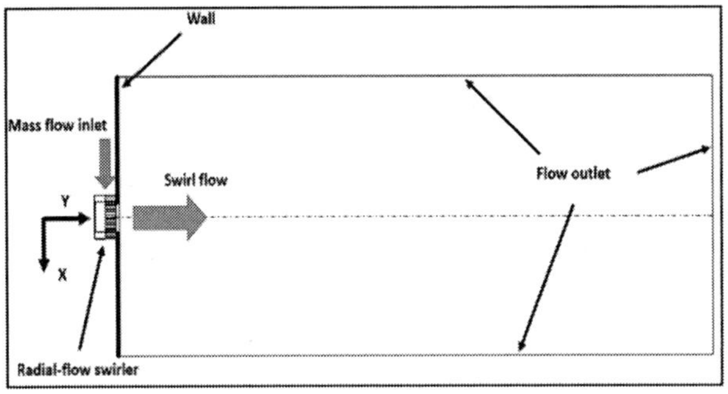

Figure 2. The control volume that is used in this study

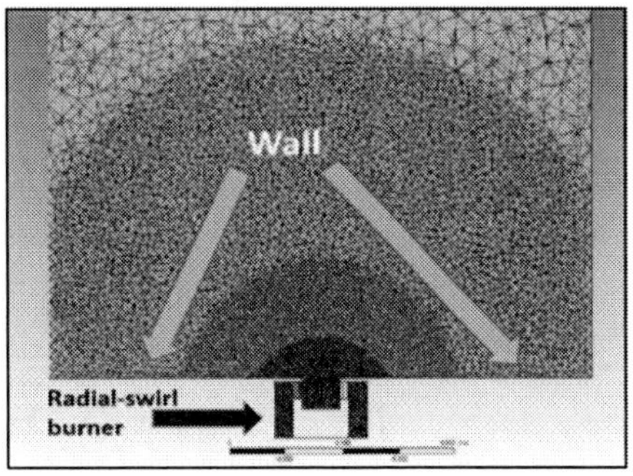

Figure 3. The mesh shape inside the control volume

4. Size of the recirculation zone

The recirculation zone is a vital for swirl burner since it is mixing process between an air-fuel mixture and hot gas that flows back to the burner occurs in this area, and it has a pivotal role in the heat release rate in combustion process [4]. As illustrated in Figure 4, the size of this zone can be determined by using this graph. Figure 4 also provides shows the value of the y-component velocity of flow along the y-axis which is located at the center of the burner. The coordinate has its origin at the end wall of the burner and the direction of this coordinate can be seen in Figure 2. The size of the recirculation zone is the distance between the two stagnation points (which velocity is zero). Between this area, the y-component velocity has a negative value which indicates that the direction of flow is inward to the burner.

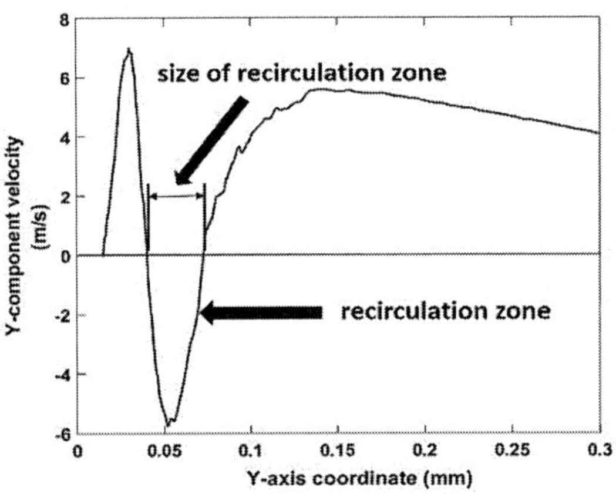

Figure 4. The y-component velocity profile at along Y-axis coordinate

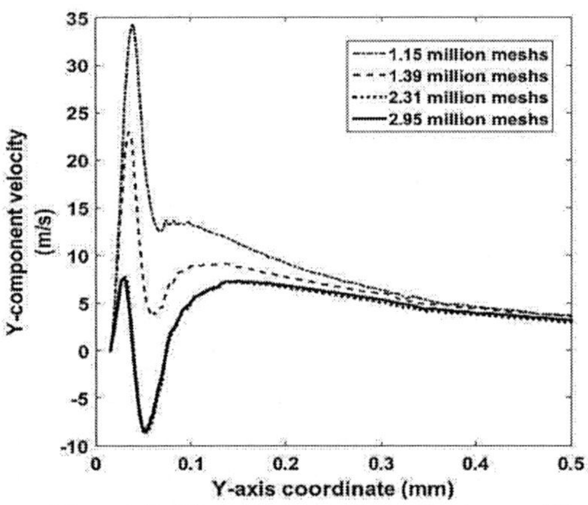

Figure 5. The y-component velocity profile each number of meshes

5. Mesh dependency testing

The mesh dependency testing is performed to determine the appropriate number of meshes that will be chosen in this study. The straight-type burner which has a swirl angle at 40° was selected for the test. The mass flow rate was set at 1.8 kg/min. The number of meshes varied from 1,153,322 to 2,953,934. The y-component velocity profiles at each number of meshes are given in Figure 5. The velocity profile drastically changed at the 1.15 and 2.31 million meshes (see also Figure 5). In the case of 1.15 and 1.39 million meshes, the recirculation zone does not develop. However, the recirculation flow occurs at 2.31 million meshes. When the number of meshes increases higher than 2.31 million, the velocity profile does not change any further (see Figure 5). Therefore, the number of meshes that has approximately 2.3 to 2.4 million was employed in this current study.

6. Results and discussion

It can be seen from the simulation data in Figure 6 that the maximum reversal flow velocity for every types of burner increases when the mass flow rate increases. In case of the straight-type burner, the maximum reversal velocity continuously increases when the angle of the inlet flow increases. For the nozzle-type burner, this value also has the same trend when the inlet angle is at 40° and 50°. However, when the inlet angle is at 60°, the maximum velocity decreases. When compare between the two types, the straight and nozzle type at the same flow inlet angle, the nozzle-type has higher maximum reversal flow velocity except at the angle 60° which the nozzle-type has a lower value.

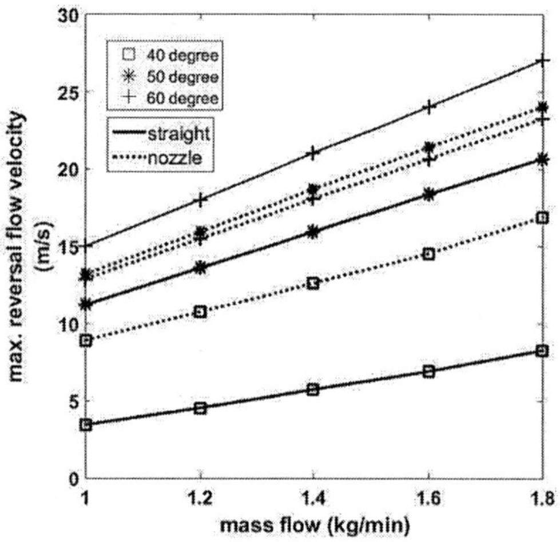

Figure 6. The maximum reversal flow velocity at each mass flow rate

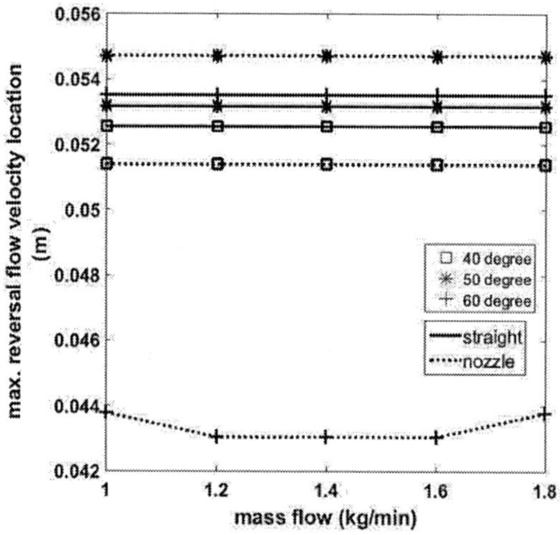

Figure 7. The location which maximum reversal velocity occurs at each

The location of the maximum reversal velocity does not change with the mass flow rate (see Figure 7) Although it changes with the value of inlet flow angle but these changes are in the narrow range between 51 mm to 55 mm from the end wall of the burner. However, for the nozzle-type at 60°, the location of maximum velocity is at 43 mm which differs very much when compares with the other cases. The relation between the size of the recirculation zone and mass flow rate shows in figure 8. The change in mass flow rate does not have any significant impact on the size of the recirculation zone. There is some change in case of 50° nozzle-type burner at a mass flow rate between 1.0 to 1.2 kg/min. However, this happens in the narrow range only. The value of the average size of the recirculation zone for each burner at a various angle can be seen in Figure 9. Regarding the straight-type burner, the size of the recirculation zone continuously increases when the inlet flow angle increases. For the nozzle-type burner, the recirculation zone size also increases when the inlet flow angle increases from 40° to 50°.

However, the size decreases when the flow angle changes from 50° and 60°. When compared between straight and nozzle type, at the same angle inlet, the recirculation zone size of the nozzle-type burner is larger except at the angle 60° which it has a smaller size.

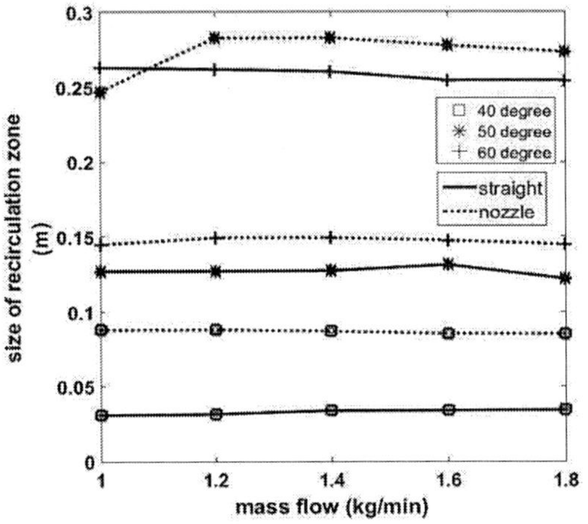

Figure 8. The relation between mass flow rate and the size of the recirculation zone

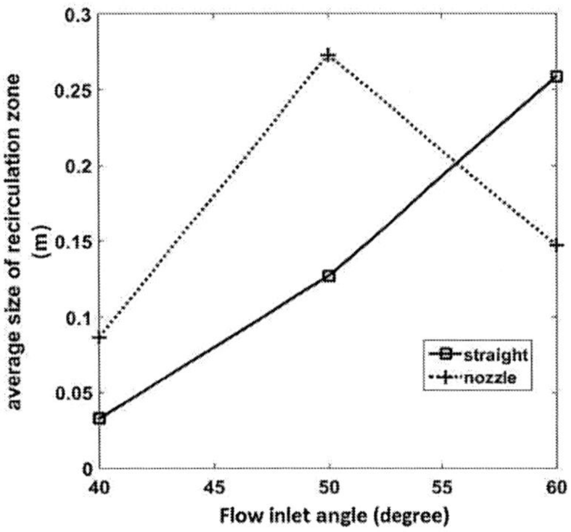

Figure 9. The average size of the recirculation zone at each inlet flow angle

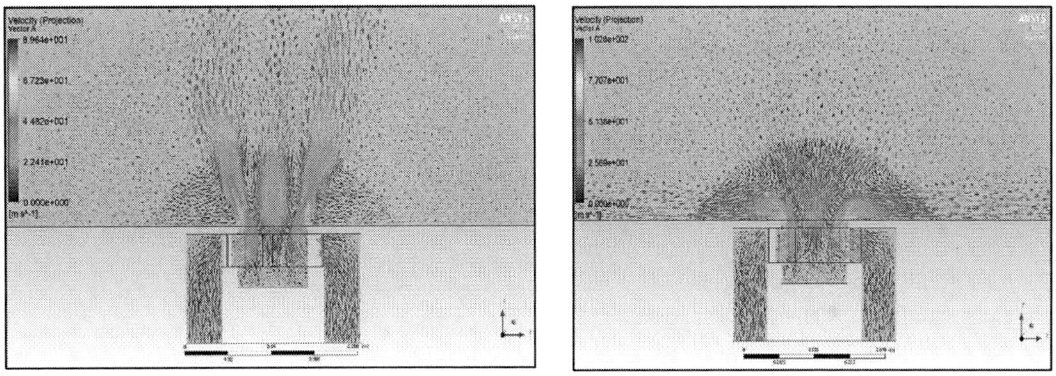

Figure 10. Comparison flow velocity vector of the 60° straight-type (left) and nozzle-type (right) passages at mass flow rate 1.8 kg/min.

Figure 10 compares of the vector flow field between the straight-type and nozzle-type burner shows in figure 10. Both of them have the inlet angle at 60° and consist of the mass flow rate at 1.8 kg/min. In regard to the straight-type, was found to cause the recirculation zone at the centreline of the burner. This pattern of the flow field also occurs with the other cases but has a difference in the size of recirculation zone and magnitude of reversal flow velocity. Whilst the nozzle-type at 60° has a difference in the flow pattern (see also Figure 10). The recirculation zone has a very small size compared to other cases. Besides, the swirling flow around the recirculation zone tends to flow outward in the radial direction or in x-axis instead of flowing upward in the y-axis direction. Another important finding was that this can cause an unwanted situation. Because when this flow pattern occured in the real combustion process, the high temperature gas flowed along the wall that burner installed. This can cause some damage to this area by the heat that might transfers to the material. Therefore, the the nozzle-type at 60° may not be suitable in the application of the combustion chamber.

7. Conclusion

This paper has argued that all type of radial-swirl burner can create recirculation. The simulation confirmed that the inlet flow angle had a direct impact on the size of the recirculation zone and maximum reversal velocity. These variables increased when the inlet flow angle increased. The nozzle-type burner created a larger size of the recirculation zone and had a higher of the reversal velocity when compared with the straight-type at inlet angle 40° to 50°. However, at inlet angle 60° these variables of nozzle-type were lower. The change of mass flow rate affected on the maximum reversal flow velocity but it did not have any significant impact on other parameters.

Acknowledgments

This research has been financially supported by King Mongkut's University of Technology North Bangkok, contract no. KMUTNB-63-DRIVE-10.

References

[1] Hesham B Nick S Phil B Agustin M and Richard M 2019 *Int. J. Energy Technol.* **1** 31-43
[2] Mohd Z Mohammad J and Tholudin L 2017 *J. Teknologi* **79**(7-3) 33-41
[3] Cheng C and Simone H 2013 *Advanced Materials Research* **622-623** 1119-1124
[4] Mohd N Jusoff K Mohammed O and Mohd I 2011 *Int. J. the Phys. Sci.* **6**(13) 3091-3098
[5] Jiri V and Jiri H 2011 *J. Phys. Conf. Series* **318** 1-6
[6] Arthur L and Dilip B 2010 *Gas Turbine Combustion Alternative Fuel and Emission* (Florida: CRC Press) 146
[7] Mohamad I and Mohammad J 2004 *J. Mekanikal* **17** 99-108
[8] David G L 1977 *AIAA J.* **15**(8) 1063-78

IOP Conf. Series: Materials Science and Engineering 1137 (2021) 012050 doi:10.1088/1757-899X/1137/1/012050

Stress analysis in upper fillet and rail-end bolt hole at rail joint

SaiKham Le[1] and Jittraporn Wongsa-Ngam[1,*]

[1]Department of Mechanical Engineering, Faculty of Engineering,
King Mongkut's Institute of Technology Ladkrabang, Bangkok, Thailand 10520

* Corresponding Author: jittraporn.wo@kmitl.ac.th

Abstract
Rail joint between two adjacent rail ends in the transportation of railway track structure is one of the weakest components leading to defects and failures. In this study, the rail joint components such as rail, joint bars, bolts, and sleeper are modelled by 3D Abaqus/CAE software and Finite-element method are used to investigate stress distribution in rail under static wheel load condition. The wheel load positions and bolt-hole locations vary as a parametric study. The results show that the critical stress occurs at the upper fillet rail end and at the edge of the rail-end bolt hole (1st hole from the rail end). The wheel load position affects stress only in case of the rail-end bolt hole located at 80 mm from the end. In addition, overall trends for the effect of rail-end bolt hole positions are similar in which the stress decreases when the distance from rail end to the rail-end bolt hole increases. Taken together, these findings suggest stress distribution is the basic information that helps understand the failure mechanism.

Keywords: Rail-end bolt hole, Upper fillet, Wheel-load positions, Bolt hole positions, Von-Mises stress distribution

1. Introduction
A rail joint is a system connecting two rails end cross-sections to provide a smooth-running surface of the rail to reduce the risk of derailment. A large number of repetitive operating cycles, on the rail joint region, was found that lead to the railhead end edge damage and bolt hole fracture, in the rail web area, because geometric discontinuities and different mechanical properties were induced by high impact loads [1, 2]. A large and growing body of literature has investigated rail joints, based on different theories and finite element software. Up to now, far too little attention has been paid to the measurements data of stresses and strains of the rail joints. Although some research has been carried out on the replacement of the rail joint gap in continuously welded rails to reduce rail end defects, no studies have been found which the rail joint is now still used in rail transit and curved track system.

Zhu et al. [1] estimated fatigue life of rail-end upper fillet with three crosstie support configurations for the joint sitting under various wheel load factors. Zhu et al. [2] analyzed the stress distributions in rail the end bolt hole and rail-head-to-web fillet from bolted joint bars and considered five crosstie support configurations under static loading by using finite element models: their model considered elastic material by evaluation with severe impact wheel load factors. Two types of rail design (100-8, 115RE) and four types of longer and thicker joint bars were modelled in ABAQUS and simulated under load applied for two positions [2]. By drawing on the results of this study, Zhu et al. [2] has been able to show that the longer joint bars did not enhance the performance, thicker joint bars noticeably reduced deflection at the rail joint. Zhu et al. [3] investigated the stress distribution in the upper fillet rail end and rail-end bolt hole by using a finite element model and considered the effect of bolt loading and missing-bolt configurations when the load directly applied on the fourth bolt hole. Their study was done with the six bolts rail joint. The maximum von Mises occurred in the contact area between the upper fillet rail end and the joint bar top and the fourth rail end bolt hole, because the load is directly applied on it. The result found that the rail end upper fillet stress is decreased when the bolt loading increased. Ding and Dhanasekar [4] used a 3D finite element model of the bonded bolted butt joints, due to bolt looseness, without considering rail shape details. Under severe loading, the loss of bolt tightness critically influenced the joint performance and life expectancy, due to increased length of the contact surface. Samantaray et al. [5] evaluated the structural integrity and joint bar deflection under static load and investigated the stress distribution in joint bars and bolt shank region under loadings. They simulated the effect of loose nuts and bolts and structure deformation behaviors by using finite element software of ANSYS and ABAQUS, for validation of the analysis results.

To date, several studies have linked the stress analysis in the rail-end bolt hole and upper fillet rail end with the rail designs of 100-8 and 115RE under static loading conditions [2]. However, the influence of the applied load positions on the railhead for UIC60 rail design and analyzed the stress in the rail-end bolt hole and upper fillet has remained unclear [5]. Therefore, this prospective study was designed to investigate the use of UIC60 rail, joint bar, bolt and nut and sleeper in ABAQUS/CAE FE software. The rail joint structure is simulated by variation in the applied load positions and variations in the distance from the rail-end bolt hole center to the rail end (65 mm and 80 mm). Consequently, the simulation results were analyzed for the stress distribution in the rail-end bolt hole and upper fillet for observation of the critical parameters which may lead to the initiation of the crack in the rail joint under loading conditions.

2. Finite element model

To study stress distribution in the rail-end bolt hole and upper fillet rail end, the components of rail joint structure such as rail, bars, bolts and sleepers are modelled in ABAQUS/CAE. Three parts of the bolt, nut and washer combination are simplified, as one component, to reduce computational cost [2]. The profile of the rail design UIC 60, joint bars and bolt were modelled by using the design of Samantaray et al. [5]. The rail joint gap between the two rail ends was 5 mm. Each solid rail element was 2.6 m long, 148 mm wide, 172 mm height and the adjacent sleeper center-to-center is 600 mm [5]. Also, the bolt hole diameter in joint bars and rail was 28 mm. The 65 mm and 80 mm distances from the rail end to the center of the rail-end bolt hole are parametric study, for verification the stress distributed in the rail-end bolt hole and upper fillet. The rail joint is placed between two adjacent sleepers as suspended crosstie support configuration. The assembly of the rail joint structure is shown in Figure 1.

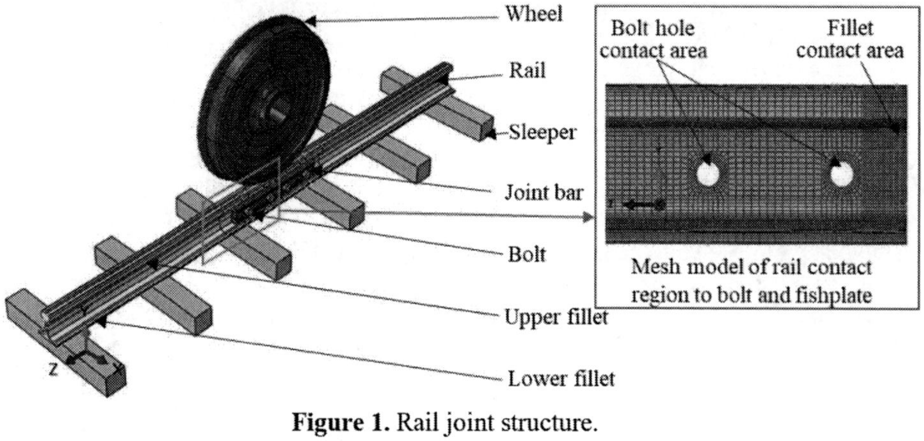

Figure 1. Rail joint structure.

The current study is assessed the stress distribution in the rail-end bolt hole and upper fillet because the stress influencing is a critical factor affecting rail web fracture in the rail joint region. The simulation steps are shown in Figure 2. From this data, we can see that the component mechanical properties are based on a linear elastic material model (see Table 1). All components were customized steel except the concrete sleepers. Each mechanical property is applied to the parts of the rail joint structure.

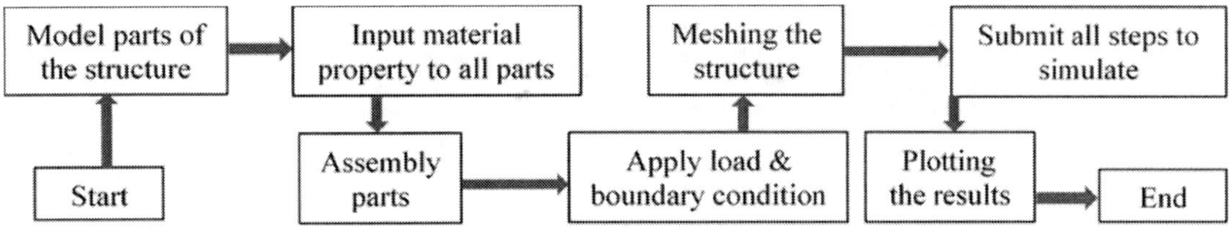

Figure 2. Steps for finite element simulation.

Table 1. Mechanical properties of the structural components

Parts	Modulus of elasticity-E (GPa)	Yield Strength (MPa)	Poisson's ratio	Density (kg/m^3)
Parts	210	540	0.3	7850
Rail (UIC 60)	210	540	0.3	7850
Joint bar	207	540	0.3	7800
Nut and bolt	30	46.6	0.18	1265

In the interaction module, Surface-to-surface contact (Standard) of element type is used in-between Contact Pairs condition for all contact components. Then, the friction coefficient between rails and fishplates (bars), between bars and bolts, between bolts and rails are chosen as 0.3 to ensure tangential behavior in the penalty formulation [6]. Likewise, the normal behavior of component contact surfaces is selected as the default "hard" contact. According to this simulation, tie constraints were used between rail and sleeper and the analysis did not consider the sleeper effects, i.e., the bottom of sleeper surface was rigidly fixed in every direction [7].

Two loading steps are created for the rail joint structure simulation: bolt preloading and applied static load on the railhead from wheel load. In the first step, the bolt is preloaded to reduce the contact surface and length of separation from a loose of bolts tightness under static load condition [4, 5], following the pretension method [3-6, 8]. The preloaded bolt load was

$$P_b = \frac{T}{K_b D} \tag{1}$$

where P_b is the bolt load evaluated from the bolt torque, D is the bolt diameter and K_b is the coefficient of the bolt torque moment. The bolt load is used a preload 184 kN and based on an M27 bolt [3-5]. The second step is the vertical wheel load, acting on the railhead surface. As Figure 3 shows, the load is 200 kN, is applied on the rail end and rail-end bolt hole when varying the distances of rail-end bolt hole center to the rail end were 65 mm and 80 mm. The wheel load applied area is considered as the contact area from wheel to rail, following with Zhu et al. [2] model.

The applied-load positions are very important to analyze crack initiation in the rail joint since the load is on the rail end and the stress has reached a maximum in the rail end upper fillet area. Similarly, the load is applied above the rail-end bolt hole when the stress distribution around the bolt hole reaches its maximum [2]. To study the stress distribution in the rail joint, the tested distances from the rail-end bolt hole center to the rail end and the applied load positions was used (see Figure 3).

The mesh size at the contact locations is chosen as the fine mesh, while the other regions are used coarse that is allowed by the software for accuracy. Using a linear hexahedral mesh, the total number of elements and nodes in the rail, bars, bolts and sleepers was 602436 elements and 701956 nodes. The structure elements are applied by using the ABAQUS `C3D3R'- an eight-node fully integrated tri-linear brick reduction - to ease integration for computational effectiveness and use hourglass control.

After the successful simulation, the results can be visualized for analyzing the stress distribution in the vicinity of the upper fillet and the rail-end bolt hole and for checking the likely service life of the rail joint.

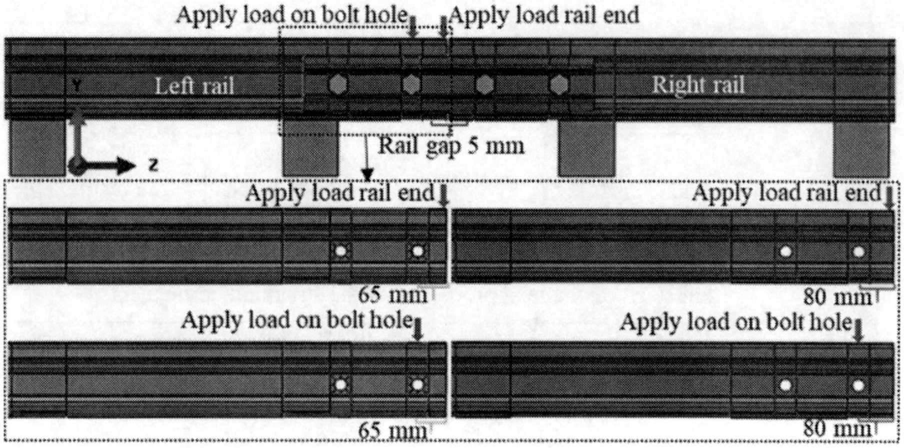

Figure 3. Rail joint of the parameters.

3. Discussion findings of simulation

The current study found that the stress distribution of a rail in a rail joint is evaluated using the finite element method. As can be seen in Figure 3, it showed the stress analysis is performed only on the left rail due to the symmetry of rail joint structure. From the data in Figure 4, we can see that there is a stress distribution in rail when load of 200 kN is applied for each condition. It is apparent that higher stresses are located in the areas of fillets and around the holes. Further analysis showed that there is a tendency of stress values of upper fillet are higher than that of lower fillet under the conditions of load applied on rail end (see Figure 4(a and c)) and load applied on top of rail-end bolt hole (see also Figure 4(b and d)). From this data, we can see that stress values around rail-end bolt hole (1st hole from the rail end) are larger than that of 2nd hole. The maximum stress measured from upper fillets, lower fillets, rail-end bolt holes and 2nd holes for each condition are summarised in Table 2. Interestingly, this correlation is related to the maximum stresses in upper fillet and rail-end bolt hole that are more critical when compared with the lower fillet and 2nd bolt hole. Therefore, the detail examination on FE analysis was concerned with stress distribution at the upper fillet area and around the rail-end bolt hole of the rail.

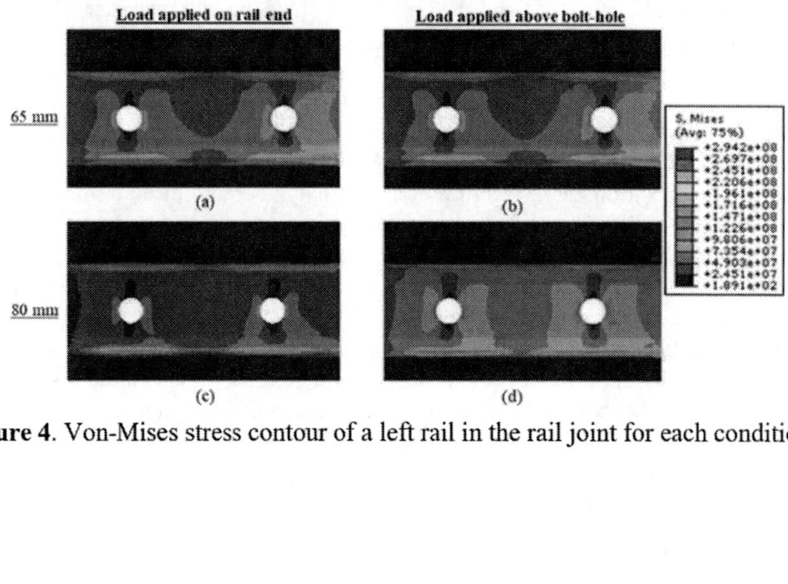

Figure 4. Von-Mises stress contour of a left rail in the rail joint for each condition.

Table 2. Maximum of Von-Mises stress (MPa) measured from upper fillets, lower fillets, rail-end bolt holes and 2nd holes for each condition.

Maximum Stress (MPa)	Load applied at rail end 65 mm	Load applied at rail end 80 mm	Load applied on the top of rail-end bolt hole 65 mm	Load applied on the top of rail-end bolt hole 80 mm
Upper fillet	269.2	210.2	268.9	102.3
Lower fillet	215.1	80.9	219.2	99.6
Rail-end bolt hole	173.1	134.3	172.2	184.4
2nd bolt hole	130.1	96.1	129.2	166.4

3.1 Stress in upper fillet region

Figure 5 represents the critical stress in the upper fillet region typically occurs at the rail end. This finding is consistent with those of other studies confirms the upper fillet crack initiates at rail-end of the rail joint leading to head-web separation failure mode [1, 3]. The influence of positions of wheel-loading and a rail-end bolt hole on stress distribution in upper fillet was also investigated. Figure 6 compares the results obtained of a single case of the local stresses along with the upper fillet, and it starts from rail end toward the bolt-hole side. Figure 7 provides the effect of position of the rail-end bolt hole (65 mm or 80 mm) on Von-Mises stress along the upper fillet. The results indicate that the local stress values reduced when the distance from the rail-end to the rail-end bolt hole (1st hole) is longer from 65 mm to 80 mm in both cases of applied loading. In particular, the reduction is higher when a load is applied on the top of the rail-end bolt hole (see also Figure 7(b)).

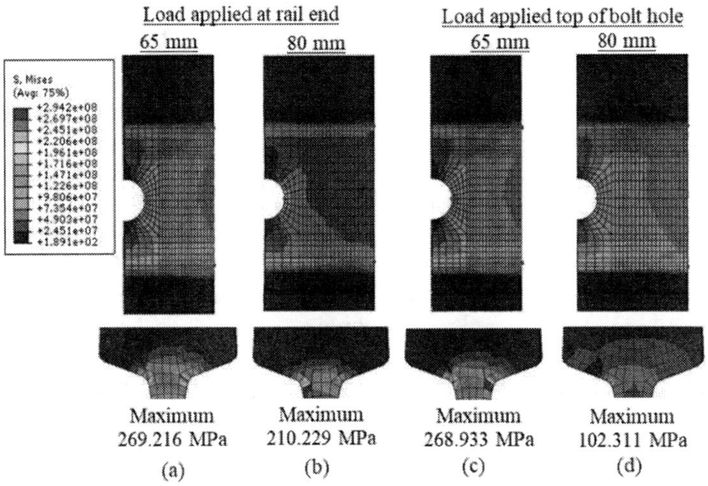

Figure 5. Von-Mises stress distribution in the upper fillet in each condition.

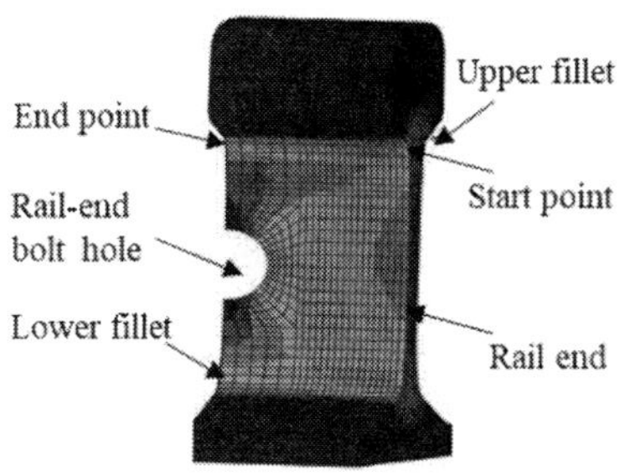

Figure 6. The local stress along the upper fillet.

The effect of wheel-load positions (applied at rail-end and the top of the rail-end bolt hole) on the upper fillet stress are shown in Figure 8. In case of rail-end bolt hole located at 65 mm from the end (see Figure 8(a)), the local stresses along the upper fillet are identical both when wheel-load is applied at the rail end and above the hole. These results suggest that the distance of 65 mm is too narrow to see the influence of location change. On the other hand, in case of rail-end bolt hole located at 80 mm, the stresses of a load applied on the rail end are higher than those of a load applied above the bolt hole. These results reflect those of Zhu et al. [3] who also found that the upper fillet stress reaches its maximum when the wheel load is at the rail end. In Table 2, there is a clear trend of increasing of 80 mm of the maximum stress of 210.2 MPa when the load applied at the rail end. However, there is a steady trend in case of 65 mm the maximum stress (~269 MPa) when the load applied at the rail end or on the top of the bolt hole.

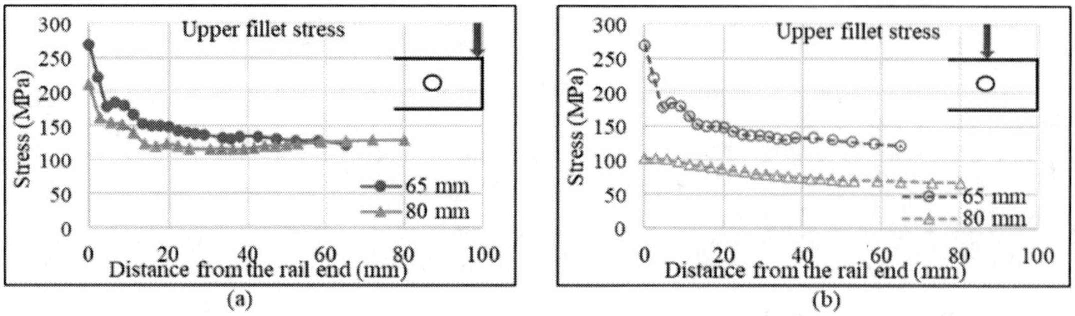

Figure 7. Effect of position of rail-end bolt hole on Von-Mises stress along the upper fillet when wheel load applied (a) at the rail end (b) on top of the rail-end bolt hole.

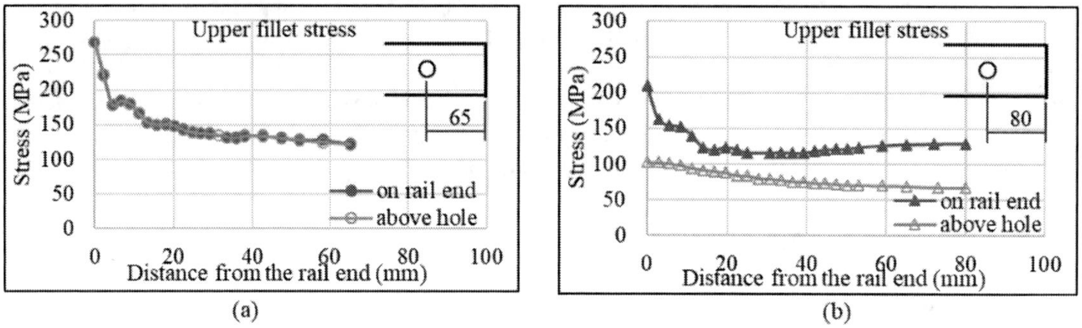

Figure 8. Effect of wheel load position on Von-Mises stress along the upper fillet when rail-end bolt hole located at (a) 65 mm (b) 80 mm from the rail end.

3.2 Stress around the rail-end bolt hole

Figure 9 presents the stress distribution around the rail-end bolt hole for each condition. The maximum Von-Mises stress of the rail-end bolt hole is located at the edge of the hole as set out in the red dot in Figure 9 These results corroborate the ideas of the defect or failure in rail joint in which bolt-hole crack commonly initiates at the very end bolt-hole and at an angle that is approximately 45° to the neutral axis of the rail [9]. The maximum stress of the rail-end bolt hole is probably dependent on the position of wheel load and position of the rail-end bolt hole. The effect of the wheel load applies on the rail-end bolt hole is similar to on the upper fillet in which the change of wheel load does not affect to the stress in rail-end bolt hole with a case of 65 mm; maximum stresses are 173.1 MPa and 172.2 MPa when the load applied at rail end and on top of the bolt hole, respectively. However, at 80 mm distance, there is a difference in stress when the load applied at rail end (134.3 MPa) and when load applied on the top of the bolt hole (184.4 MPa). In other words, the stress around the rail end bolt hole in case of 80 mm distance reaches the maximum when the load is applied on the top of the bolt hole.

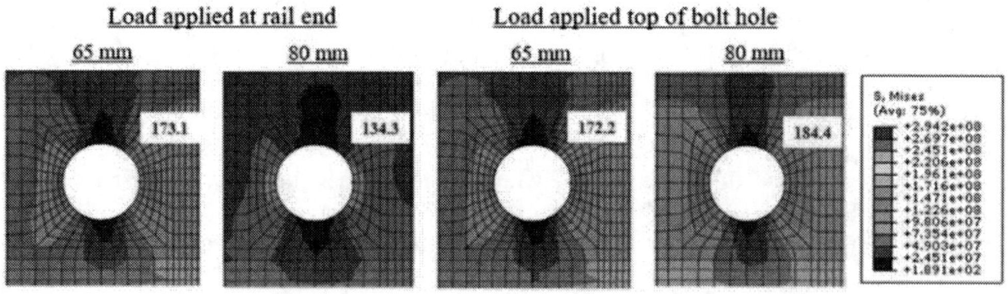

Figure 9. Von-Mises stress distribution around the rail-end bolt hole.

4. Conclusion

The purpose of the current study was to determine the performance of the FE method and to investigate the stress distribution of the UIC 60 rail in the bolt-rail joint under static loading of 200 kN with the parametric study of two positions of the applied wheel load (at the rail end and on the top of rail end bolt hole) and two rail-end bolt hole locations (65 mm and 80 mm). The summarize of FE results are shown as follows:

1) The critical locations of the bolt-rail joint are on the upper fillet rail end and the edge of the rail-end bolt where the maximum stresses are generated.
2) The wheel load position is not affected by the stress in upper fillet and rail-end bolt hole with a case of rail-end bolt hole located at 65 mm from the rail-end. However, in case of the distance from the rail end to the bolt hole is 80 mm, it is found that the stress value of upper fillet reaches the maximum when the load applied at the rail end whereas rail-end bolt hole reaches the maximum when the load applied above it.
3) Overall trends for the effect of rail-end bolt hole position are mostly similar in upper fillet and rail-end bolt hole where the Von-Mises stress of distance of 80 mm is lower than that of 65 mm except stress in rail-end bolt hole when the load applied above the hole.

These findings contribute in several ways to our understanding of stress distribution and provide a basis for stress distribution which can lead to a better understanding of structural failure.

Acknowledgments
This work was supported by King Mongkut's Institute of Technology Ladkrabang (KMITL).

5. References

[1] Zhu K Edwards J R Qian Y and Andrawes B O 2016 *Proc. 2016 ASME/IEEE Joint Rail Conf.* (Columbia, South Carolina) 1-7
[2] Zhu K Edwards J R Qian Y and Andrawes B O 2017 *Joints. Transportation Research Record.* **2607**(1) 33-42
[3] Zhu K Edwards J R Qian Y and Andrawes B O 2016 *Transportation Research Record: Journal of the Transportation Research Board* **2545** 36-45
[4] Ding K and Dhanasekar M 2007 *Advances in Engineering Software* **38**(8-9) 598-606
[5] Samantaray S K Mittal S K Mahapatra P and Kumar S 2019 *Engineering Failure Analysis* **104**(2019) 1002-1013
[6] Mohammadzadeh S Sharavi M and Keshavarzian 2013 *Engineering Failure Analysis* **29** 132-148
[7] Krishnamoorthy R R Saleheen Z Effendy A Alisibramulisi A and Awaludin A 2018 *IOP Conference Series Materials Science and Engineering* **431**(11) 11207
[8] Mandal N K Spiryagin M Berg M and Stichel S 2019 *International Journal of Fatigue* **128** 105197
[9] Mayville R A and Hilton P D 1984 *Theoretical and Applied Fracture Mechanics*

IOP Conf. Series: Materials Science and Engineering 1137 (2021) 012051 doi:10.1088/1757-899X/1137/1/012051

Effect of interference on the press fitting of railway wheel and axle assemblies

Theingi Nwe[1], Monsak Pimsarn[2,*]

[1,2]Department of Mechanical Engineering, Faculty of Engineering, King Mongkut's Institute of Technology Ladkrabang, Thailand 10520

* Corresponding Author: monsak.pi@kmitl.ac.th

Abstract

Press fitting is one of the common fastening methods used to assemble mechanical components. It has also received considerable critical many industrial applications, for example, railway wheel and axle are usually assembled by press-fitting. This process generates high stress on the axle-fillet area and fatigue is generally initiated due to the stress concentration. This paper attempts to propose the stress distribution and deformation of a railway axle. Various interference clearances were analyzed using finite element method. The contact pressure at the axle wheel seat along the axial direction was examined as it significantly influences the press-fitted components life. The analytical results from Lame's theory [1] were compared with finite element analysis results for the wheel and axle press-fitting process. The results obtained from both methods were in good agreement and the maximum relative error of the contact pressure is by 11.08 %. Additionally, the results show that the Von-Mises stresses on the axle wheel seat and fillet areas were found to be increased if the interference increased.

Keywords: Press Fitting, Railway Axle, Interference, Lame's Theory

1. Introduction

Railway axle is the vital security component in railway vehicles, and it is assembled with the wheel by the press-fitting method. The axle might be damaged by any of excessive pressure, excessive interference, or improper lubricant. The damage could evolve into fatigue cracks and these cracks are easily initiated on the axle due to fretting behaviors. Thus, the reliability of interference fits is crucial to ensure the railway wheelset is safe in service. Press fitting is a traditional method for assembly of interference fitting parts, and the quality of the assembly has been extensively studied. The contact pressure in press-fitting produces residual stresses, for example, tangential and radial stresses in the wheel and the axle. The residual stress greatly influences the occurrence of fatigue on the parts in operation.

Previous research has established that interference can have an effect to the interface between the two parts by finite element analysis. However, far too little attention has been paid to railway wheelsets. Lame's theory for thick-walled cylinder has been commonly used to predict the contact pressure of interference fit and deformation. However, the use of interference in the press-fitting has not been investigated. Partly, such an analytical theory is not adequate to predict residual stresses in press-fitting.

IOP Conf. Series: Materials Science and Engineering 1137 (2021) 012051 doi:10.1088/1757-899X/1137/1/012051

Sarabandi et al.[1] estimated the residual stress in railway wheels, due to press fitting, based on the nonlinear structural analysis and the nonlinear transient thermal analysis of the wheelset. They found that the stress distribution, due to thermal loading, is important and sensitive. Naderi [2] studied the stress intensity factor in railway wheelset by press fitting process. Hannemann and Sander [3] analyzed stress intensity factor and crack propagation, considering the effects of rotary bending and press-fitting. It was found that the press-fit load is important for the SIF solution. Marshall et al.[4] measured ultrasonic reflections to non-destructively determine contact conditions in the interference fit and investigated the wheel and axle interface with large contacts and gradual pressure variations. Zehsaz and Shahriary [5] investigated the effects of the coefficient of friction and interference on fretting fatigue strength of axles by using fretting damage parameters.

This research gives an account of the effect of interference on the wheelset press-fitting and variation of residual stresses with different interference clearances. The contact pressure and residual stresses were also evaluated analytically. The finite element analysis results confirmed that the analytic estimations can be employed to predict the stress distribution in the axle.

2. Materials and Methods

2.1 Interference Fit

Interference fits are widely used for mechanical attachment mechanisms because they are cost-effective and flexible. Concerning the wheelset assembly, shrink fitting or press-fitting can be used. However, press fitting is commonly used for assembling the wheel and axle because it is cheaper and faster. During the press-fitting, the axle wheel seat diameter is larger than that of the wheel hub bore. The assembled wheel and axle should meet the geometric requirements of some standard, e.g., EN 13261, EN 13262. The maximum and a minimum interference between the axle wheel seat and wheel hub bore are as follows [6]:

For shrink fitting,

$$0.0009dm \le 2\delta \le 0.0015dm \tag{1}$$

For press-fitting,

$$0.001dm \le 2\delta \le 0.0015dm + 0.06 \tag{2}$$

where dm is the mean diameter and δ is the radial interference in mm.

2.2 Lame's Theory for Thick-Walled Cylinder

Assuming the cylinder subjected to internal pressure is the wheel hub bore and the cylinder subjected to external pressure is the axle wheel seat, the stresses can be calculated by Lame's theory.

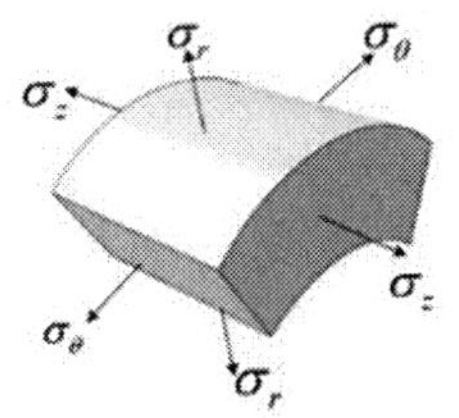

Figure 1. Stresses in a thick-walled cylinder element

Consider a small section of the cylinder, radial stresses σ_r and σ_θ tangential stresses can be achieved as a function of radius:

$$\sigma_r = \frac{r_i^2 P_i - r_o^2 P_o}{r_o^2 - r_i^2} - \frac{(P_i - P_o) r_i^2 r_o^2}{(r_o^2 - r_i^2) r^2} \tag{3}$$

$$\sigma_\theta = \frac{r_i^2 P_i - r_o^2 P_o}{r_o^2 - r_i^2} + \frac{(P_i - P_o) r_i^2 r_o^2}{(r_o^2 - r_i^2) r^2} \tag{4}$$

IOP Conf. Series: Materials Science and Engineering 1137 (2021) 012051 doi:10.1088/1757-899X/1137/1/012051

When the wheel and axle are press-fit assemblies, it can be assumed that pressure P is exerted all over the surface of the contact at the nominal radius, r. The magnitude of contact pressure, due to press-fitting assembly, can be calculated by summing the radial displacements of the inner and outer parts [7]. It was designated r_i is the inside radius of the cylinder, r_o is the outside radius, P_i is the inner pressure and P_o is the outer pressure. Where the subscripts o and i on Young's modulus E and Poisson's ratio v represent the inner and outer cylinder.

$$P=\frac{\delta}{r\left[\frac{1}{E_o}\left(\frac{(r_o^2+r^2)}{(r_o^2-r^2)}+v_o\right)+\frac{1}{E_i}\left(\frac{(r^2+r_i^2)}{(r^2-r_i^2)}-v_i\right)\right]} \qquad (5)$$

If the two members are the same material, $E_o = E_i = E,\ v_o = v_i$, equation (3) simplifies to

$$P=\frac{E\delta}{2r^3}\left[\frac{(r_o^2-r^2)(r^2-r_i^2)}{(r_o^2-r_i^2)}\right] \qquad (6)$$

3. Finite Element Analysis

In this article, finite element analysis was employed to evaluate the stress distribution and deformation of the press-fit interference on the railway wheel and axle. The geometric model of the wheel and axle assembly is shown in figure 2. The S1002 wheel profile and hollow axle were used in the analysis. The inside diameter of the axle is 60 mm and the wheel hub diameter is 190 mm. The axle length is 2300 mm and the wheel and axle contact length is 185 mm.

The wheel and axle quarter model were established using the symmetric boundary condition to reduce the running simulation time. A linear elastic material behavior with Young's modulus of 210 GPa and Poisson's ratio of 0.3 [8] is defined for both wheel and axle material properties. A three dimensional (C3D8R) 8-node linear brick (hexahedral), reduced integration, hourglass control was used for the element type. The wheel and axle contact area have meshed with a refine meshed until the stress and contact pressure changed lower than 1%. The total number of elements are 31,305 and 36,322, for the wheel and axle, respectively.

The contact surface interaction between the axle wheel seat and wheel hub was defined using a master-slave contact algorithm: the wheel hub surface was the master surface, and the axle wheel seat surface was the slave. A surface-to-surface interaction was defined with a finite sliding in the initial step. It is also required to be defined as an initial over-closure at the initial step. In the first step, the interference was resolved by automatic shrink-fit with a gradually removed slave node over closure during the step. A contact interaction property was defined as a penalty and the coefficient of friction was set to 0.6 to ensure tangential behavior. The sliding formulation was selected as "finite sliding" because the installation displacement was large. The flag indicating a non-linear geometry, Nlgeom, in the software, was set ON [9].

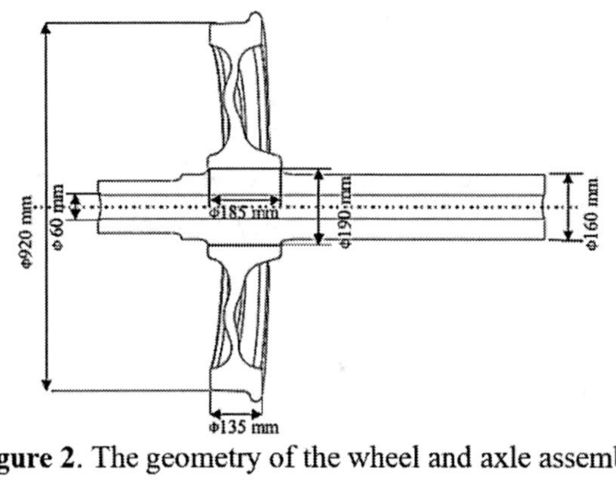

Figure 2. The geometry of the wheel and axle assembly

IOP Conf. Series: Materials Science and Engineering 1137 (2021) 012051 doi:10.1088/1757-899X/1137/1/012051

Figure 3 shows the boundary conditions of the symmetric finite element model: the symmetries of the wheel and axle assemblies used in the finite element analysis are: in the X-Y plane, $U_z, \theta_x, \theta_y = 0$,in the X-Z plane, $U_y, \theta_x, \theta_z = 0$ and at the right end of the axle YZ plane was constrained in the longitudinal direction, $U_x, \theta_y, \theta_z = 0$.

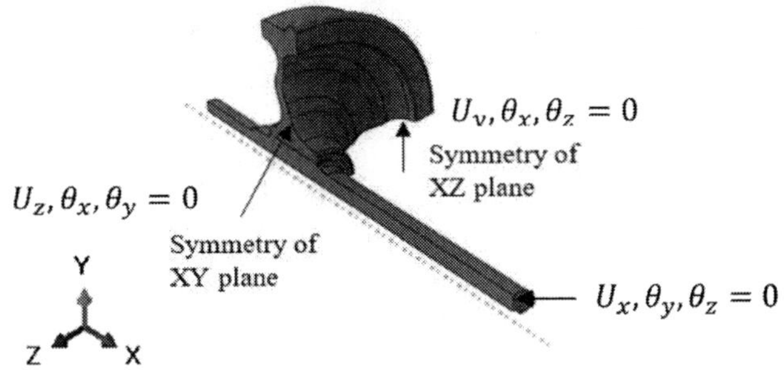

Figure 3. Boundary conditions for wheel and axle assembly

4. Results and Discussion

Several different interference clearances, ranging from 190 μm to 345 μm, obtained from Equations (1) and (2), were modeled and examined. The Equation (5) was used to determine the contact pressure distribution over the interface area. Analytical results and finite element analysis results were compared to determine whether the finite element analysis was acceptable and accurate. Table 1 shows the relative error in contact pressure for various interferences, between the finite element method and Lame's theory. It also compares the maximum deformations of the wheel and axle between modeling and theory. The axle was compressed, and the wheel was expanded since the contact pressures were equal and opposite at the interface surfaces. When the interference continuously increases, the deformation of the wheel and axle get higher. The lowest relative error of maximum contact pressure was 1.64% with the same mesh size. The Von Mises stress distribution on press-fitted wheelset between FEM and the analytical methods can be compared in Figure 4. The Von Mises stress variation along the wheelset center radial axis from both results were similar and the values rapidly increased at the wheel hub inner surface and then gradually decreased towards the wheel rim outer surface. It can be noticed that FEM shows the maximum Von Mises stress at the wheel hub inner surface, but analytical results showed it at the axle hollow inner surface. Finite element and analytical results for the radial and tangential stress distribution along the central axis of the wheelset are shown in Figures 5 and 6. The radial stresses are zero at the inner surface of the hollow axle and outer surface of the wheel because they are unpressurized surfaces. The tangential stresses on the axle are maximum at the outer surface and minimum at the inner surface. However, maximum tangential stresses were at the inner surface and minimums were at the wheel outer surface. It can be seen the axial, radial, and tangential stress distribution of the wheelset in Figures 8 to 10.

Table 1. Contact pressure and maximum deformations along the radial direction for various interferences by finite element method (FEM) and Lame's theory.

Interference (μm)	Contact Pressure (MPa)		Relative Error %	Wheel Deformation (μm)		Shaft Deformation (μm)	
	Lame's Theory	FEM		Lame's Theory	FEM	Lame's Theory	FEM
200	96	107	11.08%	60.1	68.2	-30.3	-31.7
240	115	127	10.43%	72.1	77.3	-36	-38.1
280	134	136	1.64 %	84.2	85.7	-42	-38.1
320	153	168	9.8%	96.2	103	-48	-50.8

IOP Conf. Series: Materials Science and Engineering 1137 (2021) 012051 doi:10.1088/1757-899X/1137/1/012051

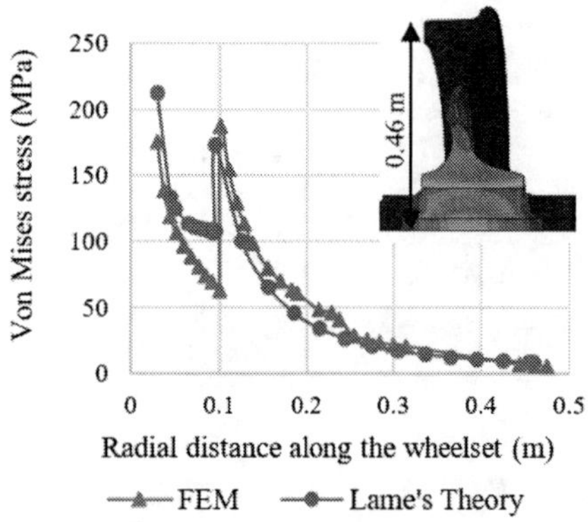

Figure 4. Von Mises stress along the centre of the wheelset radial axis.

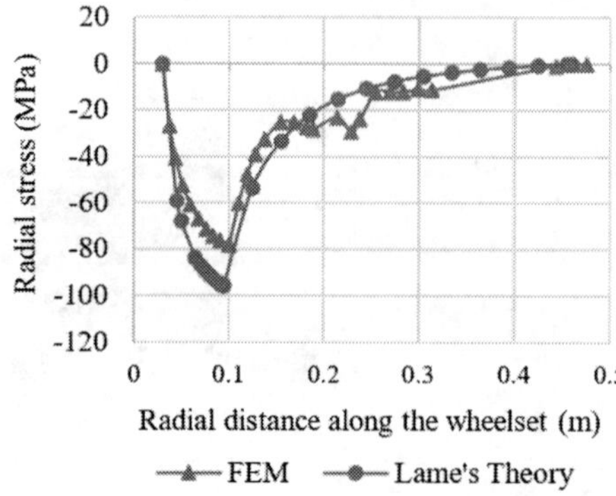

Figure 5. Radial stress distribution along the centre of the wheelset radial axis.

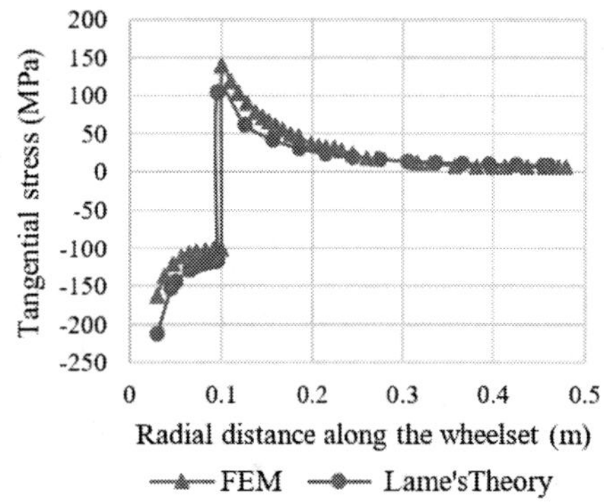

Figure 6. Tangential stress distribution along the centre of the wheelset radial axis.

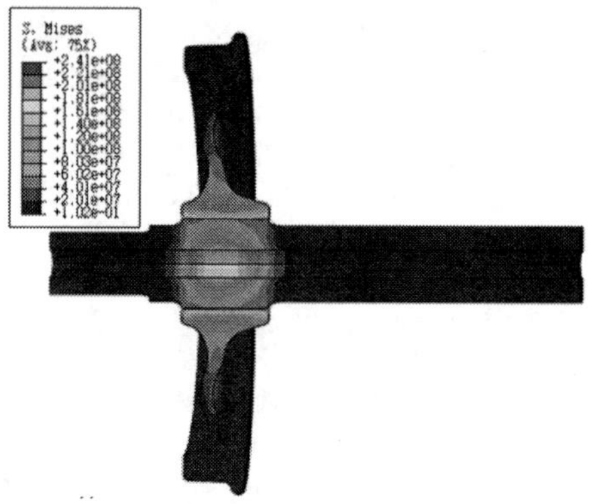

Figure 7. Von Mises stress distribution along the centre of the wheelset radial axis.

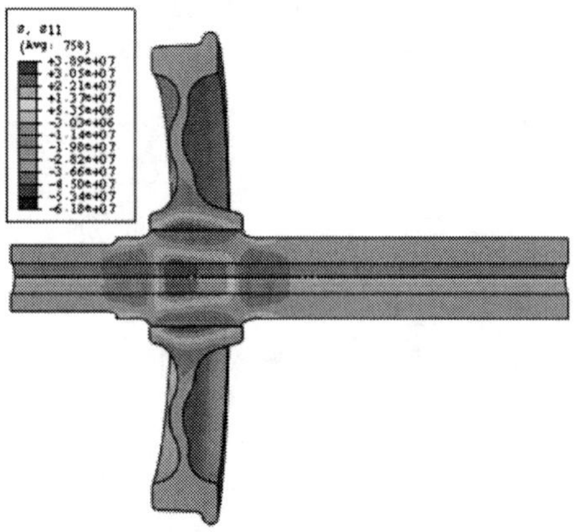

Figure 8. Axial stress distribution for wheel and axle assembly along the radial axis.

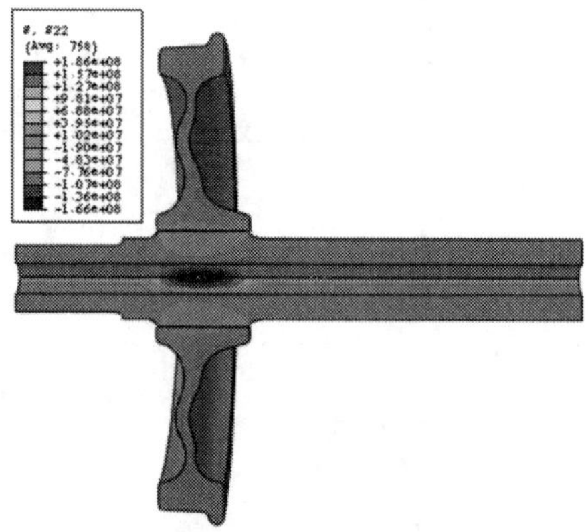

Figure 9. Radial stress distribution for wheel and axle assembly along the radial axis.

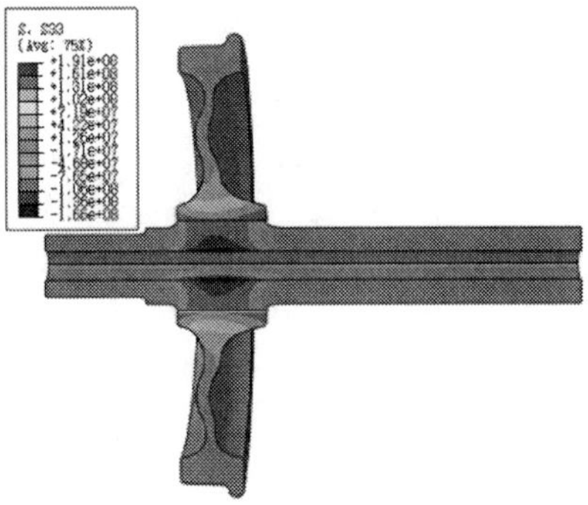

Figure 10. Tangential stress distribution for wheel and axle assembly along the radial axis.

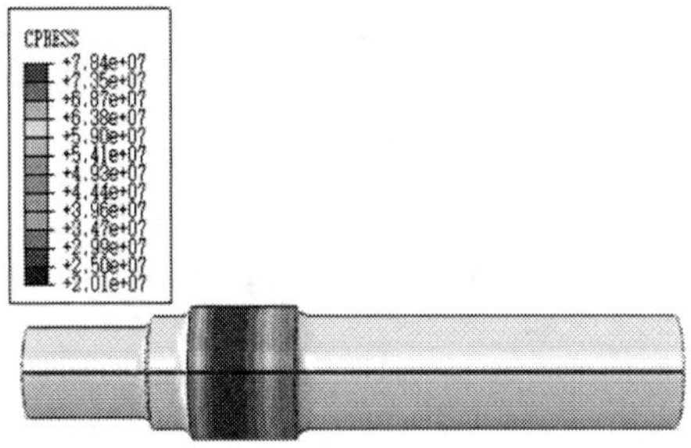

Figure 11. Contact pressure at axle wheel seat.

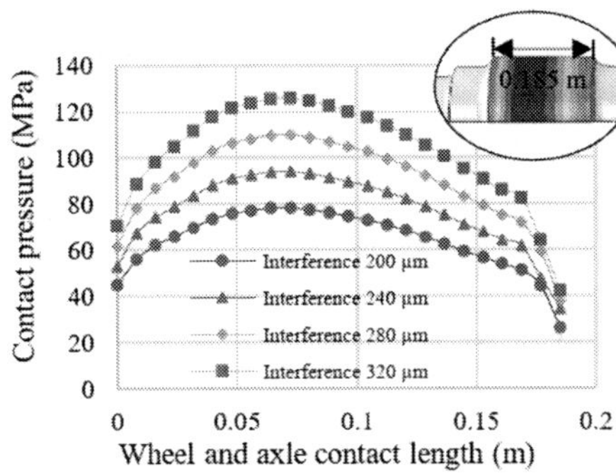

Figure 12. Contact pressure distribution along the contact length *vs* interference clearance.

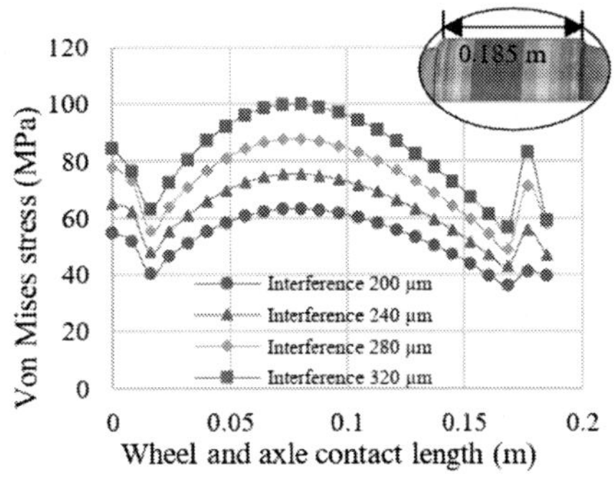

Figure 13. Von Mises stress distribution along the contact length *vs* interference clearance.

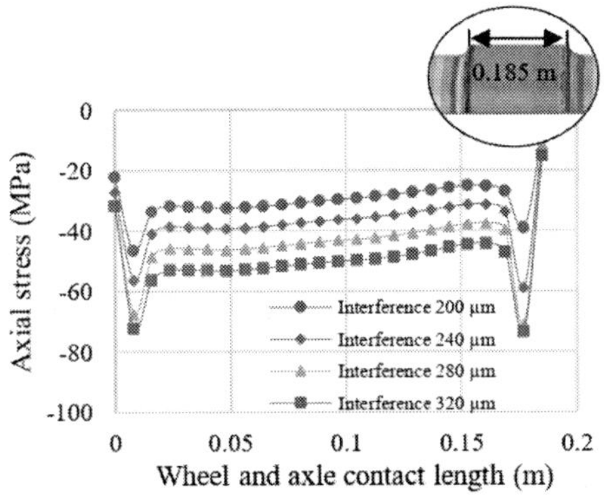

Figure 14. Axial stress distribution along the contact length *vs* interference clearance.

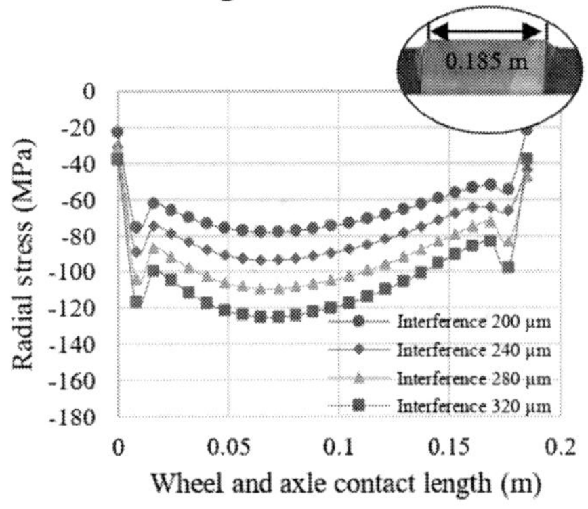

Figure 15. Radial stress distribution along the contact length *vs* interference clearance.

The variations of contact pressure along the axle wheel seat length vs interference are presented in Figure 12. By increasing the interference, the contact pressure was also increased with the same mesh size. Although the contact pressure, between the wheel and axle, was constant along the axle wheel seat length. Following Lame's theory, the contact pressure obtained from the FEM varied along the contact length. The maximum contact interference pressure was in the middle of the axle wheel seat contact length for all interferences. This was explained by the assumption that, in Lame's theory, the cylinder thickness was constant, but in the wheel model it was not. The Von Mises stress variation on the axle wheel seat is presented vs interference values in Figure 13. The axial, radial and Von Mises stresses were significantly affected by interference clearances as set out in Figures 13 to15. The results revealed that the maximum Von Mises stress occurred in the middle of the axle wheel seat contact area similar to the contact pressure.

5. Conclusion

This paper has discussed the reasons for the wheel and axle press-fitting assembly had a significant impact on the lifetime of the wheelset components. The stress distribution and deformation due to the press-fit interference effect were compared with the finite element method and Lame's theory. The following conclusion can be drawn from this research:

The FEM model agree well with the theory to within 15% relative error, the maximum relative error of contact pressure is 11.08% and the minimum value is 1.64%, which means the results is considered acceptable. Therefore, Lame's theory for a thick-walled cylinder was effective for the wheel and axle press-fitting analysis. However, the finite element method was more accurate and comprehensive than Lame's theory. The comparing results show the results obtained from the FEM are greater than the analytical results because the wheel and axle are considered ideal cylinder surface for analytical calculation and the assembly parts had a complex geometry. These results confirmed that the critical stress occurred at the contacting surfaces of the wheel hub and the tangential stress was greatest among the three principal stresses.

Deformation during press-fitting along the radial axis was unfavorably affected and strongly depended on the contact pressure at the wheel and axle interface. One of the more significant findings to emerge from this study is that increasing amount of interference caused deformation to increase, smaller clearances benefit to press-fitted parts. Taken together, these results suggest that the press-fit effect should be accounted for a railway axle design.

Acknowledgements

This work was supported by King Mongkut's Institute of Technology Ladkrabang. The author would like to thank the State Railway of Thailand for providing valuable data.

References

[1] Budynas R and Nisbett J K 2006 *Shingley's Mechanical Engineering Design 8th ed* (New York: MC Graw Hill Eduaction)
[2] Sarabandi S Soleimani H and Mahmoudi H 2019 *Sci. Iran* **26**(1) 367-374
[3] Naderi A 2018 *Eng. Fail. Anal* **94**(7) 78-86
[4] Madia M Beretta B and Zerbst U 2008 *Eng. Fract. Mech* **75**(8) 1906-1920
[5] Marhsall M B Lewis R Dwyer-Joyce R S Demilly F and Flament Y 2004 *Proc. 14th International Wheelset Congress* (Orlando, USA) **2**(41) 17-21
[6] Zehsaz M and Shahriary P 2013 *UPB Sci. Bull. Ser. D Mech. Eng* **75**(4) 71-84
[7] *EN 13260:2009+A1:2010: Railway applications - Wheelsets and bogies - Wheelsets - Product requirements.*
[8] Boresi R P 2013 *Advanced Mechanics of Material 6 th Edition* (USA: John Wiley & Sons, Inc)
[9] Yamamoto M and Ishiduka H 2017 *Int. J. Fatigue* **97** 48-55
[10] *Abaqus Analysis user's manual* vol I-IV.2010 (USA: Dassault Systèmes)

Numerical simulations of cancer treatment prediction: modeling the influence of immunity

Yuka Ezaki[1], Katsuya Nagayama[1*] and Ichiro Miura[2]

[1]Kyushu Institute of Technology Graduate School, 680-4, Kawazu, Iizuka, Fukuoka, 820-8502, Japan
[2]Department of surgical pathology, Hokkaido Institutional Society Obihiro Hospital, Japan

* Corresponding Author: nagayama@mse.kyutech.ac.jp, Tel & Fax 81-948-29-7778

Abstract
Malignant tumors, also called cancer or malignant neoplasms, are life-threatening tumors that invade the borders of other tissues or spread to other parts of the body and grow in various locations [1]. Recent evidence suggests that cancer cells generated by genetic mutations are eliminated by immune cells. Immune cells are immune cells that recognize and attack foreign substances, pathogens, and cancers in the body. A decline in immunity due to ageing or nutritional deficiencies, or due to immunosuppression is associated with increased risk of cancer cells [2]. Immunotherapy is a treatment method that activates the body's innate immune function to eliminate cancer cells. This study, therefore, set out to assess the effect of immunotherapy using numerical simulations. Parameters, such as the conditions of cancer cell death with immune cells, cell division, and angiogenesis, were theoretically set and introduced into the model.

Keywords: Cancer, Immunotherapy, Immune cell, Numerical simulation

1. Introduction
Malignant neoplasms have been thought of as a key factor in causing death. The issue of malignancies has received considerable critical attention. Recently, there has been renewed interest in FY2016. Researchers have shown an increased interest in cancer predictions with various conditions and treatment effects using numerical simulations to support the diagnoses and treatment methods of cancer [3]. The specific objective of this study was to examine the immunotherapy method as one of the cancer treatment methods. The present research explores, for the first time, the effects of the method of cancer treatments that use anticancer drugs or radiation. This treatment has proved to be useful because it has fewer side effects. After all, the patient is treated with their immune cells. Immunotherapy is expected to become increasingly widespread in the future [4]. There is still uncertainty, however, whether what kind of patients it is effective for and how long to beat the dose. The purpose of this investigation is to explore the effect of immunotherapy by numerical simulations using the particle method. Apart from the model of FY2016, we have made the model more like a real-world phenomenon by improving the accuracy of immune cell distribution and introducing nutrient-deficient cancer cell death. The results are expected to contribute to clinical research on cancer treatments and explanatory materials for cancer patients.

2. Computational Models and Analysis Targets

The analytical model is based on a particle method [5] that has previously been used for numerical simulations of tumor growth and angiogenesis. The particle method is a method for computing the behavior of a continuum by the motion of a finite number of particles. Each particle can be treated as a particle that moves around the calculation point while maintaining variables such as nutrient content and pressure. This analysis deals with cancer cell particles, vascular cell particles, and immune cell particles. Immune cell particles are placed on vascular and cancer shape data generated from cancer images and changes in cancer cell particles are observed. The cancer images used and the images processed and modeled from them are shown in Figure 1. The image in Figure 1a is binarized, and when he shading value difference is 0.4, it is recognized as cancer. The particles placed at the recognized location as cancer are free particles. The image area is 75mm wide and 75 mm high.

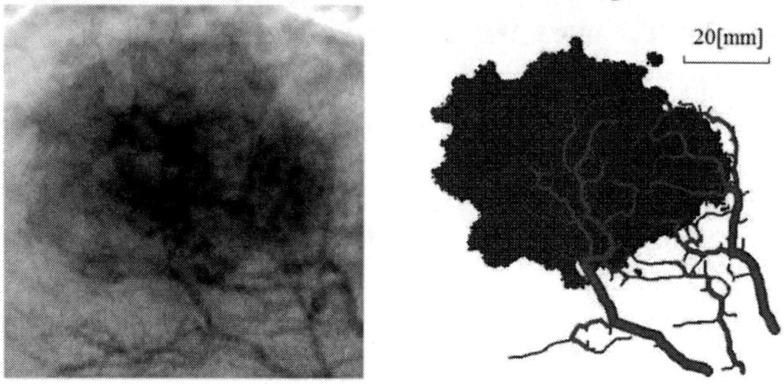

(a) Cancer imaging (b) Cancer and vascular placement

Figure 1. Physical model used in the analysis.

In this study, we focused on immune cells that recognize and attack foreign substances, pathogens, and cancers in the body. Immunotherapy is a method of treatment in which immune cells are activated outside the body and then returned to the body to attack cancer cells [6]. In this study, we compare changes in the tumors by predicting cancer growth in the absence of immunity, predicting the effects of the body's natural immunity, and predicting the therapeutic effects of immunotherapy.

For a tumor to grow, blood vessels need to be pulled into the tumor. If blood vessels cannot be pulled into the tumor, the lack of nutrients will cause the cancer cells in the tumor to suffocate or become malnourished. Accordingly, many cancer cells produce factors that promote angiogenesis [7]. These factors help the newly formed blood vessels feed the tumor and ensure that the cancer cells can continue to divide. Here, the nutrient diffusion equation between cells is shown in equation (1), where C is the amount of nutrients, D is the diffusion coefficient, and α is the nutrient consumption coefficient.

$$\frac{\partial C}{\partial t}=D\left(\frac{\partial^2 C}{\partial x^2}+\frac{\partial^2 C}{\partial y^2}+\frac{\partial^2 C}{\partial z^2}\right)-\alpha \tag{1}$$

The nutrient content of the particle of interest is given by equation (2), where i is the particle of interest and j is the target particle. X is the distance between particles. If the nutrients in the vascular cell particles are set to 1, the nutrient diffusion distance to the cancer cell particles is 200μm. Set the value of α/D that satisfies this condition. In this case, we set it to 0.1.

$$C_{\mathrm{i}}=\frac{\sum\frac{C_{\mathrm{j}}}{X_{\mathrm{ij}}^{2}}-\frac{\alpha}{D}}{\sum\frac{1}{X_{\mathrm{ij}}^{2}}} \tag{2}$$

The division of cancer cell clusters depends on the nutrient concentration within the cluster. Cancer cells with high nutrient concentrations actively divide, while nutrient-depleted cancer cells die or become dormant over time. The equation for the cell division time [8] is shown in equation (3), where T is cell division time, $T_{\min}$ is the shortest cell division time, Kc is the relaxation coefficient of the trophic term, and C is the amount of nutrient concentration. The shortest cell division time was 30 days, and the relaxation coefficient of the trophic term was 0.5.

$$\mathrm{T}=\frac{T_{min}}{1+Kc}\left(1+\frac{Kc}{C^2}\right) \tag{3}$$

3. Analysis Conditions

Effective immune cells have difficulty penetrating tumors and can only be present in approximately 20% of a tumor as transported by the blood vessels. Therefore, in the simulations, 20% of the immune cells outside the tumor were placed inside the tumor. Cancer cells were killed by the immune cells at a rate of 1% per hour. During that time, an immune cell could attack the nearest cancer cell within a range of two particles surrounding the immune cell.

Immunotherapy results in the body having approximately 10 times the normal number of immune cells [6]. In actual immunotherapy, the cytotoxic activity of the immune cells falls below the effective range in 3–4 days; therefore, new immune cells are injected in a 3–4-day cycle [9]. In this analysis, this process was represented by the probabilistic annihilation of the immune cells and their re-emergence every 4 days.

The amount of nutrients in the blood vessels, which serve as the nutrient source of the cancer cells, was set to 1; the nutrients then diffused to the surrounding particles. However, it is difficult for nutrients to reach blood vessels that are in areas of high pressure inside a tumor. Therefore, coupled with the tumor pressure distribution [3] shown in Figure 2, the nutrient content of the blood vessel cells with pressure values of 20 or more was assumed to be the value resulting from diffusion from the surrounding cells. The distribution of the nutrients is presented in Figure 3.

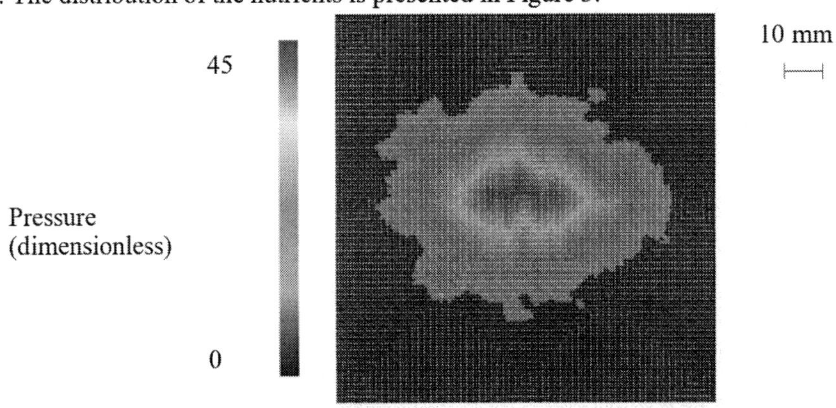

Figure 2. Pressure distribution in the modeled tumor.

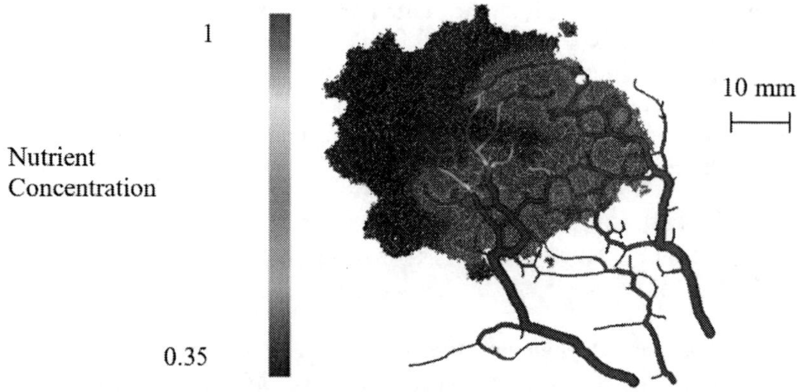

Figure 3. Nutrient distribution in the modeled tumor and blood vessels.

Malnourished cancer cells may take months to die. To represent this in the model, cells with nutrient levels below 0.42 were eliminated over an average of approximately two months.

4. Analysis Results

As can be seen from Figure 4 below, the 30-day cancer particle count predictions in the absence of immunity, with natural immunity, and with immunotherapy.

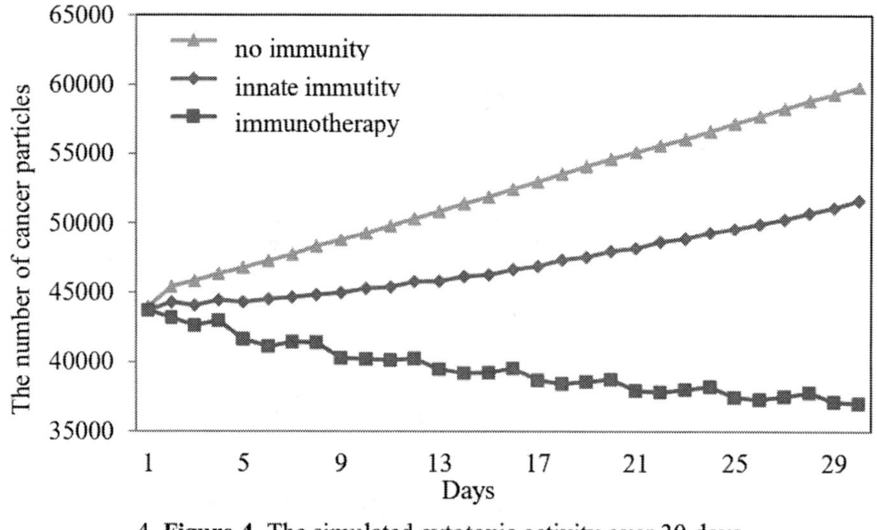

4. **Figure 4**. The simulated cytotoxic activity over 30 days.

4.1 Prediction in the absence of immunity

The purpose of this section was to provide result of the analysis of only cancer cell particles and vascular cell particles, without the placement of immune particles. As can be seen from the green triangles in Figure 4 above, the cancer growth can be represented in the case of the absence of an immune attack. Introducing cancer cell death in nutrient-deficient areas and modifying the nutrient content of the high-pressure blood vessels improved the model behavior compared to the rapid cancer growth seen in previous models [3].

4.2 Predicting the effect of natural immunity

This is the result of the analysis of the placement of immune cell particles. It can be seen from the data of the blue diamonds in Figure 4 that the number of cancer cells in the case of natural immune cells is lower than in the case of no immunity, representing the attack by immune cells on cancer cells. However, natural immunity alone is not effective in curing terminal cancers, and we can see that the number of cancer cells is increasing.

4.3 Predicting the effect of immunotherapy
The results obtained from the preliminary analysis of modeled immunotherapy by placing about 10 times as many immune cell particles as natural immunity, and making immune are set out in Figure 4. We can see that the cells disappear stochastically and reappear every 4 days. The red squares in Figure from the Figure 4 above we can see that the cancer cells are attacked and killed by periodically administered immune cells. The results indicated that the number of cancer particles was only reduced in the case of immunotherapy, which represents the treatment effect of immunotherapy. The results of an analysis of the cancer model with immunotherapy are provided in Figure 5. Figures 5a–5d compare the effects of immunotherapy on days 1, 10, 20, and 30, respectively. It is apparent from this figure that the cancer cells in the malnourished areas are gradually killed and the tumor shrinks (see Figure 5). In the same vein, regression of the blood vessels can be confirmed in response to the death of the cancer particles.

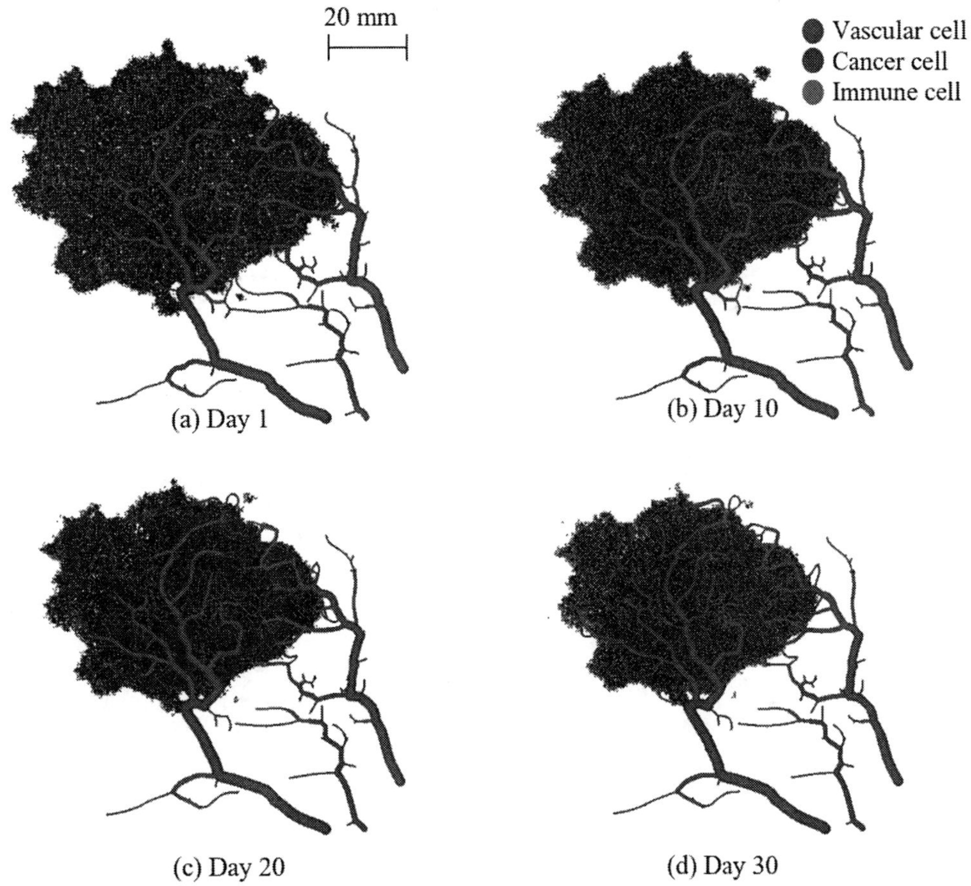

Figure 5. Modeled results of immunotherapy for days (a) 1, (b) 10, (c) 20, and (d) 30.

5. Conclusion
The present study was designed to improve the placement of the immune cells and the nutrient content of high-pressure blood vessels and to introduce the death of cancer particles in nutrient-deficient areas.

Hence, we were able to model the effect of immunity on cancer cells in a manner closer to the real phenomenon compared to conventional cancer models.

In the prediction without immunity, the current data highlight the importance of simulating a cancer model with only cancer cells and blood vessels with an appearance of the cancer growth. These findings also suggested a prediction of the effect of natural immunity so that the immune cells can be suppressed the cancer growth when compared to the prediction without immunity. Therefore, the attack on the cancer cells by the immune cells could be expressed. The prediction of the effect of immunotherapy revealed that the cancer cells were killed in the presence of periodic infusions of immune cells. The treatment effect could be observed as the tumor shrank over time.

Acknowledgments
We have received support and advice from many people in advancing this research, and we have successfully completed the paper. Dr. Ichiro Miura of Juntendo University and Dr. Sae Saito of Tohoku University gave us medical knowledge, images, experimental data, and guidance on our research, and we were able to proceed with our research in a favorable environment. We would like to thank you for this opportunity. Prof. Katsuya Nagayama has been paying close attention to our progress and has helped us to deepen our understanding of our research. We hereby express my gratitude.

References
[1] Matsubara S 1992 *Biology of Canceration and Cancerous Cells* pp 1-2
[2] Seino K 2015 *The Power of Immunity is Attractiong in Cancer Treatment* vol 1 Chapter 1
[3] Tajima K and Nagayama K 2016 *Numerical Analysis of Cancer Status from Cancer Images and Prediction of Treatment Effectiveness* pp 4-5
[4] Nakatsura T 2016 *Era of Cancer Immunotherapy has Come* pp 164-171
[5] Koshizuka S 2005 *Computational Mechanics Lecture Series 5* pp 1-8
[6] Tajima K and Nagayama K 2016 *Numerical Analysis of Cancer Status from Cancer Images and Prediction of Treatment Effectiveness* p 10
[7] Sakiyama N and Nagayama K 2019 *Numerical Simulation of Tumor Growth and Angiogenesis: Modeling from Development to End Stage* pp 8-9
[8] Shimizu H 2002 *Bioprocess Systems -Bases and Applications for Efficient Use* pp 56-62
[9] Tajima K and Nagayama K 2016 *Numerical Analysis of Cancer Status from Cancer Images and Prediction of Treatment Effectiveness* pp 44-45

Effect of audio pre-processing technique for neural network on lung sound classification

Krittaya Chaiyot[1], Supattra Plermkamon[2], Thana Radpukdee[3*]

[1,2]Department of Mechanical Engineering, Faculty of Engineering, Khon Kaen University, Khon Kaen, 40002, Thailand
[3]Department of Industrial Engineering, Faculty of Engineering, Khon Kaen University, Khon Kaen, 40002, Thailand

* Corresponding Author: tthanar@gmail.com

Abstract
In this paper, we intended to study the effect of two audio pre-processing methods to multilayer feedforward network performance on lung sound classification problem. The first one is Mel Frequency Cepstral Coefficients (MFCC) and the second is the MFCC supplement with Linear Discriminant Analysis (LDA). Datasets applied in this study came from Kaggle Respiratory Sound Database, which is the largest resource for machine learning. As a result, MFCC supplement with LDA has both good performance and significant improvement than conventional MFCC.

Keywords: Neural Network, Machine learning, Lung Sound Classification, MFCC, LDA

1. Introduction
Over the past century, physicians have diagnosed lung pathology using respiratory sounds. Those sounds can be related to respiratory health and respiratory disorders that need to have both special skill and reliable stethoscope to precision diagnose. The last two decades have seen a growing trend towards the automatic analysis of respiratory sounds. The reason is that it is not only necessary for ones to have special skills but it has also the potential to detect abnormalities in the early stages of respiratory dysfunction. Consequently, it can increase the effectiveness of decision making [1, 2], and may be of great support to the physicians.

This study set out to investigate the usefulness of automatic analysis of respiratory sounds, particularly sound classification task. This is an audio pre-processing and plays a significant role in preparation of training data on neural network. These processes can be used to identify the components of the audio signal, in the form of numeric data. It is also practical for identifying the audio content and rejecting redundant and unwanted stuff which carries information like background noise [3].

This paper can be divided into four sections. First section is introduction. Sections 2 describe the material and methods used in this study. There are four topics: Mel Frequency Cepstral Coefficient (MFCC), Linear Discriminant Analysis (LDA), Multilayer Perceptron and Data Preparation. Sections 3 include the result and discussion of the current study. The last section is the conclusion of the study.

2. Materials and Methods

Figure 1 illustrates the main framework used in this study. The lung sounds dataset in this study are preprocessed into two types. The first one, recorded lung sounds data just has been preprocessed and extracted features by MFCC so it can be used as training data for neural network, multilayer perceptron (MLP), and to perform classification task. Second, recorded lung sounds data was prepared like the first data type supplement with LDA for reducing dimension. Then, projecting the data into a subspace can be used to improve the class separability, and send as training data into multilayer perceptron. This current study applied Matlab as the main tool to implement the methods and use in the experimental for comparing classification results from different data types.

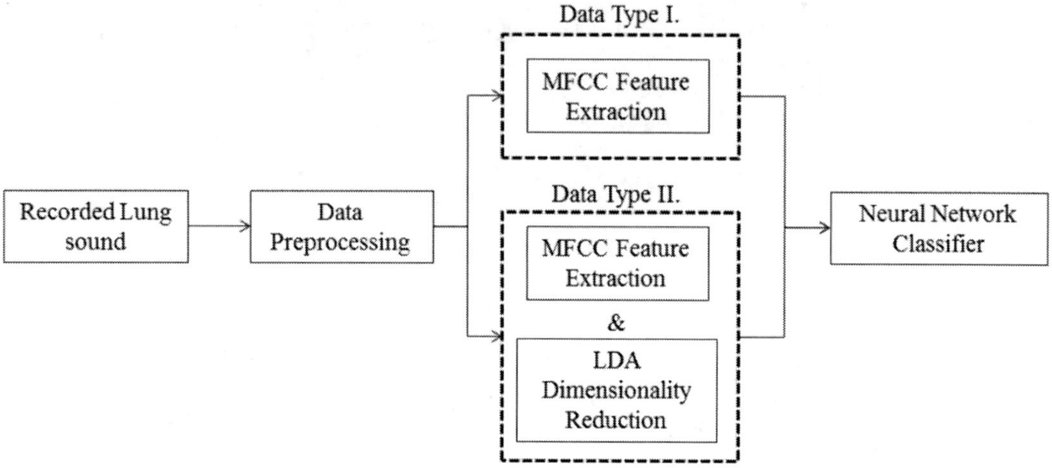

Figure 1. The framework used in this study.

2.1 Mel Frequency Cepstral Coefficient (MFCC)

MFCC is the most popular technique that is chosen to extract feature vectors in automatic speech recognition (ASR) [4-7]. The main point of MFCC is an extract of lung sound features without losing important information and rejecting redundant and unwanted stuff. The steps to perform MFCC computation process consist of Pre-Emphasis, Sampling and Windowing, Fast Fourier Transform, Mel Filter Bank, Discrete Cosine Transform, [8, 9] (see Figure 2 below).

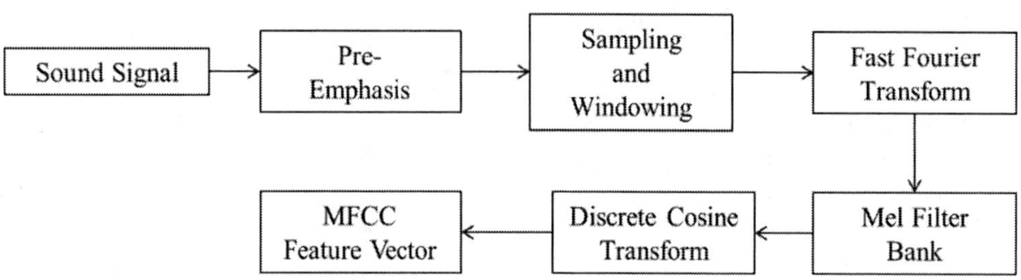

Figure 2. MFCC feature extraction process.

2.2 Linear Discriminant analysis (LDA)

Linear Discriminant Analysis is a dimensionality reduction and projecting the data into a subspace that improves the class separability technique adapted in supervised learning. This technique will project the original data matrix into low dimensional space. There are three steps involved in the LDA performance process. To calculate the between-class matrix according to the equation is used to begin the process (1), then the calculation within-class matrix as represented by equation (2). The last process includes the construction of the lower dimensional space, as set out in Equation (3), and it can maximize between-class variance and minimize within the class variance [10]. In this study, LDA function are adapted from MATLAB Central File Exchange [11].

$$(m_i - m)^2 = W^T S_{Bi} W \tag{1}$$

Where mi = the projection of the mean of the ith class, m = the projection of the total mean of all classes, W = represents the transformation matrix of LDA, SBi = between-class variance of the ith class.

$$\sum_{x_i \in \omega j, j=1,...,c} (W^T x_i - m_j)^2 = \sum_{x_i \in \omega j, j=1,...,c} W^T S_{Wi} W \tag{2}$$

Where SWi = within-class variance for each class, mj = the projection of the mean of the jth class, xi = samples of each class, W = represents the transformation matrix of LDA.

$$Y = XV_k \tag{3}$$

Where Y = new data matrix, X = original data matrix, Vk = lower dimensional space.

2.3 Multilayer Perceptron (MLP)

Multilayer Perceptron are one type of Artificial Neural Network (ANN) that use supervised training to train a network. In this study a classifier architecture (MLP type) is R-10-4 with tansig activation function in hidden layer and softmax activation function in output layer as in Figure 3. This network had been trained 10 times for each data types, and then chooses only one the best training and test result in order to compare network accuracy onward.

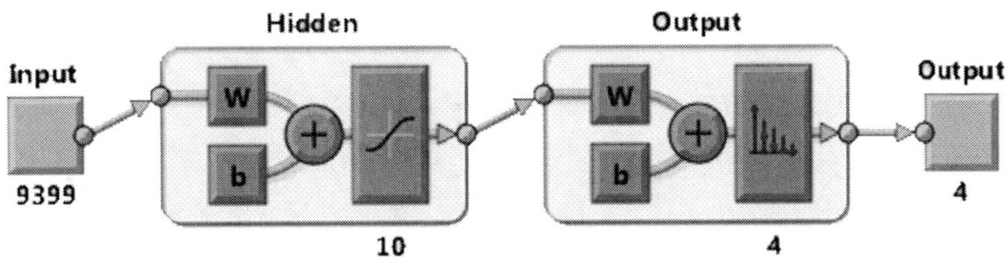

Figure 3. Classification model's architecture R-10-4 (MLP type)

2.4 Data Preparation

The dataset of the lung sound chosen in this study was from Kaggle Respiratory Sound Database [12]. This was created by two research teams in Portugal and Greece. It includes 920 annotated recordings of lung sounds. These recordings were taken from 126 patients. There is a total of 5.5 hours of recordings containing crackles, wheezes and both crackles and wheezes. Despite the limit of our computation resource, we use only 200 annotated recordings of a length 20 second each, which consist of 50 sounds

IOP Conf. Series: Materials Science and Engineering 1137 (2021) 012053 doi:10.1088/1757-899X/1137/1/012053

recorded for each class. Also, each class has frequency content (see Figure 4). (Normal < 400 Hz, Crackles < 1000 Hz, wheezes < 100 Hz and both wheezes and crackles < 1000 Hz).

However, each of 200 lung sound used in the training ANN still has a large amount of sample, 882,000 per one recorded lung sound, and it takes very high resource to computation. In order to overcome this each recorded lung sound has scaled down the sample rate from 44.1 kHz to 16 kHz [13], reference to the frequency content in Figure 4, 16 kHz can prevent an aliasing problem. As a result, sample per one recorded lung sound is 320,000 and duration is still 20 second. Then perform MFCC, which will get column vector of 25,974 elements for each recorded lung sound, so for totally, 200 recorded lung sounds we will get a matrix of 25,974 by 200. However, this matrix just still big for sending to ANN, so again all of 200 recorded lung sound have squeezed duration time from 20 seconds to 7.3 seconds by change sample rate to 44.1 kHz (Sample point just still 320,000 per each recorded lung sound). This process will increase the playback speed of recorded lung sound but still maintain the previous sample point without losing any information and also help to reduce data after performing MFCC to 9,399 by 200 matrixes, which use for train ANN so-called data type I. Figure 5 shows an audio waveform for one recorded lung sound.

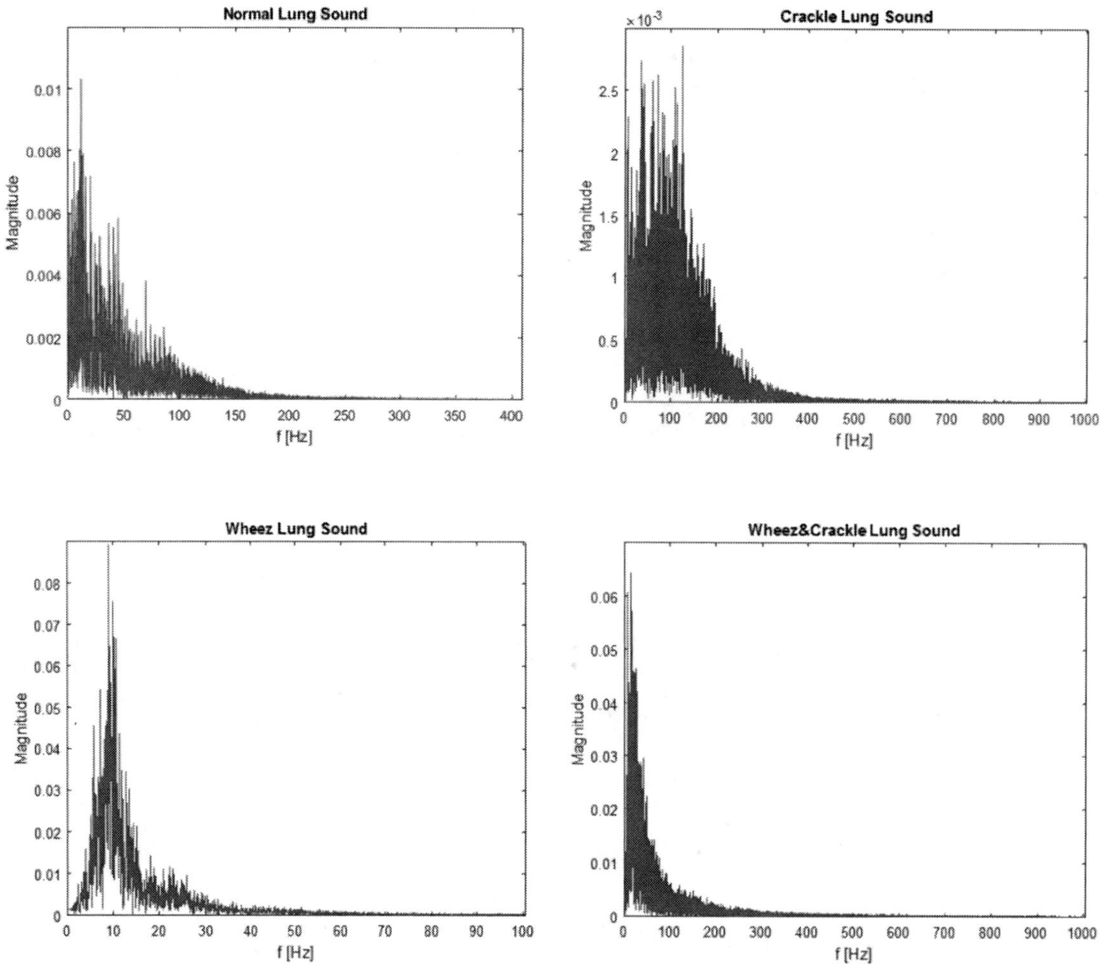

Figure 4. Frequency content for a single recorded lung sound.

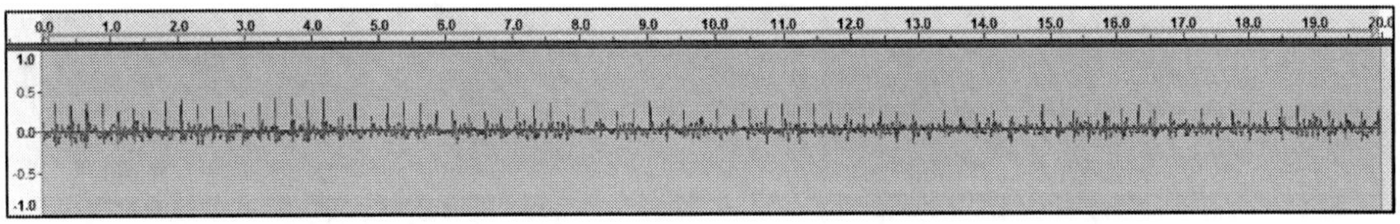

a. Recorded lung sound with 882,000 samples, Fs = 44.1 kHz, time length = 20 seconds

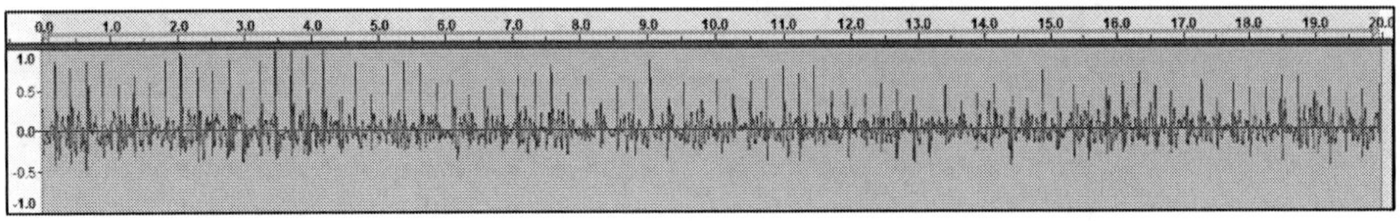

b. Recorded lung sound with 320,000 samples, Fs = 16 kHz, time length = 20 seconds

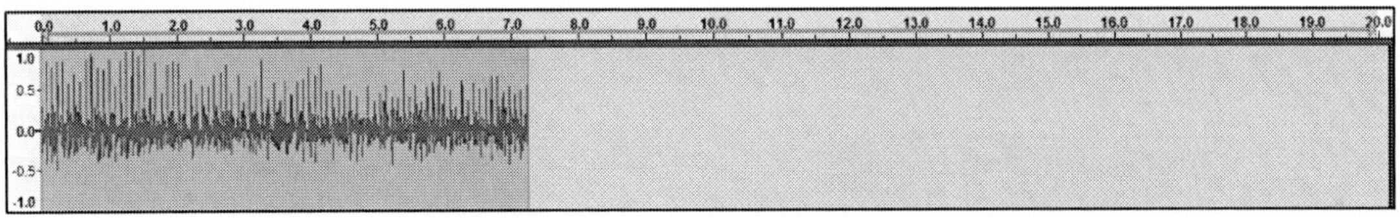

c. Recorded lung sound with 320,000 samples, Fs = 41.1 kHz, time length = 7.3 seconds

Figure 5. Compare a recorded lung sound in Audio waveform.

It should be noted that the preparation processes of data type II are the same as data type I. While the main difference in data type II is supplement with LDA after performing MFCC.

3. Result from the experiment

After preparing the training data as provided in topic 2.4, we have to train and investigate the network performance. Generally, the confusion matrix was used to measure the performance of a model in the classification task. We also use this tool to investigate the performance of the network.

To begin this training ANN process, each type of data was used to train R-10-4 network for 10 times. After training, the one that shows the best performance was indicated by a confusion matrix. The best result from data type I showed accuracy on the test set as 60% and 100% on the training set (see Figure 6). We can see that data type II indicated the best accuracy for a test set by 96.7% and for a training set is 100%, as set out in Figure 7.

The results from both types of data have good accuracy on training data, it was found that the network can learn training data quite well. Whilst the test data cannot be seen by the network never seen before. Only data type II showed a significant improvement on testing accuracy, resulting in the predominant of MFCC supplement with LDA for use in classification problem.

IOP Conf. Series: Materials Science and Engineering 1137 (2021) 012053 doi:10.1088/1757-899X/1137/1/012053

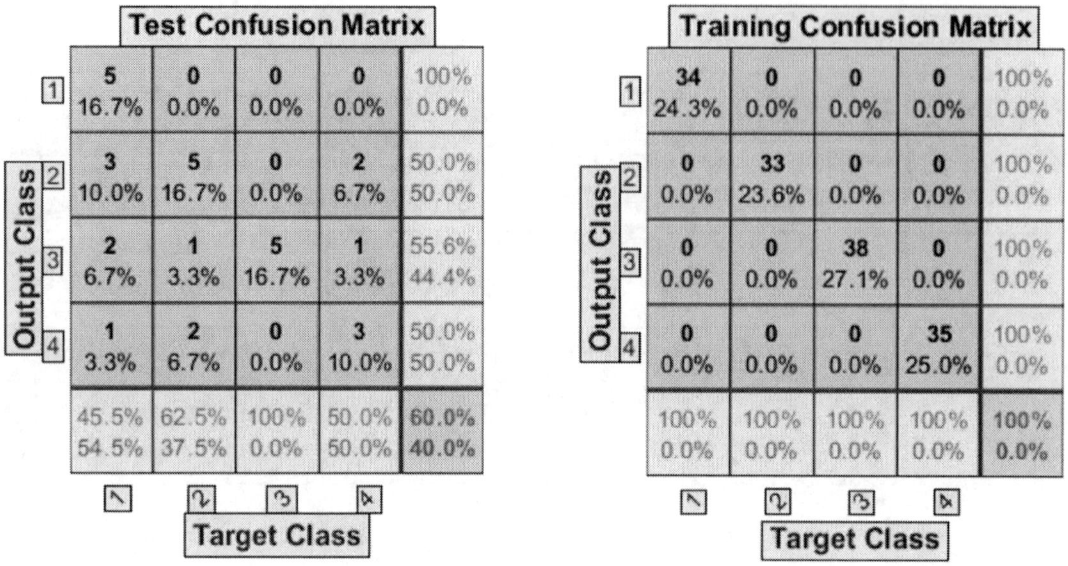

Figure 6. Test and training confusion matrix of ANN (R-10-4) on data type I.

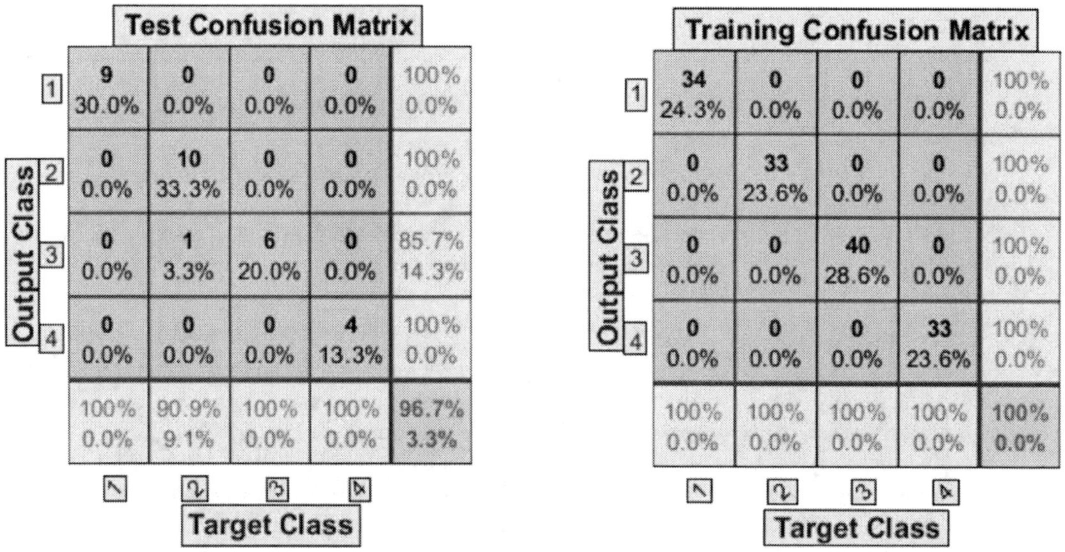

Figure 7. Test and training confusion matrix of ANN (R-10-4) on data type II.

4. Conclusion

This study has identified two input preprocessing methods that affect the classification performance. As can be seen from the result on topic 3, the MFCC supplement with LDA has the best performance for the classifier. Although this experiment has to deal with multivariable data like audio data a classifier, it can have an ability to identify each class with good performance. The findings of this study suggest that this pre-processing technique can also be applied in many similar classification tasks like a heartbeat sound etc.

References

[1] Saraiva A Santos D Francisco A Sousa J Ferreira N Soares S and Valente A 2020 *Proc. the 13th International Joint Conference on Biomedical Engineering Systems and Technologies* **3** 138-144

[2] Pasterkamp H Kraman S and Wodicka G R 1997 *American Journal of Respiratory and Critical Care Medicine* **156** 974-987

[3] Sharma D and Ali I 2015 *Proc. Int. Conf. on Advances in Computing, Communications and Informatics* (Kochi) 1052-1057

[4] Winursito A Hidayat R and Bejo A 2018 *Proc. Int. Conf. on Information and Communications Technology* (Yogyakarta) 379-383

[5] Joshi S C and Cheeran A N 2014 *MATLAB Based Feature Extraction Using Mel Frequency Cepstrum Coefficients for Automatic Speech Recognition*

[6] Davis S and Mermelstein P 1980 *IEEE Transactions on Acoustics, Speech, and Signal Processing* **28**(4) 357-366

[7] Huang X Acero A and Hon H 2001 *Spoken Language Processing: A guide to theory, algorithm, and system development* (Prentice Hall)

[8] Magre S Deshmukh R and Shrishrimal P 2013 *Proc. Int. Conf. on* Recent Advances in Statistics and Their Applications (Aurangabad, Maharashtra: Ambedkar Marathwada University)

[9] Chakraborty K Talele A and Upadhya S 2014 *International Journal of Innovative Research in Advanced Engineering* **1** 158-161

[10] Tharwat A Gaber T Ibrahim A and Hassanien A E 2017 *Ai Communications* 169-190

[11] Yarpiz 2020 Linear Discriminant Analysis (LDA) aka. Fisher Discriminant Analysis (FDA), https://www.mathworks.com/matlabcentral/fileexchange/53151-linear-discriminant-analysis-lda-aka-fisher-discriminant-analysis-fda), accessed on 17 September 2020

[12] Marsh 2019 Respiratory Sound Database, Version 2, https://www.kaggle.com/vbookshelf/respiratory-sound-database, accessed on 17 September 2020

[13] Messner E Hagmüller M Swatek P and Pernkopf F 2016 *Proc. 9th International Conference on Biomedical Electronics and Devices* (Rome, Italy) 34-39

Structural analysis of three-dimensional finite element model to design multifunction wheelchair for patients

P. Keangin*, P. Chawengwanicha, N. Wimala and T. Nakbanpotkul

Department of Mechanical Engineering, Faculty of Engineering, Mahidol University, 25/25 Phutthamonthon 4 Road, Salaya, Nakhon Pathom 73170, Thailand

* Corresponding Author: pornthip.kea@mahidol.ac.th

Abstract

The disabled and patients, especially hemiplegia and paraplegia cases and elder need to use a wheelchair for movement and admitted to the hospital or stay at home. The problem of using wheelchairs for patients and the elder is due to their different shapes. There is only a standard type of wheelchair in a hospital, therefore could not respond to differences in shape and usability. The ability of a wheelchair that does not provide enough facilities will result in the service as being ineffective in achieving and will make patients feel uncomfortable. The basis of the strength analysis, as well as stress distribution of the device structure design, is still minimal due to computer program limitations. A comprehensive understanding of the fundamentals of strength analysis will support the recommendation to design of wheelchair to handle patient for appropriate use in each situation. This research aim provides strength analysis to design the multifunction wheelchair. The multifunction wheelchair is designed to be turned into a bed, height-adjustable to use in a situation where a standard wheelchair could not be achieved due to the height is not enough and arm rest's width adjustable for the convenience of patients with different body shapes. The detail regarding concept design and strength analysis of main components are solved by the finite element method (FEM) with a three-dimensional model. The results obtained from the simulation solution are examined and compared with the analytical solution. It is found that the highest maximum stress value does not exceed the yield strength of the material. The obtained results can be helpful in determining the basis for the guideline of wheelchair design and develop medical instruments situation by using principles of engineering.

Keywords: Finite element method, Multifunction wheelchair, Stress distribution

1. Introduction

Nowadays, society becomes an aging society which means the amount of elder in the country may increase compared to other ages [1]. Since the population average age is higher than before and there are some problems from the economy effect that most people do not expect to have more heirs to reduce their expenses. Therefore, the number of patients in hospitals is higher and part of them are elder patients. Some patients need to use a wheelchair because of their health problems. However, the wheelchairs are manufactured in one standard size and therefore cannot respond to the different shapes of patients and usability. The main problem of the limitation of standard size wheelchairs can be as follows. First, the different shapes of the patient do not fit in the wheelchair, resulting in the patient feeling that the wheelchair is inconvenient to use and may cause other health problems such as aches. Second, the patients who sit in a wheelchair for a long time to wait for medical service may develop several diseases. In which case the staffs need to transfer the patient to a bed. In addition, if there is any incorrect physiological movement, it will affect the health of the patient and the staff [2]. The staffs and caregivers who have to help and support patients need to learn how to use any equipment. They need to know the technique to handle and move patients safely using medical equipment [3]. There are instructions for using any patient handling tasks associated to reduce injures. These tasks are including wheelchair and patient lifting equipment. The staffs and caregivers must know practicing for safe patients' movement [4]. Reductions of injuries associated with patient handling were studied by Wardell H. [5]. It was observed that in the 3 months following the implementation of the program, the hospital experienced only 9 reported strain and sprain injuries associated with patient handling, a 61 % reduction in the reported injury rate. Lastly, it takes a long time to transfer patients, especially those who are unable to help themselves, which may affect the time they provide medical services. The design of the multifunction wheelchair will improve patient handling patients more efficiently in health care for both patients and staffs.

There are many kinds of wheelchairs produced and available to be using nowadays. These productions were designed and developed to respond to the patient's need. The recline and tilt function were found by using a mechanic and electrical control system of the wheelchair. Some research to study the function of wheelchair by using MATLAB and to calculate the patient's average wheelchair usage time. It is found that most patients use their wheelchair 9-15 hours per day and the longest single continuous is 14.2 hours per day. The recline and tilt feature also be used together for about 5 hours per day [6]. Moreover, wheelchair also has been developed a new seat-adjustable power wheelchair to prevent pressure ulcers from long periods of sitting [7]. The lower rear portion of the power wheelchair was movable along a ball screw to change the tilting angle of the seat of the wheelchair resulted in the patient to be able to change the high-pressure area of the bottom. The mechanism can also be employed to convert the center of gravity of the wheelchair to increase the stability while the wheelchair was moving. The standard size of the wheelchair in the hospital cannot be adjusted the height of the seat but this feature is important as well. Some research described a biomechanical assessment of electric lifting chair with hip-up function. Experimental measurements of three-dimensional (3D) motion data and electromyogram (EMG) on the femoral muscle when the subject performs the standing motion on the predetermined seat height. The results showed that 15 degrees of the hip-up angle were suitable for the hip-up function and that the higher seat position was more effective to assist for standing up motion [8]. In the same year, the contour detection and localization of intelligent wheelchair for parking into and docking with U-shape bed was proposed [9]. Lines contained in the back-view image of the wheelchair were projected into the bed's superface plane and separated into several groups based on conditions of parallelism and perpendicularity.

The mechanical design that suitable for functionality and safety for a wheelchair or medical device is essential. Preliminary design of a mechanical transfer lift for the disabled patient in a Malaysian hospital was presented by Jusoh M.A. et al. [10]. It was found that the safety factor of 1.04 during static loading analysis using mild steel as the base was achieved. Mahmoud Z. et al. [11] determined the importance weights for engineering characteristics of wheelchair design by integrating a Quality Function Deployment (QFD) framework with a Fuzzy Analytic Network Process (FANP).

Although there have been studies and designed wheelchair, the wheelchairs which are available to purchasing have only one special feature for one product such as recline wheelchair will have only recline feature and height-adjustable wheelchair will have only height adjustable feature. The wheelchair features are reclined and tilt adjustable and height seat adjustable are necessary. Furthermore, comprehensive knowledge of the fundamentals of strength analysis will support the approval to design and create of the wheelchair to handle patient for appropriate use in each situation. The model of multifunction wheelchair and strength analysis is representative the basis of the model that will be used in the construction of the wheelchair.

2. Design and Simulation

This research focuses on the strength and ability to work in 3 features including recline and tilt adjustable, height-adjustable and armrest width's adjustable. The multifunction wheelchair is designed by using SolidWorks software and simulation models to analyze the strength of the main components. The size of human used in calculating is the average standard size.

2.1 Wheelchair design

The wheelchair is assembled by various parts, which are base of the wheelchair, wheel, leg rest and footrest, X-lifter, seat, armrest and backrest. Thus, the design is separated into a part and assembly to become a complete multifunction wheelchair as shown in Figures 1 and 2.

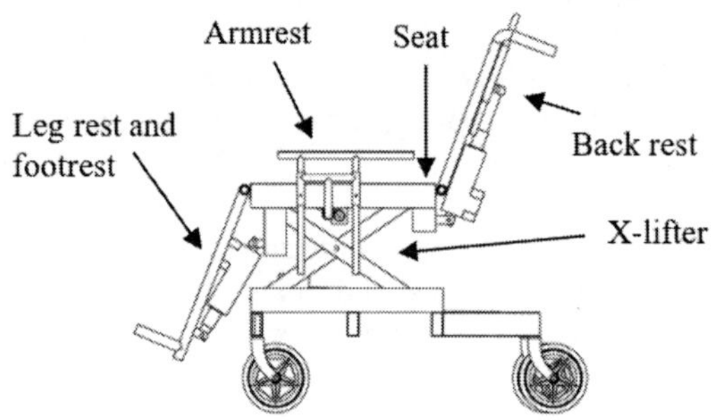

Figure 1. Side view of a multifunction wheelchair design

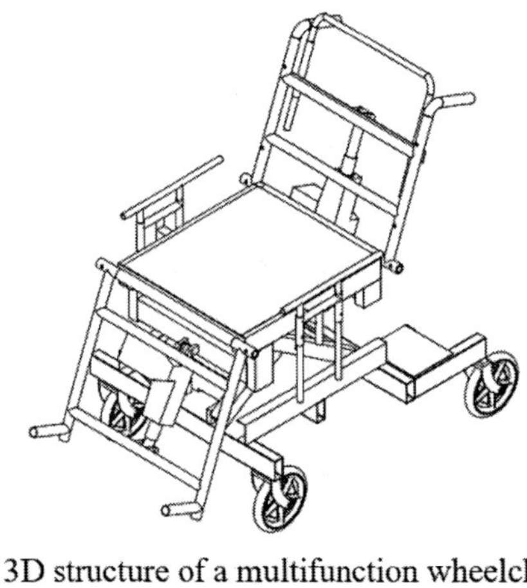

Figure 2. 3D structure of a multifunction wheelchair design

2.2 Equilibrium of wheelchair design

The conditions of design defined is user has maximum weight of 120 kg and maximum height 165 cm. Furthermore, the research of the distribution of weight on each man's proportions from human body dynamics showed that the weight distributed to each part on the wheelchair as shown in Figure 3 [12].

Consider at point A, the reaction force (F_r) is calculated using Equation (1) and found that the value can be calculated equal to 255.50 N, which the wheelchair is in balance.

$$\sum M_A = 0 \tag{1}$$

The X-lifter is used for adjusting the height of the wheelchair. The calculation for maximum load act to X-lifter uses Equations (2) and (3) with definite equal to 1200 N acts on the centroid of X-lifter as shown in Figure 4. Therefore, F_{r1} and F_{r2} are equal and can be is calculated equal to 1475 N.

IOP Conf. Series: Materials Science and Engineering 1137 (2021) 012054 doi:10.1088/1757-899X/1137/1/012054

$$\sum F_x = 0 \quad (2)$$

$$\sum F_y = 0 \quad (3)$$

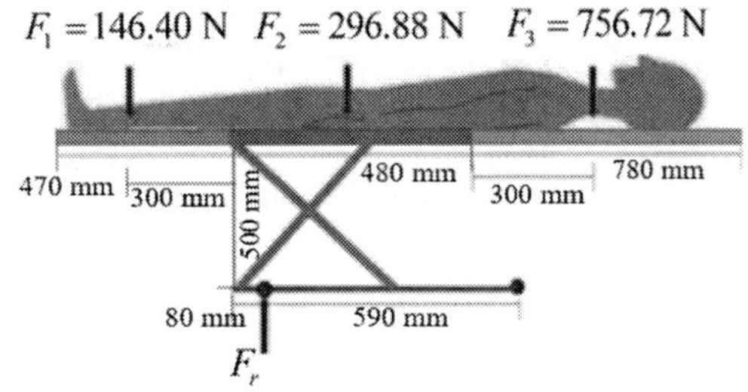

Figure 3. People's weight distribution on the wheelchair when the human weight is 120 kg and 165 cm in height

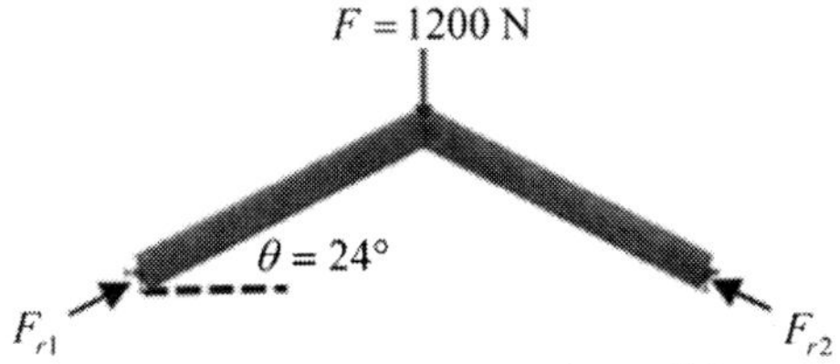

Figure 4. Forces act on the X-lifter

As can be seen in Figure 5, the rear of the backrest is used for recline and tilt features. Linear actuator at the rear of backrest gets a reaction force (F_r). Calculating for the value of reaction force (F_r) by Equation (1) The calculation result will be able to calculate the reaction force (F_r) is equal to 1991 N.

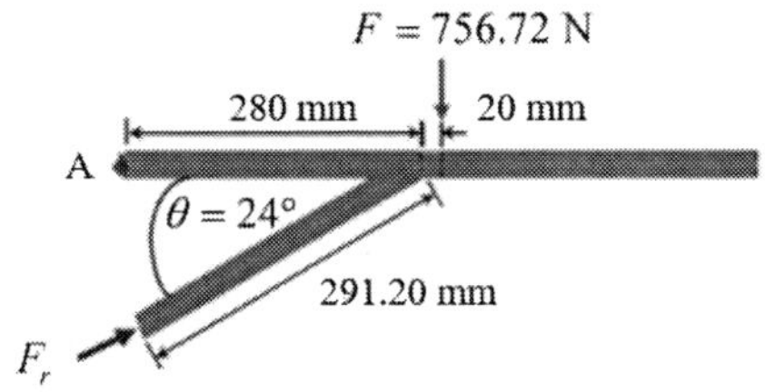

Figure 5. Forces act on the rear of the backrest

The leg rest will use a linear actuator for adjusting similar to the backrest. It can be calculated by the equation of equilibrium force and rotation as set out in Equation (1). Thus, the value of reaction force (F_r) that acts is equal to 840 N. The force distribution is given in Figure 6.

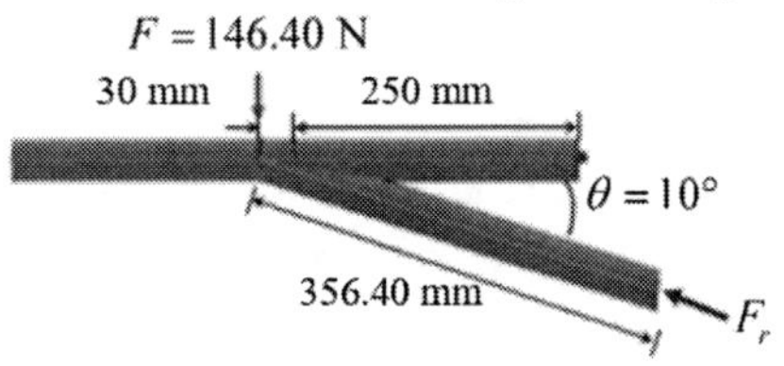

Figure 6. Forces act on the rear of leg rest

The above calculations are used to select the linear actuators for each part of the wheelchair design.

2.3 Governing equation

The equations below are calculated the solution of differential equations for the balance of solids with the flexibility in 3D [13]:

$$\frac{\partial \sigma_x}{\partial x}+\frac{\partial \tau_{xy}}{\partial y}+\frac{\partial \tau_{xz}}{\partial z}+F_x=0$$
$$\frac{\partial \tau_{xy}}{\partial x}+\frac{\partial \sigma_y}{\partial y}+\frac{\partial \tau_{yz}}{\partial z}+F_y=0 \quad (4)$$
$$\frac{\partial \tau_{xz}}{\partial x}+\frac{\partial \tau_{yz}}{\partial y}+\frac{\partial \sigma_z}{\partial z}+F_z=0$$

Where $\sigma_x,\sigma_y,\sigma_z$ are normal stress components in x, y, z direction (MPa), respectively

$\tau_{xy},\tau_{yz},\tau_{xz}$ are shearing stress components (MPa)

F_x,F_y,F_z are body force in x, y, z direction (N), respectively

2.4 Properties of material

The carbon steel SS400 is used as a material for the structural strength of wheelchair in this research as shown in Table 1. Table 1 presents the mechanical properties of the carbon steel SS400.

The mathematical models are solved by using FEM to establish the stress distribution of 4 main components of wheelchair design. The 3D model is discretized using triangular elements to approximate the stress distribution variations through each element. A grid independence test is carried out to identify the appropriate number of elements required. To obtain a good calculation, a fine mesh is specified in the sensitive areas.

Table 1. The properties of carbon steel SS400 used in the calculation procedure [14]

Property	Value	Unit
Elastic Modulus	2.1×10^{11}	N/m^2
Poisson's Ratio	0.26	-
Mass Density	7860	kg/m^3
Tensile Strength	3.0×10^8	N/m^2
Yield Strength	2.45×10^8	N/m^2

This study has discussed the design of wheelchair, the stress distributions of 4 main components namely the lower base of the wheelchair, X-lifter, the linear actuator holder at the front of the seat and the linear actuator holder at the rear of the backrest. The mathematical model, boundary condition and the number of the mesh of each component are also provided in the Results and Discussion sections.

3. Results and Discussion

3.1 Verifications of the model

To evaluate the accuracy of the present numerical solution, the reaction force (F_r) of the simulated result is validated against the analytical result in the same condition. The X-lifter is used in the validation test case. It can be seen from the data in Figure 4 that the X-lifter model and free body diagram used to compare the accuracy of the results by using a computer program. The reaction force acting along the x-axis and y-axis is calculated as provided in Equations (2) and (3). The results of the reaction force acting along x-axis and y-axis from the analytical solution are as follows:

Considering x-axis:
$$\sum F_x = 0 \quad (5)$$
$$F_{r1}\cos(24°) + F_{r2}\cos(24°) = 0 \quad (6)$$
Considering y-axis:
$$\sum F_y = 0 \quad (7)$$
$$F_{r1}\sin(24°) + F_{r2}(\sin 24°) = 1200 \quad (8)$$
$$F_{r1} = F_{r2} = 1475\ \text{N} \quad (9)$$
$$\therefore F_{r1}\cos(24°) = F_x = 1347\ \text{N} \quad (10)$$
$$\therefore F_{r1}\sin(24°) = F_y = 600\ \text{N} \quad (11)$$

The reaction force acting along with the x-axis and y-axis results from the numerical solution of the current study is given in Figure 7 and Figure 8, respectively. The comparison of the reaction force acting along the x-axis and y-axis of the X-lifter between the numerical solution and analytical solution is explained in Table 2.

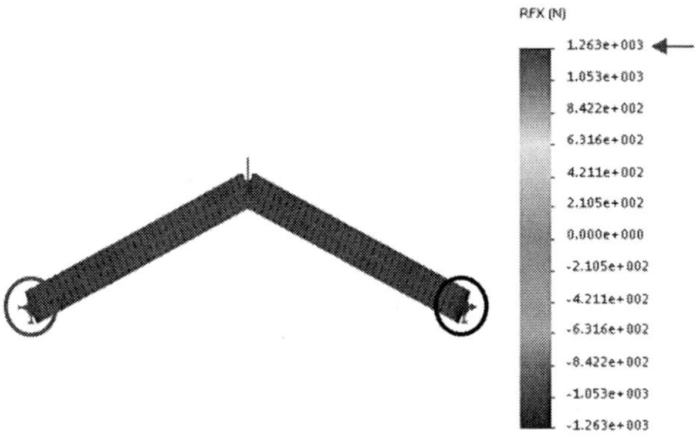

Figure 7. The reaction force acting along the x-axis from numerical solution

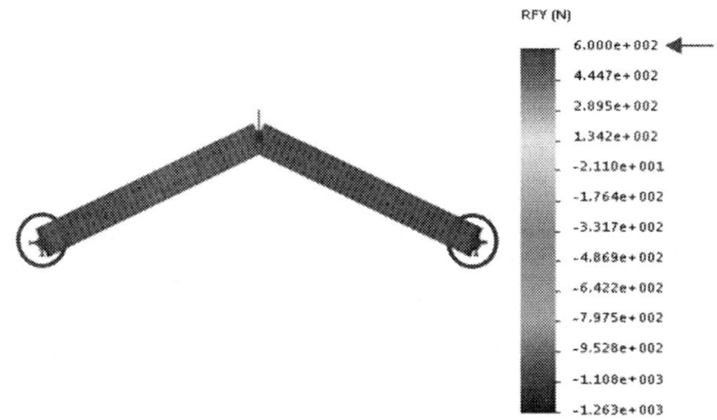

Figure 8. The reaction force acting along the y-axis from numerical solution

The error between the 2 solutions is equal to 6.24 % and 0% along the x-axis and y-axis, respectively. This favorable comparison lends confidence to the accuracy of the present numerical model.

Table 2. Comparing solution methods between the analytical solution and numerical solution

	Analytical Solution	Numerical Solution	Error
The reaction force acting along x-axis	1347 N	1263 N	6.24 %
The reaction force acting along y-axis	600 N	600 N	0 %

3.2 The lower base of the wheelchair

As can be seen from Figure 9, the lower base of the wheelchair model is considered as carbon steel SS400. The number of mesh which is suitable for analysis is about 62,000 elements. Figure 10 shows the boundary condition for analysis and Von-Mises stress distribution of the lower base of the wheelchair. It is found that the maximum Von-Mises stress value of simulation results that is equal to 117.6 MPa at the position for wearing the wheel. Likewise, considering the maximum Von-Mises stress value, the outcomes also show that in case of the lower base of the wheelchair does not exceed the yield strength of the material at 245 MPa (see also Table 1).

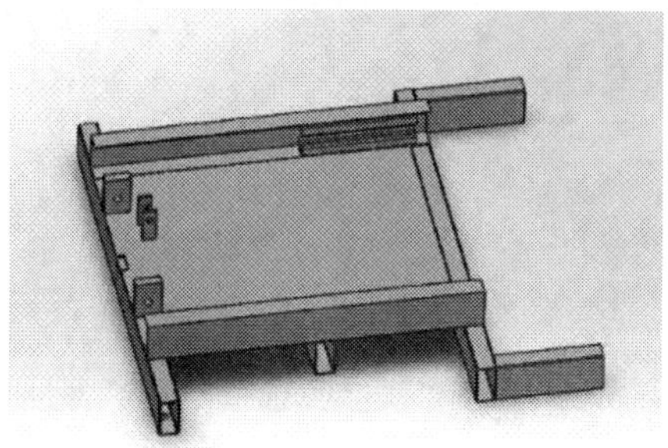

Figure 9. The lower base of the wheelchair

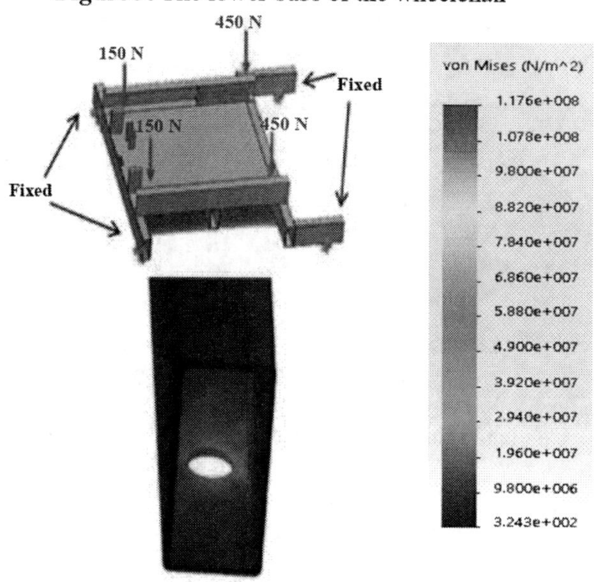

Figure 10. Results from the simulation of the lower base of the wheelchair

3.3 The X-lifter

Corresponds to the above part, the X-lifter model is considered as carbon steel SS400 as illustrated in Figure 11. The number of mesh which is suitable for analysis is about 23,350 elements. Figure 12 gives the boundary condition for analysis and Von-Mises stress distribution of the X-lifter. It revealed that the maximum Von-Mises stress value of simulation results is equal to 111.3 MPa. Besides, considering the maximum Von-Mises stress value, the outcomes also indicate that in the case of the X-lifter does not exceed the yield strength of the material at 245 MPa (refer to Table 1).

Figure 11. The X-lifter

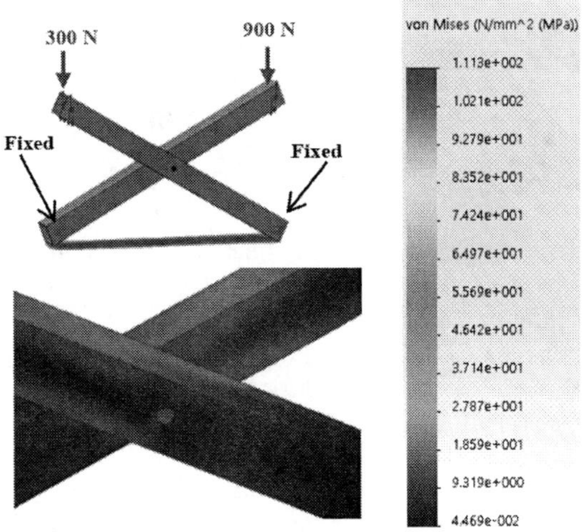

Figure 12. Results from simulation of the lower base of the X-lifter

3.4 The linear actuator holder at the front of seat

Figure 13 presents the linear actuator holder at the front of the seat model and is considered as carbon steel SS400. The number of mesh is suitable for analysis and is about 42,000 elements. Figure 14 shows the boundary condition for analysis and Von-Mises stress distribution of the linear actuator holder at the front of the seat. It is found that the maximum Von-Mises stress value of simulation results is equal to 34.82 MPa. In the same vein, considering the maximum Von-Mises stress value, it is able to explain that in the case of linear actuator holder at the front of seat does not exceed the yield strength of the material at 245 MPa (see also Table 1).

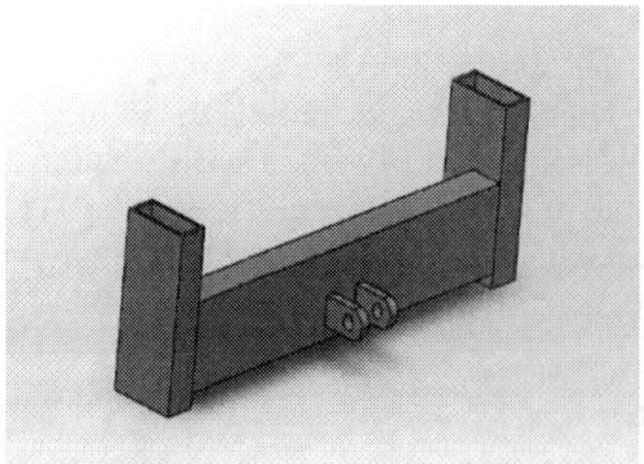

Figure 13. The linear actuator holder at the front of the seat

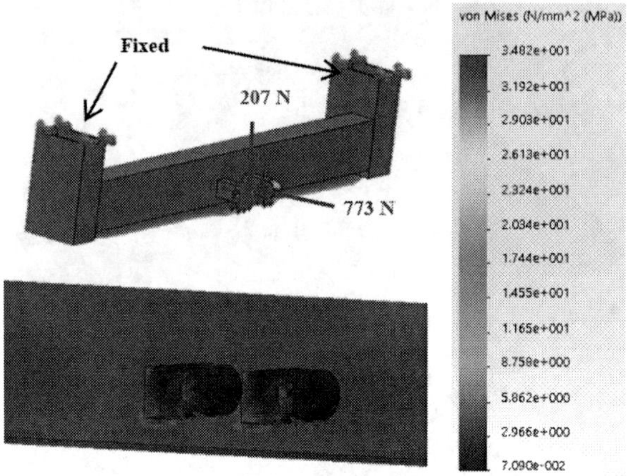

Figure 14. Results from simulation of the linear actuator holder at the front of the seat

3.5 The linear actuator holder at the rear of backrest

The last part of the analysis regards the actuator holder at the rear of the backrest. The actuator holder at the rear of the backrest model is considered as carbon steel SS400 in relation to the linear actuator holder at the front of the seat part (see Figure 15 below). The number of mesh which is suitable for analysis is about 31,000 elements. Figure 16 illustrates the boundary condition for analysis and Von-Mises stress distribution of the actuator holder at the rear of the backrest. It is observed that the maximum Von-Mises stress value of simulation results is equal to 121 MPa. In addition, considering the maximum Von-Mises stress value, it also shows that in the case of the actuator holder at the rear of backrest does not exceed the yield strength of the material at 245 MPa (refer to Table 1).

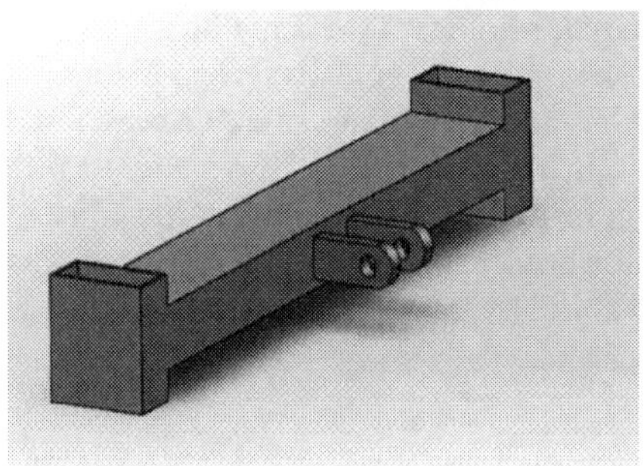

Figure 15. The actuator holder at the rear of the backrest

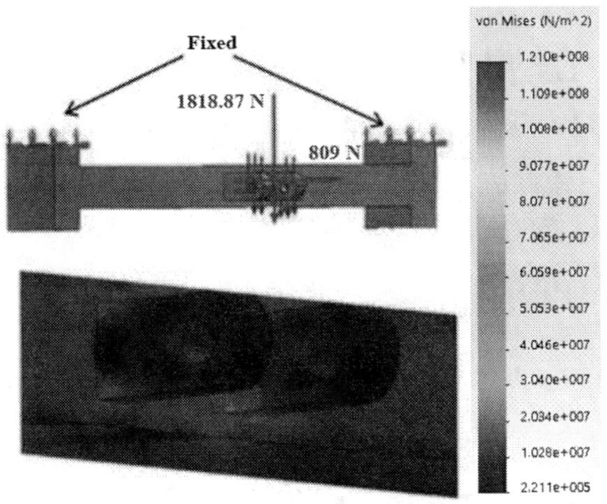

Figure 16. Results from simulation of the actuator holder at the rear of the backrest

After simulating in each part using the methods and equation above. It has found that the values which are suitable to create a wheelchair. The findings of this study suggest that the next step is to create a wheelchair according to the design.

4. Conclusion

The aim of the present research was to examine computer simulation of stress distribution to prepare the preliminary design of multifunction wheelchair. The most obvious finding to emerge from this study is that the multifunction wheelchair design has features as recline and tilt adjustable, height-adjustable and armrest width adjustable with the maximum weight equal to 120 kg. SolidWorks software is used in the design including analysis of the strength of 4 main components i.e., the lower base of the wheelchair, X-lifter, the linear actuator holder at the front of the seat and the linear actuator holder at the rear of the backrest by FEM via 3D model. The simulation results with computer programs are validated with analytical results. The results show a good agreement between the analytical results and present simulated stress distributions. It can be seen that the maximum Von-Mises stress value in all components does not exceed the yield strength of the material. The results of this research support the idea that the multifunction wheelchair will not be damaged when used. The findings reported here shed new light on the basic knowledge obtained from computer analysis can be used to improve the structure and build a prototype of the multifunction wheelchair.

References

[1] Kaufman G and Elder Jr G H 2003 *J. Aging Stud.* **17** 269-282
[2] Risør B W Casper S D Andersen L L and Sørensen J 2017 *Appl. Ergon.* **60** 74-82
[3] Nelson A Owen B Lloyd J D Fragala G Matz M W Amato M and Lentz K 2003 *Am. J. Nurs.* **103** 32-43
[4] Nelson A and Baptiste A S 2006 *Clin. Rev. Bone. Miner. Metab.* **4** 55-69
[5] Wardell H 2007 *AAOHN J.* **55** 407-412
[6] Ding D Cooper R A Cooper R and Kelleher A 2007 *Proc. 29th. Annual Int. Conf. of the IEEE Engineering in Medicine and Biology Society* 4364-4367
[7] Lee W C Yao J A and Kuo C H 2011 *Proc. Int. Conf. on System Science and Engineering* 494-497
[8] Bae J H and Moon I 2011 *Proc. IEEE Int. Conf. on Rehabilitation Robotics* 1-5
[9] Zou W Ye A Lu T Ren Y Xu Z and Yuan K 2011 *Proc. IEEE Int. Conf. on Robotics and Biomimetics* 378-383

[10] Jusoh M A Ismail M T Abdullah S C Rashid H and Hashim H 2017 *Proc. Int. Symp. on Micro-NanoMechatronics and Human Science* 1-6
[11] Mistarihi M Z Okour R A and Mumani A A 2020 *Appl. Soft Comput.* **90** 106136
[12] Tözeren A 1999 *Human Body Dynamics: Classical Mechanics and Human Movement* (Berlin: Springer Science & Business Media)
[13] Singiresu S R 2005 *The Finite Element Method in Engineering* (Amsterdam: Elsevier)
[14] Bringas J E 2004 *Handbook of Comparative World Steel Standards (*Pennsylvania: ASTM International)

IOP Conf. Series: Materials Science and Engineering 1137 (2021) 012055 doi:10.1088/1757-899X/1137/1/012055

Line tracking control of royal Thai air force nursing mobile robot using visual feedback

Jeerasak Moudpoklang [1], and Chamnan Pedchote [2*]

[1,2]Department of Mechanical Engineering, Navaminda Kasatriyadhiraj Royal Air Force Academy, 171/1 Phaholyothin Rd. Saimai District, Bangkok, 10220, Thailand.

* Corresponding Author: moudpoklang@windowslive.com, c_pedchote@hotmail.com

Abstract

The specific objective of this study was to design a Royal Thai Air Force (RTAF) nursing robot, so-called Thadloom, to assist medical personnel in delivering food, medical supplies and communicating remotely with patients during the COVID-19 pandemic. This study also seeks to develop many identical mobile robots to be used in concerned places, such as hospitals or clinics, to reduce the risk of infection to medical personnel when entering the controlled area COVID-19-patients. This paper presents the design and implementation of real-time vision systems for the RTAF nursing robots when used in the controlled area of COVID-19-patients, with a capacity of delivering food and medical supplies automatically. The robots were able to follow lines on the floor using visual feedback. A contour tracing algorithm was implemented for line identification and position estimation. Linear segments with parabolic blends were used for the robot's trajectory planning. The robot can track a line on the floor effectively by employing a PID controller. This experimental result showed that the developed vision-based state estimation and the designed PID controller. It was also found that it received a good satisfaction towards the accuracy of the position within 3.5 centimeters, i.e., The distance from the center of the robot to the center of the line.

Keywords: PID control, Mobile robot, Vision feedback, COVID-19

1. Introduction

Covid-19 impacts negatively upon a range of the global economy and daily social life [2]. It is also associated with an increased risk of infected cases and deaths. For example, more than 48 million confirmed cases and more than 1.2 million deaths (see detailed results from the World Health Organization) [1]. Of particular concern is, such a virus is easy to transmit from one person to another and can lead to a detrimental disease, with have yet provided effective vaccines.

Thailand has implemented various measures to combat the disease, such as enforcing state quarantine for arrivals, tracking and isolating possible infected patients, monitoring temperature, and promoting campaigns to keep social-distancing and encouraging wearing masks in public areas. At the beginning of the disease's spread, the country was in a panic; Thai people and health personnel were afraid and stressed. Medical staff and health workers were working at their full capacities. One of the core missions of The Royal Thai Air Force (RTAF) as a part of the Thai government is to provide supports and relief for Thai people during natural disasters and hardship. Among many other projects invented by different RTAF divisions for combating against COVID-19, developing a mobile robot was initiated at the Department of Mechanical Engineering, Navaminda Kasatriyadhiraj Royal Air Force Academy, in an attempt to help medical personnel as well as the patients. The project was selected and considered to be a potential solution as it provided a means of transportation of foods or medical supplies and remote communication between medical personnel and the patient. This is possible because a mobile robot is a vehicle that is capable of autonomous motion. In particular, autonomous robots are promising in scientific research and practical applications [3-6].

A group of researchers and air cadets were formed to carry out the design and development task. Their background in designing and implementing control for unmanned aerial vehicles was adopted to design and control this mobile robot. The objective of this current study was to develop a robot that was able to make delivery autonomously. Besides, the robot should be able to provide visual, remote communication between the medical staff and the patient and perform the additional task of measuring the patient's body temperature and then recording the data for further analysis. Regarding the assembly matter, the robot must be simple and fast to be built in order to respond to demand when the spread of the disease became wide quickly. Other requirements included easiness in operating by doctors, nurses, and being user-friendly to zero experienced users, and within an acceptable cost. Initially, it was planned to have the robots provided for government hospitals. At the moment, the developed robot has been tested and put into practical use in several hospitals and in-state quarantine operations.

2. System Architecture

2.1 Mobile robot architecture

Thadloom, The Royal Thai Air Force (RTAF) nursing robot, is an autonomous delivery vehicle used to deliver foods and medicines to patients during the COVID-19 pandemic. The robot can carry up to 20 kg of loads and moves at a maximum speed of 1 meter per second (3.6 km/hr). It can work up to 4 hours before recharging, which is sufficient for a typical working requirement. The robot was designed to have three easy-to-access levels, with an overall dimension of 55 cm. length, 45 cm. width, 120 cm. height (see Figure 1).

The body of Thadloom is made from an aluminum profile and stainless steel to provide rigidity, light-weight, and good hygiene. Thadloom is driven by two DC motors, incorporating differential technique for its movements. Maneuvering Thadloom is achieved by two controllers, comprising one high-level controller, which runs at 30 Hz operated by an onboard computer, and one low-level controller runs at 50 Hz operated by an ATmega328 microcontroller. The control routines are written in C++ language for high-level control and C language for low-level control. A webcam camera is attached to the robot and is positioned in the middle of the bottom crossbar. The webcam that provides visual information is also equipped with a laser range finder sensor. The information of the laser range finder sensor is used for obstacle avoidance. Visual information obtained by Thadloom can then be displayed in real-time on a monitor. Operating Thadloom can be done in two modes, i.e., a semi-auto mode and a full auto mode. The semi-auto mode is used when controlling the robot using a keyboard, while the full-auto mode is used when the robot tracks a line autonomously. Thadloom, when it is completely equipped, has a total weight of 28 kg, which includes a 12V DC lead-acid battery needed for power supplying to all-electric components.

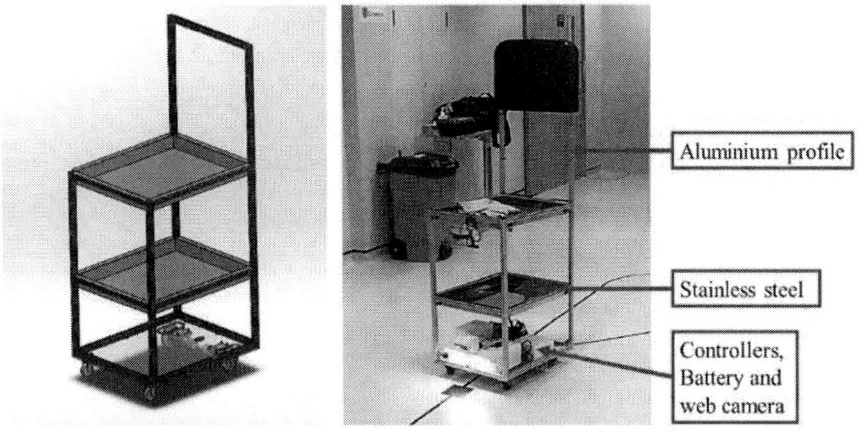

Figure 1. Thadloom's drawing and its main components.

Figure 2 demonstrates the concept of using Thadloom for delivering foods or medicines to patients infected with COVID-19. The hospital or clinic's designated area may be divided into two zones: the safety zone (not infected area) and the danger zone (infected area). An operator such as a nurse can command the robot via Remote Desktop by carrying out the following steps.

1. Select the mode in which to enter the ward; semi-automatic or fully automatic.
2. Select the patient's room from a given set of rooms.
3. Click the start button.

When the start button is pressed, the motion is initiated. The robot will follow a unique, established path by tracking and detecting the lines and shapes on the floor. The robot stops when an obstacle is detected. For example, it stops in front of the door and waits for the operator to open the door remotely. When the robot senses that there is no longer an obstacle, it then enters the patient's room. When Thadloom reaches the delivery location, the robot stops. The patient can then picks up the drinks, foods, medicines, or documents, which have been included in the delivery. The patient can also communicate with doctors, nurses, or other persons outside the patient's room visually through a teleconference system equipped with Thadloom. This allows for the patient's health inspection and also comforts the patient who has been isolated for a long time. Once the mission of delivery and communication by Thadloom is completed, the operator will command the robot to return to its starting point for cleaning and making the next round.

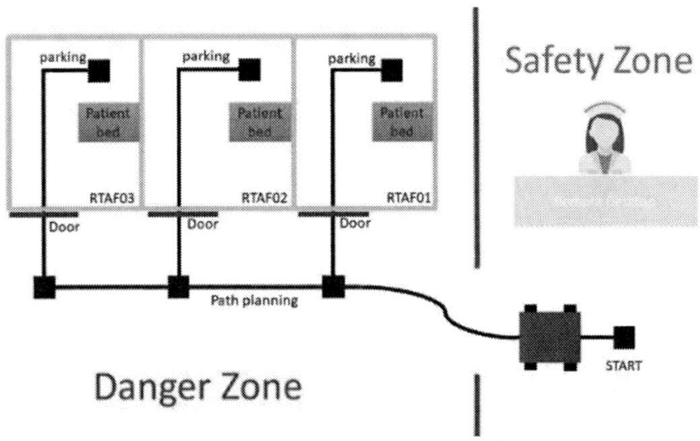

Figure 2. Concept of using Thadloom in a mission of delivery and communication.

The automatic transportation of Thadloom is achieved by employing an algorithm of line detection and target plate recognition. This current study has designed a target plate as a black square of size 12×12 cm^2, as set out in Figure 3. As can be seen in Figure 2 illustrates the target plates from both inside and outside of the patients' rooms. While the robot follows those lines of the established path, it simultaneously searches for the target plates. When the robot detects a target plate, its current location is confirmed and may move on, and then turn into a patient's room or make a stop for delivery. Once the Thadloom's delivery and communication tasks are satisfactorily done, the operator can initiate the return trip. Again, the target plates play an essential role in making the way back to the starting point.

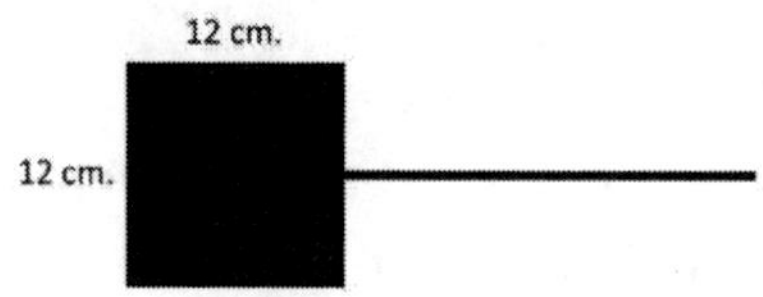

Figure 3. The designed target plate.

2.2 Line detection and Target plate recognition algorithm

In this study, the line detection and target plate recognition algorithm are programmed in C++ and based on OpenCV [7] in favor of detection. Figure 4 shows sequential steps in the designed algorithm. The image of the line with a resolution of 640×480 pixels is continuously captured. The captured image is first converted into a gray-scale image and then to a binary image using the image thresholding technique. Unwanted parts of the resultant image are eliminated. The calculation of perimeter and square-shape recognition can then be done with greater accuracy. The robot's position relative to the line on the floor is estimated and used to control the robot to make a further movement.

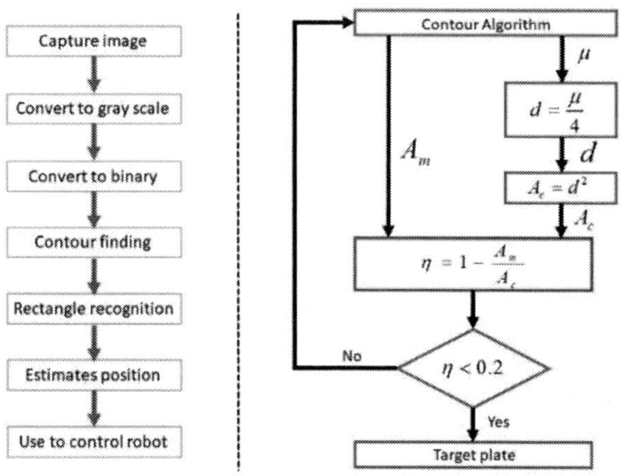

Figure 4. Line detection and Target plate recognition algorithm.

The steps taken for the target plate recognition are also represented in Figure 4. The contour length or the perimeter of a square shape from the contour algorithm is used for calculating the size of the square by using equation (1),

$$d = \frac{\mu}{4} \tag{1}$$

where d is the size of the square, and μ denotes the contour length or perimeter. The area of the square is obtained from equation (2),

$$A_c = d^2 \tag{2}$$

where A_c denotes the calculated area of the square. The error or the difference between the calculated area and the measured area obtained from the contour algorithm is then computed by using equation (3),

$$\eta = 1 - \frac{A_m}{A_c} \tag{3}$$

where η denotes the error, A_m denotes the measured area from the contour algorithm. The error η is thus used for identifying the target plate. When it η is less than 0.2, the condition indicates that a target plate is found, and the robot can make the right move for that particular path.

2.3 Mobile Robot Location Representation

Two frames can define the robot location [8], the initial coordinate (frame I), and the robot coordinate (frame R), as presented in Figure 5. The inertial coordinate is fixed at the start point of the robot movement. The robot coordinate is located at the center of the robot. The position of the robot is at a distance x and y with respect to the inertial coordinate. The positive direction of the x-axis of the robot coordinate (X_R) is in the same direction as that of the robot's heading direction. The angle θ formed by X_R and X_I is the orientation of the robot. The location of the robot can therefore be defined by a generalized coordinate vector q when $q = [x \quad y \quad \theta]^T$.

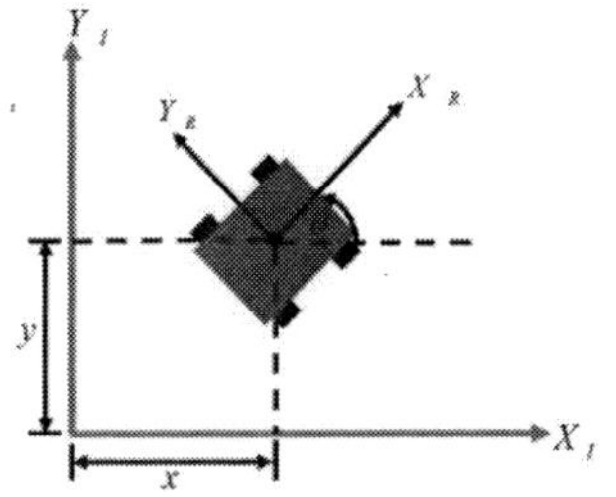

Figure. 5 Coordinate systems used for the location of the robot.

In deriving a forward kinematic of the robot, it is assumed that the robot moves without slipping. This implies that the robot cannot move laterally along its wheel axes. This assumption associates with a nonholonomic constraint described by the following equation.

$$\dot{x}\sin\theta - \dot{y}\cos\theta = 0 \tag{4}$$

Equation (4) can be rearranged into a multiplication of matrices as

$$\begin{bmatrix} \sin\theta & -\cos\theta & 0 \end{bmatrix} \begin{bmatrix} \dot{x} \\ \dot{y} \\ \dot{\theta} \end{bmatrix} = 0$$

$$A\dot{q} = 0$$

where A is called constraint matrix and $\dot{q}$ is velocities or time derivative of the vector q. A kinematic model of the robot is the null space of the constraint matrix A. This relationship is in the differential form, shown in equation 5, which cannot be integrated due to the nonholonomic constraint.

$$\begin{bmatrix} \dot{x} \\ \dot{y} \\ \dot{\theta} \end{bmatrix} = \begin{bmatrix} \cos\theta & 0 \\ \sin\theta & 0 \\ 0 & 1 \end{bmatrix} \begin{bmatrix} v \\ \omega \end{bmatrix} \tag{5}$$

In equation 5 v is the robot's heading velocity in X_R direction and ω is the robot's turning velocity. A differential-drive robot has forward kinematics representing the relationship between drive wheel rotational velocities and robot velocities as

$$\begin{bmatrix} v \\ \omega \end{bmatrix} = \frac{r}{N} \begin{bmatrix} \frac{1}{2} & \frac{1}{2} \\ \frac{1}{L} & -\frac{1}{L} \end{bmatrix} \begin{bmatrix} \dot{\varphi}_1 \\ \dot{\varphi}_2 \end{bmatrix} \tag{6}$$

Where $\dot{\varphi}_1$ and $\dot{\varphi}_2$ are rotational velocities of the right and left wheel, respectively, r is the drive wheel radius, L is the distance between wheels along their wheel axes, and N is the gear reduction ratio of a harmonic gear. The conversion of a matrix in equation (6) called inverse kinematics is described in equation (7) as

$$\begin{bmatrix} \dot{\varphi}_1 \\ \dot{\varphi}_2 \end{bmatrix} = \frac{N}{r} \begin{bmatrix} 1 & \frac{L}{2} \\ 1 & -\frac{L}{2} \end{bmatrix} \begin{bmatrix} v \\ \omega \end{bmatrix} \tag{7}$$

3. Control system

Figure 6 provides the concept implemented for controlling the Thadloom robot. The Thadloom robot is initially set up and then controlled via Remote Desktop, which has already been installed on a ground personal computer. The high-level controller performs several essential functions, including a graphic user interface (GUI), image processing, control algorithm, and sending pulse width modulation or PWM control signals to the low-level controller. The low-level controller handles tasks, such as sensor data processing, data fusion, running obstacle avoidance algorithms, and sending PWM command signals to control the motors, with a sampling rate of 50 Hz.

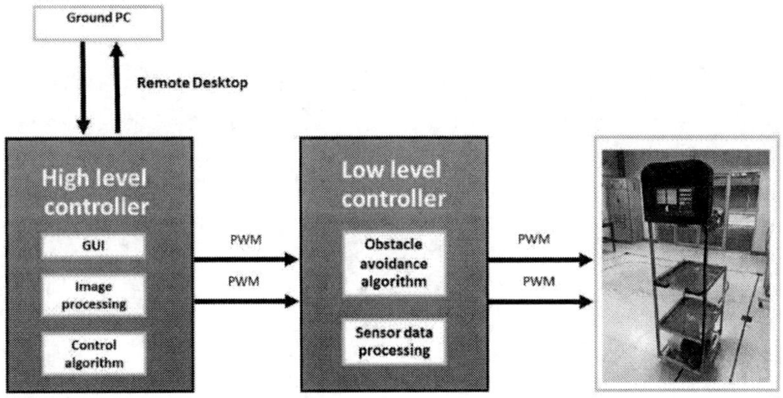

Figure. 6 The designed control concept for the Thadloom robot using a visual input.

Figure 7 illustrates the block diagram of a position control for a robot using visual input. A robot's controller has two control loops: inner loop control and outer loop control. The outer loop compares the estimated location from odometry and a camera with the reference positions and orientation. The errors and reference velocities are inputted into the tracking control to calculate the desired velocities needed to navigate the robot back to the reference trajectory. Desired wheel rotation velocities are then computed by inverse kinematics in equation (7) and passed into the motor control loop (the inner loop). Angular positions of each drive motor are then controlled with PID controllers [9].

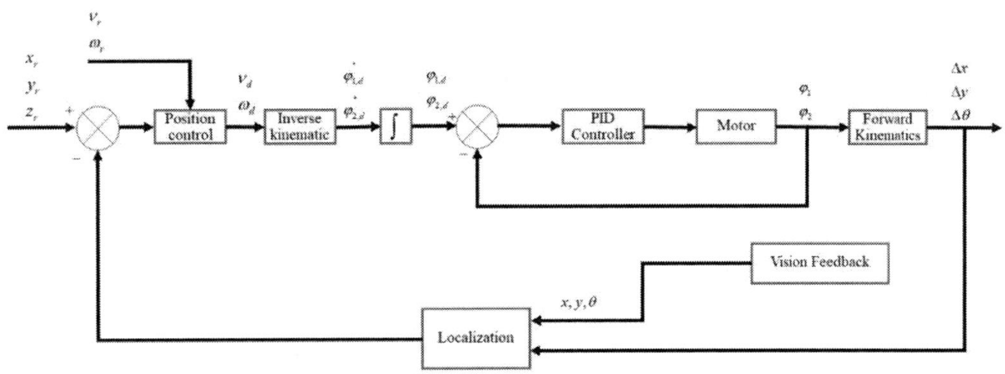

Figure. 7 The block diagram of a position control for the Thadloom robot using visual input.

4. Experiment and Result

To assess the capability of the developed robot and the performance of the designed control algorithm, experiments were set up and carried out in two stages. In the first stage, the experiment was done at the Department of Mechanical Engineering, Navaminda Kasatriyadhiraj Royal Air Force Academy. The objectives were to observe the robot's mobility and determine a good set of PID control gains and find the maximum load limit and operating time duration. Figure 8 shows the experimental setup in the first stage when the robot tracked lines on the floor with a speed of 1 m/s and continued running for 10 minutes autonomously. The differences between the center of the robot and the center of the line were analyzed and used for tunning PID control gains. The same tests were then repeated, but with a different set of PID control gains. The experimental results showed that the robot was able to continuously track the center of the lines acceptably well and with good mobility.

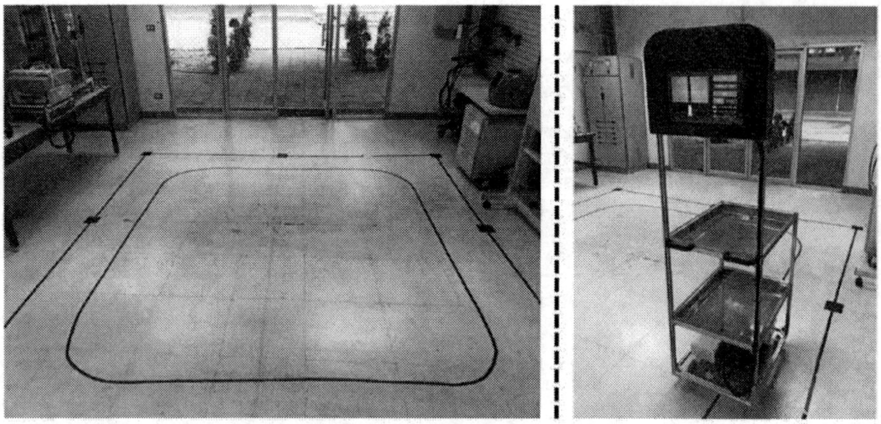

Figure. 8 Autonomous line tracking tested at a speed of 1 m/s and continued for 10 minutes.

Figure 9 provides the error or the difference between the center of the robot and the center of the line measured in centimeters for 600 seconds. These are errors of the position in the y-direction, and they are within ±3.5 cm. The best PID control gains from the tests were as follows; $K_p = 1.5$, $K_i = 0.001$ and $K_d = 10$. The robot operated continuously for 4 hours and carried a maximum load of approximately 20 kg.

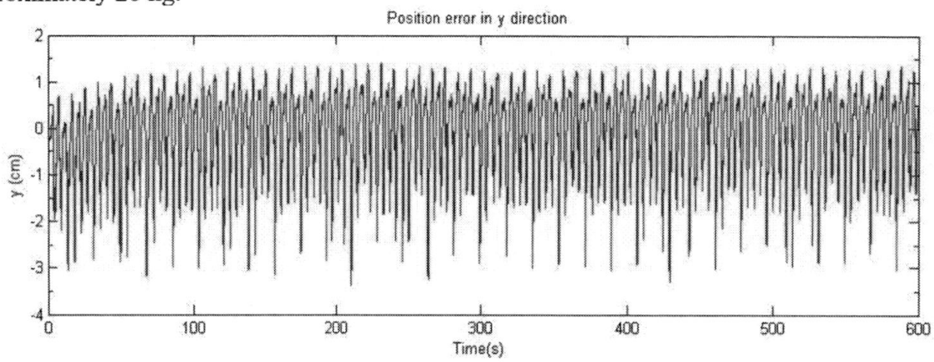

Figure. 9 The line tracking accuracy for the Thadloom robot using visual input.

The second stage of the experiments was performed in an actual environment where patients infected with COVID-19 were treated. These were carried out at two of the RTAF hospitals, including Bhumibol Adulyadej Hospital and Royal Thai Air Force Hospital, located in Don Muang District, Bangkok. The infected patients were treated in the emerging disease ward, shown in Figure 10, in which the two hospitals had Thadloom robots to help. The objectives were to assess how well medical personnel or the users operate the robot and have satisfaction feedback for further adjusting the robot and the control algorithm. After a short period of training, doctors, nurses, and operators handled the robot satisfactorily well.

Figure. 10 The emerging disease ward for treated patients with COVID-19 infections.

Figure 11 shows the Thadloom robot being tested in a COVID-19 patient room at the emerging disease ward in the Royal Thai Air Force Hospital. An operator-controlled the robot such as a nurse in a safe zone via the Remote Desktop. When the operator pressed the start button, the robot performed its task sequentially by first moving off from the starting point to the patient room at a speed of 1 m/s, as shown by number 1. The robot followed the line and automatically turned to the specified room, then it stopped and waited when encountering an obstacle, as shown in 2-3. When the operator saw the robot waiting at the front of the door through the hospital CCTV system, he/she remotely commanded the door to open. The robot then carried on moving into the patient room and stopped at the delivery point as number 4-5-6. The patient picked up and talked to a doctor or nurse via a video conference as number 7. When the mission was completed, the robot returned to the starting point as number 8-9.

Figure. 11 Thadloom mission test results; COVID-19 patient room at the emerging disease ward in the Royal Thai Air Force Hospital.

5. Conclusion

This study set out to design a mobile robot named Thadloom and develop it for food and medicines delivery or distribute necessary supplies to patients infected from the COVID-19. The target robot was designed and consider for practical use as the highest priority, with the ease to operate by a less-skilled operator. The robot was also equipped with a remote communication system. The robot control and navigation were achieved using a vision-based algorithm, which was effective and robust. The robot was first set up and tested at the Department of Mechanical Engineering, Navaminda Kasatriyadhiraj Royal Air Force Academy, before testing in the actual environment in two of the RTAF hospitals. Later, a number of identical Thadloom robots were built and sent to the RTAF hospitals and for state quarantine uses. The users were satisfied and thankful for the support made by the RTAF. These experiments confirmed that the robots prove useful in helping both medical personnel and patients. The present study appears to be the first study to have successful cooperation of the researchers, air cadets, and the RTAF.

6. References

[1] World Health Organization 2020 WHO Coronavirus Disease (COVID-19) Dashboard https://covid19.who.int, accessed 1 December 2020
[2] Farah M 2020 Journal of Taibah University Medical Sciences **15** 85-86
[3] Mester G 2006 *Proc. the 10th Intelligent Engineering Systems* (London: Metropolitan University) 282-286
[4] Mester G 2006 *Proc. the YuINFO 2006* (Kopaonik) 1-4
[5] Mester G 2006 *Proc. the Kandó Conference 2006* (Budapest) 1-3
[6] Mester G 1994 *Proc. Int. Conf. on Industrial Electronics, Control and Instrumentation* (Bologna) vol 3 1639-1644
[7] The OpenCV Library, http://opencvlibrary.sf.net, accessed 1 December 2020
[8] Siciliano B Sciavicco L Villani L and Oriolo G 2009 *Robotics Modelling, Planning and Control* (London: Springer)
[9] Sangveraphunsiri V 2004 *Control of Dynamic Systems* (Bangkok: Chulalongkorn University Press)

LTV-FIR command prefiltering for vibration reduction of a flexure-jointed X-Y motion stage with payload variation

P. Kuresangsai[1] and M. O. T. Cole[1*]

[1]Department of Mechanical Engineering Faculty of Engineering, Chiang Mai University, Chiang Mai 50200, Thailand

* Corresponding Author: motcole@dome.eng.cmu.ac.th

Abstract
This paper investigates the application of a linear-time-varying (LTV) command prefilter for vibration reduction in motion control of LTV systems. A model-based time-varying prefilter function is constructed that can be convolved with arbitrary command input signals to achieve residual vibration cancellation in a finite time. The prefilter is implemented on an experimental X-Y micro-positioning stage with a low-compliance flexure-jointed mechanism. Model coefficients within the LTV prefilter realization are varied in real-time according to measured platform position. As the dynamic model is also dependent on platform payload, a method is introduced to compensate for changes of payload mass based on the interpolation of model coefficients for extremum values. The results reveal that residual vibration can be cancelled effectively with settling time reduced from over 2 seconds to less than 0.4 seconds. Rapid point-to-point maneuvers over large travel distances, and with different known values of payload mass, can be achieved with high accuracy.

Keywords: Motion control, Vibration control, Nonlinear system, Input shaping

1. Introduction

Compliant mechanisms have certain advantages over conventional mechanism, especially for high precision motion tasks where the elimination of friction and backlash is required [1-3]. However, the performance of a compliant mechanism is often limited by undesired vibration during fast operation. One of the greatest challenges for controller design is dealing with nonlinear and/or time-varying characteristics of the mechanism dynamics [4-6].

For feedback controller synthesis, nonlinear dynamics can be treated by linear parameter varying (henceforth LPV) formulations. The main disadvantage of using LPV models for feedback control synthesis is the computational complexity and numerical fragility for high order system [7]. An alternative approach is to use feed-forward control (command prefiltering) to compensate for nonlinear effects and suppress vibration. For the FIR command prefiltering approach, a key feature is the ability to cancel residual vibration exactly within a finite time [8]. Additionally, command prefiltering can be easily implemented in practice without the risk of feedback-induced instability. To date, several studies have investigated impulse-based (time-delay) command prefilters, applied to LTI systems, where knowledge only of the natural frequency and damping ratio is required [8]. Model-based feedback may also be applied with command prefilters to improve the response of the system when subject to

IOP Conf. Series: Materials Science and Engineering 1137 (2021) 012056 doi:10.1088/1757-899X/1137/1/012056

parameter variation or uncertainty [9, 10]. A drawback of impulse-based command prefiltering is that excitation of unmodeled high-frequency modes can be amplified [11]. Therefore, in [12], an FIR command prefilter was introduced having smooth impulse response function and minimum H_2-norm that reduces excitation of high frequency modes by removing discontinuities from the command input.

This paper investigates a recently proposed time-varying FIR prefilter for settling time reduction in nonlinear motion control systems [13]. The technique is based on a second-order LTV system model and involves constructing the filter impulse response function in order to achieve exact residual vibration cancellation. An experimental study is undertaken involving an X-Y motion stage, where LTV prefiltering is applied via position-dependent scheduling of the prefilter function. A modified version of the LTV prefilter is also devised to compensate for changes in payload mass.

2. Theoretical foundation

2.1 Prototype LTI prefilter

A command prefilter for a second-order LTV system may be constructed by first considering the following undamped LTI system having infinite duration impulse response $g(t) = \sin(t)$:

$$\ddot{y}(t) + y(t) = y_d(t) \tag{1}$$

To achieve exact residual vibration cancellation, a kinematic command input $y_d(t)$, can be prefiltered by convolving with a function $h(t)$ having finite duration T_s. Zero residual vibration will be achieved with arbitrary command input if $h(t)$ satisfies a condition of orthogonality with respect to the functions $\sin t$ and $\cos t$ (see [12],[13]). Here, we consider a prefilter function with the general form $h(t) = A + Bt + C\cos t + D\sin t$. For the case with $T_s = \pi$, the solution is obtained as [14]

$$h(t) = \frac{2\pi}{(\pi^2-8)^2}\left[\pi(\pi - t) - 4(1+\cos t) - \frac{2}{\pi}(\pi^2 - 8)\sin t\right], t \in [0,\pi] \tag{2}$$

The impulse response of the overall system, including prefilter, is then finite duration and is given by

$$f(t) = \int h(\tau)\, g(t-\tau)\, d\tau$$
$$= \frac{2}{(\pi^2-8)^2}[\pi(\pi^2 - \pi t - 4) - (\pi^3 - \pi^2 t - 4\pi + 8t)\cos t - 2(\pi t - 4)\sin t], t \in [0,\pi] \tag{3}$$

1.1. LTV prefilter formulation

To apply the prefilter function given by equation (2) in the case of an LTV system described by

$$\ddot{y}(t) + C(t)\dot{y}(t) + K(t)y(t) = K(t)y_d(t) \tag{4}$$

the following construction is used. First, defining $\omega(t) = \sqrt{K(t)}$ and $v(t) = \omega(t)y_d(t)$, we obtain

$$\ddot{y}(t) + C(t)\dot{y}(t) + \omega^2(t) = \omega(t)v(t) \tag{5}$$

The prefiltering is then defined by convolution of $v(t)$ with time-varying function $\tilde{h}$ given by

$$\tilde{h}(t,q) = \omega(t)h(n(t,q)) + \big(\dot{\omega}(t)/\omega(t) + C(t)\big)f'(n(t,q)) \tag{6}$$

Here, q is the local time variable and $n(t,q)$ is a time-warping function that depends on the history of $\omega(t)$ over $(t-q,t)$:

$$n(t,q) = \int_{t-q}^{t} \omega(\tau)d\tau \tag{7}$$

The filtered input $\tilde{y}_d$ is obtained by convolving $\tilde{h}(t,q)$ with $v(t) = \omega(t)y_d(t)$, over q:

$$\tilde{y}_d(t) = \frac{1}{\omega(t)}\int v(t-q)\tilde{h}(t,q)dq \tag{8}$$

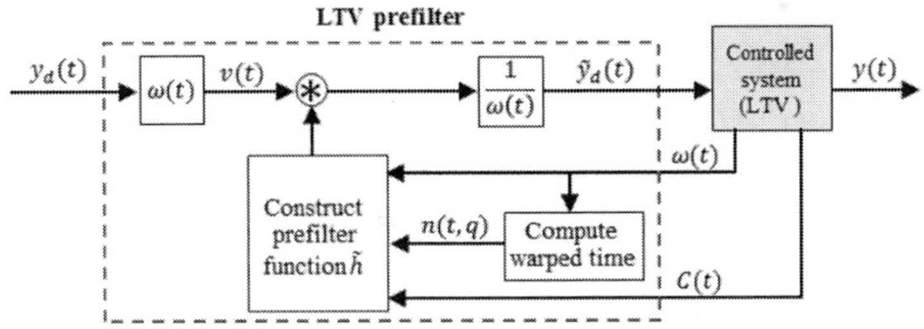

Figure 1. Implementation of LTV-FIR prefilter

It can be seen that, when substituting $v = \tilde{h}$ in equation (5), the solution must be $y = f$ because equation (5) then reduces to

$$f'' + f = h \tag{9}$$

Consequently, a finite duration input-output pair satisfying equation (1) can be used in the LTV prefilter function given by equation (6). Further details of the proof are given in [13]. The implementation of the LTV prefilter is provided in Figure 1. To see an example of the time evolution of the prefilter function, $\tilde{h}$, see Figure 9 in the experimental results section (Section 4).

3. System description and dynamic behavior

3.1 System setup

The LTV-FIR prefiltering method was applied to an experimental micro-positioning X-Y stage (see Figure 2). The system was designed based on 4 parallel RRR chains attached to the end-platform, but with the revolute joints replaced by thin flexures. The system is driven by two voice coil actuators with rotation measured by absolute magnetic encoders. The position of the end platform is measured by laser distance sensors. The overall size is approximately 250 x 250 mm and the operating space is 20 x 20 mm square. More details of the optimized design and kinematic behavior can be found in [15]. For the purpose of experiments, the motion of the system was restricted to translation along the Y-axis only. To test the system with different payloads, an additional mass is clamped at the center of the motion stage. The system dynamics are strongly dependent on platform position and this must be accounted for in the prefilter formulation, as described in Section 2. All control operations and prefiltering were implemented in discrete-time on PC-based hardware with sampling frequency 3000 Hz.

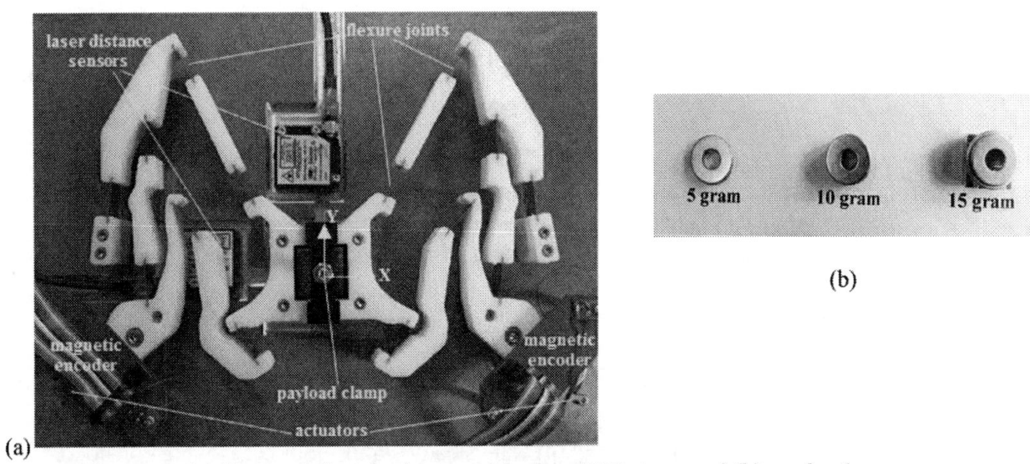

Figure 2. (a) flexure-jointed micro-positioning X-Y stage and (b) payload mass cases

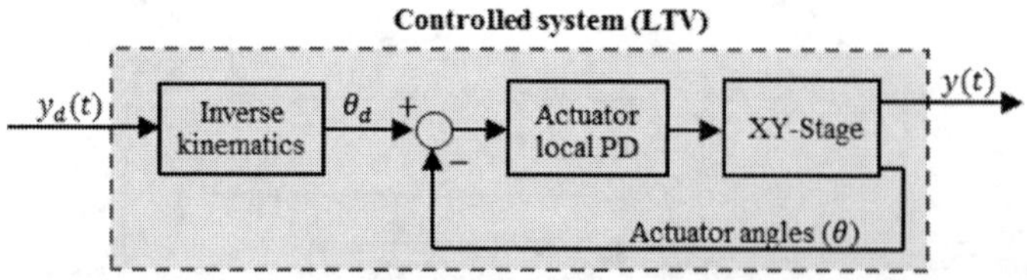

Figure 3. Controlled system with inverse kinematic model and actuator PD feedback control

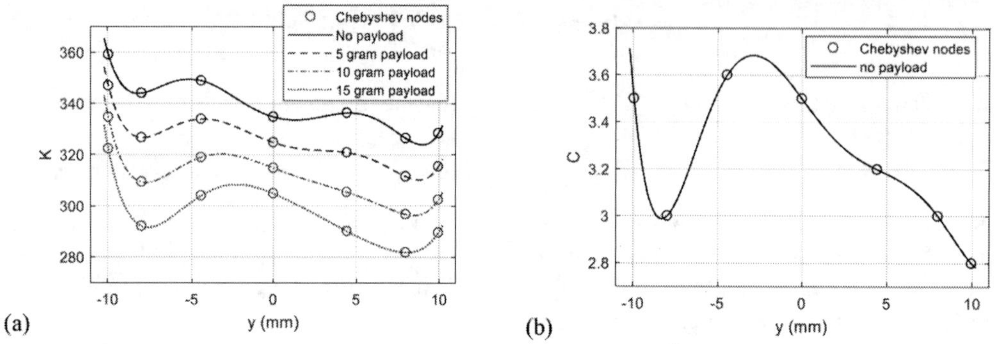

Figure 4. Normalized stiffness and damping coefficients ($K = \omega^2$ and C): Variation with platform position is shown for different platform payloads.

3.2 Inverse kinematics for control implementation

Initially, proportional-derivative (PD) feedback control was applied locally at each actuator to stabilize the system, as shown in Figure 3. It should be noted that high D gain cannot be used because flexible mechanism is prone to noise excitation and instability. Because the kinematic behavior of the mechanism is affected by joint compliance, the rigid body inverse kinematic (RBIK) model is modified by a correction function $\phi(y_d)$ so that the demand values for the actuated joints under feedback control are $\theta_d = RBIK(y_d) + \phi(y_d)$. This correction function is chosen as a 6th order summation of Chebyshev polynomials:

$$\phi(y_d) = \sum_{i=0}^{6} a_i T_i(y_d/L) \tag{10}$$

where $T_{i=0,\dots,n}$ are Chebyshev polynomials of the first kind. The constant coefficients a_i were obtained by experimental identification such that zero positioning error was achieved at the set of Chebyshev nodes given by

$$y_i = L\cos\left(\left(i + \frac{1}{2}\right)\frac{\pi}{7}\right), \quad i = 0, \dots, 6 \tag{11}$$

where $L = 10.2$ mm.

3.3 System dynamics

The dynamic properties of the system were established by applying linear system identification routines at the set of Chebyshev nodes given by Equation (11). The system identification was performed by response-matching for small step input commands. The model structure was set to match the 2nd order system model in Equation (4). Hence, the transfer function identified for each operating point (Chebyshev node) has the following form

$$\frac{Y(s)}{Y_d(s)} = \frac{\widehat{K}^i}{s^2 + C^i s + K^i} \tag{12}$$

where $i \in \{0, \dots, 6\}$. The parameter $\widehat{K}^i$ was set independently to K^i even though, with perfect kinematic behavior, the values should be equal (as $y = y_d$ in the steady state). Four cases were considered, with different platform payloads of 0, 5, 10 and 15 grams. The main results of system identification are shown in Figure 4, together with the Chebyshev interpolation curves (6th order polynomials). These are used within the prefilter implementation to set the values of $\omega = \sqrt{K}$ and C based on the current platform position, as shown by the block diagram in Figure 1. The stiffness coefficient K is additionally scheduled based on the payload mass by using linear interpolation between the two sets of extremum values (for no payload and 15-gram payload). The damping coefficient C is not affected greatly by the payload mass. Therefore, it is scheduled according to the current platform position only.

4. Experimental results and discussion

Experiments were conducted to evaluate the performance of the LTV-FIR prefilter with both position- and mass-dependent scheduling of the dynamic coefficients. Initial results for testing with no payload can be seen from Figure 5. The platform motion is produced by a change in command signal involving 20 mm/sec ramp and overall move distance of 2 mm. Clearly, the LTV-FIR prefilter was able to suppress vibration effectively and with short settling time compared with the case without prefiltering. However, with an increase in payload mass to 15 grams, significant transient oscillations occurred. These initial results show the importance of having accurate knowledge of the dynamic coefficient values. They also indicate that it is desirable to adjust the values in relation to the payload mass.

The results in Figure 5 also show that some static positioning error occurred due to residual errors in the inverse kinematic model (which is also affected by platform payload). Additional compensation for positioning error was introduced by using model reference feedback of platform position with low gain PID control according to the block diagram in Figure 6 below. The model reference feedback acts to reduce the error between the actual system output $y(t)$ and predicted output $y_p(t)$, which is calculated by convolving $y_d(t)$ with the target system impulse response ($y_p = y_d * f$). Due to the fact that f has unity steady-state gain, the use of integral feedback ensures that $y = y_d$ at steady state. The results with model reference feedback are also provided in Figure 7. These show that the model reference feedback helps to eliminate static positioning error by shifting the command input. However, an improved insensitivity to model error is also observed.

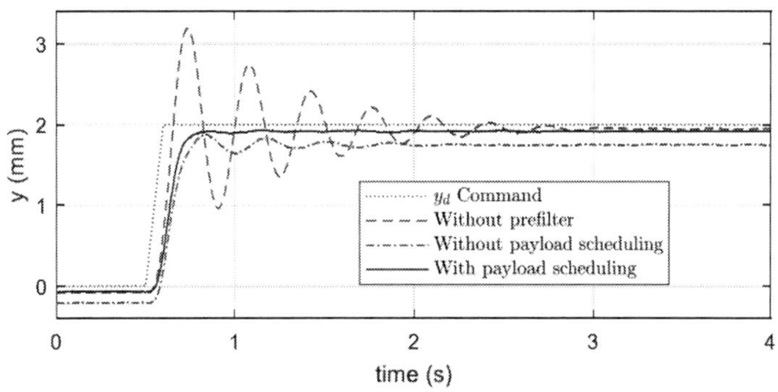

Figure 5. The result of LTV-FIR prefilter with and without payload scheduling

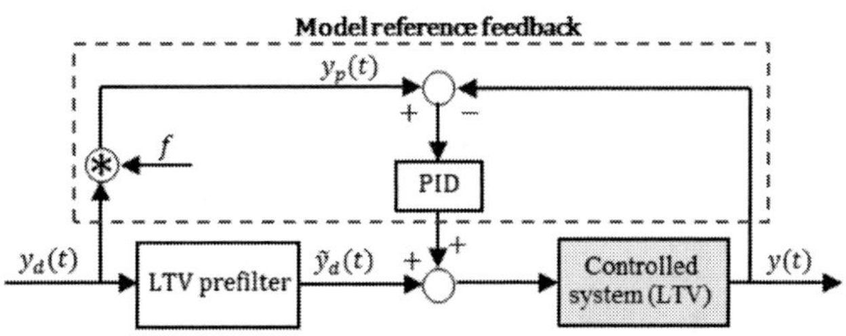

Figure 6. The control implementation with model reference feedback

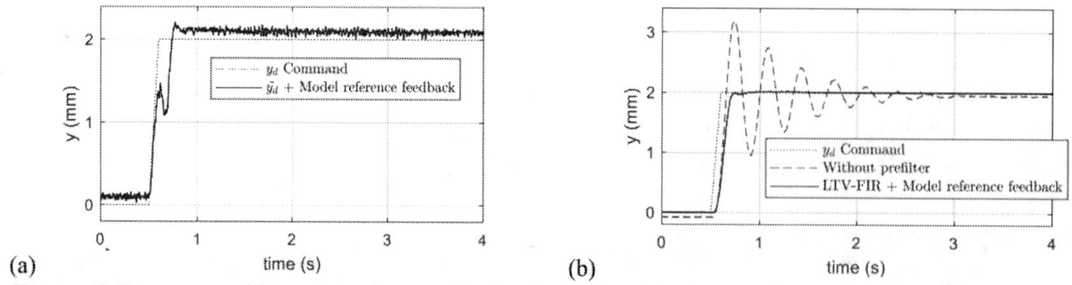

Figure 7. Response with model reference feedback (a) modified command input (b) system response

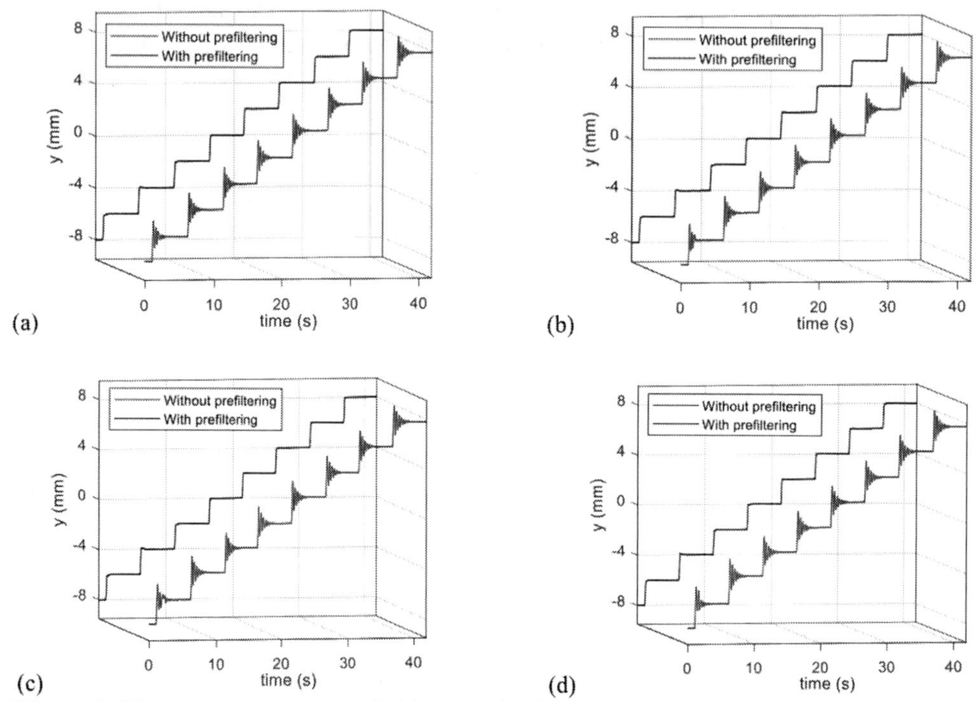

Figure 8. The system response with (a) no payload (b) 5 gram (c) 10 gram (d) 15 gram payload

To investigate the performance of the LTV-FIR prefilter with payload scheduling, tests were performed where the platform undergoes a series of motion intervals from $y = -8$ mm to $y = 8$ mm. Each change in command value involves a constant speed (ramp) of 20 mm/sec with a travel distance of 2 mm, followed by a 5 second dwell period. Figure 8 compares the results obtained with and without prefiltering for four cases with different payload mass values. Although there are significant differences in natural frequency for each payload value, the vibration cancellation effect from the prefiltering is maintained. Moreover, the position-dependent scheduling of the prefilter ensures effective control over the full range of motion.

Further results shown in Figure 9 are based on constant-speed ramp commands that produced motion of the platform between extremum points with $y = \pm 8$. It can be seen that the LTV-FIR command filtering was still effective, even though the large range maneuvers involved rapid changes in model coefficient values. The time-varying filter impulse response function $\tilde{h}$ during these maneuvers is presented in Figure 9b.

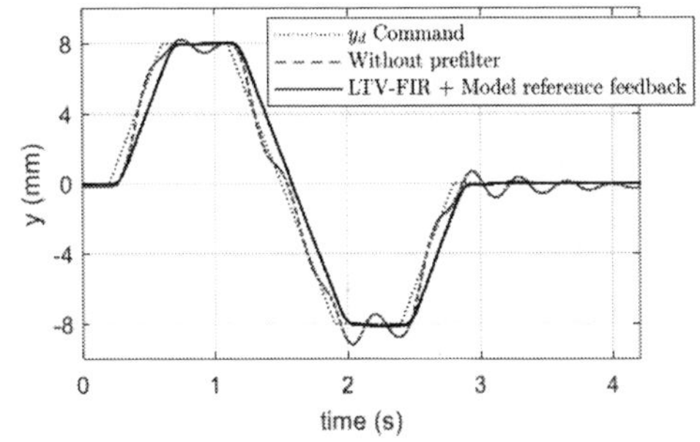

(a)

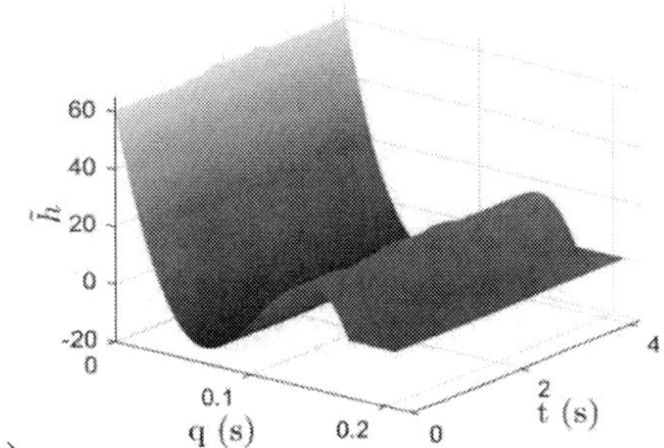

(b)

Figure 9. Large-range motion tests: (a) platform motion (b) time evolution of prefilter function $\tilde{h}$.

5. Conclusions

This paper has investigated an LTV-FIR prefiltering approach that can prevent undesired oscillations in the motion control of LTV systems. The prefiltering method was applied to a flexure-based micro-positioning stage, where the system dynamics were determined by linear identification at a set of calibration nodes. Polynomial interpolation of model coefficients based on measured platform position was used to schedule the LTV prefilter function in real-time. To deal with changes in platform payload, a linear interpolation was adopted based on extremum payload values. This gave significant improvements in vibration cancellation performance compared with cases without payload scheduling.

Acknowledgements

Pongsiri Kuresangsai was supported by the RGJ-PhD scholarship program under Thailand Research Fund (grant number PHD/0101/2561)

References

[1] Howell L L 2001 Compliant mechanisms (New York: John Wiley & Sons)
[2] Xiao S and Li Y 2014 *Journal of Microelectromechanical Systems* **23**(1) 181-190
[3] Liu X Kim K and Sun Y 2007 *Proc. IEEE International Conference on Automation Science and Engineering* 1087-1092
[4] Paijmans B Symens W Brussel H V and Swevers J 2008 *European Journal of Control* **14**(1) 16-29
[5] Rotondo D Nejjari F and Puig V 2013 *Control Engineering Practice* **21**(6) 829-846
[6] Hanifzadegan M and Nagamune R 2015 *IEEE/ASME Transactions on Mechatronics* **20**(1) 133-142
[7] Wassink M G Wal M V D Scherer C and Bosgra O 2005 *Control Engineering Practice* **13**(2) 231-245
[8] Singer N C and Seering W P 1990 *J. of Dyn. Sys. Meas. Control* **112**(1) 76-82
[9] Chatlatanagulchai W Kijdech D Benjalersyarnon T and Damyot S 2016 *J. Dyn. Sys. Meas. Control.* **138**(6) 061006-1-13
[10] Yuan D and Chang T 2008 *IEEE Transactions on Industrial Electronics* **55**(2) 842-851
[11] Mohamed Z Chee A Hashim A Tokhi M Amin S and Mamat R 2006 *Robotica* **24**(4) 499-511
[12] Cole M O T and Wongratanaphisan T 2011 *J. Dyn. Sys. Meas. Control* **133**(2) 021008-1-9
[13] Cole M O T and Kuresangsai P 2019 *Proc. 2019 IEEE Conference on Control Technology and Applications* (Hong Kong, China) 964-969
[14] Cole M O T 2012 *Automatica* **48**(9) 2377-2380
[15] Kuresangsai P and Cole M O T 2019 *Mechanism and Machine Theory* **132** 80-97

Emulation of battery discharge behaviour using programmable power supply based on LabVIEW

N. Tanruean[1], C. Charoenphonphanich[1], T. Yenwichai[2],and M. Masomtob[2*]

[1]School of Engineering, King Mongkut's Institute of Technology Ladkrabang, Chalongkrung Road, Ladkrabang, Bangkok, 10520, Thailand
[2]National Energy Technology Center (ENTEC), National Science and Technology Development Agency (NSTDA), 114 Thailand Science Park, Paholyothin Rd, Khlong Nueng, Khlong Luang, Pathum Thani, 12120, Thailand

* Corresponding Author: manop.mas@entec.or.th

Abstract
This paper presents a battery simulator based on the actual electrical behaviour of the battery cells using power supply controlled by Laboratory Virtual Instrument Engineering Workbench (LabVIEW). Firstly, the behaviours of the battery are tested under experimental conditions including constant temperatures at 25C° and two different current loads at 0.5C and 1C. Secondly, the battery behaviours are analysed and applied in LabVIEW. The algorithms which are low pass filter, PID and others are implemented in LabVIEW to reduce standard deviations and errors. The current study found that the constant load at 0.5C and 1C, which controlled by the electronic load via LabVIEW, can emulate the battery voltage and the error relative is below 5 %. Further work is required to establish the optimization of the parameters in the battery management system (BMS) since a battery simulator uses power sources before the use of the battery management system (BMS) with a real battery.

Keywords: Simulation battery management system, Battery behaviour, LabVIEW simulation

1. Introduction

A lithium-ion (Li-ion) battery plays a key role in the rapid development of technology and is also considered a vital component in electrical vehicles (EVs). There is an evidence that Li-ion batteries used for EVs plays a crucial role in the battery pack system. A battery pack system comprises multiple batteries wired together (multicell), and a Battery Management System (BMS). BMS has a pivotal role in making the battery pack safe, reliable, and efficient.[1] By using algorithms, a key aspect of BMS is being used accurately to predict the state of charge (SOC), and state of health (SOH) to protect against any short circuit or fire hazard. Additionally, the multicell in the battery pack, which consists of several batteries connected in parallel and series, requires BMS for cell balancing due to issues during discharge and charge processes.

The issue occurs when some cells reach cut off voltage first causing a fault in the battery pack. In the same manner, some cells reach full charge ahead of the other cells and results in overvoltage of those cells, and this is the reason the BMS is necessary for EVs.

The BMS has several parameters which need to be adjusted for optimal function such as value resistance in passive balancing charge, value capacitor, frequency switching, and length of adjustment time. The development of BMS depends on different algorithms which apply to different EVs. One example of this is a study in algorithm concurrent cell balancing by M. Kuer et al. [2].

Even if the batteries are made in the same factory with the same materials, their electrical characteristics are not always the same. Thus, actual battery cells should be considered for the improvement as well as the development of BMS. There are various studies about simulated characteristics of battery cells and equivalent circuit models for Li-ion batteries. There are two of the studies in line with this work. The first one is "Comparative research on RC equivalent circuit models for lithium-ion batteries of electric vehicles", L. Zhang, H. Peng, Z. Ning, Z. Mu, and C. Sun considering and discussing an equivalent circuit lithium-ion battery pack [3]. The second one is "Detailed Design of Second-Order Model of Lithium-Ion Battery Simulator Based on Experimental Measurements" by Institute of Electrical and Electronics Engineers, Universitatea Politehnică București [2] focusing on a second-order model of Li-ion battery simulator based on experiments. Souag et al. [4] working on the example of LabVIEW create power flow solution computing.

LabVIEW is used to emulate the battery behaviour based on the actual electrical discharge battery behaviours. The battery discharge behaviours under various discharge current rates are required to understand and quantify using a first-order battery equivalent circuit model. Therefore, the significant contribution of this research has two points. Firstly, a new battery cell model has been proposed to quantify the battery discharge behaviour under current load condition and provide a virtual battery instrument. Secondly, parameters in BMS can be adjusted without using the actual batteries.

2. Theory

2.1 Characterization of the battery cell

The tested cell here is a commercial 18650 lithium-ion cells with $LiNiMnCoO_2$ (NMC) and $LiNiCoAlO_2$ (NCA) cathodes. The specifications of the cell are listed in Table 1. The behaviour curves of battery during constant discharge mode shown in Figure 2.

2.2 Circuit-Based Models.

Figure 1 illustrates the Thevenin-based model [5, 6]. A series circuit consists of an open-circuit voltage (V_{oc}) as a function of the state of charge (SOC) with an ohmic internal resistor (R_0). A parallel circuit includes a resistor (R_p) and a capacitor (C_p), which used to predict battery respond to transient load evens at a particular state of charge. The terminal voltage is the battery voltage (V_b). The terminal voltage equation of the battery is given in (1):

$$V_b(t) = V_{oc} - I_b \cdot (R_0 + R_p \cdot e^{\frac{-t}{\tau}}) \tag{1}$$

Where is $V_b(t)$ is the terminal voltage battery as a function of time, I_b represents the current of pulse discharge, R_0 is the ohmic internal resistor, and R_p is resistance polarization $\tau = R_p \cdot C_p$ time constant of between R_p and C_p which are parallelly connected. The values of R_0, R_p and C_p are determined from the pulse current measurement. The more detail is mentioned in section 2.3.

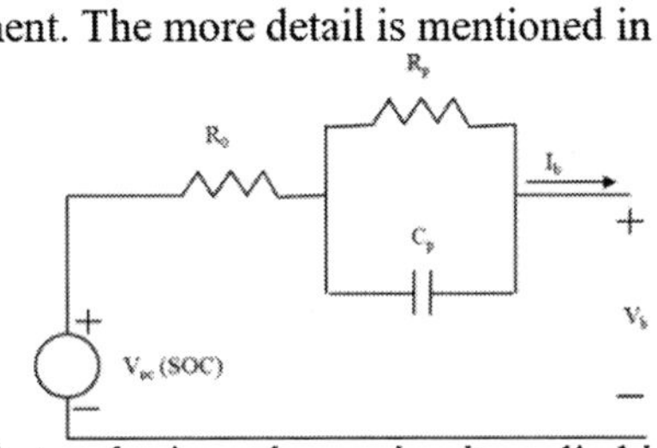

Figure 1. First-order impedance circuit applied in cell battery simulation

IOP Conf. Series: Materials Science and Engineering 1137 (2021) 012057 doi:10.1088/1757-899X/1137/1/012057

Table 1. Battery Parameters.

Item		**Specifications**
Capacity	Rate capacity (Minimum)	2750 mAh at 1C
	Nominal capacity (Typical)	2900 mAh at 0.2C
	Nominal capacity (Minimum)	2800 mAh at 0.2C
Nominal Voltage		3.6 V
Charge Voltage		4.2 V
Discharge cut-off voltage		2.5 V
Max Charge current		1C (2750mA) at 25 C°
Max Discharge current		3C (8250mA) at 25 C°

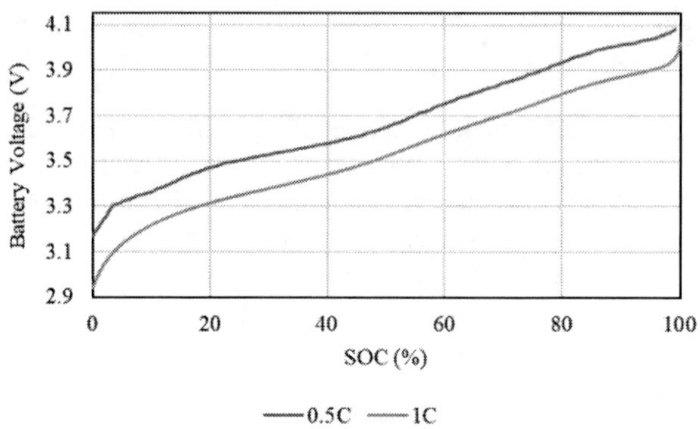

Figure 2. Characteristic curves of the battery in constant discharging mode

2.3 Model Parameter Determination.

The parameters of the Li-ion battery cell can be calculated from the experimental result, as shown in Figure 3. The calculation method is as fellow:

1) The calculation of V_{oc}, in Figure 3. $V_{oc} = V_1$
2) Calculate R_0:

$$R_0 = \frac{V_1 - V_2}{I} \tag{2}$$

In Equation (2), I is the current value of 0.5C a pulse discharge. R_0 describes the rapid voltage drop from V_1 to V_2 in Figure 3. The ohmic internal resistance of Li-ion cell batteries causes the terminal voltage to increase or decrease instantaneously during pulse charging and discharging. However, there are differences ohmic resistance between charge and discharge.

3) The formulas for calculating the RC respond are as follows, the parameters of R_p and C_p are shown in Figure 3.

$$R_p = \frac{V_2 - V_3}{I} \tag{3}$$

$$C_p = \frac{t_2 - t_1}{R_p} \tag{4}$$

The experimental details of Figure 3 can be explained as follow. Each cycle consists of

1) A 60 s pulse constant current discharge period, with 0.5C current load.
2) A 180 s pulse rest period.

IOP Conf. Series: Materials Science and Engineering 1137 (2021) 012057 doi:10.1088/1757-899X/1137/1/012057

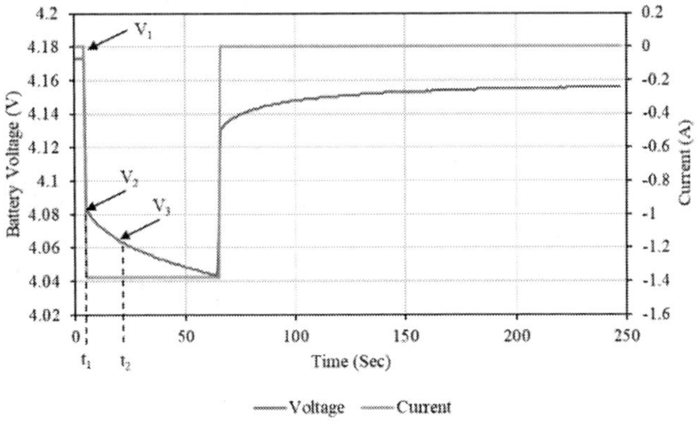

Figure 3. Current and voltage of a cell battery during discharge pulse in a cycle.

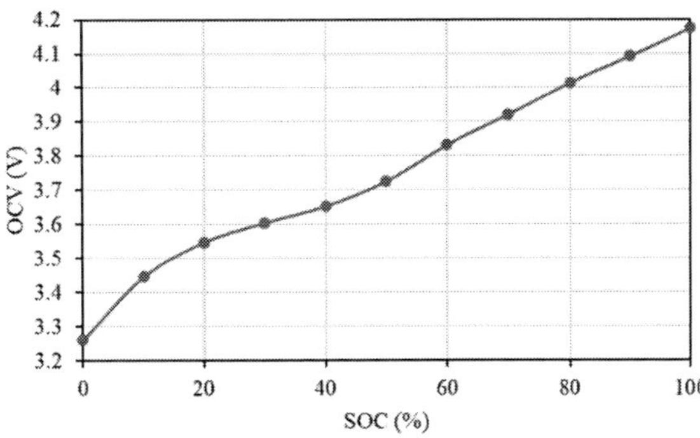

Figure 4. Experimental data and fitted curves of the relationship between OCV and SOC

Table 2. Parameters of the cell battery

DOD (%)	0.5C			1C		
	R_0 [mΩ]	R_p [mΩ]	C_p [F]	R_0 [mΩ]	R_p [mΩ]	C_p [F]
0	64.493	13.768	1016.842	57.971	14.130	990.769
10	55.072	10.870	1104.000	52.174	15.942	1442.727
20	54.348	17.391	1150.000	53.261	18.478	1461.176
30	55.797	18.116	1214.400	54.348	17.391	1552.500
40	53.623	17.391	1207.500	54.710	14.493	1794.000
50	54.348	10.145	1478.571	53.261	10.870	1748.000
60	54.348	7.246	1380.000	52.174	13.768	1888.421
70	55.797	7.971	1254.545	53.986	13.043	1916.667
80	56.522	7.246	1104.000	55.435	14.855	1750.244
90	62.319	23.913	1087.273	68.478	43.478	598.000
100	91.304	174.638	143.1535	94.203	77.536	116.0748

An accuracy of OCV is vital to improve the SOC estimate accuracy. Thus, this paper uses a ninth-order polynomial fitting method to enhance the model accuracy as shown in Equation (5), and the relationship between OCV and SOC is plotted, as shown in Figure 4.

IOP Conf. Series: Materials Science and Engineering 1137 (2021) 012057 doi:10.1088/1757-899X/1137/1/012057

$$OCV = (9.5968 \cdot 10^{-16} SOC^9) - (4.3431 \cdot 10^{-13} SOC^8) + (8.2170 \cdot 10^{-11} SOC^7) - (8.4116 \cdot 10^{-9} SOC^6) + (5.0384 \cdot 10^{-7} SOC^5) - (1.7888 \cdot 10^{-5} SOC^4) + (3.6942 \cdot 10^{-4} SOC^3) - (0.0044 \cdot SOC^2) + (0.03916 \cdot SOC) + 3.26 \quad (5)$$

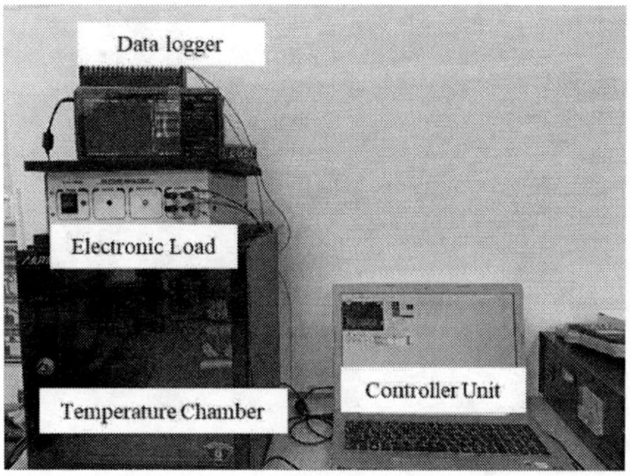

(a) Battery cells testing experiment set up.

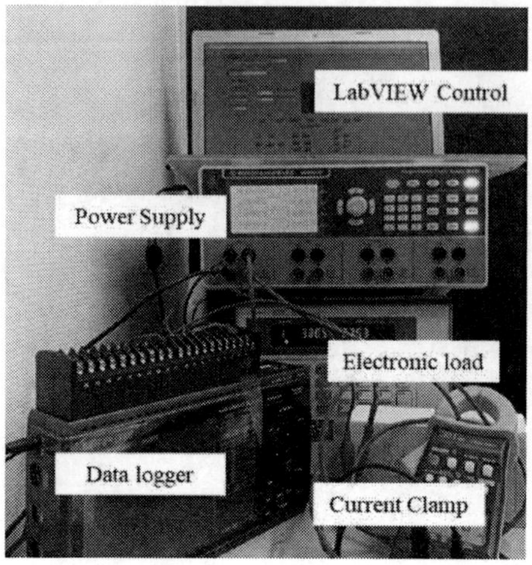

(b) LabVIEW control power supply

Figure 5. The experimental equipment setup

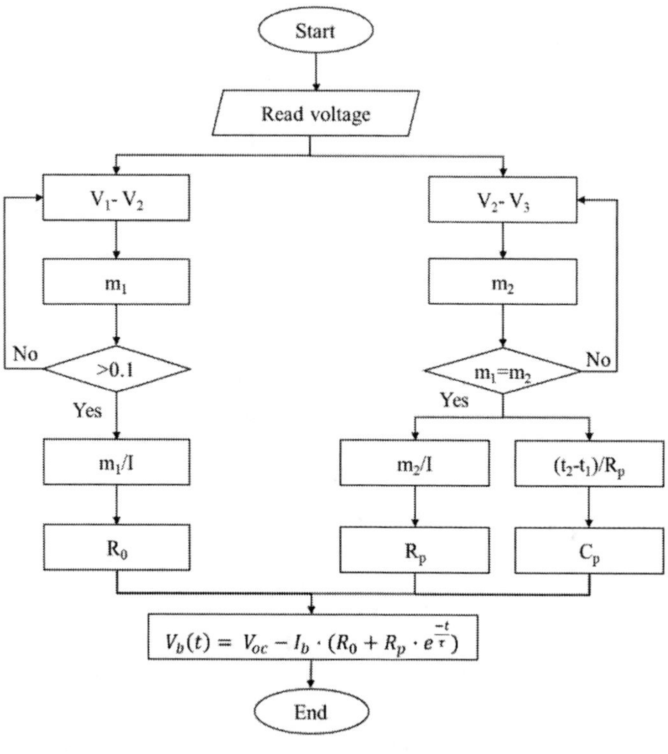

Figure 6. Flow chart of an algorithm in LabVIEW

IOP Conf. Series: Materials Science and Engineering 1137 (2021) 012057 doi:10.1088/1757-899X/1137/1/012057

3. Methodology

Figure 5a illustrates the experiment of the battery cells testing set up for tested battery behaviour with pulse current. Figure 5b provides the LabVIEW control power supply. The flow chart of an algorithm in LabVIEW is presented in Figure 6. After staring the program, the voltage is read by the controller, then the difference of voltage is calculated. If the value is higher than 0.1 V, resistance (R_0) is computed. On the other hand, if the value is less than 0.1, the difference in voltage is repeatedly calculated. After the value of resistance (R_0) is accepted, the program continuously calculates polarization resistance (R_p) and capacitor (C_p). If the program does not accept the value of resistance (R_0), the program will not calculate R_p and C_p. The program computes R_p and C_p when m_1 equals m_2. After R_p, C_p and R_0 are obtained, the voltage of battery is determined using Equation (1).

3.1. Experimental setup and process

Toriyama-N18650CL-29 is attached to a battery analyser which is a battery tester. The BA553WIN software program is applied to both charge and discharge battery. Moreover, voltage and current under specific conditions are set out in the monitor. The collected data from the pulse current condition can be seen in Figure 7. Table 2 illustrates the battery parameters.

3.2. Programming in LabVIEW

This work is to study the battery behaviour and the emulation of the battery behaviour. The program is created based on the battery equivalent circuit and the Thevenin-based model using LabVIEW. The creation of an algorithm can calculate parameters based on Equations (2) - (4), and the output of battery voltage behaviour using Equation (1). Moreover, OCV can be calculated from Equation (5).

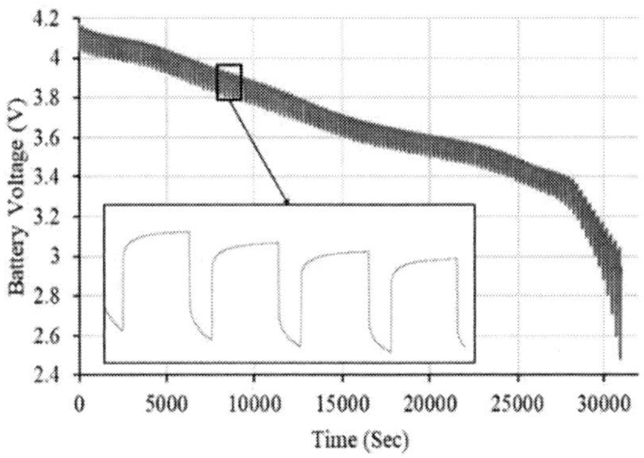

Figure 7. Experimental data of the battery voltage during the current pulse test.

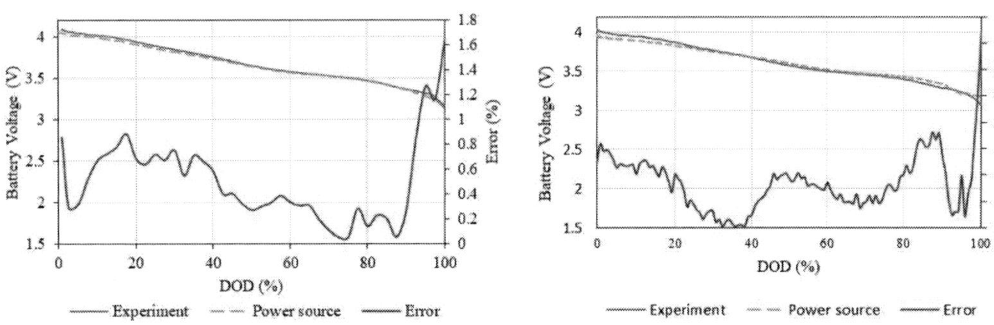

(a) 0.5C constant current (b) 1C constant current.

Figure 8. The result of the discharged battery voltage.

4. Results and discussion

The constant loads at 0.5C and 1C are controlled by the electronic load via LabVIEW to emulate the battery voltage. In Figure 8a, the findings from the constant 0.5C load reveal the occurrence of the relative error which is less than 2%. The highest error occurs during 80% to 100% of DOD which is similar to the result of the constant of the 1C load. On the other hand, the findings from the constant 1C load indicate that overall error is higher than 0.5C load due to the value resistance of 1C load, which is higher than 0.5C load, as shown Figure 8b. The peak error reaches 3.68% during 100% and 90% of DOD period. However, the output from two different constant loads is in agreement with the data of the actual battery voltage, which is below the relative error 5%.

5. Conclusion

This study identifies the performance of the algorithm due to error is below five percent. This study also shows that different loads are satisfactory. The findings of this study suggest that this algorithm can replace the actual battery, and the performance has an excellent graphical interface in which it can efficiently run the computation. These findings have significant implications for the understanding of how to adjust the parameters in the BMS as a battery simulator using power sources before the use of the BMS with a real battery. This would be a fruitful area for further work.

Acknowledgements

The research was supported by National Science and Technology Development Agency (NSTDA) and King Mongkut's Institute of Technology Ladkrabang (KMITL).

References

[1] Rahimi-Eichi H Ojha U Baronti F and Chow M Y 2013 *IEEE Industrial Electronics Magazine* **7** 4-16

[2] Ruba M Ciornei S Nemeş R and Martis C 2019 *Proc. 11th International Symposium on Advanced Topics in Electrical Engineering* (Bucharest, Romania) 1-6

[3] Zhang L Peng H Ning Z Mu Z and Sun C 2017 *Applied Sciences (Switzerland)* **7**(10) 1002

[4] Souag S Benhamida F Gherbi F Z and Graa A 2013 *Proc. 5th International Conference on Modeling, Simulation and Applied Optimization* (Hammamet) 1-6

[5] Zhang J Ci S Sharif H and Alahmad M 2010 *IEEE Transactions on Energy Conversion* **25** 1133-1141

[6] Miniguano H Barrado A Lazaro A Zumel P and Fernandez C 2020 *IEEE Transactions on Vehicular Technology* **69** 235-245

Penetration-force estimation approach for a flexure-jointed micro-injection mechanism with Lorentz force actuation

A. Sura, P. Kuresangsai, M. O. T. Cole*, T. Wongratanaphisan, and P. Puangmali

Department of Mechanical Engineering Faculty of Engineering,
Chiang Mai University, Chiang Mai 50200, Thailand

* Corresponding Author: motcole@dome.eng.ac.th

Abstract
This paper presents a novel linear-motion compliant mechanism with Lorentz force actuation and integrated force-sensing capability for automated cell micro-injection. A model-based force estimation approach is introduced such that no force sensor is required. Model identification is undertaken by applying a sinusoidal actuation signal while the mechanism contacts with objects of known stiffness. Displacement data is then used to calculate model coefficients via a least-squares optimization. By using an observer-based state estimation scheme with actuation and displacement signals as inputs, force sensing accuracy within 70 μN RMS error could be achieved within a sensing range of $0-5$ mN. This sensing capability confirms the suitability of the system for penetration force measurement in certain cases of cell micro-injection.

Keywords: Cell injection, Disturbance observer, Force sensing, Compliant mechanism

1. Introduction
The has been considerable research interest on micromanipulation systems for biological and medical procedures such as Intracytoplasmic Sperm Injection (ICSI) and In Vitro Fertilization (IVF) [1]. Manual ICSI requires skilled operators to control the manipulator and is very time-consuming. Therefore, methodologies for automating this process are currently under development [2], [3]. It is generally recognized that information about the forces acting on the micropipette can be used to better control the injection process, particularly to avoid excessive speed or force that would cause the pressure within the cell to increase and negatively affect or damage the cell [4].

To obtain real-time measurements of penetration force, Wei and Xu [5] developed a force sensor based on a parallel flexure mechanism with piezoelectric polymer (PVDF) film. The device was applied in crab egg micro-injection and could successfully measure penetration forces of approximately 30 mN. A commercial PVDF force sensor was used by Huang et al. [2] for combined position/force control in a novel batch micro-injection system. Sun et al. [6] used a capacitive force sensor with comb drives for micro-injection of mouse oocytes. This type of MEMs sensor is currently costly and extremely fragile, which makes it less practical for general use. Other micro-force sensing methods include the use of piezo-resistive elements, strain gauges, and optical methods to measure strain [7], [8]. With any type of force sensor, it is a challenge to integrate the sensing device within the position control system for the cell or micropipette, while maintaining accuracy, cost-effectiveness and ease of use in practice [3].

IOP Conf. Series: Materials Science and Engineering 1137 (2021) 012058 doi:10.1088/1757-899X/1137/1/012058

This current study considers an alternative approach to force-sensing in microinjection where interaction forces are estimated from knowledge of actuation force and measured displacement of the micropipette. To successfully implement this technique, several important system features are adopted:

1) Flexure-jointed mechanism for high fidelity and transparency of force transmission.
2) Voice coil actuation to achieve low-stiffness and high-fidelity motion.
3) Model-based micro-force estimation scheme using precise position measurements.

This paper also focuses on model calibration and initial investigations on suitable force estimation algorithms. The main results concern the influence of mechanical stiffness and issue of noise contamination.

2. System description

A motion control system for cell injection procedures has been created based on a multi-link compliant mechanism, as shown in Figure 1. In this mechanism, two flexure-jointed linkages are used to support a linear stage with a micropipette (see also Figure 1a). The mechanism is driven by rotatory voice coil actuators to produce rectilinear motion of the stage. Displacement of the stage is measured by a linear optical encoder (resolution 0.2 μm). The actuators are powered by linear drives (Sprint Electric 200 XLV) with current range $\pm$0.2 A and current control resolution of 24 μA. A manually operated XYZ stage is used to position the target object for interaction with the micropipette. The mechanism is manufactured by fused deposition modelling (3-D printing) with ABS filament. Key properties of the system are shown in Table 1, while further details on the mechanism design can be found in [9].

The flexure joints allow smooth and precise motion of the stage with low friction. In this present study, the mechanism was controlled in open-loop without feedback. Signal processing was implemented in discrete time with a sampling period of 1 ms. The intrinsic stiffness of the mechanism due to joint stiffness (as measured at the output stage) is approximately 650 N/m. To examine the effect of the mechanism stiffness on force measurements, a stiffness reduction device was attached to the linear stage. This device is a negative stiffness mechanism based on two parallel beams with compressive loading set by a preload spring. The negative stiffness property (which can be determined according to buckling theory [10]) is used to partially cancel the positive stiffness of the linear stage mechanism.

Table 1. System key properties

Property	Value	Units
Link lengths $(l_1, l_2, l_3, l_4, l_5)$	(40,20,120,50,120)	mm
Position measurement resolution	2×10^{-4}	mm
Current control resolution	2.5×10^{-2}	mA
Linear motion range (maximum)	± 3	mm

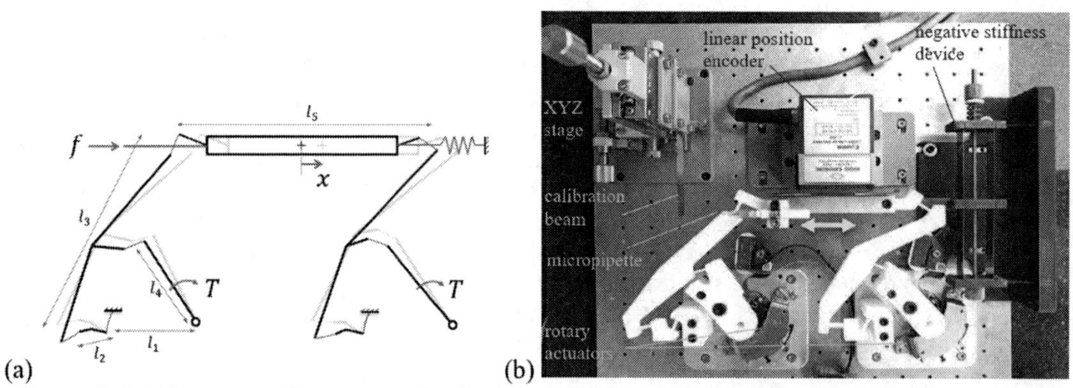

Figure 1. Motion system for automated cell micro-injection: (a) schematic diagram of flexure-jointed mechanism (b) experimental setup for calibration and testing

IOP Conf. Series: Materials Science and Engineering 1137 (2021) 012058 doi:10.1088/1757-899X/1137/1/012058

3. Mechanism model and error analysis

As a basis for force-sensing and system analysis, the mechanism can be represented by a linear dynamic model with a single positional degree of freedom, according to

$$k_x x + c_x \dot{x} + m_x \ddot{x} = k_i i + f. \tag{1}$$

Here, x is the displacement of the end link from the equilibrium position, f is the interaction force and i is the actuator coil current (see Figure 1). The constant coefficients k_x, c_x and m_x are the effective stiffness, damping and mass of the mechanism while k_i is the force/current coefficient for the actuators.

Under static conditions (with $\dot{x} = \ddot{x} = 0$), Equation (1) simplifies to $f = k_x x - k_i i$. Hence, the absolute error Δf for static force sensing, based on knowledge of x and i, may be obtained as

$$\Delta f = k_x \Delta x + k_i \Delta i, \tag{2}$$

where Δx and Δi are the absolute error/resolution for displacement measurement and actuator current, respectively. For a well-designed system, the dependency of Δf on the input/output should be balanced so that $k_x \Delta x \approx k_i \Delta i$. For example, if $k_x \Delta x > k_i \Delta i$ then reducing the mechanism stiffness will improve force-sensing precision by lowering k_x. However, if the stiffness is too low, the mechanism will be prone to positional drift and instability. In this study, the negative stiffness device is introduced so that the effect of stiffness reduction can be investigated in a systematic manner. Although force-sensing accuracy will be affected by other error sources, such as noise and vibration, as well as model error, the underlying limits of precision, as given by Equation (2), must still be examined.

4. Model calibration

4.1 Method

To perform model calibration, a sinusoidal actuation signal with frequency ω is applied to excite the system. Under steady-state conditions, the measured input-output data can be expressed as

$$i(t) = A_i \cos \omega t + B_i \sin \omega t + C_i, \qquad x(t) = A_x \cos \omega t + B_x \sin \omega t + C_x, \tag{3}$$

where the coefficients $A_{i,x}$, $B_{i,x}$ and $C_{i,x}$ are to be determined. By defining the sampled input-output data in vector form $\boldsymbol{i} = [i(t_1) \dots i(t_n)]^T$, $\boldsymbol{x} = [x(t_1) \dots x(t_n)]^T$, the errors in Equations (3) are expressed as

$$\boldsymbol{e}_i = \boldsymbol{R} \begin{bmatrix} A_i \\ B_i \\ C_i \end{bmatrix} - \boldsymbol{i}, \quad \boldsymbol{e}_x = \boldsymbol{R} \begin{bmatrix} A_x \\ B_x \\ C_x \end{bmatrix} - \boldsymbol{x}, \quad \text{where } \boldsymbol{R} = \begin{bmatrix} \cos \omega t_1 & \sin \omega t_1 & 1 \\ \cos \omega t_2 & \sin \omega t_2 & 1 \\ \vdots & \vdots & \vdots \\ \cos \omega t_n & \sin \omega t_n & 1 \end{bmatrix}. \tag{4}$$

Solutions that minimize $\boldsymbol{e}_i$ and $\boldsymbol{e}_x$ in a sum-of-squares sense can be computed as

$$\begin{bmatrix} A_i \\ B_i \\ C_i \end{bmatrix} = (\boldsymbol{R}^T \boldsymbol{R})^{-1} \boldsymbol{R}^T \boldsymbol{i}, \qquad \begin{bmatrix} A_x \\ B_x \\ C_x \end{bmatrix} = (\boldsymbol{R}^T \boldsymbol{R})^{-1} \boldsymbol{R}^T \boldsymbol{x}. \tag{5}$$

To determine the model coefficient values, a series of tests are performed where the needle contacts with a linearly elastic object of known stiffness k_b. In this case, $f = k_b x + F_0$ and so, from Equation (1), the input-output data is related by

$$k_b x + F_0 = k_x x + c_x \dot{x} + m_x \ddot{x} - k_i i. \tag{6}$$

Substituting for x and i using equation (3) and matching terms in $\cos \omega t$ and $\sin \omega t$ gives

$$\begin{bmatrix} A_x k_b \\ B_x k_b \end{bmatrix} = \begin{bmatrix} A_x \\ B_x \end{bmatrix} k_x + \begin{bmatrix} \omega B_x \\ -\omega A_x \end{bmatrix} c_x + \begin{bmatrix} -\omega^2 A_x \\ -\omega^2 B_x \end{bmatrix} m_x - \begin{bmatrix} A_i \\ B_i \end{bmatrix} k_i. \tag{7}$$

When using low-frequency excitation (small ω), inertia effects may be neglected so that the $m_x = 0$.

A number (M) of excitation tests are performed involving interaction with different objects of known stiffness: $k_b = k_b^1, k_b^2 ..., k_b^M$. For each experiment, the coefficients for the input-output data are computed using Equation (5). These are used in Equation (7) and combined in matrix form

$$\boldsymbol{f}_M - \boldsymbol{G}_M \begin{bmatrix} k_x \\ c_x \\ k_i \end{bmatrix} = \boldsymbol{0}, \text{ where } \boldsymbol{f}_M = \begin{bmatrix} k_b^1 A_x^1 \\ k_b^1 B_x^1 \\ \vdots \\ k_b^M A_x^M \\ k_b^M B_x^M \end{bmatrix}, \boldsymbol{G}_M = \begin{bmatrix} A_x^1 & \omega B_x^1 & -A_i^1 \\ B_x^1 & -\omega A_x^1 & -B_i^1 \\ \vdots & \vdots & \vdots \\ A_x^M & \omega B_x^M & -A_i^M \\ B_x^M & -\omega A_x^M & -B_i^M \end{bmatrix} \quad (8)$$

As, in general, this equation is overdetermined and cannot be exactly satisfied, the error is defined as

$$\boldsymbol{e}_F = \boldsymbol{f}_M - \boldsymbol{G}_M \,[k_x \; c_x \; k_i]^T. \quad (9)$$

Minimizing $\boldsymbol{e}_F$ in a sum-of-squares sense gives the following solution for the model coefficients:

$$\begin{bmatrix} k_x \\ c_x \\ k_i \end{bmatrix} = (\boldsymbol{G}_M^T \boldsymbol{G}_M)^{-1} \boldsymbol{G}_M^T \boldsymbol{f}_M. \quad (10)$$

Note that a minimum of two sets of data (with two different values of k_b) is required so that $rank(\boldsymbol{G}_M) \geq 3$ and the matrix $\boldsymbol{G}_M^T \boldsymbol{G}_M$ is nonsingular.

Table 2. Identified model parameters for different stiffness-reduction settings.

Case	Preload (no. of turns)	k_x (N/m)	c_x (Ns/m)	k_i (N/A)	ω_d (rad/s)	m_x (kg)
A	0	656.7	77.8	0.8196	110.4250	0.0539
B	2	329.4	50.5	0.7625	81.9724	0.0490
C	4	104.4	36.7	0.734	44.8479	0.0400

4.2 Results

To perform the model identification, a thin steel beam (thickness 0.2 mm, effective length 55 mm and width 12.7 mm) was clamped at the one end, while the other (free) end had continuous contact with the needle (see also Figure 1). The beam stiffness, calculated with Euler-Bernoulli theory, was k_b = 33.4 N/m. Displacement-current data was collected during steady motion of the mechanism under sinusoidal excitation. This was done both with and without contact of the needle with the beam. A typical data set is also illustrated in Figure 2.

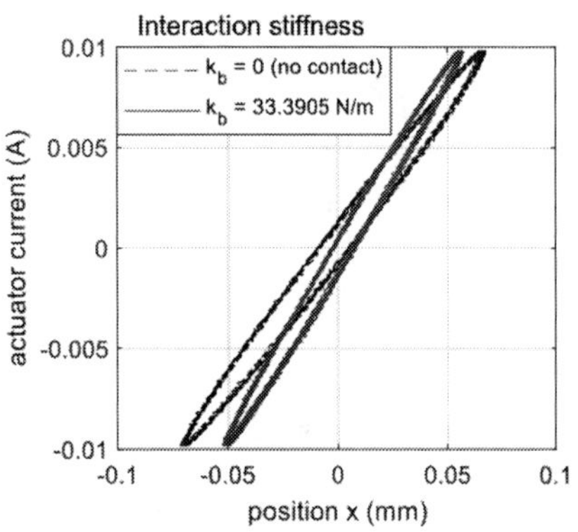

Figure 2. Model calibration data: Sinusoidal response with and without contact (case C).

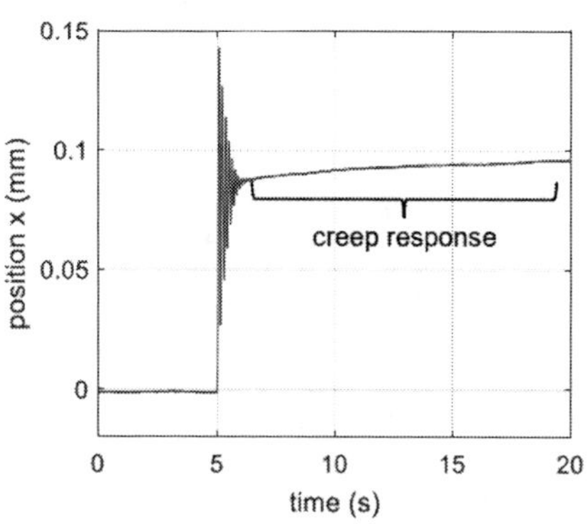

Figure 3. Mechanism response due to step-change in actuator current (case C).

Table 2 presents the identified model coefficients, calculated by the method used in this study (see Section 4.1) and for cases with different preload settings for the negative stiffness mechanism. The effective stiffness of the mechanism k_x is reduced as the preload on the negative stiffness mechanism is increased. It is observed that the effective damping coefficient c_x also decreases as the stiffness is lowered. Some variation in k_i is also observed, which is believed to be due to nonlinear effects impacting on the force transmission. Effective mass values m_x were also calculated using step response data (see example in Figure 3). By measuring the frequency of oscillation ω_d the mass value could be estimated as $m_x = k_x/\omega_d^2$, in accordance with the model defined by Equation (1).

5. Estimation of interaction forces

5.1 Inverse model approach

A simple method to estimate the object interaction force is based directly on Equation (1). If the mass effects are neglected, the force estimate can be calculated from i and x according to

$$\tilde{f} = k_x x + c_x v - k_i i, \tag{11}$$

where $v(t)$ is an estimate of the end-link velocity $\dot{x}(t)$. This may be obtained using a filtered-derivative operation (with cut-off frequency ω_c), which has the state equation $v = \dot{z} = -\omega_c z + \omega_c x$. The force estimation scheme is implemented by combining this with Equation (11):

$$\dot{z} = -\omega_c z + \omega_c x, \tag{12}$$

$$\tilde{f} = (k_x - \omega_c c_x)x - \omega_c c_x z - k_i i. \tag{13}$$

Although this approach is simple to design and implement, it does not provide an optimal estimate of f when external disturbances and noise act on the system during operation.

5.2 Disturbance observer approach

An optimal estimation of f should take account of system perturbations, such as noise and vibration excitation (e.g., due to ground vibration and acoustic noise). For stochastic disturbances, it is appropriate to consider an optimal state estimator based on a model of the system in the form

$$\frac{d}{dt}\begin{bmatrix} x \\ \dot{x} \end{bmatrix} = \begin{bmatrix} 0 & 1 \\ -\frac{k_x}{m_x} & -\frac{c_x}{m_x} \end{bmatrix}\begin{bmatrix} x \\ \dot{x} \end{bmatrix} + \begin{bmatrix} 0 \\ \frac{k_i}{m_x} \end{bmatrix} i + \begin{bmatrix} 0 \\ \frac{1}{m_x} \end{bmatrix} f + \begin{bmatrix} 0 \\ \frac{1}{m_x} \end{bmatrix} d, \tag{14}$$

$$x_m = \begin{bmatrix} 1 & 0 \end{bmatrix}\begin{bmatrix} x \\ \dot{x} \end{bmatrix} + n. \tag{15}$$

For this model, the external disturbances include mechanical excitation and measurement noise/error, as represented by d and n, respectively. An extended state estimator, which incorporates the best estimate of f as an additional state variable, may be considered in the form

$$\frac{d}{dt}\begin{bmatrix} \tilde{x} \\ \dot{\tilde{x}} \\ \tilde{f} \end{bmatrix} = \boldsymbol{A}\begin{bmatrix} \tilde{x} \\ \dot{\tilde{x}} \\ \tilde{f} \end{bmatrix} + k_i \boldsymbol{B}_d i + \boldsymbol{L}(x_m - \tilde{x}), \tag{16}$$

where $\boldsymbol{L}$ is the observer gain matrix and

$$\boldsymbol{A} = \begin{bmatrix} 0 & 1 & 0 \\ -\frac{k_x}{m_x} & -\frac{c_x}{m_x} & \frac{1}{m_x} \\ 0 & 0 & 0 \end{bmatrix}, \quad \boldsymbol{B}_d = \begin{bmatrix} 0 \\ \frac{1}{m_x} \\ 0 \end{bmatrix}.$$

Defining $\boldsymbol{e} = \begin{bmatrix} x \\ \dot{x} \\ f \end{bmatrix} - \begin{bmatrix} \tilde{x} \\ \dot{\tilde{x}} \\ \tilde{f} \end{bmatrix}$ then, from equations (14)-(16), the error dynamics are obtained as

$$\dot{\boldsymbol{e}} = (\boldsymbol{A} - \boldsymbol{LC})\boldsymbol{e} + \boldsymbol{B}_d d - \boldsymbol{L}n, \tag{17}$$

where $\boldsymbol{C} = [1\ \ 0\ \ 0]$ and the observer gain $\boldsymbol{L}$ is calculated to minimize the RMS state estimation error. If d and n are zero-mean Gaussian white-noise processes with covariances $\boldsymbol{Q}$ and $\boldsymbol{R}$, respectively, the optimal observer (Kalman filter) follows from the solution $\boldsymbol{P}$ to the following matrix Riccati equation:

$$\boldsymbol{AP} + \boldsymbol{PA}^T - \boldsymbol{PC}^T\boldsymbol{R}^{-1}\boldsymbol{CP} + \boldsymbol{B}_d\boldsymbol{Q}\boldsymbol{B}_d^T = \boldsymbol{0}. \tag{18}$$

The realization of Equation (17) with observer gain $\boldsymbol{L} = \boldsymbol{PC}^T\boldsymbol{R}^{-1}$ provides an optimal estimation of both the system state variables and the external interaction force. Note that, as $\boldsymbol{Q}$ and $\boldsymbol{R}$ are scalars, the solution for $\boldsymbol{P}$ (and hence $\boldsymbol{L}$) depends only on the ratio $\boldsymbol{Q}/\boldsymbol{R}$. Consequently, the ratio $\boldsymbol{Q}/\boldsymbol{R}$ may be treated as a design parameter that sets the bandwidth (cut-off frequency) for the force measurements. The value can be chosen empirically to give a suitable trade-off between low noise and fast dynamic response for the force estimation.

5.3 Results

Tests were conducted where the injection needle was moved sinusoidally to produce intermittent contact with a thin beam. As the stiffness of the beam is known, the actual interaction force could be calculated directly from the displacement. Figure 4 shows the model-based force estimate, calculated as described in sections 5.1 and 5.2, for cases with high and low stiffness settings.

Regarding the high stiffness case (see Figure 4a), the force estimate from the inverse model is contaminated by large errors. The main cause of the error is that step changes in the displacement signal due to quantization are amplified by the inverse model (Equations (12) and (13)). The cut-off frequency for the filtered derivative was set to $\omega_c = 4\pi$ rad/s in order to limit error amplification. The force estimate from the disturbance observer (Figure 4b) shows significant improvements in noise level and overall accuracy. The RMS error values for the inverse model and disturbance observer methods were 289 μN and 69 μN, respectively. For the low stiffness case (Figure 4c), the force estimate from the inverse model has much lower error levels. This can be attributed to the reduction in both k_x and c_x. The force estimates from the disturbance observer (see Figure 4d) show improvements in regard to less high frequency noise. However, the RMS error does not decrease significantly as the presence of low frequency disturbances causes positional drift, which is the main source of error in this case.

The results from testing with three different mechanism stiffness settings are summarized in Table 3. For the inverse model method, the effect from the stiffness reduction gives clear improvements in force-sensing accuracy, as the dominant source of error is the quantization of the displacement measurement. Overall, the disturbance observer approach gives much better force-sensing accuracy. However, the effect of stiffness reduction is less clear as a reduction in quantization effects is offset by increased low-frequency drift of the needle position, which then becomes a dominant source of error.

Table 3. Summary of force estimation results.

Case	k_x (N/m)	RMS force estimate error: $\sigma(f - \hat{f})$ (μN)	
		Inverse model	Disturbance observer
A	656.7	289	69
B	329.4	196	82
C	104.4	133	63

6. Conclusions

This study investigated a sensorless force estimation method for a flexure-based micro-injection mechanism. A model-based disturbance estimation scheme with optimal state observer was introduced to obtain real-time estimates of interaction forces. The main limitations of the approach were related to model inaccuracy, noise, and disturbance effects. In particular, this included noise and quantization error from the position sensor. Since the success of this approach relied on an accurate system model, a practical scheme was introduced to identify coefficients for a single-degree-of-freedom mechanism model. The findings reported here indicate there was good potential for using the technique in automated cell injection. Further work is needed to examine the utilization of the force-estimates within real-time control of the micropipette motion during cell micro-injection procedures.

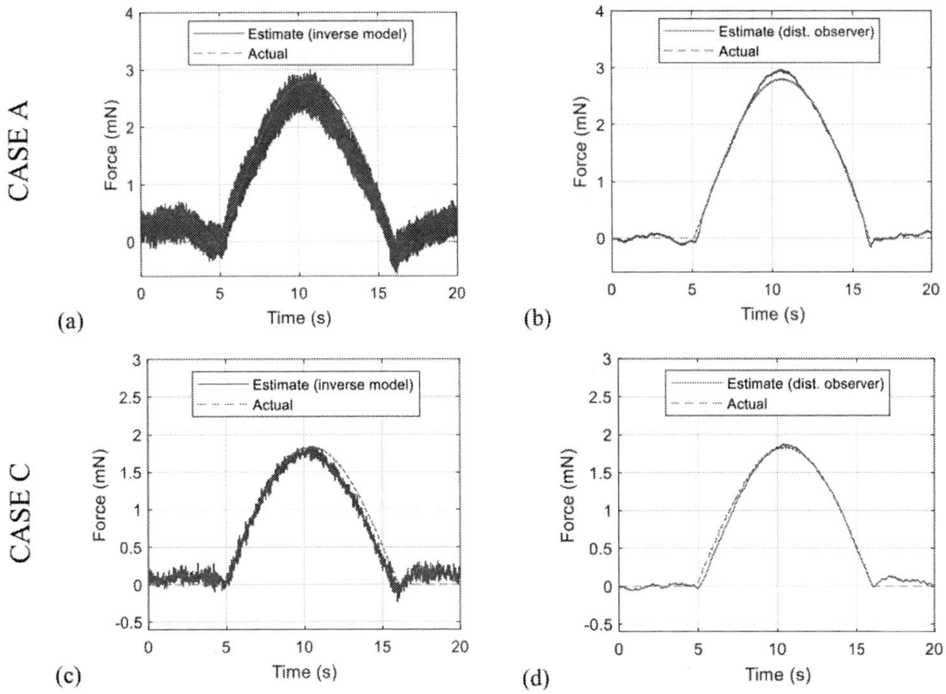

Figure 4. Comparisons of the results of the contact force estimation obtained from the inverse model and disturbance observer

References

[1] Landuyt L V Vos A D Joris H Verheyen G Devroey P and Steireghem A V 2005 *Fertility and Sterility* **83**(5) 1397-1403

[2] Huang H B Sun D Mills J K and Cheng S H 2009 *IEEE Trans. on Robotics* **25**(3) 727-737

[3] Wei Y and Xu Q 2019 *IEEE Trans. on Automation Science and Engineering*, **16**(2) 931-945

[4] Hajiyavand M Saadat M Abena A Sadek F and Sun X 2019 *Micromachines* **10**(4) 1-16

[5] Wei Y and Xu Q 2017 *IEEE Sensors Journal* **17**(13) 3975-3982

[6] Sun Y Wan K T Roberts K P Bischof J C and Nelson B J 2003 *IEEE Trans. Nanobioscience* **2**(4) 279-286

[7] Wei Y and Xu Q 2015 *Sensors and Actuator A: Physical* **234** 359-374

[8] Puangmali P Althoefer K Seneviratne L D Murphy D and Dasgupta P 2008 *IEEE Sensors Journal* **8**(4) 371-381

[9] Kuresangsai P and Cole M O T 2019 *Mechanism and Machine Theory* **132** 80-92

[10] Compos L M B C and Marta A C 2014 *Int. Journal of Mechanical Sciences* **79** 95-104

IOP Conf. Series: Materials Science and Engineering 1137 (2021) 012059 doi:10.1088/1757-899X/1137/1/012059

Orbital control for cube satellites

Chayada Thidrasamee[1], Thanapat Wanichanon[1,*]

[1]Faculty of Engineering, Mahidol University, Nakhon Pathom, 73170, Thailand

* Corresponding Author: thanapat.wan@mahidol.edu.

Abstract
Today's space technology is continuously developing so that sending satellites into space is not difficult anymore. Besides, satellites have emerged as powerful platforms for telecommunication, survey, and traffic information. For example, many countries have started to develop their satellites. Due to its compact size and uncomplicated system, a cube satellite is suitable as a starting point for the study and development of space technology. This research investigates the stability of a cube satellite with a small size, weighing less than 1 ton and orbiting in a Low-Earth Orbit (LEO) by using the Fundamental Equations of Constrained Motion (FECM) and Sliding-Mode Controller (SMC). The proposed controller guarantees motions of the satellite under the constraint that the satellite has to orbit in a circular trajectory in a Low-Earth Orbit (LEO) and able to handle unexpected changes. The MATLAB program was used to gather data from the numerical simulation. The result from the motion of the analyzing process is rather a remarkable outcome. It may help us to understand the satellite's movement in a circular orbit with less and adjustable errors. This finding, while preliminary, suggests that we can cope with changes very well particularly the external high-frequency noise. Also, the values of the acceptable errors can be determined by adjusting the control parameters in the proposed controller.

Keywords: Orbital control, Cube satellite, Sliding mode controller, Fundamental equations of constrained motion.

1. Introduction

In the history of development economics, Thailand has developed in many types of technologies, such as agricultural technology, transportation technology, medical technology, and exploration technology. Satellite is a dominant feature of exploration technology and it is of interest because it is widely used in finding natural resources, weather forecast, and explore the terrain, etc. In the same vein, space technology is one of the interesting technologies, and it can be developed continuously without the end. This current study analyses the impact of exploration technology by using a cube satellite. Recent developments in new exploration satellite have heightened the need for THEOS. The THEOS is a cube exploration satellite that has been created with cooperation between the governments of Thailand and France. This study also includes parameters of the THEOS satellite and the simulation.

The methodology developed in this work uses the sliding-mode controller together with the fundamental equations of constrained motion to calculate the satellite's orbit. Few studies have investigated the use of the sliding-mode controller to control the motion of the satellite because it takes care of an uncertainty part of the system. Concerning space, there are many uncontrollable things and satellites cannot be tuned their hardware. Consequently, a robust control system is recommended for these conditions. The system has to be managed effectively by using the fundamental equations of constrained motion in constraint following part for dealing with the ideal system and using the sliding-mode control for dealing with potential system instability.

2. Nominal system

2.1. The free body diagram of a cube satellite

The motion and moment of inertia chosen refer to the model of THEOS satellite [1]. The THEOS satellite is a cube satellite that meets the interest of this research. It can be given in Figure 1 and the property of the satellite is illustrated in Table 1.

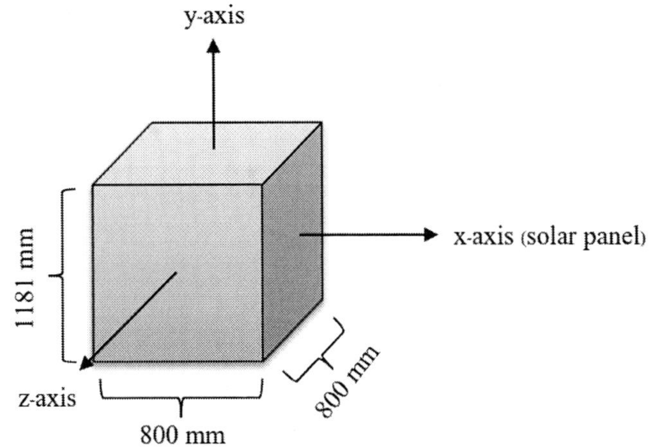

Figure 1. Model of the THEOS satellite [1]

Table 1. Property of the satellite

Property of satellite	Value
Weight	120 kg
size	800 x 800 x 1181mm

2.2 Equations of motion of the satellite

The equation used to describe the satellite's motion [5],[18] is set out in Equation (1).

$$a(t)=(M)^{-1}Q=-\frac{GM_{\oplus}}{r^3}[X \quad Y \quad Z]^T. \tag{1}$$

The acceleration's matrix shown above, is used to describe the motion of satellites, where

G is the gravitational constant, which equals to $6.67408 \text{ x } 10^{-11} \text{Nm}^2 / \text{kg}^2$

$M_{\oplus}$ is the mass of the earth, which is 5.972×10^{24} kg

r is the distance between the Earth's center and the center of the satellite that is described by

$r=\sqrt{X^2+Y^2+Z^2}$

$[X \quad Y \quad Z]^T$ is the position vector of the center of mass of the satellite in the ECI (Earth-Centred Inertial) (see Figure 2).

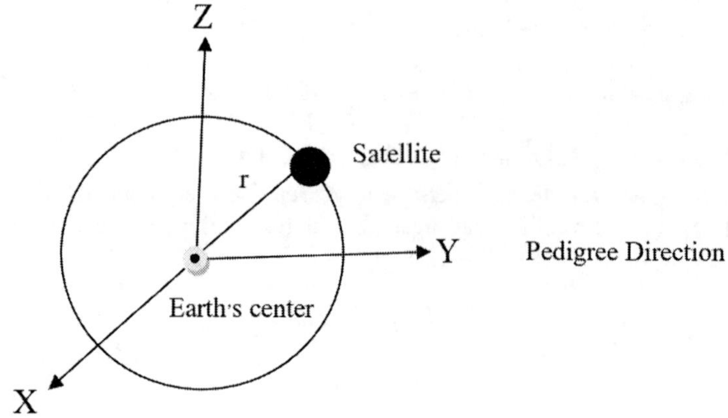

Figure 2. Orbit frame

3. Constraint equation

This section focuses on bringing the conditional equations and applying the fundamental equations of constrained motion to control the system. However, this controller is suitable for the ideal system without uncertainty.

3.1. Condition's equation

The following are the conditions that govern the system to move under the circular orbit's condition

$$2X = Z \text{ and } Y^2 + Z^2 = \rho_0^{\,2}. \tag{2}$$

Equation (2) is the circular orbit condition. The orbit should be projected on YZ frame as circular orbit with a constant radius ρ_0 when $[X \quad Y \quad Z]^T$ is in the ECI frame. Besides, X-axis of the satellite points towards the center of the Earth at all times.
To prevent the default of choosing the right initial condition assumption, Trajectory Stabilization [14], [21] is used in this paper. Thus, Equation (2) must be rewritten in the form of

$$\alpha\ddot{\phi} + \beta\dot{\phi} + \gamma\phi = 0. \tag{3}$$

This equation guarantees that $\phi \to 0$ when $t \to \infty$. Rearranging the Equation (3), then we get

$$\begin{bmatrix} 2\alpha_1 & 0 & -1\alpha_1 \\ 0 & Y\alpha_2 & Z\alpha_2 \end{bmatrix} \begin{bmatrix} \ddot{X} \\ \ddot{Y} \\ \ddot{Z} \end{bmatrix} = \begin{bmatrix} -\beta_1(2\dot{X} - \dot{Z}) - \gamma_1(2X - Z) \\ -\beta_2(Y\dot{Y} + Z\dot{Z}) - \gamma_2(Y^2 + Z^2 - \rho_0^{\,2}) - \alpha_2(\dot{Y}^2 + \dot{Z}^2) \end{bmatrix}. \tag{4}$$

3.2. Fundamental equations of constrained motion

IOP Conf. Series: Materials Science and Engineering 1137 (2021) 012059 doi:10.1088/1757-899X/1137/1/012059

The first controller, is used to control the system to move according to the specified conditions, is the fundamental equation of the constrained motion [2-4]. The first step is to model the satellite's system by using Newton's second law of motion as provided in Equation (5)

$$M(q,t)\ddot{q}=Q(q,\dot{q},t) \tag{5}$$

where

q is the generalized coordinate
t is time (second)
M is the mass matrix
Q is the given force vector.

The unconstrained acceleration of the system can be obtained by pre-multiplying of Equation (5) with M^{-1},

$$a:=M^{-1}(q,t)Q(q,\dot{q},t). \tag{6}$$

Rearranging the Equation (4) in the form of a linear function of acceleration [7]

$$A(q,\dot{q},t)\ddot{q}=b(q,\dot{q},t), \tag{7}$$

where A is according to the Equation (7) under the generalized displacement matrix 3 vector

$$q(t)=[X \quad Y \quad Z]^T, \tag{8}$$

then, we get

$$A=\begin{bmatrix}A_1\\A_2\end{bmatrix}=\begin{bmatrix}2\alpha_1 & 0 & -l\alpha_1\\ 0 & Y\alpha_2 & Z\alpha_2\end{bmatrix} \tag{9}$$

$$b=\begin{bmatrix}b_1\\b_2\end{bmatrix}=\begin{bmatrix}-\beta_1(2\dot{X}-\dot{Z})-\gamma_1(2X-Z)\\ -\beta_2(Y\dot{Y}+Z\dot{Z})-\gamma_2(Y^2+Z^2-\rho_0{}^2)-\alpha_2(\dot{Y}^2+\dot{Z}^2)\end{bmatrix} \tag{10}$$

by following the condition Equation (2).

Taking the A_1, A_2 and b_1, b_2 from the previous step to form the equation of motion and get the control force of the system that makes the satellite move following the conditions [15-16]. The formula of the control is set out in Equation (11)

$$Q^c(t)=Q^c(q(t),\dot{q}(t),t)=A^T(AM^{-1}A^T)^+(b-Aa) \tag{11}$$

where

T is the transpose of vector or matrix
+ is the Moore-Penrose inverse of a matrix.

IOP Conf. Series: Materials Science and Engineering 1137 (2021) 012059 doi:10.1088/1757-899X/1137/1/012059

4. Uncertainty control

In this section, the instability of the system is chosen. The effective controller should be able to accommodate the uncertainty of the system. Therefore, a sliding-mode controller [10-13] is selected. It is a robust controller that can compensate for instabilities and disturbances of the system as well. It is suitable for systems with highly non-linear and complicated.

Tracking- error signal of the system can be defined as follows:

$$e(t) = q_c(t) - q(t) \tag{12}$$

where

$q_c(t)$ is actual system displacement

$q(t)$ is nominal system displacement.

Differentiating Equation (12) twice, we then get

$$\ddot{e} = \ddot{q}_c - \ddot{q} \tag{13}$$

in which [10]

$$\ddot{q}_c = a_a + M_a^{-1}Q^c(t) + M_a^{-1}M\ddot{u}_c \tag{14}$$

and

$$\ddot{q} = M^{-1}Q + M^{-1}A^T(AM^{-1}A^T)^+(b - Aa) := a + M^{-1}Q^c(t). \tag{15}$$

Substituting Equation (14) and Equation (15) in Equation (13), we obtain

$$\ddot{e} = \ddot{q}_c - \ddot{q} = [a_a(q_c, \dot{q}_c, t) - a(q, \dot{q}, t)] + [M_a^{-1}(q_c, t) - M^{-1}(q, t)]Q^c(t) + M_a^{-1}M\ddot{u}_c$$

$$:= \delta\ddot{q}_c + M_a^{-1}M\ddot{u}_c = \delta\ddot{q} + [I - (I - M_a^{-1}M)]\ddot{u}_c := \delta\ddot{q} + \ddot{u}_c - \bar{M}\ddot{u}_c$$

where

$$\delta\ddot{q}(q, \dot{q}, q_c, \dot{q}_c, t) = [a_a(q_c, \dot{q}_c, t) - a(q, \dot{q}.t)] + [M_a^{-1}(q_c, t) - M^{-1}(q, t)]Q^c(t)$$

and

$$\bar{M} = I - M_a^{-1}(q_c, t)M(q, t) = I - [M(q_c, t) + \delta M(q_c, t)]^{-1}M(q, t)$$

$$= I - [M^{-1}(q, t)M(q_c, t) + M^{-1}(q, t)\delta M(q, t)]^{-1}.$$

There are uncertainties of the system being considered [20]. Mass and force are uncertain parameters that may change in time, especially, in unpredictable environments like space. Therefore, in the control system, these uncertainties must be anticipated to accommodate for future unexpected situations. Equation (16) refers to the equation of motion that adding nominal control force $Q^c(t)$ with additional control force for uncertainty $Q^u(t)$. Noting that if the system is certain, Equation (12) equals to 0.

$$M_a(q_c, t)\ddot{q}_c = Q_a(q_c, \dot{q}_c, t) + Q^c(t) + Q^u(t) \tag{16}$$

where

M_a is the actual mass, $M_a := M + \delta M > 0$

Q_a is the actual control force $Q_a := Q + \delta Q$.

The equation also contains terms of the constraint following control force [19]

$$Q^c(t) = A^T(AM^{-1}A^T)^+(b - Aa). \tag{17}$$

The discussion below is the control force that is created from the sliding-mode controller to compensate for the system's uncertainties.

$$Q^u := M\ddot{u}_c \tag{18}$$

in which

$$\ddot{u}_c = -[k\dot{e} + \beta(t) f(s)] \tag{19}$$

where

$$\|e(t)\| \le \frac{\Sigma}{2K} \text{ , } \|\dot{e}(t)\| \le \Sigma \text{, and } \Sigma = 2\varepsilon \|g^{-1}(1)\| > L_\varepsilon(t) \text{ as } t \to \infty .$$

The function $f_i(s)$ is any arbitrarily monotonically increasing [19], a continuous, odd function of s_i on the interval $(-\infty, +\infty)$ that satisfies

$$f_i(s) = g(s_i / \varepsilon), \; i = 1,2,...,n \tag{20}$$

where

$$\|f(s)\| = \|g(s/\varepsilon)\| \ge \frac{\gamma(t) + k\|\bar{M}(t)\|\|\dot{e}(t)\|}{\gamma(t) + \beta_0} . \tag{21}$$

If is outside the surface $\Omega_\varepsilon(t)$, which is defined as the surface of the n-dimensional cube around the point s=0, Equation (22) will always be satisfied when $\|f(s)\| \ge 1$ following the control law [13]. It will cause $s(t) \to \Omega_\varepsilon$ as $t \to \infty$. The value of $\beta(t)$ should be satisfied with Equation (22)

$$\beta(t) \ge \frac{n[\gamma(t) + \beta_0]}{\alpha_0} > 0 \tag{22}$$

in which

$$\beta(t) > k\|\bar{M}(t)\|\|\dot{e}(t)\| \text{ , } 0 < \alpha_0 < 1 - n\sigma\|\bar{M}(t)\| \text{ and } \gamma \le \sigma \le 1. \tag{23}$$

So, finally, we get the equation of motion of the controlled system as

$$M_a(q_c, t)\ddot{q}_c = Q_a(q_c, \dot{q}_c, t) + Q^c(t) - M[k\dot{e}(t) + \sigma\beta(t)f(s)]. \tag{24}$$

5. Results of numerical simulation

This section involves the results of the numerical simulation, which they are simulated to verify the proposed control approach. The simulation is calculated based on the THEOS satellite parameters, as provided in Table 1. It is interesting to note that the satellite was found to cause the orbiting in a circle around the Earth under the conditions of J_2 perturbation with an inclination 98.7 degrees, gravity of the Earth $G = 6.67408 \times 10^{-11} Nm^2 / kg^2$, Earth-mass $M = 5.9722 \times 10^{24}$ kg, and T (period) = 6084 seconds. In the same way, the calculation is simulated over 3 periods and the longitude of the ascending node $\Omega_L = 30°$,

$r(0) = p_0 = 7.26 \times 10^6 \mathrm{m}, \; v = \sqrt{\frac{GM}{r(0)}} = 7.41 \times 10^3 \mathrm{m/s},$

where

r is the distance from the center of the Earth to the center of the satellite
v is the satellite speed.

The initial condition used must be followed by all the constraints. Hence, the initial of the satellite [22] are

$X(0) = 2328.9694 \times 10^3 \mathrm{m}$
$Y(0) = 5995.21600 \times 10^3 \mathrm{m}$
$Z(0) = 1719.97894 \times 10^3 \mathrm{m}$
$\dot{X}(0) = 2.911101130 \mathrm{m/s}$
$\dot{Y}(0) = -0.98164053 \mathrm{m/s}$
$\dot{Z}(0) = -7.090400220 \mathrm{m/s}.$

In the actual system, there is an uncertainty that is simulated into the system, which is

$$\delta M = 10\% \text{ of its original masses and } \delta Q = \sin(100t) \text{ of its nominal parameters.} \quad (25)$$

In the sliding-mode controller the parameters are assigned as follows:

$$k = 0.1, \; \alpha = 1, \; \beta_0 = 0.1 \text{ , and } f(s) = (s/\varepsilon)^3. \quad (26)$$

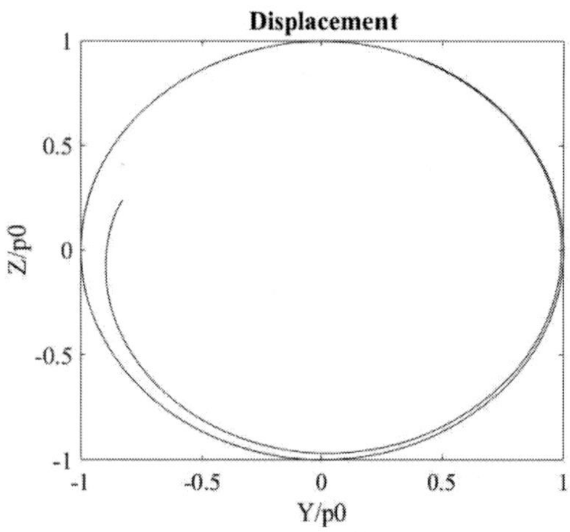

Figure 3. Satellite movement in YZ frame

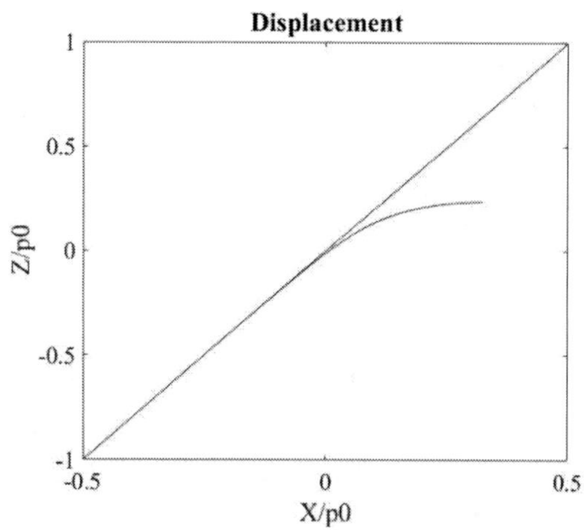

Figure 4. Satellite movement in XZ frame

The simulation results of the controller when trying to control the actual uncertain system are shown in Figure 3 and Figure 4. The satellite moves smoothly into the conditions. The diameter of the orbital radius is equal to p_0 that makes the proportion become 1, The speed of entering conditions can be increased by adjusted the value of α, γ, and β from Equations (9) and (10). In this simulation, these values are $\alpha_1 = 10, \gamma_1 = 15, \beta_1 = 0.1$, and $\alpha_2 = 1, \gamma_2 = 40, \beta_2 = 0.01$.

IOP Conf. Series: Materials Science and Engineering 1137 (2021) 012059 doi:10.1088/1757-899X/1137/1/012059

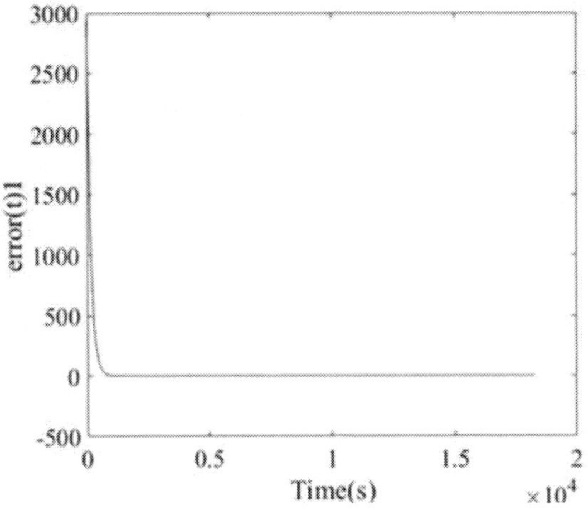

Figure 5. error $e_1(t)$ of the system

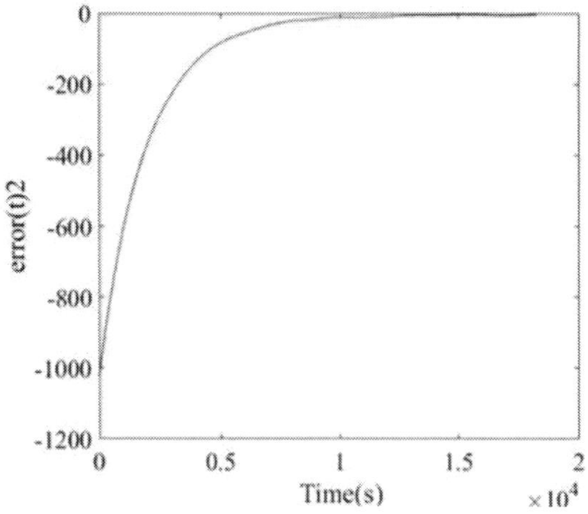

Figure 6. error $e_2(t)$ of the system

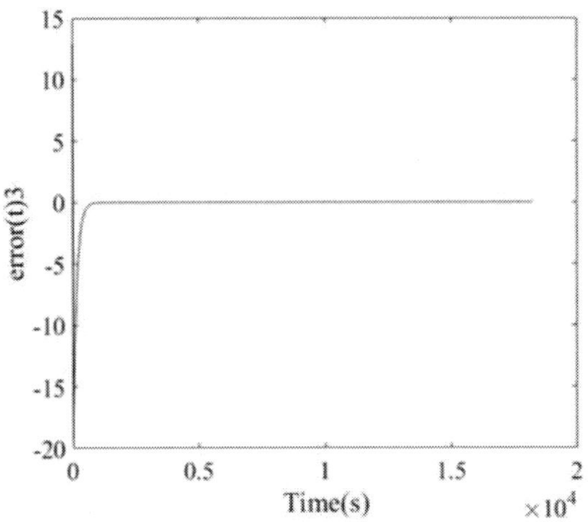

Figure 7. error $e_3(t)$ of the system

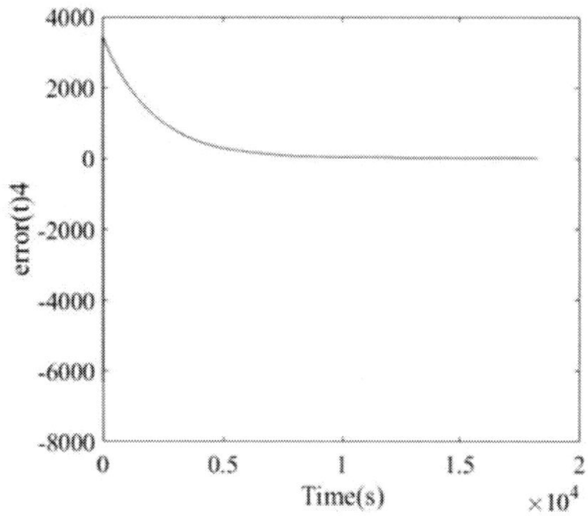

Figure 8. error $e_4(t)$ of the system

The simulation errors were measured by using constraint conditions, as set out in Equation (2) denoting $e_1(t) = 2X - Z$, $e_2(t) = \sqrt{Y^2 + Z^2} - \rho_0^2$ $e_3(t) = 2\dot{X} - \dot{Z}$, and $e_4(t) = Y\dot{Y} - Z\dot{Z}$. It can be seen that the tolerances of the systems reduced to 0 in a short period of time (see Figure 5, Figure 6, Figure 7, and Figure 8).

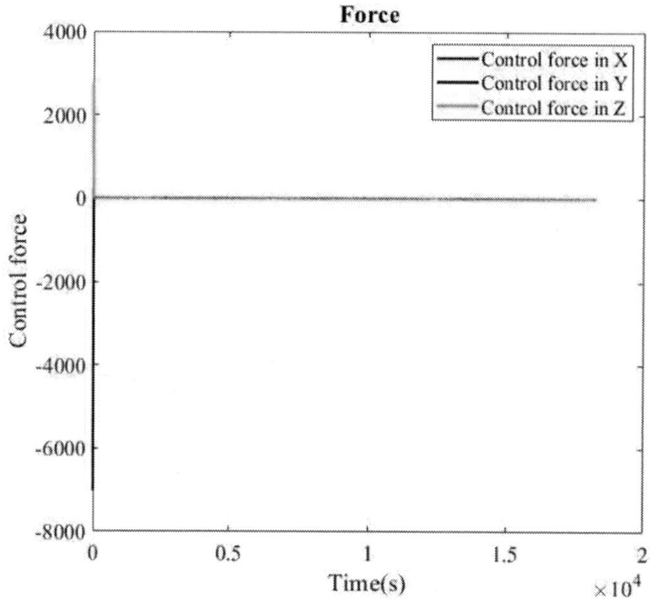

Figure 9. Required control force in XYZ frame

As can be seen in Figure 9, it illustrates the required control forces to control the satellite's movement according to the constraint conditions. These forces are referenced in the ECI frame and they vary with the control parameters used to satisfy the system's initial conditions. By starting with incorrect initial conditions, the control forces are happened to be large. The maximum control value in the X, Y, and Z axes are 1750 N, 6980 N, and 2780 N, respectively. When the satellite moves into their orbit, the control forces are gradually reduced to 0. Therefore, the whole stabilization process took place in only 200 seconds. After that, little effort (under 5 N) from the additional control force is required to handle the external high-frequency noise throughout the movement of the satellite, as represented in Equation (25).

6. Conclusion

This study sets out to develop a complete control system that can be applied in real situations. It also has discussed the reasons for adopting the value of J2 perturbation since it has an effect on the movement of the cube satellite under the condition of real orbit motion. This result of the simulation approach will prove useful in expanding our understanding of how effective the fundamental equations of constrained motion and the sliding-mode controller to control the satellite system. It can not only greatly reduce the errors in tracking the orbit of the system, but the system can also meet the stable point faster by increasing the control parameters and the values of α, γ, and β in the trajectory stabilization's equation. Being limited to the size of the control forces, this study lacks the ability of fast control to the stationary point. The results of this research support the idea that the sliding-mode controller also has the advantage of being able to accommodate instability in the system. Although this current study has set the uncertainty of the nominal values at 10% (considered as rather high values), the controlled system shows that it can handle the uncertainty terms very well. Further experimental investigations are needed to estimate efficient directional and rotational system to increase the likelihood of being implemented of the system for future practice.

References

[1] *RSDO the p200 platform 2016 USA* (NASA Goddard Space Flight Center)
[2] Udwadia F E 1996 *J. Nonlin. Mech.* **31** 951-958
[3] Udwadia F E 2000 *J. Aerosp. Eng.* **13** 17-22
[4] Udwadia F E 2003 *Proc. Math. Phys. Eng. Sci., Series A: Mathematical and Physical Sciences* **459** 1783-1800
[5] Vallado D A 2001 *Fundamentals of Astrodynamics and Applications 2nd ed* (The Space Technology Library Microcosm: Hawthorne) pp 20-23
[6] Cho H and Yu A 2010 *J. Aerosp. Eng.* **22** 445-455
[7] Cho H Wanichanon T and Udwadia FE 2014 *Proc. the 3rd Joint International Conference on Multibody System Dynamics and the 7th Asian Conference on Multibody Dynamics* (Busan, South Korea)
[8] Cho H and Udwadia F E 2010 *Acta Astronautica* **67** 369-387
[9] Udwadia F E Wanichanon T and Cho H 2014 *J. Guid. Control. Dyn.* **37** 1611-1624
[10] Udwadia F E and Wanichanon T 2014 *J.appl. Mech.* **81**(4) 041020
[11] Utkin V 1997 Transactions on Automatic Control **22**(2) 212-222
[12] Tsypkin Y Z and Releinykh T 1995 *System Avtomati Cheskogo Regulirovaniya (Theory of Switching Control Systems)* (Gostekhizdat: Moscow) p 27-84
[13] Edwards C and Spurgeon S 1999 *Sliding Mode Control: Theory and Applications Taylor and Francis* (London) p 17-74
[14] Khalil H K 2002 *Nonlinear Systems Prentice–Hall* (Upper Saddle River: NJ) p 551-589
[15] Wanichanon T Cho H and Udwadia F 2011 *AIAA SPACE 2011 Conference & Exposition* 7210
[16] Cho H Wanichanon T and Udwadia F E 2018 *Nonlinear Dynamics* **94** 2727-2747
[17] Udwadia F E and Wanichanon T 2014 *Nonlinear Approaches in Engineering Applications 2* (New York: Springer) p 101-136
[18] Cho H Wanichanon T and Udwadia F E 2015 *Proc. the 1st ECCOMAS Thematic Conference on International Conference on Uncertainty Quantification in Computational Sciences and Engineering* (Crete, Greece)
[19] Wanichanon T Udwadia F E and Cho H *COMPDYNJ. Res. Appl. Mech. Eng.* **2** 20-33
[20] Wanichanon T Cho H and Udwadia F E 2015 *Proc. the 23rd ABCM International Congress of Mechanical Engineering* (Rio de Janeiro, Brazil)
[21] Udwadia F E and Wanichanon T 2012 *ASME Int Mech Eng Congress Expo.* **45202** 921-940
[22] Sharaf M A and Selim H H 2014 *NRIAG J. Astron. Geophys.* **2** 134-138

IOP Conf. Series: Materials Science and Engineering 1137 (2021) 012060 doi:10.1088/1757-899X/1137/1/012060

Automatic precision docking for autonomous mobile robot in hospital logistics - case-study: battery charging

S Vongbunyong[1*], K Thamrongaphichartkul, N Worrasittichai, and A Takutruea

[1]Innovation and Advanced Manufacturing Research Group (I AM)
Institute of Field Robotics, King Mongkut's University of Technology Thonburi.
126 Pracha Uthit Rd, Bang Mot, Thung Khru, Bangkok, Thailand 10140

* Corresponding Author: supachai.von@kmutt.ac.th

Abstract
Autonomous Mobile Robots (AMRs) plays a vital role in various logistic applications, especially in the health industry. Precise localization is required for automated navigation of the robot throughout the operations. Docking is one of the key processes that the robot moves and physically attaches to the station. Higher precision in position and orientation is required for this process to ensure the secured connection. This article presents an implementation of precise docking maker technique for localization of the robot. The robot is equipped with affordable Lidar sensor while a geometrical marker is on the docking station. The proposed method should help the robot achieve better than ±20 mm error and 0.05rad error in position and orientation.

Keywords: Autonomous Mobile Robots, Localization, Automatic Docking, Precision Marker, Hospital Logistics

1. Introduction - AMRs in hospital logistics during COVID-19

Logistics management in hospital is involved with effective and efficient processes to transport various items, e.g. food, medicine, medical devices, clothes, trash, and other items in hospitals [1]. In general, these items are manually transported by health workers. Some safety and security issues are involved, especially for restricted items, e.g., controlled medicine, infected items. Robotic technology can be concerned one of the possible solutions to improve the transportation process. Automated Guidance Vehicles (AGV) and Autonomous Mobile Robot (AMR) are concerned the possible solutions for automating transportation system. They convey items between specific locations automatically. They are widely used in industry for ages and recently applied in other working environments, e.g., hospitals.

During pandemic situation, especially COVID-19, avoiding physical contact between patients and health workers is one of the major practices to prevent infection. Many AGVs and AMRs are applied to help keeping distance between them in many activities, e.g., food, medical supplies, medicine delivery, tele-medicine, remoted symptom diagnosis, disinfection, general communication, etc. [2]. Apart from the level of accuracy that these robots need to meet, the contamination control of these machines remains a major challenge.

In general, "*zone*" is allocated precisely in COVID-19 patients wards. (a) "*Dirty zone*" refers to the area having chance of contamination where the patients are situated; while (b) "*Clean zone*" mostly refers to other working areas of the hospitals. The *dirty zone* includes the negative pressure patients' room and the area nearby the entrance of room. The health workers normally need to put on personal protective equipment (PPE) suits before entering the room. The changeover of the equipment must be done outside the room in the *dirty zone* area before crossing to the *clean zone*. This is cumbersome for the health workers and reduces the efficiency of their operations.

Improving this activity, AMRs can be assigned to work in the dirty zone to reduce the needs of the health workers to enter those areas. AMRs must stay within the dirty zone at all times to prevent the contamination in the clean zone. Therefore, monitoring, control, and services of AMRs, must be able to achieve on-line.

In this research, an AMR system, named "CARVER" [3], is developed to transport items in hospitals automatically; and in this case, it is used in the dirty zone of COVID ward. Remote monitoring and control is enabled via an *Internet-of-Things* (IoT) platform [4]. However, physical operations such as battery charging and items transfer are still required to complete the tasks. Since the health workers cannot get into the dirty zone, those operation must be done automatically. In this case, the automatic docking ability needs to be developed as it will be used in many applications. Automatic docking for battery charging will be used as a case-study in this article.

2. Literature review

2.1 Applications of AGV and AMR in hospitals

AGV and AMR are used for transportation of items in various environment, e.g., hospital, office building, residential area, factory, etc. The AMRs can navigate through space by using data from various sensors to sense the internal and external factors that affect the motion; for example [5] combined detection and control architecture for detecting payload containers with horizontally mounted 2D laser scanner and a 3D camera for transport heavy payload. In hospitals, many systems are researched and developed. [6] provided technical specifications of an AGV designed for the healthcare facility. The simulation was used to evaluate the system from technical and economic points of view. The analysis on investment was done and the results showed that, currently, the AGV technology is economically feasible only for large scale facilities. [7] developed a path finder of AGV for transportation of material in hospital environments. On-site tests were done in a local hospital and the results confirmed that the robot can move along with the global path and reach the goal without colliding with static and moving objects.

2.2 Docking for autonomous mobile robot

AMRs and AGVs should be able to work automatically throughout the working process, such as scheduled tasks, power recharging, localization, path planning, etc. The docking process is one of the important processes that allow fully automatic operations.

To date, several studies have investigated the docking system by combining Lidars and various types of sensors, markers, and mechanisms to find position and orientation for the robot and the station for precisely docking. [8] proposed automatic docking system for recharging a surveillance robot. This docking method is based on self-localization of the robot and the infrared detectors of the docking station. The structure of the robots and docking station were designed based on the arc-shaped docking interface. [9] proposed a system combining the use of QR codes and infrared distance sensors. The QR code is used as a landmark therefore the robot can perceive its position to the dock. The experiments were done from different starting points and showed the robustness of the proposed solution. [10] proposed the use of a portable effector docking mechanism for path planning and localization of the robot. The portable effector was placed on the floor for correcting the alignment at the time of docking. In this research, the docking process is developed based on Lidars and a precision marker on the station to avoid further modification of the hospital floor.

IOP Conf. Series: Materials Science and Engineering 1137 (2021) 012060 doi:10.1088/1757-899X/1137/1/012060

3. Methodology

3.1 Concept of automatic docking

Having a physical connection to particular working stations, the AMR needs to be able to move to a specific location precisely. This current study employed a "*battery recharging station*" as a case study. The AMR needs to automatically dock at the station within ±20mm error to achieve the task. According to the hospital environment, extra installation, such as magnetic tape, is restricted in most cases due to many reasons; e.g., modification of the premise is prohibited; access for service and maintenance is limited, etc. Therefore, the *precision marker* located on the station is used in this research (Figure 1).

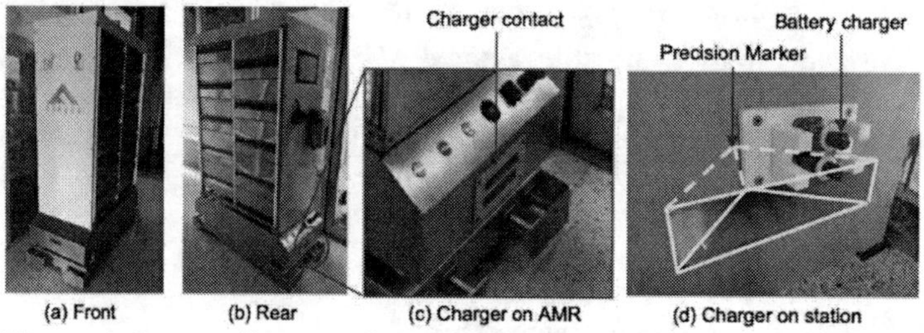

Figure 1. System diagram of CARVER platform with Docking Section

The system architecture of AMR "CARVER" with IoT platform is shown in the diagram in Figure 2 where the further detail of the system is described in [4]. In this article, the extension from the system in [4] is the "Docking section", consisting of a *docking module* and a *charging station*.

The AMR works in two navigation modes: (a) *normal operation* and (b) *docking operation*. The navigation scheme will be switched between these two navigation modes. To perform docking, the AMR will automatically navigate to the docking zone with normal operation scheme; and then switch to the docking operation scheme. The *normal operation* allows a larger position error which is in the range of ±80mm, while *docking operation* limits the error within ±20mm. An RPLidar sensor [11], a low-cost Lidar sensor for indoor localization, is used to detect the marker with a specific shape. The marker is simply an angle with two connected straight lines. The Lidar can detect data points and the lines will be recognized. As a result, the position of the docking station can be located relative to the marker position.

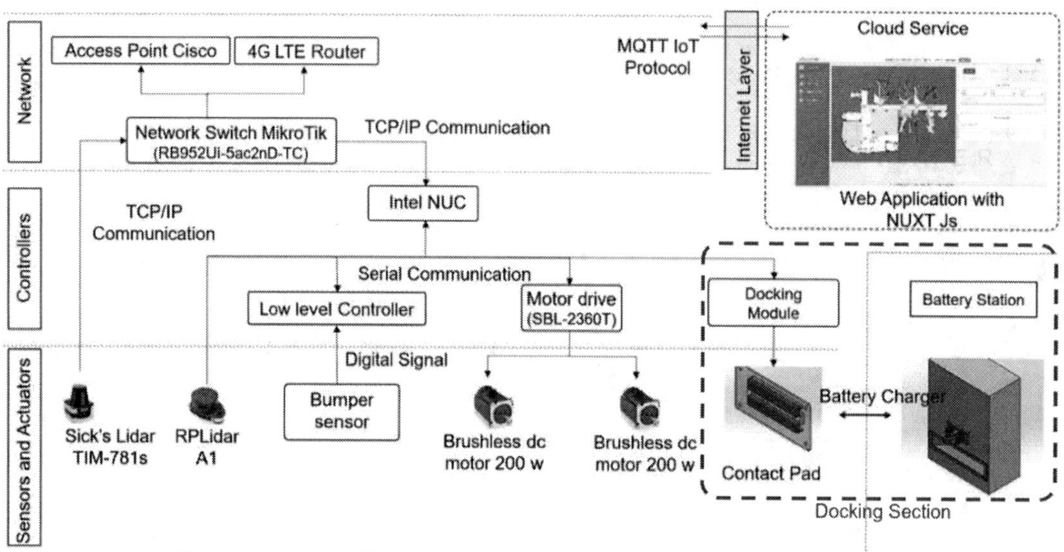

Figure 2. System diagram of CARVER platform with docking Section

3.2 Algorithm for localization of the marker

The algorithm of the docking process is operated as a state machine running in the Robot Operation System (ROS). Several ROS packages [12] are used in the docking process and the system diagram is set out in

Figure **3**a. (a) "*RPLidar*" [11] connects the sensor and convert to *sensor_msgs/LaserScan* message format. (b) "*Laser Line Extraction*" [13] extracts the line from *sensor_msgs/LaserScan* message. The output of this package is to provide two endpoints of the line. (c) "*Localization Marker*" uses the output from *Laser Line Extraction* to find the lines which are used to identify the location of the marker according to Equations (1-3). (d) "*Base Controller*" controls the linear velocity and angular velocity of the robot. (e) *"State Machine"* is the algorithm to dock at the charger station. The docking procedure is run based on the pseudocode in

Figure **3**b. Further explanation can be seen in Section 3.3.

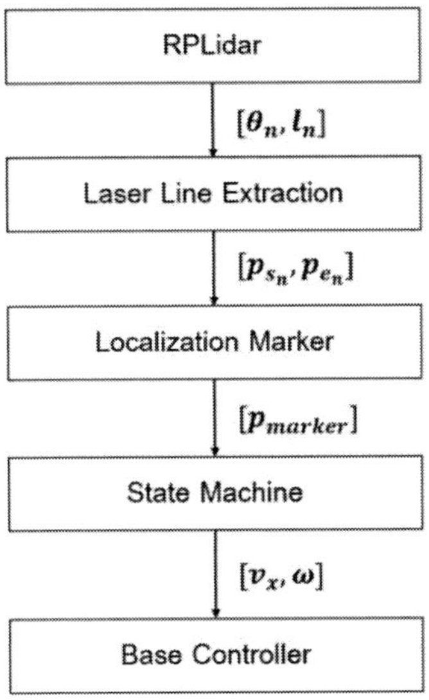

(a)

```
Pseudo Code: State Machine
-----------------------------------------------
ReadCurrentPosition()
findError()
if state == Go_To_Pre_Contact:
    if (absolute_distance > distanc_pre_contact):
        if heading_error > d_e:
            forward and turnright
        elif heading_erro < -d_e:
            forward and turnleft
        else:
            forward
    else:
        state = Aliment_Pre_Contact
elif state == Aliment_Pre_Contact:
    if heading_error > d_e:
        turnright
    elif heading_error < - d_e:
        turnleft
    else:
        state = Go_To_Contact
elif state == Go_To_Contact:
    if (absolute_distance > distanc_contact):
        if heading_error > d_e:
            forward and turnright
        elif heading_error <  -d_e:
            forward and turnleft
        else:
            forward
    else:
        state = End
```

(b)

Figure 3. Algorithm for the docking procedure (a) Overall diagram (b) Pseudocode of state machine

The output of the *Laser Line Extraction* package is $\boldsymbol{l_n} = [\boldsymbol{p_{sn}}, \boldsymbol{p_{en}}]$. Where $\boldsymbol{l_n}$ is line extraction; $\boldsymbol{p_{sn}}$ is the start point of line extraction where the position on *xy*-plane is (x_{sn}, y_{sn}); and $\boldsymbol{p_{en}}$ is the endpoint of line extraction where the position on *xy*-plane is (x_{sn}, y_{sn});

The start and endpoints are used to create a linear equation. Two conditions are used for identifying the marker (see Figure 4(c)). The *first condition* is the delta angle ($\Delta\theta_{23}$) between two lines which is between ±5 degree. The *second condition* is the distance (Δd_{23}) between end point ($\boldsymbol{p_{en}}$) and start point ($\boldsymbol{p_{sn+1}}$) which is less than 0.05 mm. If both conditions are satisfied, the intersection between two lines, $(x_{intersect}, y_{intersect})$ can be identified as the position of the maker by using equations (1-3).

$$y_n = m_n x_n + c_n \quad (1)$$

$$x_{intersect} = \frac{(c_{n+1} - c_n)}{(m_{n+1} - m_n)} \quad (2)$$

$$y_{intersect} = m_n x_{intersect} + c_n \quad (3)$$

IOP Conf. Series: Materials Science and Engineering 1137 (2021) 012060 doi:10.1088/1757-899X/1137/1/012060

As can be seen in Figure 4d, the angle of the *marker frame* with *base link frame* (the reference coordinate frame at the base of the robot) can be obtained with the geometric approach as the three following steps:

1. Create a circle equation at the point $p_{intersect}$ with radian r mm;
2. Find intersection point p_{cl} and p_{cr} between the line equation and circle equation; and,
3. Find the angle of marker θ_{marker} by $Atan2(\mathbf{p_{cl}}, \mathbf{p_{cr}})$.

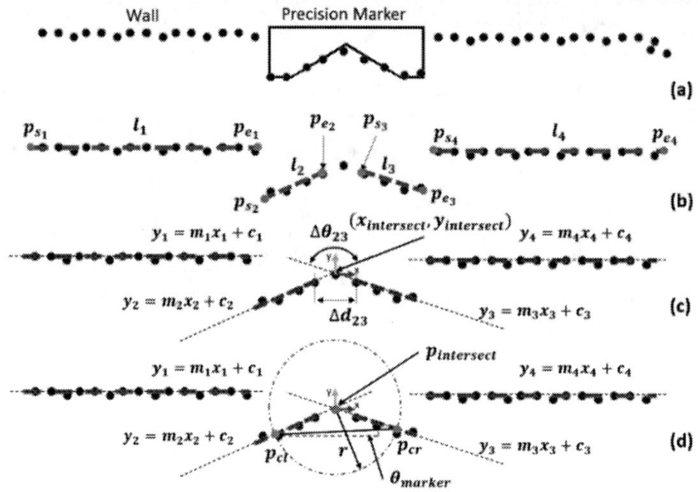

Figure 4. Identification of the markers according to the data from lidar

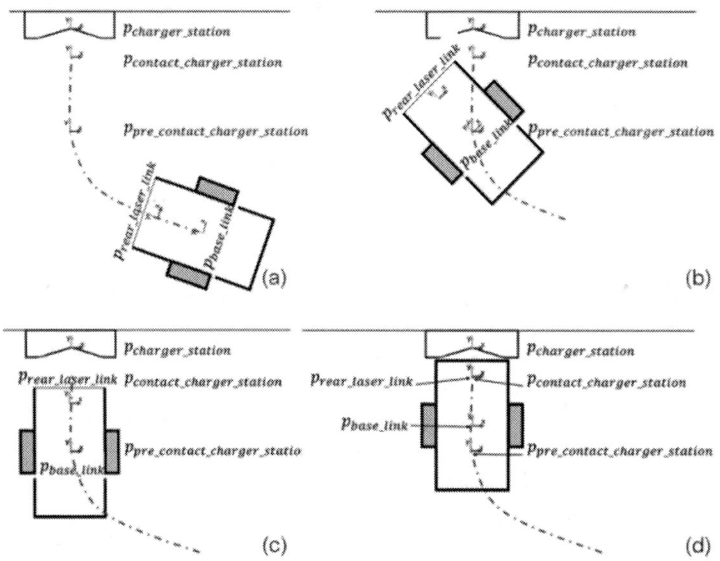

Figure 5. Docking states
(a) Go to pre-contact point (b) Align with pre-contact (c) Go to contact point (d) Final state

3.3 Algorithm for docking

IOP Conf. Series: Materials Science and Engineering 1137 (2021) 012060 doi:10.1088/1757-899X/1137/1/012060

After the marker location $\boldsymbol{p}_{intersect}$ is obtained, this position is hereby called the *position of charger station* ($\boldsymbol{p}_{charger\_station}$). Transformation matrices are applied to find other reference points for docking (see Figure 5), namely, *contact charger station point* ($\boldsymbol{p}_{contact\_charger\_station}$) and *pre-contact charger station point* ($\boldsymbol{p}_{pre\_contact\_charger\_station}$).

The docking process uses $\boldsymbol{p}_{base\_link}$ and $\boldsymbol{p}_{pre\_contact\_charger\_station}$ as references for the positions of robot and charger, respectively. Hysteresis control is used to control the position and velocity of the robot. From the basic docking algorithm provided in
Figure **3**b, the robot operates in five states:

1. "*Initial state*": the robot locates the marker and knows to obtain the distance between them;
2. "*Go to pre-contact point state*": the robot uses the *base link* to move to the pre-contact charger station;
3. "*Align with pre-contact state*": the robot adjusts itself to align with the pre-contact charger station;
4. "*Go to contact point state*": the robot moves to the contact charger station point; and
5. "*Final state*": the robot is at the charger station position and recharges the battery.

4. Implementation and experiment

4.1 Experiment method

The purpose of the experiment is to test the performance of the proposed docking procedure. The test-bed is set up in the lab environment to perform automatic docking at the battery recharging station (see Figure 6a). The experiments were done in 2 cases with 10 times each. *Case-1* represents a basic docking operation since the path is straight in front of the station. *Case-2* represents actual movement when the starting point can be anywhere in the space (see Figure 6b).

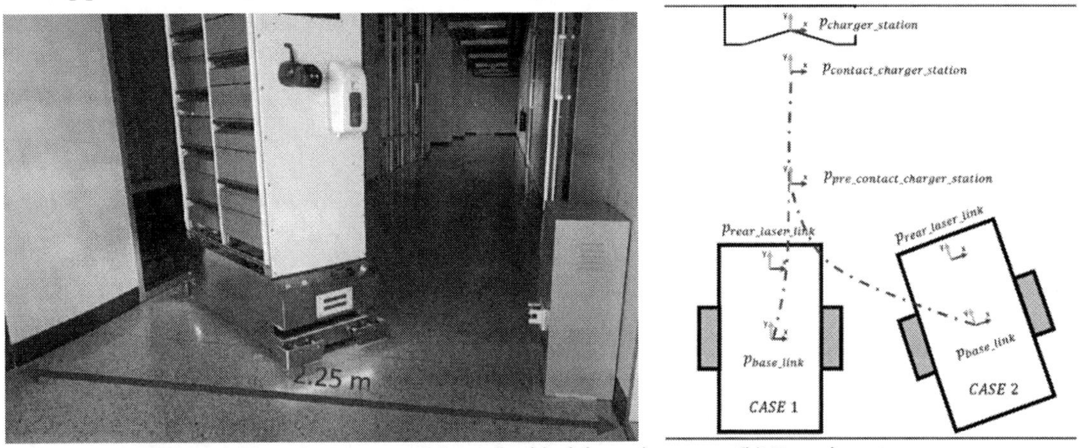

Figure 6. Experiment setup (a) test-bed in lab environment (b) Experiment cases

4.2 Experiment results

The experiment result of the position and orientation of the robot at the final state are summarized in Table 1. There was an insignificant difference between both cases. The average position errors were within 13.0 mm on the x-axis and 10.9 mm on the y-axis, while the orientation is represented by heading where the error was within 0.01 rad. Based on the battery charging application, this proposed docking process was able to accomplish the task with compliant design at the charging contact points which allow this level of position and orientation error.

Table 1. Results of the summary – errors in the position and the orientation at the final state

	Case-1	**Case-2**

(n = 10)	x (mm)	y (mm)	heading (rad)	x (mm)	y (mm)	heading (rad)
Min	0.5	1.4	0.005	2.8	6.6	0.001
Max	26.2	19.1	0.015	17.2	16.7	0.009
Average	13.0	10.9	0.010	9.3	12.8	0.004
S.D.	9.0	6.3	0.003	5.0	3.0	0.002

5. Discussion and conclusion

AMRs have been currently used in many applications. The second aim of this study was to investigate the effects of AMR operation in hospital applications. To enable fully autonomous ability, automatic docking is one of the important functions as it allows the AMRs to physically interact with other items and equipment. Battery recharging is one of the main features to ensure that the AMR can be continuously operated without the aid of human operators. In this article, automatic docking is developed based on the precision marker and RPLidar. A geometrical approach is used to locate and identify the precision marker. The robot can plan the path and navigate to the station accurately. From the experiment results, the position and orientation errors at the final state are in the allowable range and the AMR was able to achieve the recharging task.

However, from observation, there was some deviation of heading direction along the way between the starting point and contact position. More advanced controller and trajectory control will be investigated and developed to increase docking precision in the future. In a few cases that the deviation was over the allowable range, the size of the charger contact is considered to enlarge to prevent the operation failure.

Acknowledgement

This project is partially funded and supported by (a) Project “Mod Boriraksa” FACO: FIBO Against COVID-19, (b) NXPO Office of Higher Education Science Research and Innovation Policy Council as part of the project “Promotion of Robotics for All & Development of Robotics Innovators, Researchers, Engineers, and Startups”, and (c) “Petchra Pra Jom Klao” Research Scholarship from King Mongkut’s University of Technology Thonburi, Thailand.

References

[1] Gunjan M and Pankaj L 2016 *Global Opportunity Analysis and Industry Forecast* 2014-2022
[2] Mišeikis J Caroni P Duchamp P Gasser A Marko R Mišeikienė N Zwilling F de Castelbajac C Eicher L Früh M and Früh H 2020 *IEEE Robot. Autom. Lett.* 5(4) 5339-5346
[3] Vongbunyong S Worrasittichai N Tripathi S P Takutruea A Thamrongaphichartkul K and Prayongrak T 2020 *Int. Jt. Symp. Artif. Intell. Nat. Lang. Process. and Int. Conf. Artif. Intell. Internet Things*
[4] Thamrongaphichartkul K Worrasittichai N Prayongrak T and Vongbunyong S 2020 Proc. *Int. Jt. Symp. Artif. Intell. Nat. Lang. Process. and Int. Conf. Artif. Intell. Internet Things*
[5] Buck S Hanten R Bohlmann K and Zell A 2017 *Proc. - IEEE Int. Conf. Robot. Autom.* 718-723
[6] Pedan M Gregor M and Plinta D 2017 *Procedia Eng.* **192** 665-670
[7] Bacik J Durovsky F Biros M Kyslan K Perdukova D and Padmanaban S 2017 *IEEE Access* **5** 26892-26900
[8] Song G Wang H Zhang J and Meng T 2011 *IEEE Transactions on Consumer Electronics* **57**(2) 428-435
[9] Quilez R Zeeman A Mitton N and Vandaele J 2015 *Proc. EAI Int. Conf. Testbeds Res. Infrastructures Dev. Networks Communities*
[10] Takeuchi E and Tsubouchi T 2006 *Proc. - IEEE Int. Conf. Robot. Autom.* **2006**(May) 3380-3386
[11] GitHub, Inc. 2020 RPLIDAR ROS package, github.com/robopeak/rplidar_ros, accessed on 1 December 2020
[12] Robot Operating System, www.ros.org, accessed on 1 December 2020
[13] Laser Line Extraction, github.com/kam3k/laser_line_extraction

Development of learning materials for automatic control course

S. Phetsong[1,*], and N. Weerayuth[2]

[1]Department of Mechanical Engineering,
Silpakorn University, Nakhon Pathom, Thailand
[2]Department of Mechanical Engineering,
Ubonratchathani University, Ubonratchathani, Thailand

* Corresponding Author: phetsong_s@su.ac.th

Abstract
In the new global economy, automation technologies play an important role in the industry, agriculture, transportation and daily life in the 4.0 era. Generally, studying the theory of automatic control involves the design of the dynamic control system. Students have to understand the system and be able to create mathematical models or models of control systems, both open-loop control system and closed-loop control systems, based on the knowledge of physics, mathematics, and differential equations. They have to use all knowledge to design a control system that provides the desired results by simulating the system with computer software, which is difficult to understand and practice their skills for real implementation. By using the teaching materials that emphasize hands-on practice along with theoretical studies, is very important in creating practical experience and skills. This research has developed learning materials in designing a basic automatic control system by systematically designing and building a controller with the appropriate actuators and sensors. The designing steps are clear and easy to understand include of application of control laws, programming the controller via microcontrollers, and using a Graphic User Interface (GUI) to display the response of the system. The learning materials can compare the results from the different control systems by adjusting the parameters of the dynamics system. Students can experiment with adjusting the parameters of the controller, which makes them understand the automatic control system through actual hands-on operations with the low cost and easy to produce teaching materials.

Keywords: Automatic control, Learning material, Microcontroller, GUI

1. Introduction

In the history of development economics, automatic control technology has been thought of as a key factor in today's industry, agriculture, transportation, as well as daily life in the 4.0 era. Generally, studying the theory of automatic control involves the design of the dynamic control system. Students have to understand the system and be able to create mathematical models or models of control systems, both open-loop control systems and closed-loop control systems, based on the knowledge of physics, mathematics, and differential equations. They have to use all knowledge to design a control system that provides the desired results by simulating the system with computer software, which is difficult to understand and practice their skills for real implementation. The teaching materials that emphasize a hands-on practice along with theoretical studies is fundamental to create practical experience and skills. [1], [2]

In this research, we have developed the learning materials that the students can design and tune the simple automatic control system. [3] They can practice in designing both open-loop and closed-loop control system. [4] The learning materials consist of the hardware system that they can set up parameters and the software GUI for interface computer with the hardware sets and also display the signal outputs on-screen.

2. Background and Theory

There are several types of automatic control system in common use. The simple system that can act as On-Off to the complicated system includes P, PI, PID, and fuzzy. The basic system that is suitable for basic learning is a closed-loop control system and it is associated with the use of a control mechanism such as P, PI, PID. The PID control system considers the output error relative compare with the setpoint (desired output) by using the three relevant terms P, I, and D, which in the control system P, PI, PID will consider the error from the related term differently. [5]

The P-Term shows only the error that the output is different from the setpoint only, according to Equation *1*. The P-Control system will result in the remaining steady-state error (SSE). The system output cannot touch the set point. If the P-Term is increased, the SSE value may be reduced, and it results in an unstable system and output oscillation.

$$P-Term = K_p \cdot e(t) \tag{1}$$

When compared with a PI Control system that has I-Term terms in the system, the system will sum all output error of the system. The value of this term will result in the system to compensate for the error that occurred and achieve the desired output. Consequently, the high magnitude overshoot may occur, and it can also affect the system.

$$I-Term = \int_0^T K_i \cdot e(t)dt \tag{2}$$

Regarding the PID Control system, there is also a D-Term. It indicates the rate of change of the error value (see Equation 3). This can benefit the system, and it has less overshoot than the PI-Control system, or it may not happen.

$$D-Term = K_d \cdot \frac{de(t)}{dt} \tag{3}$$

The D-Term indicates a rapid change in output. Whilst it can cause the noise to the output and can easily lead to instability of the system. That system is formed by the values of all three terms, P, I, and D-Term, as set out in Equation 4. The system model can be seen in Figure 1.

$$output = K_p \cdot e(t) + \int_0^T K_i \cdot e(t)dt + K_d \cdot \frac{de(t)}{dt} \tag{4}$$

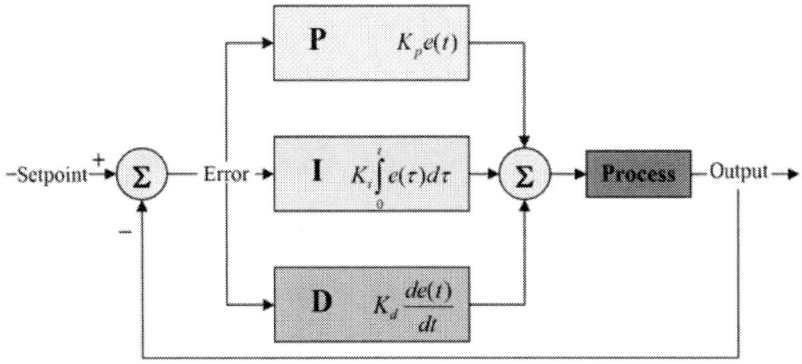

Figure 1. PID Control System model. [6]

In designing a feedback control system, we can use Equation 4 directly in both P, PI, and PID control systems. Also, we can set appropriate Kp, Ki, and Kd values, or set unnecessary parameters to zero. [7]

3. Design and component

In this project, the learning material was developed from a simple device and interfacing software to work together. The Arduino microcontroller is the key to make all components work properly.

Arduino firmware includes all functions required, such as output control, sensor reading, Software interfacing, and control law calculation. The Interfacing Software was used to select the weather of open-loop and closed-loop. It allows users to set up directly.

The developed material was also associated with the use of acrylic as the main component. It accepts the installation of other equipment in the form of tipping arms. It is equipped with a force-generating device at various positions on the arm with a height of 100 mm. The arm length from the pivot point to the motor is 120 mm. The weight position is at 60, 80, and 100 mm on the opposite side. Users can install mass or setup a DC Motor for an adaptive load. With a more complex design, learning material can be set up as a simple feedback control system, or install spring and damper at those positions. The simple method is to put weights at different distances to create a vertical force on arm. Users can also change the setup by installing a motor to be able to vary the force as needed. at the other end, a DC motor is installed to generate a lifting force to keep the system in equilibrium. With complex systems or applied to mechanical vibration systems, additional springs or dampers have installed in the system.

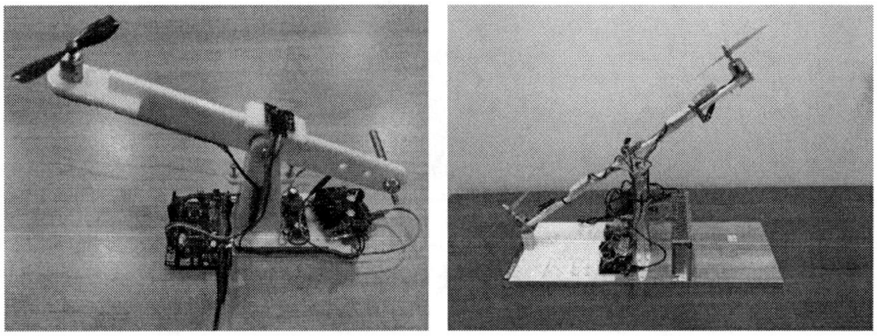

Figure 2. The learning material.

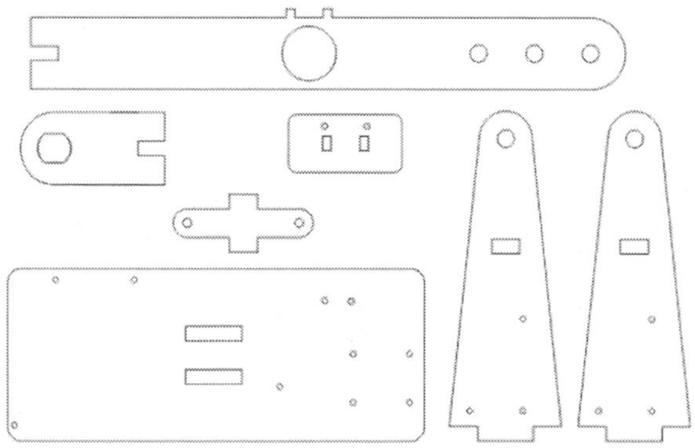

Figure 3. 5mm acrylic model.

Figure 4. Arduino UNO has been used in this learning material.

Figure 5. L298N DC Motor driver.

About the electronic parts, the system has been designed by using an Arduino Uno microcontroller. Likewise, it can connect to the DC motor unit to generate lift force using the L298N driver. The microcontroller sends the PWM signal to control motor direction and speed through the analog. Then, it can write a function from the Arduino platform. There is an accelerometer for measuring the tilt angle. The MPU6050 sensor is installed to read the angle of inclination calculating from the z-axis acceleration force. It is also connected to the microcontroller via the I2C bus. The measurement of the value is ranged between -2 to + 2g. The resolution depends on the range of maximum acceleration setting. Regarding programming in Arduino, it is not necessary to convert to acceleration value, except the direct use of the raw sensor data in the calculation. [8]

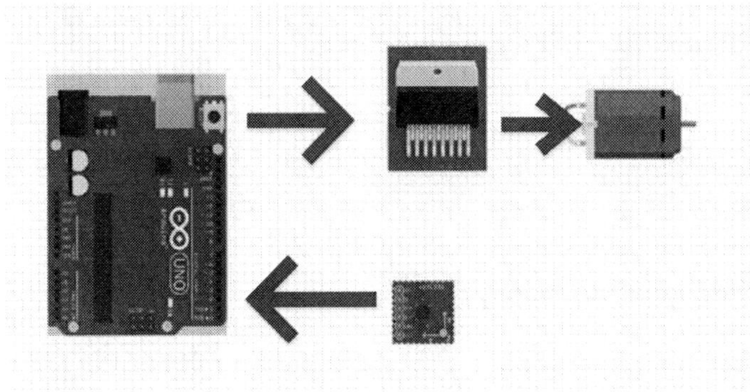

Figure 6. Electronic parts connection.

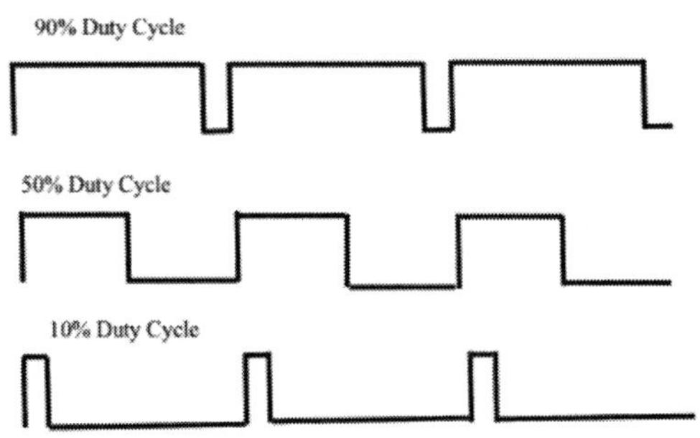

Figure 7. PWM Signal from microcontroller can vary DC Motor speed by varying duty cycle.

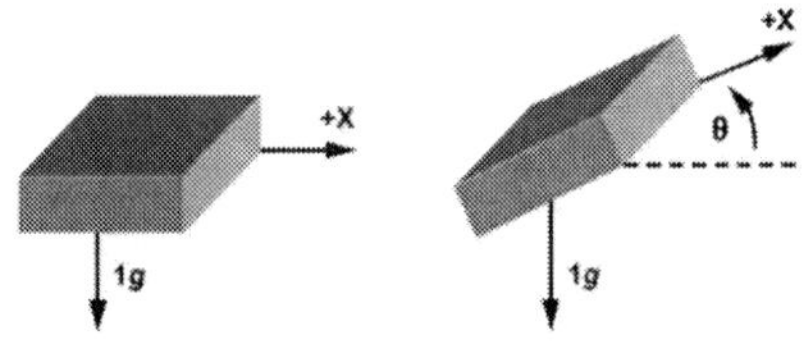

Figure 8. Calculating tilting angle by sum all force in vertical direction. [8]

Another important part of the learning material is user interface software. The software will be used for parameter setup, function control, and display parameter and output value in both value and graph display. [9] The system is connected with MegunoLink software which is an interface development tool for Arduino. The interface display consists of the setting parameter sections on the left. Users can set PID parameters, Start-Stop control, setup the output angle which is usually at 0, but it can be set to others if necessary. In the case that incorporates with another DC motor as vary load, user can setup percent of load output, and change on-the-fly. On the right side, there is the graph area which can be set to display

IOP Conf. Series: Materials Science and Engineering 1137 (2021) 012061 doi:10.1088/1757-899X/1137/1/012061

output or set point as well as the PWM level signal output. To integrate MegunoLink with Arduino, MegunoLink has a set of built-in functions prepared to communicate between PC Software and Arduino. By pressing a button on the Interface screen, it will activate a sub-function in the Arduino to configure various parameters or control the microcontroller according to the specified parameters. On the other side, Arduino sends plotting data via MegunoLink proprietary protocol to display in the graphing area.

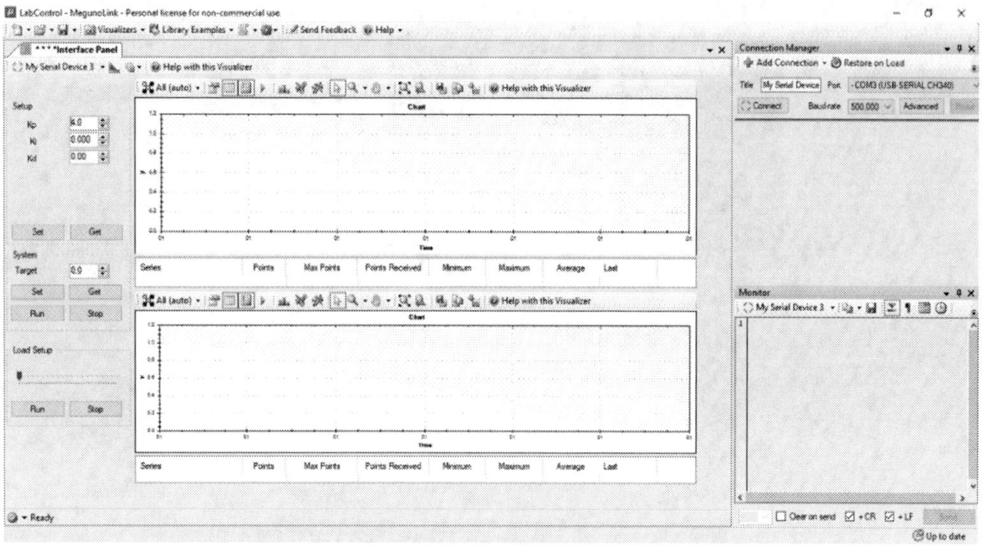

Figure 9. User interface created by MegunoLink.

The control system uses Arduino as a controller. Arduino process the value received from the MPU6050 sensor and send the signal to control motor speed according to the result of the control law which can be the P, PI, or PID control through the setting of the coefficients Kp, Ki, and Kd.

4. Results of the study

Concerning learning material, the system can accept control from the user. Users can set up parameters for the learning material and monitor the result of the system. Users can compare the result from settings by export data to other software or can compare graphs or display in various plot styles. The rising time, overshoot, settling time and steady-state error can be easily noticed from the graph. Users can configure the parameters Ki, Kp, and Kd for the experiment, which affects the output of the system. If there is an improper configuration, the system will show different behaviors, such as oscillation, divergence. When the system is convergence with different control system parameters, the user can notice the changes from the graph that is displayed.

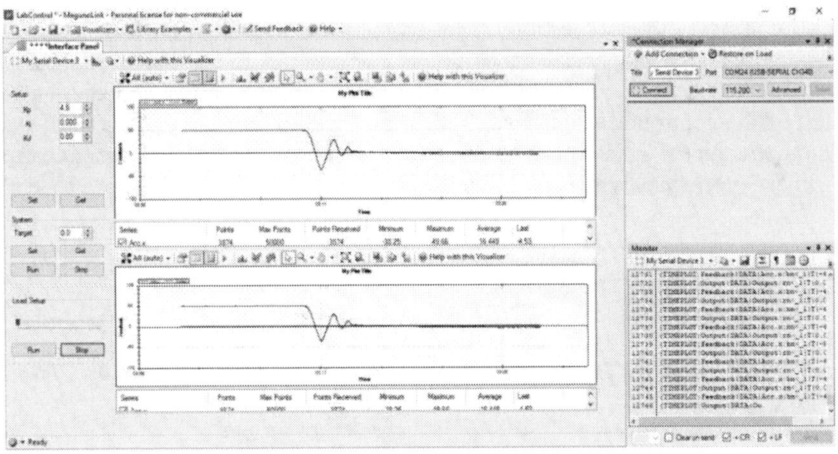

Figure 10. User Interface in MegunoLink when connecting with the microcontroller.

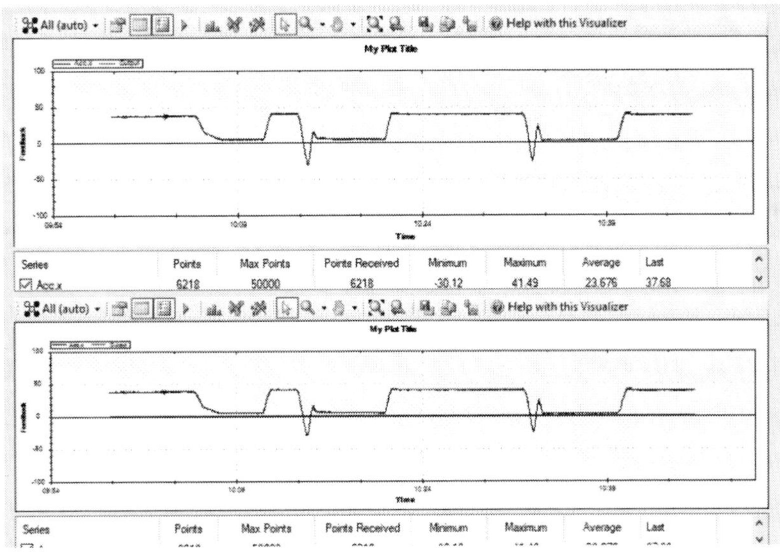

Figure 11. Graphing data send from Microcontroller

Regarding microcontroller parts, the user can select the sampling rate for error calculation. System response has changed when the sampling rate is changed. There are some more steady-state errors and oscillation. Since the size of the learning material is small in both structure and motor power, the result error and disturbance to the system is quite noticeable compared to the control signal. The slight increase in external force from the internal friction of the device or the wind from the air conditioner affects the experimental set at the observed level.

5. Conclusion

The investigation of the use of learning material in the course has shown that users have to use all knowledge to design a control system. In the current study, the results showed a significant increase by simulating the system with computer software, which is difficult to understand and practice their skills.

The most obvious finding to emerge from the analysis is that creating practical experience and skills using the teaching materials emphasize hands-on practice along with theoretical studies.

With the low cost and easy to produce teaching materials, students can do the experiment by adjusting the parameters of the controller, which makes them understand the automatic control system through actual hands-on operations. This finding has important implications for developing learning materials for the preliminary mechanical vibration system.

References

[1] Liu W Hu J and Pan H 2019 *IFAC Paper online* **52** 202-206
Clayton S. Semenzin, Martin, Eric Wu, Jo P. Pauls, Benjamin Simpson, Shaun D. Gregory, Geoff Tansley
[2] Semenzin C S Mapley M Wu E Pauls J P Simpson B Gregory S D and Tansley G 2020 *HardwareX* **8** e00140
[3] Losoya E Z Gildin E and Noynaert S F 2018 *IFAC Paper online* **51** 56-61
[4] Omar M El-Kassaby M M and Abdelghaffar W 2017 *Alexandria Engineering Journal* **56** 359-370
[5] James L 2020 *The working principles of a PID controller*, https://www.mes-insights.com/amp/the-working-principles-of-a-pid-controller-a-929190, accessed on 1 December 2020
[6] Avery P 2009 *Introduction to PID control*, https://www.machinedesign.com/automation-iiot/sensors/article/21831887/introduction-to-pid-control, accessed on 10 September 2020
[7] Bista D 2016 M.S. Thesis (Virginia: Virginia Commonwealth University)
[8] Fisher C J 2011 *Using An Accelerometer for Inclination Sensing*, https://www.digikey.at/en/articles/using-an-accelerometer-for-inclination-sensing, accessed on 10 September 2020
[9] Rossiter J A 2018 *IFAC Paper online* **51** 322-327

Enhanced thermal performance in a square-duct heat exchanger with inclined square-rings

Chitakorn Khanoknaiyakarn[1], Pitak Promthaisong[2], Panuwat Hoonpong[3], Pongjet Promvonge[4] and Sompol Skullong[5,*]

[1]Division of Mechanical and Automotive Engineering Technology, Faculty of Engineering and Technology, King Mongkut's University of Technology North Bangkok, Rayong Campus, 19 moo 11 Nonglalok, Bankhai, Rayong 21120, Thailand
[2]Department of Mechanical Engineering, Faculty of Engineering, Mahanakorn University of Technology, Bangkok 10530, Thailand
[3]Department of Industrial Technology, Faculty of Industrial Technology, Thepsatri Rajabhat University, Talaychubsorn, Lopburi 15000, Thailand
[4]Department of Mechanical Engineering, Faculty of Engineering, King Mongkut's Institute of Technology Ladkrabang, Bangkok 10520, Thailand
[5]Department of Mechanical Engineering, Faculty of Engineering at Sriracha, Kasetsart University Sriracha Campus, 199 M.6, Sukhumvit Rd., Sriracha, Chonburi 20230, Thailand

* Corresponding Author E-mail: sfengsps@src.ku.ac.th, sompol@eng.src.ku.ac.th

Abstract
An experimental work has been conducted to explore thermal and turbulent flow friction characteristics in a uniform heat-fluxed square-duct contained with inclined square-rings (SR). Experiments were performed for different geometric parameters of the SR elements, focused on their detailed effects on the optimal thermal performance. Influences of five different ring-to-duct height or blockage ratios (BR=b/H= 0.1, 0.15, 0.2, 0.25 and 0.3) on friction factor (f), Nusselt number (Nu), and thermal enhancement factor (TEF) in the square duct are investigated for Reynolds number (Re) from 4210 to 25,800. The inclination angle (α) and the ratio of axial pitch length of rings to the height of duct (called pitch ratio, PR=P/H) were kept constant at 60° and 3.0, respectively. The experimental result has shown that the SR insert gives greater heat transfer and friction loss than the plain duct acting alone. Increasing the blockage ratio (BR) results in the drastic enhancement in Nu and f values. The inserted duct with BR = 0.25 has the maximum TEF around 1.57. Comparing thermal performance at identical operating conditions, the SR insert at BR = 0.15 gives around 1.1%, 2.3%, 3.2 and 4.5% higher than the one at BR = 0.2, 0.25, 0.3 and 0.1, respectively.

Keywords: Square duct, Vortex generator, Heat exchanger, Square-ring, Thermal performance

IOP Conf. Series: Materials Science and Engineering 1137 (2021) 012062 doi:10.1088/1757-899X/1137/1/012062

1. Introduction

A heat exchanger is a crucial link in the management of the industrial system and energy conservation, which usually determines the heat transfer rate and thermal performance of such a system. Therefore, a heat transfer enhancement technique is a subject of considerable interests to researchers, especially for the passive heat transfer enhancement techniques [1]. One of such techniques is the use of vortex flow devices. The tube/duct is one of the most commonly used heat exchanger components, which are often found in energy-related systems such as thermoelectric power generation, thermal power plant, waste energy conservation and solar energy collection. Vortex flow devices (so-called "vortex generators", VGs) such as twisted tape [2], coiled wire [3], conical/diamond ring [4,5], baffle [6], and winglet [7] which all belong to a vital group of the VGs by insertion into the heat exchanger ducts/tubes, have been extensively adopted in enhancing the thermal performance of those systems. There have been many attempts to enhance the convection heat transfer rate in the tubes/ducts. Recently, García et al. [8] investigated the heat transfer augmentation in a tube using three types: corrugated tubes, dimpled tubes and wire coils. A new design of a twisted-tape (double V-ribbed twisted-tape) inserted in a heat exchanger tube was reported by Tamna et al. [9] for turbulent flow regime. They showed that TEF for the V-ribbed twisted- tape rises around 22% above that for the typical twisted-tape. A comparison of the thermo-hydraulic performances between twisted tapes and wire coils was suggested by Wang and Sunden [10] for both turbulent and laminar flows. They reported that the wire coils give high performance in augmenting heat transfer in a higher turbulent flow regime while the twisted tapes provide a poorer overall performance. Promvonge et al. [11] experimentally examined the heat transfer enhancement in a round tube using the angled ring inserts and suggested the ring conditions giving the highest thermal-hydraulic performance. Influences of using a vortex-induced impinging-jet phenomenon on enhancing thermal-hydraulic performance in a duct heat exchanger with oblique horseshoe baffles (HBs) were numerically and experimentally explored by Skullong et al. [12]. They found that the increase of thermal-hydraulic performance for the HBs is considerably higher than that for the wire coils, due to the effect of vortex-induced impinging-flow (VI), resulting in the drastic rise in the convection heat transfer within the duct.

Most investigations stated earlier have emphasized on enhancing thermal-hydraulic performance in many heat exchangers by methods of vortex generator (VG) inserts with various pitches, heights and different arrangements: angled/inclined, transverse. The experimental study on turbulent flow through the tandem inclined square-rings (SR) contained in a square duct has rarely been reported so far. The current SR device has been developed by the combined advantages of winglet, twisted-tape and baffle turbulators. This implies that the inserted SR device is expected to have a vortex flow like winglets, ease of practical use like twisted-tapes and drastically greater heat transfer rate like baffles. Thus, a newly 60° SR element is introduced to enhance convection heat transfer better than the twisted tape/coiled wire as a result of the VI effect as mentioned earlier. Hence, this article aims to examine the effect of the SR insert at various BR values on the rate of heat transfer, pressure loss and thermo-hydraulic performance in the square-duct heat exchanger. Experimental setup and formulas for measuring the Nusselt number, friction factor and thermal enhancement factor are presented. The test range of the flow in the duct is for Reynolds number (Re) between 4210 and 25,800.

2. Experimental program and procedure

The square duct in the present work had a cross-section of 45-mm height (H) × 45-mm width (W) with a 3500-mm total length included the 1000-mm test section (L). The SR elements made of a 0.8-mm aluminium sheet to shape them as shown in Figure 1 were fastened each other by putting four small straight wires into small holes on the four corners of the SRs. The 60° inclined SRs at five different ring-to-duct height or blockage ratios (BR=b/H= 0.1, 0.15, 0.2, 0.25 and 0.3) were fitted in the square-duct by surface-attached position.

The schematic of the experimental setup in this study is depicted in Figure 2. The setup mainly consists of a calm section, test section and exit section. The test duct was made of aluminium plates in which air flowed through it. To heat the duct uniformly, an electrical heating wire was continually winding around the duct to give a constant wall heatflux. The outermost duct-wall surface was covered entirely by a 50-mm thick ceramic insulation to avert the convective heat loss to the surrounding. The 2-kW blower directed the air with T_{in} = 25°C through an orifice flow meter. An inverter was utilized to adjust the airflow rate by controlling the motor speed of the blower to achieve the desired Reynolds number between 4210 and 25,800. For measuring temperatures of the air at the outlet and inlet (T_{out} and T_{in}), RTD-typed temperature sensors were employed at certain positions upstream and downstream of the test duct using a multi-channel temperature measurement unit whereas 28 T-type thermocouples for measuring the wall temperatures (T_w) were placed equally on each of the side and upper walls along the test duct. The thermocouple voltage outputs were fed into a data acquisition system (Fluke 2680A) and then recorded via a personal computer. A digital manometer (Dwyer Mark III) was employed to obtain the pressure drop across the air duct. More details on the experimental set-up, method and uncertainty analysis were already mentioned in another paper of the authors [12].

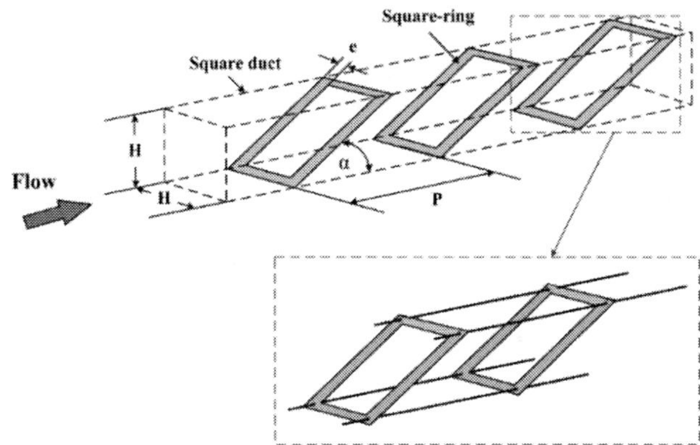

Figure 1. Test square-duct with inclined 60° SRs.

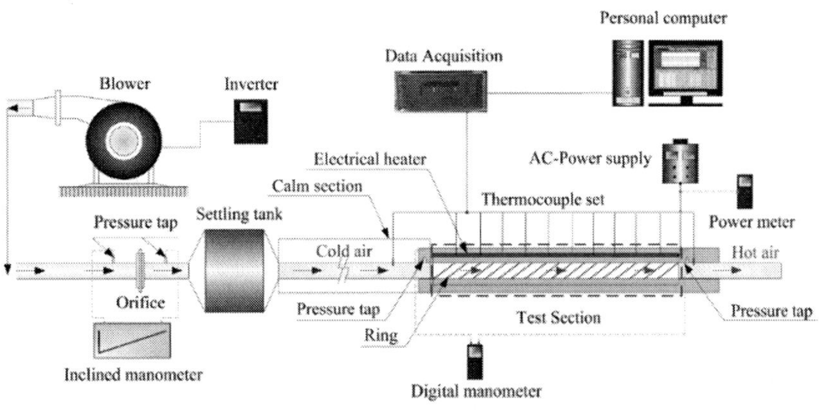

Figure 2. Schematic sketch of the experimental set-up.

3. Data Processing

The current experimental work aims to examine the pressure loss (friction factor, f), heat transfer (Nusselt number, Nu) and thermal-hydraulic performance in a square duct fitted with SRs. The data processing is as follows:

The parameters involved include the flow blockage ratio (BR) and Reynolds number (Re). The Reynolds number is written as

$$\mathrm{Re} = \frac{\rho \times U \times D}{\mu} \tag{1}$$

The f evaluated using the pressure drop of airflow along the test duct is defined as

$$f = \frac{2}{(L/D)} \frac{\Delta P}{\rho \times U^2} \tag{2}$$

where U is the average velocity of the fluid in the duct.

In this work, air entered the test duct having a constant wall heat-flux. At a steady-state, the rate of heat transfer to the air can be presumed to equate the convection heat loss of the tested duct expressed as:

$$Q_{\mathrm{air}} = Q_{\mathrm{conv}} \tag{3}$$

where

$$Q_{\mathrm{air}} = \dot{m} \times C_{\mathrm{p,air}} \times (T_{\mathrm{out}} - T_{\mathrm{in}}) \tag{4}$$

The heat transfer within the test duct can be estimated by

$$Q_{\mathrm{conv}} = h \times A \times (\tilde{T}_{\mathrm{w}} - T_{\mathrm{b}}) \tag{5}$$

in which

$$\tilde{T}_{\mathrm{w}} = \sum T_{\mathrm{w}} / 28 \tag{6}$$

and

$$T_{\mathrm{b}} = (T_{\mathrm{out}} + T_{\mathrm{in}})/2 \tag{7}$$

where T_{w} is the local wall-temperature positioned equally along the test duct. The average wall temperature, $\tilde{T}_{\mathrm{w}}$ was estimated from 28 points of the local surface temperatures. Thus, the evaluations of the mean heat transfer coefficient (h) and the average Nu are as follows:

$$h = \dot{m} \times C_{\mathrm{p,air}} \times (T_{\mathrm{out}} - T_{\mathrm{in}}) / A \times (\tilde{T}_{\mathrm{w}} - T_{\mathrm{b}}) \tag{8}$$

Then, the Nu is achieved by

$$\mathrm{Nu} = h \times D / k \tag{9}$$

All properties of the air are determined at the bulk mean air temperature (T_{b}) by Eq. (7).

A novel thermal enhancement factor (TEF) is employed as a thermal-hydraulic performance indicator of the inserted duct which its h is relative to the smooth duct at the same pumping power as suggested by Refs. [13,14] and expressed by

$$\mathrm{TEF} = \left.\frac{h}{h_0}\right|_{\mathrm{pp}} = \left.\frac{\mathrm{Nu}}{\mathrm{Nu}_0}\right|_{\mathrm{pp}} = \left(\frac{\mathrm{Nu}}{\mathrm{Nu}_0}\right) \Big/ \left(\frac{f}{f_0}\right)^{\frac{0.8}{2.75}} \tag{10}$$

where subscript "0" stands for the smooth duct only.

4. Results and Discussion

4.1. Validation test

The smooth square-duct was tested to verify the experimental setup. The comparison of the f and Nu results of the current smooth square-duct with the results of the published correlations [15] of Blasius and Dittus-Boelter, as prescribed by Eqs. (11) to (12), is shown in Figure 3 and 4, respectively.

Correlation of Blasius,

$$f = 0.316\ \mathrm{Re}^{-0.25} \tag{11}$$

Correlation of Dittus-Boelter:

$$\mathrm{Nu} = 0.023 \mathrm{Re}^{0.8} \mathrm{Pr}^{0.4} \tag{12}$$

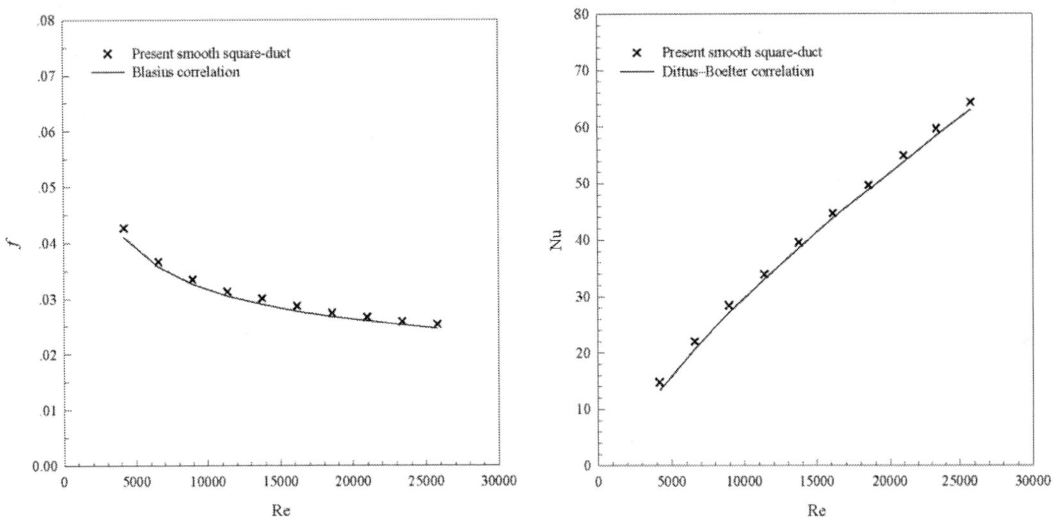

(a) (b)

Figure 3. Validation test of (a) *f* and (b) Nu for smooth square-duct.

Figure 3a and 3b show the comparison between the present experimental work and the standard correlations of Blasius (Eq.11) and Dittus-Boelter (Eq.12). In the figures, the results of the present work agree reasonably well within ±7.6% for *f* of Blasius and ±6.8% for Nu of Dittus–Boelter correlations. Hence, the experimental setup is reliable.

4.2. Influence of SR on heat transfer

Figures 4a, b and c display the plots of the heat transfer coefficient (h), Nusselt number (Nu) and Nusselt number ratio (Nu/Nu_0) against Re for various BR values of the 60° SR. The results in the figure show that h of the SR inserts is increased considerably with the rise of Re and BR (see Figure 4a). The Nu for the inclined SR insert is also seen to be higher than of the smooth duct around 262–436%, and the greatest Nu is found at the highest BR (BR=0.3) as shown in Figure 4b. This is because the presence of the SRs can help induce faster flow mixing, destruct the boundary layer and create vortex-induced impingement (VI) apart from prolonging the residence time in the duct. The variation of a ratio of enhanced Nu to smooth duct Nu, (Nu/Nu_0) with Re is exhibited in Figure 4c. As observed in the figure, Nu/Nu_0 has a slightly reducing tendency with rising Re. It is obvious that the largest Nu/Nu_0 can be found at BR = 0.3. This is because using the highest blockage ratio, BR = 0.3 leads to stronger vortex strength behind the SR and then, higher the convection heat transfer. In the current experimental data, the values of Nu/Nu_0 from the 60° SR insert are about 5.13–5.36, 4.93–5.17, 4.66–4.92, 4.35–4.63 and 3.61–3.94 times at BR = 0.3, 0.25, 0.2, 0.15 and 0.1, respectively. The BR = 0.3 yields the average Nu/Nu_0 around 4%, 9.8%, 17.3% and 40.5% higher than the BR = 0.25, 0.2, 0.15 and 0.1, respectively.

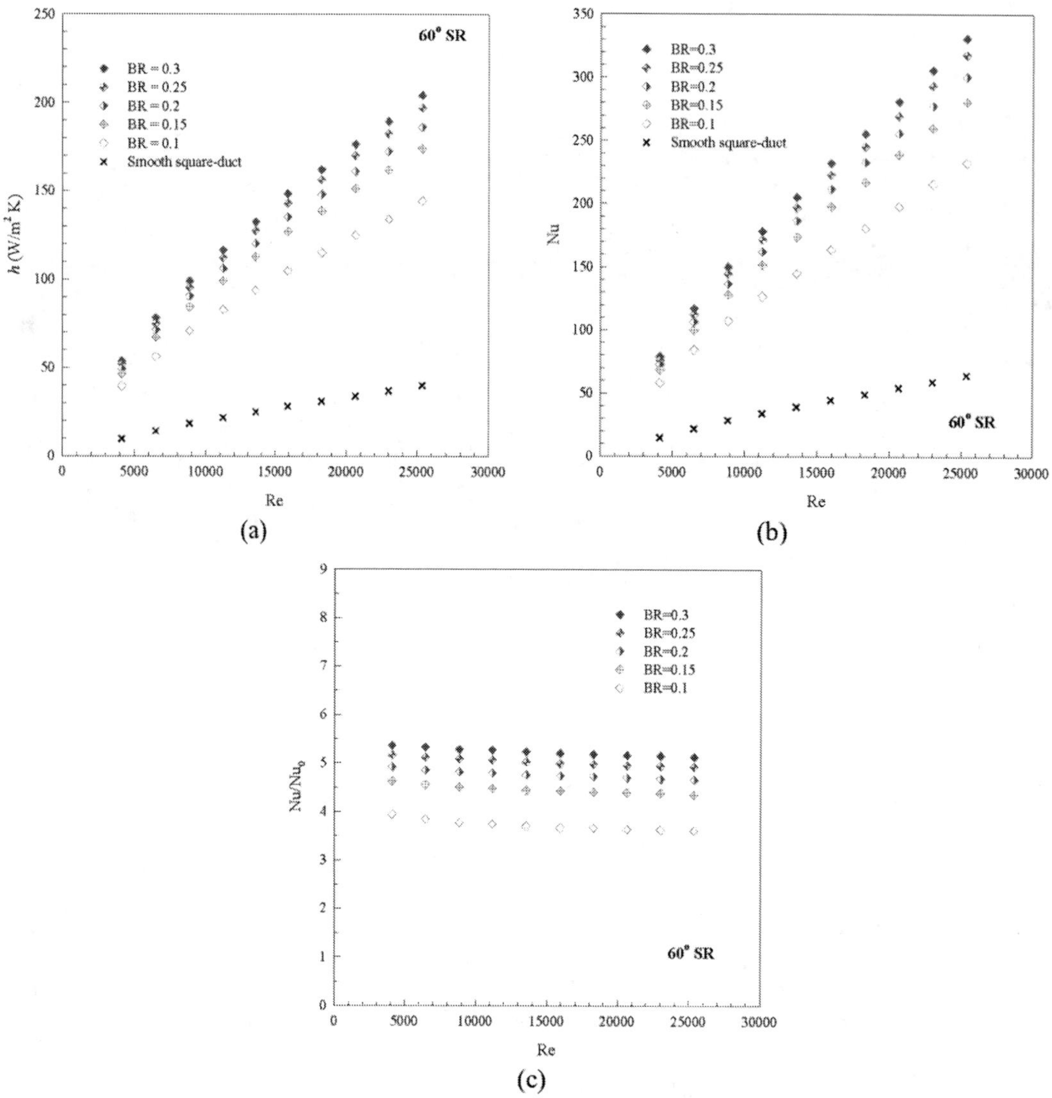

Figure 4. Variations of (a) h, (b) Nu and (c) Nu/Nu_0 with Re for SR inserts.

4.3. Influence on friction loss

Effects of the SR insert on pressure drop (ΔP), friction factor (f) and friction factor ratio (f/f_0) are examined by plotting against Re as displayed in Figures 5a, b and c, respectively. In Figure 5a, the variations of ΔP with Re have the same trends for all cases, ΔP gradually increases with increasing Re until Re ≈ 10,000 and then, rises rapidly for Re > 10,000. ΔP of the inserted duct is found to be considerably higher than the smooth duct alone. It can be seen in Figure 5b that the SR insert has an extreme rise in f above the smooth square-duct owing to higher flow blockage of larger BR value aside from the pressure dissipation due to increasing surface area of the SR. The f of the SR inserts increases around 2361–6645% above that of the smooth duct. Figure 5c shows the distributions of f/f_0 with Re for different BRs. As seen in the figure, f/f_0 has the increasing tendency with Re for all insert cases. The mean values of f/f_0 at BR = 0.3, 0.25, 0.2, 0.15 and 0.1 are, respectively, about 57.8, 49.6, 41.1 and 32.6 times. The maximum f/f_0 of about 67.45 times is found at the largest blockage ratio (BR = 0.3). The

IOP Conf. Series: Materials Science and Engineering 1137 (2021) 012062 doi:10.1088/1757-899X/1137/1/012062

average increase of f/f_0 for BR = 0.3 is, respectively, about 14.2%, 40.6%, 77.2% and 168% higher than that for BR = 0.25, 0.2, 0.15 and 0.1.

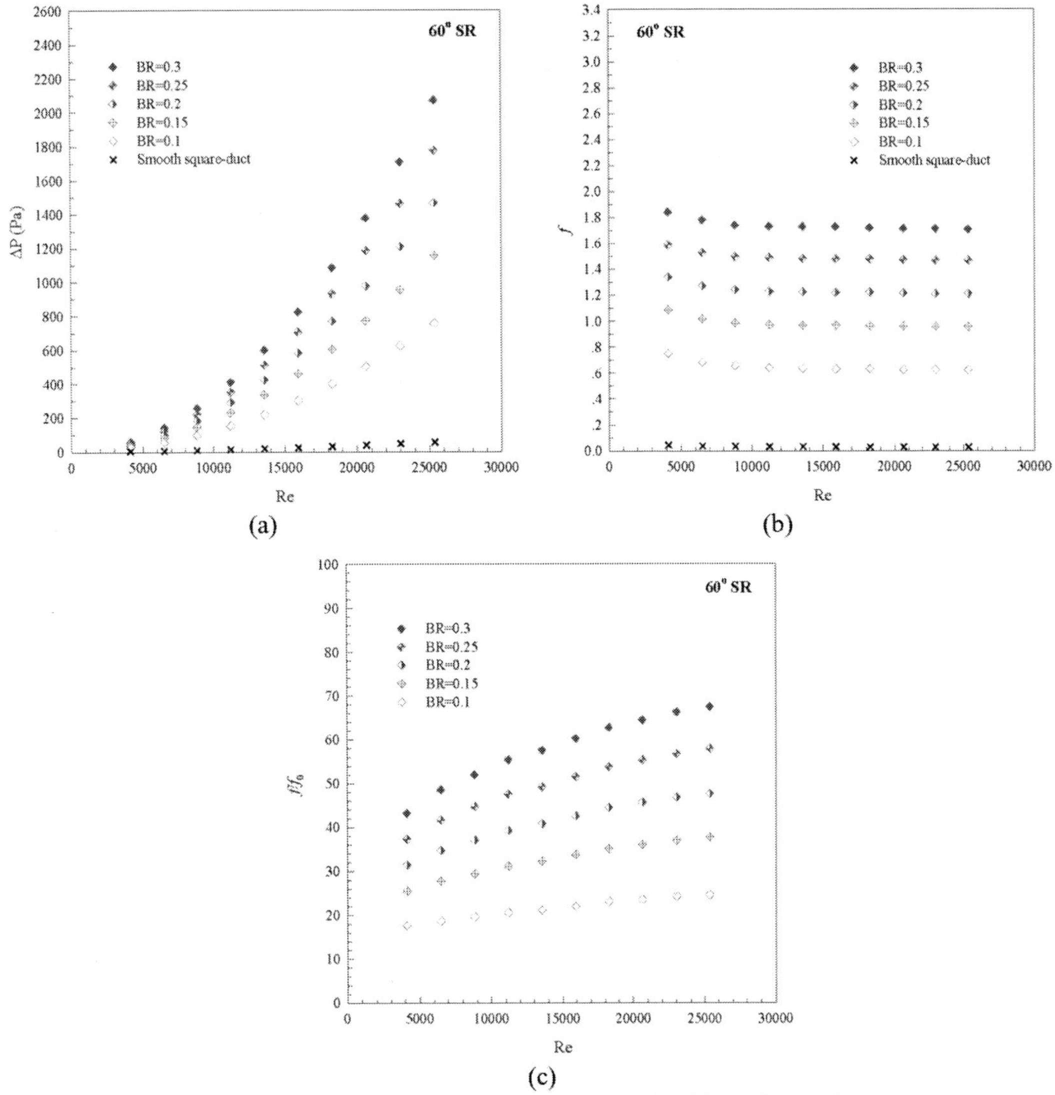

Figure 5. Variations of (a) ΔP, (b) f and (c) f/f_0 with Re for SR inserts.

4.4. Effect on thermal performance

The potential assessment from using the SR insert for a real practice is performed in the form of TEF. In general, TEF can be achieved by simultaneously evaluating the f and Nu of the duct with SR inserts and of the smooth duct at identical pumping power conditions as denoted by Equation (10). The distribution of TEF with Re is portrayed in Figure 6. It is visible from the figure that TEF has a declining trend with rising Re for all insert cases. The values of TEF for BR = 0.3, 0.25, 0.2, 0.15 and 0.1 are, respectively, around 1.26–1.53, 1.27–1.55, 1.28–1.56, 1.23–1.57 and 1.24–1.51, depending on Re. The optimal TEF of about 1.57 is seen for BR = 0.15 at the lowest Re. Hence, if the choice of a vortex flow device is the SR, the BR = 0.15 condition should be used to achieve superior thermal performance.

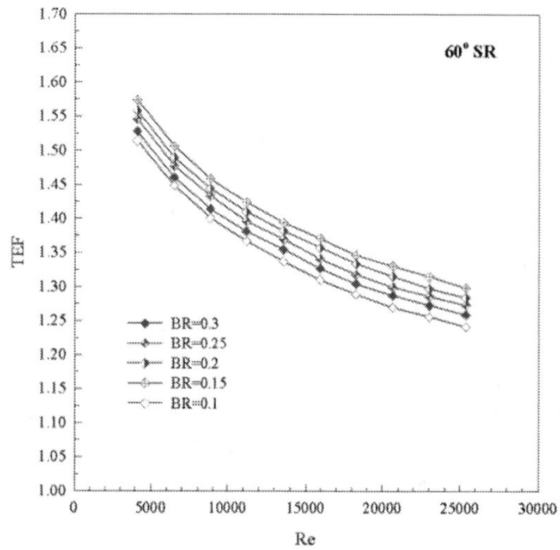

Figure 6. Variation of TEF with Re for SR inserts.

5. Conclusions

An experimental procedure has been conducted to investigate a single-phase turbulent flow and heat transfer behaviors in a square-duct heat exchanger equipped with inclined square-ring vortex generators. Air is employed as the working fluid. The test range of the rate of fluid flow is Reynolds number between 4210–25,800 at a uniform wall heat-flux. The experimental results show that the f and Nu increase considerably with rising Re and BR, especially for BR=0.3. The 60° SR at BR = 0.3 gives the maximum Nu/Nu_0 and f/f_0 of about 5.36 and 67.45, respectively. The SR insert with BR = 0.15 has the maximum TEF around 1.57 and shows the declining trend with increasing Re.

References

[1] Liu S and Sakr M A 2013 *Renew. Sustain. Energy Rev.* **19** 64-81
[2] Eiamsa-ard S and Chuwattanakul V 2020 *Int. J. Therm. Sci.* **153** 106358 1-25
[3] Keklikcioglu O and Ozceyhan V 2018 *Appl. Therm. Eng.* **131** 686-695
[4] Ibrahim M M Essa M A and Mostafa N H 2019 *Int. J. Therm. Sci.* **137** 138-160
[5] Promthaisong P and Skullong S 2019 *J. Res. Appl. Mech. Eng.* **7** 1-10
[6] Tamna S Skullong S Thianpong C and Promvonge P 2014 *Sol. Energy* **110** 720-735
[7] Baissia M T Brima A Aoues K Khanniche R and Moummi N 2020 *Appl. Mech. Eng.* **2020** 165 113563 1-13
[8] García A Solano J P Vicente P G and Viedma A 2012 *Appl. Mech. Eng.* **35** 196-201
[9] Tamna S Kaewkohkiat Y Skullong S and Promvonge P 2016 *Case Stud. Therm. Eng.* **7** 14-24
[10] Wang L and Sunden B 2002 *Int. Commun. Heat Mass Transf.* **29** 45-56
[11] Promvonge P Koolnapadol N Pimsarn M and Thianpong C 2014 *Appl. Therm. Eng.* **62** 285-292
[12] Skullong S Thianpong C Jayranaiwachira N and Promvonge P 2016 *Chem. Eng. and Process.* **99** 58-71
[13] Promvonge P and Skullong S 2020 *Appl. Therm. Eng.* **164** 114424 1-11
[14] Promvonge P Promthaisong P and Skullong S 2020 *Int. J. Heat Mass Transf.* **151** 119351 1-18
[15] Incropera F P Witt P D Bergman T L and Lavine A 2006 *Fundamentals of Heat and Mass Transfer* (John-Wiley & Sons)

Experimental investigation of pressure drop of water flow in a double vertical coils and shell heat exchanger induced by 28 kHz ultrasound

Suphakorn Tuntarungsri[1], Wanarat Rakpakdee[1], Mongkol Pornnattawut[2], Masaaki Motozawa[3], Mitsuhiro Fukuta[3], and Weerachai Chaiworapuek[1,*]

[1] Department of Mechanical Engineering, Faculty of Engineering, Kasetsart University, Bangkok Thailand 10900
[2]Ample Cool Co., Ltd., Samutprakarn Thailand 10280
[3] Department of Mechanical Engineering, Faculty of Engineering, Shizuoka University, Hamamatsu Japan 432-8561

* Corresponding Author: fengwcc@ku.ac.th

Abstract
This research experimentally investigated the pressure drop characteristics induced by 28 kHz ultrasound in a double vertical coils and shell heat exchanger, typically used in 3 – 10 tons water chiller. Also, the water inlet was positioning as 3 configurations, including 1: configuration 1 is the original model from a commercial type (2 ports to the coil center), 2: configuration 2 is the modified model to make a swirl in the heat exchanger (4 ports on 4 side walls), and 3: configuration 3 is the modified model that vertically injects water to the coil (4 ports on a top wall). The results showed that the pressure drop increased in accordance with the flow rate from 0.25 to 0.75 l/s. The second inlet configuration was found to be a promising setup due to its lowest occurring flow friction. Especially in the operation under ultrasonic waves, this configuration provided a lower pressure drop of approximately 20%. Therefore, the maximum reduction of pressure drop was at 86.6% from the original configuration at the lowest flow rate. These results lead to the better design of the heat exchanger and the consequence of higher efficiency water chiller in the future.

Keywords: Ultrasound, Heat exchanger, Vertical coil, Swirl flow, Pressure drop

1. Introduction
The water chiller is a machine that used to chill water for many applications such as in buildings, heavy machines, and manufacturing processes, etc. For the typically small size chiller (1-10 tons chiller), it employs a coil and shell heat exchanger to exchange the heat from water to refrigerant. This type of heat exchanger gives benefit in term of the price and maintenance [1]. This heat exchanger also has high reliability and excellent adaptability. Following the advancement of numerical simulation, this heat exchanger has been studied numerically and experimentally [1-6]. In order to augment the efficiency of heat exchanger, the heat transfer enhancement techniques such as the use of twisted tape [7], extended area [8-9], turbulent spot [10-13], baffle [14], water jet [15], and ultrasound [16-18] have

been employed. However, the use of ultrasound may cause a negative effect on pressure drop of liquid flow across the heat exchanger.

The ultrasound is the sound waves, having frequencies greater than 20 kHz and can classify as low, high, and very high frequency types. Low frequency ultrasound is defined when the waves have frequency between 20 kHz and 100 kHz. Under these frequencies, the acoustic cavitation dominates the fluid domain when the waves are released. This phenomenon occurs as the gas bubbles, generated during the wave rarefaction. When these bubbles collapse, they create micro jets in the fluid domain and lead to the initiation of turbulence. When the frequency is above 1 MHz, the acoustic waves are defined as very high frequency ultrasound. The induced flow phenomenon under this frequency range is dominated by the acoustic streaming effect. This effect causes a driven flow by momentum gradient from the waves. When the wave power is high, the turbulent flow can be created via this effect. This effect is strengthened when the cavitation is observed in the system [19]. This information clearly shows that both acoustic cavitation and streaming directly affect the flow, so, this paper focuses on the investigation of pressure drop across a double vertical coils and shell heat exchanger induced by 28 kHz ultrasound. The obtained results lead to the understanding of ultrasonic effects on the flow past the heat exchanger and they are a key to improve the heat exchanger's efficiency in the future.

2. Experimental Method

The experimental apparatus had 3 main parts that were chiller, test section, and the load tank as depicted in figure 1. In this figure, T and PD refer thermocouple and pressure difference sensor, respectively. The water was cooled by the 3 tons chiller, supplied by Ample Cool Co., Ltd. and it flowed through the coils in the heat exchanger. This chilling coil is responsible for the temperature decrease of shell-side water. When shell-side water was cooled by the chilling coil, it was circulated to the load tank, installing 2 coil heaters, having power of 9,000 W. Its heating power was adjustable by a dimmer, so, the water flowed back into the shell of heat exchanger with higher temperature. The inlet and outlet temperatures of shell- and coil-side waters were detected by thermocouples and the flow rates were detected by SITRANS F M MAG 5100 W flow meter.

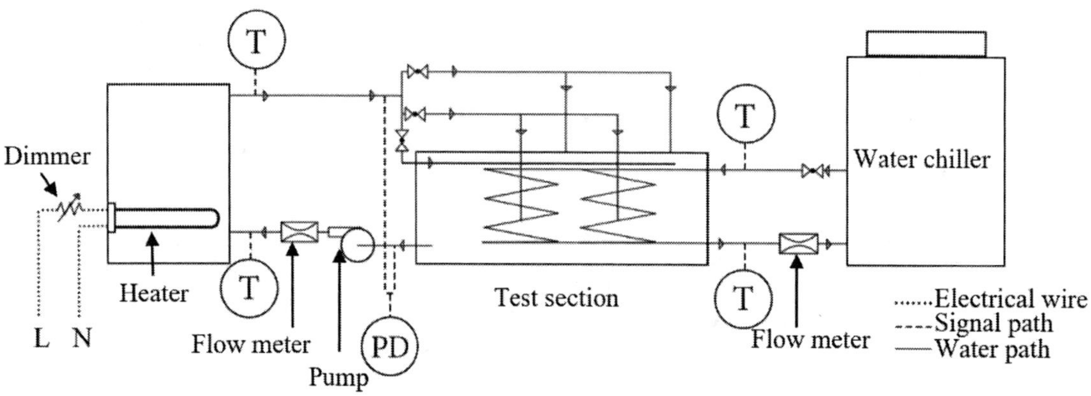

Figure 1. The schematic diagram of experiment

The test section was modeled from a heat exchanger in a 5 tons commercial chiller of Ample Cool Co., Ltd. This heat exchanger had a width of 0.390 m, height of 0.385 m, and length of 0.8 m and made from 3 mm stainless steel. The double coils, having a nominal diameter of 0.304 m were produced from copper tube, having a diameter tube of 0.013 m as shown in figure 2. The outlet port, having a diameter of 0.028 m was installed at bottom corner of tank, while an air vent valve was utilized to release the air from the system. The test system was covered by thermal insulation to prevent the heat loss from the system to surroundings. In this experiment, there were 3 inlet

IOP Conf. Series: Materials Science and Engineering 1137 (2021) 012063 doi:10.1088/1757-899X/1137/1/012063

configurations, including 1: configuration 1 was the original model from a commercial type (2 ports to the coil center), 2: configuration 2 was the modified model to make a swirl in the heat exchanger (4 ports on 4 side walls), and 3: configuration 3 was the modified model that vertically injected water to the coil (4 ports on a top wall) as depicted in figure 3. The inlet of configuration 1 and 3 had a diameter of 0.028 m, while the inlet diameter of configuration 2 was 0.013 m to increase the water velocity. Besides, twelve 28 kHz ultrasonic transducers, having power of approximately 8 W were installed around the heat exchanger to emit the ultrasound into the water flow in shell side.

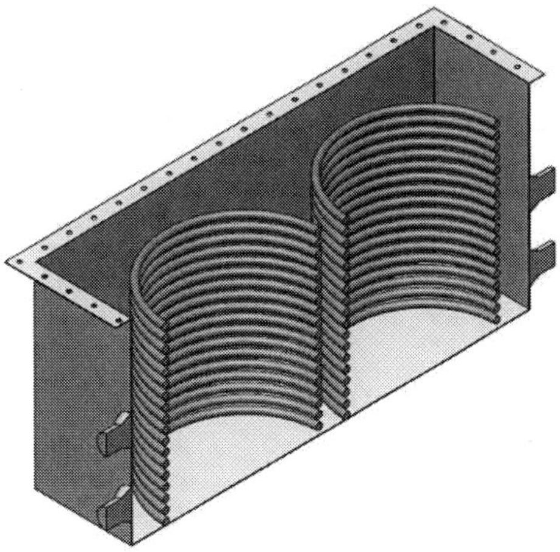

Figure 2. Double coils and shell heat exchanger

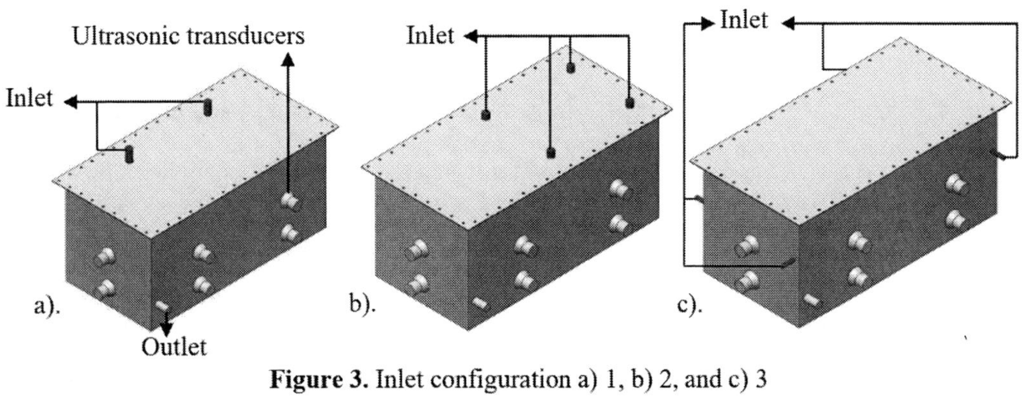

Figure 3. Inlet configuration a) 1, b) 2, and c) 3

In this research, the inlet temperature of water in shell side and coil side was controlled at 22°C. 15.5°C, respectively. The flow rate of shell side water was set at 0.25, 0.5, and 0.75 l/s to investigate the pressure drop behaviour under the change of flow rate and ultrasonic condition. The friction factor, used to present a pressure loss in dimensionless form was determined in accordance with [20] as follows:

$$f=(2d_h\Delta p)/(\rho H_s u^2_{eff}) \quad (1)$$

where Δp is pressure drop (Pa)
ρ is water density (kg/m³)
H_s is shell height (m)
u_{eff} is effective velocity (m/s)

The hydraulic diameter was defined as follows:

$$d_h=4((\text{shell side volume})/(\text{surface contact with fluid})) \quad (2)$$

The effective velocity was evaluated as follows:

$$u_{eff}=(\dot{m}_s)/(\rho A) \quad (3)$$

where $\dot{m}_s$ is mass flow rate of water (kg/s)
A is cross sectional area of pipe (m^2)

3. Results

Figure 4 shows the example of pressure drop signals, directly measured from the pressure difference sensor from the case of Config. 1 and flow rate of 0.75 l/s. The signals show that the magnitudes of pressure drop fluctuate within 100 Pa due to the turbulence. Thus, these signals were averaged and employed for friction factor calculation.

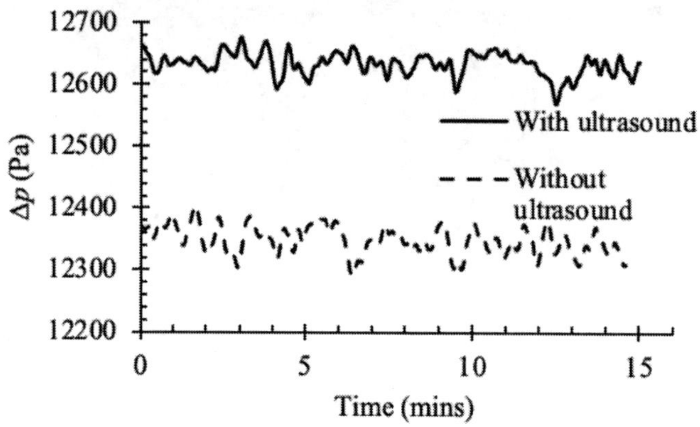

Figure 4. Example of pressure drop signals from the case of configuration 1 and flow rate of 0.75 l/s

During the experiment, 6 data sets of results were acquired at each test condition. The conditions with and without ultrasound are abbreviated as w and w/o, respectively. When the signals from 6 data sets were averaged, the mean pressure drop across the shell side of heat exchanger were achieved as shown in figure 5. They were found to increase with an increment of the flow rate in shell side. Config. 1 provided the maximum pressure drop, while Config. 2 gave the relatively lowest pressure drop. Moreover, the ultrasound, that is abbreviated as US in this paper was found to slightly affect the pressure drop characteristics.

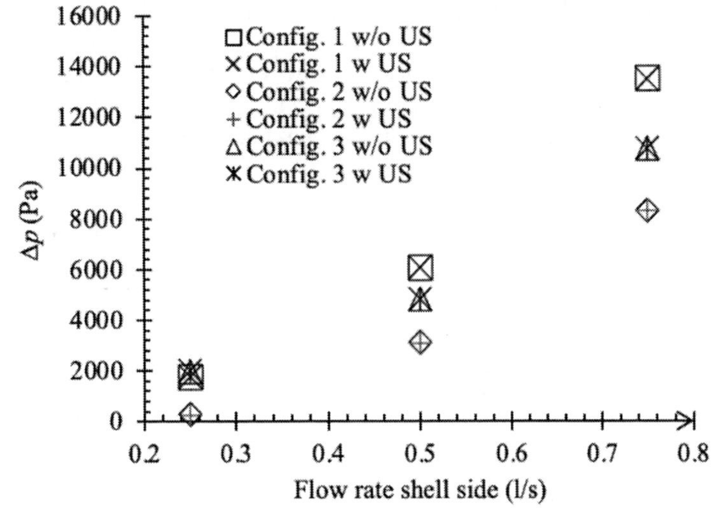

Figure 5. Pressure drops induced by 3 inlet configurations and ultrasonic conditions

Figure 6 shows the pressure drop ratio and friction factor ratio under the condition with and without ultrasound. In this paper, Δp_{us}, Δp_o, f_{us}, and f_o refer pressure drop with ultrasound, pressure drop without ultrasound, friction factor with ultrasound, and friction factor without ultrasound, respectively. The result revealed that the ultrasound slightly affected the water flow in the heat exchanger. There were only 1-2% difference of pressure drops and friction factors for Config. 1 and 3. However, at the relatively low flow rate, Config. 2 showed the potential of ultrasound usages in order to reduce pressure drop of water flow in heat exchanger. About 20 % reduction in pressure drop and friction factor were obtained at the flow rate of 0.25 l/s, while the water flow at relatively higher flow rate was not affected much by ultrasound.

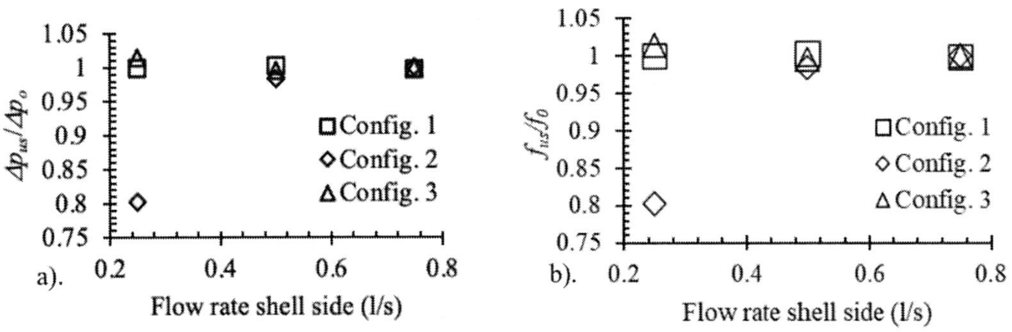

Figure 6. a) pressure drop ratio and b) friction factor ratio under the condition with and without ultrasound

4. Conclusion

This research focused on the determination of pressure drop and friction factor characteristics of water flow across a double coils and shell heat exchanger. The effect of 3 inlet configurations and ultrasound were investigated during the flow rate of 0.25 - 0.75 l/s. The results show that pressure drop and friction factor increased with an increment of the flow rate in shell side. Config. 2 provided the relatively lowest pressure drop and friction factor. Meanwhile, the ultrasound reduced the pressure drop and friction factor approximately 20 % at the flow rate of 0.25 l/s. When the pressure drop was compared with the original model in a commercial chiller, it decreased 86.6 %. Hence, the obtained results show that the ultrasound can significantly decrease the pressure loss in the heat exchanger.

Acknowledgement
The authors gratefully thanks to Research and Researchers for Industries (RRI) and Ample Cool Co., Ltd. under project no. 60-1-2013 for financial support and Prof. Dr. Tanongkiat Kiatsiriroat for valuable suggestions.

References

[1] Ghorbani N, Taherian H, Gorji M and Mirgolbabaei H 2010 *Int. Commun. Heat Mass* **37** 775
[2] Suvanjumrat C and Loskupapaiboon K 2020 *Eng. J.* **24**(2) 109
[3] Suvanjumrat C 2017 *Eng. J.* **21**(5) 225
[4] Suvanjumrat C 2017 *Eng. J.* **21**(3) 207
[5] Chaiworapuek W, Champagne JY, El Hajem M and Kittichaikarn C 2010 *AIP Conf. Proc.* **1225** 562
[6] Thungthong T, Chalearmwattananon N, Mongkolkitngam T and Chaiworapuek W 2020 *Int. J. Heat Mass Transfer* **163** 1
[7] Mashoofi N, Pourahmad S and Pesteei SM 2017 *Case. Stud. Therm. Eng.* **10** 161
[8] Wang J, hashemi S, Alahgholi S and Mehri M 2018 *Int. J. Refrig.* **94** 11
[9] Bozzoli F, Cattani L and Rainieri S 2015 *Int. J. Heat Mass Transfer.* **101** 76
[10] Chaiworapuek W, Nongnoi S and Kittichaikarn C 2014 *Kasetsart J. (Nat. Sci.)* **48** 637
[11] Chaiworapuek W and Kittichaikarn C 2016 *Int. J. Heat Mass Transfer* **92** 850
[12] Chaiworapuek W and Kittichaikarn C 2016 *Exp. Therm Fluid Sci.* **77** 179
[13] Srichan S, Rakpakdee W, Kiatsiriroat T and Chaiworapuek W 2019 *Exp. Heat Transfer* **32** 488
[14] Cao X, Dub T, Liu Z, Zhai H and Duan Z 2019 *Int. J. Heat Mass Transfer* **142** 118437
[15] Rakpakdee W, Kittichaikarn C and Chaiworapuek W 2019 *Int. J. Heat Mass Transfer* **135** 491
[16] Mongkolkitngam T, Mingbunjerdsuk J, Chaiworapuek W and Priyadumkol J 2020 *Kasetsart J.* **108** 45
[17] Kitboonkrailers W, Tongkratoke A, Khaothong K and Chaiworapuek W 2019 *ME-NETT.* vol 33 (Thailand: Udon-Thani)
[18] Inworn N and Chaiworapuek W 2018 *Int. J. Heat Mass Transfer* **122** 1153
[19] Sakamoto S and Watanabe Y 1999 *Jpn. J. Appl. Phys.* **38** 3050
[20] Sepehr M, Hashemi S, Rahjoo M, Farhangmehr V and Alimoradi A 2017 *Chem. Eng. Res. Des.* **134** 277

Investigation of pressure loss in a circular pipe under ultrasonic waves released along main stream flow

Pongsapak Treegosol[1], Jetsadaporn Priyadumkol[2], Kanet Katchasuwanmanee[1], and Weerachai Chaiworapuek[1,*]

[1]Department of Mechanical Engineering, Faculty of Engineering, Kasetsart University, Bangkok, 10900, Thailand
[2]Department of Mechanical Engineering, Faculty of Engineering, Mahidol University, Nakhon Pathom, 73170, Thailand

* Corresponding Author: fengwcc@ku.ac.th

Abstract
Recently, ultrasound was discovered to have great potential in enhancing heat transfer from thermal systems., However, there is little information of the flow resistance, induced by its waves. Therefore, this research investigates the pressure loss of turbulent water flow in a circular pipe disturbed by 25 kHz ultrasound at the Reynolds number of 10,000, 12,500, 15,000, 17,500, 20,000, 22,500, and 25,000. The size of test section was modeled in accordance with the power plant's condenser and the flow had been verified by the comparison of friction factor with theory. The maximum and minimum differences were 9.79 and 0.33%, respectively, confirming the reliability of the test setup. In this study, ultrasound was released along the mainstream flow from 3 and 15 transducers, installed upstream of the pipe. The results confirmed that the ultrasound was slightly affected on the friction loss of turbulent pipe flow. The average increase of pressure loss was 0.077% and 0.833% by the waves when they were emitted from 3 and 15 transducers, respectively. With very low inducement of flow resistance, ultrasound can be accounted as one of the most promising techniques for heat transfer enhancement, applicable to many thermal systems such as condensers, in the future.

Keywords: Condenser, Ultrasonic waves, Turbulent flow, Heat transfer enhancement, Pressure drop

1. Introduction

The heat transfer of a pipe flow is a basic phenomenon that is applicable to many heat exchangers. However, a heat transfer enhancement is needed for the reduction of operating cost. Following the advancement of numerical method and devices, this heat exchanger has been studied numerically and experimentally [1-6]. In an attempt to augment the heat exchanger's performance, the heat transfer enhancement techniques such as the use of twisted tape [7], extended area [8], turbulent spot [9-12], porous media [13], water jet [14], etc. have been utilized. The use of ultrasound is one of those techniques, giving high performance to enhance the heat transfer [15-16]. When the ultrasonic waves are released into a liquid, 2 important phenomena occur and these typically use for heat transfer enhancement [17]. The first phenomenon is acoustic cavitation that appears during the compression and rarefaction by the variation of the local pressure [18]. When the local pressure is lower than the vapor pressure, gas bubbles are formed, oscillate, grow, and then collapse violently [19]. The bubble implosion around a solid-liquid interface disrupts thermal and velocity boundary layers and leads to the enhancement of heat transfer. Moreover, it was found that the process of acoustic cavitation reduced the turbulence intensity, leading to a reduction in wall friction loss [20]. Another phenomenon is acoustic streaming which arises from the dissipation of acoustic energy into the liquid [21]. When liquid streaming occurs due to the increase of momentum energy, the heat transfer rate around the surface can be enhanced. However, the information of not only heat transfer, but also the pressure loss of pipe flow under the ultrasound, emitted in the direction of the main flow is very little. Thus, this paper focuses on the measurement of pressure drop of water turbulent flow in a circular pipe, disturbed by the 28 kHz ultrasound. The obtained results in this research will clarify the pressure drop characteristics of a water pipe flow, under the interference of ultrasound.

2. Experimental methods

2.1 Experimental setup

The investigation was carried out in a closed-loop system, as shown in Figure 1. The apparatus consisted of a 1.65 kW centrifugal pump, supplying the water from a water tank to the ultrasound generating zone, installing fifteen 28kHz and 120 W ultrasonic transducers. Then, the water flowed through a copper tube that was the test section of this research and the tube had an outer diameter of 19.05 mm. The water was heated by heaters to maintain the constant temperature condition at 45°C on the pipe surface. This condition was modelled from the operating condition of a condenser in the power plant. The heaters were controlled using proportional-integral-derivative (PID) controllers that received local input signals from eight stations of type-K thin leaf thermocouples which have an uncertainty of 0.75%. These thermocouples were installed between the surface of the copper tube and heaters. This setup was wrapped by a high thermal resistance insulator to prevent heat losses to the environment as shown in Figure 2. The temperature of the inlet water was controlled at 25°C by water chiller. The water flow rate was measured and adjusted at the most downstream of the test section using an ultrasonic flow meter (TUF-2000H) and the gate valve, respectively. Finally, the water circulated back to the water tank and cooled down using a cooling coil.

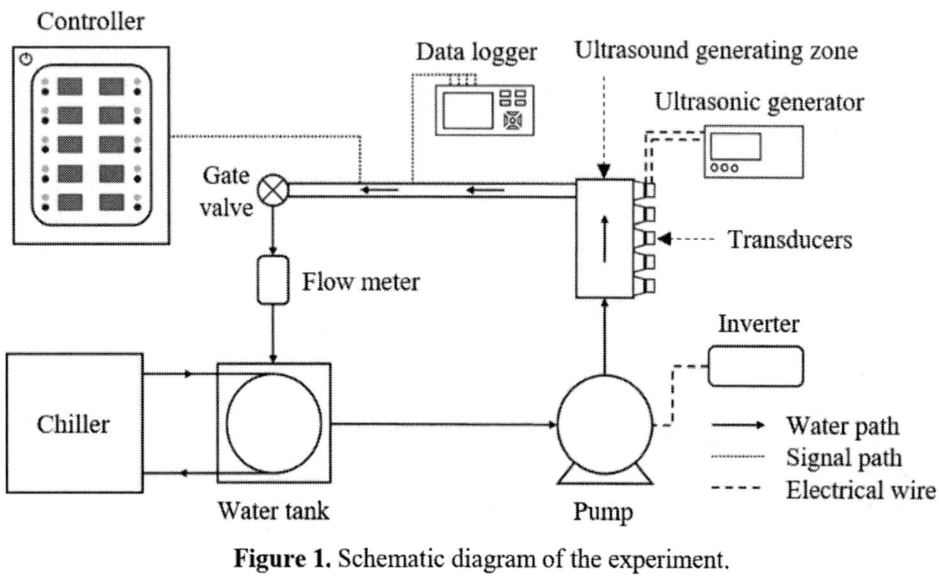

Figure 1. Schematic diagram of the experiment.

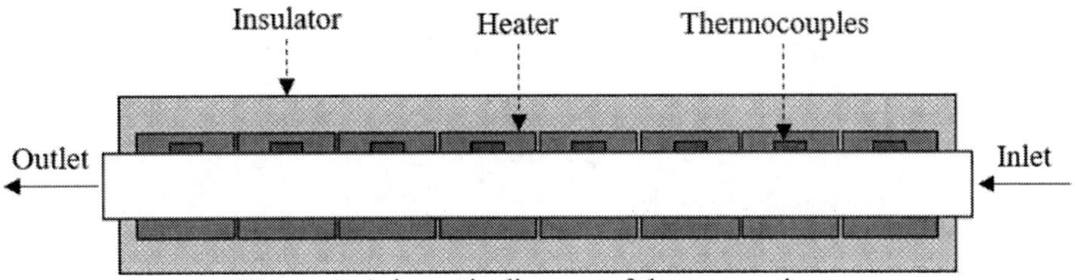

Figure 2. Schematic diagram of the test section.

In the test section, the pressure difference was measured using the differential pressure sensor (Siemens Sitrans P DS III) which has an uncertainty of 0.5% between the streamwise distances of 0.02m and 1.02m from the pipe entrance. This sensor received power from the power supply and sent the signal of the pressure difference to the transmitter. Then, the pressure difference data were achieved using the data logger (Graphtec midi LOCKER GL240), as shown in Figure 3. Before the tests, the differential pressure sensor would be calibrated for the accuracy of a measurement. After entering the steady-state condition, the data of pressure difference was recorded by the data logger for 10 minutes without the interference of ultrasound. After that, the system was disturbed by ultrasound and the data was grabbed for another 10 minutes. In order to provide reliable data, all experiments were repeated 3 times and the data was averaged to minimize the experimental error. In this research, the ultrasound would be released from 3 most top transducers and 15 transducers to see the effect of the waves, emitted from different number of transducers. Also, the measurement would be performed during the Reynolds number of 10,000 – 25,000 with an increment of 2,500.

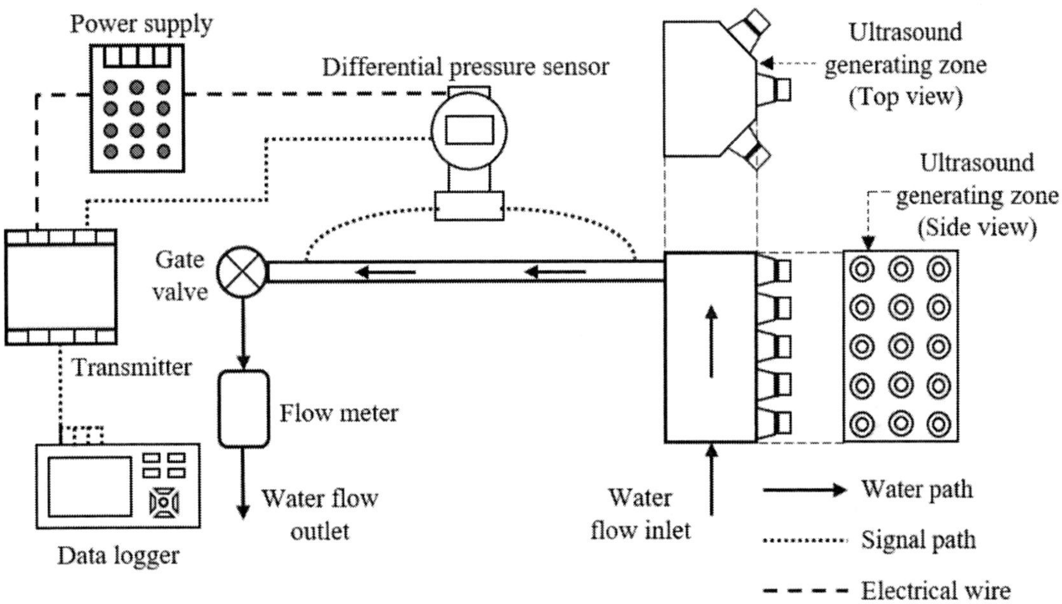

Figure 3. Setup of differential pressure sensor.

2.2 Data analysis

The Reynolds number describes the relationship between density, velocity, diameter and viscosity. It can be calculated as follows:

$$Re = \frac{\rho v D}{\mu} \tag{1}$$

where ρ is fluid density [kg/m^3]
μ is dynamic viscosity of fluid [Pa·s]

Also, the friction factor is defined as:

$$f = \frac{2L_c \Delta P}{L \rho v^2} \tag{2}$$

where L_c is the characteristic length [m]
ΔP is the pressure difference [Pa]
L is the distance to measure differential pressure [m]

2.3 The verification of friction factor

In this research, the friction factor under the condition without ultrasound was verified with the friction factor of turbulent flow in a smooth tube following Petukhov [22]. It could be calculated as follows:

$$f = \left(0.79\,ln\left(Re\right) - 1.642\right)^{-2} \text{ for } 3000 < Re < 5\times10^6 \tag{3}$$

The percentage error of the comparison between the results from the experiment and Petukhov's correlation was defined as follows:

$$\%Error = \left[\left(f_{exp} - f_{eqn}\right) / f_{eqn}\right] \times 100 \tag{4}$$

where f_{exp} is the experimental friction factor [-]
f_{eqn} is the friction factor from [22] [-]

Table 1 shows the percentage errors of the comparison between the friction factor from the experiment and Petukhov's correlation during the *Re* of 10,000 – 25,000. It is found that they are in the same tendency that the friction factor is decreasing with the increase in the *Re* as depicted in Figure 4. The maximum and minimum percentage errors are 9.79% and 0.33% at the *Re* of 25,000 and 17,500, respectively. These results shown that this experimental setup is reliable and able to conduct an experiment of ultrasound.

Table 1. Friction factor averaged from experiment and calculated from Petukhov's correlation.

Re	f_{eqn}	f_{exp}	$\%Error$
10,000	0.0315	0.0304	-3.47
12,500	0.0296	0.0307	3.64
15,000	0.0282	0.0273	-3.24
17,500	0.0271	0.0272	0.33
20,000	0.0262	0.0259	-0.99
22,500	0.0254	0.0246	-3.27
25,000	0.0247	0.0271	9.79

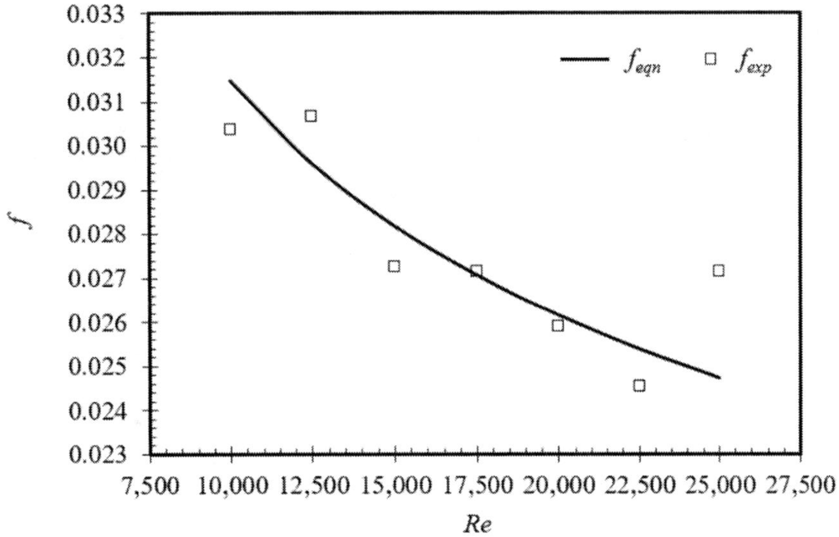

Figure 4. Friction factor averaged from experiment and calculated from Petukhov's correlation.

3. Results and discussion

Figure 5 shows an example of pressure difference signals, recorded during the experiment by a data logger. The signals were obtained when the experiment was conducted under the conditions without and with ultrasound from 3 ultrasonic transducers (3 UTs) and 15 ultrasonic transducers (15 UTs). The data show that the magnitudes of pressure difference fluctuate because of the turbulent flow, and the magnitude of pressure drop can be decreased, due to the effects of ultrasound. Therefore, these data were averaged and used in the evaluation of pressure difference and friction factor.

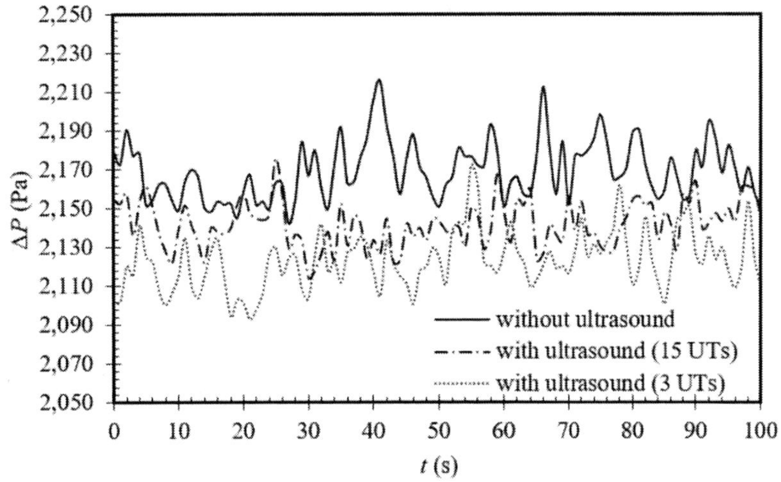

Figure 5. The example of pressure difference signals.

Figure 6 shows the pressure difference ratio and the friction factor ratio under the condition with and without the disturbance of 28 kHz ultrasound versus the Reynolds number during 10,000 – 25,000. In this figure, the subscript w and nw represent the condition with and without interference of ultrasound, respectively. From Figure 6a, it is found that the pressure difference changed by ultrasound and the ratio was in the range between 0.9805 - 1.0188 and 0.9883 – 1.0378 when the system was induced by 3 and 15 transducers, respectively. The maximum pressure difference ratio under the effect of 3 transducers is 1.0188 at the *Re* of 20,000 while the case of 15 transducers shows the ratio of 1.0387 at the *Re* of 17,500. Furthermore, the average ratio of pressure difference increased about 0.077% and 0.833% under the waves from 3 and 15 transducers, respectively. The value of friction ratio was similar to the pressure difference as depicted in Figure 6b. Finally, the results showed that the range of Reynolds number that mostly affected the pressure difference or friction factor was during 15,000 – 20,000 for the case of 15 UTs. Meanwhile, the pressure difference and friction factor under the case of 3 UTs did not show a significant change by the waves.

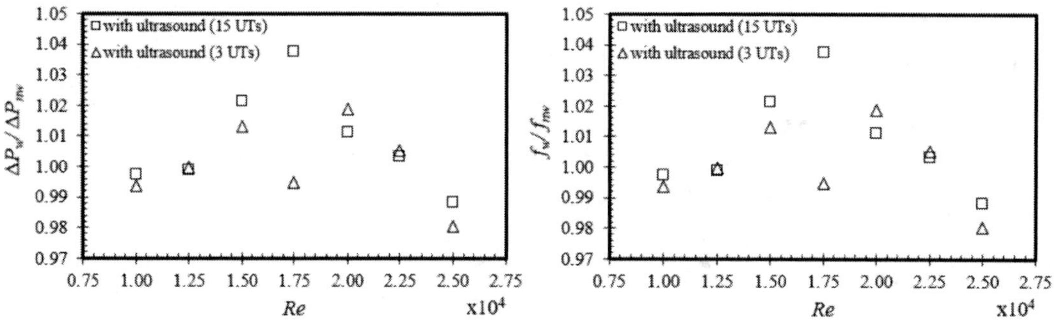

Figure 6. (a) pressure difference and (b) friction factor ratios under ultrasound, released from 3 and 15 transducers.

4. Conclusion

This research investigated the pressure difference and friction factor of turbulent water flow, in a circular pipe under the 28kHz ultrasound released along the mainstream flow, at the Reynolds numbers of 10,000, 12,500, 15,000, 17,500, 20,000, 22,500, and 25,000. The water flow in the pipe was heated to maintain the constant surface temperature of 45°C. The reliability of experimental setup was verified by comparing friction factor with those calculated from theory. The maximum and minimum percentage errors were 9.79% and 0.33% at the *Re* of 25,000 and 17,500, respectively. The results were presented as the ratio of pressure difference and the friction factor under the condition with and without the interference of 28 kHz ultrasound. The maximum ratios are 1.019 and 1.038 under the ultrasonic waves from 3 and 15 transducers, respectively. Meanwhile, the ratios increased averagely about 0.077% and 0.833% when the system was induced by waves from 3 and 15 transducers, respectively. Thus, these results show that the ultrasound slightly affects the pressure drop and friction factor of the flow system. It can be accounted as one of the promising techniques for heat transfer enhancement, applicable to many thermal systems, such as condensers in the future.

Acknowledgements
The authors gratefully acknowledge the financial support from the Electricity Generating Authority of Thailand (EGAT).

References

[1] Suvanjumrat C and Loskupapaiboon K 2020 *Eng. J.* **24(2)** 109
[2] Suvanjumrat C 2017 *Eng. J.* **21(5)** 225
[3] Suvanjumrat C 2017 *Eng. J.* **21(3)** 207
[4] Thungthong T Chalearmwattananon N Mongkolkitngam T and Chaiworapuek W 2020 *Int. J. Heat Mass Transfer* **163** 1
[5] Li X Zhu D Sun J Mo X and Yin Y 2018 *Exp. Therm. Fluid Sci.* **96** 146

[6] Chaiworapuek W Champagne J Y El Hajem M and Kittichaikarn C 2010 *AIP Conf. Proc.* **1225** 249
[7] Mashoofi N Pourahmad S and Pesteei S M 2017 *Case. Stud. Therm. Eng.* **10** 161
[8] Wang J hashemi S Alahgholi S and Mehri M 2018 *Int. J. Refrig.* **94** 11
[9] Chaiworapuek W Nongnoi S and Kittichaikarn C 2014 *Kasetsart J. (Nat. Sci.)* **48** 637
[10] Chaiworapuek W and Kittichaikarn C 2016 *Int. J. Heat Mass Transfer* **92** 850
[11] Chaiworapuek W and Kittichaikarn C 2016 *Exp. Therm Fluid Sci.* **77** 179
[12] Srichan S Rakpakdee W Kiatsiriroat T and Chaiworapuek W 2019 *Exp. Heat Transfer* **32** 488
[13] Thongjamroon J Techapiesancharoenkij R Chaiworapuek W 2019 *J. Phys: Conf. Ser.* **1380** 1
[14] Rakpakdee W Kittichaikarn C and Chaiworapuek W 2019 *Int. J. Heat Mass Transfer* **135** 491
[15] Mongkolkitngam T Mingbunjerdsuk J Chaiworapuek W and Priyadumkol J 2020 *Kasetsart Eng. J.* **108** 45
[16] Inworn N and Chaiworapuek W 2018 *Int. J. Heat Mass Transfer* **122** 1153
[17] Legay M Gondrexon N Person E Boldo P and Bontemps A. 2011 *Int. J. Chem. Eng.* **17** 1
[18] Monnot A Boldo P Gondrexon N and Bontemps A 2007 *Heat Transf. Eng.* **28** 3
[19] Neppiras E 1984 *Ultrasonics.* **22** 25
[20] Nomura S Murakami K and Sasaki Y 2000 *Jpn. J. Appl. Phys.* **39** 3636
[21] Lighthill J 1978 *J. Sound Vib.* **61** 391
[22] Petukhov B 1970 *Adv. Heat Transf.* **6** 496

IOP Conf. Series: Materials Science and Engineering 1137 (2021) 012065 doi:10.1088/1757-899X/1137/1/012065

Energy efficiency evaluation from micro trip database of HDV bus tracking system

Kampanat Thapmanee[1], Phumanan Niyomna[1], Peerawat Saisirirat[1*], Kampanart Silva[1], and Nuwong Chollacoop[1]

[1]Renewable Energy and Energy Efficiency Research Team,
National Energy Technology Center (ENTEC), Pathum Thani, Thailand 12120

* Corresponding Author: peerawat.sai@entec.or.th

Abstract
Fuel consumption (L/100 km) is one of the most important parameters for assessing fuel economy and energy efficiency in the road transport sector. However, the fuel consumption database for Thailand heavy duty vehicles (HDVs) is usually collected from road-side surveys, as whole trip summaries, e.g., total consumed fuel quantity, and measured driving distances. Therefore, the effects of trip details, e.g., topography, route curvature, and driving style on fuel consumption, cannot be calculated. Nowadays, the modern vehicle tracking system, which can acquire vehicle traveling information at every second, including real-time fuel consumption, provides numerous sets of micro trip data. Therefore, this study aims to indicate important parameters effected on the fuel consumption of HDV buses, from the analysis of micro trip databases. Out of a number of fixed HDV buses routes, the Bangkok – Nakhon-Ratchasima route (about 300 km) was selected from seven HDV bus routes because this route has sufficient elevation gain (uphill) and elevation loss (downhill) for the analysis. From the analysis results, the correlation between fuel consumption and elevation gain was distinctively observed, while the relationship with other energy-related parameters were not observable. During the vehicle ascending, fuel consumption increased linearly as 0.75 L/100km by an elevation gain in m/km ($y = 0.75x + 29.71$). However, from elevation gain above 50 m/km fuel consumption fluctuated around a constant value of 59.03 ± 7.25 L/100km.

Keywords: Micro trip database, Fuel consumption, Topography, Energy efficiency

1. Introduction

Proportional to national economic developments, the energy consumption of Thailand gradually increases. Consequently, the transport sector shares the largest contribution, as shown in Table 1 and Figure 1 [1]. In addition, the majority of energy consumption in the transport sector, depends only on fossil fuel. This causes a national trade deficit for imported fossil fuel, inevitable Greenhouse Gas (GHG) emissions and insecurity for our national energy resources.

IOP Conf. Series: Materials Science and Engineering 1137 (2021) 012065 doi:10.1088/1757-899X/1137/1/012065

Table 1. Thailand historical records of final energy consumption by economic sector [1]

	Final energy consumption [ktoe]					Increment	
	2015	2016	2017	2018	2019	2018	2019
Agricultural	3,082	2,987	2,652	2,876	2,940	8.4%	2.2%
Industrial	21,083	29,475	28,459	30,440	31,144	7.0%	2.3%
Household	8,935	11,071	10,870	11,001	11,171	1.2%	1.5%
Commercial	4,270	6,215	6,452	6,549	6,846	1.5%	4.5%
Transport	21,714	30,181	32,319	33,086	33,607	2.4%	1.6%
Total	59,084	79,929	80,752	83,952	85,708	4.0%	2.1%

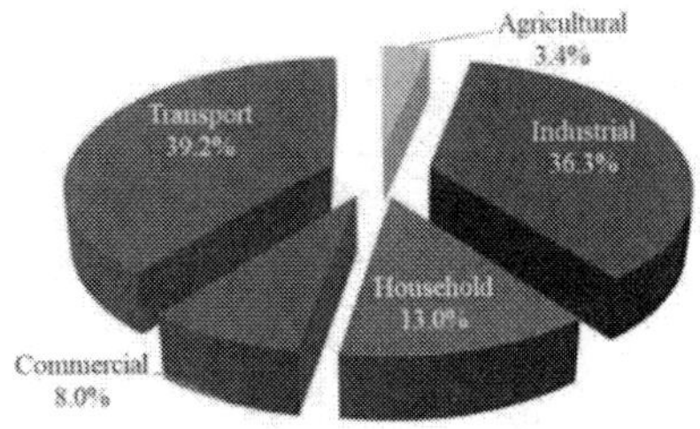

Figure 1. Contribution of energy consumption by economic sector [1]

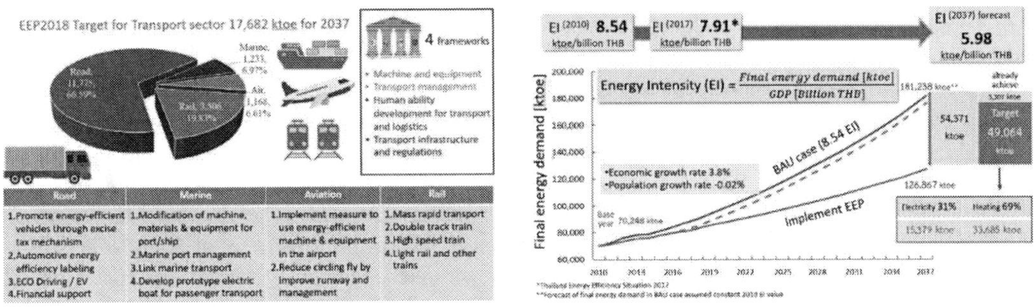

(a) (b)

Figure 2. (a) Projection of final energy demand and target of EEP2018 plan and (b) EEP2018 target for transportation sector [4]

Consider on these issues, the government has announced several national energy plans including the Energy Efficiency Plan (EEP2018) [2, 3, 4]. The final target is to achieve 30% reduction of Energy Intensity (EI, energy demand per economic activity) within 2037 as shown in Figure 2(a). The EEP measures for transport sector contribute to 36% out of this target, composed of four policy frameworks implemented on four transport modes including marine, aviation, rail and the road transport. The most important mode is the road transport, which is accounted for 66% of total transport EEP target as shown in Figure 2(b).

Road transport comprises of light- and heavy- duty vehicles (LDVs and HDVs) for transporting passengers and goods. One of the most important parameters involved in EI analysis, is the vehicle fuel consumption, defined as a consumed fuel quantity in a unit of driving distance, e.g., liter/100km, kgCNG/100km, (informal recognized as fuel economy, km/liter or km/kgCNG), apart from the average vehicle driving distance (vehicle kilometer of travels, VKT), vehicle number, type of fuel used etc. The fuel consumption is required for evaluating fuel cost from a vehicle owner viewpoint [5, 6] (beside of vehicle purchasing cost, maintenance cost, etc.) and for the policymakers to project policy impacts [7, 8, 9, 10]. Fuel consumption depends on many parameters, e.g. vehicle deterioration, load factor, driving behavior (driver), and road conditions, etc. Therefore, the aim of this study is to identify and evaluate the impacts of those parameters on the vehicle fuel consumption of bus transport.

2. Energy Efficiency Observation for Heavy Duty Vehicles

As mentioned above, the fuel consumption and other energy-related parameters are important for road transport EI analysis. Therefore, the Energy Policy and Planning Office (EPPO) has been continuously conducting field research projects to update those important parameters as the Thailand transport database [11, 12, 13, 14]. In the report, the fuel consumption data was collected as well as average vehicle driving distance (vehicle kilometer of travels, VKT), share of various vehicles by regions and fuel systems, and also vehicle owner information from the collected sample group. The data was collected from various vehicle types, according to registration types under the motor vehicle act and land transport vehicle act. The fuel consumptions of diesel HDV buses are 12.5, 21.6 and 15.2 L/100km for fixed route-, none fixed route- and private buses in the latest report (8.0, 4.6 and 6.6 km/L). This information included all legislation vehicles in the passenger bus category, which include passenger vans, light buses, mini-buses, coaches, mini-coaches, and double-decker buses etc. [15].

On the other hand, the Department of Land Transport (DLT) has implemented the vehicle location-tracking project for safety reason. Every public vehicle, which is used for carrying passengers or goods, must install vehicle-tracking equipment [16, 17] and this equipment must be able to transfer the vehicle GPS (Global Positioning System) location to DLT – GPS tracking system management center (DLT GPS) in real time. The current technology of vehicle tracking equipment has been further developed to obtain larger information such as vehicle identification, driver identification, as well as all energy-related parameters. Referring to advance vehicle tracking technology and requirement for annual vehicle registration, the Department of Alternative Energy Development and Efficiency (DEDE) initiates the on-going research project entitled Feasibility Study on Energy Efficiency in HDV Transport, which composes four sub projects including various HDV: the agricultural product trucks, container trucks, oil tanker trucks, and heavy-duty buses for passenger transport. The major goal of this project is to initiate a new survey approach for acquiring energy efficiency indicators in real time. The project offers a partial subsidizing for the HDV operator companies, for upgrading their GPS location tracking systems into these new advanced systems. In this initial state, the target of sample size is set at 400 vehicles in each group of HDVs, equivalent to totally 1,600 sample vehicles. This new survey approach provides advantage in acquiring energy-related parameters in HDVs, for example:

- The vehicle and driver identification can be used to categorize database into various types of HDVs which is useful for the policymaker to analyze their implementation of energy efficiency measures.
- In the most successful case, if this advance vehicle tracking is widely accepted, there will be many energy-related databases available from a number of HDV sample volumes.
- Moreover, the real-time energy-related database can help the energy-policymakers or the traffic authorizers to gain efficient-control over transport infrastructure.

This study uses the acquired data from this DEDE project.

3. Methodology

This section explains the vehicle details, data collection method and data analysis in this study. The collected data from GPS tracking equipment used in this study composes three functional modules, i.e., vehicle position tracking, driver identification and the vehicle operational information which are tapped from the vehicle CAN bus system. On the other hand, the considered driving route are selected from all operational routes of 400 samples of HDV buses which serve passenger transport all over the country. The Bangkok – Nakhon-Ratchasima route is considered for this study because it has appropriate length (around 400 km), has sufficient elevation gain (uphill) and elevation loss (downhill) and there are 7 HDV buses operated on this route more than the others. The operational driving routes of all HDV buses and information of selected route are shown in Figure 3. This route starts from Bangkok and ends at Mittraphap road, Donwai, Nonsung, Nakhon-Ratchasima. The highest altitude is in Pakchong district measured about 400 m above sea level and 172 km far from the starting point.

3.1. Considered HDV bus

All the selected HDV buses in this study were built up from SCANIA buses. These vehicles provide both CAN bus systems and OBD-II connectible ports. The specifications are shown in Table 2.

3.2. Data collection

The data of seven HDV buses were collected for 30 days. All the collected information was acquired, transmitted through the mobile network and recorded in the cloud database system. Then the administrators, bus operators, vehicle owners can achieve the data via a web-based system. Figure 4 shows the schematic diagram of data and information flow.

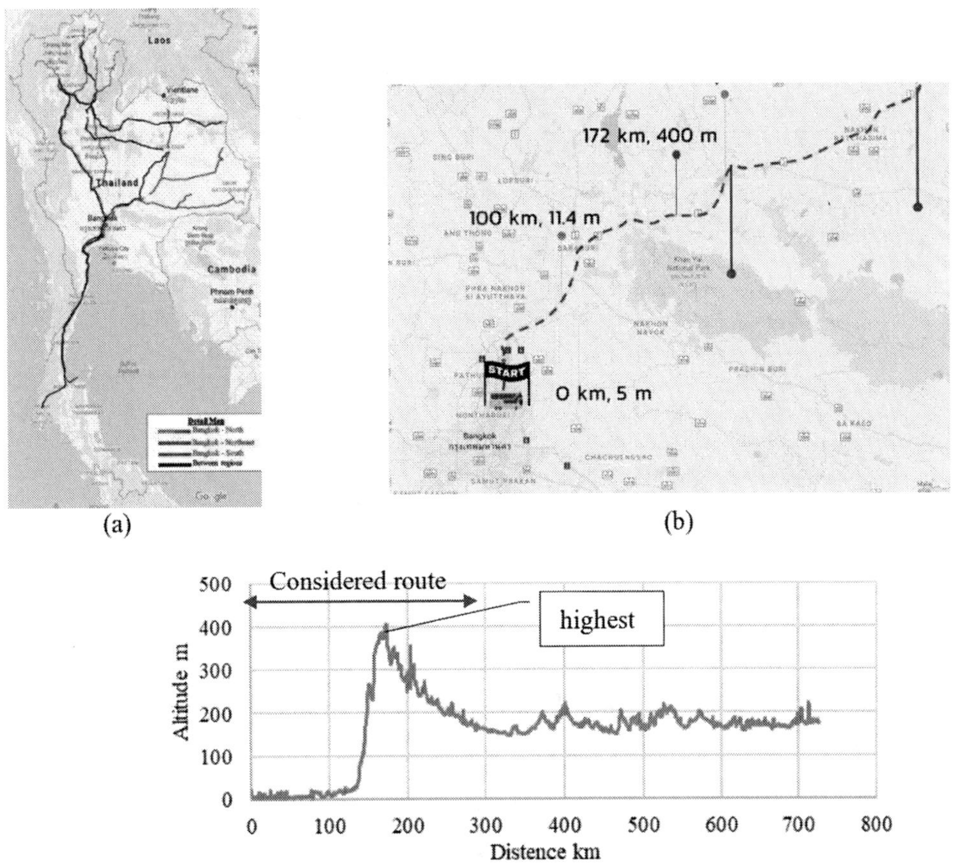

(a) (b) (c)

Figure 3. Driving routes of the sample vehicles (a) Driving route of all 400 sample vehicles, (b) selected driving route in this study (Bangkok – Nakhon-Ratchasima) and (c) altitude and distance from Bangkok

Table 2. Specification of considered HDV buses

Origin of engine and chassis part	SCANIA
Engine size (BHP)	310 – 420
Fuel use	Diesel fuel
Seat number	44 (max)
Total weight (ton)	18 – 24 (21 – 27 included passenger)
Load factor (of maximum passenger number)	80% – 100%
Wheel number	8 wheels
Sample photo (taken from bus company)	

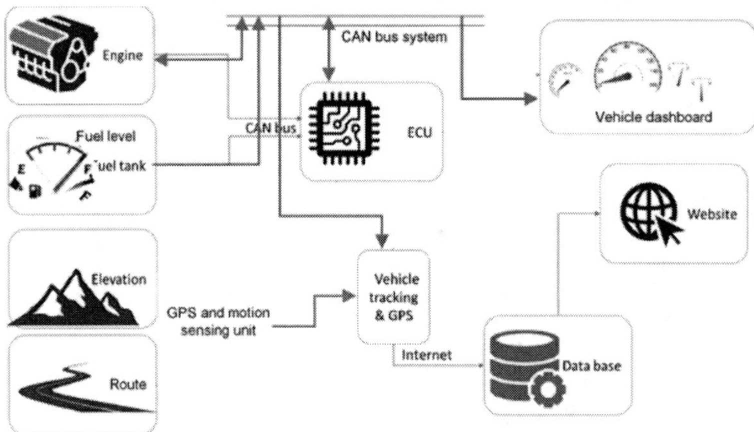

Figure 4. Schematic diagram of data collection methodology

The data flow consists of four parts. First, all vehicle operational information was read and transmitted with the vehicle Engine Control Unit (ECU). Then the information was acquired and shown on the vehicle user interface (vehicle dashboard). The vehicle tracking equipment also collected all CAN bus information as well as observing the vehicle GPS position and motion from the GPS, and motion sensing functional module. Finally, all the information was preliminary analyzed, recorded in the database and provided on an internet website. The preliminary real-time data, which was taken from the internet web site went through pre-processing, are shown in Figure 5. Then the daily data was divided into small trips between Bangkok – Nakhon-Ratchasima, for departing- and returning trips separately. To ensure that all the information was correctly observed, the validated routines were performed for all considered vehicles. The road altitude information was randomly checked with Google Earth software [18]. The driving distance was validated between GPS information and recorded vehicle odometer. The fuel consumption results which were determined from fuel consumed quantity of real- time vehicle CAN bus were validated with the recorded fuel expense given from the HDV bus company.

Table 3. List of considered energy-related parameters to be analyzed with fuel consumption

Parameters	Unit
Averaged speed	km/h
Percentage of idling operation	%
Percentage of moving acceleration	%
Percentage of moving deceleration	%
Load factor	%
Elevation gain ratio	m/km
Elevation loss ratio	m/km

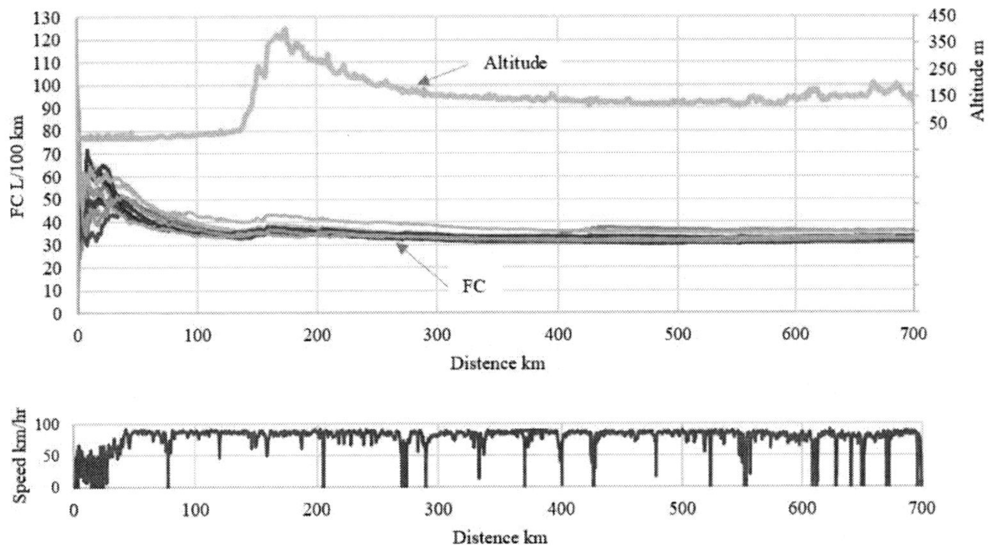

Figure 5. Preliminary analyzed data from the web based system

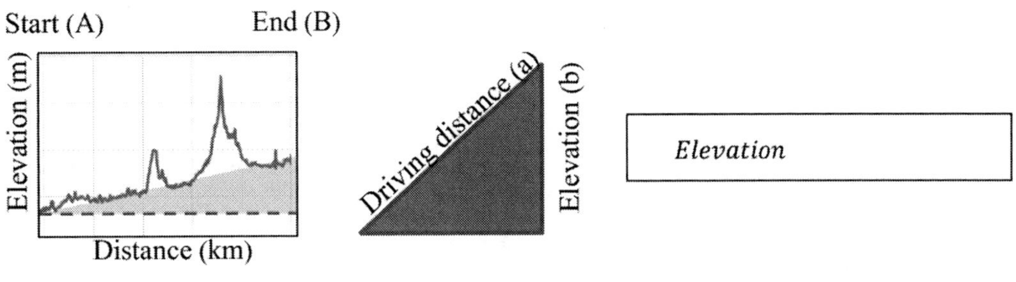

Figure 6. Definition of elevation gain and elevation loss ratios

3.3. Data analysis

From the preliminary results shown in Figure 5, the energy-related parameters were analyzed to identify and correlate the ones that effect fuel consumption (FC). The considered parameters are shown in Table 3. The situations of vehicle operational status were collected in the percentage unit. The boundary limits of vehicle accelerated and decelerated situations were set at +2 and -2 m/s2. The elevation gain and elevation loss ratios are calculated as shown in Figure 6.

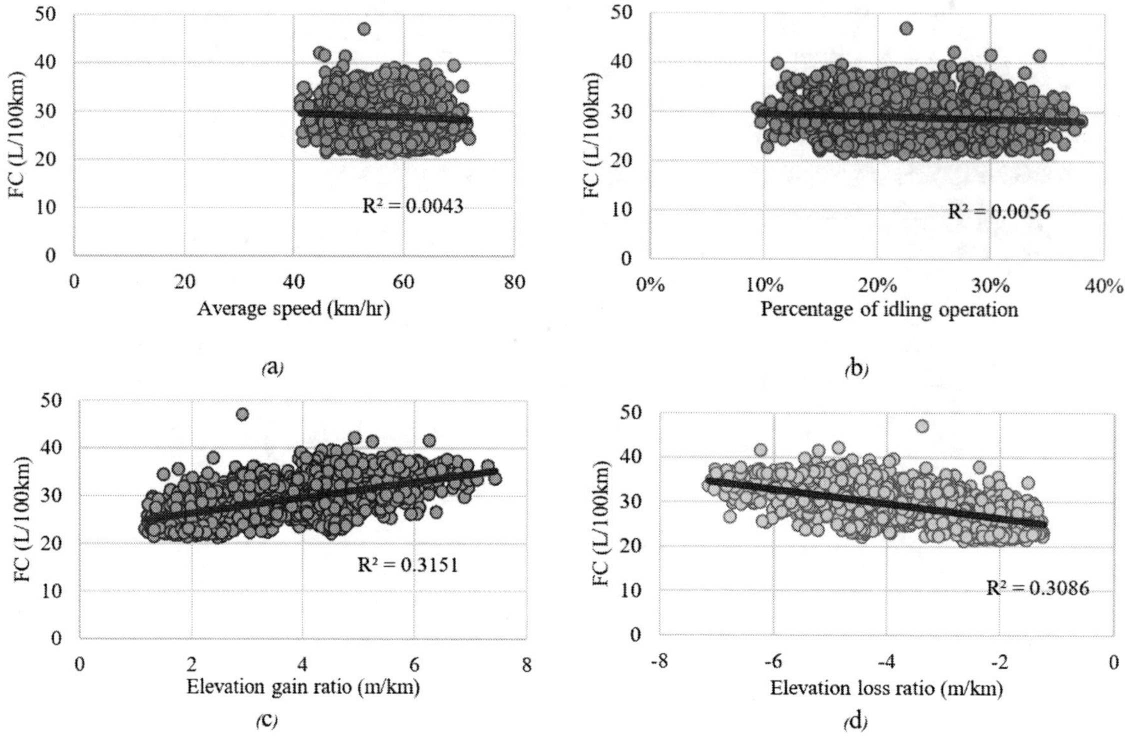

Figure 7. Observed linear relationship between influential parameters and fuel consumption (a) Averaged speed (b) Percentage of idling operation (c) Elevation gain ratio and (d) Elevation loss ratio

4. Results and discussion

The relationship between considered energy-related parameters and fuel consumption (FC) were analyzed by using linear regression analysis. Then the influence parameters were indicated according to the regression results. The 30 points of daily-averaged data were analyzed first. The results show that there are only four measured parameters in Table 3 exhibit observable linear relationship with the fuel consumption data as shown in Figure 7

Next, the real-time micro trip data of four energy-related parameters previously found linked with fuel consumption was used to perform further analysis. The micro trip results show fuel consumption consistently increases with elevation gain ratio as shown in Figure 8, while the impacts of other parameters do not show observable evidence. By dividing elevation gain ratio into intervals of 10, the results show a different trend at 50 m/km as shown in Figure 9. For the range lower than 50 m/km, the fuel consumption increases linearly with a slope of 0.75 (L/100km by m/km, $y = 0.75x + 29.71$). After that, the fuel consumption fluctuates around a constant value of about 59.03 ± 7.25 L/100km.

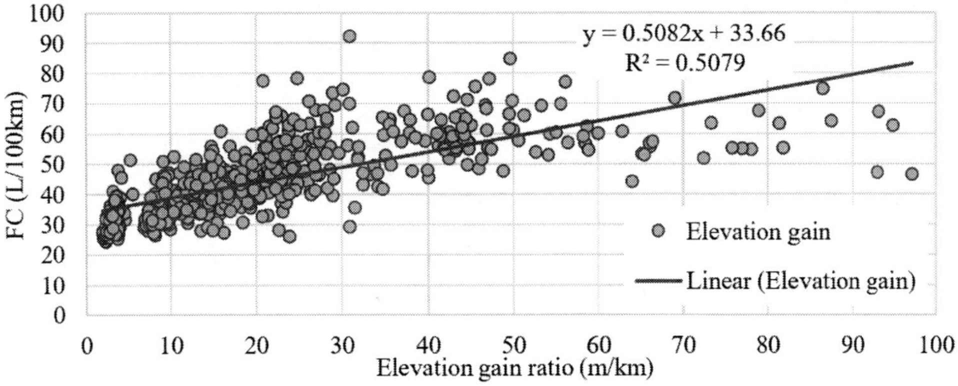

Figure 8. Relationship between elevation gain ratio and measured fuel consumption

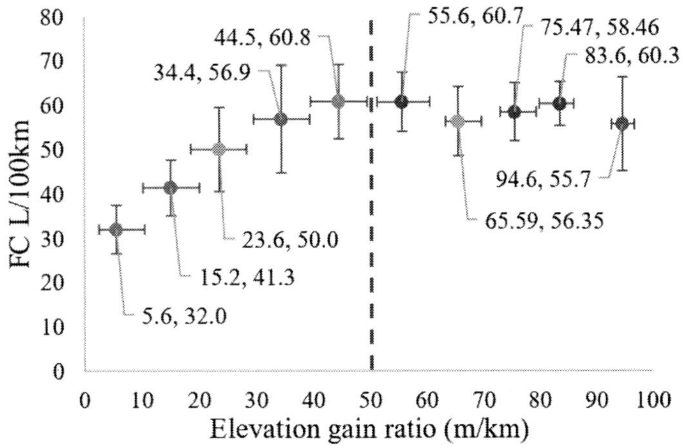

Figure 9. Different relationship for low and high elevation gain ratio

5. Conclusion

In this study, the fuel consumption of HDV buses was evaluated with recourse to micro trip database obtained from a vehicle positioning and energy consumption tracking system. The results of this study are a part of a research project which aims to conduct a feasibility study on including energy consumption tracking with the vehicle tracking device which has become the safety installation requirement of the Department of Land Transport. The real time database was analyzed to identify and evaluate the impacts of considered energy-related parameters on the fuel consumption of HDV buses. The results proved that fuel consumption depends on energy-related parameters such as vehicle deteriorated situation, load factor, driving behavior, and road condition, including the changing of vehicle altitude. However, in this study, the impacts of other parameters are in a minor evidence, except for the elevation gain ratio. The results show that fuel consumption linearly increases with elevation gain in a slope of 0.75 L/100km by m/km of elevation gain ratio. However, the fuel consumption achieves a maximum constant around 59.03 L/100km when the elevation gain ratio reaches 50 m/km.

Acknowledgments

The authors would like to acknowledge the Department of Alternative Energy and Efficiency and the Energy Conservation Fund (ENCON fund) for the financial support on conducting the project entitle "Feasibility Study on Energy Efficiency in HDV Transport". All the results of this study are a part of this project.

References

[1] Department of Alternative Energy Development and Efficiency, Ministry of Energy 2020 *Energy Situation of Thailand*, https://www.dede.go.th/ewt_news.php?nid=47349, accessed on 1 December 2020
[2] Energy Policy and Planning Office, Ministry of Energy 2020 *Thailand Energy Efficiency Development Plan*, 2016, 28 March [Online] http://www.eppo.go.th/index.php/en/policy-and-plan/en-tieb/tieb-eep, accessed on 1 December 2020
[3] Energy Policy and Planning Office, Ministry of Energy 2020 *Resolutions of National Energy Policy Council,* http://www.eppo.go.th/epposite/index.php/th/component/k2/itemlist/category /429-resolutionsnepc#s5, accessed on 1 December 2020
[4] Department of Alternative Energy Development and Efficiency, Ministry of Energy 2020 *News and Announcements – Executive News, Public Hearing Seminar on EEP2018,* https://www.dede.go.th/ewt_news.php?nid=52418&filename=index, accessed on 1 December 2020
[5] Kerdlap P and Gheewala S H 2016 *J. Ind. Ecol.* **20**(6) 1399-1411
[6] Thananusak T Rakthin S Tavewatanaphan T and Punnakitikashem P 2018 *Int. J. Electr. Hybrid Veh.* **9**(4) 361-381
[7] Tanatvanit S Limmeechokchai B and Chungpaibulpatana S 2003 *Renew. Sust. Energ. Rev.* **7**(5) 367-395
[8] Pongthanaisawan J, Sorapipatana C and Limmeechokchai B 2010 *Renew. Sust. Energ. Rev.* **14**(9) 2966-2975
[9] Limanond T Jomnonkwao S and Srikaew A 2011 *Energy Policy* **39**(5) 2754-2763
[10] Chollacoop N Saisirirat P Fukuda T and Fukuda A 2011 *Energies* **4**(1)108-125
[11] National Energy Policy Office 1997 *The Study of Energy Conservation Guidelines for Vehicles*
[12] Energy Policy and Planning Office 2008 *The Study and Survey of Energy Consumption in Thai Transport Sector*
[13] Energy Policy and Planning Office 2013 *The Study on Development of Forecasting Model and Survey of Energy Consumption in Thai Transport Sector*
[14] Energy Policy and Planning Office 2019 *The Study and Survey of Energy Consumption in Thai Transport Sector: Executive Summary*
[15] Department of Land Transport 2020 *Definition of Passenger Carried Vehicle,* https://web.dlt.go.th/statistics/, accessed on 1 December 2020
[16] Head of the National Council for Peach and Order, Measures to Increase Safety in Public Passenger Vehicles 2020 *Government Gazette 134:85D (in Thai),* http://gps.dlt.go.th/?page_id=136, accessed on 1 December 2020
[17] Department of Land Transport 2012 Requirements, Specifications and Functionality of the Vehicle Tracking for the Dangerous Goods-Transport Vehicles, *Government Gazette 129:194D (in Thai),* http://gps.dlt.go.th/?page_id=136, accessed on 1 December 2020
[18] Google Earth Pro 2017 https://www.google.com/earth/

IOP Conf. Series: Materials Science and Engineering 1137 (2021) 012066 doi:10.1088/1757-899X/1137/1/012066

Numerical investigation of air flow in a vertically paddy bed dryer using computational fluid dynamics

Weerachai Chaiworapuek[1], Nuttakorn Chuenboonma[1,*], Teerapat Thungthong[1], Ratiporn Munprom[2], Sudathip Sae-tan[3], and Damrongvudhi Onwimol[4]

[1]Department of Mechanical Engineering, Faculty of Engineering, Kasetsart University, Bangkok Thailand 10900
[2]Department of Materials Engineering, Faculty of Engineering, Kasetsart University, Bangkok Thailand 10900
[3]Department of Food Science & Technology, Faculty of Agro-Industry, Kasetsart University, Bangkok Thailand 10900
[4]Department of Agronomy, Faculty of Agriculture, Kasetsart University, Bangkok Thailand 10900

* Corresponding Author: fengwcc@ku.ac.th

Abstract
This research focuses on the numerical investigation of air flow in a vertically paddy bed dryer using computational fluid dynamics - CFD. In this study, the Tubtim Chumphae paddy and perforated plates were assumed to be porous media and their properties were characterized experimentally in a 0.05 m × 0.05 m × 1 m duct. In this paper, the obtained porosity and permeability were applied to a numerical model, the air flow through a 100 kg paddy, stored in a vertical bed dryer. The results from the simulation and experiment agreed well with each other via the calibration process. The pressure drop across the dryer was found to increase when the inlet velocity gained from 0.025 to 0.1 m/s with an increment of 0.025 m/s. In this study, the results showed that air flow was controlled past the paddy with the deviation of less than 2.82% to achieve a high rice quality after the drying process. Because this drying technique used low temperature air and had low energy consumption, therefore, it is found to be one of the promising techniques and truly benefits the farmer in the future.

Keywords: CFD simulation, Bed grain drying, Energy, Paddy, Porous media

1. Introduction

In the last decade, more than 15,000 papers have been published to present experimental and mathematical models, which enhance the drying efficiency of grain dryers [1, 2]. Paddy is one of those grains, and it is the main food in many countries around the world [3]. After the paddy is harvested in rice fields, it has very high moisture content, so the drying process is needed before storage and milling operations. One of the simplest but effective dryers is the shallow-bed dryer, because the bed thickness is less than 20 cm. In this dryer, paddy depth does not affect the change of flow rate [4]. In industrial applications, paddy drying is an energy-intensive process, consuming specific primary energy between 1.44 and 74.73 MJ/kg. It could affect the odour and nutrition of paddy rice [5]. Meanwhile, the dryer that operates using air with temperature below 40°C and flow rates less than 0.3 m^3/s can preserve rice quality [6]. Following the advancement of numerical simulation, the computational fluid dynamics-CFD has been employed to investigate the flow in many applications including the air flow through the paddy [7-12]. The paddy is typically modelled as porous media, having porosity between 0.462 and 0.542 [13]. Mathematical modelling and simulation in 3D based on momentum and continuity equations with viscous resistance, inertial resistance and porosity of paddy porous zone are 10.626×10^7, 18,605 and 0.57, respectively showed 10% error. This was studied under the case of 25 cm bed depth and air flow rate between 0.10 and 0.22 kg m^{-2} s^{-1} [14]. Some simulation results revealed that the shallow-bed dryer, having 4 ft bed height had a higher porosity when the bed height increased and humidity decreased [15]. The improvement of air distribution in a woodchip fixed-bed dryer, producing almost perfectly uniform air distribution could give the better performance for drying process [16]. Thus, this research focuses on the numerical investigation of air flow in a vertically paddy bed dryer using computational fluid dynamics – CFD to predict the uniformity of air flow that enters the paddy zone. The Tubtim Chumphae paddy and perforated plates, employed in the dryer were assumed to be porous media and their properties were characterized experimentally. The results obtained in this research would lead to the design of high performance vertical-bed dryer in the future.

2. Porous media investigations

2.1 Experimental setup

The permeability and porosity of Tubtim Chumphae paddy and 4 different types of perforated plate were investigated in the test section as depicted in Figure 1 to perform as porous media in the simulation. The test section had a size of 5 cm width × 5 cm height × 1 m length and the porous media was installed at the middle of the test section as depicted in the figure. For the paddy, the bed thickness *(Δx)* was 30 mm.

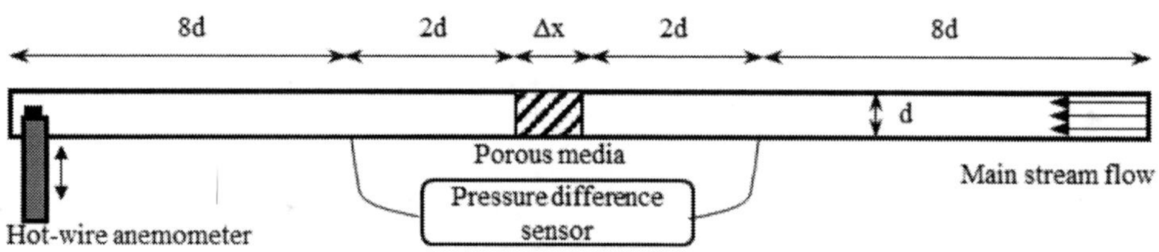

Figure 1. Experimental setup for permeability and porosity measurement.

For the test of perforate plate, the plate configurations were presented as shown in Table 1 and Figure 2. In the experiment of both paddy and perforated plate, the inlet velocity was set during 0.2 – 3 m/s, corresponding to the Reynolds number in the range of 340 – 10,300.

Table 1. The dimensions of perforate plate.

Hole diameter *(ø)*	Porosity *(ε)*	Hole pitch *(p)*	Thickness *(Δx)*
5.0 mm	0.36	8.0 mm	1.0 mm
8.0 mm	0.77	12.0 mm	1.0 mm
9.5 mm	0.53	12.5 mm	1.5 mm
10.0 mm	0.47	14.0 mm	1.3 mm

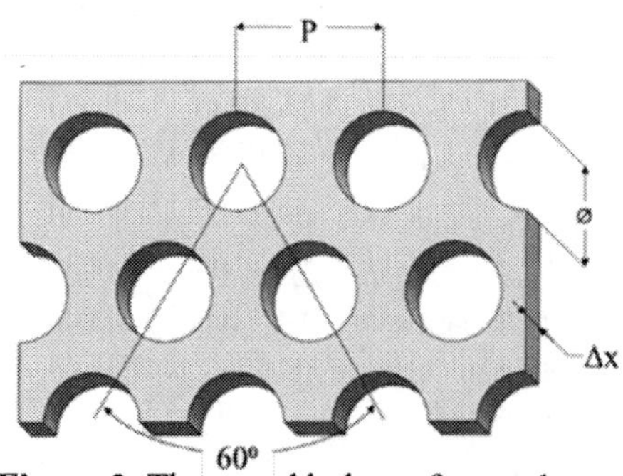

Figure 2. The round hole perforate plate.

2.2 Relating equations

For the calculation using computational fluid dynamics, the continuity and momentum equations are mutually utilized and they are as follows:

$$\partial \rho/\partial t + \nabla\cdot(\rho \boldsymbol{v}) = 0 \tag{1}$$

$$\partial \rho \boldsymbol{v}/\partial t + \nabla\cdot(\rho \boldsymbol{v}\boldsymbol{v}) = -\nabla p + \mu\nabla^2\cdot \boldsymbol{v} + \rho\mathbf{g} + \boldsymbol{F} \tag{2}$$

The fluid flow in porous media is modelled via a simple modification of the Navier-Stokes equations. The source term is added, responsible for an additional resistance or pressure drop to the flow exerted by the porous material in Equation 2. The source term of porous media was added as an external body force and the Darcy-Forchheimer equation or Equation 3 was employed. In the zone of porous media, the air flowed in one specific direction, shown as subscript i. Thus, the Darcy-Forchheimer equation can be described as follows:

$$\Delta p/\Delta x_i = S_i = -\mu\cdot v_i / \alpha + 0.5 \cdot C_2 \cdot \rho \cdot |v| \cdot v_i \tag{3}$$

$$\Delta p = a v_i^2 + b v_i \tag{4}$$

$$1/\alpha = b/(\mu\cdot\Delta x_i) \tag{5}$$

$$C_2 = 2a/(\rho\Delta x_i) \tag{6}$$

Where $1/\alpha$, C_2 and Δx_i represent the viscous resistance coefficient, inertial resistance coefficient and thickness of porous material, respectively.

3. Numerical method

In this paper, a validation process was conducted in the flow domain as the experiment, explained previously. It was a rectangular domain having 5 cm width × 5 cm height × 1 m length and the porous medias were set at the middle of the domain as depicted in Figure 3a. The permeability and porosity of porous medias were obtained from the experiment and the velocity inlet was set to be during 0.2 – 3 m/s as the experiment. The pressure difference obtained from the experiment and simulation would be compared to check the accuracy of numerical model. Then, a fixed-bed dryer was designed to have 2 vertical paddy beds and each had a cross-sectional area of 1.2 m × 0.6 m and a bed thickness of 0.2 m. The air domain inside the dryer is shown as Figure 3b. The inlet was set as velocity inlet of 1, 2, 3 and 4 m/s, corresponding to the velocity through the paddy of 0.025, 0.050, 0.075 and 0.100 m/s, respectively. The air from the inlet flowed pass 3 perforated plates and fins to decrease the turbulence and control the flow to have a uniformity of velocity distribution. A symmetry boundary condition was employed to reduce the computing resources and times. The mesh sensitivity method was conducted at the element number of 1,205,834, 2,899,632, 4,836,546, 6,905,562, 9,605,356 and 12,658,956. The pressure drop across the dryer was found to converge at the element number of 6,905,562 element model. Thus, this mesh size was selected and utilized through this study. Besides, the uniformity of air flows was checked by 2 standard deviation *(2SD)* and 2 relative standard deviation *(2RSD)* of air velocity on the paddy surface. The velocity on the paddy surface was sampling on the 4-vertical lines on the middle of each plenums of Z-axis coordinated at z = 0.125(150 mm), 0.375(450 mm), 0.625(750 mm) and 0.875(1,050 mm) from the bed width of 1,200 mm.

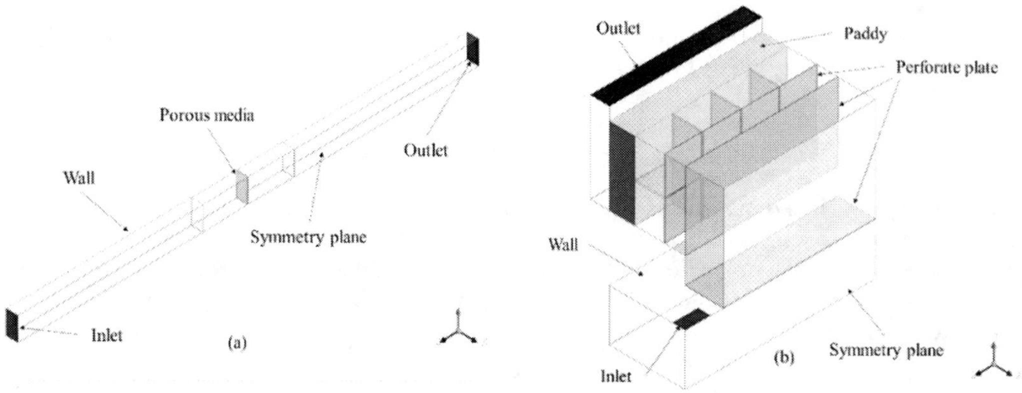

Figure 3. Air flow domain of (a) test section for validation process and (b) dryer.

4. Results and discussion

4.1 Porous media properties and validation results

From the experiment, the properties of porous medias were measured as shown in Table 2. They were input to the numerical model as explained previously in the topic of validation process. The coefficient of determinations *(R^2)* between the pressure differences from experiment and simulation were obtained. The results showed that most cases except the case of a perforated plate with a diameter of 10 mm had the R^2 above 0.95, confirming the accuracy of the numerical model. And, this model will be employed to predict the air flow in a dryer.

Table 2. Porous media properties and the R^2 from validation process.

Porous media	Porosity	Permeability	Viscous resistance coefficient	Inertial resistance coefficient	R^2
Tubtim Chumphae paddy	0.41	2.265×10^{-9}	4.435×10^{-8}	1884.0	0.9956
Perforate plate: ⌀5 mm	0.36	5.965×10^{-7}	1.677×10^{-6}	11183.7	0.9883
Perforate plate: ⌀8 mm	0.77	5.965×10^{-7}	1.677×10^{-6}	9575.5	0.9687
Perforate plate: ⌀9.5 mm	0.53	8.947×10^{-7}	1.118×10^{-6}	3907.5	0.9504
Perforate plate: ⌀10 mm	0.47	7.754×10^{-7}	1.290×10^{-6}	3579.3	0.8865

4.2 Air flow results in a dryer

Figure 4a shows the air flow inside the dryer domain installing a perforated plate with 8 mm diameter when the velocity inlet is 4 m/s. The results present that the swirl flow appears at the entrance region and it decayed after passing 3 perforated plates. The velocity contour of air flow on the entering paddy surface was presented in Figure 4b. On this plane, the mean velocity was 0.112 m/s and 2 standard deviations is 0.0032 m/s or 2.825% for the case of the perforate plate with 8 mm diameter. This showed that the uniformity of velocity distribution performed at the entrance of the paddy chamber.

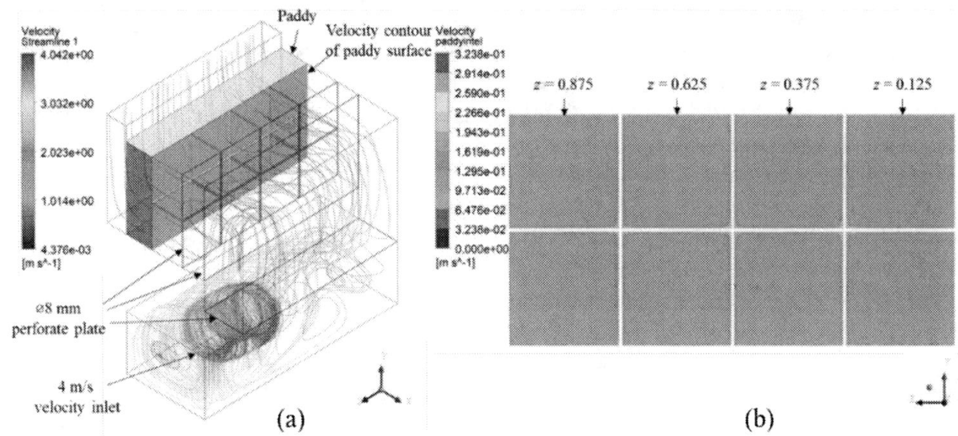

Figure 4. (a) Air path lines in the dryer (b) velocity contour at the entrance of paddy chamber.

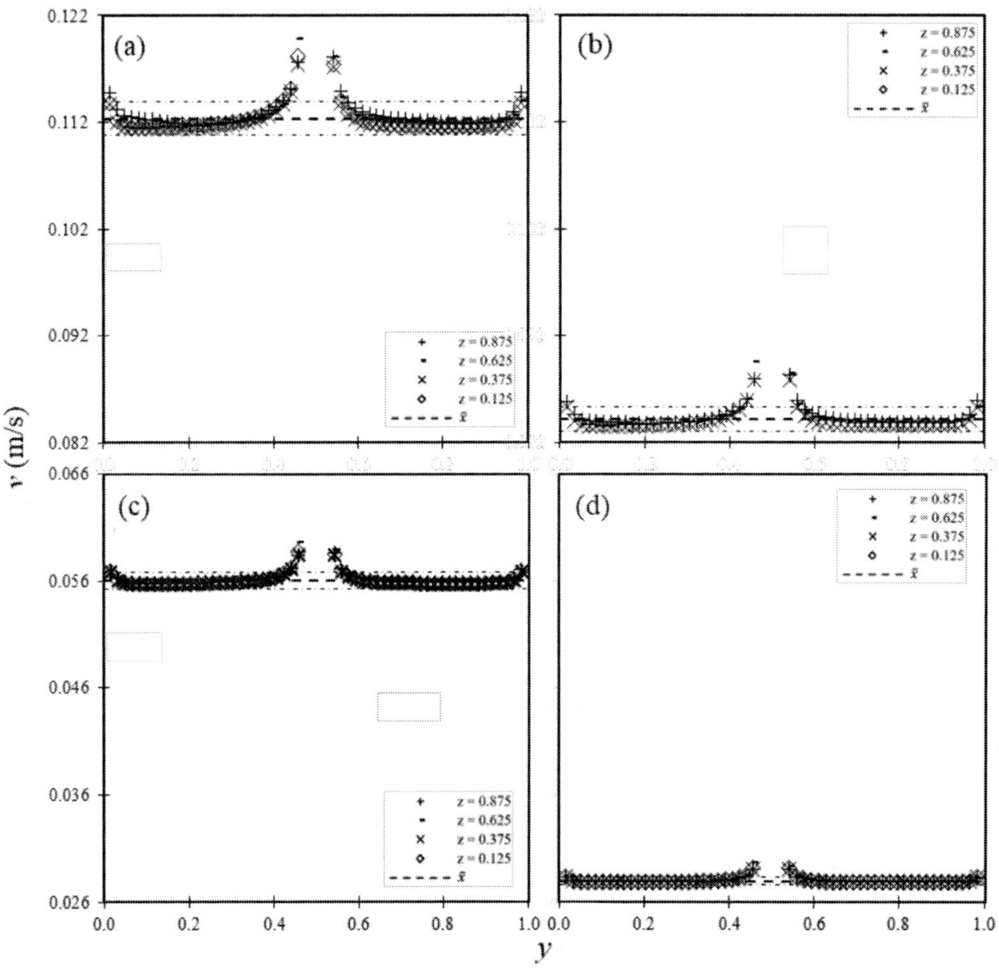

Figure 5. Velocity profiles in the case of perforated plate with diameter of 8 mm and inlet velocity of (a) 0.1 m/s, (b) 0.075 m/s, (c) 0.05 m/s, and (d) 0.025 m/s.

The velocity profiles on the plane at the entrance of the paddy chamber are presented as shown in Figure 5. The results show that the velocity along the line of $z = 0.125$, 0.375, 0.625, and 0.875 are in the same tendency, meaning that the velocity is quite uniform over paddy chamber. Only in region of $y = 0.5$, the profiles behave differently with other regions due to the separation flow from the horizontal fin before the air flow enters the paddy zone.

Moreover, the sizing effect of the hole diameter of perforated plate on the uniformity of air flow in dryer was also investigated. Figure 6 shows the velocity distribution error (*2RSD*) and pressure drop across the dryer (ΔP) at 4 different hole diameters and 4 inlet velocities. The error as shown in Figure 6a gains with the increase in inlet velocity and it characterizes similarly when the diameter is 5 and 8 mm. Meanwhile, if the diameter is greater than 8 mm, the error is found to be increased when the hole diameter is bigger. The pressure loss as depicted in Figure 6b is slightly affected by the change in the hole diameter. Hence, the 8 mm diameter plate was recommended in attempt to employ in the dryer because it provided a uniformity of velocity distribution as the 5 mm diameter plate but it had a relatively lighter.

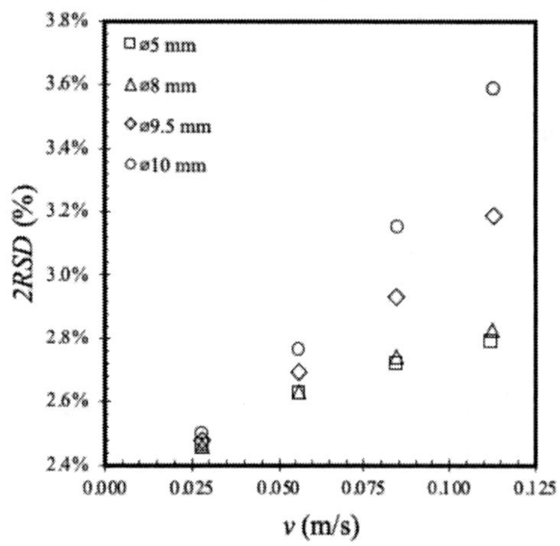

(a)

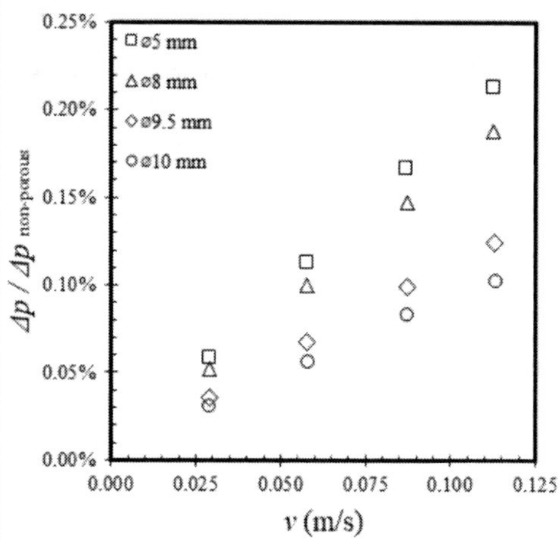

(b)

Figure 6. (a)The deviation of velocity profiles and (b) ratio of pressure drop across the dryer.

5. Conclusion

The uniformity of velocity distribution in a fixed-bed dryer, over the entry plane of paddy was investigated over 4 perforated plates, having different diameter size using computational fluid dynamics. The results showed that the 8 mm diameter model provided the relatively highest uniformity of velocity distribution with the error less than 2.82%. Besides, this paper also revealed the porous characteristics of Tubtim Chumphae paddy and perforated plate, having diameter of 5, 8, 9.5, and 10 mm. The obtained results in this research will offer key information for designing a vertical-bed dryer for paddy in the future.

Acknowledgments

The authors would to thank Kasetsart University, Bangkok, Thailand and National Research Council of Thailand-NRCT for financial support.

References

[1] Sivakumar R Saravanan R Perumal A and Iniyan S 2016 *Renew. Sust. Energ. Rev.* **61** 280
[2] Tohidia M Sadeghib M and Harcheganic M 2017 *Renew. Sust. Energ. Rev.* **70** 519
[3] Vaughan D Morishima H and Kadowaki K 2003 *Curr. Opin. Plant Biol.* **6-2** 139
[4] Doungporna S Poomsa-adb N and Wiset L 2012 *Food Bioprod. Process* **90-2** 187

[5] Jittanit W Saeteaw N and Charoenchaisri A 2010 *J. Stored Prod. Res.* **46** 209
[6] Sarker M Ibrahim M Aziz N and Salleh P 2014 *Energy Convers. Manag.* **77** 389
[7] Suvanjumrat C and Loskupapaiboon K. 2020 *Eng. J.* **24(2)** 2
[8] Suvanjumrat C 2017 *Eng. J.* **21(5)** 225
[9] Suvanjumrat C 2016 *Eng. J.* **21(3)** 207
[10] Chaiworapuek W Champagne J Y Hajem M E Kittichaikarn C 2010 *AIP Conf. Proc.* **1225** 249
[11] Thungthong T Chalearmwattananon N Mongkolkitngam T and Chaiworapuek W 2020 *Int. J. Heat Mass Transfer* **163** 1
[12] Zare D Naderi H and Ranjbaran M 2015 *Dry Technol.* **33** 570
[13] Bhattacharya K Sowbhagya C and Swamy Y 1972 *J. Sci. Food Agric.* **23** 171
[14] Ranjbaran M Emadi B and Zare D 2014 *Dry Technol.* **32** 919
[15] Srivastava V and John J 2002 *Energy Convers. Manag.* **43** 1689
[16] Roman F Schaefer V and Hensel O 2012 *Biosystems Engineering* **112**(4) 359

Energy management of precooling process for green cabbages

Thanwit Naemsai[1,*], Banyat Niyomvas[1], and Jaruwat Jareanjit[1]

[1]Department of Mechanical Engineering, Faculty of Engineering, Rajamangala University of Technology Srivijaya, Songkhla, Thailand 90000

* Corresponding Author: Thanwit.n@rmutsv.ac.th

Abstract
Cold storage is one of the most widely used and effective processes to preserve shelf-life harvested agricultural products during long periods. Research to date has not yet determined the energy management of pre-cooling methods of green cabbage. This study seeks to investigate the cooling performance of the cold storage, keeping green cabbages with good quality and to report different pre-cooling strategies. The cold storage specifications included 15 HP compressor with a voltage of 380 V, a temperature range of -50 to 50°C, and R22 refrigerant. The mass of green cabbage was examined with three values of 30, 40, and 50 kg. Moreover, the pre-cooling strategies of the green cabbages were divided into four scenarios including the parallel pattern in the front zone, staggered pattern in the front zone, parallel pattern in the back zone, and staggered pattern in the back zone. The experimental results revealed that the coefficient of performance (henceforth COP) was significantly affected by the mass of the cabbage and cooling zone of the cold storage. While the pattern in case of 30 kg cabbage precooling had a slight effect on cooling performance due to the lower density arrangement and heat extraction. The highest COP appeared at the 30 kg cabbage precooling with the strategy of staggered pattern in the front zone. The COP value was 2.61.

Keywords: Energy, Management, Precooling, Green cabbages

1. Introduction

Recent evidence suggests that cabbages provide people with a variety of essential vitamins, antioxidants, and dietary fiber, and benefit to many aspects of health [1]. Post-harvest losses and waste in the supply chain of cabbage are considerably high before the product even reaches the consumer. Industrial production is faced with the challenge of minimizing post-harvest losses. Hence, research efforts are needed to prolong the shelf-life of vegetables and subsequently reduce postharvest losses. The solution is a rapid pre-cooling process in cold storage because it can enhance keeping nutrition ingredients and freshness, improving coldness, and prevent chilling injury [2]. Generally, the capacity of the cold storage must be based on a thorough heat load calculation for each project. Refrigeration load can vary widely for stores of the same capacity depending on the design, local conditions, product mix, etc. In the practice, a safety margin of some 50 percent of the theoretical calculation has been applied [3].

Several attempts have been made to investigate the variables affecting the cooling performance. The type of cold storage greatly influenced the freezing time. For example, the immersion freezer froze the product faster than the air blast freezer operating at the same temperature due to improved surface heat transfer [4]. An appropriate operating temperature was the main factor of cold storage with energy savings [5]. Of particular concern is the temperature of the product. To give an example, the storage temperature of fish meat and the chicken breast was approximate -30±0.5°C [6,7]. The strawberry and green bean were frozen at -27±0.5°C [8]. For the cabbage, the storage temperature was 4±0.5°C [2]. However, the lower temperature of the freezer may cause an increase in the freezing cost. Since it has been observed that most freezers are designed to operate a few degrees below the required storage temperature of the product [9]. Tian et al. [10] proposed that the freezing time was reduced as the air speed was increased by the heat pipe-assisted cooling supply. Moreover, Chauhan et al. [11] indicated that the heat transfer coefficient was changed by the velocity of each shelf. Thus, the cooling performance of each shelf was different.

Recently, in view of the increasing demand for energy and the escalating costs of energy from all sources, energy management, and energy conservation have become increasingly important in all industries and also in the food systems. Good energy management and energy conservation depend on the operation of cold stores, such as product pattern and cooling zone. Mukama et al. [12] reported the product pattern (liner-packaged stacks), and they found that the most decisive factor affecting the cooling rate and energy consumption for the forced-air precooling of pomegranate fruit inside ventilated cartons. Gruyters et al. [13] demonstrated that the apple precooling condition with the using the reusable plastic crate inside the front zone provided the best cooling rate, although the high airflow rates might cause chilling injury to the apples. Han et al. [14] found that the numerical model with computational fluid dynamics (CFD) was well suited to evaluate package performance. The aerodynamic resistance and energy consumption of a certain package design provides a more comprehensive analysis of the overall forced-air cooling process. Wu et al. [15] investigated cooling time and uniformity of the packed produce, and its resistance to airflow. They found that it could often provide key information about the overall cooling efficiency of a specific package design in full-scale experiments of forced-air precoolers for citrus fruit.

Previous studies of energy management and energy conservation have not dealt with the case of green cabbage. Besides, no research has been found that surveyed the effects of the cooling zone and product pattern on the cooling performance during precooling. This paper attempts to show the effective way for energy management and energy conservation of the green cabbage inside the cold storage.

2. Materials and Methods

2.1 Experimental Setup and Procedure

The samples of green cabbage (comprising several cultivars of Brassica oleracea) with the average size (diameter of 12.5 cm and height of 8 cm) were obtained from the cabbage farm in Songkhla, Thailand. The samples were cleaned and all extraneous materials were removed. Cold storage specifications included 15 Hp (11.18 kW) compressor with a power supply voltage of 380 V, refrigerant of R22, storage space with the dimension of 2.4 m x 3.5 m x 2m (Width x Length x Height), and stainless-steel trolley with 9 tray capacity (see Figure 1). For the cold storage feature, the cooler unit was mounted at the roof level. There were no special means of directing the air over the product and therefore it generally tended to swirl about in the empty spaces in the front zone of the room and slightly flowed between the shelves or trays.

In the experiment, the mass of green cabbage was examined with three values of 30, 40 and 50 kg. Moreover, the pre-cooling strategies of the green cabbages were divided into four scenarios including the parallel pattern in the front zone (F_PL), staggered pattern in the front zone (F_ST), parallel pattern in the back zone (B_PL), and staggered pattern in the back zone (B_ST). For the initial condition, the distance between cabbages was controlled at 5±0.5 cm (25% of product hydraulic diameter). The temperature before entering the cold storage was controlled at 28±0.5°C. The temperature differential setting was also controlled at 2°C (from -1°C to 1°C) to keep the cabbage temperature range of 4°C to 5°C. This range was the storage value [2]. Moreover, the distance between the thermal center and the outer surface of the cabbages was given by 6 cm. The finish time was the time taken to lower the temperature of the cabbage from its initial temperature to a given temperature at its thermal center. In addition, the temperature trend of the thermal center must be hardly drooped or almost constant.

The data acquisition module consisted of a data logger (Yokogawa MV1000), Chromel-Alumel thermocouple (type K), 3 phase power meter (PM2230), and relative humidity sensor (PCE-P18-3). The temperature and humidity points were located on the experimental apparatus as depicted in Figure 2. All of the tests, measuring the same position at least 3 times and the recorded parameter points were monitored every minute. The air-cooling temperature was calculated from the average of two temperature points (T_{17} to T_{18}). The cabbage temperatures at the upper tray were calculated from the average of eight temperature points (T_5 to T_8) and (T_9 to T_{12}) for the back zone. Also, the cabbage temperatures at the lower tray were determined by the average temperature points from T_1 to T_4 (front zone) and from T_{10} to T_{16} (back zone). Whereas, the relative humidity of the evaporator was measured from two points (RH_{19} and RH_{20}).

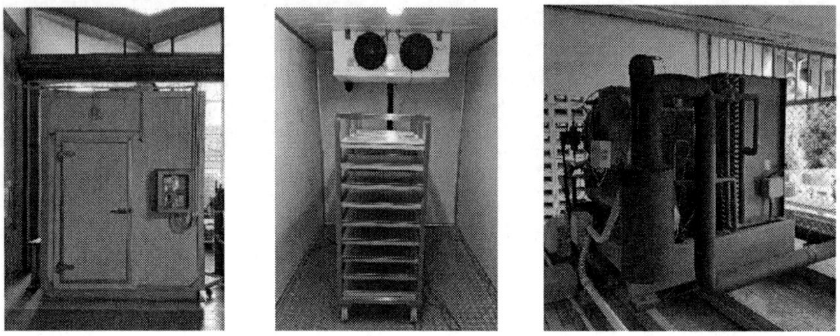

Figure 1. Cold storage

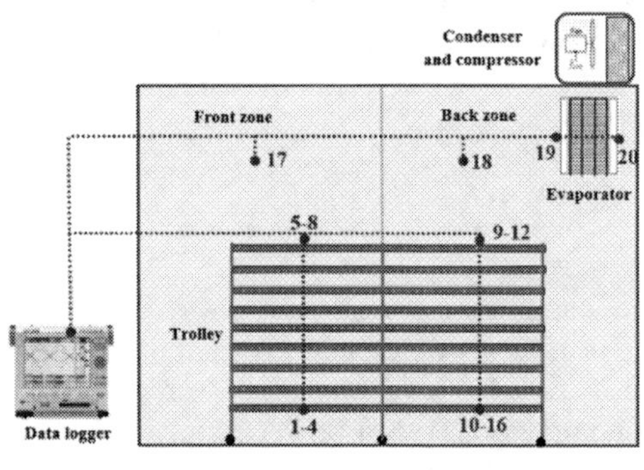

Figure 2 Experimental apparatus

2.2 Performance analysis

The experimental test was performed to investigate the cooling performance of cold storage under different pre-cooling strategies. The efficiency of a refrigerator or heat pump is given by a parameter called the coefficient of performance (COP) [4-5]. The COP is determined by the ratio between the energy usage of the compressor (W) and the amount of useful cooling at the evaporator (Q_c). COP can be calculated from Equation (1).

$$\mathrm{COP} = \frac{Q_C}{W} \tag{1}$$

The useful cooling at the evaporator was evaluated by the rate of heat transferred to the ambient air and applying the conservation of energy equation for a heating process at constant specific humidity [10], which was shown as Equation (2)

$$Q_C = \dot{m}_{air}\left(h_{in} - h_{out}\right) \tag{2}$$

where $\dot{m}_{air}$ is the mass flow rate of dry air and h_{out} and h_{in} are the enthalpies per unit mass of dry air at the exit and the inlet of the evaporator section, respectively.

All experiments were replicated three times and the uncertainties of parameters throughout the experiments were presented in Table 1. The result R is a given function in terms of the independent variables. W_R is the uncertainty in the result and $w_1, w_2, \ldots, w_n$ are the uncertainties in the independent variables ($x_1, x_2, \ldots x_n$). If the uncertainties in the independent variables are all given with the same odds, then the uncertainty in the result having these odds is given as Equation (3) [16].

$$W_R = \left[\left(\frac{\partial R}{\partial x_1} w_1\right)^2 + \left(\frac{\partial R}{\partial x_2} w_2\right)^2 + \ldots + \left(\frac{\partial R}{\partial x_n} w_n\right)^2\right], \tag{3}$$

Table 1 Uncertainties of parameters throughout the experiments

Parameters	Uncertainty	Unit
Temperature	±0.5	°C
Relative humidity	±0.1	%
Time measurement	±0.1	min
Air velocity	±0.1	m/s

3. Results and discussion

3.1 Effect of the cooling zone

The cabbages inside both front and back zones were simultaneously examined to cooling performance. From Figure 3 below, we can see that the condition of parallel pattern, the temperature and relative humidity variations over the cooling time. It found that as the cabbage temperatures gradually decreased to their equilibrium temperatures which were about 4.8°C, the air temperature and relative humidity were fluctuated by the compressor operation. The results of the study indicate that the temperature difference between the upper rack and the lower rack of 30_F_PL was around 2°C. It was found to cause higher temperature than 30_B_PL approximately 1°C, as provided in figure 3(a). Table 2 above illustrates COP of 30_F_PL, which was 2.58, was almost equal to COP of 30_B_PL. Whereas, the cabbage mass of 50 kg, it showed that there is a difference of the temperature between the upper rack and lower rack the 50_F_PL and 50_B_PL were around 4°C and 2°C, respectively. As can be seen as from Figure 3(b), the front zone had a lower temperature than the back zone due to the better air ventilation in the front zone. Likewise, the front zone had higher forced convection than the back zone, and these results match those observed in earlier study [13].

Regarding the cooling performance, the COP showed the difference between 50_F_PL and 50_B_PL (11%). The results revealed that the cooling zone of the lower cabbage mass had a slight effect on the COP. However, it has been suggested that the cooling zone of the higher cabbage mass had a more effect on COP (see Section 3.3 for more details).

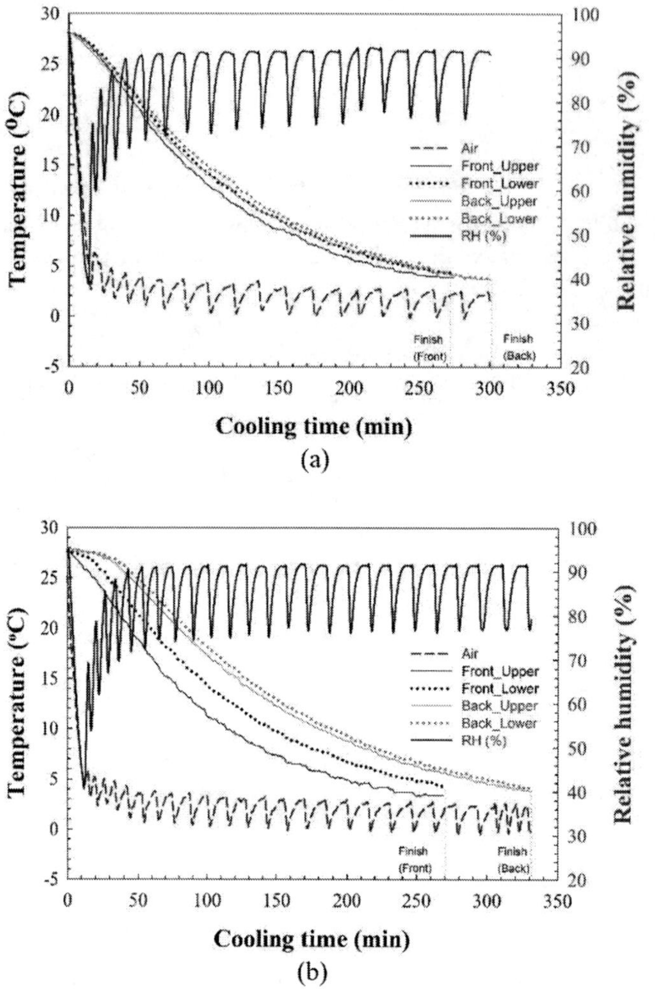

Figure 3. Temperature and relative humidity variations of the cabbages over the cooling time: (a) 30 kg; (b) 50 kg

3.2 Effect of product pattern

Figure 4 presents average temperature variations over the cooling time for different patterns. At the front zone, the difference of average temperature between both patterns of 30 kg and 50 kg were approximately 1.5°C and 0.3°C, respectively. Both COP patterns of 30 kg showed no differences. However, it can be seen that COP of 50_F_ST (2.27) was 10% higher than the 50_F_PL (see Table 2). A possible explanation for this might be that an airflow resistance and limited obstructions in the cooling process possibly occurred in the parallel pattern. These results are likely to be related to previous work [12], turning now to the experimental evidence on the effect of the cooling load on the cooling performance.

Table 2 Performances of the cold storage

Cooling conditions	COP
No cooling load	3.50
30_F_PL	2.58
30_B_PL	2.53
30_F_ST	2.61
30_B_ST	2.55
40_F_PL	2.38
40_B_PL	2.31
40_F_ST	2.49
40_B_ST	2.36
50_F_PL	2.13
50_B_PL	2.00
50_F_ST	2.33
50_B_ST	2.12

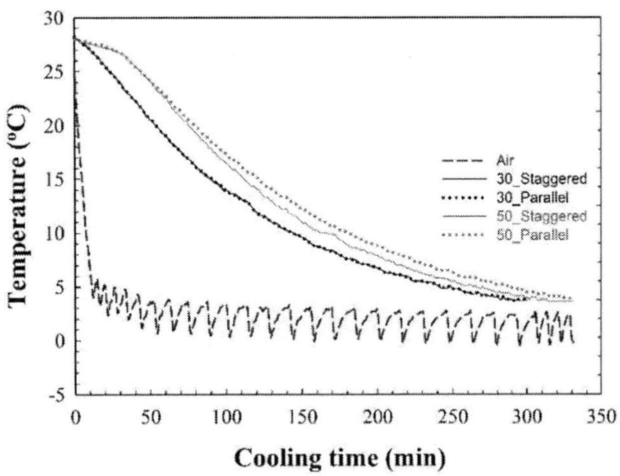

Figure 4. Average temperature variations over the cooling time for different patterns

3.3 Effect of cooling load

Figure 5 provided an effect of cooling loads on COP. It can be seen that the more increase of the COP's cooling load, the steady decrease of the COP was. Interestingly, the COP was observed to have higher heat extraction inside the evaporator. While the cooling load contained 30 kg, the cooling zone and pattern had less effect on the COP. Nevertheless, for the cooling load of 50 kg, the COP was significantly changed by both cooling zones and patterns. Additionally, the maximum COP with the cooling load appeared at 30_F_ST. It can be seen from the data that the higher COP provided the faster pre-cooling time of the cabbages to approach their average or equilibrium temperature, as illustrated in Figure 4. After that, the cabbages were transferred to keep in the freezer under their storage temperature. The results indicate that the cabbage pre-cooling strategy included the product arrangement with a staggered pattern and the front zone due to more energy savings. Together, these results provided important insights into the different heat transfer enhancements of the cold storage, particularly the appropriate precooling strategy.

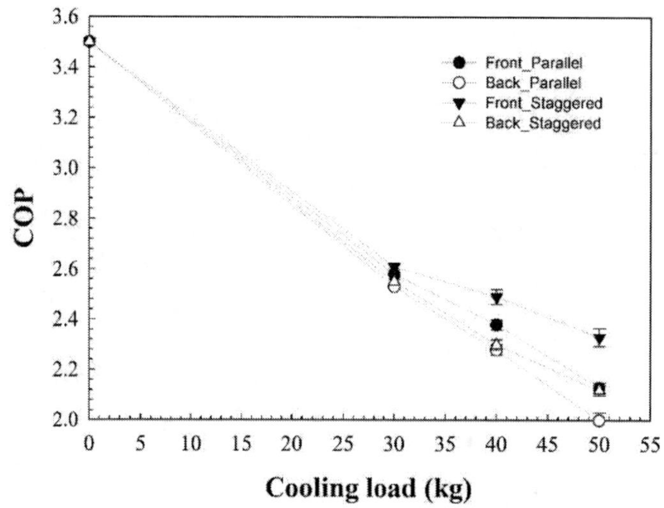

Figure 5. Effect of cooling loads on COP

4. Conclusion

The aim of the present research was to examine cooling temperature profile, cooling time, and COP of the forced-air pre-cooling of the green cabbages using four different cooling strategies. The second aim of this study was to investigate the effects of product mass under different cooling strategies on the COP. This study was designed to experiment obtained cooling curves to estimate the finish time of the cabbage pre-cooling process. The evaporator fan and the compressor of the cooling unit were also used to calculate the energy usage of the pre-cooling process. The results of this investigation show that the COP was inversely proportional to the cooling load. On average values, the maximum COP occurred in the case of 30_F_ST. The pre-cooling condition of product arrangement with a staggered pattern inside the front zone can be greatly improved using the cooling performance because of the higher heat transfer rate. The findings of this research provided insights for using pre-cooling strategy in the energy management of the green cabbages.

Acknowledgments

This research was supported by the Department of Mechanical Engineering, Faculty of Engineering, Rajamangala University of Technology Srivijaya, and granted by university research fund (2020).

References

[1] Cao B Li N and Xu K 2020 *Environmental and Experimental Botany* **179** 104209
[2] Elansari A M Fenton L D and Callahan C W 2019 *Postharvest Technology of Perishable Horticultural Commodities* **6** 167-207
[3] Duan Y Wang G B Fawole O A Verboven P Zhang X R Wu D Opara U L Nicolai B and Chen K 2020 *Trends in Food Science & Technology* **100** 278-291
[4] Xin Y Zhang M Xu B Adhikari B and Sun J 2015 *International journal of refrigeration* **57** 11-25
[5] Biglia A Comba L Fabrizio F Gaya P and Aimonino D R 2016 *Energy Procedia* **101** 305-312
[6] Sepahvandi F Heravi H M and Saleh S R 2017 *International journal of heat and technology* **35** 75-81
[7] Zilio C Righetti G Pernigotto G and Longo G A 2018 *International Journal of Refrigeration* **95** 38-50

[8] Bulut M Bayer O Kırtıl E and Bayındırlı A 2018 *International Journal of Refrigeration* **88** 360-369
[9] Brito P Lopes P Reis P and Alves O 2014 *International Journal of Energy and Environmental Engineering* **5** 88-103
[10] Tian S Yang Q Hui N Bai H Shao S and Liu S 2020 *Energy* **205** 117987
[11] Chauhan A Trembley J Wrobel L and Jouhara H 2019 *Energy* **171** 77-94
[12] Mukama M Ambaw A Berry T M and Opara U L 2020 *Journal of Food Engineering* **215** 126-133
[13] Gruyters W Defraeye T Verboven P Berry T Ambaw A Opara U L and Nicolai B 2019 *International Journal of Refrigeration* **106** 338-349
[14] Han J W Zhao C J Qian J P Ruiz-Garcia L and Zhang X 2018 *International Journal of Refrigeration* **89** 131-141
[15] Wu W Häller P Cronjé P and Defraeye T 2018 *Biosystems Engineering* **169** 115-125
[16] Sevik S 2014 *Solar Energy* **105** 190-205

IOP Conf. Series: Materials Science and Engineering 1137 (2021) 012068 doi:10.1088/1757-899X/1137/1/012068

Photovoltaic forecasting model in Thailand: case study solar farm at Nakhon Ratchasima province

Withaya Yongchareon[1,*], Sumlee Thongthew[2], and Pensiri Juncharoen[3]

[1]Faculty of Engineering. Chulalongkorn University

[2]Faculty of Education. Chulalongkorn University

[3]Graduate School, Chulalongkorn University

* Corresponding Author: Withaya Yongchareon: yongchareon@gmail.com

Abstract

The photovoltaic module performance was normally tested at the standard conditions. The use of the actual field performance has not been investigated. One purpose of this study was to find the predicted equation for forecasting the energy produced from photovoltaic modules. Data for this study were collected using the data from the solar farm in Nakhon Ratchasima province: solar irradiation, ambient temperature, module temperature and wind velocity. The daily composite data were 1,461 from 1st January 2015 – 31st December 2018. The five predicted equations, including both the main effect and the interaction effect, were statistically analysed by regression method using the Minitab software program for multiple linear regression. The requiring dependent variable (y) was the energy produced from the photovoltaic module. The four independent variables were solar irradiation (x_1), ambient temperature (x_2), module temperature (x_3) and wind velocity (x_4). For prediction preciscion, all the predicted equations were validated with the new gathered data from 1st January 2019 – 30th June 2019. The analysis results showed, the forecasting model together with the appropriate predicted equation with the highest coefficient of determination (R^2 of 0.9873) and the standard deviation of prediction (S = 2.67%). This predicted equation, $\hat{y}_2 = 5258 + 5310.0x_1 - 100.31x_2 + 66.2x_4$, consisted of the three external independent variables; solar irradiation, ambient temperature, and wind velocity. The independent variable sensitivities were also determined. The solar irradiation was the most sensitive to the predicted equation. Moreover, this predicted equation would be suggested to utilize as the estimator for the energy produced from the new photovoltaic power plant to be installed at the north-eastern part of Thailand.

Keywords: Photovoltaic, Forecasting Model, Regression Method, Solar Farm

1. Introduction

There is evidence that a solar panel play a crucial in generating clean, emission free electricity. One of the greatest challenges is it produces only direct current electricity (DC) and is limited by appliances use. Solar photovoltaic systems (solar PV systems) are often made of solar PV panels (modules) and inverter (changing DC to AC). Solar PV panels are of interest because they are mainly made of solar photovoltaic cells. As much module performance is generally rated under standard test conditions (STC): irradiance of 1,000 W/m^2, the solar spectrum of AM 1.5 and module temperature at 25°C.

So far, however, there has been little discussion about, the actual field performance. Evidence suggests that the actual voltage and current output of the module changes according to lighting, temperature, and load conditions. Therefore, there is never one specific voltage, current, or power at which the module operates. Performance varies depending on the time of the day, amount of solar insolation, direction and tilt of modules, cloud cover, shading, state of charge, temperature, geographic location, and day of the year. Concerning the photovoltaic power plant, the electrical power output, according to the production license, must be guaranteed. Since the lifetime of the solar panels is generally last at approximately 25 years, the electrical power output produced must be forecasted before actual use to make sure it meets the guaranteed power output. Preliminary work on multiple regression analysis technique was under taken by Pichet Vongkiem [1]. In an analysis of multiple regression analysis, he showed that it can forecast the electrical energy needs for the PEA (Provincial Electricity Authority). This paper purposes a new methodology of the predicted equation for forecasting the energy produced from photovoltaic modules using the past actual data from the solar farm at Nakhon Ratchasima province. As can be seen from Table 1 below, electrical solar cell module performance at standard conditions include maximum power (P_{MAX}), open circuit voltage (V_{OC}), short circuit current (I_{SC}), maximum power voltage (V_{MPP}), maximum power current (I_{MPP}), and module efficiency (%).

Table 1. Solar Cell Module Performance

Electrical Performance	Unit	Value
Irradiation	W/m2	1000
Module Temperature	°C	25
Maximum Power	W	210
Maximum Power Voltage	V	26.6
Maximum Power Current	A	7.9
Open Circuit Voltage	V	33.2
Short Circuit	A	8.58
Module Efficiency	%	12.1

Table 1. also shows that the module temperature increased when electrical power and efficiency decreased. It was found that the module temperature depended on the balance of heat gain and heat loss from the module, in the environment.

2. Regression method [4]

The regression method has been proposed to analyse the forecasting model, and the predicted equation containing various parameters. The regression method, through the statistic test can eliminate the unnecessary parameter in the predicted equation and it has less contribution to the dependent variable, which is the solar power output in this study.

To demonstrate regression method, the simple (1st order) linear regression was selected as an example. A set of observed data consisted of independent variable y and independent variable x. As shown in the following procedures.

1) The assumed model was expressed as in the equation (1)

$$y_i = a + bx_i + \varepsilon_i \quad (1)$$

where y_i and x_i were the observed variables

a and b were constant parameter

ε_i was the random error variable of y and normally distributed with the mean of 0 and variance of 1 and also independent on the level of x

Then, the predicted equation could be expressed as in the equation (2)

$$\hat{y}_i = a + bx_i \tag{2}$$

2) The least square of error was applied as in the equation (3)

$$\sum \varepsilon_i^2 = \sum (y_i - (a + bx_i))^2 = \text{mininum} \tag{3}$$

To satisfy equation (3) the partial derivative of parameters should be set to zero as in the equation (4)

$$\frac{\partial \sum \varepsilon_i^2}{\partial a} = 0 \text{ and } \frac{\partial \sum \varepsilon_i^2}{\partial b} = 0 \tag{4}$$

From Equation (4) the parameter a and b could be mathematically solved as in equation (5)

$$a = \bar{y} - b\bar{x} \quad \text{and} \quad b = \frac{\sum (x_i - \bar{x})(y_i - \bar{y})}{\sum (x_i - \bar{x})^2} \tag{5}$$

where $\bar{y}$ was the mean value of y_i and $\bar{x}$ was the mean value of x_i

3) Finding the goodness of fit by using the coefficient of determination R^2 as in the equation (6)

$$R^2 = \frac{SS_{REG}}{SS_T} = \frac{\sum (\hat{y}_i - \bar{y})^2}{\sum (y_i - \bar{y})^2} = 1 - \frac{\sum (y_i - \hat{y}_i)^2}{\sum (y_i - \bar{y})^2} \tag{6}$$

where SS_{REG} was sum square due to regression and SS_T was the total sum square

$R^2=1$ was the perfect fit, all data points of y were lined on the regression line. However, in the actual situation there were errors due to measurement uncertainties and the assumed model was not the true model. Thus $R^2<1$

4) Testing the hypothesis by using statisticT for individual parameter and statistics F for all parameters. If the individual parameter was significant, then it was in the predicted equation.

3. Solar farm data [2]

The recorded field data used in this study were solar irradiation, ambient temperature, module temperature wind velocity and energy production. The daily composite data were 1,461 from 1 January 2015 - 31 December 2018 for construction of models. The data were 183 from 1 January 2019 - 30 June 2019 for validation of the model. The example of recorded data was given in Table 2. The statistic data of 1461 was shown in Table 3.

Table 2. The example of recorded data on the solar farm

Date	Irr (kWh/m2)	Amb temp. (°C)	Module temp. (°C)	Wind velocity (m/sec)	Energy Production (kWh)
1/1/2015	6.397	25.130	33.283	2.092	38,400
2/1/2015	6.308	25.240	33.797	1.833	37,600
3/1/2015	6.081	26.281	35.484	1.097	35,920
4/1/2015	6.405	30.114	41.766	0.541	36,880
5/1/2015	6.104	32.507	43.484	0.463	34,800
6/1/2015	5.885	33.141	43.285	0.636	33,440
7/1/2015	4.454	32.072	39.764	0.918	25,360
8/1/2015	1.192	25.660	26.341	1.282	7,280
9/1/2015	4.675	28.787	36.584	1.412	27,360
10/1/2015	5.396	29.456	38.765	1.326	31,520

Table 3. The statistic data of solar farm

List	Irr (kWh/m2)	Amb temp. (°C)	Module temp. (°C)	Wind velocity (m/sec)	Energy Production (kWh)
Mean	5.484	32.9	42.3	1.1	31156
Minimum	0.620	2.4	16.9	0.0	3680
Maximum	7.517	41.9	54.3	3.1	42800
Range	6.896	39.5	37.4	3.0	39120
Standard Deviation	1.200	3.7	5.4	0.5	6464

4. Selection of Models

The linear polynomial model method is particulary useful in studying since most of the engineering data could be fitted with a popular polynomial equation, and the required model was a simple approximation model with acceptable precision. However, the higher-order term such as the radiation loss term which is the function of (temperature)^4 was not included in the model because the value of temperature variables was less than 100°C.

All recorded data of solar farm were used. Energy Production was selected as the dependent variable (y), while solar irradiation (x_1), ambient temperature (x_2), module temperature (x_3), and wind velocity (x_4) were independent variable. The forecasting approximation model could be many models that represent the field data set. First, the used full model consisted of all field variables and only the main effect of x_1, x_2, x_3, and x_4 (see Equation (7)). The predicted equation was also given in Equation (8). The model included an error to represent the data points.

$$y_i = a + bx_{1i} + cx_{2i} + dx_{3i} + ex_{4i} + \varepsilon_i \tag{7}$$

$$\hat{y}_i = a + bx_{1i} + cx_{2i} + dx_{3i} + ex_{4i} \tag{8}$$

Where ε_i was the random error in y, it is normally distributed and independ to the level of x. It can be seen from the data in Table 4., the regression analysis results obtained from the Minitab software program.

Table 4. Analysis of Variance (ANOVA)

Source	Degree of freedom	Sum Square	Mean Square	F-Value	P_Value
Regression	4	60255940916	15063985229	29006.63	0.000
Solar Irradiation	1	13607620571	13607620571	26202.31	0.000
Amb.Temp.	1	1541163	1541163	2.97	0.085
Module Temp.	1	20156982	20156982	38.81	0.000
Wind velocity	1	9096653	9096653	17.52	0.000
Error	1455	755623775	519329		
Total (corrected)	1459	61011564690			

The predicted equation was also provided in Equation (9)

$$\hat{y}_1 = 4594 + 5494.6x_1 + 40.2x_2 - 118.0x_3 + 90.8x_4 \quad (9)$$

with coefficient of determination $R^2 = 0.9876$ and standard deviation S = 720.6 kWh. In this current study the predicted equation represented the field data with the precision of 98.76%. Table 4 illustrates the ambient temperature (x_2). From this data, it was not significant because the probability of F (P_ value) in the last column, was greater than the significant at the $p = 0.05$ level . It can thus be suggested that the ambient temperature would not existed in the predicted equation. Secondly, there were many approximation models and predicted equations which consisted of the linear combinations of x-variables including only the main effect, for example $\hat{y}_2$ and $\hat{y}_3$, or including both the main effect and the interaction effect, for example $\hat{y}_4$ and $\hat{y}_5$. The predicted equations fitted to the observed field data (see Table 5.). Table 5. presents the results of the regression analysis of some predicted equations. It reviewed that these four selected predicted equations was a high value of R^2 showing more than 0.97. Further analysis showed that these approximation predicted equations could represent the observed solar farm data set. It also could explain the variation of data, showing more than 97 %. However, the equation $\hat{y}_2$ had the high value of R^2 of 0.9873 and contained only the three external independent variables; solar irradiation, ambient temperature, and wind velocity. While x_3, module temperature, depends on solar irradiation, ambient temperature, and wind velocity. Therefore, this predicted equation $\hat{y}_2$ was appropriately predicted equation representing the forecasting equation.

Table 5. Analysis Results of Regression

Predicted Equation	R^2
$\hat{y}_2 = 5258 + 5310.0x_1 - 100.31x_2 + 66.2x_4$	0.9873
$\hat{y}_3 = 4849 + 5448.1x_1 - 86.51x_3 + 80.7x_4$	0.9876
$\hat{y}_4 = 2471.7 + 5165.7x_1 + 0.2382x_2x_3x_4$	0.9847
$\hat{y}_5 = 2241.5 + 5175.4x_1 + 15.07x_2x_4$	0.9786

The result of the residual ($y_i - \hat{y}_i$) composed of the lack of fit and the random noise ε_i. The state of lack of fit could be observed from the residual plots. If the distribution of the residual was nearly the normal distribution then it shows a small state of lack of fit. The residual plot of the predicted equation $\hat{y}_2$ was nearly normal distribution because of its high value of R^2 (see Figure 1.)

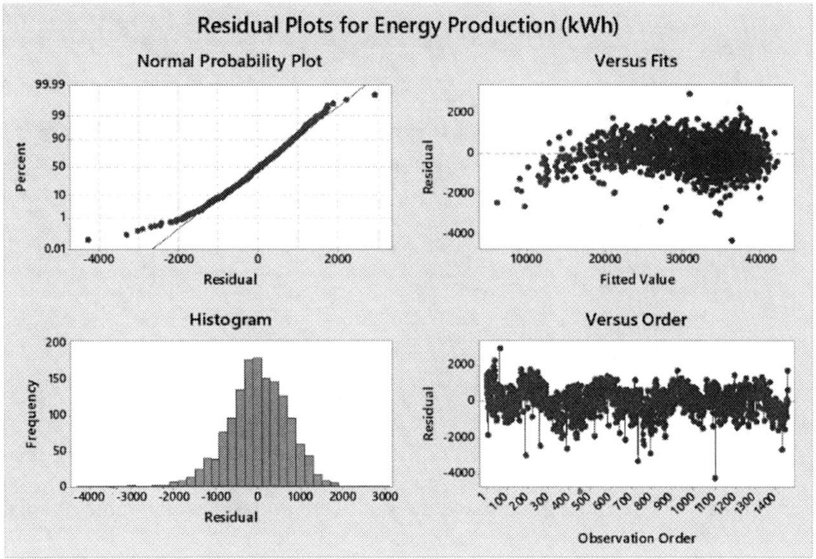

Figure 1. The residual plots of the predicted equation $\hat{y}_2$

In general, therefore, it seems that the predicted equation could be used to represent the observed solar farm data set. It is possible to use this predicted equation as the estimator for the energy produced from the exiting photovoltaic power plant or the new one to be installed in the future. This predicted equation need to be validated with the new data set from 1 January 2019 - 30 June 2019. The predicted values from the predicted equation were used to compare with the observed data set. As shown in Figure 2., the straight line passed to the origin. The equation was $y = 1.0063\,x$ with the R^2 of 0.9749. These findings suggest that the predicted equation $\hat{y}_2$ could be used as the etimator for energy production.

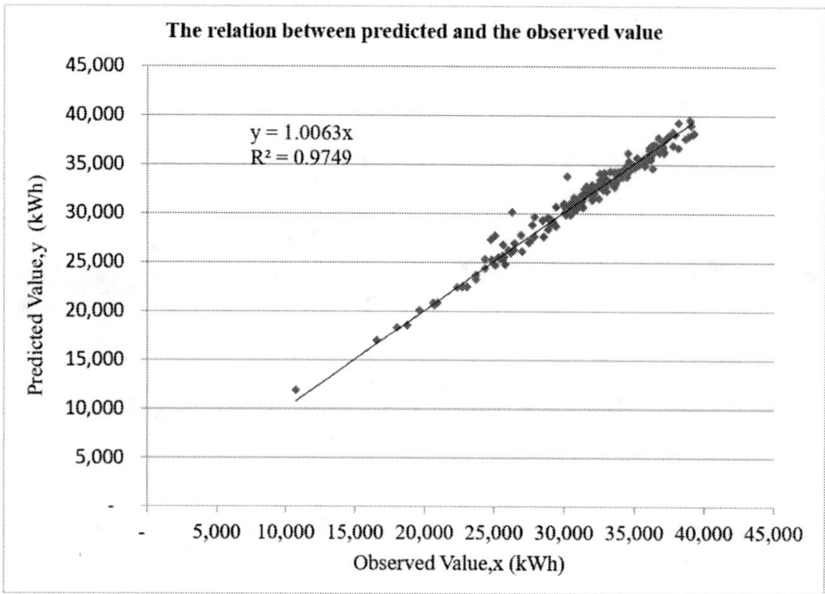

Figure 2. The relation between the predicted and the actual value of solar farm

5. Sensitivity Analysis

The energy production $\hat{y}_2$ depended on the solar irradiation x_1, ambient temperature x_2, and wind velocity x_4. To find how much each variable had an influence on the energy production, the variable sensitivity had been analyzed by changing each variable by 10 % while keeping other variables constant and obtained the change of the energy production. The sensitivity results were given in Table 6.

Table 5 provides the solar irradiation and it was the most sensitive variable. Whilst the wind velocity was the least sensitive variable. Also, the ambient temperature was a negative sensitivity. When the ambient temperature increased the energy production decreased because of the increased in the module temperature. This finding is consistent with that of Keattisak Phunjumpa [3]

Table 6. Analysis results of Sensitivity of the predicted Equation

List	Daily Actual Average				% Diff. $(y_{+10\%} - \hat{y})/\hat{y}$
	Irradiation kWh/m2	Amb temp. °C	Wind velocity m/s	Energy Production kWh	
	x_1	x_2	x_4	$\hat{y}_2$	
Actual	5.484	32.9	1.1	31151	
+10% by variable	6.032	32.9	1.1	34063	9.35%
	5.484	36.2	1.1	30821	-1.06%
	5.484	32.9	1.2	31158	0.02%
Note:	$\hat{y}_2 = 5258 + 5310x_1 - 100.31x_2 + 66.2x_4$				

IOP Conf. Series: Materials Science and Engineering 1137 (2021) 012068 doi:10.1088/1757-899X/1137/1/012068

6. Conclusion

This study has identify the data of solar farm that comprised of solar irradiation, ambient temperature, module temperature and wind velocity. This study had found that generally daily composite data were 1,461 from 1st January 2015 – 31st December 2018 for construction of the model. The data were 183 from 1 January 2019 - 30 June 2019 for validation of the model.The method used in this current study was a regression and it is chosen to forecast the model along with the appropriate predicted equation with the highest coefficient of determination (R^2 = 0.9873) and the standard deviation of prediction (S = 2.67%). This predicted equation contained only the three external independent variables; solar irradiation, ambient temperature, and wind velocity. The independent variable sensitivities were also determined. The solar irradiation was the most sensitive to the predicted equation. The finding reported here shed new light on the estimator for the energy produced from the new photovoltaic power plant that will be installed in the north-east of Thailand.

Acknowledgments

The researcher was grateful to obtain the data of the solar farm from the solar power company.

References

[1] Vongkiem P and Rerkpreedapong D 2015 *J. Kasetsart Engineering* **91** 31-40

[2] Juncharoen P and Yongchareon W 2019 *Proc. Conf. on Solar Energy Forecasting Model in Thailand (Thai Version)* (Phitsanulok: Thailand) 1-6

[3] Phunjumpa K Kongnok K and Plangklang B 2017 *Proc. Conf. on Study on the Effect of Ambient Temperature to Electrical Energy Production from Solar Cells (Thai Version)* (Chanthaburi: Thailand) p 217

[4] Draper N.R., Smitch.H, 1966 *Applied Regression Analysis* (London: John Wiley & Sons) Chapter1 pp 1-33

IOP Conf. Series: Materials Science and Engineering 1137 (2021) 012069 doi:10.1088/1757-899X/1137/1/012069

Numerical study of performance augmentation of a vertical coil and shell heat exchanger using high frequency ultrasonic waves

Weerachai Chaiworapuek[1], Jirameth Chungprempree[1,*], Jaytawee Pukrushpan[1], Kunthakorn Khaothong[2],

[1]Department of Mechanical Engineering, Faculty of Engineering, Kasetsart University, Bangkok Thailand 10900
[2]Department of Mechanical Engineering, Faculty of Engineering, Kasetsart University, Nakhon Pathom Thailand 73140

* Corresponding Author: jirameth.ch@ku.th

Abstract
This research focuses on the performance augmentation of a coil and shell heat exchanger using high frequency ultrasonic waves by the commercial computational fluid dynamics program FLUENT. The model was validated both heat transfer and ultrasound propagation with the results from other researchers. Then, the simulation of the flow and heat transfer enhancement in the heat exchanger under 1.7 MHz ultrasound. Four ultrasonic transducers were installed at the corners of the shell at the middle height of the coil. The ultrasound from transducers was emitted to induce the vortex flow in the shell with the angle of 0, 15, and 30 degrees with the horizontal line. The results show that the high frequency ultrasound is applicable for heat transfer enhancement of a shell and coil heat exchanger. Therefore, the obtained results can be used to innovate an ultrasonic coil and shell heat exchanger in the future.

Keywords: Ultrasound, Heat exchanger, Heat transfer, Vertically coil.

1. Introduction

A coil and shell heat exchanger are important equipment, used in many thermal applications such as chillers, oil cooling, SDHW systems, power plant cooling systems, etc. This heat exchanger type provides high thermal efficiency, low installation space, and low maintenance cost [1-4]. However, performance augmentation is still needed for the reduction of operating cost. Following the advancement of numerical simulation, this heat exchanger has been studied numerically and experimentally [2-8]. In an attempt to augment the heat exchanger performance, the heat transfer enhancement techniques such as the use of twisted tape [9], extended area [10], turbulent spot [11-14], baffle [15], water jet [16], etc. are typically employed. The use of ultrasound is one of those techniques, giving high performance to enhance the heat transfer [17-18]. Besides, ultrasonic waves, having high frequency or above 1 MHz can initiate the streaming by momentum gradient [8]. They should significantly increase heat transfer of thermal systems. Thus, this research study's the heat transfer augmentation of a coil and shell heat exchanger using 1.7 MHz ultrasonic waves by the commercial computational fluid dynamics program

FLUENT. The continuity, momentum, and energy equations are mutually utilized and they are as follows:

$$\partial\rho/\partial t + \nabla\cdot(\rho \boldsymbol{v}) = 0 \tag{1}$$

$$\partial\rho \boldsymbol{v}/\partial t + \nabla\cdot(\rho \boldsymbol{v}\boldsymbol{v}) = -\nabla p + \mu\nabla^2\cdot \boldsymbol{v} + \boldsymbol{F} \tag{2}$$

$$\rho c_p(\boldsymbol{v}\cdot\nabla T) = k\nabla^2 T + S \tag{3}$$

In this study, 4 ultrasonic transducers were installed in the heat exchanger and the waves were released in 3 directions. The results, including pumping power factor, heat transfer performance factor, and thermal performance factor will be presented to evaluate the performance of the heat exchanger.

2. Numerical validation

2.1 Heat transfer validation

The heat transfer of numerical models was validated by comparing the results from the simulation of water flow in a vertical coil and shell heat exchanger and the experiment of [2]. The fluid model was created in accordance with their test section. The water flow rate and temperature inlet were 0.05 kg/s and 20°C, respectively, while the outlet pressure was 0 Pa. In the simulation, coil surface was set to have a constant heat flux of 23.16 kW/m². The Reynolds number in shell-side [8] was calculated as follows:

$$Re = \rho U D_{eq}/\mu \tag{4}$$

where ρ is density, μ is dynamic viscosity, U is inlet velocity. The D_{eq} is equivalent diameter [8], calculated as follows:

$$D_{eq} = 4V_{sh}/\Pi dL \tag{5}$$

where V_{sh} is fluid volume in shell-side, d is coiled tube diameter, and L is coiled tube diameter length.

For the simulation, the standard k-epsilon model was used and the element number of 2147164, 1,380,417, 913,022, 739,016, 398,540, 126,247, 90,485 and 66,575 were employed for the mesh sensitivity and validation process as depicted in Figure 1.

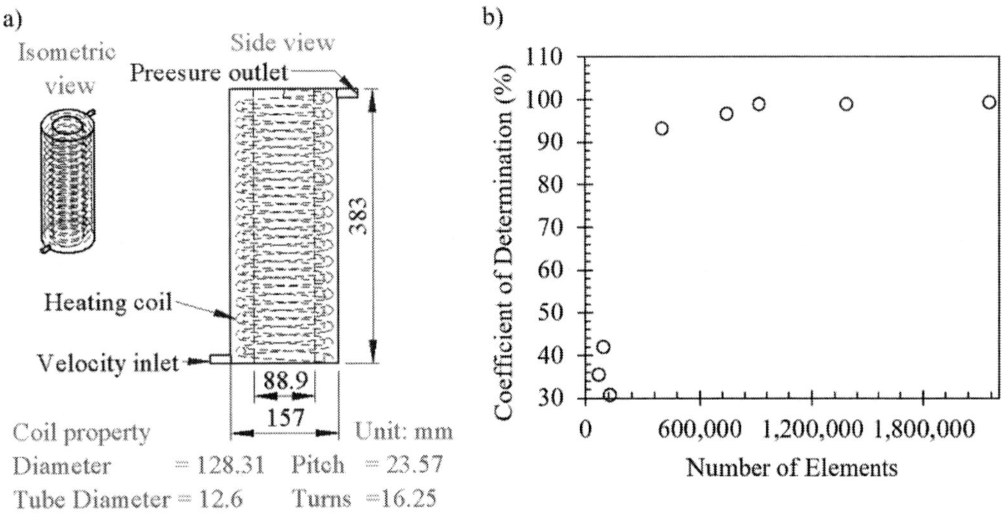

Figure 1. a) Details of fluid model of [2] and b) results of heat transfer validation

Form the figure, the coefficient of determination, R^2 was determined from the comparison between numerical and experimental coil temperatures. It converged at 98.90% and the element number of 913,022. Thus, this model shows a very high reliability for heat transfer prediction in a coil and shell heat exchanger.

2.2 Ultrasound propagation validation

The numerical validation for ultrasound propagation was conducted as the experiment of [19]. For the numerical setup, the momentum energy was added to the cylindrical volume, having the length of the near field distance [8] as follows:

$$N = d_s^2/4\lambda \tag{6}$$

where N is Fresnel length or near field region length, λ is acoustic wavelength, and d_s is sound source diameter. The numerical validation for ultrasound propagation was conducted as the experiment of [19]. For the numerical setup following [8], the momentum energy was added to the force term in acoustic streaming equation as follows:

$$\rho(\boldsymbol{v}\cdot\nabla\boldsymbol{v}) = -\nabla p + \mu\nabla^2\boldsymbol{v} + \boldsymbol{F} \tag{7}$$

In the near field region, the acoustic force was added as 73.62 N/m^3 in the direction of ultrasonic waves. The heater surface was set as a constant heat flux of 6.9, 16.8, 34.1, and 55.5 kW/m^2, respectively. The water surface was set to be atmospheric condition or gauge pressure of 0 Pa. Other wall surfaces were constant temperature of 0°C. Also, the standard k-epsilon turbulence model was used in this work. The mesh sensitivity method was conducted at the element number of 1,503,744, 1,181,604, 734,713, 592,794, 377,757, 302,845, 271,793, and 122,556. It was found that the element number of 592,794 gave the convergence of surface temperature. At this mesh aspect, the validation process was done at the all-heat flux conditions and the results showed that the percentage error is below 6.57%, confirming the reliability of the numerical model of ultrasound propagation.

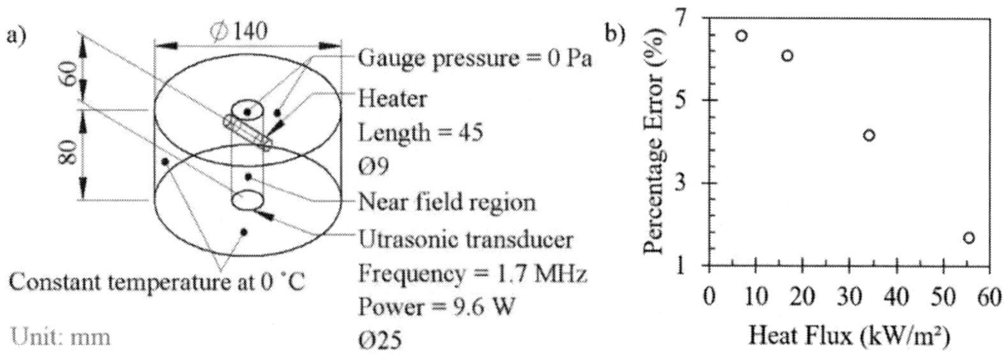

Figure 2. a) Details of fluid model of [8] and b) results of ultrasound propagation validation

3. Numerical Method

This heat exchanger type, studied in this research was the model that was utilized in a commercial 1-ton water chiller of Ample Cool Co., Ltd. Its shell dimensions were width of 280 mm x length of 280 mm x height of 391 mm. The outlet was set near the bottom corner and the inlet was installed on the top wall. Both diameter sizes were 25 mm. The inlet was defined to have constant temperature of 25°C and volume flow rate of 10, 15, 20 and 25 L/min. Meanwhile, the temperature and gauge pressure at the outlet was 13°C and 0 Pa, respectively. The vertical coil was mounted at middle of the shell and it was lifted from the bottom wall about 23 mm. The coiled tube diameter, coil diameter, pitch, and coil height were 10 mm, 337.5 mm, 30 mm, and 225 mm, respectively as shown in Figure 3.

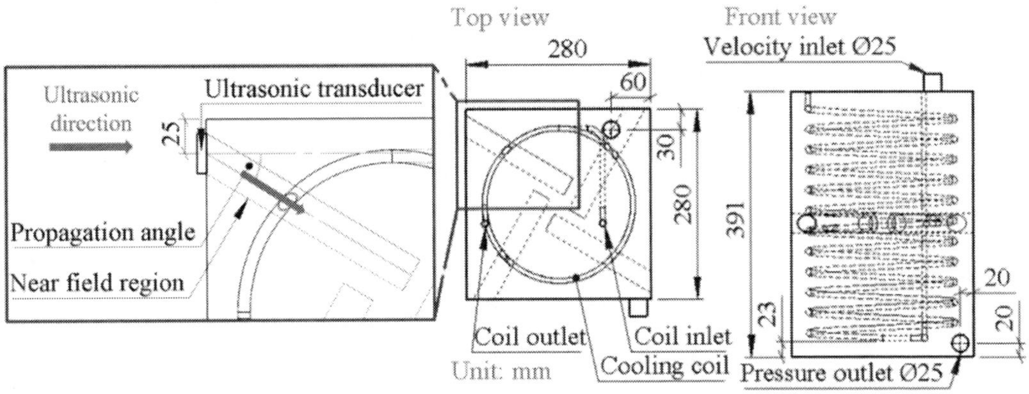

Figure 3. Details of shell side water in the heat exchanger

An ultrasonic source, having diameter of 25 mm was installed at each corner of the shell on the middle height level of heat exchanger to emit the high frequency ultrasound in clockwise direction. The angles with respect to each horizontal line were 0, 15, and 30 degrees, respectively. The ultrasonic waves were set to have frequency and power of 1.7 MHz and 9.6 W, respectively. To generate the ultrasound propagation, the acoustic force was added as 73.62 N/m^3 in the direction of ultrasonic waves in equation 7. The standard k-epsilon model was used as the turbulence model in this part. The spatial discretization of all convection terms was second order upwind except the gradient that was set as the least squares cell based. The calculation stopped when all scaled residuals were less than 10^{-5}. For mesh sensitivity check, the element number of 12,053,443, 9,529,035, 7,823,081, 5,634,562, 4,201,682, 2,224,615 and 1,322,985 were investigated. The results show that the mesh number of 5,634,562 was proper for the calculation as depicted in Figure 4 due to the convergence of temperature between inlet and outlet of the heat exchanger. Thus, this meshing method was employed through this research.

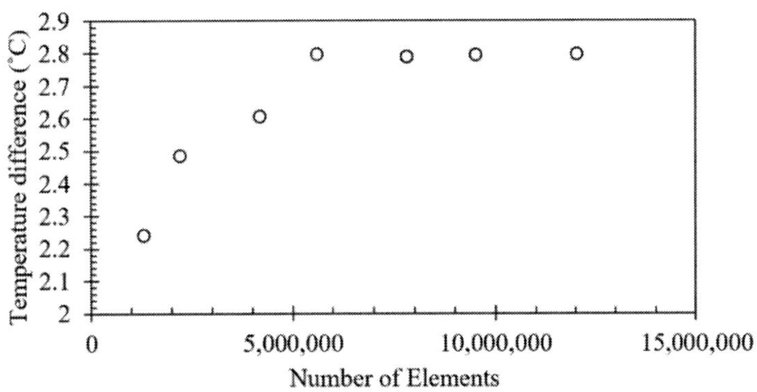

Figure 4. Mesh sensitivity results from the simulation of water flow in a coil and shell heat exchanger

This work investigated the effect of flow rate change and the angle of ultrasound propagation on heat transfer and pressure drop in a vertical coil and shell heat exchanger. In attempt to evaluate the performance of a heat exchanger, the pumping power factor, *F*, the heat transfer performance factor, *J*, and the thermal performance factor, *TP*, were presented and the pumping power factor [8] was determined as follows:

$$F = f \times Re^3 \tag{8}$$

where f is friction factor [8]. It was calculated as follows:

$$f = \Delta P(D_{eq}/l_s)(2/\rho U^2) \tag{9}$$

where ΔP is pressure drop, l_s is heat exchanger length.

The heat transfer performance factor [8] was determined as follows:

$$J = j \times Re \tag{10}$$

where j is Colburn j-factor [8], evaluated as follows:

$$j = hPr^{2/3}/\rho Uc_p \tag{11}$$

where h is heat transfer coefficient. Pr is Prandtl number. c_p is specific heat.

Finally, the thermal performance factor [8] was defined as follows:

$$TF = (j/j_{ref})/(f/f_{ref})^{1/3} \tag{12}$$

where the subscript ref means the condition without the interference of ultrasound.

4. Results and discussion

Figure 5 presents the J, the F and the TP, depending on the Re at 4.21×10^5, 3.37×10^5, 2.53×10^5, and 1.69×10^5. The results show that when the Re was increased, the values of J and F gained too. In addition, Figure 5a shows that the application of ultrasound can enhance the heat transfer of exchanger. At the lowest Re, the J was clearly increased by ultrasonic waves. The strength of this effect declined with the increase in Re. In the meantime, Figure 5b shows that the F is slightly affected by the ultrasonic disturbance and the most effect from ultrasound is less than 3.5%. Figure 5c reveals the overview effect of ultrasound on the flow and heat transfer in the heat exchanger. In the figure, the maximum TP of 1.1345 was yielded when ultrasound was emitted at angles of 0° and the lowest Re. It also decreased with the increase in the ultrasound propagation angle or the Re.

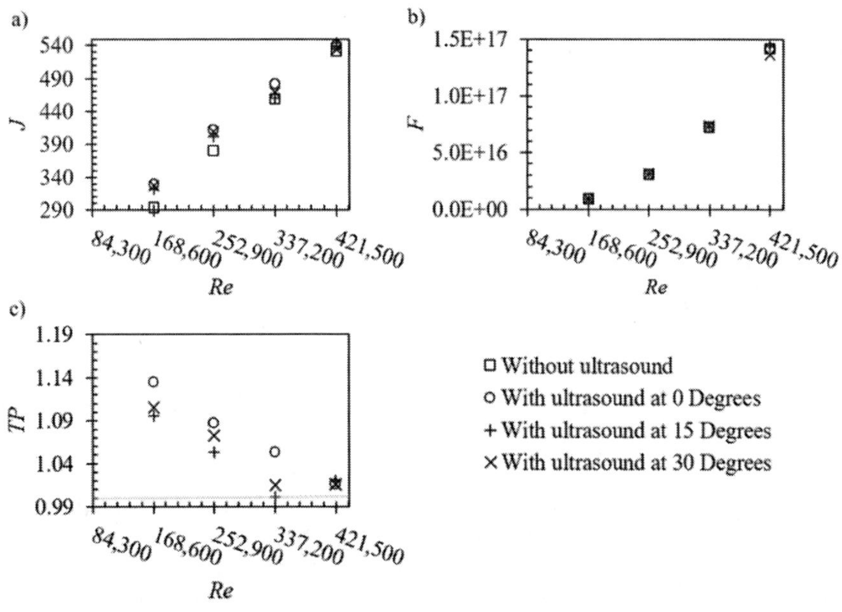

Figure 5. a) J, b) F, and c) TP vs Re under the effect of ultrasound propagation angle

Figure 6 shows the distribution of velocity vectors on the plane of middle height heat exchangers under conditions with and without ultrasound at the *Re* of 1.65×10^5. Figure 6a shows that the water velocity on this plane is mostly below 0.05 m/s, while a relatively higher velocity region occurs at the lower left corner due to the inducement of inlet water. The velocity around the coil clearly increased when ultrasound was released into the heat exchanger as depicted in Figure 6b-d. The figures show that a large vortex has been created by ultrasound and different vortex behavior is obtained. At the propagation degree of 0°, the vortex was formed at outer side of the coil and this gives the relatively highest performance as shown in Figure 6b. At the higher propagation angle of the waves, not only the size of vortex, but also the performance of heat exchanger is reducing as illustrated in Figure 6c-d. This clearly shows that the size of the induced vortex directly depends on the performance of the ultrasonic heat exchanger.

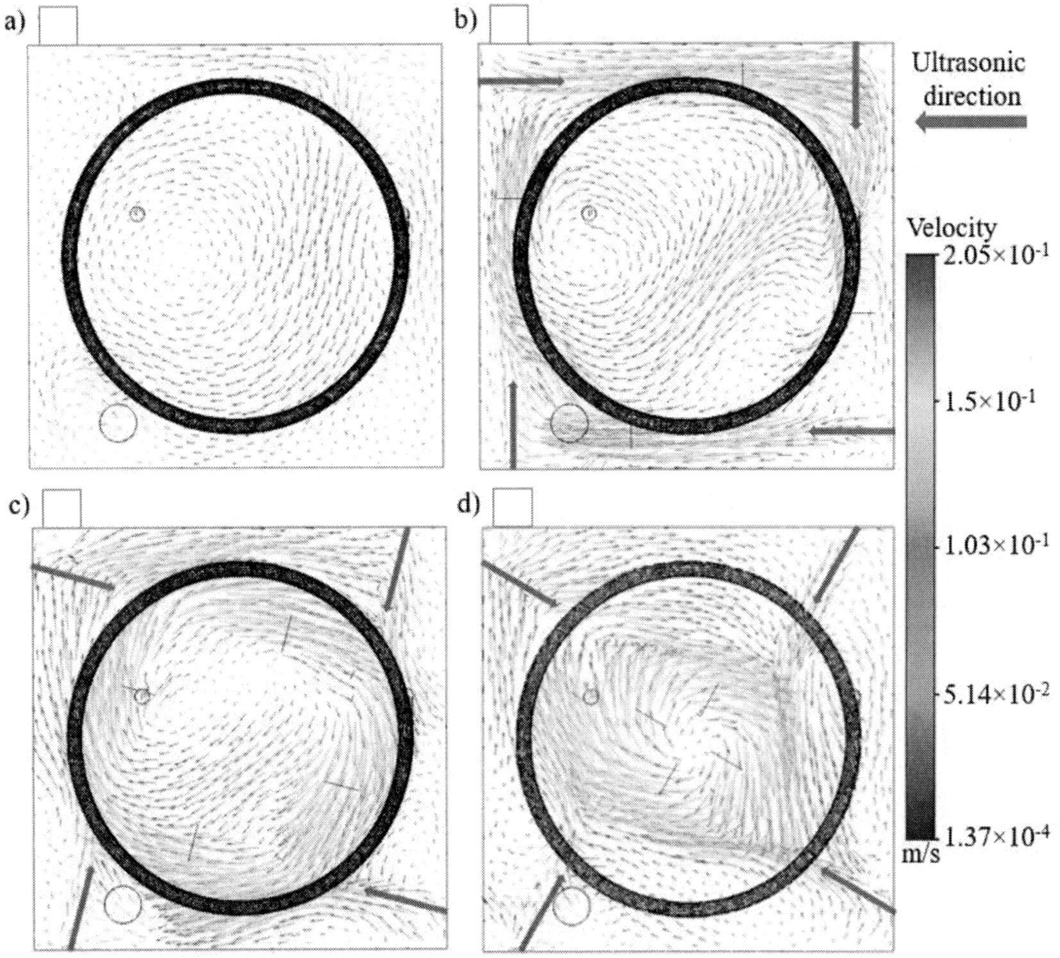

Figure 6. Distribution of velocity vectors on the plane of middle height of heat exchanger at the *Re* of 1.65×10^5 under the condition a) without and with ultrasound at b) 0°, c) 15°, and d) 30°

5. Conclusion

The performance of a vertical coil and shell heat exchanger under the ultrasound, was investigated at the flow rate effects of 10, 15, 20 and 25 L/min and ultrasound propagation angles of 0°, 15°, and 30°. The results showed that the high frequency ultrasonic waves slightly affected the pumping power factor. Meanwhile, this ultrasound increased the thermal performance factor up to 13.45% at the relatively lowest *Re* and the angles of 0°. At this angle, the vortex flow was initiated at the outer side of the coil. The higher angle gave the smaller size of induced vortex and also the performance of heat exchanger under the ultrasound.

Acknowledgments

The authors would like to acknowledge Ample Cool Co., Ltd. for the supports and Prof. Dr. Tanongkiat Kiatsiriroat for valuable suggestions.

References

[1] Sikandor M 2019 *Int. J. Eng. Res.* **10** 303
[2] Ghorbani N Taherian H Gorji M and Mirgolbabaei H 2010 *Int. Commun. Heat Mass* **37** 775
[3] Ghorbani N Taherian H Gorji M and Mirgolbabaei H 2010 *Exp. Therm. Fluid Sci.* **34** 900
[4] Alimoradi A and Veysi F 2016 *Int. J. Therm. Sci* .**107** 196
[5] Suvanjumrat C and Loskupapaiboon K 2020 *Eng. J.* **24(2)** 109
[6] Suvanjumrat C 2017 *Eng. J.* **21(5)** 225
[7] Suvanjumrat C 2017 *Eng. J.* **21(3)** 207
[8] Thungthong T Chalearmwattananon N Mongkolkitngam T and Chaiworapuek W 2020 *Int. J. Heat Mass Transfer* **163** 1
[9] Mashoofi N Pourahmad S and Pesteei SM 2017 *Case. Stud. Therm. Eng.* **10** 161
[10] Wang J hashemi S Alahgholi S and Mehri M 2018 *Int. J. Refrig.* **94** 11
[11] Chaiworapuek W Nongnoi S and Kittichaikarn C 2014 *Kasetsart J. (Nat. Sci.)* **48** 637
[12] Chaiworapuek W and Kittichaikarn C 2016 *Int. J. Heat Mass Transfer* **92** 850
[13] Chaiworapuek W and Kittichaikarn C 2016 *Exp. Therm Fluid Sci.* **77** 179
[14] Srichan S Rakpakdee W Kiatsiriroat T and Chaiworapuek W 2019 *Exp. Heat Transfer* **32** 488
[15] Cao X Dub T Liu Z Zhai H and Duan Z 2019 *Int. J. Heat Mass Transfer* **142** 118437
[16] Rakpakdee W Kittichaikarn C and Chaiworapuek W 2019 *Int. J. Heat Mass Transfer* **135** 491
[17] Mongkolkitngam T Mingbunjerdsuk J Chaiworapuek W and Priyadumkol J 2020 *Kasetsart J.* **108** 45
[18] Inworn N and Chaiworapuek W 2018 *Int. J. Heat Mass Transfer* **122** 1153
[19] Abolhasani M Rahimi M Dehbani M and Alsairafi A 2012 *Numer. Heat Transf. A* **62** 822

Effect of low-frequency ultrasonic waves on heat transfer of laminar water flow over a heating flat plate

Satanon Phetchoo, Jirachai Mingbunjerdsuk*, Kanet Katchasuwanmanee, and Weerachai Chaiworapuek

Department of Mechanical Engineering, Faculty of Engineering, Kasetsart University, Bangkok Thailand 10900

* Corresponding Author: fengjim@ku.ac.th

Abstract
This research investigates the effect of 40, 80, and 120 kHz ultrasound on heat transfer of laminar water flow over a heating flat plate using thin leaf thermocouples in a low turbulence water tunnel. The mainstream velocity was set at 0.127 m/s, corresponding to the local Reynolds number during 65,181 – 148,390 in the test section. The ultrasonic waves were released perpendicular to the mainstream and vertically from the upper wall at the distance of 0.15 m to the heating plate having constant heat flux of 1,040 W/m^2. The results show that the ultrasound, having frequencies of 40 and 120 kHz provides the maximum Nusselt number up to 15% and 31%, respectively at the most downstream position of the test region. In the meantime, the 80 kHz waves slightly affect the heat transfer over a flat plate. Hence, in an attempt to employ the ultrasound for heat transfer enhancement over a flat plate, the distance and ultrasonic frequency must be well considered.

Keywords: Ultrasound, Thin leaf thermocouple, Heat transfer, Laminar flow

1. Introduction

The active techniques of heat transfer enhancement have emerged as powerful platforms in increasing in the efficiency of the thermal system such as the use of jet, vibration, magnetic field [1-6]. The use of ultrasonic waves including the acoustic waves is one of the most widely used techniques that has a frequency above 20 kHz. The frequency range is one of the main parameters, inducing different ultrasonic phenomena such as acoustic cavitation and acoustic streaming. This leads to a different level of heat transfer augmentation [7]. On a flat plate, the 2 MHz ultrasound was found to enhance the heat transfer by increasing turbulence into the liquid flow and caused the higher convective heat transfer coefficients [8]. Meanwhile, the maximum increase in Nusselt number of laminar flows over a flat plate was 10%, 16.5%, and 20.8% under 25, 33, and 40 kHz [9]. Following the advancement of numerical simulation, the computational fluid dynamics-CFD has been employed to investigate the flow and heat transfer for many applications as well as the flow under the ultrasound [10-15]. However, the information about heat transfer under 40 – 120 kHz ultrasound is very little comparing with those from 20 – 40 kHz waves. Thus, this research investigates the heat transfer of laminar flow over a flat plate induced by 40, 80, and 120 kHz ultrasonic waves. The obtained results will be presented in form of surface temperature and Nusselt number distributions. This work will generate fresh insight into improvement of the thermal system, and can be applied to thermal systems such as food, power plant, cleaning industries, etc.

2. Experimental setup

This experiment was carried out in a low free stream turbulence water tunnel, as shown in Figure 1. This test section had a size of 0.2 m width × 1.5 m length × 0.15 m height and was made of 10 mm acrylic plate. A flat plate was fabricated from 3 mm aluminium plate and it was covered by a thin PVC sheet to dampen the heat from a plate heater. The heater was utilized to generate the heat flux of 1,040 W/m^2 to the water flow. Between the heater and 3 mm plate, a 10 mm aluminium plate was inserted for the better heat distribution. Underneath the heater, a thermal insulator was installed to block the heat flux in the downward direction and then a steel plate was mounted to strengthen the flat plate installation. Above a PVC sheet, 7 type-K thin leaf thermocouples, having an uncertainty of 0.75% were well glued at the center of section width and at the distance of 0.47 - 1.07 m from the plate leading edge with an increment of 0.1 m. A new boundary layer was created at the plate leading edge by the aid of the bleeding channel. In this research, the mainstream velocity was set at 0.127 m/s, corresponding to the local Reynolds number between 60,000 and 150,000 at the test area. This velocity was measured with a Nixon Streamflow velocity meter type 403 having an uncertainty of 0.35%. Also, the water temperature was maintained at 24 °C. An ultrasonic transducer was attached on a 1 mm stainless steel plate above the test surface at the streamwise distance of 0.47 m or $Re_x = 65{,}181$. It could release the 40, 80, and 120 kHz ultrasound in the downward direction with the power of 60 W. With these setups, the heat transfer capability of water laminar flow over a heating plate, induced by ultrasound could be characterized.

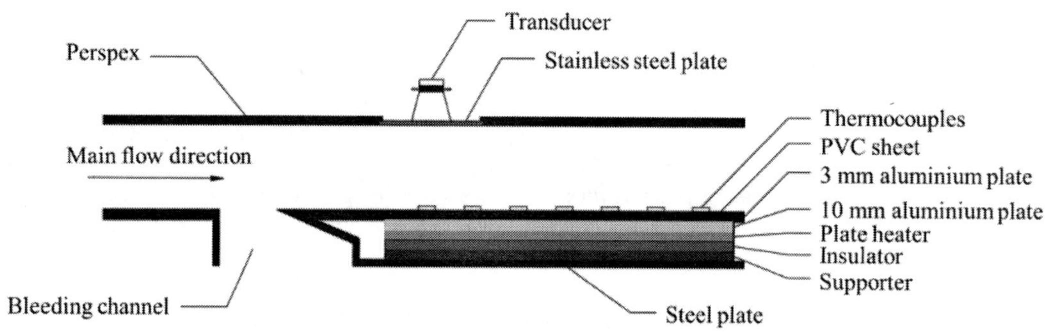

Figure 1. Schematic diagram of the test section

In this study, the local Nusselt number, Nu_x in the streamwise direction was determined as follows:

$$Nu_x = h_x x / k \tag{1}$$

where h_x = local heat transfer coefficient ($W/m^2{\cdot}K$)
x = streamwise distance from the plate leading edge (m)
k = thermal conductivity ($W/m{\cdot}K$)

From the experiment, the local heat transfer coefficient was evaluated as follows:

$$q = h_x(T_s - T_\infty) \tag{2}$$

where q = heat flux (W/m^2)
T_s = surface temperature (K)
T_∞ = free stream temperature (K)

3. Results and discussions

The experimental setup was verified by comparing the local Nusselt number, obtained from experiment and [16] during Re_x = 65,181 to 148,390 as depicted in Figure 2. The local Nusselt number from the experiment was obtained from Equations (1) and (2), while the value from [16] was evaluated as follows:

$$Nu_{x,uh} = Nu_x / (1 - (\xi / x)^{3/4})^{1/3} \quad (3)$$

where ξ = unheated starting length (m)

In this equation, the Nu_x was calculated as follows [16]:

$$Nu_x = 0.453Re_x^{0.5}Pr^{1/3} \quad (4)$$

where Pr = Prandtl number

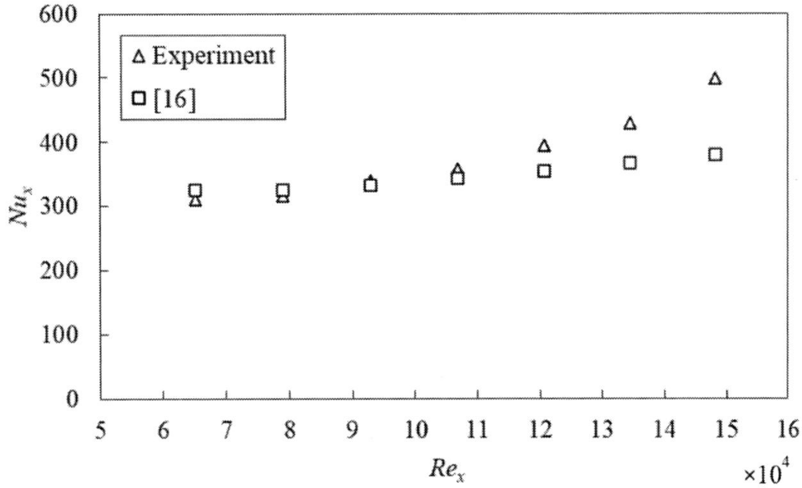

Figure 2. Comparison of local Nusselt numbers from experiment and [16]

As shown in Figure 2, these results are consistent with data obtained in [16]. This study has been unable to demonstrate that the test setup is reliable. Figure 3 shows the surface temperature at the Re_x of 65,181, 79,049, 92,917, 106,785, 120,653, 134,522, and 148,390 under the condition with and without the interference of 40, 80, and 120 kHz ultrasound. It shows that the surface temperature gains with the increasing streamwise distance. When the ultrasound released from the transducer, the surface temperature under the 40 and 120 kHz waves increased at the location of Re_x = 65,181. The waves decreased the surface temperature when Rex was greater than 92,917. The 80 kHz ultrasound was found to have a slight effect on the distribution of surface temperature along the test surface. At each wave frequency, the maximum decrease in temperature was at the farthest position in the downstream direction.

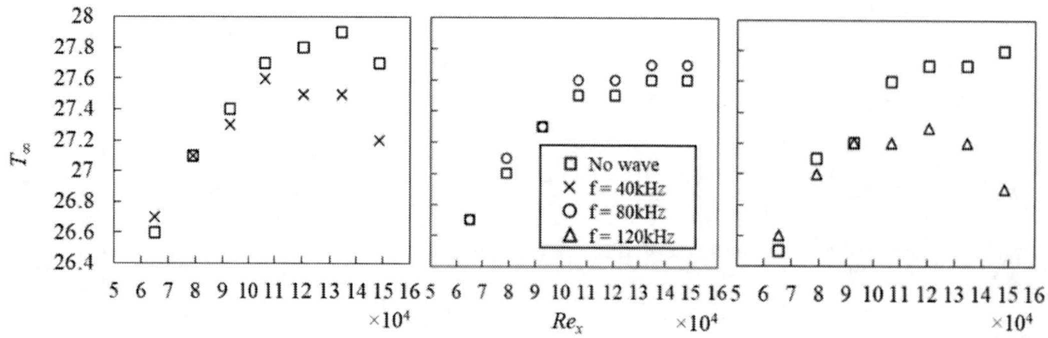

Figure 3. Surface temperature distribution under 40, 80, and 120 kHz ultrasound

In this research, the temperature ratio variable, θ_x induced by ultrasound was determined as follows:

$$\theta_x = (T_{w,x} - T_\infty) / (T_{nw,x} - T_\infty) \tag{5}$$

where $T_{w,x}$ = local surface temperature under ultrasound (°C)
$T_{nw,x}$ = local surface temperature without ultrasound (°C)
T_∞ = free stream temperature (°C)

From Figure 4 below we can see that the distribution of the θ_x represented the surface temperature of a heating plate under the 40 and 120 kHz. It also revealed that the ultrasound decreased in the streamwise direction. Further analysis showed that the heat transfer between water flow and the heating surface is enhanced by the ultrasonic waves. At the farthest position or the Re_x of 148,390, the 40 and 120 kHz waves reduced the θ_x to 0.87 and 0.78, respectively. In the meantime, the θ_x induced by 80 kHz was around 1 over the entire test surface. It means that 80 kHz ultrasound has a small effect on heat transfer enhancement. The lowest θ_x under these waves was at 0.97.

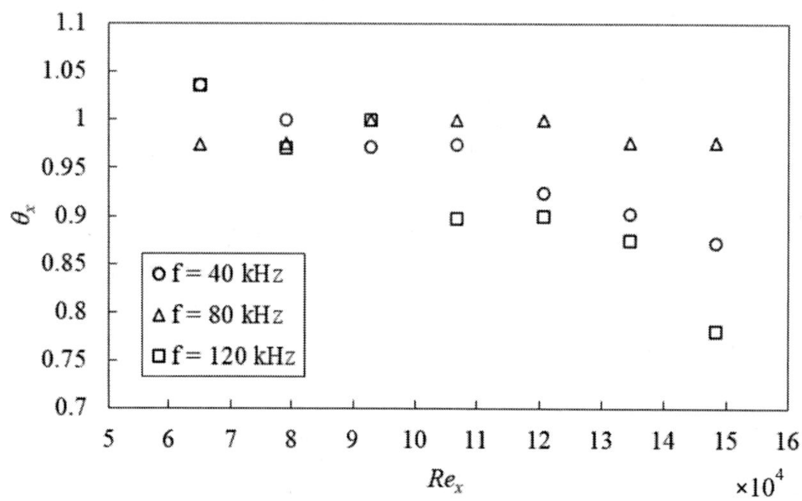

Figure 4. Dimensionless surface temperature under 40, 80, and 120 kHz ultrasound

Figure 5 provides the ratio between the local Nusselt number under the condition with and without ultrasound. At the location underneath the ultrasonic transducer, the Nusselt number was reduced approximately 4% by 40 and 120 kHz ultrasonic waves, while it gained about 3% by 80 kHz ultrasound. The Nusselt number increased up to 15 and 31% at the Re_x of 148,390 by 40 and 120 kHz ultrasound, respectively. The results indicated that the 120 kHz ultrasonic waves are attributed to for the heat transfer enhancement over a flat plate.

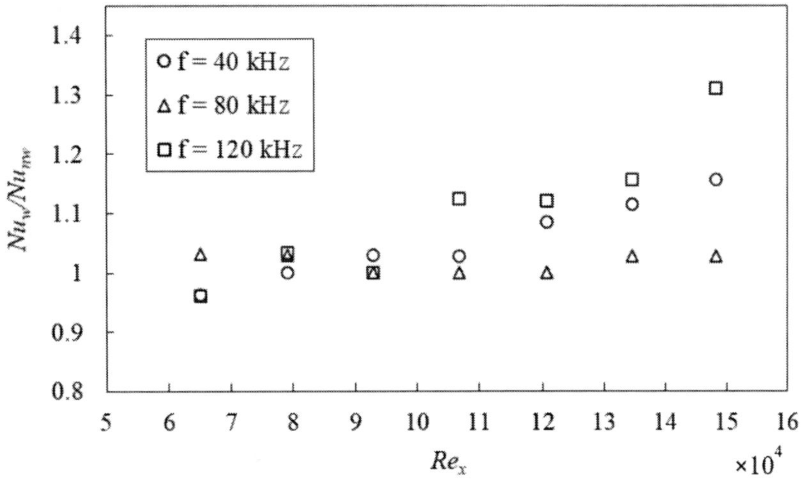

Figure 5. Nusselt number ratio under 40, 80, and 120 kHz ultrasonic waves

4. Conclusion

The present study was designed to determine the effect of ultrasound on heat transfer of laminar water flow over a heating flat plate when the waves have the frequency of 40, 80, and 120 kHz during the Re_x of 65,181 – 148,390. This study has found that generally the ultrasound, having frequencies of 40 and 120 kHz, provides the maximum Nusselt number up to 15% and 31%, respectively at the most downstream position of the test area. These results are likely to be related to the effect of acoustic cavitation and streaming. A possible explanation for this might be that the ultrasound frequency of 80 kHz, was found to have very low effect on heat transfer at the experimental condition. The present study raises the possibility that the distance and ultrasonic frequency must be well--considered to use the ultrasound to enhance the heat transfer of a flat plate.

References

[1] Rakpakdee W Kittichaikarn C and Chaiworapuek W 2019 *Int. J. Heat Mass Transfer* **135** 491
[2] Srichan S Rakpakdee W Kiatsiriroat T and Chaiworapuek W 2019 *Exp. Heat Transfer* **32** 488
[3] Chaiworapuek W and Kittichaikarn C 2016 *Exp. Therm Fluid Sci.* **77** 179
[4] Chaiworapuek W and Kittichaikarn C 2016 *Int. J. Heat Mass Transfer* **92** 850
[5] Chaiworapuek W Nongnoi S and Kittichaikarn C 2014 *Kasetsart J. (Nat. Sci.)* **48** 637
[6] Motozawa M Chang J Sawada T and Kawaguchi Y 2010 *J. Phys.Procedia.* **9** 190
[7] Mongkolkitngam T Mingbunjerdsuk J Chaiworapuek W and Priyadumkol J 2020 *Kasetsart J.* **108** 45
[8] Bulliard-Sauret O Ferrouillat S Vignal L Memponteil A and Gondrexon N 2017 *J. Ultrason Sonochem.* **39** 262
[9] Inworn N and Chaiworapuek W 2018 *Int. J. Heat Mass Transfer* **122** 1153
[10] Suvanjumrat C and Loskupapaiboon K 2020 *Eng. J.* **24**(2) 109
[11] Suvanjumrat C 2017 *Eng. J.* **21**(5) 225
[12] Suvanjumrat C 2017 *Eng. J.* **21**(3) 207

[13] Chaiworapuek W Champagne J Y Hajem M E and Kittichaikarn C 2010 *AIP Conf. Proc.* **1225** 249
[14] Thungthong T Chalearmwattananon N Mongkolkitngam T and Chaiworapuek W 2020 *Int. J. Heat Mass Transfer.* **163** 1
[15] Tang Q and Hu J 2015 *J. Ultrasonics.* **58** 27
[16] Cengel Y A and Ghajar A J 2015 *Heat and Mass transfer: Fundamentals & Applications, 5th ed* (New York: McGraw-Hill)

The energy harvester investigation in exercising by using the piezoelectric effect

Settakorn Kamuang[2], Nutchanok Pramuanjaroenkij[1,*]
Wayuda Wachum[1], Amarin Tongkratoke[2]

[1]Sakolrajwittayanukul School, Sakon Nakhon, Thailand
[2]Faculty of Science and Engineering, Kasetsart University, Chalermpharakiat Sakon Nakhon Province Campus, Thailand

* Corresponding Author: nutchanokp@gmail.com

Abstract
Energy is important in everyday human life, for that reason, energy harvesting is one of the most important issues. Many countries have aging societies, exercise is promoted to keep people in good health in old age. This work focused on an integration of the energy harvesting and exercise. Therefore, this work was aimed at producing electrical energy from exercise shoes by using the piezoelectric effect. Different piezoelectric circuits; series, parallel and combination circuits, and types; Lead Zirconate Titanate (PZT) and ceramic or buzzer types, were investigated. Then, an energy harvesting prototype was built to charge ions to a capacitor by attaching the prototype to shoe pads. From the circuit comparison results, the parallel piezoelectric circuit provided the highest current density and this circuit was carried out in other experiments. Next, the comparison was made by applying both piezoelectric types into two separate parallel circuits with 2 pieces per circuit. The PZT circuit harvested the same amount of electrical power as the buzzer circuit, but the energy harvesting by the PZT circuit took less time at 15 mins to complete its harvest. Finally, the PZT circuit was installed on the shoe pads as a prototype and investigated for its capability. The experiment was carried out by an experimenter who wore the shoes with the prototype and walked at a constant speed of 1 km per hr. The experimental distances were 200, 400 and 600 m and the electric power were calculated accordingly. From the experimental results, walking with the constant speed could enable the PZT piezoelectric materials to produce electric power of electrical energy constantly because of its elastic ability which could tolerate mechanical energy applied to it and because the time used to harvest energy by its circuit is short. It could be concluded from the experiments that the more the distance, the more electric power. Therefore, this prototype proved to produce electrical energy from the piezoelectric effect in harvesting energy and could be further developed as a renewable energy harvester.

Keywords: Piezoelectric, Energy, Energy harvesting, Exercising

1. Introduction

There are many energy harvesting methods such as from electrochemistry [1] and mechanical forces. Piezoelectricity is the appearance of 2 charges on opposite sides of particular non-conducting crystals. The 2 charges consist of the positive electric charge on one side of the crystal and the negative electric charge on the opposite side [2]. Piezoelectricity naturally occurs in ferroelectric crystals such as quartz with not very strong effects but is very stable. The Piezoelectric effect can be divided into 2 types; the direct effect which results from a distortion of the crystal lattice caused by the applied mechanical stress resulting in the appearance of electrical dipoles, and the inverse effect which happens by applying an electric field to the crystal. The electric field applied causes a distortion of the lattice resulting in an induced mechanical strain. Piezoelectricity can also be induced through poling in other materials which are ferroelectric crystals, ceramics or polymers. A piezoelectric ceramic is produced by pressing ferroelectric material grains (typically a few micrometers in diameter) together. After the ceramic powder is heated above Curie temperature and has cooled down, the Perovskite ceramic undergoes phase transformation from the paraelectric state to the ferroelectric state, resulting in the formation of randomly oriented ferroelectric domains. These domains are arranged in grains, containing either 90° or 180° domains. This random orientation leads to zero (or negligible) net polarization and piezoelectric coefficients. The application of a sufficiently high electric field to the ceramic causes the domains to reorient in the direction of the applied electric field. However, after removing the applied electric field, the ferroelectric domains do not return in their initial orientation and a permanent polarization remains in the direction of the applied electric field [3].

Thainiramit [4] designed a beam system and a circuit to harvest electrical energy using the piezoelectric effect. The PZT type was used in this project by attaching two PZTs onto both sides of a cantilever beam, resulting in parallel bimorph harvesting structure. The PZTs were connected by the parallel circuit. The frequency of the vibration applied was approximately 17-18 Hz and at a constant displacement of the beam tip. He found that the beam length affected the frequency the most and a pressure-controlled circuit to prevent overload. The generated power was enough to be used with devices that consumes 1 mW and below.

Chotwattanakankul and Khunboa [5] developed a mechanical energy harvesting system using piezoelectric materials to support a wireless sensor node. The system composed of piezoelectric materials, a power management circuit and an energy storage device attached to a sensor node. The results showed that a strum of 12.7 x 31.8 x 0.51 mm^3 piezoelectric plate was able to produce a maximum voltage 21.3 V and an average energy of 25.94 μJ which could be used as energy for a wireless sensor node when entered sleep mode in 1.44 seconds or in data transformation across wireless sensor nodes by 1 package per 32.29 strumming.

Putthongchai [6] studied the energy harvesting from PZT piezoelectric material with a bulk shape for its effect on mechanical energy acquisition when a mechanical force of 100-500 N was applied to the PZT with low mechanical frequency of 0.5-0.8 Hz. The different electrical circuits were also studied and a prototype was also implemented. The results showed that the diameter of the substance determines the electrical energy the PZT material may acquire. The external resistance had influence on the voltage and power of the piezoelectric material by connecting the circuit with an external resistance of 25 MΩ provided the most electric power. The half-wave rectifiers charging circuit which included a filter is appropriate for energy harvesting from piezoelectric materials under low frequency conditions (less than 1 Hz). Due to less energy loss, the voltage was enough to supply a 1.2 V battery.

Casimiro [7] invented a device which could generate electricity by walking. The piezoelectric materials were applied in a shoe sole and were experimented that it could produce electrical energy. Although this product was not in the stage to be sold, his design was a great benefit to many people worldwide.

This work aimed to find the optimal circuit to develop a prototype which could produce electrical energy from the piezoelectric effect. The different piezoelectric circuits; series, parallel and combination circuits, and types; Lead Zirconate Titanate (PZT) and ceramic or buzzer types, were investigated. Then,

IOP Conf. Series: Materials Science and Engineering 1137 (2021) 012071 doi:10.1088/1757-899X/1137/1/012071

the piezoelectric circuit most suitable was installed on a shoe pad and was connected to an energy harvesting circuit to charge ions to a capacitor.

2. Experimental Setup

2.1 The study of piezoelectric circuits

Three piezoelectric circuits were studied; a series circuit, a parallel circuit and a combination circuit. A circuit which does not refute each other and produces the most electric current would be applied to connect the piezoelectric materials. The electric current can be calculated from Ohm's law as following:

$$V=IR \quad (1)$$

Each circuit in this study had resistors with an exact value of resistance. After acquiring the suitable circuit, the electric circuit with a purpose to harvest energy would be designed. The three circuits; the series circuit (Figure 1), the parallel circuit (Figure 2) and the combination circuit (Figure 3) were studied. The formula used for calculation from the series circuit as following:

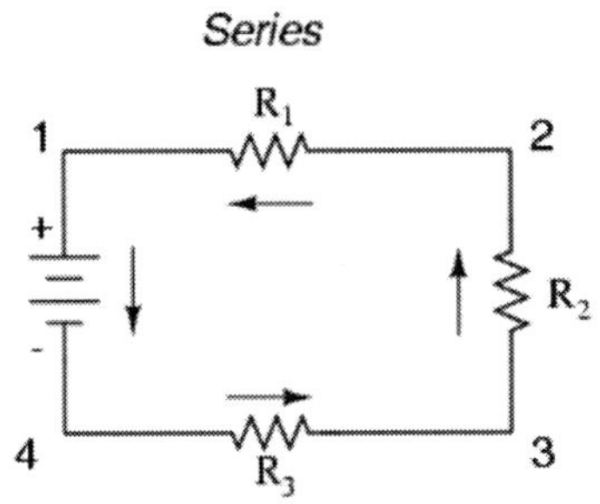

Figure 1. The series circuit [8].

$$VT=V1+V2+V3+\cdots+Vn \quad (2)$$

$$IT=I1=I2=I3=\cdots=In \quad (3)$$

$$RT=R1+R2+R3+\cdots+Rn \quad (4)$$

$$PT=P1+P2+P3+\cdots+Pn \quad (5)$$

The formula used for calculation from the parallel circuit as following:

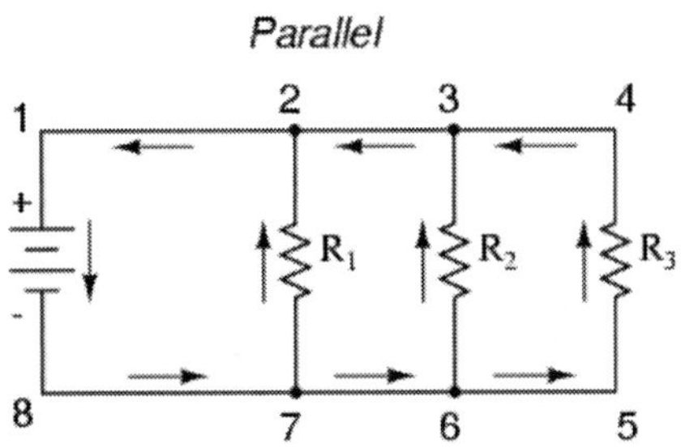

Figure 2. The parallel circuit [8].

$$VT=V1=V2=V3=\cdots=Vn \quad (6)$$

$$IT=I1+I2+I3+\cdots+In \quad (7)$$

$$1RT=1R1+1R2+1R3+\cdots+1Rn \quad (8)$$

$$PT=P1+P2+P3+\cdots+Pn \quad (9)$$

The formula used for calculation from the combination circuit as following:

IOP Conf. Series: Materials Science and Engineering 1137 (2021) 012071 doi:10.1088/1757-899X/1137/1/012071

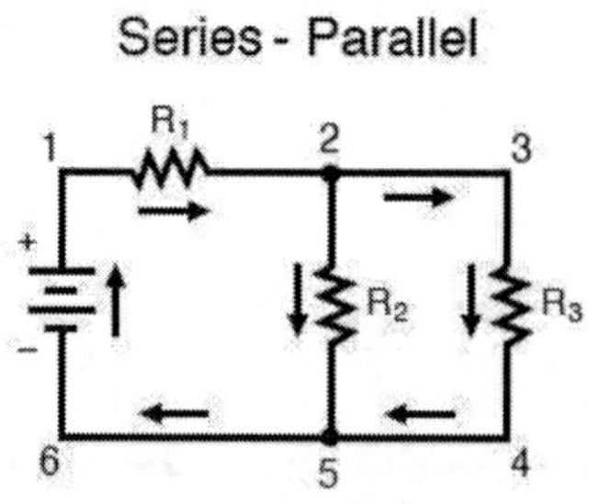

Figure 3. The combination circuit [9].

$$IT=ETRT \quad (10)$$

$$RT=R1+R2,3 \text{ where } R2,3 \text{ could be calculated from } R2,3=R2R3R2+R3 \quad (11)$$

$$PT=ETIT \quad (12)$$

2.2 The study of piezoelectric materials

The two piezoelectric materials; which were the buzzer type with a diameter of 3.5 cm and the PZT type (Figure 4) with dimensions as 5.54 cm in length, 2.34 cm in width and a thickness of 0.46 cm, were connected in two separated circuits. The piezoelectric circuits chosen in the study contained 2 pieces of each piezoelectric material. The voltage of the two circuits were obtained and recorded in a constant period by using an oscilloscope. The frequency used in vibrating the piezoelectric materials; 0.55 Hz, 3.33 Hz and 5 Hz, were studied. The electric current was calculated after connecting each piezoelectric circuit with a capacitor with a capacity of 1000 mF and voltage of 5 V by assessing the time in seconds that a light bulb could maintain its brightness. After recording the time, the electric current can be calculated as following:

$$I=qt \quad (13)$$

When the electric current was calculated, the time used by each circuit in harvesting the current was compared. The electric power can be calculated after attaining both voltage and electric current by Ohm's law as following:

$$P=IV \quad (14)$$

Each of the piezoelectric circuits was connected to a capacitor of 1000 μF and 100 V. As a consequence, the voltage obtained from the capacitor was measured. The PZT parallel circuit was investigated by deforming the piezoelectric with a torsion displacement of 1 cm at a constant frequency of 3 Hz. The buzzer parallel circuit was investigated by tapping with a frequency of 3 Hz. The voltage obtained from the capacitor was recorded in different periods; 10, 20, 30, 40, 50 and 60 s. The electric power obtained from the capacitors of both circuits were calculated subsequently. After achieving the suitable piezoelectric type for harvesting energy, the piezoelectric circuit was installed on a shoe pad.

Figure 4. The piezoelectric transducer; the PZT type, used in this work.

2.3 The experiment and analyzing the results of the prototype
The circuit with a higher level of the electric power installed on the shoe pad was connected to an energy harvesting circuit containing an energy harvester and a boost converter. The circuit was an experimental prototype and investigated for its capability. The experiment was carried out by a volunteer who had a constant weight wearing shoes with the prototype installed and walking at a constant speed of 1 km per hr on a treadmill. The factors studied was the distance. 3 different distances; 200 m, 400 m, and 600 m, were investigated.

3. Results and discussions

3.1 Results and discussions of the study of piezoelectric circuits
It was found that to produce electric current the total voltage has to be equalled; hence, the series circuit is not suitable because it would lead to the piezoelectric materials refuting each other. From the Equations (6) to (12), we had made a comparison between the parallel circuit and the combination circuit as shown in Figure 5 by assuming the exact values of the resistance (5Ω) and the voltage (10V):

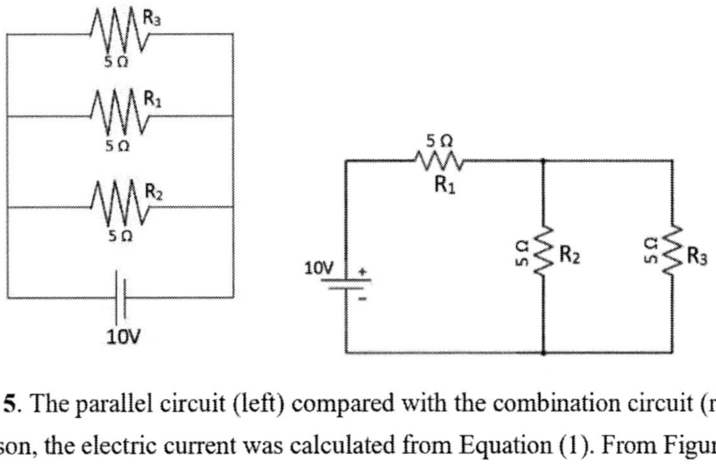

Figure 5. The parallel circuit (left) compared with the combination circuit (right).

To make comparison, the electric current was calculated from Equation (1). From Figure 6 both circuits had the same voltage. In the parallel circuit, the electric current was calculated resulting in 6 A. The electric current from the combination circuit was calculated resulting in 1.33 A. From the results, when studying circuits theoretically, it influenced decision making to make it less difficult and more precise because the decision was made by the theory supported in Figures 1 to 3. Hence, the parallel circuit was the most suitable for harvesting energy because it provided the most current at 6 A. The difference between the parallel circuit and the combination circuit in percentage was 77.83%. Therefore, the further experiments were conducted by using the parallel circuit. The design for the piezoelectric circuits to be experimented on further were made as shown in Figures 6 and 7:

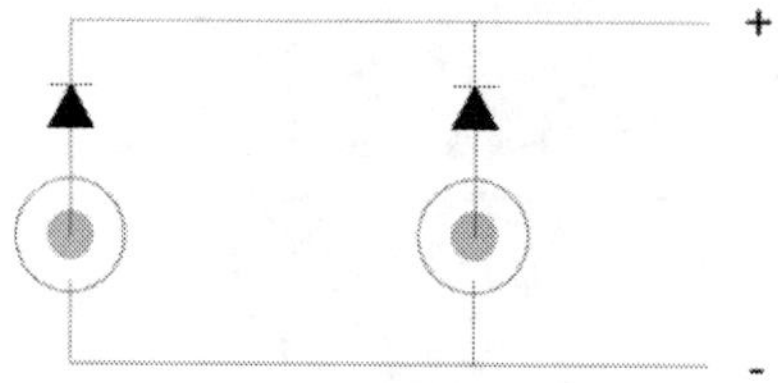

Figure 6. The buzzer type parallel circuit.

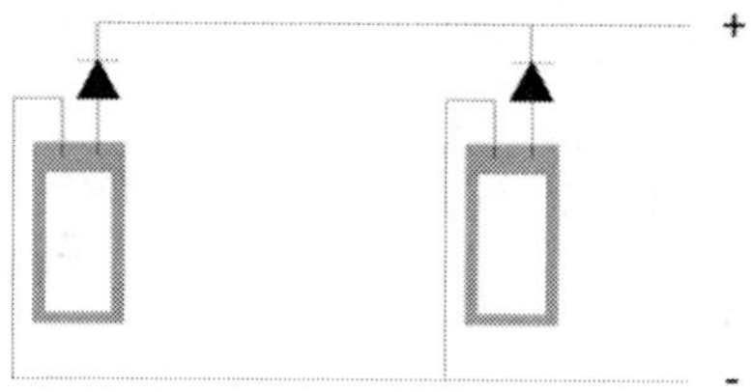

Figure 7. The PZT type parallel circuit.

3.2 Results of the study of piezoelectric materials
From the experiments, the voltage from the parallel circuit of the PZT type was greater than the voltage produced from the parallel circuit of the buzzer type as shown in Figure 8. In a constant period, it could be elaborated from Figure 8 that, from all experimental tests, the voltage from each piezoelectric material fluctuated. The average values of the voltage from the PZT type at 0.55, 3.33, and 5 Hz were 1.14, 2.56 and 3.7 V, respectively. The average voltage of the buzzer type at 0.55, 3.33, and 5 Hz was approximately 0.64, 1.08 and 1.98 V, respectively. As a consequence, by comparing the two types, the PZT type could produce greater voltage. Next, it was found that the electric current from both circuits were the same as shown in Figure 9. From Figure 9, the electric current harvested from the PZT type and the buzzer type were shown with a constant value at 50 μA. Even though the amount of electric current produced was the same, the time used to harvest energy of the PZT parallel circuit was less than the buzzer parallel circuit. The PZT circuit took 15 minutes to finish its harvest while the buzzer circuit took 60 minutes to finish its harvest.

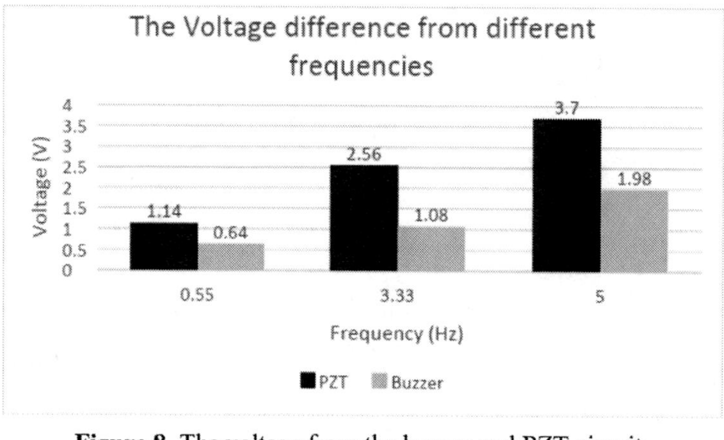

Figure 8. The voltage from the buzzer and PZT circuit.

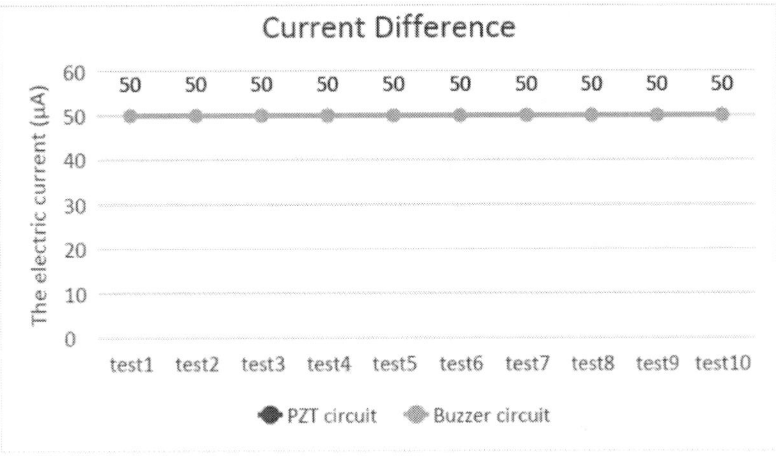

Figure 9. The electric current from the buzzer and PZT circuit.

After having the values of voltage and the electric current, the electric power was calculated using equation (14), resulting in the PZT circuit having electric power of 57 µW, 128 µW and 185 µW at an average when conducted with a frequency of 0.55 Hz, 3.33 Hz and 5 Hz, respectively. The buzzer circuit produced electric power at 32 µW, 54 µW and 99 µW at average when conducted with a frequency of 0.55 Hz, 3.33 Hz and 5 Hz, respectively, as shown in Figure 10.

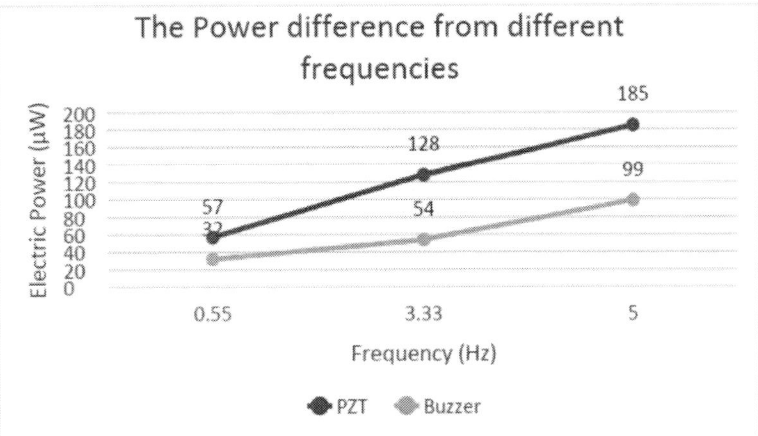

Figure 10. The electric power from the buzzer and PZT circuit.

When the voltage obtained from the capacitor in each circuit was investigated, as shown in Figure 1, it could be seen that the voltage of the capacitor connected to the PZT circuit and the buzzer circuit increased during 6 different periods; 10 s, 20 s, 30 s, 40 s, 50 s and 60 s. The voltage obtained from the capacitor connected with the PZT circuit was the same as the buzzer circuit during 10 s. However, the voltage of both circuits began to differentiate at 20 s with the PZT circuit producing greater voltage than the buzzer circuit.

IOP Conf. Series: Materials Science and Engineering 1137 (2021) 012071 doi:10.1088/1757-899X/1137/1/012071

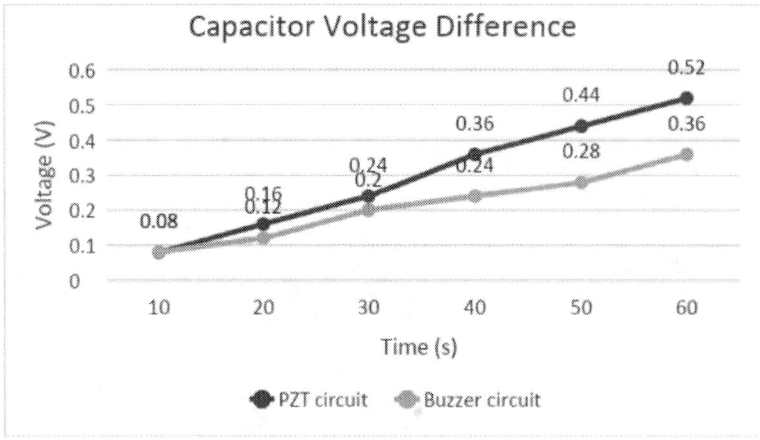

Figure 11. The capacitor voltage from the buzzer and PZT circuit.

After having the values of voltage from the capacitor of both circuits, the electric power was calculated using the equation (14), resulting in the capacitor connected to the PZT circuit having electric power of 4 µW, 8 µW, 12 µW, 18 µW, 22 µW, and 26 µW when conducted for 10 s, 20 s, 30 s, 40 s, 50 s, and 60 s, respectively. The capacitor connected to the buzzer circuit produced electric power at 4 µW, 6 µW, 10 µW, 12 µW, 14 µW, and 18 µW when conducted for 10 s, 20 s, 30 s, 40 s, 50 s, and 60 s, respectively. Thus, as shown in Figure 12, it could be noticed that the electric power from the capacitor connected to the PZT circuit was greater than the capacitor connected to the buzzer circuit.

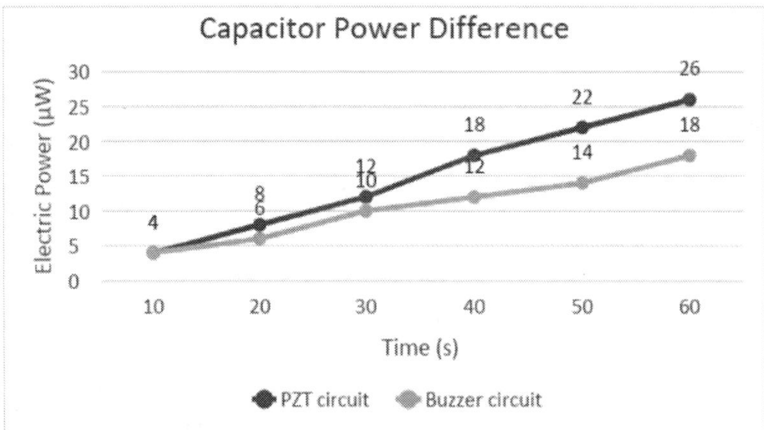

Figure 12. The capacitor electric power from the buzzer and PZT circuit.

3.3 Results of the experiment and analyzing the results of the prototype

From the experiment's results, it was found that walking further distances could produce more electric power as shown in Figure 13. The average electric power calculated was 85.9 µW followed by 135.3 µW and 214.15 µW which represented the average electric power from the distances of 200 m, 400 m and 600 m consecutively.

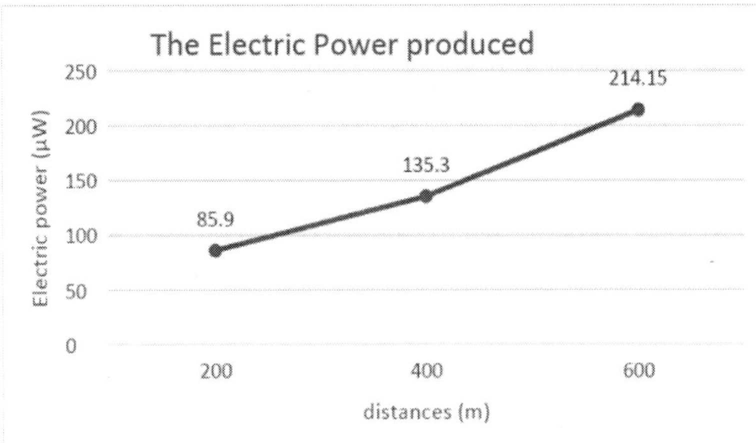

Figure 13. The electric power produced from the piezoelectric circuit.

4. Conclusion

This project aimed to produce a prototype which could harvest energy from the piezoelectric effect. Different circuits: series, parallel and combination were studied. The study resulted in the parallel circuit producing the most electric current without refuting each other. The latter experiment was to compare piezoelectric material types which are Lead Zirconate Titanate (PZT) and ceramic or buzzer types. The PZT type produced greater electric power than the buzzer type. Hence, the energy harvesting circuit was designed using the PZT type. The PZT piezoelectric circuit was installed onto a shoe pad and connected to the energy harvesting circuit resulting in the prototype which was experimented with afterwards. The results elaborated that the prototype could be used as an energy harvester. Therefore the energy harvested from the piezoelectric effect installed on shoe pads could be applied as a renewable energy and developed further.

Acknowledgements

We would like to thank our families and Faculty of Science and Engineering, Kasetsart University, Chalermphrakiat Sakon Nakhon Province Campus, for giving full support to this project.

References

[1] Huang H Wang Z Zhou X Liu H Wei Y Pramuanjaroenkij A Bordas A Page M Cai S and Zhang X 2013 *ECS Transactions* **58**(2) 243-254
[2] The Editors of Encyclopaedia Britannica 2020 *Piezoelectricity*, https://www.britannica.com/science/piezoelectricity, accessed on 10 August 2020
[3] SDTools 2020 *Basics of piezoelectricity,* https://www.sdtools.com/help/pz_basics, accessed on 10 August 2020
[4] Thainiramit P 2013 *Master Thesis* (Songkla: Prince of Songkla University)
[5] Chotwattanakankul T and Khunboa C 2013 *GRC* (Khon Kaen, Thailand) p 195
[6] Putthongchai N 2009 *Master Thesis* Suranaree (Nakhon Ratchasima: University of Technology)
[7] The Editors of Josh Bianc 2020 *Angelo Casimiro: Filipino Teen Invents Electricity-Generating Shoes*, https://philnews.ph/2014/06/05/angelo-casimir, accessed on 10 August 2020
[8] An example of a series circuit 2020 https://sites.google.com/site/lamsnc2dvella/Home/grade-9e-science/unit-3---electricity/series-and-parallel-circuit, accessed on 10 August 2020
[9] Series-Parallel Configuration Circuit 2020 https://www.technocrazed.com/5-1-what-are-series-and-parallel-circuits, accessed on 10 August 2020

Potentials of the direct use of ethanol fuel cell as backup electricity for renewable energy system

Penyarat Saisirirat[1], Peerawat Saisirirat[2,*], Papopchote Kongdai[3], and Brodindech Joommanee[4]

[1]Department of Mechanical Engineering Technology College of Industrial Technology, King's Mongkut University of Technology North Bangkok (KMUTNB),
1518 Pracharat 1 Rd., Wongsawang, Bangsue, Bangkok, 10800, Thailand
[2]National Energy Technology Center (ENTEC),
National Science and Technology Development Agency (NSTDA),
114 Thailand Science Park, Phahonyothin Rd., Khlong Nueng,
Khlong Luang, Pathum Thani, 12120, Thailand
[3]Bureau of Machinery, The Royal Irrigation Department,
200, Tiwanon Rd., Bangtarad, Pakkred, Nonthburi 11120
[4]Department of Tool and Die, Faculty of Engineering and Architecture,
Rajamangala University of Technology Suvarnabhumi,
221 Nonthaburi 1 Rd., Suan-Yai, Muang, Nonthaburi, 11000, Thailand

* Corresponding Author: Peerawat.sai@entec.or.th

Abstract

The issue of renewable energy has received considerable critical attention. This has led to a decline in the population of Greenhouse Gas (henceforth GHG) emissions, health-risk related emissions and energy security. The main challenge faced by many researchers is major issues of low reliability for producing electricity from solar PV. Wind energy has accentuated the problem of sharing of these renewable energies down in the National Energy plans. The solar PV and wind energy may cause fluctuation depending on the daytime and season. The backup electricity is required for increasing reliability of these renewable energy sources. It is now well established that the Direct Ethanol Fuel Cell (henceforth DEFC) is a high bio-fuel production country. Ethanol also has many advantages such as, high energy density, easy for storage and transport, less dangerous to handle compared to hydrogen gas. In addition, the energy efficiency of DEFC is comparable to that of other fuel cell technology. The low operating temperature means less startup period and thermal input energy. This study aims to analyze the potentials to use DEFC as backup electricity for the renewable energy system. The performance characteristics were examined according to updated experimental results. The advantages and challenges were also discussed in this study.

Keywords: Reliability, Backup electricity, Ethanol, Renewable.

1. Introduction

Depending on global economic development and growth of energy resource demand, the GHG emissions are inevitably produced leading to worldwide climate change crises, e.g., increasing average ambient temperature, fluctuating ultimate rainfall -amount, extreme and unpredictable weather condition. The change in average global temperature may accelerate genetic changing [1] and contribute to newly emerging diseases [2], including the ongoing global pandemic of corona virus disease 2019 so-called COVID-19. One of the greatest challenges for Thailand economy is rapid development that may lead to GHG emission production and with high fossil energy dependence. The trends of economic activity development (GDP and population) and GHG emission are shown in Figure 1(a), resulting from various economic sectors and primary energy consumption (see Figure 1(b) and Figure 1(c)) [3]. Electricity produced from industrial activities and transport sectors is the leading cause of the highest GHG emission. The energy consumption is insecurely dominated by fossil energy, more than 80% depending on natural gas and crude oil. It is now well established that the economic activity can impair global pandemic of corona virus disease 2019.

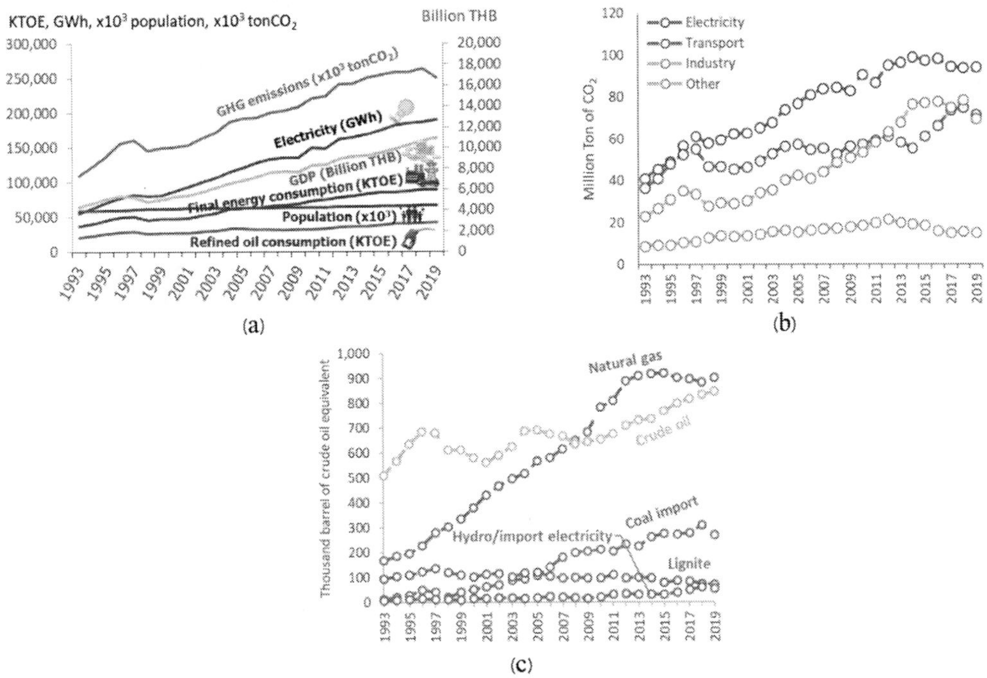

Figure 1. Energy and GHG emission in Thailand [3]
(a) Historical record of economic activities and GHG emission,
(b) GHG emission by economic sectors and (c) Primary energy consumption

To mitigate GHG emission and improve energy security, renewable energy is supported in Thailand through the Alternative Energy and Development Plan (AEDP [4,5]). Beside of biodiesel and ethanol for suppressing transport fuels and other primary energies for heat generation, the natural energy sources, such as hydropower, biomass, photovoltaics and wind turbines, are widely regarded for electricity. However, some kinds of these renewable sources are defined as the variable renewable energy (henceforth VRE) [6] because those are non-dispatchable due to its fluctuating and low reliable nature, such as wind power and solar power. The hydropower also relies on seasonal rainfall while the biomass, which is generally a byproduct of the agricultural sector, depends on the overall planting season. Expanding the share of these renewable energies requires essential supply-demand management

IOP Conf. Series: Materials Science and Engineering 1137 (2021) 012072 doi:10.1088/1757-899X/1137/1/012072

and induces interesting on energy storage technology [7]. The energy storage system absorbs energy and stores it for some time, before releasing supply energy when renewable power down. The energy storage technology is the necessary components in the most energy system, especially the VRE to achieve a low carbon future. However, there are some disadvantages and limitations in applying energy storage technology especially the investment cost and short-life time. Therefore, the backup generator is required to optimize investment cost depending on the size of backup electricity. The fuel cell technology is another alternative environmental-friendly option of backup electricity generation in coupling with the VRE or other renewable energy. Supplied fuels can be the carbon-free hydrogen or the carbon-neutral ethanol from a cultivated plant. In a reason that the ethanol supplied-chain has already completed in Thailand [8]. It can be produced from agricultural products such as molasses, sugarcane juice and cassava. Comparing to methanol, ethanol has a greater advantage in mass-energy density and less toxic. Moreover, it is easy for storage and distribution compared to hydrogen fuel. Therefore, the Direct Ethanol Fuel Cell (DEFC) is focused in this study. The main objective is to highlight potentials to use DEFC for backup electricity application and describe challenge issues.

2. Direct ethanol fuel cell reviews

The direct ethanol fuel cell (DEFC) is an electrolyte class of the lowest operating temperature, ranging from low to high order of 1) polymer electrolyte or proton-exchange membrane fuel cell (PEFC or PEMFC), 2) alkaline fuel cell (AFC), 3) phosphoric acid fuel cell (PAFC), 4) molten carbonate fuel cell (MCFC), and 5) solid oxide fuel cell (SOFC). Some details of each fuel cell categorize are shown in Table 1 [9].

Table 1. Summary of major differences in fuel cell types [9]

	PEFC (PEMFC)	AFC	PAFC	MCFC	SOFC
Electrolyte	Hydrated polymeric ion exchange membranes	Mobilized or immobilized potassium hydroxide in asbestos matrix	Immobilized liquid phosphoric acid in SiC	Immobilized liquid molten carbonate in $LiAlO_2$	Perovskites (Ceramics)
Electrodes	Carbon	Transition metals	Carbon	Nickel and Nickel oxide	Perovskite and perovskite metal cermet
Catalyst	Platinum	Platinum	Platinum	Electrode materials	Electrode materials
Interconnect	Carbon or metal	Metal	Graphite	Stainless steel or Nickel	Nickel, ceramic or steel
Operating temperature	40 - 80 °C	65 - 220 °C	205 °C	650 °C	600 - 1000 °C
Prime cell components	H^+	OH^-	H^+	$CO_3^=$	$O^=$
Product water management	Evaporative	Evaporative	Evaporative	Gaseous Product	Gaseous Product

As can be seen from Table 1 above, we consider using alcohol as one of the energy sources of the fuel cell. The alcohol solution can be directly fed into the anode side while oxygen or air is fed in another cathode side (see also Table 2) [10]. At the Anode side, the active fuel solution breaks down and provides active protons diffused through the separated electrolyte layer. The active electrons flow through the electric junction and release the electricity. At another Cathode side, the oxidant receives inactive

IOP Conf. Series: Materials Science and Engineering 1137 (2021) 012072 doi:10.1088/1757-899X/1137/1/012072

protons and the electrons form water as the fuel cell products. The performance of this alcohol fuel cell can be maintained by compromised proton diffusing rate with the alcohol crossing over characteristics.

Table 2. The stoichiometric chemical reaction in PEMFC and schematic diagram

	Anode	Cathode
PEMFC: H_2 and oxygen/air	$2H_2 \rightarrow 4H^+ + 4e^-$	$O_2 + 4H^+ + 4e^- \rightarrow 2H_2O$
DMFC:methanol and oxygen/air	$CH_3OH + H_2 \rightarrow CO_2 + 6H^+ + 6e^-$	$\frac{3}{2}O_2 + 6H^+ + 6e^- \rightarrow 3H_2O$
DEFC:ethanol and oxygen/air	$C_2H_5OH + 3H_2O \rightarrow 2CO_2 + 12H^+ + 12e^-$	$3O_2 + 12H^+ + 12e^- \rightarrow 6H_2O$
	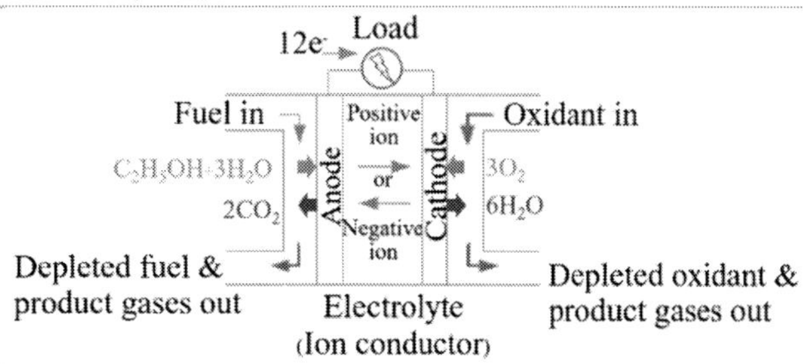	

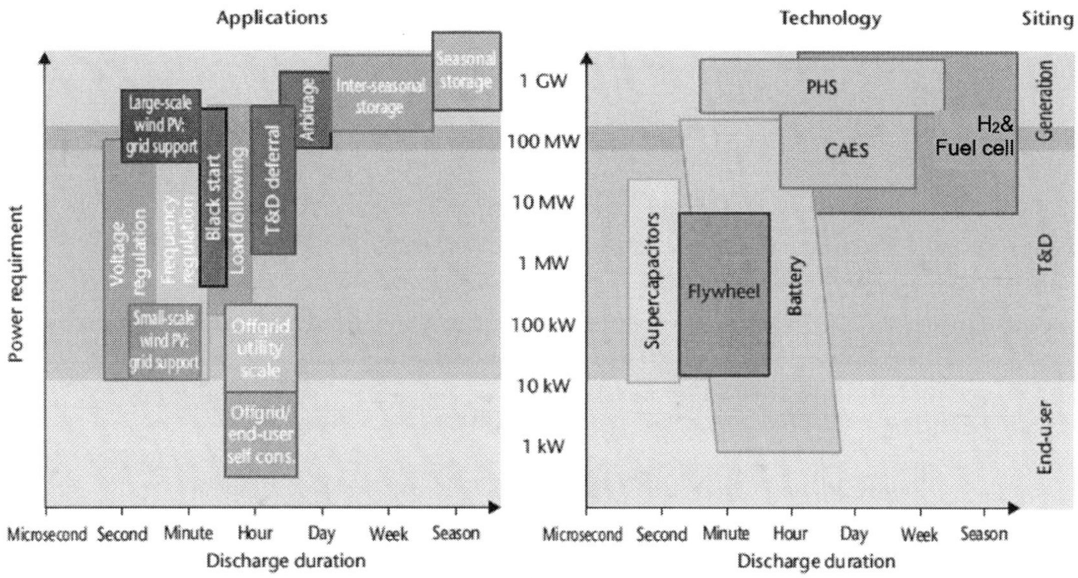

Note: CAES = compressed air energy storage; PHS = pumped hydro energy storage.

Figure 2. Electricity storage applications and technologies [15]

For the direct alcohol fuel cell, the DEFC uses ethanol feeding to fuel cell instead of the more toxic methanol as the Direct Methanol Fuel Cell (DMFC). Ethanol is a more promising option in Thailand than methanol because it comes with a supply chain that is already in place. Ethanol is a hydrogen-rich liquid and it has a higher specific energy (8.0 kW/kg) compared to methanol (6.1 kWh/kg) or better energy density (6.28 kWh/liter) compared to compressed hydrogen (0.18 kWh/liter, for 70 bar 25°C

IOP Conf. Series: Materials Science and Engineering 1137 (2021) 012072 doi:10.1088/1757-899X/1137/1/012072

hydrogen). Ethanol can be obtained in great quantity through a fermentation process of renewable energy resources such as molasses, sugarcane juice and cassava produced in Thailand or from wheat, corn and straw. Bio-generated ethanol (or bio-ethanol) is thus attractive since growing crops for bio-fuel production absorbs much of carbon dioxide emitted into the atmosphere by burning the bio-fuels themselves, as the carbon-neutral fuel. This is in sharp contrast to the use of fossil fuels. Ethanol also remains the easiest fuel to work with and is important for a wide range of use to the consumers. There is evidence that ethanol plays a crucial role in regulating the storage and infrastructure challenges of hydrogen for fuel cell application. To use DEFC as backup electricity for the renewable energy system, ethanol can be stored in a long time without additional cost. Otherwise, it can provide continuous electricity as long as fuel is providing. Figure 2 [11] shows the electricity providing duration (similar to battery discharge duration) from various backup electricity (or electricity storage) applications and technologies.

The hydrogen and fuel cell can provide long electric supply in a wide range of output duration rather than other technologies. Likewise, the DEFC, which is a PEMFC type, has the lowest operating temperature (see also Table 1). It theoretically requires the shortest start-up period [12, 13, 14], among other fuel cell system as well as shorter than internal combustion engines. For prolong discharge duration, the battery energy storage needs additional battery capacity, so investment cost increases. However, the hydrogen and fuel cell can continuously supply electricity as long as the fuel is fed. Figure 3 [15] presents the relationship of investment cost risen and discharge duration for various backup electricity technologies. The DEFC and other fuel cell system provide a reduction of investment cost to prolong discharge duration than the others.

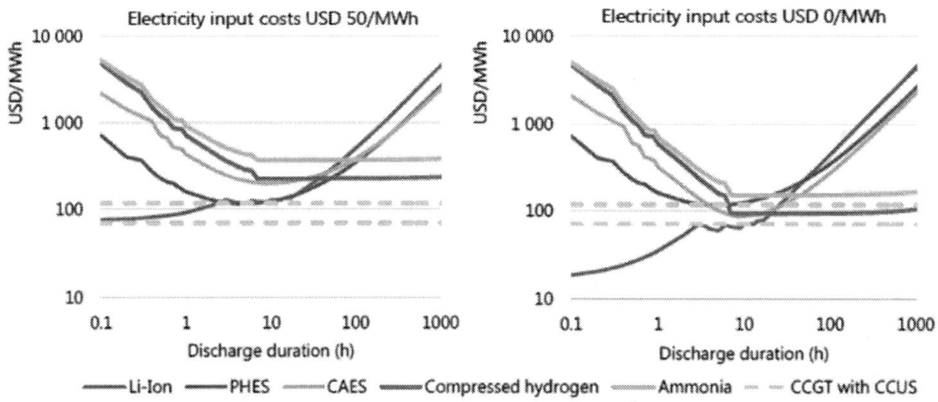

Notes: PHES = pumped-hydro energy storage; CAES = compressed air energy storage; Li-Ion = lithium-ion battery. Compressed hydrogen storage refers to compressed gaseous storage in salt caverns, ammonia storage to storage in tanks.
Source: IEA 2019. All rights reserved.

Figure 3. Energy storage cost risen as a function of discharge duration [11]

3. DEFC performance study

This section presents progress of DEFC performance study followed the experimental investigation and theoretical calculation in the previous works [14, 16,17]. By using a micro-DEFC demonstration kit (4.0 cm^2 active area, 100/1,400/177.8 μm of Catalyst/Diffusion/Membrane layer thickness [18]), electrical performances were measured by varying ethanol quantity (95% purity) in 60 mL Anode-side active liquid mixture (blending with water) coupled to the atmospheric air Cathode side.

IOP Conf. Series: Materials Science and Engineering 1137 (2021) 012072 doi:10.1088/1757-899X/1137/1/012072

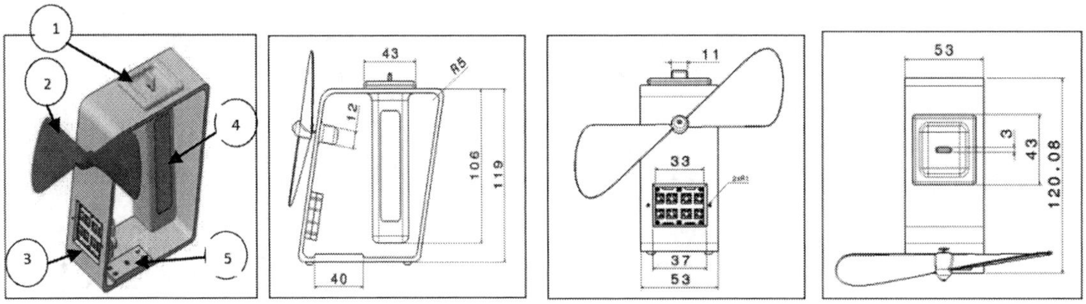

(a) (b) (c) (d)

Figure 4. (a) direct ethanol fuel cell (b) drawing front view (c) drawingside view (d) drawing top view (all dimensions are in millimeters)

The details on the Figure 4 are: 1. Lid set (Micro-DEFC), 2. Vane set, 3. Membrane (MEA), 4. Set structure (Micro-DEFC) and 5. Cylinder solution. Bio-energy feed ethanol set is fuel direct demonstration for changing the ethanol fuel to electrical energy.

A small heater with temperature controller was dipped in active ethanol-water mixture tank so the operating temperature can be maintained. The output electric junctions were attached to a small fan as an electric load, so the cell performance can be observed. The operating parameters were measured using a high-speed data acquisition from National Instrument. Table 3 provides the calculated ethanol concentration according to the ethanol quantity in the mixture. The temperature of micro-DEFC demonstration kit was varied between 30 – 65°C.

Table 3. Ethanol quantity in 60 mL Anode-side active liquid mixture [14, 16, 17]

95% ethanol quantity (mL)	Ethanol concentration (%)
5	7.92%
7	11.08%
9	14.25%
11	17.42%
13	20.58%
15	23.75%

The results of DEFC performance (Current density, Power density and Cell Voltage) in various conditions are shown in Figure 5. It was found that the performance of the demonstration DEFC kit increases with operating temperature as cell voltage, current density and power density increasing. However, the cell performance tends to fall with ethanol concentration achieving some condition. On the other hand, the optimum operation can be obtained at the cell temperature of 51°C. The current density, power density and cell voltage increase continuously with ethanol concentration at this operating temperature. The current density can be varied between 3.35 – 4.83 mA/cm^2 and the power density can be varied between 3.19 – 4.76 mW/cm^2, at a nearly constant cell voltage of 0.97±0.01 V. The maximum power of this DEFC is 76 mW according to 2 cm × 2 cm of cell active area at the maximum ethanol concentration of 23.75 %vol.

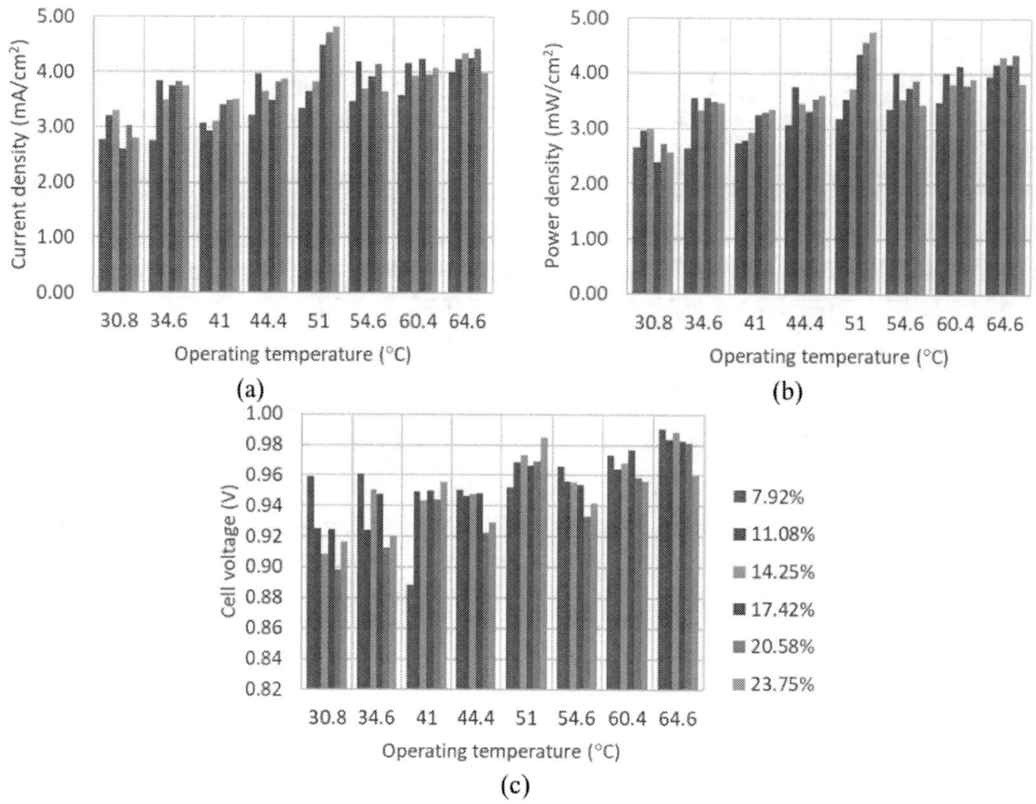

Figure 5. Results of DEFC test in various ethanol concentration (a) Current density, (b) Power density, and (c) Cell voltage

4. Advantages and challenged issues of the DEFC

The performance characteristics of the DEFC were observed in this study. The alcohol crossover the cell is a specific characteristic of the direct alcohol fuel cell. The alcohol crossover rate is generally increasing with temperature, current density, and feed concentration [12, 16, 17, 19, 20]. These characteristics have a more influence in the DMFC (Direct Methanol Fuel Cell) type. However, this alcohol crossover can be minimized within the operating temperature range. This current study revealed that the cell performance fell with ethanol concentration for mostly operating temperature conditions, excepted at the observed optimum temperature of 51°C where the crossover can be maintained and the ethanol concentration can be further increased (see in Figure 5). Therefore, the advantages and challenging issues of the DEFC can be discussed below.

4.1 Reviewed of key advantages of DEFC

- **More sustainable energy source:** The DEFC uses ethanol as the energy input, which is the better attractive than other fuels. Ethanol can be obtained from sustain-renewable resources such as molasses, sugarcane juice and cassava produced in Thailand or from wheat, corn and straw. This gives the ethanol in sharp contrast to fossil fuel since growing crops for biofuel production absorbs much of carbon dioxide emitted into the atmosphere. It has higher energy content compared to hydrogen and methanol, which also has higher toxicity. The infrastructure and storage of ethanol ease to be maintained compared to hydrogen. Besides, the ethanol can be stored in a long time without additional cost and can provide

IOP Conf. Series: Materials Science and Engineering 1137 (2021) 012072 doi:10.1088/1757-899X/1137/1/012072

continuous electricity as long as fuel is providing. This gives much more attractive than the battery energy storage to prolong continuous electricity supply.

- **Low operating temperature and less start-up period**: Low optimal operating temperature has been confirmed from the observed results (51°C), as shown in Table 1. Therefore, the DEFC (and all PEFC) requires the shortest theoretical start-up period compared to other fuel cell technologies, as well as shorter than internal combustion engines.
- **Other advantages**: Comparing to the electric generator with internal combustion engines, the DEFC has better thermal efficiency in both theoretical and practical viewpoints. The current retail-fuel station is applicable for ethanol supply. The vehicle technology shifts toward electric vehicles will give more opportunity on DEFC because there will be excessive ethanol production from available ethanol supply-chain.

4.2 Challenged issues reported in the literature

- **DEFC up-scaling**: The major weakness of DEFC is that there is currently not any successive large-scale DEFC power production, for both cells up-sizing or prototype stack such as a 100 kW power production. The current DEFC are only micro-FCs. The available 10 kW DEFC [21] which success in laboratory prototype composes of ten 36 cell-stacks (27.7 W each with 18 cm × 18 cm active area). The major up-scaling issue depends on several weakness such as in-complete reaction of feed ethanol, low ethanol concentration in practical operation, a difficulty of carbon-carbon bond breaking [20].
- **High cost of catalyst materials:** Like many fuel cells types, the major drawback of DEFC is the cost related to catalysts. However, the cost related issue may become minor due to demand development with renewable power plant expansion. Furthermore, one research direction considers the Anion-Exchange Membrane (AEM), which involve on hydroxide ion conduction in electrolyte instead of the proton. The AEM concept allows the possibility to use non-platinum, less expensive, metal catalysts.
- **Matching with electricity load requirement:** Like other fuel cell or any electric generation technology, the supplied electricity must balance with load demand. The over voltage or power shortage issues will cause widely-damage for many electric devices, reduce technology confidence and slow down the production volume to achieve mass-production levels. This issue requires more effort on research and development.

5. Conclusion

This study has shown that the Direct Ethanol Fuel Cell (DEFC) is considered as another option of backup fuel cell generator for Variable Renewable Energy (VRE). It is selected instead of other active fuel cell reactant (e.g., hydrogen, methanol, natural gas) because the ethanol is one of the most sustainable energy in Thailand. This study has found that generally, the DEFC is one fuel cell technology which operates in the lowest operating temperature so it requires shorter start-up period so appropriates in backup generator application. The experimental investigation is performed in a DEFC demonstration kit. The updated results confirm that the optimum operating temperature of DEFC is in low-temperature range where the ethanol crossing over the cell is compromised with the active-ions diffused process. The optimum operating temperature of 51°C is observed whereas cell temperature increases continuously with ethanol concentration. The DEFC demonstration kit can achieve maximum current and power densities of 4.83 mA/cm^2 and 4.76 mW/cm^2, as respectively at a maximum ethanol concentration of 23.75 %vol. The results of this study indicate that there are some potentials on using DEFC as a backup generator of VRE in Thailand. The DEFC requires the shortest start-up period regarding its lowest operating temperature range (lowest warm–up period). Furthermore, the DEFC uses sustainable renewable bio-ethanol, which is available in Thailand. In the same vein, the thermal efficiency of DEFC is comparable to hydrogen fuel cell but provide easier to handle with (considered on storage, transport and distribution). The major limitation of this study is the up-scaling methodology, high cost of catalyst materials and management unit to matching the DEFC with variable electricity load. Further work needs to be done to establish whether the DEFC technology can be applied in Thailand.

Acknowledgments
This research was supported by College of Industrial Technology, King Mongkut's University of Technology North Bangkok, National Energy Technology Center (ENTEC), National Science and Technology Development Agency (NSTDA), Bureau of Machinery, The Royal Irrigation Department and Department of Tool and Die, Faculty of Engineering and Architecture, Rajamangala University of Technology Suvarnabhumi.

References

]1[World Health Organization 2002 *Climate Change and Human Healt- Risks and Responses : Summar,* https//:www.who.int/globalchange/summary/en/index5.html, accessed on 1 December 2020
]2[Bernstein A 2020 *Coronavirus, Climate Change, and the Environment,* https//:www.hsph.harvard.edu/c-change/subtopics/coronavirus-and-climate-change/, accessed on 1 December 2020
]3[Energy Policy and Planning Office 2020 Economic Data and Analysis", http//:www.eppo.go.th/index.php/en/en-energystatistics/energy-economy-static, accessed on 1 December 2020
[4] Energy Policy and Planning Office 2015 *Alternative Energy Development Plan* http//:www.eppo.go.th/images/POLICY/ENG/AEDP2015ENG.pdf, accessed on 1 December 2020
[5] Energy Policy and Planning Office 2020 N*EPC Approved for Four Main Energy Plans, Energy Fund Management Plan, Acceleration of Nationwide EV Charging Station Distribution and Criteria for Community Power Plants,* http//:www.eppo.go.th/index.php/th/?option=com_k2&view=item&id=15611:news-190363-2, accessed on 1 December 2020
[6] International Renewable Energy Agency 2017 *Variable Renewable Energy* https//:www.irena.org/power/Variable-Renewable-Energy, accessed on 1 December 2020
[7] International Energy Agency 2014 *IEA Technology Roadmap :Energy Storage,* https//:www.iea.org/reports/technology-roadmap-energy-storage, accessed on 1 December 2020
[8] Tunpaiboon N 2020 Ethanol Industry, https://www.krungsri.com/bank/getmedia/0c42d6fd-18d7-41c1-9369-96dded234800/IO_Ethanol_190710_EN_EX.aspx, accessed on 1 December 2020
[9] EG&G Technical Services 2004 *Fuel Cell Handbook, 7th ed* (U.S. Department of Energy, DE-AM26-99FT40575)
[10] Fordi 2011 *Direct Ethanol Fuel Cell,* https://en.wikipedia.org/wiki/Direct-ethanol_fuel_cell, accessed on 1 December 2020
[11] International Energy Agency 2015 *Technology Roadmap :Hydrogen and Fuel Cells,* https//:www.iea.org/reports/technology-roadmap-hydrogen-and-fuel-cells, accessed on 1 December 2020
[12] Nutthapon W Apichai T Supaporn T Kumar S M S and Keith S 2013 *Int. J. Hydrog. Energy* **38**(22) 9464-9473
[13] Liu H Li P Robles D J Wang K and Guerrero H 2014 *Front. Energy Res.* **2** 2
[14] Penyarat S and Bordindech J 2017 *Energy Procedia* **138** 187-192
[15] International Energy Agency 2019 *The Future of Hydrogen,* https//:www.iea.org/reports/the-future-of-hydrogen, accessed on 1 December 2020
[16] Penyarat S and Bordindech J 2018 *IOP Conf. Ser. Mater. Sci. Eng.* **297** 012002

[17] Penyarat S and Papopchote K 2019 *Journal of Mahasarakham University, Research & Knowledge* **5**(1) 1-5
[18] NDCPower EOS vs Conventioinal Fuel Cell Technology)Horizon Fuel Cell Technologies Pte Ltd] (1 February 2012 [http//:www.ndcpower.com/green-power.php
[19] Pereira JP Falcao DS Oliveira VB and Pinto AMFR 2014 J. Power Sources 256 14-9
[20] Helena Berg Joakim Nyman Per Erlandsson Partrik Johansson and Aleksandar Matic "Direct Ethanol Fuel Cells :Ethanol for Our Future Fuel Cells? "Energiforsk, Report 2015:137, ISBN 978-91-7673-137-6
[21] Badwal SPS Giddey S Kulkarni A Goel J and Basu S 2015 Appl. Energy 145 80-103

Simulation of the variable assembly of MGT-SOFC systems to optimize the mechanical efficiency

R. Dückershoff[1],*, H.P. Berg[2], M. Kleissl[2], A. Himmelberg[2]

[1]University of Applied Sciences Mittelhessen (THM), Friedberg, Department of Mechanical Engineering, Mechatronics and Material Technology, ZEuUS, Wilhelm-Leuschner-Str. 13, 61169 Friedberg, Germany
[2]BTU Cottbus-Senftenberg, Chair of Combustion Engines and Flight Propulsion, Siemens-Halske-Ring 14, D-03046 Cottbus, Germany

* Corresponding Author: Roland.Dueckershoff@m.thm.de

Abstract
Simulations of various modifications and technical solutions to improve the energy yield when connecting micro gas turbines and SOFC high-temperature fuel cell systems are presented. Above all, heat integration measures enable an increase in system efficiency and a reduction in primary energy input. On the basis of analysis of efficiencies and the heat output it is shown, which high influence the modifications for the improvement of the hybrid process have on the energy yield. Finally, an optimal configuration is worked out, which enables primary energy to be used with the highest possible efficiency.

Particular emphasis is placed on the use of gaseous fuels from renewable energy sources. Preparation mechanisms, e.g., the reforming of used methane, and the improvement of the heat insulation materials was also taken into account. All in all, electrical total efficiencies for the optimized hybrid system of more than 65 % could be proven and additional thermal energy for heating purposes was extracted. Overall efficiency for the utilization of the energy content of the primary energy of up to 95 % was then achieved.

Entire systems with a maximum system output of around 270kW were examined. In the near future, these plants can replace existing fossil-fuel power plants. Due to the achievable flexibility, the previous security of supply and load fluctuation compensation can also be performed.

Keywords: MGT-SOFC, Hybrid System, Simulation, Fuel Cell, Micro gas turbine.

1. Introduction

One task to protect the climate is to reduce primary energy consumption. At the same time, the switch to renewable energy sources must be tackled [13]. The big problem here is the time difference between production and consumption. For example, the so-called dark lull is critical for the operation of a supply network. A method must be found to store energy generated from renewable sources safely and effectively for later conversion into electrical energy. For efficient storage, power-to-gas systems are

IOP Conf. Series: Materials Science and Engineering 1137 (2021) 012073 doi:10.1088/1757-899X/1137/1/012073

also an option if there is an efficient possibility of regenerating electricity. Efficiencies of more than 75 % for the production of fuel gas (CH_4) and more than 80 % for the production of pure hydrogen are already being achieved. There is no procedure for the efficient conversion of electricity back into electricity [1],[2],[4],[5].

In fuel cells, the direct conversion of chemical energy into electricity takes place through electrochemical processes. Direct conversion usually results in a higher system efficiency [7]. Among the highly efficient fuel cells are SOFCs (Solid Oxide Fuel Cells). These high-temperature fuel cells use membranes made of yttrium-stabilised zirconium oxide as solid electrolyte. This type of fuel cell can achieve an electrical efficiency of more than 50 % at operating temperatures of more than 700°C under atmospheric conditions.

Simulations also show that an increase in the operating pressure of SOFC fuel cells can increase cell performance. It may therefore make sense to couple an energy converter that can convert energy under pressure and high operating temperatures with such cells. Efficient, recuperated micro gas turbines are a suitable solution for this purpose. They supply compressed combustion air which, after preheating, can be made available as cathode air for fuel cell conversion. The use of the components can be coupled in different ways to make better use of the primary energy input. With conversion of the residual fuel in an afterburner chamber, the electrical yield and the electrical efficiency of the entire system can be further increased with a turbine.

Various hybrid interconnection variants of micro gas turbines and SOFC stack clusters allow heat flows and energy conversions to be coupled in different ways. Selected arrangements are to be compared with each other using Ebsilon simulations [10] and heat analyses [8]. A circuit variant of the individual apparatuses will be worked out, which promises a maximum electrical efficiency for an entire machine. The simulation results will be presented and compared with each other and the influence of the selected individual components on the overall system will be discussed.

2. Determination of performance and efficiency of hybrid coupling of Micro gas turbines and SOFC fuel cells

To investigate the performance of hybrid systems consisting of SOFC high-temperature fuel cells and micro gas turbines, the electrical efficiency in relation to the lower heating value (LHV) of individual circuit variants is compared with each other. The overall efficiency of the hybrid system in terms of heat output and electrical efficiency also plays a role. Hybrid supply systems MGT-SOFC can be used both for local heating and electrical supplies [1].

In an SOFC fuel cell, the chemical energy of the fuel is directly converted into electricity by electrochemical processes ("cold combustion"). For their electrochemical reactions, such fuel cells use hydrogen as fuel, which is supplied to them either directly or as a fossil precursor. Air is fed to the cell as an oxidizer, separated by a solid oxide membrane. The oxygen contained there can diffuse ionised as O2- at high temperatures through the membrane. On the anode side, these ions then react with the hydrogen to form water, releasing electrons and heat. The electron migration can be used as electric current via the connected electrodes [7].

Anode: $$2\ H_2 + 2\ O^{2\ominus} \rightleftharpoons 2\ H_2O(g) + 4\ e^{\ominus} \qquad E_0 = 0\ V \qquad (1)$$
$$CO + O^{2\ominus} \rightleftharpoons CO_2 + 2\ e^{\ominus}$$

Cathode: $$O_2 + 4\ e^{\ominus} \rightleftharpoons 2\ O^{2\ominus} \qquad E_0 = 1{,}23\ V \qquad (2)$$

$$2\ H_2 + O_2 \rightleftharpoons 2\ H_2O$$

Since a free enthalpy of the reaction of $\Delta G_0 = 237.13\ kJ/mol$ and an enthalpy under standard conditions of $\Delta H = 285.83\ kJ/mol$ can be determined, regarding the theoretical efficiency follows:

$$\eta_{SOFC}^{rev} = \frac{\Delta G_0}{\Delta H} = \frac{237.13\ kJ/mol}{285.83\ kJ/mol} = 82.96\% \qquad (3)$$

IOP Conf. Series: Materials Science and Engineering 1137 (2021) 012073 doi:10.1088/1757-899X/1137/1/012073

These 82.96% represent the theoretical maximum which can be achieved with the reacting substances H_2 and O_2. The efficiency of the fuel cell (η_{FC}) shows the ratio of supplied chemical energy and generated electrical energy, which summarizes the losses inside the fuel cell.

$$\eta_{FC} = \frac{electrical\ power}{chemical\ performance\ in\ used\ hydrogen} \tag{4}$$

This can be represented in an efficiency chain, so that the internal losses can be more specifically attributed to their origins and the Fuel Utilization of Hydrogen in the FuelCell F_U:

$$\eta_{FC} = \eta_{SOFC}^{rev} \cdot \eta_U \cdot \eta_I \cdot \eta_H \cdot F_U \tag{5}$$

In the system under investigation, the energy expenditure of the transformation from CH_4 to H_2 in the reformer reaction is taken into account for evaluation. In the turbo system, efficiency can be assessed on the basis of the individual efficiencies of compressor, turbine and mechanical coupling. To determine the el. efficiency, the generator efficiency is also required. All in all, this results in an electrical efficiency for the turbo machinery part [8],[11]:

$$\eta_{el,MGT} = \eta_{Comp} \cdot \eta_m \cdot \eta_{Turb} \cdot \eta_{Gen} \tag{6}$$

$$\eta_{el,Turb} = \frac{electrical\ power}{chemical\ performance\ in\ used\ methane} \tag{7}$$

2.1 Efficiencies for hybrid systems made of MGT and SOFC

In order to be able to compare different system circuits with each other, it is necessary to analyze the overall electrical efficiency with η_C as the efficiency of the fuel conversion:

$$\eta_{SOFC-MGT}^{real} = \left(\eta_{MGT}^{real} + \eta_{SOFC}^{rev} \cdot \eta_U \cdot \eta_I \cdot \eta_C \cdot \left(1 - \eta_{MGT}^{real}\right)\right) \cdot \eta_{System} \tag{8}$$

For comparison, the power output of the fuel cell was kept constant. Therefore, the total mass flow of combustion air approximately was taken to be constant. These fixed points were used to compare electrical efficiency, overall efficiency and power output. This was used to determine which subsystem configuration has a positive influence on the performance of the overall system.

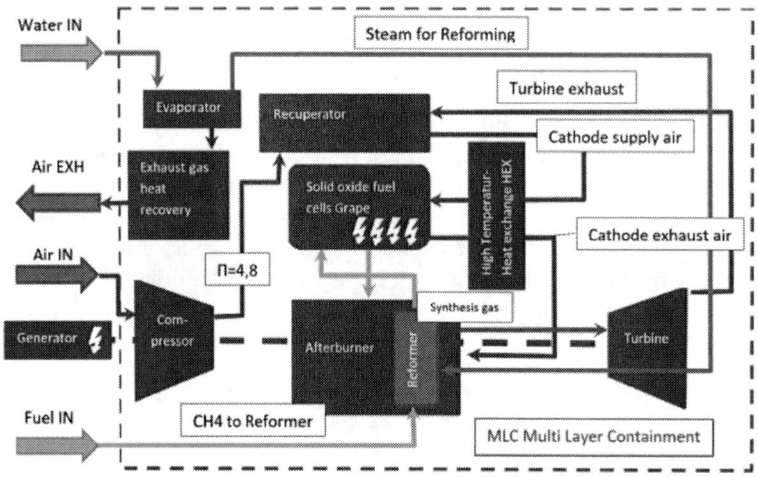

Figure 1: MGT-SOFC-Hybrid Energy Converter

IOP Conf. Series: Materials Science and Engineering 1137 (2021) 012073 doi:10.1088/1757-899X/1137/1/012073

3. Researched systems

First, a reference simulation for a MGT SOFC system with single-stage compression and turbine is presented. This system operates at 4.8bar abs. operating pressure and can convert a high degree of primary energy into electrical energy. SOFC stacks are used, which show their optimum operating point at 750°C. These are the operating conditions of an MGT SOFC of the TurboFuelCell 1.0 type, which is currently under development in the authors' working group [1],[2],[5],[6],[14].

For comparison purposes, the number of stacks was kept constant and the power output of the entire system was kept similar. Care was taken to ensure that all the load cases examined had approximately the same primary energy input. Furthermore, both full load and partial load cases with a machine load of 50% were compared with each other.

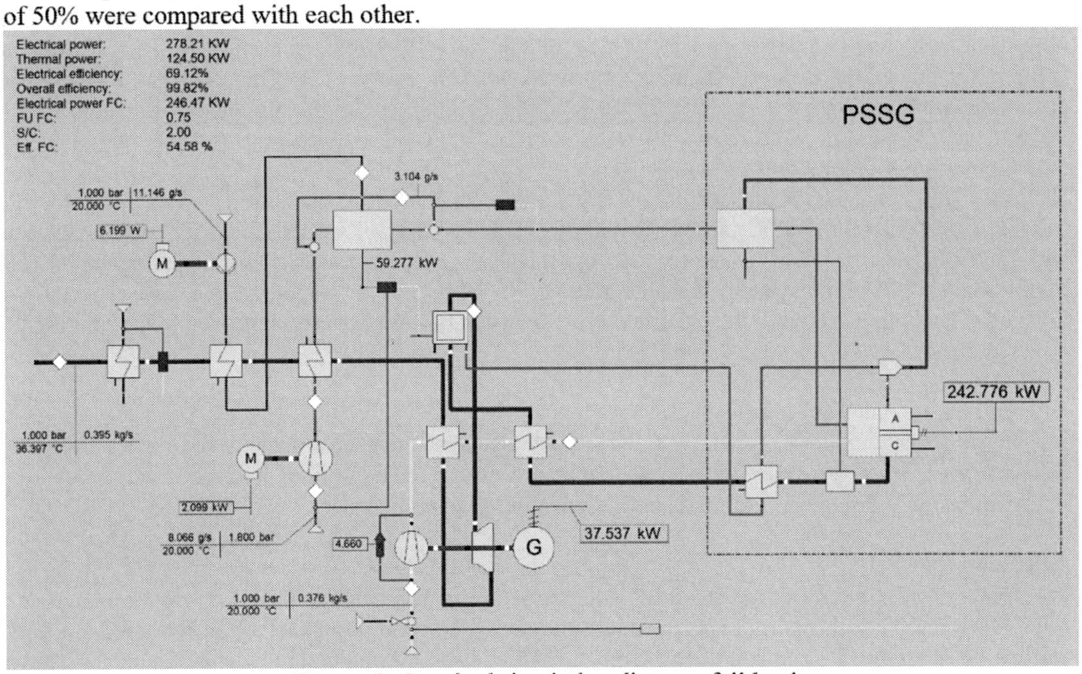

Figure 2: Standard simulation diagram full load

3.1 Two-stage turbo system

In particular, if the compressor system is two-stage, impellers with a higher efficiency can be used. Above all, the gap losses can be reduced by lower pressure stages in the respective impellers. It will be investigated whether a uniform pressure staging of two impellers can be improved compared to compression with only one compressor impeller, but a larger pressure staging. A turbine wheel is to be used as the drive in the evaluation. A distinction is made between compression with intercooling and without such an extension. The influence of the heat deficit due to the changed compression stage on the overall process is also evaluated. The heat exchange for the intercooling process is done by a heat exchanger. An investigation of the injection of water for direct cooling can be found in [3].

3.2 Electrical separation of turbine and compressor

Especially for an optimization of the partial load behavior it is necessary to operate the turbine wheel and compressor wheel in a hybrid system at the optimum operating point. It may be necessary to operate both wheels in different load configurations and at different speeds. One possibility is to drive both systems with electric motors. For this purpose, it is being investigated whether the advantage of decoupled operation allows an increase in efficiency for the overall system. The efficiency of the electrical operation and the evaluation of an independent control system are of particular importance.

3.3 Optimization of heat management

Various circuits for optimizing the heat flow are being investigated for their influence on the overall system of an MGT-SOFC. The aim is to show that highly integrative circuitry improves the utilisation of the primary energy input. Furthermore, it will be shown under which conditions low-temperature energy can be made available.

3.4 Use of a turbine system with VTG guide vanes

By using a VTG (variable turbine geometry) guide grid in the turbine section, the efficiency and response of a radial turbine furthermore can be further increased. Thus, at the design point under full load, an optimum flow to the turbine wheel is set. By adjusting the inflow conditions in the partial load range, the response behaviour and efficiency of the turbine system is improved. This allows the compressor to operate in a significantly improved working range [12]. Thus, the gas flow can be optimally introduced into the machine for the required load range with the highest possible efficiency.

3.5 Optimization of the flow of the supply gases

Heat exchange, pipe losses and insulation expenditure for the entire machine can be optimized with suitable flow control. In particular, MLC airflow (MLC Multi-layer Containment for system isolation and media packaging) and pressure loss optimization can have a significant influence on the performance of the entire system. Variable guide grids are also used to optimize the load, especially in the partial load range. Overall, the electrical efficiency can be significantly improved by optimizing the pipe systems.

3.6 Influence of the stack load

It is shown to what extent the stack load can lead to degeneration and a change in the service life of the cell systems used. In particular, it is shown that an increase in the number of stacks certainly leads to an improvement in the stability and performance yield of the overall system.

4. Results

For comparison, a circuit corresponding to the structure of a TurboFuelCell 1.0 was simulated as a reference circuit. For this purpose, various adaptations of the plant schematic were examined in the 1D simulation program Ebsilon from Steag GmbH. For the standard model, the calculation data as shown in Table 1 were determined for an optimum efficiency and a system output of about 270kW.

TurboFuelCell 1.0-Standard	
Electrical Power delivered to grid	278.21kW
Thermal Power	124.50kW
Electrical Efficiency	69.12%
Overall Efficiency	99.82%
Electrical Power Fuel Cell/Turbine	246.47kW/37.537kW
Fuel utilization/Efficiency Fuel Cell	75%/54.58%
Reformer Steam/Carbon	2.0
Air in	0.376kg/s
Fuel in	8.066g/s CH_4
Pressure ratio π	4.65

Table 1: Performance data of the TurboFuelCell 1.0 Standard simulation

Due to the high electrical efficiency a very good utilization of the primary energy carrier methane is achieved. Heat recovery measures also help to achieve a high overall degree of utilization, whereby the thermal output can be used for heating purposes and low-temperature heat.

IOP Conf. Series: Materials Science and Engineering 1137 (2021) 012073 doi:10.1088/1757-899X/1137/1/012073

description	el. Efficiency %	el. Power kW	air flow g/s	fuel flow g/s	thermal power KW	overall efficiency %
Standard simulation full load	69.12	278.21	376	8.066	124.5	99.82
standard simulation partial load	71.05	163.08	188	4.599	68.62	100.73
standard simulation partial load, VTG	72.14	169.62	188	4.710	68.17	100.93

Table 2: Performance data Standard configuration

4.1 Study of the two-stage turbo system

The use of a two-stage compressor system reduces the required pressure step per impeller to the baseline having one pressure step [12]. This reduces the pressure ratio to 2.16 per stage while maintaining the same gap loss. Thus, the isentropic efficiency of each individual stage can be slightly increased. For the simulation, an increase in the isentropic efficiency per stage of 4% was assumed.

A two-stage compressor system results in a noticeable increase in the output of the entire machine. This shows that such a system can certainly provide advantages in terms of primary energy consumption, irrespective of more complicated flow control.

description	el. Efficiency %	el. Power kW	air flow g/s	fuel flow g/s	thermal power KW	overall efficiency %
two stage compressor, full load	69.64	280.26	375	8.064	123.05	99.99
two stage compressor, partial load	71.45	164.32	188	4.608	68.19	100.88
two stage compressor, full load, intercooler	71.28	286.96	356	8.066	93.71	101.2
two stage compressor, partial load, intercooler	73.2	169.52	177	4.639	52.41	102.05

Table 3: Performance data of two stage compressor system, one shaft

A further improvement can be achieved by the use of intercooling (Table 3, line 3+4). This reduces the energy required for compression and increases the electrical efficiency. However, the dissipatable thermal power is also significantly reduced. The heat dissipated by the intercooling is missing in the heat balance to reach the stack operating temperature and has to be removed elsewhere. This means that part of the low-temperature heat is no longer available for heating purposes.

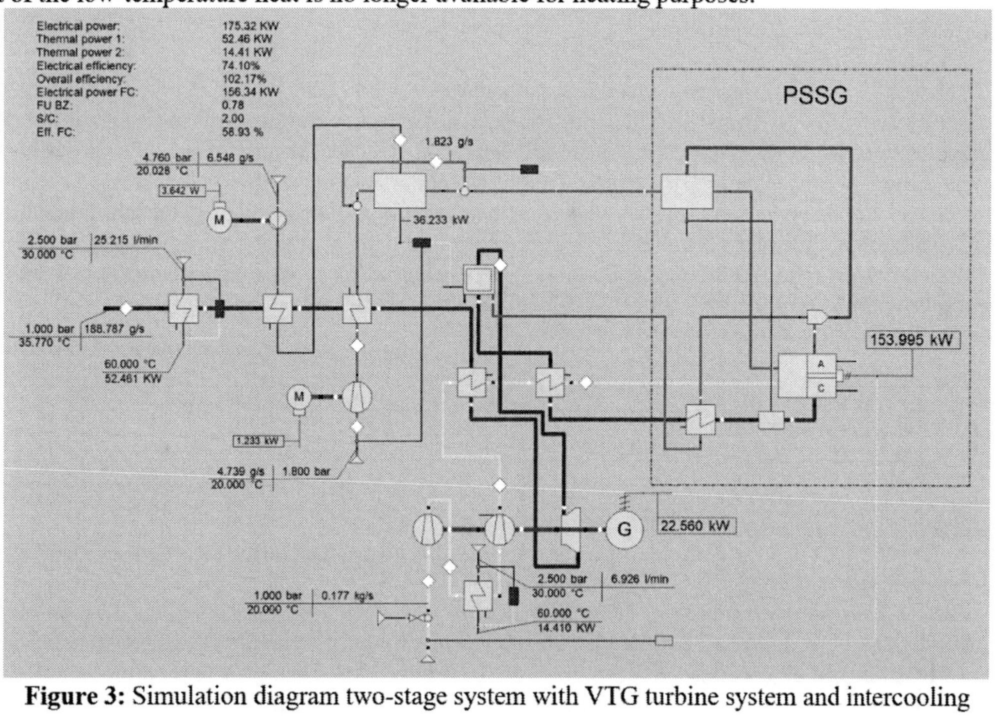

Figure 3: Simulation diagram two-stage system with VTG turbine system and intercooling

4.2 Investigation of the electrical separation of turbine and compressor

For better utilization of the impeller performance, especially in the partial load range, systems were investigated which allow the speeds of turbine and compressor to be varied differently. In particular, there is the possibility to strengthen the electrical generator of the turbine section and to drive the compressor wheel with the additionally gained energy. Thus, both systems can be operated in better efficiency ranges. The standard configuration and a two-stage compression were investigated, as well as the use of adjustable turbine geometries to improve the partial load behavior.

4.2.1 Standard Configuration with electrical separation

description	el. Efficiency %	el. Power kW	air flow g/s	fuel flow g/s	thermal power KW	overall efficiency %
Standard simulation full load, electrically isolated	67.54	271.79	376	8.066	124.5	98.24
standard simulation partial load, el. isolated	69.57	159.64	188	4.599	68.62	99.24
standard simulation partial load, VTG, el. isolated	70.73	166.28	188	4.710	68.17	99.52

Table 4: Performance data of standard turbo system, compressor/turbine on two shafts, electrical isolated

If the compressor and turbine are separated in the standard configuration, the additional loss of efficiency due to the electrical conversion will come more in focus. Table 4 shows the calculation regarding this configuration. In spite of high efficiencies in high frequency electric motors and generators, the electrical efficiency in particular will decrease due to the additional double energy conversion under full load conditions. It is also to be expected that the electrical output of the overall system will be reduced, since the additional conversion loss requires more drive energy. Furthermore, a higher degree of efficiency can be expected under partial load than under full load, since the fuel cells achieve higher efficiency under lower load. However, the losses due to electrical separation cannot be compensated for by this increase.

By using a VTG system, an increase in turbine efficiency can be achieved, especially in the partial load range, through improved compressor delivery. Thus, the electrical output can be improved at the expense of the thermal power output.

4.2.2 Two-stage turbo system with electrical separation

Electrically separated systems, as listed in Table 5, do not provide any improvement in maximum efficiencies even in the two-stage variant. In all cases, the electrical results of the overall machine are below the respective mechanically coupled systems (impellers on one shaft). The decrease in efficiency due to the electrical separation is clearly visible in the electrical yield as well as in the additional thermal power available at full and partial load. A further problem for a timely realisation is the additional development effort required for the air bearing system required for a two-part system. This would then have to allow for a significantly improved controllability for different speed ranges.

description	el. Efficiency %	el. Power kW	air flow g/s	fuel flow g/s	thermal power KW	overall efficiency %
two stage compressor, full load, el. Isolated	67.96	274.85	377	8.106	123.72	98.32
two stage compressor, partial load, el. Isolated	69.85	161.46	189	4.632	68.56	99.29
two stage compressor, partial load, VTG, el. Isolated	71.05	168.26	189	4.745	68.06	99.57
two stage compressor, full load, intercooler, el. Isol.	69.87	282.33	356	8.098	122.22	99.89
two stage compressor, partial load, intercooler, el. Isol.	71.89	167.21	177	4.660	67.78	100.83
two stage compressor, partial load, intercooler, VTG, el. Isol.	72.85	173.08	177	4.759	67.37	101.01

Table 5: Two stage compressors, compressor/turbine on two shafts, electrical isolated

4.3 Results from the optimization of heat management

Heat losses can be reduced by insulation measures. For this purpose, improved heat transfer can be realized by integrative measures and nesting of the apparatus parts. Heat transfer and pressure losses in pipe systems can also be improved by equalization of the flow and lowering the flowspeed. A decoupling of low-temperature heat for local heating concepts can cause an increase in the recoverable thermal output at a flow temperature of 40°C compared to 80°C. Such a system can be used directly for feeding into underfloor heating systems and provides an additional heat quantity of 63.5 kW at full load. Modern low temperature heating systems operate with a 60°C/30°C flow/return interval. Even for such systems an additional heating cable of approx. 31.5 kW can be taken from the system. However, the utilization of the condensation of water, which is relevant to the calorific value, plays a considerably greater role in the gain in thermal output. The condensation water obtained in this way can also be used for the extraction of low-temperature heat.

4.4 Use of a turbine system with VTG guide grid (variable turbine geometry)

description	el. Efficiency %	el. Power kW	air flow g/s	fuel flow g/s	thermal power KW	overall efficiency %
two stage compressor, partial load, VTG	72.55	170.97	188	4.721	67.69	101.07
two stage compressor, partial load, VTG, intercooler	74.10	175.32	177	4.739	52.46	102.17
standard simulation partial load, VTG, el. isolated	70.73	166.28	188	4.710	68.17	99.52
two stage compressor, partial load, VTG, el. Isolated	71.05	168.26	189	4.745	68.06	99.57
two stage compressor, partial load, intercooler, VTG, el. Isol.	72.85	173.08	177	4.759	67.37	101.01
standard simulation partial load, VTG	72.14	169.62	188	4.710	68.17	100.93

Table 6: turbine system with VTG (variable turbine inlet geometry)

By using adjustable turbine geometry, an additional increase in performance can be determined for both single-stage and two-stage systems, especially under partial load. Assuming that the maximum output can be achieved at full load with a suitable setting, the electrical efficiency of two-stage compressor systems increases to up to 74.1 % in the investigated partial load range of 50 % mass flow (see table 6). At the same time, however, the usable thermal output decreases for reasons already described above. This can be an advantage especially in the case of load requirements of low thermal power.

4.5 Optimization of the flow of the supply gases

The effort required to reach the cathode air temperature can be reduced by an additional heat input in the fuel gas flow and water heating already with the exhaust air flow. This results in a slight increase in the electrical efficiency. By converting the heating of the cathode exhaust air in counterflow compared to co-current flow, an improved heat exchange between cathode exhaust gas and cathode supply air can be achieved. If the preheating of the cathode supply air is forced by increasing the recuperator surface area, an improvement in heat coupling can be achieved and the surface area of the high-temperature heat exchanger can be reduced.

4.6 Influence of the stack load

By reducing the stack load, the fuel cells can be operated in a higher electrical efficiency range. This also results in longer service lives of the systems, as the degradation mechanisms are slowed down. However, this requires a higher number of fuel cell stacks to raise the power output to the standard value. The necessary, more complex distribution of the combustion air means that the distribution system must be further adapted. A marginal loss can be expected due to higher pipe resistances, but this will be compensated by the improved efficiency of the fuel cells.

5. Analysis

Various thermodynamic and technical measures are available to increase the efficiency of thermal turbo machines. These include above all cooling mechanisms and the increase in the number of stages of a turbo-compression. Here, a standard simulation was first carried out with a single-stage compressor and turbine and generator on the same shaft. Performance data were determined and for a standard system of a MGT SOFC (TurboFuelCell 1.0) having a maximum Power at start of operation of 278 kW an electrical efficiency of 69.12 % was determined. Derived from this, simulations were performed to estimate the influence of a two-stage compressor system, with and without intercooling, the efficiency improvement by a VTG system, and an electrical separation of the compressor and turbine systems.

IOP Conf. Series: Materials Science and Engineering 1137 (2021) 012073 doi:10.1088/1757-899X/1137/1/012073

Electrical and overall efficiency were then compared, as well as the additional thermal energy extraction for heating purposes and low-temperature heat.

A two-stage compressor system with intercooling can significantly increase system efficiency. When using a VTG system, higher efficiency can be achieved in some cases, especially in the partial load range, than under full load. This is due to improved compressor operation as well as improved efficiency of the fuel cell systems under lower load. The further improved stack efficiency through load reduction has a clear effect on system efficiency, but leads to more complex and larger systems, as more stacks are required for the same output. If intercooling is used, however, more exhaust gas energy must be used to reach the stack working temperature. This reduces the usable thermal output and the overall efficiency.

Electrical separation of turbine and compressor does not lead to any increase in efficiency in the systems investigated. An additional reduction in efficiency due to losses in the electrical drive technology cannot be compensated for by the efficiency increases of the overall system achieved as a result. This applies to both single-stage and two-stage systems and also systems with variable turbine geometry lead to a decrease in efficiency due to runner separation.

The most effective increase in overall efficiency was achieved by the VTG system. This enabled performance increases of up to 2% to be achieved, particularly in the partial load range. A coupled, two-stage system with VTG delivered the best performance in the partial load range. The compressor efficiency is stabilized in part load and less energy is required for compression, which can be taken from the turbine system as additional electrical energy. Based on the knowledge of the advantages of the VTG system, this system will also provide further advantages in turbine response and possibly enable an even more flexible system.

By reducing the working temperature of the exhaust gas heat exchanger, heat can also be extracted at the level of a 60°C/30°C low temperature heating system. There the overall efficiency will increase. A further use of waste heat for local heating concepts means an increased expenditure on equipment, but it can meet additional requirements if a corresponding reduction of the offered performance form is possible. Further concepts for cooling systems and direct heating systems with condensed exhaust air water are conceivable and will be dealt with in further work.

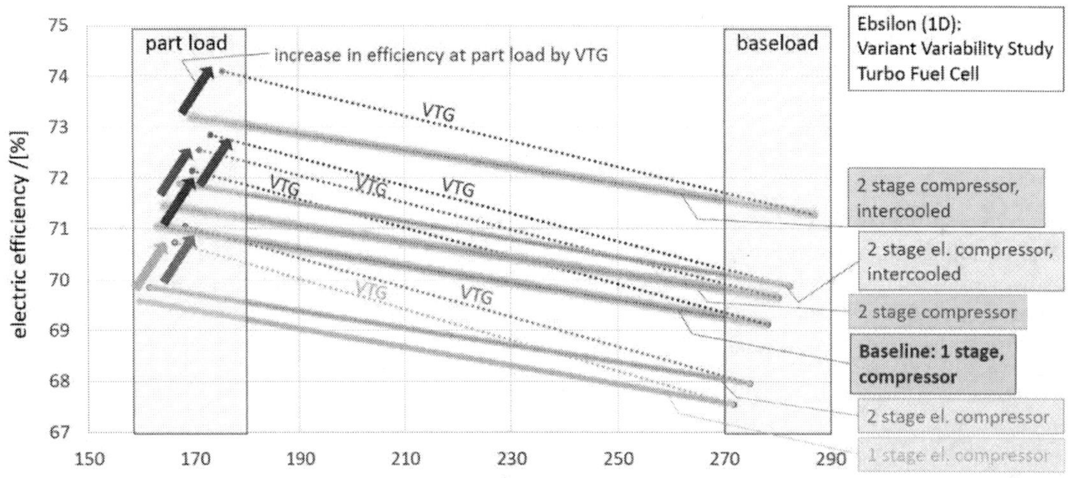

Figure 4: Simulation of the variable assembly of MGT-SOFC-Systems to optimize the mechanical efficiency

At least, Figure 4 shows a comparison of different variants of the simulated configurations. A system with a two-stage turbine system, i.e., also with two-stage compression, as well as intercooling shows the greatest increase in the electrical efficiency of the overall system compared to a system with a simple micro gas turbine (baseline). The simulations show an efficiency increase of more than 2%. In the partial load case, all configurations show a further increase in efficiency of about 1%, cause of the increasing efficiency of the Fuel Cell Stacks. For all the variants considered, this can be additionally increased by a further percentage point by the use of a VTG system. In contrast, decoupling the turbine and compressor sections with the help of electric drives and generator systems leads to a decrease in efficiency in all load cases considered.

6. Conclusion

Various simulations were carried out to investigate the influence of system additions to an MGT-SOFC (System TurboFuelCell 1.0). After comparing the variants, an increase in efficiency and performance could be achieved, starting from the standard system, mainly by VTG systems. However, two-stage compressor systems also achieved an improvement in electrical efficiency, especially when using intercooling. VTG systems can improve the response behavior and performance of the system, especially at partial load operating points.

References

[1] Berg H P Himmelberg A Lehmann M Dückershoff R and Neumann M 2018 *IOP Conf. Series: Materials Science and Engineering* **297** 012004
[2] Berg H P Kleissl M Himmelberg A Lehmann M Prechavut N and Vorpahl M 2019 *IOP Conf. Series: Materials Science and Engineering* **501** 012007
[3] Dückershoff R Berg H P Himmelberg A Lehmann M and Kleissl M 2020 *IOP Conf. Series: Materials Science and Engineering* **886** 012041
[4] Henke M Willich C Kallo J and Friedrich K A 2014 *International Journal of Hydrogen Energy* **39** 12434
[5] Berg H P and Krienke C 2015 *12. Magdeburger Maschinenbautage* CD-ROM (UB Magdeburg)
[6] Berg H P Dückershoff R Lehmann M and Prechavut N 2017 *Proc. 12th European Conference on Turbomachinery Fluid dynamics & Thermodynamics* (Stockholm, Sweden) ETC2017-266
[7] Kurzweil P 2016 *Brennstoffzellentechnik* (Wiesbaden: Springer Fachmedien Wiesbaden GmbH)
[8] Cerbe W and Wilhems G 2017 *Technische Thermodynamik* (Hanser)
[9] Marek R and Nitzsche K 2015 *Praxis der Wärmeübertragung* (München: Hanser)
[10] STEAG System Technologies 2018 *EBSILON®Professional-User Manual* (Zwingenberg, STEAG Energy Services GmbH)
[11] Vorpahl M 2020 *Master Thesis* (BTU Cottbus-Senftenberg, Germany)
[12] Pucher H and Zinner K 2012 *Aufladung von Verbrennungsmotoren* (Springer)
[13] Droste-Franke B 2009 *Brennstoffzellen und Virtuelle Kraftwerke* (Ethics of Science and Technology Assessment, Vol. 36) ed Gethmann CF (Berlin: Springer) chapter 3 pp 43-133
[14] Berg H P 2020 *Die T-Cell als Energieerzeuger der Zukunft*, https://www.b-tu.de/T-Cell/, accessed on 14 September 2020

Analyze open channel for continuously low speed used carbon dioxide bubbles

L. Kirasamutranon[1,*], K. Kirasamutranon[2], and S. Saimek[3]

[1]Departments of Mechanical Engineering, Faculty of Technical Education, Rajamangala University of Technology Krungthep, Bangkok,10120 Thailand
[2]Departments of Mechanical Engineering Technology, College of Industrial Technology, King Mongkut's University of Technology North Bangkok, Bangkok, 10800, Thailand
[3]Department of Mechanical Engineering, King Mongkut's University of Technology Thonburi, Bangkok, 10140, Thailand

* Corresponding Author: lerchai.k@mail.rmutk.ac.th

Abstract
Designing a recirculating open-channel for continuously low speed in place of a towing tank required checking on its flow characteristics. There are many methods for checking the flow characteristics such as food coloring techniques, using buoys and hydrogen gas bubbles, etc. For this research, we use carbon dioxide bubbles which are caused by the reaction between sodium bicarbonate (NaHCO3) and hydrochloric acid (HCl) that is diluted in water until the acidity is equal to pH 3. Our experiments were carried at the area near the wall and the center of the test range at depths 6, 8 and 10 centimeters from the free surface level respectively. From the experimental results, the trend of water flow patterns in the center of the channel using food coloring needles technique tends to form the fluid flow close to the computational fluid dynamics (CFD) while the results from the carbon dioxide bubble technique are different compared with the two previous techniques by less than 20 percent, and the trends of the water flow pattern results of the numerical method, the food coloring needle, and the carbon dioxide bubble techniques. The tendency of the flow pattern in the same direction.

Keywords: Carbon dioxide bubble technique, Food coloring technique, Open-channel for continuously low speed

1. Introduction

Testing a specimen to know the pattern of flow through that specimen can be done in various ways, such as wind tunnels or water tunnels. However, we will encounter problems with noise and relatively large wind tunnel sizes to resolve the issue where the boundary layer thickness will increase as the test distance increases. As for the water tunnels, the barriers are mostly structural and usually encounter leakage problems. This includes the water tunnels that have to be closed systems [1] while the hulls are formed like an open gutter and the specimen must be dragged back and forth. The problem with this test suite is if the test length is insufficient, deviation of the actual flow pattern on the object to be tested or even obtained data extremely deviates from the actual operating conditions [2]. In this research [3], the concept of a low cost, small, recirculating open-channel, continuous water flow was developed. The system can be simply set up and maintained. The flow patterns of the sample specimen can also be easily observed. However, this type of water tunnel is suitable for relatively low speed flow. Other than the development of a recirculating open-channel flow tunnel, we also developed various relatively low-cost techniques to obtain flow patterns for the tunnel. The paper starts from theory and measuring procedures related to open-channel flow pattern. There are two methods of flow velocity measurements, i.e., food coloring and carbon dioxide bubble techniques. Next section, we provide details of our experiments from various methods. Details of the experiments' set up such as the position of instruments and substance with respect to the tunnel, experimental procedures and results are also provided. We then compare our experimental results with numerical simulations. Finally, we conclude our results from the different methods.

2. Theory

2.1 Measuring fluid velocity in an open-channel flow apparatus

To find appropriate fluid measurement and flow visualization methods, the research focused on flow visualization in irrigation engineering. There are three main groups for measuring water currents:

1. Mechanical measuring instruments, such as water flow meters rotated around a vertical axis, etc.
2. Electronic water flow meters such as a tidal speed meter, electromagnet (Electromagnetic Velocity Meters) etc.
3. Special tools, which are not in the first 2 groups, for example, float method, Dilution (Dye) method, etc.

In this paper, we measured flow pattern of the water flow using dilution method and carbon dioxide bubble comparing to numerical calculations.

2.2 Measuring fluid velocity using dilution method

Measuring flow velocity in different depths can be done via timing specific distance that released paint moved. Before experiments started, we placed a few hypodermic syringes in different depths in the open-channel areas. Food coloring was then injected through the syringes. The movement of the paint was recorded and processed as follows. When the colored water header reached the first landmark the timer was started and when the colored water header reached the second landmark the timer was stopped. The capture time(t) was obtained from the timer. The following equations were utilized to calculate velocity and flow rate of the flow:

$$Q=AV \tag{1}$$

And

$$V=\frac{L_d}{t} \tag{2}$$

Where Q = volume flow rate ($m.^3 / s$)
A = flow cross-section area ($m.^2$)
V = velocity of flow (m / s)
L_d = distance of 2 measuring positions (m.)
t = time of water flow through both measurement positions (s)

2.3 Numerical Model

It is very important to predict fluid flow through computer calculations. Since it can reduce the time of trial and error in design, save costs, resources and time to build in the research facilities for open-channel

IOP Conf. Series: Materials Science and Engineering 1137 (2021) 012074 doi:10.1088/1757-899X/1137/1/012074

flow. A computational method was applied using the ANSYS CFX Release 15.0 program. The flow was a multi-state flow type which was selected in the mode of a homogeneous flow. Initial conditions such as height variables for upstream and downstream, and output pressure were also specified. The main equation related to our calculation was a homogenous flow equation given as follows:

$$\frac{\partial}{\partial t}(r_\alpha \rho_\alpha \varphi_\alpha)+\nabla\cdot(r_\alpha \rho_\alpha U_\alpha \varphi_\alpha)-\nabla\cdot\left(r_\alpha\left(\rho_\alpha D_\alpha^{(\varphi)}+\frac{\mu_{t\alpha}}{Sc_{\alpha t}}\right)\nabla\varphi_\alpha\right)=S_\alpha^{(\phi)}+T_\alpha^{(\phi)} \tag{3}$$

Where ρ_α = density of state a

φ_α = Mass of state a

ϕ_α = Volume of state ($\phi_\alpha = \rho_\alpha \varphi_\alpha$)

$D_\alpha^{(\phi)}$ = Kinematic viscosity of the state a

$S_\alpha^{(\phi)}$ = External volume from state a

$T_\alpha^{(\phi)}$ = Sum of φ_α

ϕ = Exit of State a

a = Volume Fraction

Equation 3 [5] is a Eulerian-Eulerian multiphase equation which is used in ANSYS CFX Release 15.0. The equation is a single state flow with the density and viscosity of the combined states added. In our research, two types of fluids were used, water and air, and the turbulent homogeneous flow equation were used to solve the problem in the case of water and air mixtures. We also defined various constants ranging from water inflow rate, outlet pressure, pressure equivalent to the atmosphere in the open surface, moreover, all losses that were proportional by the volume of air and water in different areas for example air and water inlet and an outlet of pipes were added.

2.4 Measuring fluid velocity using carbon dioxide bubbles method

Using carbon dioxide bubbles in the measurement started from adding acid to the water in the open-channel flow until the mixture reached a pH of 3. We then added sodium bicarbonate (NaHCO3) to produce carbon dioxide bubbles. The carbon dioxide foaming mechanism could be described from the chemical equation as follows:

$$NaHCO_{3(s)} + HCl_{(aq)} \rightarrow NaCl_{(aq)} + H_2O_{(l)} + CO_{2(g)} \tag{4}$$

It can be seen that when sodium bicarbonate reacts with hydrochloric acid (HCl), the resulting product contains salt, water, and carbon dioxide as shown in equation 4. The resulting product is carbon dioxide which forms as spherical bubbles. Very small bubbles which have low buoyancy move along the flow with a relatively low floating speed. However, with a clear boundary of the bubble, we could then achieve more accurate measurements of the flow with relatively low cost. The experiment to measure the velocity of the fluid in open-channel flow could be performed as shown in Figure 1, the arrangement of sodium bicarbonate in water mixed with hydrochloric acid to prepare for the flow test.

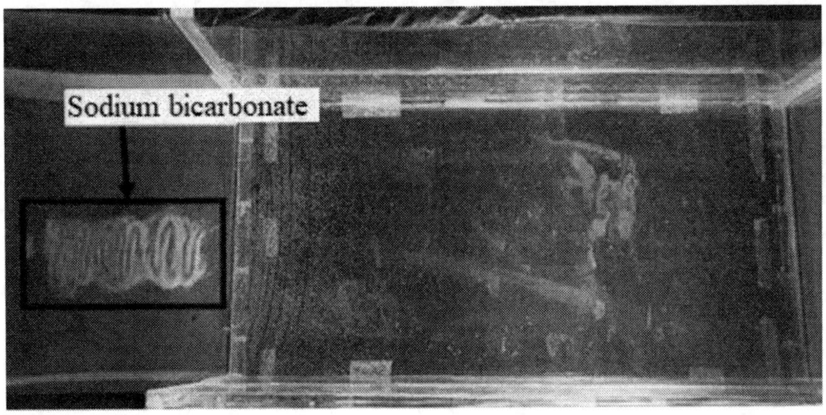

Figure 1. Sodium bicarbonate arrangement in open-channel for continuously low speed

3. Experiment

3.1 Experiment model in open-channel for continuously low speed experiment set-up using carbon dioxide bubble and food coloring techniques.

Figure 2 shows the position setting for both experiments in the open-channel flow, using food coloring and carbon dioxide bubble techniques. The experiment was conducted in 2 areas: the middle of a low vortex open gutter and the area near the wall. The red dot areas were the places where the calcium carbonate located to generate the bubbles. The green dot areas were the location of the syringes for releasing the dilution for food coloring techniques. We monitored the movement of the bubbles at water levels 10, 8, and 6 cm measuring from the bottom of the channel respectively. Notice that in the technique of carbon dioxide bubbles, the subsequent addition of water caused the pH to change, which directly affected the rate of the chemical reaction, resulting in the change in the size of the carbon dioxide bubbles. For the technique of food coloring, abruptly disturb the system, e.g., adding water too quickly, might result in variation of released positioning depth.

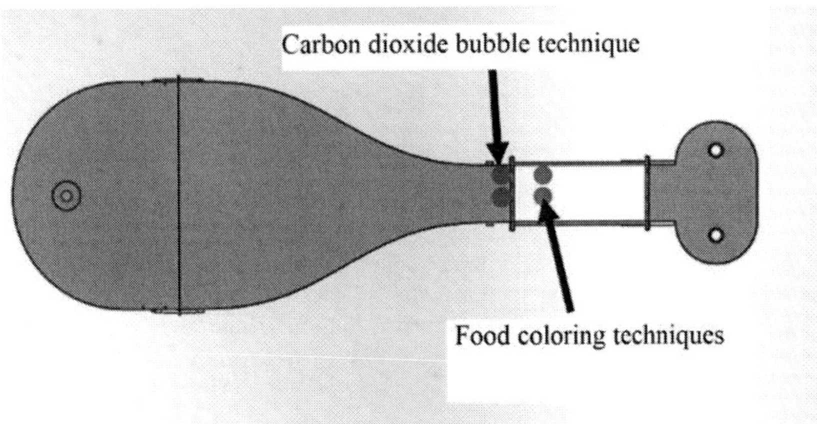

Figure 2. Equipment installation point for open-channel for continuously low speed

In this experiment, the needle was installed in the center and near the wall of the channel with three level of depth for each location as shown in Figure 3.

Figure 3. Pin installation in both areas of a low-flow open water spout.

As mentioned in the section 2, the experiments were recorded and processed. Since the color was very thin at the starting point, it therefore was difficult to identify its position correctly. In order for us to obtain better information, we waited for a second after we saw the color from our video before started collecting data for calculations. The following figure was an example of information obtained from the video capture.

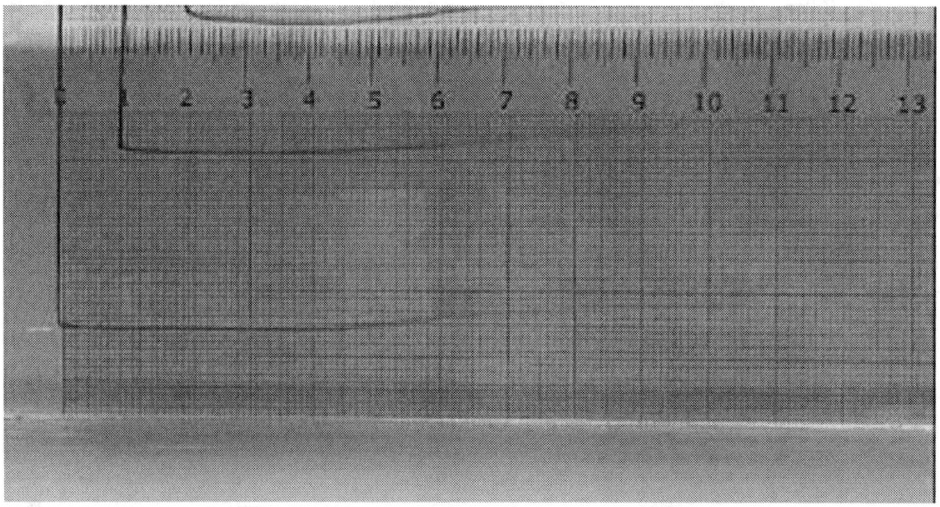

Figure 4. Example of using food coloring technique

For the carbon dioxide gas bubble technique, sodium bicarbonate powder would be arranged in the area before reading of the carbon dioxide bubble about 3-5 centimeters prior to the test section in order to make the carbon dioxide bubbles flow with the water. Figure 5 showed the arrangement of the sodium bicarbonate powder on the wall of the open-channel flow.

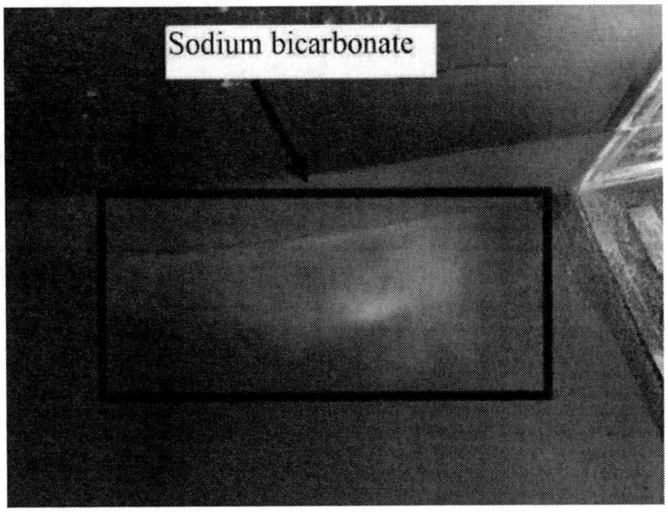

Figure 5. Equipment installation point for open-channel for continuously low speed

The size of the bubbles in the carbon dioxide bubble technique was the most important parameter to achieve accurate testing results, since the floating velocity component would contribute significantly to total velocity to the flow. We therefore had to limit the bubble size in order to limit effect of the floating velocity. The size of the bubble used to measure the fluid velocity is about 0.6 mm in spherical diameter. Figure 6 shows our set-up measurement axes. The bubble was measured with respect to y-axis only where at the top 105.6 mm to 106.2 mm at the bottom edge of the bubble as shown in figure 7. Notice that we did not consider the z-axis in this experiment. Using Adobe Photoshop CS 6, the measurements could be taken from the top left corner of the image as the origin, as shown in Figure 6-7

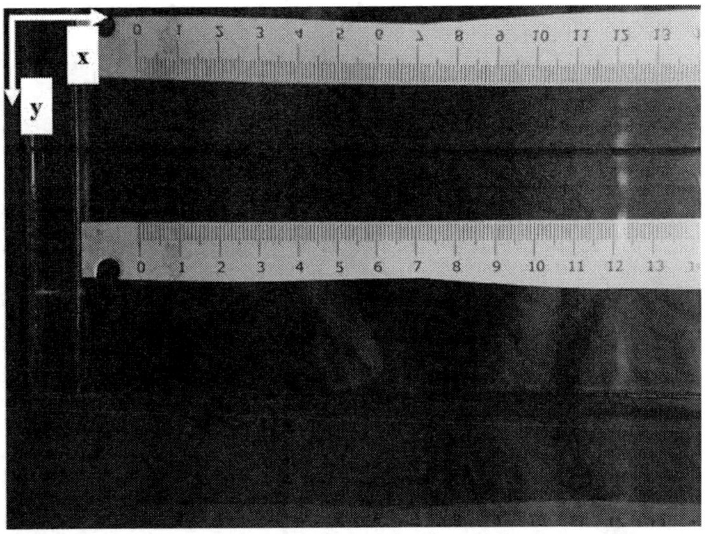

Figure 6. Figure used to locate the resulting carbon dioxide bubbles.

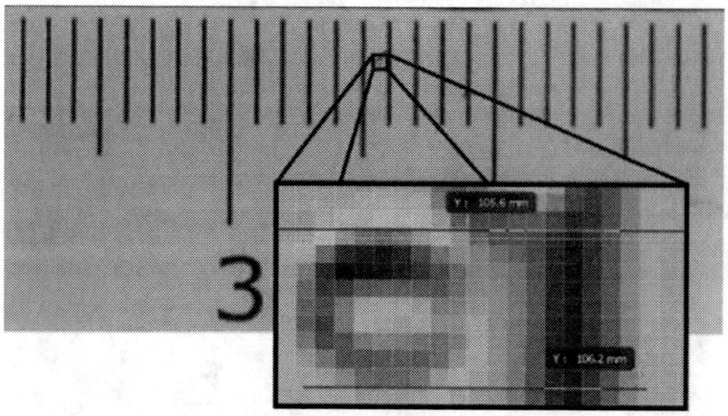

Figure 7. Size of the carbon dioxide bubble that is formed

The experimental conditions led to the flow of food coloring or carbon dioxide bubbles. To calculate the velocity at any particular point, we relied on the time in the video. The point of interest was tracked for 1 second in case of food coloring or carbon dioxide bubbles. Floatation of more than 5 mm from the point of interest was not calculated in order to avoid the influence of the buoyancy force and the expansion of carbon dioxide bubbles, including the excessive spread of diffusion and the velocity of water at different depths were not the same.

3.2 Method for testing open-channel for continuously low speed using food coloring technique.
Before testing begun, we needed to turn on the small water pump to expel all the air in the syringe. The following steps were performed in the food coloring method:

1) Turn on the small pump and open the food coloring liquid valve. To let the liquid flow from the tip of the syringe, gently open the valve so that the liquid overcomes the loss of the hose, syringe system and the syringe exit pressure.
2) When the valve position according to the first procedure is found, mark the valve position. Turn off the small water pump and valve and wait for the color to fade away.
3) Turn on the main pump for a few minutes to allow the flow to reach its steady condition. We then turn on the liquid valve to position marked from the second procedure and record the flow. Close the liquid valve and turn off the small pump and then the main pump.
4) Post processing the data we obtain from the third procedure.

Notice that if water level in the open-channel flow is altered, we need to restart from the first step. If the water level is the same, we can start from third step by using the same liquid valve position since the pressure relating to the food coloring system is relatively unchanged.

3.3 Method for testing open-channel for continuously low speed using carbon dioxide gas bubbles
The following steps were performed in carbon dioxide gas bubbles method:

1) Mix approximately 100 cc of 12 mol hydrochloric acid with water in the open-channel flow to achieve a pH value of 3 using pH paper 0 -14 to read the value.
2) Place the sodium bicarbonate powder in the desired location using paper. By placing the sodium bicarbonate powder before the area to be measured about 3-5 cm, with a width of 2-3 mm longer than the measurement and 5 cm in length of the sodium bicarbonate powder as shown in figure 8 to allow gas bubbles to form and flow in the area to be measured.
3) Turn the pump to the desired flow rate and beams of light to the area of interest to make the gas bubbles more prominent. Turn off all surrounding lights. Record the video of the area you want to record.
4) Post processing the data we obtain form the third procedure.

There are various techniques for measuring fluid velocity for open-channel for continuously low speed. The proposed techniques have many advantages such as clear visualization and the point of interest can be specified for the food coloring technique. However, there are some disadvantages such as color diffusion and an uncontrolled diffusion pattern for this food coloring technique. The advantage for the carbon dioxide bubble technique is the ability to keep track of the carbon dioxide bubbles throughout the flow without confusion. The carbon dioxide bubbles are tiny, resulting in a fine flow path. The disadvantage of this technique is that we cannot specify any particular points needed. Due to buoyancy force on the bubbles, the bubbles rise to the surface of the flow. Moreover, the bubbles expand as they rise due to diminishing of surrounding pressure, this process could disturb the overall flow field in a way that we did not anticipate.

4. Results

4.1 Water flow patterns at 10 cm-depth in the middle of open-channel for continuously low speed.
A reverse calculation was performed from the water meter installed under the channel. The water flow rate in the test set was 53 L/hr. Therefore, reverse calculations were made to find the model's inlet velocity and set up the calculations, found that the result was as shown in Figure 8, and plotted the point of the flow velocity from both techniques are shown in Table 1

Table 1. Comparison table of fluid velocity of each technique against Numerical calculation results at the water flow pattern at 10 cm-depth in the middle of open-channel for continuously low speed

	Food coloring	Carbon dioxide bubbles
Point 1	Faster (11.11 percent)	Slower (1.11 percent)
Point 2	Equal to the calculation result	Faster (2.78 percent)
Point 3	Equal to the calculation result	Equal to the calculation result

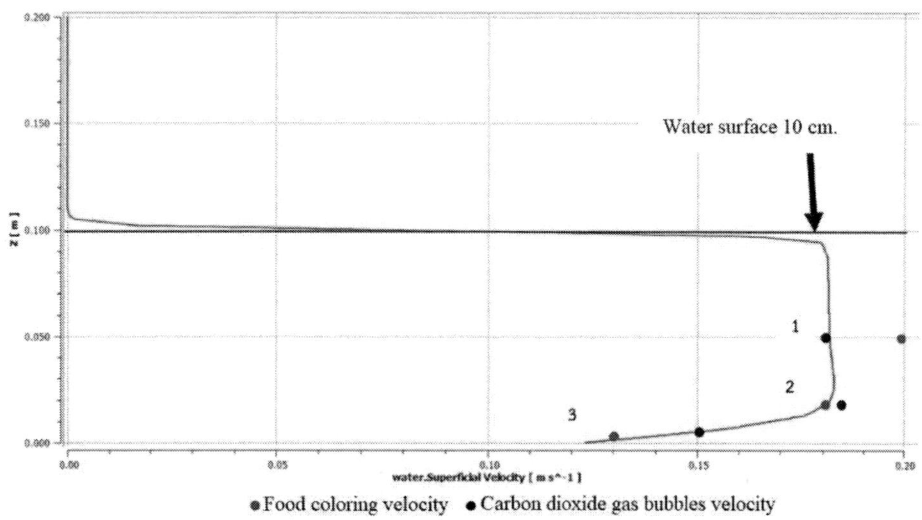

Figure 8. Numerical calculation results at the water level of 1 0 cm in the center of open-channel for continuously low speed

4.2 Water flow patterns at 10 cm-depth at the wall of open-channel for continuously low speed.

The reverse calculation was performed from the water meter installed under the channel. The water flow rate in the test set was 53 L/hr. Therefore, reverse calculations were made to find the model's inlet velocity and set up the calculations, found that the result was as shown in Figure 9, and plotted the point of the flow velocity from both techniques are shown in table 2

Table 2. Comparison table of fluid velocity of each technique against Numerical calculation results at the water flow pattern at the water level 10 cm is on the wall of open-channel for continuously low speed.

	Food coloring	Carbon dioxide bubbles
Point 1	Faster (1.58 percent)	Slower (6.35 percent)
Point 2	Equal to the calculation result	Faster (1.52 percent)
Point 3	Equal to the calculation result	Faster (3.06 percent)

In all three techniques were performed in center and wall open-channel a continuously low speed was used. The velocity obtained from the techniques was calculated from the video recording data. The velocity of the flow at the middle of the channel was faster than the velocity of the flow near the wall.

5 Conclusion

From the experiments at different water levels, it was found that the fluid flow patterns for both techniques deviated from the numerical calculation results. By using depths of 10, 8, and 6 centimeters from the water surface in the middle of open-channel for continuously low speed. Techniques for using food coloring tend to have flow patterns close to numerical calculation results. While the carbon dioxide bubble technique tends to be faster and slower at different points, the plane and the height from the floor of the carbon dioxide bubble could not be determined exactly. Therefore, the experimental results of the carbon dioxide bubble technique were highly variable. Despite being highly variable, the results showed less than 20% error rate compared to the food coloring and the numerical calculation. For the bubble technique at 6 cm-depth, it was not possible to interpret the data at some point due to the limitations of the camcorder. The maximum frame rate of 30 frames per second making it impossible to clearly see the carbon dioxide bubbles moving fast and the expanding of the bubbles. At the wall of the low vortex open gutter, it was found that the results of the food coloring technique and the carbon dioxide gas bubble technique tended to agree with each other. The results from the numerical calculation showed less than 9% error rate when compared with the food coloring and the numerical calculation. Only at the depth of 6 cm provided the results of the carbon dioxide bubble flowing at a speed of 0.125 m/s over the 5 cm flow pattern have a 23.80% error rate which could be explained by the expanding and floating of the bubbles.

References

[1] The American Society of Mechanical Engineers 1996 *A Historic Mechanical Engineering Landmark Dedication* (Pennsylvania: State College) p 21

[2] Saimek S and Thongnong W 2014 *Subject to the design and development of the back-to-back water turbine* (Bangkok, Thailand: King Mongkut's University of Technology Thonburi)

[3] Kirasamutranon L and Saimek S 2019 *Thesis* (Bangkok, Thailand: King Mongkut's University of Technology Thonburi)

[4] Polapanavani P 2019 Principle of calculating the amount of water passing through the hydraulic structures, http://ridceo.rid.go.th/buriram/WMM-RID8.html, accessed on 15 September 2019

[5] ANSYS, Inc. 2018 ANSYS CFX-Solver Theory Guide, http://read.pudn.com/downloads500/ebook/2077964/cfx_thry.pdf, accessed on 15 July 2019

IOP Conf. Series: Materials Science and Engineering 1137 (2021) 012075 doi:10.1088/1757-899X/1137/1/012075

Effect of sodium hydroxide solution concentration on liquid entrainment in a steam boiler

Piyatida Trinuruk[1,*], Somjate Meekaew[1], Ratchayothin Thong-in[1], Nut Satsananun[1], Thanapol Laklaem[1]

[1]Department of Mechanical Engineering, Faculty of Engineering, King Mongkut's University of Technology Thonburi, Bangkok Thailand 10140

* Corresponding Author: piyatida.tri@mail.kmutt.ac.th

Abstract

Liquid entrainment or carryover is a phenomenon that liquid droplets are carried out from a boiler or steam drum during steam generation. The droplets may convey some dissolved solids and cause damage to steam equipment such as scaling deposits in valves and steam piping and erosion of turbine blades. The appearance of dissolved solids in boiler water mainly comes from the internal water treatment process by dosing some chemical substances into the boiler in order to prevent corrosion and scale formation. Trisodium phosphate (Na_3PO_4) is one of chemical substances commonly added during boiler operation to control the pH of boiler water. However, it can cause the occurrence of sodium hydroxide (NaOH) solution which results in the increase of total dissolved solid (TDS) concentration in boiler water. Therefore, this study is aimed to develop a test section in order to investigate the parameters affected by liquid entrainment in the boiler. The experiment is performed in ambient conditions under an air-water system. Three variable factors are taken into consideration, including the concentration of NaOH solution, the height of the water level in the test tank, and the gas superficial velocity which represents the flow velocity of steam during steam generation. The results show that the increase of the gas superficial velocity and the water level height in the tank resulted in the enhancement of the entrainment rate. The increase of the water level height played an important role on the liquid entrainment. The increase of NaOH concentration directly impacted the existing film bubbles and the longer stability of foam bubbles at the water surface which resulted in a higher entrainment rate. Therefore, TDS control was one of the most important factors which could be incorporated with the appropriate height of the water level and the consistency rate of steam generation in order to minimize the liquid entrainment.

Keywords: Carryover, Gas superficial velocity, Height of water level, Liquid entrainment, NaOH solution

1. Introduction

In the process of steam production in a boiler, saturated steam is preferred for heating processes due to advantages in heat transfer, while superheated steam is mostly used in power generation to prevent wetted steam which can erode turbine blades and decrease thermal efficiency. If the boiler is operated under inappropriate or abnormal conditions, it may introduce some water droplets, solid and vaporous escaped from the water surface and carried out with the steam. This phenomenon is called carryover or liquid entrainment which is defined as the amount of liquid compared to the steam quantity that leaves the boiler. Carryover or liquid entrainment is an undesirable phenomenon because it can cause the damage to the equipment, especially turbine blades, and by depositing solids in the piping system and valves [1], as shown in Figure 1.

Priming and foaming are two main causes of carryover. When the contamination in water droplets leaves the steam drum during the heating process, it is called 'priming'. Priming can be caused by improper boiler design, rapid changes in steam demand, and high levels of boiler water during the operation. It can say that priming is caused by mechanical factors. Another cause of carryover is called foaming. Foaming occurs when bubble film is built up on the surface of the boiler water and then released as the bubbles burst and went out with the steam [1]. Foaming can be caused by high solid contamination such as oil, fat, grease, organic matter and suspended solids in the boiler water and excess alkalinity which are effects of chemical factors.

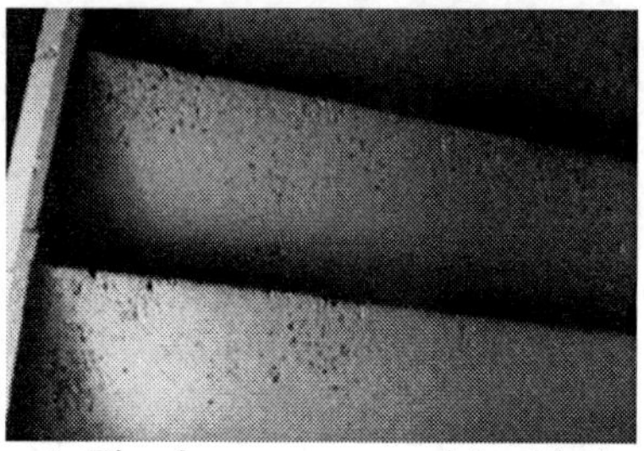

Figure 1. The damage on a turbine blade

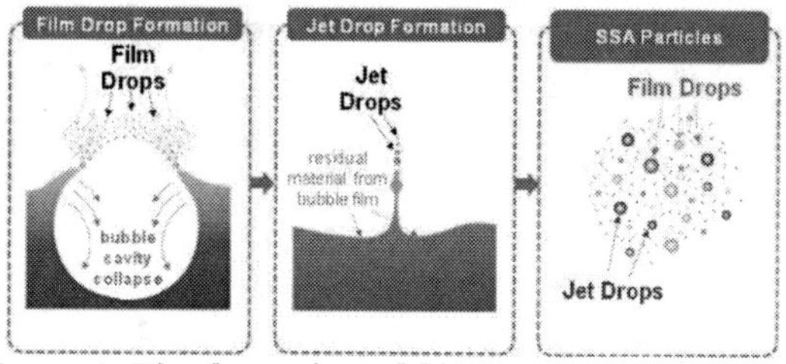

Figure 2. The formation of film drops and jet drops

Davis R F [2] studied the mechanisms of water droplets detachment. The first mechanism was called "Film drop" which was the bursting of steam bubble and then some small droplets were carried out of the free surface. Another one was called "Jet drop" when the bubble cavity collapsed, the liquid rim had enough momentum to rise up the center of depression, then the ejection of a tiny liquid jet occurred. The formation of a film drop and a jet drop are shown in Figure 2.

To prevent the liquid entrainment phenomenon, the quality of boiler water is needed to be controlled following Thailand's laws and regulations. The total dissolved solid (TDS) content in boiler water is required to be controlled within the permissible limit of lower than 3,500 ppm and the pH value is required in the range of 8.5-11.8 [3]. To maintain the pH of boiler water within the regulations, trisodium phosphate (Na_3PO_4) is generally added into the boiler for caustic treatment purposes. It can interact with the water and produces disodium phosphate (Na_2HPO_4) and sodium hydroxide (NaOH) as the by-products [4]. Equation (1) illustrates the chemical reaction of those substances. However, if a solid alkali is not applied appropriately, it can cause the scale deposition on the steam turbine and risks caustic gouging.

$$Na_3PO_4 + H_2O \rightarrow Na_2HPO_4 + NaOH \quad (1)$$

Therefore, this study is aimed to investigate the effect of NaOH concentrations on the liquid entrainment in conjunction with the variation of the water level height and the flow velocity of fluids in a horizontal drum used in the experiment.

2. Entrainment rate

The movement of steam inside the boiler is one of the characteristics of two-phase flow. The term of "*gas superficial velocity*" (j_g) is commonly used to present the velocity of multiphase fluids. Gas superficial velocity is a hypothetical flow velocity which is referred to as the volumetric flow rate of gas

IOP Conf. Series: Materials Science and Engineering 1137 (2021) 012075 doi:10.1088/1757-899X/1137/1/012075

flow through the pipe divided by the cross-sectional area of that pipe. It can be calculated as given in Equation (2):

$$j_g = \frac{r_m \overline{\rho}_{in} A_{in} \overline{V}_{in}}{\rho_{atm} A_{sw}} \tag{2}$$

where r_m is the discharge coefficient of mass flow rate. A_{in} and A_{sw} are the cross-section area of an inlet air pipe at flow meter and the average cross-section area at swell water level. $\bar{\rho}_{in}$ is the inlet air density at the flow meter, ρ_{atm} is air density at atmosphere.

Variation of the gas superficial velocity was performed during the experiment to investigate its effect on the liquid entrainment. Therefore, a dimensionless number of entrainment rate (E_{fg}) was employed as a qualitative analysis of entrained liquid which is defined as a ratio of entrained liquid mass flux to the gas mass flux. The entrainment rate is given in Equation (3).

$$E_{fg} = \frac{M}{\rho_g j_g A_{sw} t} \tag{3}$$

where j_g is the gas superficial velocity referred at swell water level. M is the mass of entrained liquid, ρ_g is the density of gas phase in the tank and t is time.

3. Design of experiment

To achieve the objective of this study, a horizontal cylindrical drum with transparent flat heads was designed and fabricated to visualize the turbulence of water surface and the detachment of water droplets. The test tank was designed to be similar to a steam drum in a water tube boiler or fire tube boiler. An air-water system was adopted in this experiment by controlling conditions under ambient pressure and a temperature of 25°C. A method of geometric similarity was applied to determine the dimension of the test tank; the pipe size and the connections were based on the actual scale of the steam drum in the real power plant. Figure 3 shows the model of the drum and the scaling factor of dimension and Figure 4 presents the set of the experiment.

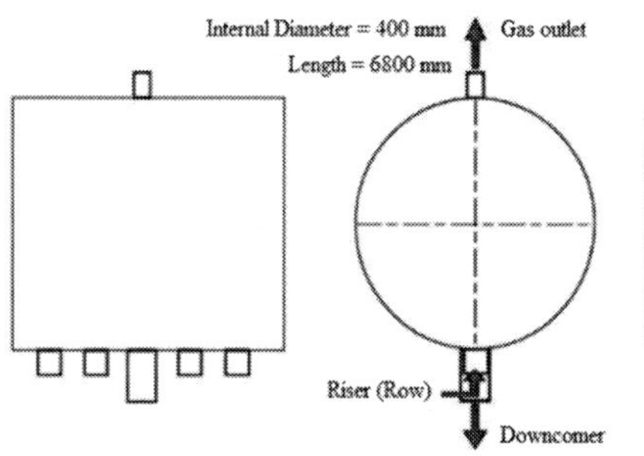

Parameter	Prototype	Model
Diameter (mm)	1,480	400
(L/D)	1.149	
(A_M/A_R)	32.675	
(A_M/A_O)	242	

Figure 3. The drum model and the scaling factor of dimension

Figure 4. The set of experiment

The pipe connected with the tank included four risers and one balancing pipe at the tank bottom and one outlet pipe at the top. A 1¼ inch diameter pipe was designed as a riser, however, a larger cross-section area of pipe can affect the entrainment phenomenon. Therefore, a grill with 3 mm diameter holes was installed on each riser to enhance uniform air flow distribution [5]. Four generations of low-cost capacity level sensors were developed to continuously measure the swell water level in the test tank. To select a high-quality sensor, it can be evaluated from the linearization between the frequency and the water level [6]. The pattern of square waves succeeded to provide a linear graph rather than other sensors as shown in Figure 5. This method was adopted in this study to overcome the human error from the image visualization for measuring the swell water level.

The initial water level height (H) of 0.24 and 0.28 m, the gas superficial velocity (j_g) of 0.3, 0.4 and 0.5 m/s and the NaOH concentration of 0, 1200, 2400 and 3500 ppm, which was presented in term of TDS were varied as parametric studies to evaluate the amount of liquid entrainment (M).

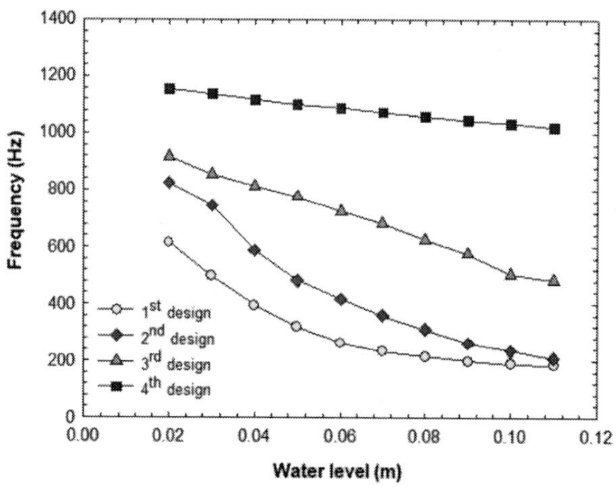

Figure 5. The frequency of capacity level sensors with respect to water level

4. Results

To distinguish the liquid entrainment phenomenon, the amount of entrained water (M) into the gas phase as the droplets were collected and measured the weight at the end of 10-min experiment. The preliminary result reported the effects of the initial water level height and the gas superficial velocity on the amount of entrained liquid under the condition of pure water. The amount of entrained waters were plotted with respect to the gas superficial velocity in the different of initial water level height. Figure 6 shows that the increase of gas superficial velocity resulted in larger amounts of entrained liquid. The enlargement of gas superficial velocity which means higher entrance on the mass flow rate of air can contribute turbulence on the liquid surface and the generation of numerous foam bubbles. If the flow velocity was high enough to overcome the surface tension, the breakup of air bubbles into small droplets can occur as expressed in Figure 7. It also can clearly notice that the initial water level height played as a key parameter affected on the rapid increase of liquid entrainment rather than the enlargement of gas superficial velocity.

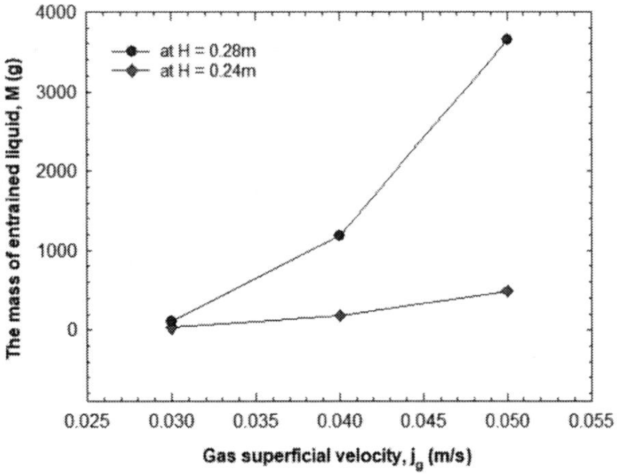

Figure 6. The entrainment of pure water in various conditions

Figure 7. The liquid entrainment phenomenon in the pure water

The impurity of water may affect the amount of entrained liquid. In this study, NaOH which is commonly added to maintain boiler water at proper pH and alkalinity levels was emphasized its effect on the liquid entrainment. Excessive NaOH compound can cause the increase of water concentration or TDS which affects the increase of mechanical properties of liquid i.e., density, surface tension and dynamic viscosity [7]. The increase of NaOH concentration tended to enhance the swell water level. Foaming existed on the liquid surface with longer time stability and larger bubble coalescence which contributed to the liquid entrainment. Figure 8 presents the effect of NaOH concertation on the liquid entrainment which observes two distinctive results.

Firstly, the increase of TDS can cause a gradual decrease of entrained water which occurs at a low initial water level height of 0.24 m for all velocity cases. A second observation is at the condition of high velocity and high water level, an increase of NaOH solution displays enlargement of liquid entrainment, but the phenomenon declined significantly when the TDS excess was 2,400 ppm even if the gas superficial velocity become very high. This was because the flow velocity of air cannot succeed in dealing with the increment of surface tension when the NaOH solution increased. It resulted in a lower amount of entrained liquid carried out with the air even the initial water level height and gas superficial velocity become higher.

Using the mass of entrained water as an indicator of the liquid entrainment phenomenon cannot represent the phenomenon under the different conditions. Therefore, a dimensionless number of entrainment rate (E_{fg}) was introduced for comparative analysis. The entrainment rates of all cases were plotted with respect to the ratio of dimensionless number between the gas superficial velocity and the height above water interface (j_g^*/h^*). The experimental results were compared with the calculation results from Ishii-Kataoka's correlation as demonstrated in Figure 9. The results show that the experimental data have a tendency to agree with the calculation results. There were small deviations occurring between the calculation and the experiment since this correlation had been developed from the experimental results of a vertical cylindrical drum which was different from this study.

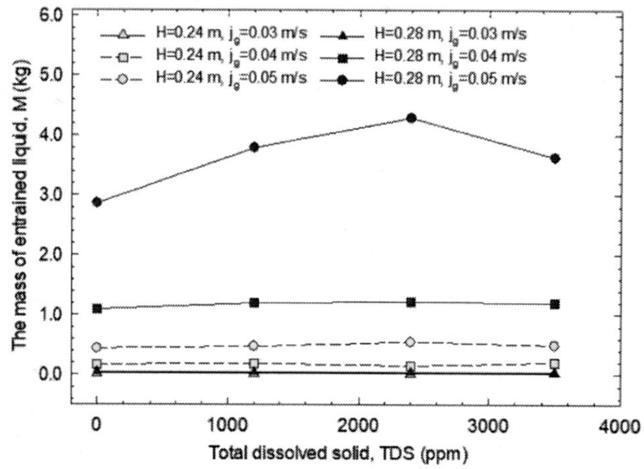

Figure 8. Influence of NaOH solution in various water level heights and gas superficial velocities on the liquid entrainment

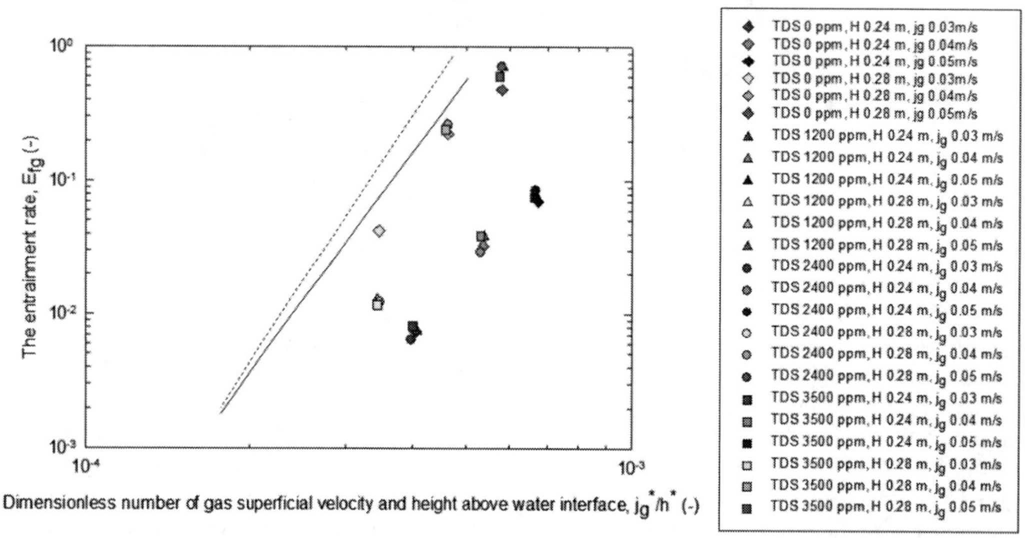

Figure 9. Comparison of the experimental data with Ishii-Kataoka's correlation when solid line and dash line are the trendline of calculation results from Ishii-Kataoka's model at H = 0.24 m and 0.28 m, respectively.

5. Conclusions

The visualization of liquid entrainment phenomenon can successfully be observed via a horizontal drum with air-water system. The increase of three parameters, including with the initial water level height, the gas superficial velocity and the NaOH concentration or TDS, can impact on the enhancement of liquid entrainment but in a different magnitude. In the case of pure water without any effect on the water impurity, the increase of the initial water level height had a significant impact on the liquid entrainment rather than the rise of gas superficial velocity. When the effect of total dissolved solids was included, it seemed that the increase of NaOH concentration resulted in a larger number of bubbles and longer foam stability which impacted on the increase of liquid entrainment. For the case of high initial water level and high gas superficial velocity, it can be noticed that excessive increase of NaOH concentrations above some certain concentrations cannot increase the liquid entrainment because the NaOH solution can cause an increase of fluid properties, particularly the surface tension and the density. The increase of mechanical properties of fluids tend to obstruct the liquid entrainment.

From this study, it can be concluded that improper conditions such as a high water level operation, a rapid change of steam demand, and the accumulation of total dissolved solids in boiler water, should be avoided during the boiler operation because they may cause water droplets carrying out some solids and liquids with the steam. That can cause damage to the steam equipment.

Acknowledgments

The authors sincerely acknowledge the support provided by the "Research Chair Grant" National Science and Technology Development Agency (NSTDA), King Mongkut's University of Technology Thonburi through the "KMUTT 55th Anniversary Commemorative Fund".

References

[1] Bagul R K Pilkhwal D S Vijayan P K and Joshi J B 2013 *Sadhana* **38**(6) 1173-1217
[2] Davis R F 1940 *Proc. Inst. Mech. Eng.* **144**(1) 198-216
[3] Announcement of Ministry of Industry Thailand 2006 *The properties of boiler water* **123** 29
[4] Cooper J R and Dooley R B 2008 *The International Association for the Properties of Water and Steam* 1-11
[5] Prabhudharwardkar D M More R Z and Iyer K N 2010 *Nucl. Eng. Des.* **240** 76-83
[6] Chetpattananondh K Tapoanoi T Phukpattaranont P Jindapetcha N 2014 *Sens. Actuators A-Phys.* **209** 175-182
[7] Wen J Shi K Sun Q Sun Z and Gu H 2018 *Front. Energy Res.* **6**(12) 1-7

IOP Conf. Series: Materials Science and Engineering 1137 (2021) 012076 doi:10.1088/1757-899X/1137/1/012076

Investigation of thermal stratification in a thermal storage tank with a curve edge obstacle plate

Papangkorn Jenyongsak[1], Piyatida Trinuruk [1,*]

[1]Department of Mechanical Engineering, Faculty of Engineering, King Mongkut's University of Technology Thonburi, Bangkok, 10140, Thailand

* Corresponding Author: piyatida.tri@mail.kmutt.ac.th

Abstract

Nowadays, thermal storage tanks have been widely installed because of high demanding for hot water in the production process or daily activities in a household. Numerous energy has been consumed to produce hot water. To reduce the energy consumption and improve the energy efficiency of the hot water system, the design of thermal stratification in a storage tank can play as an important parameter on its thermal efficiency. The basic concept of the thermally stratified tank is attempted to separate a layer of hot and cold water by means of density variation and gravitational effects which cold water is aimed to existing at the bottom and hot water is at the upper part of the tank. The level of temperature stratification significantly impacts on the thermal efficiency. The higher degree of thermal stratification, the better thermal efficiency of the hot water system. However, there are many factors strongly affected the thermal stratification such as height-to-diameter ratio, the inner structure of thermal storage tank, the water flow rate, and etc. Therefore, the purpose of this study is to design a thermally stratified tank by improvement the inner structural design of the tank. Two influenced factors are focused: i) the different configuration of inlet pipes which can impact on the flow pattern and flow velocity of the water and ii) the different design of obstacle plates which can help control a recirculation area of the water in the tank. To obtain a high thermal efficiency of the stratified tank, the computer simulation program Ansys Fluent is applied for the analysis. The results show that uniform distribution of water flow and lower flow velocity due to the appropriate design of inlet pipe and proper buffer plate design can enhance the temperature stratification in the storage tank which increase the thermal efficiency during discharging processes.

Keywords: Obstacle Plate, Thermal Stratification, Thermal Storage Tank

1. Introduction

A key aspect of the thermal stratified tank is designed to encourage the separation regions of the hot water and the cold water by using the fundamental properties of fluid on the different temperatures represented different densities. The interface between hot water and cold-water layers is called the thermocline. Studies over the past two decades have provided important information on the development of the thermal stratification in the storage tank, and many parameters applied such as the height to diameter ratio, the water inlet velocity, the shape of the storage tank and the inlet water pipe [1, 2, 3, 4,

5]. Another way to improve the thermal stratification is the implement of obstacle plates inside the tank to reduce the water recirculation in the whole section.

Some research has been carried out on the obstacle plate utilization where Computational Fluid Dynamics (henceforth CFD) has been adopted on many thermal investigations. Shah and Furbo [6] studied the shape of the obstacle plate where was placed over the inlet pipe during the discharging process. The baffle plate provided the best thermal stratification whereas the metro pipe and the bare pipe were inferior, respectively. Bouhal et al. [7] has studied the variation of the position, the number, and the rotating angle of the flat obstacle plates to investigate their effects on the thermal stratification. The results showed that the change of single flat plate positions could not represent the improvement of temperature stratification distinctively. The best stratification occurred on the obstacle plate with 30-degree rotating angle. Altuntop et al. [8] analysed the effect of the obstacle shape on the thermal stratification where it located at the bottom part of the tank. It was found that the single hole flat plate was the greatest model. Ali et al. [9] has tended to focus on the different types of curved baffles as stratifier in the solar thermal storage tank by varying the radius of curved baffles. The result revealed that a single quarter ellipse curve can minimize the water recirculation over the baffle plate resulted in the enhancement of thermal stratification. It has previously been observed that the flat and curved obstacle plate had different advantages and drawbacks as well as the inlet pipe configurations which effected on the thermal stratification.

The purpose of this paper is to consider the implications of the combination of each dominance. It also attempts to show that the deviation of inlet pipe designs and the different arrangement of obstacle plates has an effect on the thermal stratification during the discharging process.

2. Dimensionless parameters

In this study, the performance of thermal stratification was analyzed via the temperature profile and the velocity profile. Additionally, dimensionless numbers shown in Table 1, including with Richardson Number (Ri), Stratification Number (Str), Dimensionless time (τ_{use}), and Discharging Efficiency ($\eta(\tau_{use})$) were also used to characterize the thermal stratification in the storage tank. To comparatively evaluate the different storage tank configurations, dimensionless time is given, and it refers to the mass flow rate multiply with time step divided by the whole replacement of water in the tank. Richardson number is used to identify the stratified temperature which is defined as the ratio of buoyancy forces to mixing force. A larger Richardson Number identifies a better stratification. While Stratification Number is the ratio of the average of the temperature gradient at that time interval at the initial state. Besides, discharging Efficiency which is the comparative value between the amount of useful heat at a specific temperature and the total heat recoverable. The temperature of 45°C was determined as the temperature limit of useful heat in the tank.

Table 1. The equation of dimensionless number

Dimensionless Number	Equations
Dimensionless time [5]	$\tau_{use} = \frac{\tau}{\sigma}$ where $\sigma = \frac{V_T}{\dot{m}}$
Richardson Number [3]	$Ri = \frac{g\,\beta\,H\,(T_{top} - T_{bottom})}{{v_s}^2}$
Stratification Number [7]	$Str(t) = \frac{(dT/dy)_t}{(dT/dy)_{max}}$ where $\left(\frac{dT}{dy}\right)_t = \frac{1}{J-1}\left[\sum_{j=1}^{J-1}\left(\frac{T_{j+1}-T_j}{\Delta y}\right)\right]$, $\left(\frac{dT}{dy}\right)_{max} = \frac{T_{max}-T_{in}}{(J-1)\Delta y}$
Discharging Efficiency [3]	$\eta(\tau_{use}) = \frac{E(\tau_{use})}{E_{st}(\tau_{use}=0)}$ where $E(\tau_{use}) = Mc_p\tau_{use}\left\{\frac{\sum_{n=1}^{k} T_{o-use}(\tau_{use})}{k} - T_{in}\right\}$ and $E_{st}(\tau_{use}) = mc_p\sum_{j=1}^{J}(T_{j-use}(\tau_{use}=0) - T_{in})$

3. Numerical investigations

3.1 The geometry model and meshing

The shape of the thermal stratified tank at a 520-liter vertical cylindrical with elliptical end caps was chosen. Figure 1 illustrates the tank dimension had a height of 1,480 mm (H), a diameter of 690 mm (D). The inlet pipes and outlet pipes had the diameter of 30 mm and located in the positions of a = 170 mm, b = 190 mm, c = 130 mm, and d = 700 mm, respectively.

Two modification concepts were adopted to evaluate the thermal stratification in the storage tank: modified inlet pipes and using obstacle plates. Figure 2 provides the three different models of the inlet pipe, i.e., direct downward, annular downward, and annular upward, while the original design was non-insertion of the inlet pipe into the tank. The annular inlet pipe had the radius of 172.5 mm included the six water outlets equivalent. Regarding the obstacle plate, four models of plate feature and various in its position (y) were used as the obstacle plate. Model 1, model 2, and model 3 were normal plate, whereas, model 4 was a perforated plate with 30 mm in diameter. The mesh quality is very good where the skewness and orthogonal quality were about 0.23 and 0.77. The minimum volume of meshing is 6.677×10^{-10} m^3. The number of mesh elements was approximately 1,300,000 elements.

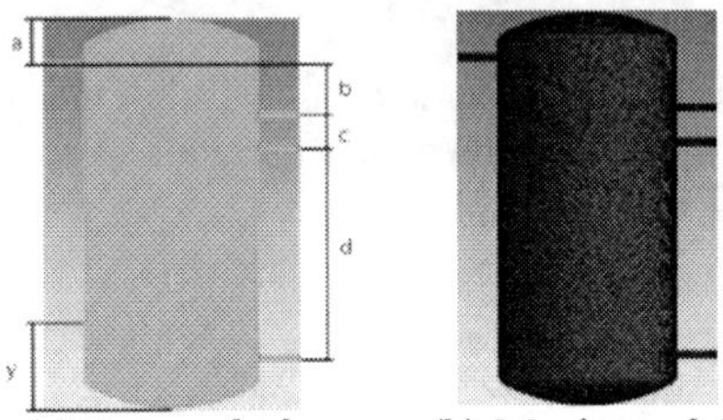

(a) External view (b) Mesh model

Figure 1. The geometry and meshing

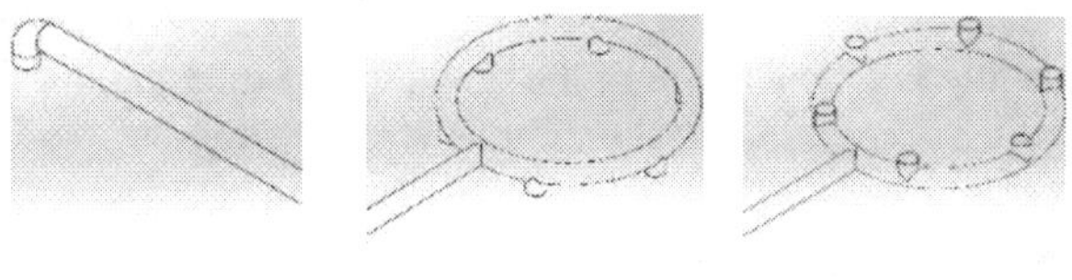

Direct Downward Annular Downward Annular Upward

Figure 2. The modification of inlet pipes

Table 2. The dimensions of obstacle plates

Obstacle model	L (mm)	y (mm)	Thickness (mm)	Normal obstacle plate (Model 1,2,3)	Perforate obstacle plate (Model 4)
Model 1	345	240	2	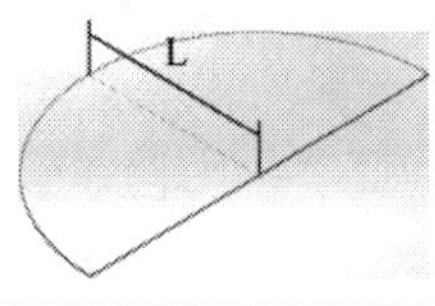	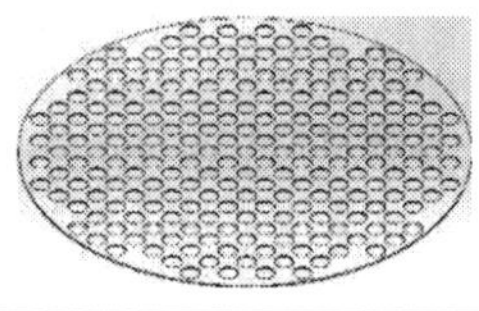
Model 2	517.5	240	2		
Model 3	345	240, 120	2		
Model 4	690	240	2		

3.2. Governing equations

The simulation model was currently chosen on the flowing fluid and the energy transfer between fluid. Therefore, Computational Fluid Dynamics (CFD) Ansys Fluent was applied to solve the temperature and velocity distributions in each cell through 3 conservative models (see Table 3).

Table 3. The governing equations

Contents	Equations
Conservation of mass	$\frac{D\rho_f}{Dt} + \rho_f \nabla \cdot \vec{V} = 0$
Conservation of momentum	$\rho_f \frac{D\vec{V}}{Dt} = -\nabla p + \mu_f \nabla^2 \cdot \vec{V} + (\mu_f + \lambda)\nabla(\nabla \cdot \vec{V}) + \vec{F}$
Conservation of energy	$\rho_f c_{p,f} \frac{DT_f}{Dt} = \rho_f c_{p,f} \left[\frac{\partial T_f}{\partial t} + (\vec{V} \cdot \nabla) T_f \right] = \nabla(k_f \nabla T_f) + \beta_f T_f \frac{Dp}{Dt} + \mu_f \Phi$

3.3 Boundary conditions

In Ansys Fluent model, the water was adopted as a single fluid domain with the application of the turbulent model of laminar flow. The gravitational effect was also included. Three-dimensional flow with transient calculation was set. The mass flow rate of inlet water was fixed at 0.259 kg/s and its density was changed depending on its temperature. The PISO and body force weight formulations were used as calculation algorithm.

4. Experimental design

The dimension and configuration of the storage tank were used to model the actual storage tank, and it was then carried out using the performance test of the heat pump at KMUTT, Thailand. It was made by 2 mm thickness of stainless steel and insulated with 45 mm thickness of polyurethane. The simulation was started from the full existence of 55°C hot water in the tank. The inlet pipe connected to the 25°C water supply tank. During the discharging process, the hot water was discharged from the storage tank while the cold water of 25°C simultaneously flowed into the tank with the same flow rate. Five probes of type K thermocouples were installed at the real tank to acquire the transient water temperature which was used to validate the accuracy of the simulation model.

5. Results

5.1. Model validation

The accuracy of the simulation model was validated with the experimental data during the discharging period. The numerical results were used to identify the relative errors of temperature at the thermocouple probe which located at the top, middle and bottom of the storage tank were 2.17%, 2.91%, and 1.55%, respectively.

5.2 Richardson Number analysis

The Richardson Number of three different inlet pipe configurations were plotted respect to Dimensionless time as shown in Figure 3. The results found that all pipe modifications can achieve to provide higher Richardson Number than the original model. It also can observe the tendency of Richardson Number of all models were similar by promptly increased at the beginning, then gradually increased or constant depended on the pipe models and decreased rapidly at the end of the process. It can be noticed that the annular downward pipe can offer the best Richardson Number while the annular upward has been inferior, and the worst model was the direct downward. Higher Richardson Number represents a larger temperature difference between the upper part and the lower part of the storage tank.

Figure 4 illustrates the Richardson Number of each obstacle plate model. The result cannot notice the deviation of the Richardson Number in each model. Nevertheless, the use of obstacle plates can provide larger Richardson Number than the original one.

It is possible, therefore, that the Richardson Number can only indicate the temperature difference between the upper part and the lower part of the tank, but it is not applicable to evaluate the temperature distributions at the middle part of the storage tank. Thus, other parameters (i.e., Stratification Number) are required to specify the performance of the stratification in the storage tank.

IOP Conf. Series: Materials Science and Engineering 1137 (2021) 012076 doi:10.1088/1757-899X/1137/1/012076

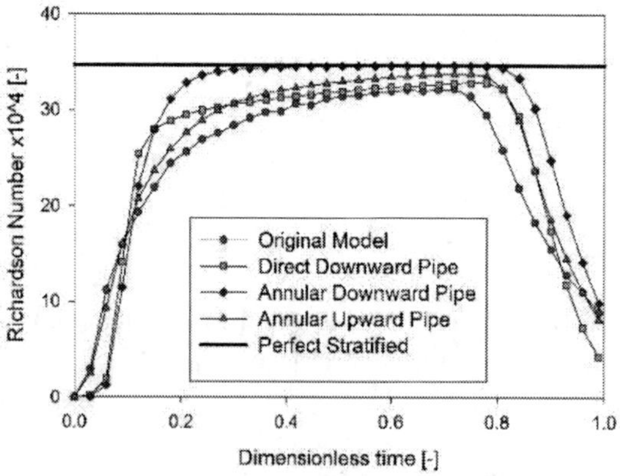

Figure 3. *Ri* of the modified pipes models

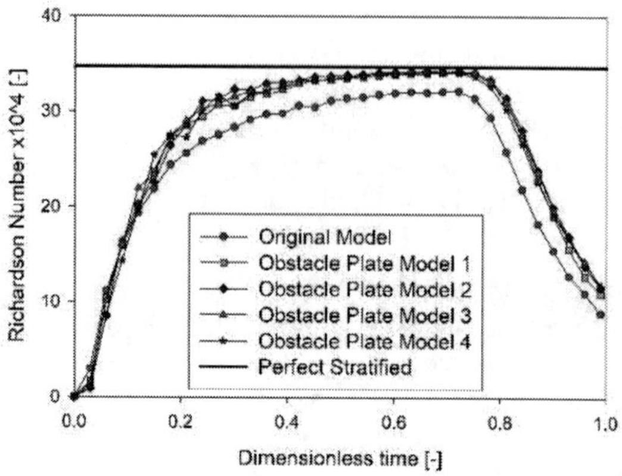

Figure 4. *Ri* of the obstacle plate models

5.3 Stratification Number analysis

Figure 5 provides the Stratification Number of each modify pipe model gives explicitly difference. The annular downward pipe can approach the highest Stratification Number ($Str = 1$) and faster than other models even direct downward pipe model. It also provides higher Stratification Number at low dimensionless time. The current study found that the use of an annular downward pipe can be effectively applied to improve the thermal stratification in the storage tank compared to the other models.

As can be seen in Figure 6, the use of the obstacle plate in the tank did not evidently show the difference from the Stratification Number. However, the increasing length of the obstacle plate (see model 2) can notice small improvement of the Stratification Number rather than adding the number of obstacle plates (see model 3). Figure 7 compares the highest Stratification Number of each modification. This study confirms that the modified inlet pipe is associated with the improvement of thermal stratification rather than the obstacle plate.

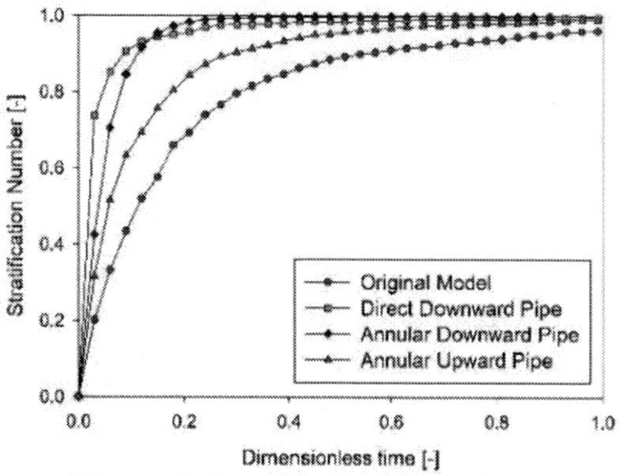

Figure 5. Str of the modified pipes models

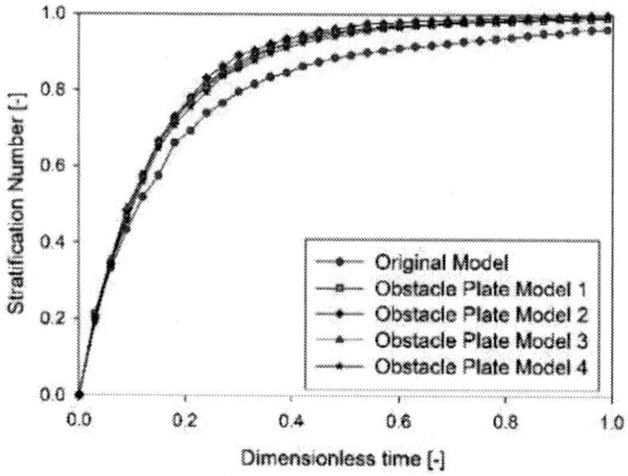

Figure 6. Str of the obstacle plate models

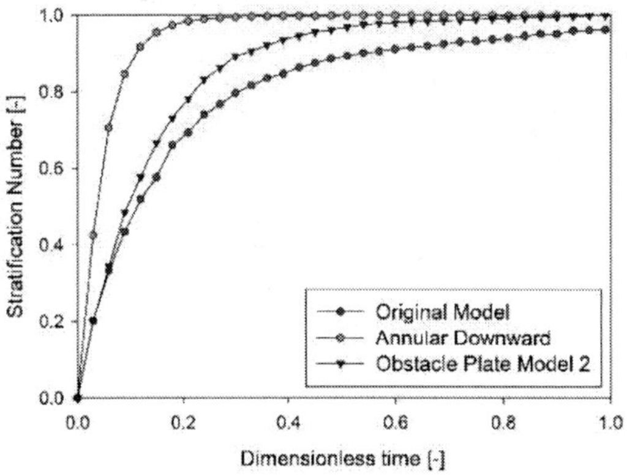

Figure 7. *Str* of the best models of both modifications

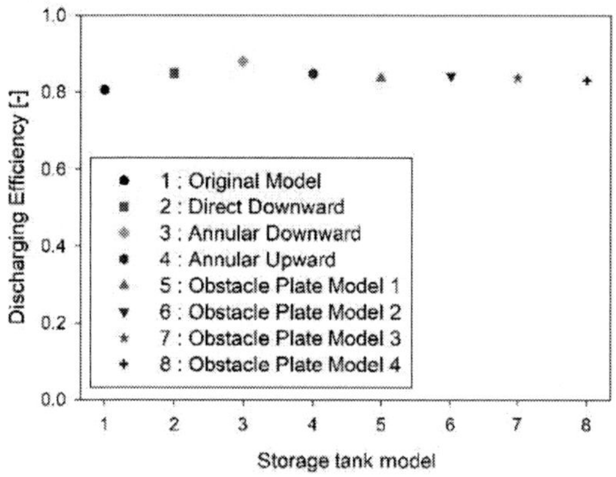

Figure 8. Discharging efficiency

5.4. Discharging Efficiency

A high value of the Discharging Efficiency represents a larger volume of useful energy which reveals a better thermal stratification of the storage tank. Figure 8 summarizes the Discharging Efficiency for all models when the useful temperature of hot water is limited at 45°C. The results indicate that the implementation of either the water inlet pipes or the obstacle plates can achieve toward the enhancement of the discharging efficiency, and it was higher than the conventional storage tank. However, all tank models cannot reach the unity of Discharging Efficiency as much as the location of the hot water outlet pipe is not being at the top of the tank. Therefore, some hot water cannot be drawn off, especially the hot water in the area of an elliptical end cap.

5.5. Temperature profiles and Velocity profiles

Temperature profiles are the best visualization method to explain stratified behaviour in the tank. The original tank was used as a base case to compare the performance of stratification with the other modifications.

Figure 9 (a) illustrates the temperature distribution of water in the storage tank at τ_{use} = 0.3 for all models. The model of annular downward pipe can perform outstandingly on the excellent thermal stratification with clearly thermocline expression rather than other models. These were the results from a lower inlet water velocity by increasing the cross-section area of water inlet and the downward flow direction of inlet water which effectively reduces the water recirculation in the tank as represented the velocity profile in figure 10.

The implementation of the obstacle plate can be noticeable as a smaller control flow area of the water recirculation which is limited only in the bottom part of the storage tank. It was also found that the thermocline was in a broad region. The use of the obstacle plate can achieve towards the control of the recirculation area whereas it is unable to reduce the velocity of the water recirculation resulting in the dispersal of the temperature profiles.

Figure 9 (b) shows the end of the discharging process at τ_{use} = 0.9. The water temperature of the obstacle plate models is still able to distinguish a broad temperature gradient which represents a poor performance of thermal stratification. However, all modification models can improve the performance of the thermal stratification as compared to the original model.

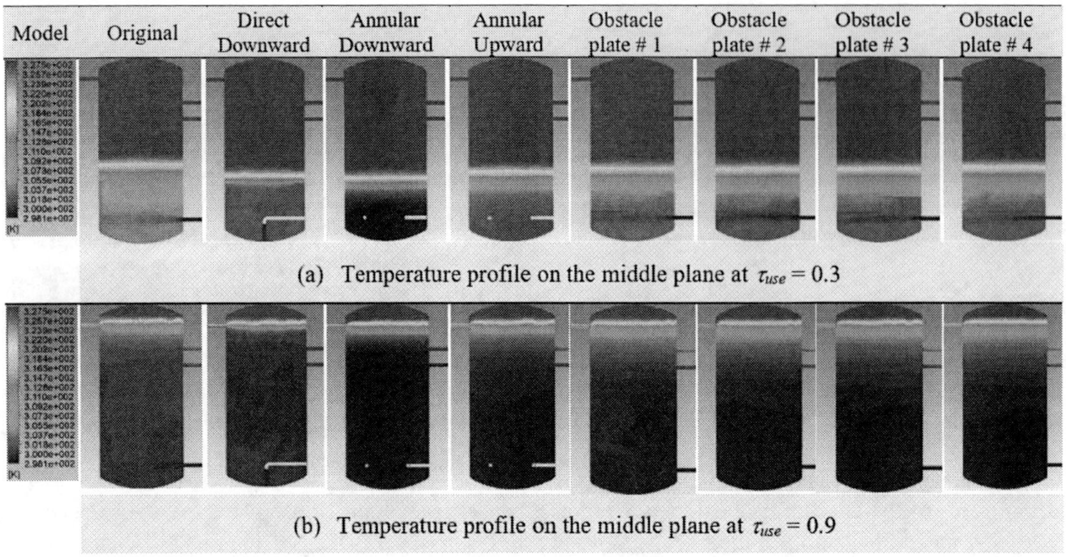

(a) Temperature profile on the middle plane at $\tau_{use} = 0.3$

(b) Temperature profile on the middle plane at $\tau_{use} = 0.9$

Figure 9. Temperature profile on the middle plane at the dimensionless time of 0.3 and 0.9

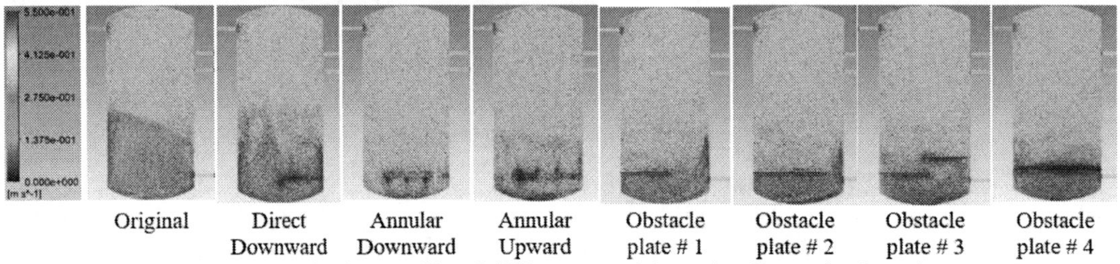

Figure 10. Velocity profile of different models at the dimensionless time = 0.5

6. Conclusions

This study has identified the modification of inlet water pipes efficiently performs the improvement of the thermal stratification in the storage tank as compared to the installation of the obstacle plate. It also set out to identify the significant decrease of inlet water velocity and the controlling flow direction of inlet water. The findings from this study make several contributions to the current literature. First, the design of obstacle plate models can be done by enlarging the plate length, increasing in the obstacle plate numbers, or using perforated plate design. It seems that it is able to control the recirculation area of the water. Second, the results of this investigation show that this approach cannot achieve to reduce the inlet water velocity which is a key parameter that has an effect on the stratification. These findings suggest that in general Richardson Number, the Stratification Number, and Discharging Efficiency can have a pivotal role in the significant indicators of the thermal stratification performance. Finally, the use of well-design thermal stratified storage tank can have a benefit in obtaining greater withdrawal high quality and quantity of useful heat. These findings have significant implications for the understanding of how to improve the efficiency of the heat production systems such as a heat pump and a solar collector which can result in high energy efficiency.

Nomenclature

τ_{use}	Dimensionless time [-]	M	Total mass of water in tank [kg]
τ	Discharge time [s]	c_p	Specific heat capacity [kJ/kg·K]
σ	The unit replacement time [s]	k	The number of data acquired
V_T	Volume of water tank [m^3]	T_{o-use}	Outlet water temperature [K]

IOP Conf. Series: Materials Science and Engineering 1137 (2021) 012076 doi:10.1088/1757-899X/1137/1/012076

$\dot{m}$	Water volume flow rate [L/min]	T_{j-use}	The temperature on each water layer [K]
g	Gravitational acceleration [m/s^2]	m	Mass of each water layer [kg]
β	Thermal expansion coefficient [1/K]	ρ_f	Density of fluid [kg/m^3]
H	Height of water tank [m]	$\vec{V}$	The velocity vector [m/s]
T_{top}	Temperature on top of the tank [K]	p	The local pressure [Pa]
T_{bottom}	Temperature on bottom of the tank [K]	$\mu_f \Phi$	The viscous dissipation term [-]
v_s	Inlet water velocity [m/s]	λ	The second viscosity coefficient [Pa·s]
J	Number of thermocouples	$\vec{F}$	The body force vector per unit volume [N]
T_j	Temperature at thermocouple [K]	$c_{p,f}$	Specific heat capacity of fluid [kJ/kg·K]
T_{max}	Maximum water temperature in tank [K]	T_f	The temperature of fluid [K]
T_{in}	Inlet water temperature [K]	k_f	Fluid thermal conductivity [W/m·K]
Δy	Distance between thermocouples [m]	β_f	Thermal expansion coefficient of fluid [1/K]
$E(\tau_{use})$	Total energy discharge [kJ]	ρ_w	Density of water [kg/m^3]
$E_{st}(\tau_{use})$	Total energy stored in tank [kJ]	T_w	Water temperature [K]

Acknowledgments
The authors are grateful to be supported by the "Research Chair Grant" National Science and Technology Development Agency (NSTDA), the King Mongkut's University of Technology Thonburi through the "KMUTT 55th Anniversary Commemorative Fund" and Thermal Engineering Centre, Department of Mechanical Engineering, KMUTT.

References
[1] Lavan Z and Thompson J 1977 *Sol. Energy* **19**(5) 519-524
[2] Khalifa A Mustafa A T and Khammas F A 2011 *J. Eng. Appl. Sci.***6** 53-60
[3] Castell A Medrano M Solé C and Cabeza L F 2010 *Renew. Energy* **35**(10) 2192-2199
[4] Li A Cao F Zhang W Shi B and Li H 2018 *Sol. Energy* **173** 882-892
[5] Li S Li Y Zhang X and Wen C 2013 *Int. J. Low-Carbon Tech.* **8**(3) 203-209
[6] Shah L J and Furbo S 2003 *Sol. Energy* **75** 337-348
[7] Bouhal T Fertahi S Agrouaz Y Rhafiki T Kousksou T and Jamil A 2017 *Sol. Energy* **157** 441-455
[8] Altuntop N Arslan M Ozceyhan V and Kanoglu M 2005 *Appl. Therm. Eng.* **25** 2285-2298
[9] Ali A Hashim A and Abdulrasool A 2019 *IOP Conf. Ser.: Mater. Sci.*

IOP Conf. Series: Materials Science and Engineering 1137 (2021) 012077 doi:10.1088/1757-899X/1137/1/012077

A Mathematical Model of Laser Drilling with Laser Absorption in the Plasma

M Suchatawat*
Faculty of Engineering, King Mongkut's Institute of Technology Ladkrabang
1 Soi Chalongkrung 1, Ladkrabang, Bangkok, Thailand, 10520

* Corresponding Author: maturose.su@kmitl.ac.th

Abstract
Despite being an accurate process with high processing speed, laser drilling suffers from several drawbacks including wall taper, recast layer and spatter. Hence, in order to improve the process efficiency and achieve the good hole quality, it is crucial to have a precise model which can be used as a tool to relate the process parameters to the process output. In this paper, an analytical model of laser drilling is developed. The governing equations are established from the energy equations at the solid-liquid and liquid-vapor interfaces. Absorption of laser energy in the plasma and an additional energy generated from the exothermic reaction are also included in the model. Aiming at simplifying the model, a constant value of the plasma absorption coefficient of the laser beam is employed. Validation of the model is done by comparing with the available experimental data. It is concluded that the recommended value of the plasma absorption coefficient for low carbon steel drilling is 0.2.

Keywords: laser drilling, mathematical model, plasma absorption

1. Introduction

Laser drilling of metal can be performed by means of trepanning drilling, single pulse drilling or percussion drilling. In trepanning drilling, the process begins by drilling a central hole followed by contour cutting to enlarge the hole diameter. It typically generates the hole with diameter larger than 25 mm in the material with thickness ranging from 0.25 to 20 mm [1]. For single pulse drilling, a single shot of high energy laser pulse is used to drill a hole on a thin substrate. This technique is recommended when the production time is more significant than the hole quality. In percussion drilling, multiple laser pulses impinge on the same spot to produce a hole. Between each successive pulse is the pulse off period which solidification may take place. This technique has recently become an attractive candidate for a wide range of industrial applications due to its ability to produce a large number of small diameter holes with tight tolerance in a short time period.

Despite continuous improvement over the decades, laser drilling has not yet achieved its maximum performance. This is due to the fact that there are numerous complex phenomena and parameters involving in the process. During laser drilling of the metals, the irradiated laser beam heats and melts the solid substrate. At high laser beam intensity, vaporization occurs almost immediately. Vaporization of the melt not only remove material in the form of vapour, it also generates the recoil force which acts on the melt surface and pushes the melt up along the cavity wall. The melt expelled from the cavity may adhere around the hole entrance or may be ejected as a group of small droplets. Once the cavity depth reaches the exit of the workpiece material, the remaining melt is flushed down through the hole bottom. In laser drilling, oxygen is generally supplied coaxially with the laser beam. In addition to protecting the optical lens from the debris, it plays three roles in the process [2]. Firstly, the molten metal may oxidize with oxygen gas and the exothermic energy produced will enhance the drilling rate. Secondly, the oxygen gas convects heat away from the interaction zone, hence lowering the material temperature. Lastly, the assist gas pressure blows the melt away from the cavity resulting in increasing of the material removal rate.

In some cases, further absorption of the laser energy by metal vapour may lead to formation of laser induced plasma (LIP) which reflects the laser beam and traps a large amount of laser energy in it. Sankaranarayanan et al. [3] examined the effects of plasma plume on laser attenuation above the workpiece surface. They found that at low laser intensity, the etch rate per pulse increased as the laser energy increased. However, there was a critical laser intensity in which the etch rate per pulse decreased. At incident laser intensity higher than 4.4×10^8 W/cm^2, less than 20% of the laser energy was delivered to the workpiece surface. Pandey et al. [4] proposed a numerical model of percussion laser drilling. The numerical results obtained were compared with those obtained from their experiments. They found that for laser drilling of SUPERNI 163A, laser induced plasma was observed in all experiments conducted. Therefore, only 5% of the laser energy was assumed to reach the workpiece surface in their model. Hoffman et al. [5] studied effects of the laser wavelength on ablated carbon plume. Experiments were conducted using the first, the second and the third harmonic of Nd:YAG laser. They found that the plasma temperature was highest for the case of 1064 nm laser wavelength compared to those of 532 and 355 nm wavelengths. This implied that the largest amount of laser energy was trapped in the plasma when 1064 nm laser beam is employed. Tokarev et al. [6] pointed out that up to 80% of the laser energy was transformed into plasma energy, hence slowed down the drilling speed. Moreover, the plasma formed can be treated as a secondary drilling tool which enlarged the hole diameter.

Numerous laser drilling models have been proposed to date [7-15]. During the initial period, most of the work emphasized on analytical approach of the one-dimensional heat conduction problem [7-11]. The predicted drilling rates obtained from these models were in reasonable agreement with the experimental data. With the recent availability of the high performance computer, the current trends of the laser drilling models have been shifted to the numerical solutions [12-15]. Drilling speed, hole shape and temperature distribution in the substrate were predicted and effects of the relevant parameters were evaluated. Although the accuracy of the laser drilling models have been greatly improved, there are plenty of gaps to be completed. This is due to the fact that a number of assumptions was typically made in order to simplify the model. For examples, effects of the laser-induced plasma, multiple reflections inside the cavity and the exothermic reaction, which in fact have considerable influence on the laser drilling mechanisms, were often disregarded in the model.

In this paper, a mathematical model of single pulse laser drilling is developed. Laser absorption in the plasma plume is also included. Values of the laser absorption in the plasma are assumed and the appropriate value is determined from comparison with the experimental data. Effects of the oxygen assist gas are also taken into consideration.

2. Mathematical Formulation

Irradiation of the high intensity laser beam on solid substrate, which is initially at temperature T_0, leads to formation of the solid-liquid and liquid-vapour interfaces at $z_m(t)$ and $z_v(t)$ as shown in Figure 1. Because the laser intensity used for laser drilling is typically high, it can be assumed that melting and vaporization occur instantly. Moreover, the hole diameter is normally much larger than the melt layer thickness, hence the one-dimensional heat transfer is applicable.

In addition to the above, followings are assumptions made for the present model.

1) The laser beam has uniform spatial profile and rectangular temporal distribution. This assumption is reasonable especially for the Nd:YAG laser, which is widely used for metal drilling.
2) The changes in laser absorptivity, melting and boiling temperatures of the substrate due to oxidation are ignored.
3) Not all of the metal oxidises with the oxygen assist gas. The oxidation efficiency is introduced in the model to account for the portion of the metal that actually oxidises.
4) Thermophysical properties of the materials and oxygen assist gas are constant i.e. these properties are independent of the temperature.
5) Multiple reflections inside the cavity is ignored.

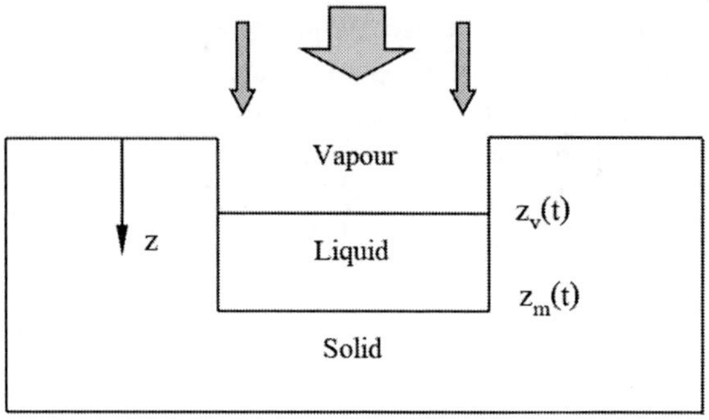

Figure 1. Schematic diagram of the model

At the liquid-vapour interface, the energy equation can be expressed as:

$$I_{abs} + \rho_l H_{ox} \eta_{ox} \frac{\partial z_m}{\partial t} - h_g (T_{l0} - T_a) + k_l \frac{\partial T_l}{\partial z} = \rho_l L_v \frac{\partial z_v}{\partial t} \tag{1}$$

At the solid-liquid interface, the energy balance can be written as:

$$k_s \frac{\partial T_s}{\partial z} - k_l \frac{\partial T_l}{\partial z} = \rho_s L_m \frac{\partial z_m}{\partial t} \tag{2}$$

where I_{abs} is the laser energy absorbed by the liquid surface,
k_l, k_s are the thermal conductivities of liquid and solid metal,
ρ_l, ρ_s are the densities of liquid and solid,
L_m, L_v are the latent heat of melting and the latent heat of vaporization,
H_{ox} is the enthalpy of oxidation,
η_{ox} is the oxidation efficiency,
z_m, z_v are the locations of the solid-liquid and liquid-vapour interfaces,
h_g is the convection heat transfer coefficient at the liquid surface, and
T_l, T_s, T_a, T_{lo} are the temperatures in the liquid and solid, temperature of the surrounded gas and the temperature of the liquid surface which can be estimated from:

$$T_{l0} = I_{abs} \sqrt{\frac{4t}{\pi \rho_l c_{eff} k_l}} \,. \tag{3}$$

Note that t is the laser heating time and c_{eff} is the effective specific heat defined by $c_{eff} = c_{ps} + \frac{L_m}{T_m}$, where c_{ps} is the specific heat of the solid and T_m is the melting point of material.

The second term on the left hand side of equation (1) is the energy added due to exothermic reaction. Because not all of the metal oxidises, the oxidation efficiency η_{ox} is included in the model. Ng et al.[2] laser drilled the mild steel plate and deduced the oxidation percentage from the cross-sectional area of the melt droplets. The oxidation percentage was found to be in the range of 26±18%. Hence, the oxidation efficiency η_{ox} is assumed to be 0.26 in this present model.

As the laser beam passes through the plasma, a fraction of the laser energy is absorbed in the plasma. If the laser intensity entering the plasma is I_i, the laser intensity exiting the plasma I_{po} can be approximated by:

$$I_{po} = (1 - A_{ib})I_i \tag{4}$$

where A_{ib} is the absorption coefficient.

Sankaranarayanan et al. [3] found that the fraction of laser energy that was absorbed by the plasma plume depends on the incoming laser intensity, I_i. At $I_i > 40$ MW/cm², up to 80% of the laser energy might be trapped in the plasma. However, the results from their model which was based on this value of the absorption coefficient showed an overestimation of the laser absorption in the plasma. Mahdavi and Ghazizadeh [18], however, reported that for 1.064 μm-wavelength laser beam generated by a Nd:YAG laser, there was only 10% of the laser energy absorbed in the plasma. A large difference in the absorption coefficient values used in these studies indicated the uncertainty in approximating the plasma absorption coefficient. In this present model, values of the plasma absorption coefficient are assumed. An appropriate value is subsequently selected by comparing with the available experimental data.

After the laser beam leaving the plasma, it irradiates on the liquid substrate. The laser energy absorbed by the liquid layer can be determined from:

$$I_{abs} = A_l I_{po} = A_l (1 - A_{ib})I_i \tag{5}$$

Oxygen assist gas plays several roles in laser drilling process. It protects the optical lens from debris, blows the melt away from the cavity, and adds exothermic energy to the system. Moreover, employing assist gas in the process increases the convection heat transfer from the interaction zone. The heat transfer coefficient h_g for this case can be determined from [14]:

$$h_g = \frac{k_g}{2r_b}\left(C_c \mathrm{Re}^{n_c}\, \mathrm{Pr}^{1/3}\right) \tag{6}$$

where r_b is the beam radius, k_g is thermal conductivity of the assist gas, Re and Pr are Reynolds number and Prandtl number, respectively. The constants C_c and n_c for the forced convection heat transfer normal to the liquid surface are $C_c = 0.228$ and $n_c = 0.731$.

The Reynolds number of the gas flow is given by

$$\mathrm{Re} = \frac{\rho_g v_g 2r_b}{\mu_g} \tag{7}$$

where ρ_g, v_g and μ_g are the density, velocity and viscosity of the oxygen assist gas.

Thickness of the melt and the heat penetration depth in the solid layer are very thin compared with the material thickness. Therefore, it may be further assumed that the temperature profiles in the melt and penetration depth of the solid layers are linear. The energy equation at the two interfaces can now be rewritten as follows:

$$I_{abs} + \rho_l H_{ox} \eta_{ox} \dot{z}_m - h_g (T_{l0} - T_a) - k_l \frac{T_{l0} - T_m}{(\dot{z}_m - \dot{z}_v)\tau} = \rho_l L_v \dot{z}_v \quad (8)$$

$$k_l \frac{T_{l0} - T_m}{(\dot{z}_m - \dot{z}_v)\tau} - k_s \frac{T_m - T_a}{2\sqrt{\alpha_s \tau}} = \rho_s L_m \dot{z}_m \quad (9)$$

where α_s is the thermal diffusivity of the solid, $\dot{z}_m$ and $\dot{z}_v$ are the propagating velocity of the solid-liquid and liquid-vapour interfaces, τ is the pulse duration and T_m is the melting temperature of the material.

Solving of equations (8) and (9) gives the velocity of the two interfaces $\dot{z}_m$ and $\dot{z}_v$.

3. Material properties

The thermophysical properties of low steel are listed in Table 1. Table 2 shows the laser absorptivity of 1.06 μm laser wavelength. Laser absorptivity of the liquid metal is approximated by $A_l \approx 1.05 A_s$ [16].

Table 1. Thermophysical properties of the low carbon steel [17-19]

	Low carbon steel
ρ_s (kg/m³)	7800
ρ_l (kg/m³)	6980
c_{ps} (J/kg K)	628
c_{pl} (J/kg K)	748
α_s (m²/s)	0.014×10^{-3}
α_l (m²/s)	0.007×10^{-3}
T_m (K)	1808
T_v (K)	3100
L_m (J/kg)	276×10^{3}
L_v (J/kg)	6088×10^{3}

Table 2. Laser absorptivity of low carbon steel at 1.06 μm laser wavelength [16-17].

	Low carbon steel
A_s	0.37
A_l	.0389

4. Results and discussion

In order to validate the model, the calculated results are compared with the available experimental data. In Low et al. [11]'s experiment, a hole was drilled on low carbon steel plate using a single pulse laser generated from the Nd:YAG fiber laser. Figures 2 and 3 show the comparison of the predicted drilling velocity with the measured data at various absorbed laser intensity for the pulse duration of 0.5 and 1.0 ms, respectively. Values of the absorption coefficient are varied from 0 to 0.2 with the incremental step of 0.1 in the model. It is found that $A_{ib}=0$, which is the case of no absorption in the plasma, gives the best prediction for both cases. For the case of 0.5 ms pulse width as shown in Figure 2, average errors are found to be 11.4 and 16.0% for $A_{ib}=0$ and $A_{ib}=0.1$. However, for the case of 1.0 ms pulse duration, average errors are quite close i.e. the average errors are 16.1 and 16.9% for $A_{ib}=0$ and $A_{ib}=0.1$, respectively. These errors may be due to the assumptions used in the model. For example, the beam multiple reflections which normally occur in the actual drilling process are not accounted in the model. If the beam multiple reflections are included, more laser energy will be absorbed by the cavity wall which will subsequently leads to the higher predicted drilling speed. Moreover, the present model assumes a constant value of the material absorptivity whereas the actual value varies with temperature and material properties.

Although the model with $A_{ib}=0$ gives good prediction of the drilling velocity, it is important to note that to estimate the absorbed laser intensity from the impinging laser beam intensity, Low et al. [11] assumed the reflectivity of the low carbon steel to be 0.21. In other words, the material absorptivity used in their calculation was 0.79. However, in this present model, the absorptivity at the liquid surface is $A_l=0.389$ which is much lower than that employed by Low et al. This results in less laser energy absorbed by the liquid. In addition to that, the present model also accounts for absorption of laser energy in the plasma which lessens the amount of laser energy reaching the liquid surface. Thus, the drilling velocity predicted by the present model tends to lower than that obtained experimentally.

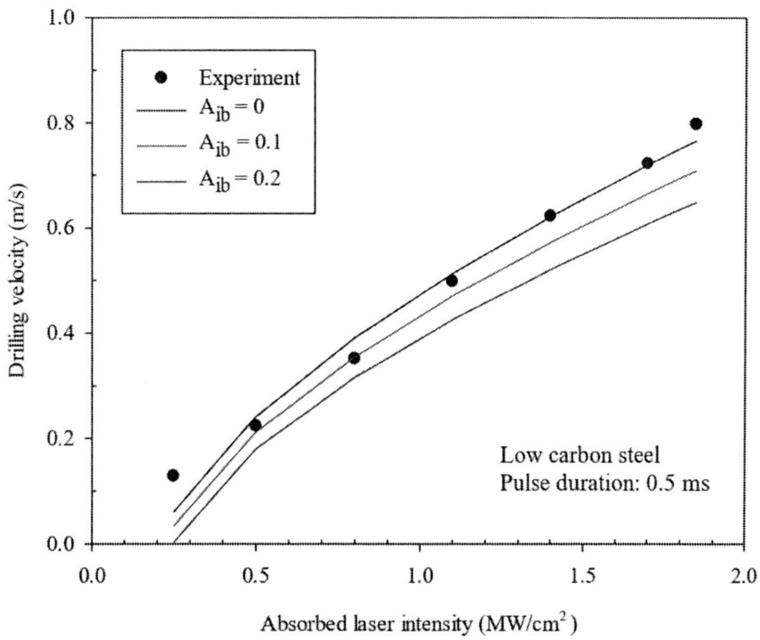

Figure 2. Comparison of the predicted drilling velocity with the experimental data of Low et al. [11]. Pulse duration is 0.5 ms.

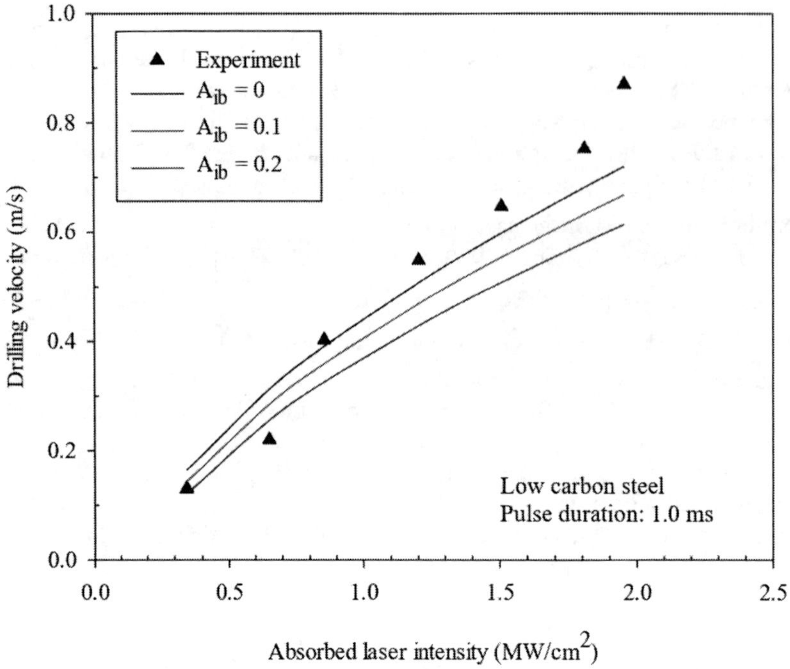

Figure 3. Comparison of the predicted drilling velocity with the experimental data of Low et al. [11]. Pulse duration is 1.0 ms.

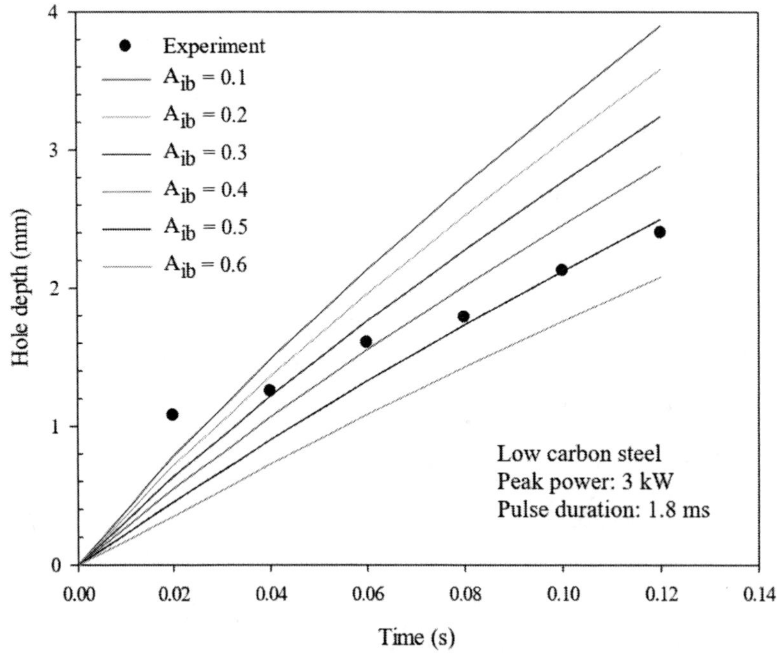

Figure 4. Comparison of the predicted hole depth with the experimental data of Maturose and Sheikh [20]

Figure 4 shows the predicted hole depth in comparison with the experimental data of Maturose and Sheikh [20]. The laser beam used in their experiment was Nd:YAG laser which produced a laser beam at 1.06 μm wavelength. The predicted results in this case show fair agreement with the measured hole depth. The smallest average error is 18.5% which is obtained using $A_{ib} = 0.5$. It can also be seen from Figure 4 that during the initial time of laser drilling, the measured hole depth is much deeper than the calculated ones. This is probably because there is only small amount of plasma formed at the beginning of the actual drilling process. Thus, more laser energy is allowed to irradiate on the material surface and hence the deeper hole.

All in all, according to Figures 2 and 3, the proper value of the plasma absorption coefficient is $A_{ib} = 0$ while in Figure 4, $A_{ib} = 0.5$ is recommended. Therefore, the approximated value of plasma absorption coefficient $A_{ib} = 0.2$ is used throughout this present model.

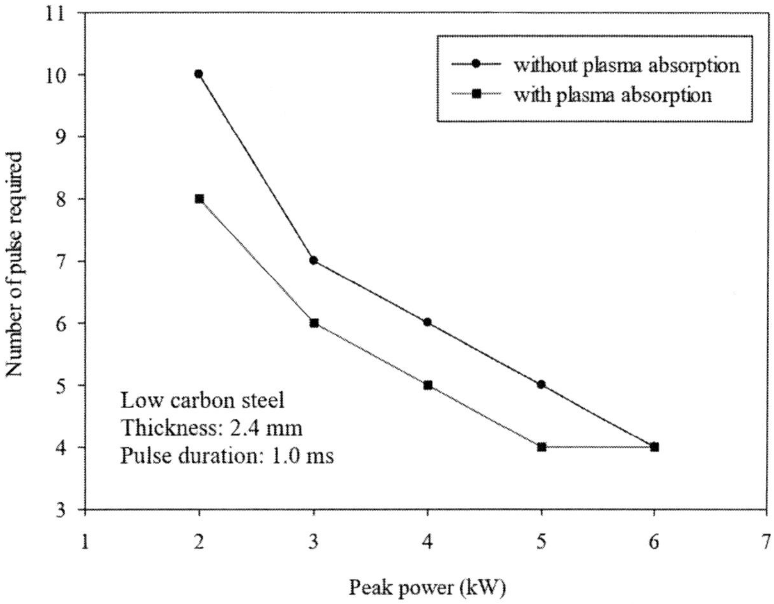

Figure 5. Number of pulses required to produce a through hole

Figure 5 illustrates number of pulses required to initiate breakthrough. It is obvious that including of the plasma absorption results in more pulses required. This is because less laser energy is delivered to the workpiece surface, therefore less material is melted, vaporized and removed from the interaction zone. This subsequently leads to a shallower hole depth. At 6 kW peak power, although 4 laser pulses are required for both cases, the hole depths obtained are different. By including the plasma absorption, the calculated hole depth is 2.53 mm whereas excluding the plasma absorption predicts 2.98 mm hole depth.

Figure 5 also shows that as the pulse peak power increases, less pulse is required to produce a through hole. This is not surprised because the higher the peak power, the more laser energy impinges on the surface. Consequently, more material is removed resulting in the formation of the deeper hole.

5. Conclusion

This paper presents a mathematical model of laser drilling with laser beam absorption in the plasma. A range of plasma absorption coefficients is assumed and the suitable value is selected from comparison with the available experimental data. It is found that the absorption coefficient of 0.2 gives the most compromising prediction. However, it is crucial to note that this recommended value of the absorption coefficient is based on the laser drilling of low carbon steel using 1.06 μm wavelength Nd:YAG laser.

Acknowledgements

The author would like to thank the National Science and Technology Development Agency for providing financial support to this work.

References

[1] Corcoran A, Sexton L, Seaman B, Ryan P and Byrne G 2002 *J. Mater. Process. Technol.* **123** 100-6
[2] Ng G K L, Crouse P L and Li L 2006 *Int. J. Heat Mass Transf.* **49** 1358-74
[3] Sankaranarayanan S, Emminger H and Kar A 1999 *J. Phys.* **D 32** 1605-11
[4] Pandey N D, Shan H S and Mohandas T 2006 *Int. J. Comput. Appl. Technol* **6** 218-32
[5] Hoffman J, Moscicki T and Szymanski Z 2011 *Appl. Phys.* A **104** 815-9
[6] Tokarev V N, Cheshev E A, Bezotosnyi V V, Khomich V Yu, Mikolutskiy S I and Vasil'yeva N V 2015 *Laser Phys.* **25** 056003
[7] Chan C L, Mazumder J 1987 *J. Appl. Phys.* **62** 4579-86
[8] Yilbas B S 1995 *J. Laser Appl.* **7** 147
[9] Xie J and Kar A 1997 *J. Appl. Phys.* **81** 3015-22
[10] Shen Z H, Zhang S Y, Lu J and Ni X W 2001 *Opt. Laser Technol.* **33** 533-7
[11] Low D K Y, Li L and Byrd P J 2002 *J. Manuf. Sci. Eng.* **124** 852-62
[12] Ganesh R K, Bowley W W, Bellantone R R, and Hahn Y 1996 *J. Comput. Phys.* **125** 161-76
[13] Collins J and Gremaud P 2011 *Math. Comput. Simul* **81** 1541-52
[14] Zhang Y, Shen Z and Ni X 2014 *Int. J. Heat Mass Transfer* **73** 429-37
[15] Abidou D, Yusoff N, Nazri N, Awang M A O, Hassan M A, Sarhan A A D 2017 *Precis. Eng.* 49 69-77
[16] Schneider M, Berthe L, Fabbro R and Muller M 2008 *J. Phys.* **D 41** 6
[17] Allmen M v 1987 *Laser-Beam Interactions with Materials* (Berlin: Springer)
[18] Mahdavi M and Ghazizadeh S F 2012 *J. Appl. Sci.* **12** 12-21
[19] Rohsenow W M, Hartnett J P and Gamic E N 1985 *Handbook of Heat Transfer Fundamentals, 2nd ed* (New York: McGraw-Hill Book Company)
[20] Suchatawat M and Sheikh M 2013 Mathematical modeling of laser drilling *Laser in Manufacturing* ed J P Davim (London: ISTE Ltd) chapter 4 pp 139-77

AUTHOR INDEX